The Kinin
System

THE HANDBOOK OF IMMUNOPHARMACOLOGY

Series Editor: Clive Page
King's College London, UK

Titles in this series

Cells and Mediators

Immunopharmacology of Eosinophils
(edited by H. Smith and R. Cook)

The Immunopharmacology of Mast
Cells and Basophils
(edited by J.C. Foreman)

Lipid Mediators
(edited by F. Cunningham)

Immunopharmacology of Neutrophils
(edited by P.G. Hellewell and
T.J. Williams)

Immunopharmacology of
Macrophages and other
Antigen-Presenting Cells
(edited by
C.A.F.M. Bruijnzeel-Koomen and
E.C.M. Hoefsmit)

Adhesion Molecules
(edited by C.D. Wegner)

Immunopharmacology of Lymphocytes
(edited by M. Rola-Pleszczynski)

Immunopharmacology of Platelets
(edited by M. Joseph)

Immunopharmacology of Free Radical
Species
(edited by D. Blake and
P.G. Winyard)

Cytokines
(edited by A. Mire-Sluis, forthcoming)

Systems

Immunopharmacology of the
Gastrointestinal System
(edited by J.L. Wallace)

Immunopharmacology of Joints
and Connective Tissue
(edited by M.E. Davies and J. Dingle)

Immunopharmacology of the Heart
(edited by M.J. Curtis)

Immunopharmacology of Epithelial
Barriers
(edited by R. Goldie)

Immunopharmacology of the
Renal System
(edited by C. Tetta)

Immunopharmacology of the
Microcirculation
(edited by S. Brain)

Immunopharmacology of the
Respiratory System
(edited by S.T. Holgate)

The Kinin System
(edited by S. Farmer)

Drugs

Immunotherapy for Immune-
related Diseases
(edited by W.J. Metzger,
forthcoming)

Phosphodiesterase Inhibitors
(edited by C. Schudt, G. Dent and
K. Rabe)

Immunopharmacology of AIDS
(forthcoming)

Immunosuppressive Drugs
(forthcoming)

Glucocorticosteroids
(forthcoming)

Angiogenesis
(forthcoming)

The Kinin System

edited by

Stephen G. Farmer

Bioscience, Zeneca Pharmaceuticals
1800 Concord Pike, Wilmington
DE 19850, USA

ACADEMIC PRESS
San Diego   London   Boston
New York   Sydney   Tokyo   Toronto

Academic Press Inc.
525 B Street, Suite 1900, San Diego, California 92101-4495, USA
http://www.apnet.com

Academic Press Limited
24–28 Oval Road, London NW1 7DX, UK
http://www.hbuk.co.uk/ap/

ISBN 0-12-249340-0

A catalogue record for this book is available from the British Library

Typeset by Phoenix Photosetting, Chatham, Kent
Printed in Great Britain by The Bath Press, Bath
97 98 99 00 01 02 EB 9 8 7 6 5 4 3 2 1

Contents

6.   *Immunological Probes for the Bradykinin B$_2$ Receptor. A Toolbox*   99
Werner Müller-Esterl

7.   *Metabolism of Bradykinin by Peptidases in Health and Disease*   111
Ervin G. Erdös *and* Randal A. Skidgel

18. Bradykinin as a Growth Factor   301

David A. Walsh *and* Tai-Ping D. Fan

19. The Role of Kinins in the Cardiac Effects of ACE Inhibitors and Myocardial Ischemia   315

Wolfgang Linz, Gabriele Wiemer, Klaus Wirth *and* Bernward A. Schölkens

Contributors

S.P. Andrade
Department of Physiology and Biophysics,
Institute of Biological Sciences,
Federal University of Minas Gerais,
Av. Antônio Carlos, 6627 – C.P. 486,
31270 Belo Horizonte, MG,
Brazil

W.T. Beraldo
Department of Physiology and Biophysics,
Institute of Biological Sciences,
Federal University of Minas Gerais,
Av. Antônio Carlos, 6627 – C.P. 486,
31270 Belo Horizonte, MG,
Brazil

K.D. Bhoola
Department of Experimental and Clinical Pharmacology,
Medical School,
University of Natal,
Private Bag 7,
Congella 4013,
Durban
South Africa

J.A. Clements
Centre for Molecular Biotechnology
School of Life Sciences,
Queensland University of Technology,
2 George Street,
GPO Box 2434,
Brisbane, QLD 4001,
Australia

R.W. Colman
Thrombosis Research Center,
The Sol Sherry School of Medicine,
Temple University School of Medicine,
3400 N. Broad Street,
Philadelphia,
PA 19140,
USA

J.W. Dear
Department of Pharmacology,
University College London,
Gower Street,
London, WC1E 6BT,
UK

A. Dray
Astra Pain Research Unit,
275 boul. Armand-Frappier,
Edifice 3000,
Laval,
Quebec, H7V 4A7,
Canada

E.G. Erdös
Department of Pharmacology,
University of Illinois College of Medicine,
835 S. Wolcott Avenue,
M/C 868,
Chicago,
IL 60612-3796,
USA

T.-P.D. Fan
Department of Pharmacology,
University of Cambridge,
Tennis Court Road,
Cambridge CB2 1QJ,
UK

S.G. Farmer
Bioscience,
Zeneca Pharmaceuticals,
1800 Concord Pike,
Wilmington,
DE 19850,
USA

J.C. Foreman
Department of Pharmacology,
University College London,
Gower Street,
London WC1E 6BT,
UK

T. Griesbacher
Department of Experimental and Clinical Pharmacology,
Universitätsplatz 4,
A–8010 Graz,
Austria

J.M. Hall
Receptors and Cellular Regulation Group,
School of Biological Sciences,
University of Surrey,
Guildford GU2 5XH,
UK

J.F. Hess
Human Genetics,
Merck Research Laboratories,
WP26-265,
PO Box 4,
West Point,
PA 19486,
USA

A.P. Kaplan
Division of Allergy and Clinical Immunology,
Department of Medicine,
State University of New York,
Stony Brook,
NY 11794-8161,
USA

F. Lembeck
Department of Experimental and Clinical Pharmacology,
Universitätsplatz 4,
A–8010 Graz,
Austria

U.H. Lerner
Department of Oral Cell Biology,
Umeå University,
S–901 87 Umeå,
Sweden

W. Linz
Hoechst-Marion-Roussel,
DG Cardiovascular (H 821),
D–65926 Frankfurt/Main,
Germany

C.A. Maggi
Department of Pharmacology,
A Menarini Pharmaceuticals,
Via Sette Santi 3,
Florence, S0131,
I–50131,
Italy

F. Marceau
Centre de Recherche,
L'Hôtel-Dieu de Québec,
11, Côte-du-Palais,
Québec, G1R 2J6,
Canada

I.K.M. Morton
Pharmacology Group,
King's College London,
London SW3 6LX,
UK

W. Müller-Esterl
Institute for Physiological Chemistry and
Pathobiochemistry,
The University of Mainz,
Duesbergweg 6,
D–55099 Mainz 1,
Germany

Y. Naidoo
Department of Experimental and Clinical Pharmacology,
Medical School,
University of Natal,
Private Bag 7,
Congella 4013,
Durban,
South Africa

M.N. Perkins
Sandoz Institute for Medical Research,
5 Gower Place,
London WC1E 6BN,
UK

R.A. Pixley
Thrombosis Research Center,
The Sol Sherry School of Medicine,
Temple University School of Medicine,
3400 N. Broad Street,
Philadelphia,
PA 19140,
USA

S.R. Reddigari
Division of Allergy and Clinical Immunology,
Department of Medicine,
State University of New York,
Stony Brook,
NY 11794-8161,
USA

A.H. Schmaier
Division of Hematology and Oncology,
Department of Internal Medicine,
University of Michigan,
102 Observatory Street,
Ann Arbor,
MI 48109-0724,
USA

B.A. Schölkens
Hoechst-Marion-Roussel,
DG Cardiovascular (H821),
D–65926 Frankfurt/Main,
Germany

M. Silverberg
Division of Allergy and Clinical Immunology,
Department of Medicine,
State University of New York,
Stony Brook,
NY 11794-8161,
USA

R.A. Skidgel
Department of Pharmacology,
University of Illinois College of Medicine,
835 S. Wolcott Avenue,
M/C 868,
Chicago,
IL 60612-3796,
USA

D.A. Walsh
Inflammation Group,
Arthritis and Rheumatism Council Building,
The London Hospital Medical College,
Whitechapel,
London E1 2AD,
UK

G. Wiemer
Hoechst-Marion-Roussel,
DG Cardiovascular (H821),
D–65926 Frankfurt/Main,
Germany

K. Wirth
Hoechst-Marion-Roussel,
TD Cardiovascular Agents (H821),
D–65926 Frankfurt/Main,
Germany

Series Preface

The consequences of diseases involving the immune system such as AIDS, and chronic inflammatory diseases such as bronchial asthma, rheumatoid arthritis and atherosclerosis, now account for a considerable economic burden to governments worldwide. In response to this, there has been a massive research effort investigating the basic mechanisms underlying such diseases, and a tremendous drive to identify novel therapeutic applications for the prevention and treatment of such diseases. Despite this effort, however, much of it within the pharmaceutical industries, this area of medical research has not gained the prominence of cardiovascular pharmacology or neuropharmacology. Over the last decade there has been a plethora of research papers and publications on immunology, but comparatively little written about the implications of such research for drug development. There is also no focal information source for pharmacologists with an interest in disease affecting the immune system or the inflammatory response to consult, whether as a teaching aid or as a research reference. The main impetus behind the creation of this series was to provide such a source by commissioning a comprehensive collection of volumes on all aspects of immunopharmacology. It has been a deliberate policy to seek editors for each volume who are not only active in their respective areas of expertise, but who also have a distinctly *pharmacological* bias to their research. My hope is that *The Handbook of Immunopharmacology* will become indispensable to researchers and teachers for many years to come, with volumes being regularly updated.

The series follows three main themes, each theme represented by volumes on individual component topics.

The first covers each of the major cell types and classes of inflammatory mediators. The second covers each of the major organ systems and the diseases involving the immune and inflammatory responses that can affect them. The series will thus include clinical aspects along with basic science. The third covers different classes of drugs that are currently being used to treat inflammatory disease or diseases involving the immune system, as well as novel classes of drugs under development for the treatment of such diseases.

To enhance the usefulness of the series as a reference and teaching aid, a standardized artwork policy has been adopted. A particular cell type, for instance, is represented identically throughout the series. An appendix of these standard drawings is published in each volume. Likewise, a standardized system of abbreviations of terms has been implemented and will be developed by the editors involved in individual volumes as the series grows. A glossary of abbreviated terms is also published in each volume. This should facilitate cross-referencing between volumes. In time, it is hoped that the glossary will be regarded as a source of standard terms.

While the series has been developed to be an integrated whole, each volume is complete in itself and may be used as an authoritative review of its designated topic.

I am extremely grateful to the officers of Academic Press, and in particular to Dr Carey Chapman, for their vision in agreeing to collaborate on such a venture, and greatly hope that the series does indeed prove to be invaluable to the medical and scientific community.

C.P. Page

Preface

Immunopharmacology and the Kinin System. What's in a Name?

In the preface to this series, Clive Page notes several points that are especially relevant to this volume on the kinin system. Thus, he has sought editors of individual volumes who have a "... distinctly pharmacological bias ...". As I was trained in Professor John Gillespie's Department in Glasgow University, I hope I can claim a pharmacological tendency. Nevertheless, a pharmacologist who develops an interest in inflammation can only begin to understand and expand his or her vocation by learning and applying the nuances of immunology. Perhaps this is how some of us realize an "immuno-" prefix to our discipline. Professor Page also notes the three major themes of this series: (1) cell types and inflammatory mediators; (2) organ systems and their diseases; and (3) drugs used to treat inflammation and diseases of the immune system. The last represents immunopharmacology (or maybe just plain old pharmacology) in action. The kallikrein–kinin system has a major impact on each of the first two and, with the future discovery and development of safe, selective and potent nonpeptide kinin receptor antagonists, the third theme will assuredly be impacted significantly.

The kallikrein–kinin system is ancient on the evolutionary scale and is not, perhaps, as voguish as newer discoveries in biomedical research. All vertebrates and many nonvertebrate animals and arthropods have a kinin-generating apparatus, not least of which is *Bothrops jararaca*, an extremely venomous South American pit viper. As described in the opening chapter, the venom of this snake was reported in 1949 by Rocha e Silva and colleagues (two of whom have written this chapter) to release from mammalian blood a substance that caused contraction of guinea pig ileum – surely the prototype tissue preparation on which all pharmacologists cut their experimental teeth. It is from its ability to elicit slow movements of the gut that "bradykinin" derives its name. Perhaps because of these original studies on kinin pharmacology, for decades, most kinin research focused

on effects on smooth muscle derived from a bewildering array of tissues, organs and species. Indeed, kinin receptors were first named "B_1" and "B_2" by Regoli and colleagues following extensive experiments utilizing vascular smooth muscle.

In contrast, the aim of the present volume is to have a distinctly *immuno*pharmacological bias to the kinin system, particularly with a view to findings indicating inflammatory effects or, indeed, protective effects of kinins. The cloning, sequencing and expression of the kininogens and kallikreins has recently facilitated our understanding of the regulation of the components of this inflammatory cascade. Furthermore, molecular biology has confirmed the existence of at least two subtypes of receptor molecules. Then again, pharmacology had already done that.

I find it fascinating that vascular endothelial and other cell types, neutrophils, for example, express on their surfaces some or all members of the protein armamentarium with which to generate, degrade and respond to endogenous kinins both in an autocrine and paracrine fashion. Given that the half-life of circulating kinins is a few seconds only, it is hardly surprising that they are generated by tissues, and that their physiological and pathophysiological actions are likely to be mediated on a local level. Locally produced bradykinin has an important role in regulating blood flow in the microcirculation, including that of the heart. Indeed, there is evidence that the clinical effects of angiotensin-converting enzyme (ACE) inhibitors in hypertension and myocardial ischemia are mediated via inhibition of kinin degradation in addition to and perhaps in some instances, rather than, preventing formation of angiotensin II. ACE is also known as "kininase II", and is one of many substances named from one (or more) of its biological activities. However, names that describe a biological effect can be confusing and can mask other, often more important effects of a given substance *in vivo*. Take enkephalinase, platelet-activating factor and tumor necrosis factor, for example – or bradykinin!

Numerous investigations on kinin receptor signal transduction are conducted in fibroblasts, probably due partly to their availability and the relative ease of maintaining them in tissue culture. Yet, relatively few studies have addressed the question of the role(s) played by fibroblast kinin receptor-mediated responses in the body. In fibroblasts, as well as endothelial and smooth muscle cells, and lymphocytes, kinins exert quite potent mitogenic or antimitogenic activity, and synergize with other growth factors. Thus, a growing awareness of the role of the kinin system in angiogenesis, tissue repair and cancer is apparent, and should continue.

Although many of us tend to consider the kinins and other so-called "pro-inflammatory" substances as solely noxious, Nature bestowed upon us cytokines, chemokines and biogenic lipids, amines and peptides as regulators of homeostasis, and usually their release is necessary, normal and healthy. It is only during exaggerated and/or protracted immune responses following trauma, severe infection, allergy or during autoimmune processes that these mediators can become harmful to the host. There has been a dramatic increase in knowledge of the seemingly pivotal roles played by the kinin system in animal models of severe, often life-threatening human diseases such as pancreatitis, sepsis, asthma and acute respiratory distress syndrome, as well as not-so-lethal, but chronically debilitating diseases such as arthritis and many other painful conditions. As noted earlier, the development of kinin receptor antagonists and kallikrein inhibitors as drugs is eagerly awaited. As therapeutic agents, they cannot fail to improve the human condition.

Finally, in a book such as this one, it is inevitable that chapters overlap in some subject matter. While every effort was taken to minimize repetition, and each chapter cross-references others, it would have been impossible and unnecessary to completely eliminate "redundancy". Different researchers observe and interpret the same phenomena often quite uniquely and that, too, is normal and healthy. I thank wholeheartedly each of the contributors to what, perhaps immodestly, we should consider to be a timely and valuable book on a significant component of our innate immunopharmacology.

Stephen G. Farmer

1. Discovery of Bradykinin and the Kallikrein–Kinin System

Wilson T. Beraldo *and* Silvia P. Andrade

1. Introduction

The existence of the kallikrein–kinin system was first disclosed nearly 50 years ago when a series of coincidental discoveries by two independent groups came together revealing the major biological roles of the system. The present chapter aims to present the pioneering work in the field which led to our current understanding of the assembly of the various components of the kallikrein–kinin system. The discovery of the physiological and pathological roles of kinins is highlighted showing their involvement in vasodilation, regulation of blood pressure, inflammation, and the production of pain. Thus, through knowledge gained from these pioneering basic studies, the actions of kinins were unveiled, leading to the beginnings of our understanding of many diseases associated with kinin formation.

1.1 THE DISCOVERY OF BRADYKININ AND ITS RELEASE BY SNAKE VENOM

The venom of *Bothrops jararaca*, an extremely poisonous pit viper occurring only in South America, was long known to cause vascular shock in the dog and other species, including man. A sample of the venom was brought to the Instituto Biológico in São Paulo, Brazil, where we set out to determine the role of venom in releasing histamine, and examine the mechanisms underlying venom-induced vascular shock. The assumption, that the effects of *B. jararaca* venom were mediated via histamine, was based on earlier work by Feldberg and Kellaway (1938) who had shown the release of histamine by the venoms of many Australian and Indian snakes.

The blood from a dog that was shocked with *B. jararaca* venom was applied to guinea-pig gut segments suspended in Tyrode's solution containing an antihistamine and atropine. To our surprise, we observed that the venom-treated blood had a stimulating effect upon the isolated smooth muscle. The effect was not due to direct action of the venom, since the preparation had previously been made refractory to it. No desensitization was observed after several additions of the serum to the bath containing the guinea-pig ileum. The contractions produced by the active principle were slow in nature when compared to those induced by histamine and acetyl-choline, and exhibited a short latent interval. Rocha e Silva and collaborators (1949) coined a name derived from Greek for this substance, using the word *kinin* (indicating movement) with the prefix *brady* (indicating slow) to describe the slow effect of the substance on the guinea-pig ileum. Thus the new principle was christened "bradykinin".

Following these initial observations, experiments were performed to study the release of bradykinin (BK) by the venom in a more direct way. Canine liver was perfused with defibrinated blood to which the venom had been added. After passage through the liver, considerable amounts of the smooth muscle-contracting substance were found. The effects of the new factor were not blocked by atropine or antihistaminics, and were rapidly inactivated after being released. The effects could not, therefore, be due to acetylcholine, histamine or, indeed, any other factors then identified as endogenous active agents. In the sequence of our investigations it was soon demonstrated that addition of venom or trypsin directly to defibrinated blood *in vitro* was sufficient to produce activity which disappeared rapidly. The active principle, BK, was defined as a polypeptide, originating from an inactive precursor, bradykininogen, released from the pseudoglobulin fraction of normal plasma, by the action of venom or trypsin. BK had a stimulating effect on isolated smooth muscle preparations of guinea-pig ileum and uterus, or rabbit duodenum, and produced hypotension when administered intravenously in dogs, rabbits and cats.

In terms of its chemical properties, BK was defined as thermostable, dialysable through cellophane, resistant to prolonged boiling in 0.1–1.0 N HCl solution, but rapidly destroyed if heated in alkaline solution (Rocha e Silva *et al.*, 1949). Significantly, also in the seminal publications by Rocha e Silva and Beraldo (1949), was the association of BK with the expression "autopharmacology", coined by Sir Henry Dale in the 1930s to indicate the important phenomenon of endogenous chemical mediation of physiological and pathophysiological events in the body.

2.   *The Discovery of Kallikrein and Kallidin*

2.1   KALLIKREIN

In 1926, while researching at the Munich University Hospital, Frey first noted a considerable reduction in blood pressure when he injected human or animal urine into dogs. This effect was attributed to a specific substance, designated "F-substance", which was isolated from human urine and from diverse other tissues (Frey and Kraut, 1926). F-substance was later called "kallikrein"

after a Greek synonym for the pancreas, since the substance existed in that organ in such high concentrations that it was originally thought to be produced mainly in this gland (Kraut *et al.*, 1930).

Soon after this discovery, a kallikrein inactivator was found (Kraut *et al.*, 1928). Urine from patients with postoperative reflex anuria was thought to contain very high concentrations of F-substance. When Frey injected such urine into a dog, it failed to produce any effect on blood pressure. The F-substance had been inactivated by a principle in the blood which heavily contaminated urine from anuric patients. This principle was termed "serum inactivator" (Kraut *et al.*, 1928).

2.2   KALLIDIN (LYS-BRADYKININ)

Werle (1936) was convinced that he was dealing with an enzyme during the course of his researches with kallikrein, but the physiological importance of the enzyme was not clear. He found that kallikrein preparations from human urine elicited contractions in isolated dog intestine. Thus kallikrein had been shown to possess two activities: induction of hypotension *in vivo*, and contraction of isolated intestinal smooth muscle. In the course of his experiments, Werle found that kallikrein alone did not contract the intestine but, rather, it had to be mixed with serum before spasmogenic activity was observed *in vitro*. Thus, the possible *de novo* formation of an active substance in serum-treated kallikrein had to be considered. The pharmacological actions of this substance were the same as those of kallikrein. Furthermore, this low molecular weight, thermostable substance could be concentrated. The substance, which was thought to be a peptide, was not a split product of kallikrein but, rather, was derived from a serum protein. It was coined "substance DK" from the German, *darmkontrahierender stoff* (intestine-contracting substance; Werle *et al.*, 1937). Shortly thereafter, substance DK was renamed "kallidin" (KD) and its precursor was named "kallidinogen". Owing to the fact that substance DK migrated in a broad pH range toward the cathode, it was established that basic amino acids were present in its composition (Werle and Däumer, 1940).

3.   *Isolation and Purification of Bradykinin and Kallidin*

3.1   ISOLATION

Rocha e Silva *et al.* (1949) showed that, if appropriate doses of *B. jararaca* venom or trypsin were incubated with bovine pseudoglobulin, maximum release of BK was observed after 2–3 minutes. When the incubation period was prolonged for 20–30 minutes, the liberated BK was inactivated completely. In initial attempts to isolate and purify BK, purification assays using different organic solvents and chromatographic methods were described (Prado *et al.*, 1950). We noted that BK was soluble and could be concentrated in glacial acetic acid. BK was also found to be soluble in ethyl alcohol and trichloroacetic acid, but completely insoluble in ethyl ester, acetone and chloroform. The first significant advance towards the isolation of a pure substance was the work of Andrade and Rocha e Silva (1956), who utilized the protein precipitate from ox plasma plus 25% ammonium sulfate as a substrate. The enzyme source was whole snake venom, and preliminary purification was on cellulose and aluminium oxide. The final step was ion-exchange chromatography on Amberlite IRC-50. An overall increase in biological activity was achieved, and we produced the first evidence that BK had been isolated successfully. It was claimed that BK was among the most active smooth muscle spasmogens yet to be isolated from mammals, even when compared to well-known tissue hormones such as acetylcholine, histamine or 5-hydroxytryptamine. Its potency in rat isolated uterus was of the order of nanograms, similar to the order of activity of angiotensin or oxytocin. In guinea-pig ileum, BK's potency was at least ten- and seven-fold higher than molar equivalents of histamine and acetylcholine, respectively (Andrade and Rocha e Silva, 1956). Similar observations were made for the hypotensive or vasodilator effects of BK *in vivo*. Injections of 1–5 µg caused profound falls in blood pressure in rabbits, cats and dogs (Pernow and Rocha e Silva 1955; Rocha e Silva, 1960).

3.2   PURIFICATION

It took almost a decade for the complete purification of BK. Elliott *et al.* (1959a) obtained BK utilizing a column of carboxymethylcellulose combined with chromatography and paper electrophoresis. They used as a substrate the protein precipitate formed from ox serum between 33% and 45% of saturation with ammonium sulfate. It was heated in solution at pH 1.37 for 30 min to destroy kininase and was then digested with trypsin at pH 7.5 for 6 h. Purification of the peptide mixture that resulted from alcoholic precipitation of the digest was performed in a butanol:water:trichloroacetic acid system followed by gradient elution ion-exchange chromato-

graphy on carboxymethyl-cellulose (CM-cellulose) with volatile ammonium acetate buffers. These were then relatively new techniques developed by Peterson and Sober (1956). Five constituent amino acids, arginine, phenylalanine, proline, glycine and serine, were identified in the following molar proportions 2:2:2:1:1. Further experiments by Boissonnas *et al.* (1960a) indicated the lack of a proline residue in the above composition, and yielded the correct sequence of BK to be:

$$\begin{array}{ccccccccc} 1 & 2 & 3 & 4 & 5 & 6 & 7 & 8 & 9 \end{array}$$
H-Arg-Pro-Pro-Gly-Phe-Ser-Pro-Phe-Arg-OH

The absence of proline in position 7 abolished the activity, although substitution with hydroxyproline in position 3 accounts for a fraction of the peptide's natural activity. The identity of pure BK as a nonapeptide was thus obtained by the action of trypsin upon the plasma globulins (Boissonnas *et al.*, 1960b,c). Purification of BK was reported by Zuber and Jaques (1960) from material obtained by the action of the venom of *B. jararaca* on plasma globulins.

Full characterization of BK, performed in a variety of tissue preparations, showed that the biological actions of both crude and pure material were comparable. Thus, Elliott *et al.* (1959b, 1960) showed that purified natural BK stimulated smooth muscle, produced vasodilation, increased capillary permeability and evoked pain. In the 1960s it was also possible to compare the actions of pure natural BK with those of synthetic BK, produced by Sandoz and Parke-Davis Laboratories. Lewis (1960) studied the effects of pure and synthetic BK on 11 different biological preparations. Elucidation of the structure of BK was decisive in the investigation of the kinin system. Table 1.1 summarizes the pioneering events on the discovery of BK.

The first syntheses of KD were accomplished by Pless *et al.* (1962) after it had been ascertained that it was BK with a lysine residue at the amino end (Werle *et al.*, 1961). (Editor's note: For consistency, "KD" will be referred to as "Lys-BK" throughout this volume). The two major kinins, BK and Lys-BK, were thus identified as the effectors of the system, being chemically closely related and comparable in their physiological and pharmacological actions.

At that time, it was also revealed that, apart from the release by trypsin or snake venom, BK could be formed

Table 1.1   Pioneering events in the history of bradykinin (BK). From discovery to purification

Year	Event	Authors
1949	Release of BK by snake venom. Association of BK with autopharmacology	Rocha e Silva *et al.*
1950	First attempts to purify BK using organic solvents and chromatography	Prado *et al.*
1956	Partial purification of BK using ion-exchange chromatography on Amberlite IRC-50	Andrade and Rocha e Silva
1959	Isolation and purification of BK identified as an octapeptide	Elliot *et al.*
1960	Identification of BK as the nonapeptide, Arg-Pro-Pro-Gly-Phe-Ser-Pro-Phe-Arg	Boisonnas *et al.*

Table 1.2   Pioneering events in the history of kallidin. From discovery to purification

Year	Event	Authors
1926	Identification of a hypotensive substance in human urine (F substance)	Frey
1928	Discovery of a kallikrein inactivator, "serum inactivator"	Kraut *et al.*
1930	Substance F designated as kallikrein	Kraut *et al.*
1937	Identification of Substance DK (*darmkontrahierender stoff*)	Werle *et al.*
1937	Substance DK designated as kallidin	Werle *et al.*
1961	Identification of kallidin's structure as Lys-Arg-Pro-Pro-Gly-Phe-Ser-Pro-Phe-Arg	Werle *et al.*
1962	Synthesis of kallidin	Pless *et al.*

by contact activation (Armstrong *et al.*, 1954), by the action of human (Pierce and Webster, 1961) or bovine (Werle *et al.*, 1961) kallikrein on the homologous plasma fraction, or by spontaneous generation from human and ox plasma (Hamberg *et al.*, 1961; Elliott *et al.*, 1963).

Many analogs and BK derivatives have since been synthesized, and their pharmacological and physiological actions tested. Table 1.2 summarizes the pioneering events on the discovery of Lys-BK.

4.   The Bradykinin Precursor – Kininogen

From a historical point of view, two independent groups detected kinin activity only in combination with the kinin precursor. Kallikrein by itself failed to cause contraction of isolated smooth muscles, and the addition of serum was necessary (Werle *et al.*, 1937). The second observation was made by Rocha e Silva *et al.* (1949), who described an ileum-stimulating factor in blood after application of snake venom *in vivo* or *in vitro*. The name "bradykininogen" given to the precursor of BK, was inspired by the nomenclature proposed by Braun-Menendez for the precursor of hypertensin and hypertensinogen (i.e. angiotensin/angiotensinogen) (Rocha e Silva, 1964).

Localization of the precise source of BK or its precursor, bradykininogen, in plasma and other biological fluids was of obvious importance in understanding the physiological role of this peptide in health and disease. The principal difficulties encountered in developing methods for determining kininogen were the elimination of trypsin inhibitors, which interfere with the release of BK, and also the presence of enzymes, which inactivate it in the plasma (Diniz *et al.*, 1961). These difficulties were overcome by heating the plasma in boiling water in the presence of dilute acetic acid. This procedure yielded a denatured substrate with the trypsin inhibitors and bradykininolytic enzymes inactivated. A satisfactory method based on the above procedure was developed by Diniz and Carvalho (1963). Numerous modifications of this method were published afterwards leading to the

identification of two types of kininogen. Habermann (1963) named the monomer and polymer of bovine kininogen as I and II, respectively. Highly purified kininogens of various species with a low molecular weight (50,000), the low molecular weight kininogen (LMWK), and higher molecular weight kininogen (HMWK) were soon discovered. These molecules are the multifunctional proteins which release endogenous kinin upon cleavage by kallikrein (Müller-Esterl, 1989).

5.   Kinin Inactivators – Kininases

Rapid disappearance of endogenous BK soon after its release was noted simultaneously with its discovery. The peptide was initially assumed to be destroyed by the same trypsin that had previously been used to liberate BK from the dog plasma globulin (Rocha e Silva *et al.*, 1949). Subsequently, however, it was found that liberation and inactivation of BK were independent processes. The term "kininase" was coined to define the enzymes that inactivate kinins. The sources of kininases ranged from blood to microorganisms (Erdös and Yang, 1970). The mechanism of inactivation of kinins in blood was began to be characterized after the availability of carboxypeptidase N (kininase I), isolated from human plasma by Erdös (1961), which cleaves the carboxyl terminal Arg from many peptides. In the process, Yang and Erdös (1967) also isolated kininase II, which was shown to inactivate BK by cleaving the link between Pro^7 and Phe^8, removing the dipeptide, Phe-Arg, from the carboxyl terminal end of BK. Kininase II later proved to be identical with the angiotensin I-converting enzyme (ACE; Erdös, 1979). This work led to the development of synthetic inhibitors of kininase II, also known as the ACE inhibitors.

5.1.   KININASE INHIBITORS AND KININ POTENTIATORS

The rapid inactivation of kinins in the circulation prompted the search for endogenous kininase inhibitors. Rocha e Silva and collaborators were the first to demonstrate that 2,3-dimercaptopropanol potentiated the effects of kinin *in vivo* (Ferreira *et al.*, 1962; Ferreira and Rocha e Silva, 1962, Rocha e Silva, 1963). This inhibitor, however, exhibited toxic effects. Erdös and Wohler (1963) showed that several thiol compounds and other agents such as Ca-ethylenediamine tetraacetic EDTA and 8-hydroxy-5-quinoline sulfonic acid potentiated responses to BK in guinea-pig ileum. While examining the ability of these kininase inhibitors to prolong BK action, Ferreira (1965) found that the venom of *B. jararaca* and other snakes had a potent potentiating effect upon BK-induced contractions of guinea-pig isolated ileum, or on hypotension *in vivo*. He demonstrated the presence of a BK-potentiating factor (BPF) in the venom and suggested that the actions of the factor were related to inhibition of enzymatic inactivation of BK (Ferreira, 1965; Ferreira and Vane, 1967). The factor was isolated and synthesized by Stewart *et al.* (1971), and shown not only to inhibit kininases but also to block ACE activity (Bakhle, 1968; Ng and Vane, 1970). This discovery stimulated vigorous work on these peptides, resulting a year later, in the determination of their structures (Ondetti *et al.*, 1971).

In 1977, the Squibb Institute for Medical Research, used BPF to develop captopril and other ACE inhibitors, which act by blocking the conversion of angiotensin I to II. They are widely employed clinically in hypertension and congestive heart failure (Ondetti *et al.*, 1977). A summary of the pioneering work on the components of the system is presented in Table 1.3.

The discovery of the individual components resulted in the elucidation of the kallikrein–kinin system, which is represented diagramatically in Fig. 1.1. Kinins are released from precursors (kininogens) by kinin-forming enzymes (the kininogenases or kallikreins). The kinins (BK and Lys-BK) are rapidly destroyed by enzymes (kininases), which in turn can be inactivated by kininases inhibitors.

6. Discovery of the Physiological and Pathological Roles of Kinins

Some of the major physiological roles of kinins were described in parallel with their discovery. A valuable clue was found by Frey (1926) when he injected human urine into dogs while searching for the substance inducing anuria in surgery patients. Urine injection caused a dose-dependent decrease in blood pressure, and Frey claimed the discovery of a new substance having a physiological action in the heart (Frey and Kraut, 1926; Kraut *et al.*, 1928). Werle (1936) demonstrated that kallikrein preparations from human urine contracted dog isolated intestine and lowered blood pressure. Rocha e Silva and Beraldo (1949) and Rocha e Silva *et al.* (1949) reported the existence of BK as an autopharmacologic agent, released from the pseudoglobulin fraction of normal plasma by snake venom. BK had been shown to stimulate smooth muscle of the guinea-pig ileum and uterus, or rabbit duodenum, and to produce hypotension when injected intravenously in dogs. Confirming pioneering studies with natural BK, Elliott *et al.* (1959b, 1960) demonstrated activity of the pure peptide on rats' uterus and cat blood pressure, and vascular permeability. The

Table 1.3   Pioneering events in identification of the components of the kallikrein–kinin system

Year	Event	Authors
Kininogen		
1937	Discovery of kallidinogen	Werle *et al.*
1949	Discovery of bradykininogen	Rocha e Silva *et al.*
1961	Development of a method for determining bradykininogen	Diniz *et al.*
1963	Identification of kininogens I (LMWK[a]) and II (HMWK[a])	Habermann
Kininases		
1961	Kininase I (carboxypeptidase N)	Erdös
1967	Kininase II (angiotensin I-converting enzyme, ACE)	Yang and Erdös
Natural bradykinin potentiators		
1965	Bradykinin-potentiating factor (BPF) in snake venoms	Ferreira
1971	Synthesis of BPF	Stewart *et al.*
Synthetic kininase inhibitors		
1962	2,3-dimercaptopropanol	Ferreira *et al.*
1963	Thiol compounds	Erdös and Wohler
1977	Captopril (first ACE inhibitor)	Ondetti *et al.*
Bradykinin receptor antagonists		
1985	Analogs of [DPhe[7]]-BK	Vavrek and Stewart

[a] Low molecular weight and high molecular weight kininogens.

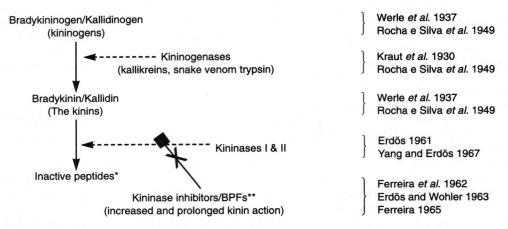

Figure 1.1   The history of the discovery of the kallikrein–kinin system. Kinins are released from precursors (kininogens) by kininogenases (kallikreins). The kinins [BK and KD (Lys-BK)] are rapidly destroyed by enzymes (kininases), which in turn can be inhibited by kininases inhibitors. *The peptides resulting from kininase I degradation of kinins, desArg^9-BK and Lys-desArg^9-BK, are specific agonists for B$_1$ receptors (see Chapter 8). **BPFs, bradykinin-potentiating factors.

bronchoconstrictor action of intravenous (i.v.) BK in guinea-pigs was first described by Collier and colleagues (1960). In experiments in humans, local intra-arterial (i.a.) injection of BK produced an increase in blood flow, and intense vasodilation in the hand and forearm (Fox *et al.*, 1960). Similarly, in a series of studies on kallikrein, the involvement of kinins in mediating physiological vasodilation in glands, and regulating blood flow and electrolyte composition in the kidney was revealed (Hilton and Lewis, 1956; Heidenreich *et al.*, 1964; Webster and Gilmore, 1964). Evidence for the participation of kinins in pregnancy and labor was provided by Martinez *et al.* (1962) and others, who showed formation of plasma kinins during labor, along with a decrease in bradykininogen and increased kininase activity.

In 1959, Chapman and Wolff established that BK-like activity was detected in the spinal fluid of individuals with inflammatory or degenerative diseases of the central nervous system, or in cases of migraine and chronic schizophrenia. The pain-producing actions of BK were first described by Armstrong *et al.* (1952), and Lewis (1964) noted that the kinins elicit the cardinal signs of inflammation: swelling (edema), redness and heat (vasodilatation) and pain. This discovery increased the number of studies on the role of kinins in inflammation, tissue damage, antigen–antibody reactions and shock. Evidence of kinins participation in cell proliferation and tissue regeneration through stimulation of DNA synthesis and mitosis was demonstrated by Rixon and Whitfield (1973). The involvement of the kallikrein–kinin system in angiogenesis has also recently been shown (Ferreira *et al.*, 1992). The biological effects of kinins are mediated by receptors subtypes B$_1$ and B$_2$ (Regoli and Barabé, 1980), and possibly B$_3$ (Farmer *et al.*, 1989). B$_1$ receptors are

Table 1.4   Pioneering events in identification of the physiological and pathological roles of the kallikrein–kinin system

Year	Physiological effect	Authors
1926	Hypotension induced by kallikrein	Frey
1936	Stimulation of smooth muscle by kallikrein	Werle
1949	Hypotension and stimulation of smooth muscle by bradykinin	Rocha e Silva *et al.*
1952	Stimulation of pain	Armstrong *et al.*
1956	Vasodilatation in glands	Hilton and Lewis
1959	Disturbances of the central nervous system	Chapman and Wolff
1959	Increased vascular permeability	Elliot *et al.*
1960	Bronchoconstriction	Collier *et al.*
1962	Increased during pregnancy and labor	Martinez *et al.*
1964	Role in inflammation	Lewis *et al.*
1973	Cell proliferation and tissue regeneration	Rixon and Whitfield

generally thought to be expressed *de novo* in pathological conditions (Marceau *et al.*, 1983), whereas B_2 receptors seem to be involved in mediating most effects of BK and Lys-BK (Regoli and Barabé, 1980). Table 1.4 summarizes the pioneering work on the identification of physiological and pathological roles of the kinin system.

7. From Discovery to Clinical Application

The first member of the kinin family to be employed clinically was aprotinin, a kallikrein inhibitor discovered by Werle (1936). Werle *et al.* (1958) used this drug (as Trasylol from Bayer, AG) for treating acute pancreatitis in humans, as a result of the finding that aprotinin inhibited uncontrolled proteolytic activities. The second major contribution of the kinin research to clinical application was the development, in 1977, of the ACE inhibitor, captopril (from Squibb). The structure of captopril was based on the BK potentiating factor isolated from the venom of *B. jararaca* by Stewart *et al.* (1971). Captopril and other ACE inhibitors are used successfully in the treatment of hypertension. Owing to the involvement of kinins in pain, the search for BK receptor antagonists has been very active for over a decade. Vavrek and Stewart (1985) announced the discovery of the first BK antagonist, [DPhe^7]-BK and others. Since then, more potent, competitive and selective kinin receptor antagonists have been described (Stewart, 1992), and these have made major contributions to our understanding of the kinin system. It is apparent that kinins participate in the regulation of most major physiological systems and that they are the main initiators of many inflammatory responses in the body. Thus, the story that began as a simple laboratory observation, has resulted in major advances of our understanding of inflammation and potential novel therapies for its alleviation.

8. Acknowledgements

The authors are supported by grants from CNPq, PRPQ and FAPEMIG – Brazil.

9. References

Andrade, S.O. and Rocha e Silva, M. (1956). Purification of bradykinin by ion-exchange chromatography. Biochem. J. 64, 701–705.

Armstrong, D., Dry, R.M.L., Keele, C.A. and Markham, J.M. (1952). Pain-producing substances in blister fluid and serum. J. Physiol. 117, 4P.

Armstrong, D., Keele, C.A., Jepson, J.B. and Stewart, J.W. (1954). Development of pain producing substance in human plasma. Nature (Lond.) 174, 791–792.

Bakhle, Y.S. (1968). Conversion of angiotensin I to angiotensin II by cell-free extracts of dog lung. Nature 220, 919–921.

Boissonnas, R.A., Guttmann, S.T. and Jaquenoud, P.A. (1960a). Synthèse de la L-arginyl-L-prolyl-L-prolyl-glycyl-L-phénylalanyl-L-séryl-L-phényl-alanyl-L-arginine. Distinction entre cet octapeptide et la bradykinine. Helv. Chim. Acta 43, 1481–1487.

Boissonnas, R.A., Guttmann, S. and Jaquenoud, P.A. (1960b). Synthèse de la L-arginyl-L-prolyl-L-prolyl-glycyl-L-phénylalanyl-L-séryl-L-prolyl-L-phénylalanyl-L-arginine, un nonapeptide présentant les propriétés de la bradykinine. Helv. Chim. Acta 43,1349–1358.

Boissonnas, R.A., Guttmann, S.T., Jaquenoud, P.A., Konzett, H. and Stümer, E. (1960c). Synthesis and biological activity of peptides related to bradykinin. Experientia 16, 326.

Chapman, L.F. and Wolff, H.G. (1959). Studies of the proteolytic enzymes in cerebrospinal fluid. Arch. Intern. Med. 103, 86–93.

Collier, H.O.J., Holgate, J.A., Schachter, M., Shorley, P.G. (1960). The bronchoconstrictor action of bradykinin in the guinea-pig. Br. J. Pharmacol. 15, 290–297.

Diniz, C.R. and Carvalho, I.F. (1963). A micromethod for determination of bradykininogen under several conditions. Ann. N.Y. Acad. Sci. 104, 77–88.

Diniz, C.R., Carvalho, I.F., Ryan, J. and Rocha e Silva, M. (1961). A micromethod for determination of bradykininogen in blood plasma. Nature (Lond.) 192, 1194–1195.

Elliott, D.F., Lewis, G.P. and Horton, E.W. (1959a). The isolation of bradykinin: a plasma kinin from ox blood. Biochem. J. 74, 15P.

Elliott, D.F., Horton, E.W. and Lewis, G.P. (1959b). Biological activity of pure bradykinin – a plasma kinin from ox blood. J. Physiol. 150, 6P.

Elliott, D.F., Horton, E.W. and Lewis, G.P. (1960). Actions of pure bradykinin. J. Physiol. 153, 473–477.

Elliott, D.F., Lewis, G.P. and Smyth, D.G. (1963). A new kinin from ox blood. Biochem. J. 87, 21P.

Erdös, E.G. (1961). Enzymes that inactivate active polypeptides. Biochem. Pharmacol. 8, 112P.

Erdös, E.G. (1979) In "Handbook of Experimental Pharmacology", Vol. 25 suppl. (ed. E.R. Erdös), pp. 427–448. Springer-Verlag, Heidelberg.

Erdös, E.G. and Wohler, J.R. (1963). Inhibition *in vivo* of the enzymatic inactivation of bradykinin and kallidin. Biochem. Pharmacol. 12, 1193–1199.

Erdös, E.G. and Yang, H.Y.T. (1970). In "Handbook of Experimental Pharmacology" (eds. E.G. Erdös and A.F. Wilde), pp. 288–323. Springer-Verlag, Berlin.

Farmer, S.G., Burch, R.M., Meeker, S.N. and Wilkins, D. (1989). Evidence for a pulmonary B_3 receptor. Mol. Pharmacol. 36, 1–8.

Feldberg, W. and Kellaway, C.H. (1938). Liberation of histamine and formation of lysolecithin-like substances by snake venom. J. Physiol. 94, 187–226.

Ferreira, M.A.N.D., Andrade, S.P., Pesquero, J.L., Feitosa, M.H., Oliveira, G.M.R., Rogana, E., Nogueira, J.C. and Beraldo, W.T. (1992). Kallikrein–kinin system in the angiogenesis. Agents Actions Suppl. 38/II, 165–174.

Ferreira, S.H. (1965). A bradykinin-potentiating factor (BPF) present in the venom of *Bothrops jararaca*. Br. J. Pharmacol. 24, 163–169.

Ferreira, S.H. and Rocha e Silva, M. (1962). Potentiation of bradykinin by dimercaptopropanol (BAL) and other inhibitors of its destroying enzyme in plasma. Biochem. Pharmacol. 11, 1123–1128.

Ferreira, S.H. and Vane, J.R. (1967). The disappearance of bradykinin and eledoisin in the circulation and vascular beds of the cat. Br. J. Pharmacol. 30, 417–421.

Ferreira, S.H., Corrado, A.P. and Rocha e Silva, M. (1962). Potenciação do efeito hipotensor da bradicinina por inibição da enzima bradicininolítica plasmática. Cien. Cult. 14, 238.

Fox, R.H., Goldsmith, R., Kidd, D.J. and Lewis, G.P. (1960). Bradykinin as a vasodilator in man. J. Physiol. 154, 16P.

Frey, E.K. (1926). Zusammenhänge zwischen Herzarbeit und Nierentätigkeit. Arch. Klin. Chir. 142, 663–669.

Frey, E.K. and Kraut, H. (1926). Ein neues Kreislaufhormon und seine Wirkung. Arch. Exp. Pathol. Pharmakol. 133, 1–56.

Habermann, E. (1963). Über pH-bedingte Modifikationen des Kininliefernden α-globulins (Kininogen) aus Rinderserum und das Molekulargewicht von Kininogen I. Biochem. Z. 337, 440–448.

Hamberg, V., Bumpus, F.M. and Page, I.H. (1961). Isolation and amino acid composition of bradykinin released by venom of *Bothrops jararaca* from bovine plasma. Biochim. Biophys. Acta 52, 533–545.

Heidenreich, O., Keller, O. and Kook, Y. (1964). Die Wirkungen von Bradykinin und Kallidin auf die Nierenfunktion des hundes. Arch. Exp. Pathol. Pharmakol. 247, 243–253.

Hilton, S.M. and Lewis, G.P. (1956). The relationship between glandular activity, bradykinin formation and functional vasodilation in the submandibular salivary gland. J. Physiol. 134, 471–483.

Kraut, H., Frey, E.K. and Bauer, E. (1928). Über ein neues kreislaufhormon II. Mitteilung. Hoppe-Seylers Z. Physiol. Chem. 189, 97–106.

Kraut, H., Frey, E. and Werle, E. (1930). Der Nachweis eines Kreslaufhormons in der Pankreasdrüse. Hoppe-Seylers Z. Physiol. Chem. 175, 97–114.

Lewis, G.P. (1960). Natural and synthetic bradykinin. Nature 188, 999P.

Lewis, G.P. (1964). Plasma kinin and other vasoactive compounds in acute inflammation. Ann. N.Y. Acad. Sci. 116, 847–854.

Marceau, F., Lussier, A., Regoli, D. and Giroud, J.P. (1983). Pharmacology of kinins: their relevance to tissue injury and inflammation. Gen. Pharmacol. 14, 209–229.

Martinez, A.R., Carvalho, I.F. and Diniz, C.R. (1962). The bradykininogen content of the plasma from women in labour. J. Obstet Gynaecol. Br. Commonwealth 69, 1011–1014.

Müller-Esterl, W. (1989). Kininogens, kinins and kinships. Thromb. Haemostas. 61, 2–6.

Ng, K.K.F. and Vane, J.R. (1970). Some properties of angiotensin converting enzyme in the lung *in vivo*. Nature 225, 1142–1144.

Ondetti, M.A., Williams, N.J., Sabo, E.F., Pluscec, J., Weaver, E.R. and Kogy, O. (1971). Angiotensin-converting enzyme inhibitors from the venom of *Bothrops jararaca*. Biochemistry 10, 4033–4039.

Ondetti, M.A., Cushman, D.W. and Rubin, B. (1977). Design of specific inhibitors of angiotensin-converting enzyme: a new class of orally-active anti-hypertensive agents. Science 196, 441–444.

Pernow, B. and Rocha e Silva, M. (1955). A comparative study of bradykinin and substance P. Acta Physiol. Scand. 34, 59–66.

Peterson, E.A. and Sober, H.A. (1956). Chromatography of proteins I Cellulose ion-exchange adsorbents. J. Amer. Chem. Soc. 78, 751–755.

Pierce, J.V. and Webster, M.E. (1961). Human plasma kallidins: isolation and chemical studies. Biochem. Biophys. Res. Commun. 5, 353–357.

Pless, J., Stümer, E., Guttmann, S. and Boissonnas, R.A. (1962). Kallidin, Synthese und Eigenschaften. Helv. Chim. Acta 45, 394–396.

Prado, E.S., Beraldo, W.T. and Rocha e Silva, M. (1950). Bradykinin, assay of purification. Arch. Biochem. 27, 410–417.

Regoli, D. and Barabé, J. (1980). Pharmacology of bradykinin and related kinins. Pharmacol. Rev. 32, 1–46.

Rixon, R.H. and Whitfield, J.F. (1973). In "Kininogenases, Kallikrein" (eds. G.L. Haberland and J.W. Rohen), pp. 131–145. F.K. Schattauer-Verlag, Stuttgart.

Rocha e Silva, M. (1960). In "Polypeptides Which Affect Smooth Muscle and Blood Vessels" (ed. M. Schachter), pp. 210–252. Pergamon Press, Oxford.

Rocha e Silva, M. (1963). The physiological significance of bradykinin. Ann. N.Y. Acad. Sci. 104, 190–210.

Rocha e Silva, M. (1964). Angiotensin and bradykinin: a study in contrasts. Can. Med. Assoc. J. 90, 307–311.

Rocha e Silva, M. and Beraldo, W.T. (1949). Um novo princípio autofarmacológico (bradicinina) liberado do plasma sob a ação de venenos de cobra e da tripsina. Cien. Cult. 1, 32–35.

Rocha e Silva, M., Beraldo, W.T. and Rosenfeld, G. (1949). Bradykinin, a hypotensive and smooth muscle stimulating factor released from plasma globulin by snake venoms and by trypsin. Am. J. Physiol. 156, 261–273.

Stewart, J.M. (1992). Bradykinin antagonists: The state of the art. Agents Actions Suppl. 38/I, 546–550.

Stewart, J.M., Ferreira, S.H. and Greene, L.J. (1971). Bradykinin-potentiating peptide Pca-Lys-Tryp-Ala-Pro. Biochem. Pharmacol. 20, 1557–1567.

Vavrek, R.J. and Stewart, J.M. (1985). Competitive antagonists of bradykinin. Peptides 6, 161–164.

Webster, M.E. and Gilmore, J.P. (1964). Influence of kallidin-10 on renal function. Am. J. Physiol. 206, 714–718.

Werle, E. (1936). Über kallikrein aus blut. Biochem. J. 287, 235–261.

Werle, E. and Däumer, J. (1940). Über das Verhalten des kallikreins des Inaktivators aus Lymphdrüsen, der substanz DK, sowie der Histidindecarboxylase bei der katophorese. Biochem. Z. 304, 377–386.

Werle, E., Götze, W. and Keppler, A. (1937). Über die Wirkung des Kallikreins auf den isolierten darm und über eine neue darmkontrahierende Substanz. Biochem. Z. 289, 217–233.

Werle, E., Tauber, K., Hartenbach, W. and Forell, M.M. (1958). Zur Frage der Therapie der Pandreatitis. Munch. Med. Wochenschr. 100, 1265–1267.

Werle, E., Trautschold, I. and Leysath, G. (1961). Isolierung und struktur des kallidin. Hoppe-Seylers Z. Physiol. Chem. 326, 174–176.

Yang, H.Y.T. and Erdös, E.G. (1967). Second kininase in human blood plasma. Nature 215, 1402–1403.

Zuber, H. and Jaques, R. (1960). Isolierung von Bradykinin aus Rinderplasma nach Einwirkung von Schlangengift (*Bothrops jararaca*). Helv. Chim. Acta 43, 1128–1130.

2. The Pharmacology and Immunopharmacology of Kinin Receptors

Judith M. Hall *and* Ian K.M. Morton

1. Introduction

For more than 40 years, it has been appreciated that the kinins play a pivotal role in the pathophysiology of inflammatory disease. It was only in the last decade or so, however, that convincing evidence emerged establishing that most of the actions of kinins are mediated via cell surface receptors. As a consequence of the discovery of potent, subtype-selective receptor antagonists, and through the cloning and expression of bradykinin (BK) receptor genes, it is known that at least two major G protein-coupled receptor subtypes, B_1 and B_2, mediate most of the kinins' actions.

In this chapter, we first discuss the kinin receptor

agonists and antagonists, as well as evidence for receptor subtypes. Second, we discuss the pro-inflammatory actions mediated by kinin receptors and, in particular, critically analyze the evidence for important roles for kinins in inflammation.

2. Kinin Receptor Classification. B_1 and B_2 Receptors

In the late 1970s, the existence of two types of BK receptor, designated B_1 and B_2, was proposed (Regoli and Barabé, 1980). The classification of the subtypes has been confirmed subsequently with selective antagonists, and gene cloning and receptor expression studies. B_1 and B_2 receptors were defined by the potency order of kinin receptor agonists (see Fig. 2.1 for structures) in isolated tissues as shown below:

B_1 receptors:

desArg^9-BK > Tyr(Me)^8-BK > BK

B_2 receptors:

Tyr(Me)^8-BK > BK > desArg^9-BK

Several recent reviews have discussed BK and kinin receptor ligands in detail (Bathon and Proud, 1991; Burch and Kyle, 1992; Farmer and Burch, 1992; Hall, 1992; Gobeil and Regoli, 1994; Marceau, 1995; Regoli et al., 1995; Stewart, 1995).

2.1 B_1 RECEPTORS

B_1 receptors are in some ways atypical, in as much as their cell surface expression is often inducible rather than being constitutive. Further, and in contrast to B_2 receptors, B_1 receptors have been demonstrated only in certain species. The induction of B_1 receptors in immunopathology is discussed in Section 3.2, and in Chapter 8 of this volume. As mentioned, B_1 receptors were originally defined in terms of agonist rank order of potencies, desArg^9-BK being the most potent relative to BK and this, along with an appreciable affinity for an antagonist, desArg^9-[Leu^8]-BK, was taken to demonstrate B_1 receptors in various cells and tissues. The original studies were conducted in rabbit isolated aorta (Regoli and Barabé, 1980), although B_1 receptors have since been confirmed in numerous preparations of the cardiovascular system (Regoli and Barabé, 1980; Churchill and Ward, 1986; Chahine et al., 1993; Campos and Calixto, 1994; DeWitt et al., 1994; Pruneau et al., 1995a), the urinary (Marceau et al., 1980; Butt et al., 1995) and intestinal tracts (Couture et al., 1981, 1982; Brown et al., 1992; Rhaleb and Carretero, 1994). Further, B_1 receptor-mediated effects have been reported in various cultured vascular cells (Cahill et al.,

1988; Issandou and Darbon, 1991; Wiemer and Wirth, 1992; Tropea et al., 1993; Schneck et al., 1994), embryonic calvarium bones (Ljunggren and Lerner, 1990), tracheal smooth muscle (Marsh and Hill, 1994) and fibroblasts (Goldstein and Wall, 1984; Marceau and Tremblay, 1986). In studies in vivo, B_1 receptors have been implicated to mediate effects of kinins on blood pressure (Marceau et al., 1980; Regoli et al., 1981; Nakhostine et al., 1993; Tokumasu et al., 1995), in persistent hyperalgesia (Perkins et al., 1993; see Chapter 9), and in plasma extravasation (Cruwys et al., 1994).

B_1 receptors are G protein-coupled and signal via phosphatidylinositol (PI) hydrolysis in smooth muscle (Schneck et al., 1994; Butt et al., 1995). The characteristics of B_1 receptors and their ligands, together with recent radioligand binding and molecular biology studies, are described below. A detailed analysis of the pharmacology of B_1 receptors has recently been published (Marceau, 1995), and their regulation of expression and induction are covered in Chapter 8.

2.1.1 Selective Agonists at B_1 Receptors

In contrast to most other peptide mediators, the study of BK receptors has been hampered by the lack of subtype-selective agonist ligands. In the case of B_1 receptors, the first selective agonist described was desArg^9-BK, in which the C-terminal arginine of BK is deleted (Regoli et al., 1977; Fig. 2.1). DesArg^9-BK has proven to be invaluable in the study of B_1 receptors, and no other agonist has been described with improved selectivity or potency (**Editor's note.** Recent studies indicate that the most potent agonist at human B_1 receptors is Lys-desArg^9-BK, i.e. desArg^{10}-kallidin; see Chapter 3). In isolated tissues, desArg^9-BK has a potency around ten-fold greater than BK, but is over 100-fold less active than BK in B_2 receptor-containing preparations (Regoli et al., 1977; Drouin et al., 1979a). Both desArg^9-BK and Lys-desArg^9-BK, can be formed in normal blood and serum in the circulation as a result of enzymatic cleavage of BK or Lys-BK (kallidin) by carboxypeptidase enzymes, including kininase I (Erdös and Sloane, 1962; Marceau, 1995). Indeed, effects of BK in tissues expressing B_1 receptors are due almost entirely to its conversion by endogenous peptidases to desArg^9-BK (Babiuk et al., 1982; Regoli et al., 1986), and this has led to speculation that this peptide may be the physiological ligand for B_1 receptors (Drapeau et al., 1991). Importantly, desArg^9-BK production can be increased in coagulating blood (Marceau, 1995), and this finding is particularly relevant to the role of kinin receptors in inflammation, as addressed in Section 3.2. Other kinin analogs including Lys-desArg^9-BK (Gaudreau et al., 1981b; Regoli et al., 1990a) and Lys-[Hyp^3]-BK ([Hyp^4]-kallidin; Rhaleb et al., 1990) have been proposed to show a degree of selectivity for B_1 receptors, although the pharmacology of these agents has not been documented in any great detail. Sar-[DPhe^8,desArg^9]-BK is

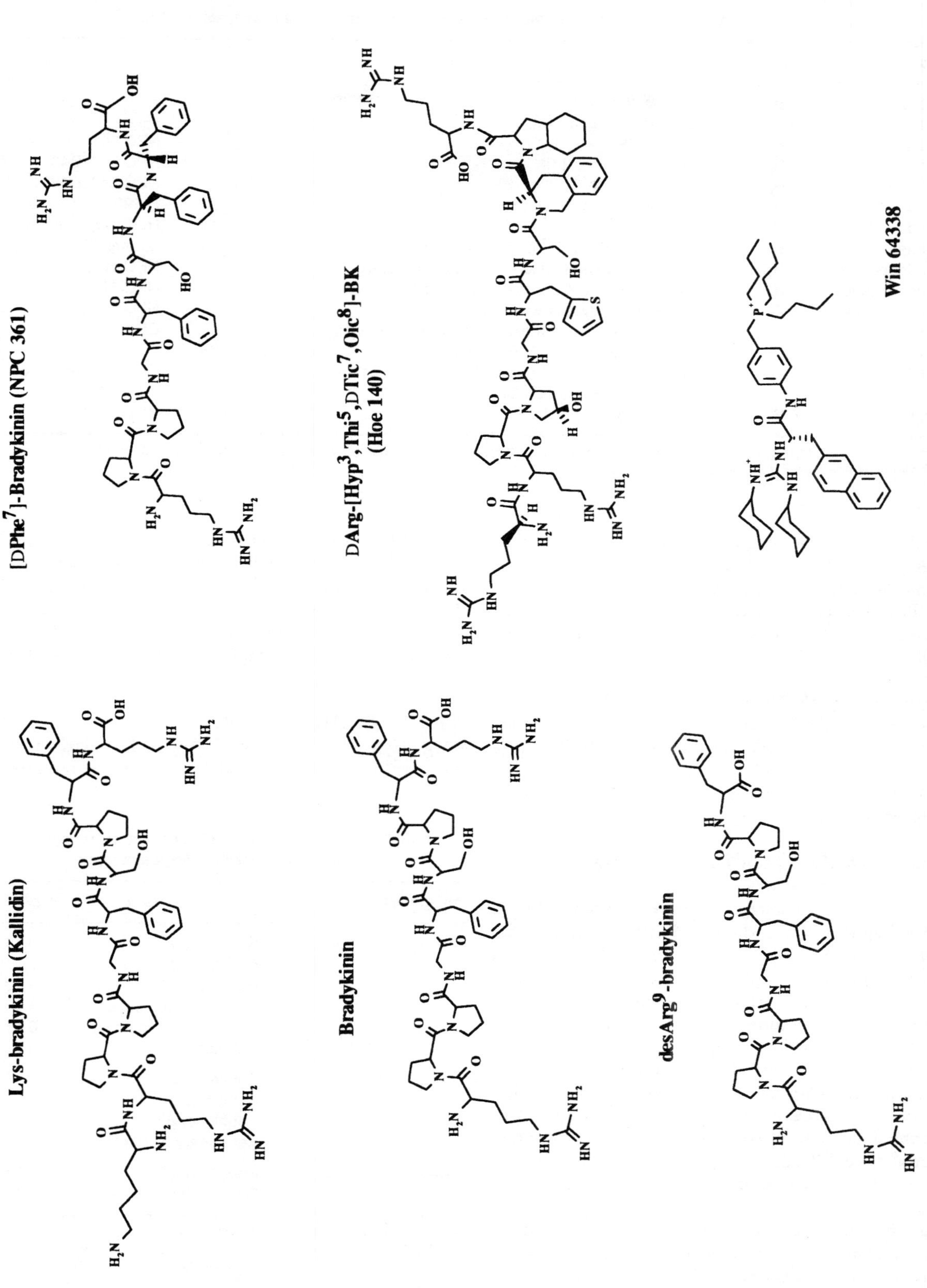

[DPhe⁷]-Bradykinin (NPC 361)

DArg-[Hyp³,Thi⁵,DTic⁷,Oic⁸]-BK (Hoe 140)

Win 64338

Lys-bradykinin (Kallidin)

Bradykinin

desArg⁹-bradykinin

Figure 2.1   Structures of some kinin receptor agonists (left) and antagonists (right).

Table 2.1   Sequences and pseudonyms of typical antagonists of kinin receptor subtypes[a]

Type	Subtype	Antagonist	Pseudonym(s)
Peptide	B_1	desArg^9-[Leu^8]-BK	—
Peptide	B_1	Lys-desArg^9-[Leu^8]-BK	—
Peptide	B_2	[DPhe^7]-BK	NPC 361
Peptide	B_2	DArg-[Hyp^3,DPhe^7]-BK	NPC 567, B4801
Peptide	B_2	DArg-[Hyp^3,Thi^{5,8},DPhe^7]-BK	NPC 349, B4881
Peptide	B_2	DArg-[Hyp^3,DPhe^7,Leu^8]-BK	R-493
Peptide	B_2	DArg-[Hyp^3,Thi^5,DTic^7,Oic^8]-BK	Hoe 140, Icatibant
Peptide	B_1	DArg-[Hyp^3,Thi^5,DTic^7,Oic^8,desArg^8]-BK	desArg^{10}-[Hoe 140]
Pseudopeptide	B_2	DArg,Arg-(aminotridecanoyl)-Ser,DTic,Oic,Arg	NPC 18325
Pseudopeptide	B_2	DArg-[Hyp^3,DHypE(*trans* propyl)^7,Oic^8]-BK	NPC 17731
Pseudopeptide	B_2	DArg-[Hyp^3,DHypS(*trans* phenyl)^7,Oic^8]-BK	NPC 17761
Peptide dimer	B_1/B_2	1,6-*Bis* (succinimido)hexane heterodimer of DArg-[Hyp^3,Cys^6,DPhe^7,Leu^8]-BK and DArg-[Cys^1,Hyp^3,Leu^8,desArg^9]-BK	CP-0364
Peptide dimer	B_2	*Bis* succinimidoalkane (LCys^6) homodimer of DArg-[Hyp^3,Cys^6,DPhe^7,Leu^8]-BK	CP-0127, Bradycor
Nonpeptide	B_2	Win 64338[a]	Win 64338[a]

[a] Phosphonium, [[4-[[2-[[*bis*(cyclohexylamino)methylene]amino]-3-(2-naphthalenyl)-1-oxopropyl]-amino]-phenyl]-methyl]-tributyl, chloride, monohydrochloride (see Fig. 2.1 for antagonist structures).

metabolically stable and is ten-fold more potent than desArg^9-BK (Drapeau *et al.*, 1991, 1993). A weak B_1 receptor-selective agonist, "ovakinin" (Phe-Arg-Ala-Asp-His-Pro-Phe-Leu), derived from synthetic proteolytic cleavage of ovalbumin is of some interest (Fujita *et al.*, 1995).

2.1.2   Selective Antagonists at B_1 Receptors

To date, all of the B_1 receptor antagonists described are peptides which were developed through substitution of the amino acids in desArg^9-BK. Table 2.1 shows some structures and pseudonyms of B_1 receptor (and B_2 receptor) antagonists.

2.1.2.1   *DesArg^9-[Leu^8]-BK and Related Antagonists*

The first B_1 receptor antagonists were reported by Regoli and colleagues (1977), who found that substitution with aliphatic amino acids in place of the phenylalanine at position eight in the sequence of desArg^9-BK resulted in the formation of B_1 receptor-selective antagonists in rabbit aorta. Substitution with leucine at position 8, resulted in a competitive antagonist with a relatively high affinity (Regoli *et al.*, 1977; Drouin *et al.*, 1979a,b). In most studies, desArg^9-[Leu^8]-BK is selective (see Regoli *et al.*, 1977, 1990a) and is still the most widely utilized B_1 receptor antagonist. A recent report, however, noted that desArg^9-[Leu^8]-BK inhibited contractions of rabbit aorta evoked by angiotensin, and bound with relatively high affinity to angiotensin AT_1 and AT_2 receptors in radio-ligand-binding studies (Pruneau *et al.*, 1995a). This is interesting in light of the observation that the B_1 receptor shares as much sequence homology with the angiotensin AT_1 receptor as it does with the kinin B_2 receptor (Menke

et al., 1994). Several other antagonists based on desArg^9-[Leu^8]-BK, such as Lys-desArg^9-[Leu^8]-BK, have been reported to have about ten-fold higher affinity and longer duration of action (Drouin *et al.*, 1979a,b,c; see Regoli *et al.*, 1990a). More recently, peptidase-resistant, acetylated peptide B_1 receptor antagonists have been described (Drapeau *et al.*, 1993).

2.1.2.2   *DesArg^{10}-Hoe 140*

DArg-[Hyp^3,Thi^5,DTic^7,Oic^8,desArg^9]-BK (referred to as "desArg^{10}-Hoe 140") is a modification of the B_2 receptor-

Table 2.2   Affinity estimates[a] for some B_1 receptor antagonists in pharmacological assays

Antagonist	Tissue	pA_2
DesArg^9-[Leu^8]-BK	Rabbit aorta[7]	6.75
	Rabbit aorta[5]	7.27
	Rabbit aorta[1]	6.33
	Rabbit mesenteric artery[3]	6.46
	Rabbit urinary bladder[1]	6.06
	Rat portal vein[2]	6.7
Lys-desArg^9-[Leu^8]-BK	Rabbit aorta[5]	8.37
	Human colon[4]	8.2
	Rabbit aorta[8]	7.6
	Rat urinary bladder[6]	8.1
DesArg^{10}-[Hoe 140]	Rabbit aorta[8]	7.57
	Rabbit urinary bladder[1]	7.18

[a]Affinity estimates, in terms of pA_2 values, are shown only where desArg^9-BK was used as the agonist. References: [1]Butt *et al.* (1995); [2]Campos and Calixto (1994); [3]Churchill and Ward (1986); [4]Couture *et al.* (1981); [5]Drouin *et al.* (1979a); [6]Marceau *et al.* (1980); [7]Regoli *et al.* (1977); [8]Rhaleb *et al.* (1992a).

selective antagonist, DArg-[Hyp^3,Thi^5,DTic^7,Oic^8]-BK (Hoe 140; Wirth *et al.*, 1991a, 1992b; Section 2.2.2.2). The IC$_{50}$ of desArg^{10}-Hoe 140 at B$_1$ receptors in rabbit isolated aorta (12 nM) is approximately ten-fold lower than that for desArg^9-[Leu^8]-BK. In contrast, desArg^{10}-Hoe 140 was 1,000-fold less potent than Hoe 140 when tested in B$_2$ receptor preparations (Wirth *et al.*, 1991a). As with Hoe 140 (Hock *et al.*, 1991; Wirth *et al.*, 1991b), desArg^{10}-Hoe 140 has previously unparalleled metabolic stability. Both the high affinity and metabolic stability have been attributed to the additional basic N-terminal amino acids (Wirth *et al.*, 1991a). Affinity estimates obtained in isolated tissue preparations for this antagonist are shown in Table 2.2.

Rhaleb *et al.* (1992a) recently reported that desArg^{10}-Hoe 140 inhibits B$_2$ receptor-mediated contractions of the guinea-pig ileum and rabbit jugular vein (pA$_2$s, respectively, of 6.1 and 8.3). Nevertheless, neither of these preparations respond to desArg^9-BK and the antagonist was tested against BK, a very poor B$_1$ receptor agonist. The data are, therefore, difficult to interpret in terms of receptor subtype selectivity.

2.1.2.3 Bis*succiminidoalkane Peptide Dimers at B$_1$ receptors*

These compounds are produced by substituting cysteine residues into the sequence of receptor antagonists, followed by formation of dimers through the use of *bis*maleimidoalkane linkers (Cheronis *et al.*, 1992a,b; Whalley *et al.*, 1992; see Section 2.2.2.3). B$_1$ receptor antagonists of this type include a heterodimer composed of desArg^9-[Leu^8]-BK, cross-linked with 1,6-Cys-Cys-*bis*(succinimido)hexane to a B$_2$ receptor antagonist, DArg-[Hyp^3,DPhe^7,Leu^8]-BK to yield CP-0364. This compound antagonizes both B$_1$ and B$_2$ receptor-mediated effects *in vitro* and *in vivo* (Table 2.1).

CP-0364 has an IC$_{50}$ of 7.5 in rabbit isolated aorta (Whalley *et al.*, 1992).

2.1.3 Radioligand-binding Studies with B$_1$ Receptors

Some characteristics of B$_1$ receptor binding sites are shown in Table 2.3. Specific binding of [^{3}H]-desArg^9-BK to B$_1$ receptors was first described in rabbit mesenteric vein by Barabé *et al.* (1982). Binding was displaced with the rank order,

$$\text{Lys-BK} > \text{desArg}^9\text{-[Leu}^8\text{]-BK} > \text{desArg}^9\text{-BK} > \text{BK} > [\text{Tyr(Me)}^8]\text{-BK},$$

and unlabeled desArg^9-BK was one order of magnitude more potent in displacing [^{3}H]-desArg^9-BK than the B$_2$-selective agonist [Tyr(Me)^8]-BK (Section 2.2.1). This profile is compatible with B$_1$ receptor pharmacology. More recently, [^{3}H]-desArg^9-BK was utilized to label rat mesangial cell B$_1$ receptors, which also express B$_2$ receptors (Bascands *et al.*, 1993). Scatchard analysis revealed a single class of B$_1$ receptor sites with an affinity of 8.7 nM, and specific binding was displaced by B$_1$, but not B$_2$ receptor antagonists.

More recent binding studies have utilized [^{3}H]-Lys-desArg^9-BK and [^{125}I]-Tyr,Gly,Lys,Aca,Lys-desArg^9-BK (Aca is amino-caproic acid) to label B$_1$ receptors (Table 2.3). Similarly, Lys-([^{3}H]-Pro^1),desArg^9-BK ([^{3}H]-desArg^{10}-kallidin) labels B$_1$ receptors in RAW264.7 cells (Burch and Kyle, 1992), exhibiting a K_d of 2.4 nM, and specific binding was displaced by several B$_1$ receptor ligands, but not by B$_2$ receptor ligands such as NPC 567 (structure in Table 2.1; Burch and Kyle, 1992). More recently, [^{3}H]-Lys-desArg^9-BK was used to label B$_1$ receptors cloned from human lung fibroblasts (Menke *et al.*, 1994), and rabbit vascular smooth muscle (Galizzi *et al.*, 1994; Schneck *et al.*, 1994). The presence of B$_1$ receptors was concluded from the rank order of potencies

---

Table 2.3   Identification and characteristics of kinin B$_1$ receptor binding sites

Species/tissue	Ligand	Affinity (nM)[a]	Density
Rabbit			
Aorta smooth muscle[6]	[^{3}H]-Lys-desArg^9-BK	0.3	680 sites/cell
Aorta smooth muscle[4]	[^{125}I]-Tyr,Gly,Lys,Aca,Lys-desArg^9-BK	0.2	660 sites/cell
Mesenteric artery[3]	[^{3}H]-Lys-desArg^9-BK	0.4	20,000 sites/cell
Mouse			
RAW264.7 macrophages[2]	[^{3}H]-Lys-desArg^9-BK	2.4	700 sites/cell
Rat			
Mesangial cells[1]	[^{3}H]-desArg^9-BK	8.7	15 fmol/mg
Human			
Transfected receptor[5]	[^{3}H]-Lys-desArg^9-BK	0.4	100 fmol/mg

[a] Affinity of ligand. References: [1]Bascands *et al.* (1993); [2]Burch and Kyle (1992); [3]Galizzi *et al.* (1994); [4]Levesque *et al.* (1995); [5]Menke *et al.* (1994); [6]Schneck *et al.* (1994).

for inhibition of binding by kinin analogs, and the observation that B_1 receptor antagonists potently displaced binding.

A B_1 receptor binding assay has also been developed utilizing the extended analog $[^{125}I]$-Tyr,Gly,Lys,Aca,Lys-desArg^9-BK (affinity, 0.2 nM; Levesque et al., 1995), and has been used to label a single B_1 receptor binding site in rabbit aorta smooth muscle cells. BK itself appears to have virtually no affinity at this B_1 site. Moreover, in experiments showing displacement of specific B_1 receptor radiolabels, it is necessary to incubate for prolonged periods or high temperatures (see Marceau, 1995, for discussion). Binding experiments allowed confirmation of upregulation of B_1 receptor sites in rabbit aorta by cytokines or growth factors (Schneck et al., 1994). In rabbit mesenteric artery smooth muscle cells, pre-exposed to interleukin 1 (IL-1), there was a five-fold increase of B_{max} with no change in affinity (Galizzi et al., 1994).

2.1.4   Molecular Biology Studies with B_1 Receptors

An account of the molecular pharmacology of kinin receptors can be found in Chapter 3 of this volume, so some aspects only are touched on here. In 1992, co-expression of B_1 receptors with B_2 receptors in Xenopus oocytes was revealed using mRNA from a human fibroblast cell line (Phillips et al., 1992). Subsequent studies demonstrated that the receptor cloned from this source is encoded by mRNA distinct from that encoding the B_2 receptor, being 2 kilobases (kb) shorter and differing in sequence (Webb et al., 1994). The human B_1 receptor gene was also cloned from human embryonic lung fibroblasts (Menke et al., 1994), and had the structure typical of seven transmembrane-domained G protein-coupled receptors. This receptor of 353 amino acids had three potential sites for N-glycosylation, and showed an overall amino-acid sequence 36% identical to the cloned B_2 receptor. In binding studies, the expressed receptor exhibited typical B_1 receptor pharmacology (Menke et al., 1994). The mouse BK receptor gene has also been cloned, and the expressed receptor comprised two populations, one with pharmacology of B_1 receptors, and the second, larger population having properties of B_2 receptors (McIntyre et al., 1993). Recently, the cloning and characterization of a rabbit B_1 receptor was described (MacNeil et al., 1995), and is discussed in Chapter 3 of this volume.

2.2 B_2 Receptors

As noted earlier, B_2 receptors were defined from the agonist rank order of potency, [Tyr(Me)^8]-BK > BK > desArg^9-BK. It was not until the mid-1980s that competitive antagonists became available, paving the way for investigating the role of B_2 receptors in pathophysiology.

Peptide and nonpeptide antagonists, along with application of molecular biology, have led to an explosive growth in our knowledge of B_2 receptors. B_2 receptors have a ubiquitous distribution and have been identified in most species. B_2 receptor-mediated effects have been demonstrated in the gastrointestinal tract (Couture et al., 1982; Hall and Morton, 1991; Field et al., 1992a), genitourinary system (Donoso and Huidobro-Toro, 1989; Maggi et al., 1989; Butt et al., 1995), respiratory tract (Rajakulasingam et al., 1991; Field et al., 1992a), the eye (Everett et al., 1992), in neuronal tissue (Fujiwara et al., 1988; Sharif and Whiting, 1991; Babbedge et al., 1995), and in the cardiovascular system (Gaudreau et al., 1981a). In binding studies, specific B_2 receptor sites have been identified in most tissues (Section 2.2.3). Further, the majority of the effects of kinins in vivo have been attributed to B_2 receptors. These include bronchoconstriction (Lembeck et al., 1991; Wirth et al., 1991b), hypotension (Benetos et al., 1986), and edema formation (Schachter et al., 1987; Sakamoto et al., 1992). In some instances, B_2 receptors in the CNS have been proposed to mediate effects of systemically administered kinins (Walker et al., 1995; Hall and Geppetti, 1995). The regulation of B_2 receptor expression, internalization and signal transduction are covered in Chapter 6.

2.2.1   Selective Agonists at B_2 Receptors

The search for B_2 receptor-selective agonists has not been fruitful, although [Tyr(Me)^8]-BK, which was used in classifying B_2 receptors, was reported originally to be a relatively selective, full agonist for B_2 receptors (Barabé et al., 1977; Gaudreau et al., 1981a,b). This analog displaces B_2 receptor binding (e.g. Emond et al., 1990), and is inactive at B_1 receptors (Gaudreau et al., 1981a). Another B_2 receptor-selective agonist, [Hyp^3,Tyr(Me)^8]-BK, is inactive at B_1 receptors in the rabbit aorta, but equally active or more potent than BK in several B_2 receptor preparations (Rhaleb et al., 1990; Regoli et al., 1991) including human bronchus (Molimard et al., 1994). Several other, less widely characterized agonists with apparent B_2 receptor selectivity have been developed, and include reduced bond "pseudopeptide" (Ψ) analogs (Rhaleb et al., 1990; Vavrek et al., 1992) and peptides based on Arg-Pro-Hyp-Gly-Ser-Pro-4-(Me)-Tyr-(CH$_2$NH)-Arg (RMP-7; Doctrow et al., 1994).

2.2.2   Selective Antagonists at B_2 Receptors

Table 2.1 and Fig. 2.1 show the structures and pseudonyms of some B_2 receptor antagonists. The first antagonists were not particularly potent, but the development of analogs containing natural amino acid residues led to the synthesis of more potent, "second-generation" antagonists, such as Hoe 140 (Hock et al., 1991; Wirth et al., 1991b), and hybrid antagonists constructed from bissuccinimidoalkane cross-linked dimers (Cheronis et al., 1992a,b). More recently, nonpeptide B_2 receptor antagonists have been disclosed, notably Win 64338 (Salvino *et al.*, 1993; Fig. 2.1). With the advent of stable and selective antagonists, the role of B_2 receptors in pathophysiology began to be explored, and examples of antagonists are currently in some stage of clinical trials.

2.2.2.1 [DPhe⁷]-Bradykinin Analogs

Since detailed accounts of the development of these agents are in earlier publications (Stewart, 1979; Vavrek and Stewart, 1985) and more recent reviews (Steranka *et al.*, 1989; Burch *et al.*, 1990; Bathon and Proud, 1991; Stewart and Vavrek, 1991; Farmer and Burch, 1992; Hall, 1992), they are not discussed in detail here. Structural conformations of B_2 receptor antagonists are discussed by Kyle (1994) and Kyle *et al.* (1991a,b).

The first competitive antagonists of B_2 receptors were BK analogs (Vavrek and Stewart, 1985). The key modification was the replacement of proline at position seven with D-phenylalanine (DPhe), and the prototype, [DPhe⁷]-BK, proved to be a moderately potent and selective antagonist in guinea-pig ileum (pA_2 ca 5) (Vavrek and Stewart, 1985). [DPhe⁷]-BK represented a milestone in the design of B_2 receptor antagonists, and led to the synthesis of many similar compounds, some of which are shown in Table 2.1. Substitution of (2-thienyl)-alanine (Thi) at position 5 of [DPhe⁷]-BK (Dunn and Stewart, 1971; Claeson *et al.*, 1979) led to [Thi⁵,⁸,DPhe⁷]-BK, an antagonist in guinea-pig ileum (pA_2, 5.2–6.4) and rat uterus (pA_2, 6.4) (Vavrek and Stewart, 1985). Other antagonists in this series include analogs with hydroxyproline residues substituted at positions 2 and/or 3 (see Stewart and Vavrek, 1991). Addition of basic residues such as two lysine residues, or DArg, at the N-terminus was also used to confer protection from degradation (see Ward, 1991). More recently, Regoli and co-workers described modified [DPhe⁷]-BK analogs where leucine has been substituted at position eight of the BK sequence (Regoli *et al.*, 1990a,b, 1991; see Table 2.1). Acetyl-DArg-[Hyp³,DPhe⁷,Leu⁸]-BK is reported to have a reduced tendency to cause histamine release (Section 3.1). B_2 receptor antagonists that have proved to be the most popular are noted in Table 2.1, and the affinities of some of these are shown in Table 2.4.

2.2.2.2 [DTic⁷]-Bradykinin Analogs

The next stage in the development of B_2 receptor antagonists was DArg-[Hyp³,Thi⁵,DTic⁷,Oic⁸]-BK (Hoe 140; Table 2.1; Hock *et al.*, 1991; Wirth *et al.*, 1991b). The residue substituted at position 7, DTic, is D-(1,2,3,4-tetrahydroisoquinolin-2-yl-carboxylic acid), a cyclized form of phenylalanine, and Oic at position 8 is l-[(3aS,7aS)-octahydroindol-2-yl-carboxylic acid], a proline analog. At much the same time as the disclosure of Hoe 140, there were revealed similar compounds including DArg-[Hyp³,Thi⁵,DTic⁷,Tic⁸]-BK (NPC 16731) and DArg-[Hyp³,DHypE(*trans*propyl)⁷,Oic⁸]-BK (NPC 17731) (see Kyle *et al.*, 1991b, and Table 2.1).

Hoe 140, which is orders of magnitude more potent than earlier B_2 receptor antagonists, has been tested in numerous assays *in vitro* (Hock *et al.*, 1991; Lembeck *et al.*, 1991; Louittit and Coleman, 1993; Perkins *et al.*, 1991; Field *et al.*, 1992a; Rhaleb *et al.*, 1992b; Wiemer and Wirth, 1992; Santiago *et al.*, 1994) and *in vivo* (Bao *et al.*, 1991; Lembeck *et al.*, 1991; Wirth *et al.*, 1991b; Damas and Remacle-Volon, 1992; Linz and Schölkens, 1992; Madeddu *et al.*, 1992; Sakamoto *et al.*, 1992). Affinity estimates for Hoe 140, including results obtained in human tissues, are summarized in Table 2.4. The pharmacology of the LTic⁷ stereoisomer of Hoe 140 was recently described, exhibiting a 200-fold lower binding affinity than Hoe 140 at B_2 receptors in IMR-90 human fibroblasts (Sawutz *et al.*, 1994b). In studies with Hoe 140, the increased affinity and metabolic stability imparted by the DTic⁷,Oic⁸ substitution was evident, and the long pharmacological duration of action was unprecedented.

Although Hoe 140 has proved to be invaluable in investigating the pharmacology of B_2 receptors and their pathophysiological roles (Section 3), some concern is warranted. This includes the apparently noncompetitive antagonism exhibited by Hoe 140 in some studies, and methodological complications arising from its extremely long duration of its action (Field *et al.*, 1992a; Griesbacher and Lembeck, 1992a; Rhaleb *et al.*, 1992b). However, in view of the very high affinity of Hoe 140 for B_2 receptors, its enhanced stability compared to the [DPhe⁷]-BK series, and its lipophilicity, it is possible that the anomalous kinetics might be explained in terms of failure to reach an equilibrium state in some assays. This has been discussed previously (Hall, 1992; Griesbacher and Lembeck, 1992a). Briefly, if the antagonist dissociates very slowly from the receptor, it would be anticipated that, in normal functional competition experiments, agonist and antagonist would fail to reach equilibrium, and that this departure from simple competitive kinetics would show itself in the form of a depression of maximum responses to agonist, with possible departure of Schild plots from slopes of unity. If this interpretation is correct, then antagonism should be regarded as "pseudo-irreversible competitive" rather than "noncompetitive" (Kenakin, 1993). This point is important, because if the antagonism is truly noncompetitive, experimental estimates of its affinity cannot easily be made. However, if it is a slowly equilibrating competitive antagonist, functional estimates of its affinity are likely to be biased. Nevertheless, there should be no great difficulty in using radioligand methods to estimate its affinity. Indeed, binding studies using radiolabeled Hoe 140 ([¹²⁵I]-PIP-Hoe 140; Section 2.2.3) showed competitive kinetics in guinea-pig ileum (Brenner *et al.*, 1993).

Recently, Hoe 140 was reported to be a relatively potent agonist, both in rabbit jugular vein and sheep femoral artery. In the latter tissue, the effect of Hoe 140 seemed to involve BK receptors, in as much as another B_2

Table 2.4   Affinity estimates for some B_2 receptor antagonists in pharmacological assays

	NPC 567[a]	NPC 349[b]	R-493[c]	Win 64338[d]	Hoe 140[e]	References[f]
Rabbit						
Jugular vein	7.2, 8.0, 8.67	7.9	8.86	5.66, 6.14	9.04, 9.19, 9.9	12, 22, 6; 22; 22; 22, 17; 6, 22, 12
Iris sphincter	8.1	8.5		6.61	10.46	12; 12; 13; 12
Vena cava			8.50	5.43	9.24	22; 22; 22
Colon				5.83		22
Urinary bladder				6.38		22
Rat						
Uterus	6.8, 7.2	6.9			9.7, 9.7	1, 11; 1; 11, 19
Urinary bladder			6.96	5.41	8.51	22; 22; 22
Vas deferens	6.11	6.46	6.87	5.33	8.47	21; 21; 21; 22; 21
Duodenum	7.3	7.3			10.21	12; 11; 14
Portal vein			6.55	5.24	8.8, 9.3	22; 22; 22, 3
Stomach			6.98	5.46	8.05	22; 22; 22
Human						
IMR-90 fibroblasts (^{45}Ca efflux)[g]				7.1		25
Ileum					8.36	23
Bladder			7.15		8.81	23; 23
Stomach				<4.9		22
Umbilical vein				5.99	8.00, 8.2	17, 7; 17
Umbilical artery					8.16	7
Pulmonary artery					7.97	7
Bronchus					8.19	7
Guinea-pig						
Ileum: longitudinal muscle	5.41, 5.5, 5.57	5.8, 6.3, 6.34	6.31, 6.77	7.57, 7.79, 7.97, 8.2	8.42, 8.43, 8.6, 8.74, 8.8, 8.94	21, 1, 20; 1, 2, 21; 20, 21; 20, 22, 5, 25; 15, 20, 10, 18, 19, 21
Taenia caeci (functional)	5.89	5.81			8.42	8; 8; 8
Taenia caeci (PI turnover)[h]	5.4				8.4	9; 9
Trachea	5.58, 5.94	5.87	6.02	6.43, 7.36	7.42, 8.13, 8.94	20, 8; 8; 20; 26, 20; 23, 20, 8
Pulmonary artery			6.12	7.1	8.44	22; 22; 22
Stomach			6.53		8.81	22; 22
Colon			6.60	7.43	8.32	22; 22; 22
Gall bladder	5.05				8.54	4; 4
Lung parenchyma	5.55		6.06	7.51	8.52	20; 20; 20; 20
Jugular vein	5.65		5.81	6.89	8.05	20; 20; 20; 20
Urinary bladder			6.42	7.55	8.50, 8.79	22; 22; 14, 22
Mouse						
Superior cervical ganglion[i]					9.65	27
Vas deferens (nerve)					9.65	16
Hamster						
Bladder	7.11	6.46	7.16		8.81	21; 21; 21; 21
Dog						
Carotid artery	7.93	7.86	8.29		9.42	21; 21; 21; 21
Pig						
Iris sphincter					8.40	14

[a] NPC567 is DArg-[Hyp³,DPhe⁷]BK.
[b] NPC349 is DArg-[Hyp³,Thi⁵‚⁸,DPhe⁷]BK.
[c] R-493 is DArg-[Hyp³,DPhe⁷,Leu⁸]BK; (Rhaleb et al., 1992c)
[d] Win 64338 is phosphonium, [[4-[[2-[[bis(cyclohexylamino)methylene]amino]-3-(2-naphthalenyl)-1-oxopropyl]amino]phenyl]methyl]tributyl, chloride, monohydrochloride.
[e] Hoe 140 is DArg-[Hyp³,Thi⁵,DTic⁷,Oic⁸]-BK.
[f] References: 1, Birch et al. (1991); 2, Braas et al. (1988); 3, Campos and Calixto (1994); 4, Falcone et al. (1993); 5, Farmer and DeSiato (1994); 6, Félétou et al. (1994); 7, Félétou et al. (1995b); 8, Field et al. (1992a); 9, Field et al. (1994); 10, Griesbacher and Lembeck, (1992a); 11, Hall (1992); 12, Hall et al. (1992); 13, Hall et al. (1995); 14, see Fig. 2.3; 15, Hock et al. (1991); 16, Maas et al. (1995); 17, Marceau et al. (1994); 18, Medeiros and Calixto (1993); 19, Perkins et al. (1991); 20, Pruneau et al. (1995b); 21, Regoli et al. (1993); 22, Regoli et al. (1994); 23, Rhaleb et al. (1992c); 24, Rifo et al. (1987); 25, Sawutz et al. (1994a); 26, Scherrer et al. (1995); 27, Seabrook et al. (1995). (Reference numbers refer to each affinity estimate. Different columns are separated by semicolons).
Most responses shown are for contraction of smooth muscle preparations. Other responses measured include: [g] increase in loss of ^{45}Ca; [h] increase in phosphatidylinositol hydrolysis; [i] and electrical depolarization.
Values are shown as pK_B values which, depending on the work cited, include exact pK_B values with tests for competition, apparent pK_B values assuming competition, and pA_2 values from Schild plot intercept estimates.

receptor antagonist, NPC 567, blocked contractile responses to Hoe 140 (Félétou *et al.*, 1994).

Other antagonists that represent elaborations of the [DTic^7]-BK series have recently been developed, and include DArg-Arg-(amino-tridecanoyl)-Ser-DTic-Oic-Arg (NPC 18325; Chakravarty *et al.*, 1995) (see Table 2.1). Some of these pseudopeptides are reported to have a higher affinity at the human, as compared to the guinea-pig B$_2$ receptor (Kyle, 1994; Chakravarty *et al.*, 1995). Also, moderately active B$_2$ receptor antagonists have been derived by part cyclization of the DArg-[DTic^7]-BK motif (Chakravarty *et al.*, 1993; Kyle, 1994). A further variation of Hoe 140, which also has a very prolonged duration of action in several tissues, is *p*-guanidinobenzoyl-[Hyp^3,Thi^5,DTic^7,Tic^8]-BK (S 16118; Félétou *et al.*, 1995a,c).

2.2.2.3 Bissucciminidoalkane Peptide Dimers at B$_2$ Receptors

These compounds result from incorporation of cysteine residues into the sequence of known B$_2$ receptor antagonists, followed by dimerization with hexamethylene *bis*-maleimidoalkanes (BSH), as cross-linkers. The BSH homodimer derived through a 1,6-Cys-Cys-*bis*-(succinimido)hexane cross-linking two molecules of DArg-[Hyp^3,DPhe^7,Leu^8]-BK (CP-0127), has a 50–100-fold higher affinity than DArg-[Hyp^3,DPhe^7,Leu^8]-BK in several isolated tissues (Cheronis *et al.*, 1992a,b; Whalley *et al.*, 1992). Several of these antagonists maintain their high affinity and long duration of action *in vivo* (Whalley *et al.*, 1992; Christopher *et al.*, 1994). A heterodimer composed of the B$_1$ receptor-selective antagonist desArg^9-[Leu^8]-BK-1,6-Cys-Cys-*bis*(succinimido)hexane) cross-linked to DArg-[Hyp^3,DPhe^7,Leu^8]-BK (CP-0364), was discussed in Section 2.1.2.3. Against B$_2$ receptor-mediated responses in rat uterus, CP-0364 has a pA$_2$ value of 8.3 (Cheronis *et al.*, 1994).

2.2.2.4 Nonpeptide Antagonists

Putative nonpeptide BK receptor antagonists, some derived from plant extracts, have been reviewed (Burch *et al.*, 1990; Calixto *et al.*, 1991). The first selective, relatively high-affinity, nonpeptide B$_2$ receptor antagonist disclosed was Win 64338 (Salvino *et al.*, 1993; see Table 2.1 and Fig. 2.1). This agent is a competitive antagonist of B$_2$ receptor-mediated responses in guinea-pig ileum (pA$_2$, 6.9) (Salvino *et al.*, 1993), and has been tested in other preparations including human tissues (Marceau *et al.*, 1994; Sawutz *et al.*, 1994a). Affinity estimates for Win 64338 are shown in Table 2.4.

In vivo, Win 64338 potently inhibits plasma extravasation evoked by trigeminal nerve stimulation in guinea-pigs (Hall *et al.*, 1995). High concentrations of the antagonist do not alter responses to nonkinin agonists (Farmer and DeSiato, 1994; Marceau *et al.*, 1994; Hall *et al.*, 1995) or to sensory nerve stimulation evoked by capsaicin in rabbit isolated iris sphincter (Hall

et al., 1995), suggesting a relative degree of selectivity. **Editor's note.** A recent study reported that i.v. Win 64338 was active *in vivo* in conscious guinea-pigs, in that it inhibited aerosol BK-induced bronchoconstriction (Sawutz *et al.*, 1995). In contrast, this antagonist was not active against BK-induced hypotension in rats, and proved quite toxic in this species. Win 64338 also inhibited the increase in cyclic GMP evoked by desArg^9-BK (but not BK) in bovine cultured aortic endothelial cells (Wirth *et al.*, 1994), although the significance of this observation is not clear.

2.2.3 Radioligand-binding Studies with B$_2$ Receptors

[^{3}H]-BK (Innis *et al.*, 1981) has proved to be a popular ligand for detecting saturable, high-affinity B$_2$ receptor binding, although recent studies have used higher specific activity ligands including [^{125}I]-[*p*-Phe^5]-BK (Liebmann *et al.*, 1994a) (Table 2.5). In general, B$_2$ receptors are indicated when binding is not displaced by B$_1$ receptor ligands, but displaced by B$_2$ receptor antagonists. Saturable, high-affinity B$_2$ receptor binding sites listed in Table 2.5 have been demonstrated in membrane preparations of a variety of tissues, cells and cloned receptors.

Although direct comparisons of data obtained in different studies are compromised by differences in methodology, particularly the radiolabel used, some generalizations may be discussed. Most studies using radiolabeled BK identify a single, saturable binding site, although several report a second site. Furthermore, the affinity of BK between tissues and species is variable (Table 2.5). Two binding sites have been identified in tissues including guinea-pig ileum (Manning *et al.*, 1986; Liebmann *et al.*, 1994b), kidney and heart (Manning *et al.*, 1986), guinea-pig lung (Trifilieff *et al.*, 1991), and rat myometrium (Liebmann *et al.*, 1991) and glomeruli (Emond *et al.*, 1992). A human fibroblast line was reported to constitutively express a single population of B$_2$ receptors, but at 60% of the cell defined life-span in culture, to then spontaneously induce expression of a second, lower affinity site (Baenziger *et al.*, 1992). There are also reports describing three binding sites in a cloned neuroblastoma line (Snider and Richelson, 1984) and in guinea-pig ileum (Seguin *et al.*, 1992; Seguin and Widdowson, 1993) (Table 2.5). Whether these multiple binding affinities represent different receptors (see Section 2.3), or different affinity states of the same receptor, is not clear and may be tissue dependent. In NG108-15 cells, one line of evidence suggests that apparent multiple binding sites represent two affinity states of the same receptor (Osugi *et al.*, 1987), a phenomenon demonstrated in other G protein-coupled receptors. However, other evidence indicates the existence of both rat and mouse receptor genomic material in this cell line (Yokoyama *et al.*, 1994) (see Section 2.3.2.2). In intestinal smooth muscle using [^{125}I]-[*p*-Phe^5]-BK as

Table 2.5   Identification and characteristics of bradykinin B_2-receptor binding sites

Tissue type	Ligand	Affinity[a] (pM)	Receptor density (pmol/g)	References
Rat				
Renal mesangial cell (cultured)	[^{125}I]-[Tyr^0]-BK	2,000	88	6
Cortical astrocytes (cultured)	[^{3}H]-BK	16,600	352	3
Brain (embryonic) (cultured)	[^{125}I]-[Tyr^0]-BK	1,000	100	18
Duodenum (whole)	[^{3}H]-BK	1,000	43	19
Cardiac myocytes (cultured)	[^{125}I]-[Tyr^8]-BK	240	18.4	24
Cardiac fibroblasts (cultured)	[^{125}I]-[Tyr^8]-BK	32,400	248	24
Uterine myometrium	[^{3}H]-BK	16	58	20
Uterine myometrium	[^{3}H]-BK	1,800	—	36
Uterine myometrium	[^{3}H]-NPC 17731[b]	900	—	17
PC12 (clone expressed in COS cells)	[^{3}H]-BK	4,400	—	25
Guinea-pig				
Ileum (whole)(TES buffer)[c]	[^{3}H]-BK	16	244	27
Ileum (whole) (physiological buffer)	[^{3}H]-BK	289	266	27
Ileum (whole)	[^{3}H]-BK	92	38	9
Ileum (whole)	[^{3}H]-BK	13	8	23
Ileum (whole)	[^{3}H]-BK	20	6	31
Ileum (whole)	[^{3}H]-BK	500	25	14
Ileum (whole)	[^{125}I]-PIPHoe 140[d]	15	193	2
Ileum (smooth muscle)	[^{125}I]-[pPhe^5]-BK	3	22	21
Ileum (smooth muscle)	[^{125}I]-[pPhe^5]-BK	192	245	21
Ileum (smooth muscle)	[^{125}I]-[Tyr^8]-BK	1,800	58	33
Ileum (epithelium)	[^{125}I]-[Tyr^8]-BK	1,600	156	33
Ileum (whole)	[^{3}H]-BK	18	220	28
Ileum (longitudinal smooth muscle)	[^{3}H]-BK	172	341	28
Ileum (circular smooth muscle)	[^{3}H]-BK	191	391	28
Iluem (mucosa)	[^{3}H]-BK	196	318	28
Ileum (smooth muscle)	[^{125}I]-[X]-BK[e]	16,800	2,080	34
Lung	[^{3}H]-BK	500	35	22
Lung	[^{3}H]-BK	302	8	9
Lung	[^{3}H]-BK	15	12	36
Lung	[^{3}H]-BK	570	45	36
Trachea	[^{3}H]-BK	284	1	9
Trachea (smooth muscle)	[^{3}H]-BK	496	221	10
Trachea (epithelium)	[^{3}H]-BK	440	—	26
Trachea (epithelium)	[^{3}H]-BK	10,000	—	26
Brain (whole)	[^{3}H]-BK	100	8	11
Gall bladder	[^{3}H]-BK	4.5	546	8
Nasal turbinates	[^{3}H]-BK	60	13	12
Bovine				
Aortic endothelial cells (cultured)	[^{3}H]-BK	152	5	15
Uterine myometrium	[^{3}H]-BK	8	378	17
Pulmonary artery endothelial cells (cultured)	[^{3}H]-BK	1,280	111	32
Pulmonary artery endothelial cells (cultured)	[^{3}H]-BK	>0.5 μM	>5,000	32
Ovine				
Lung	[^{3}H]-BK	128	5.2	9
Trachea	[^{3}H]-BK	342	1	9
Human				
Fibroblasts (cultured)	[^{3}H]-BK	4,000	250	29
Fibroblasts (cultured)	[^{3}H]-BK	2,200	80	7
Synovial cells (cultured)	[^{3}H]-BK	2,300	—	1
Fibroblast (clone transfected into COS cells)	[^{3}H]-BK	130	—	13
Receptor (HG10 clone transfected into CHO cells)	[^{3}H]-BK	150	—	5
A431 epidermoid carcinoma cells	[^{3}H]-BK	3,000	—	30
Others				
NG108-15 neuroblastoma × glioma (cultured)	[^{3}H]-BK	309	242	37
Canine tracheal epithelium (cultured)	[^{3}H]-BK	257	19	4
Canine tracheal epithelium (cultured)	[^{3}H]-BK	2,500	25	38

[a] Affinity of ligand.
[b] NPC 17731 is (DArg-[Hyp^3,DHypE(*trans*propyl)^7,Oic^8]-BK.
[c] TES, is trimethylaminoethanesulphonic acid.
[d] [^{125}I]-PIP-Hoe 140 is [^{125}I]-*p*-iodophenyl-Hoe 140.
[e] [^{125}I]-[X]-BK denotes [^{125}I]-[Tyr-DArg-Hyp^3,DPhe^7,Leu^8]-BK.

References: 1, Bathon *et al.* (1992); 2, Brenner *et al.* (1993); 3, Cholewinski *et al.* (1991); 4, Denning and Welsh (1991); 5, Eggerickx *et al.* (1992); 6, Emond *et al.* (1990); 7, Etscheid *et al.* (1991); 8, Falcone *et al.* (1993); 9, Farmer *et al.* (1989); 10, Farmer *et al.* (1991b); 11, Fujiwara *et al.* (1988); 12, Fujiwara *et al.* (1989); 13, Hess *et al.* (1992); 14, Innis *et al.* (1981); 15, Keravis *et al.* (1991); 16, Leeb-Lundberg and Mathis (1990); 17, Leeb-Lundberg *et al.* (1994); 18, Lewis *et al.* (1985); 19, Liebmann *et al.* (1987); 20, Liebmann *et al.* (1990); 21, Liebmann *et al.* (1994); 22, Mak and Barnes (1991); 23, Manning *et al.* (1986); 24, Minshall *et al.* (1995); 25, Nardone *et al.* (1994); 26, Proud *et al.* (1993); 27, Ransom *et al.* (1992a); 28, Ransom *et al.* (1992b); 29, Roscher *et al.* (1990); 30, Sawutz *et al.* (1992); 31, Sharif and Whiting (1991); 32, Sung *et al.* (1988); 33, Tousignant *et al.* (1991); 34, Tousignant *et al.* (1992); 35, Trifilieff *et al.* (1991); 36, Tropea *et al.* (1992); 37, Wolsing and Rosenbaum (1991); 38, Yang *et al.* (1995).

ligand, 5′-guanylylimidodiphosphate did not change the high-affinity site into one of low affinity (Liebmann et al., 1994a), suggesting that the two sites represented distinct B_2 receptors. In the same study, two B_2 receptor antagonists, Hoe 140 and NPC 349 (Table 2.1), displaced BK binding from both sites, supporting the proposal for two distinct B_2 receptors in this preparation (Liebmann et al., 1994a). Seguin et al. (1992) and Seguin and Widdowson (1993), using [^{3}H]-BK, demonstrated three binding sites in guinea-pig ileum, but suggested only one of these may be G protein-coupled on the basis of the effects of guanine nucleotides. The proposed existence of B_2 receptor subtypes is discussed in Section 2.3.

Displacement of B_2 receptor binding by [DPhe^7]-BK analogs has been demonstrated in guinea-pig lung (Trifilieff et al., 1991), trachea (Field et al., 1992b), gall bladder (Falcone et al., 1993), nasal turbinate (Fujiwara et al., 1989), ileum (Steranka et al., 1988; Tousignant et al., 1991; Ransom et al., 1992a; Liebmann et al., 1994a), brain (Sharif and Whiting, 1991), spinal cord (Lopes et al., 1993), and epithelial cells (Proud et al., 1993), and in rat uterus (Liebmann et al., 1991), mesangial cells (Emond et al., 1990), cerebral cortical astrocytes (Cholewinski et al., 1991), bovine aortic endothelial cells (Keravis et al., 1991), and canine tracheal smooth muscle (Yang et al., 1995).

Binding studies with radiolabeled B_2 receptor antagonists such as [^{125}I]-Tyr-DArg-[Hyp^3,DPhe^7,Leu^8]-BK), showed two sites in guinea-pig ileum epithelium (Tousignant et al., 1991). At the second, low-affinity site, BK, Lys-BK, and various B_1- and B_2-selective agonists and antagonists were inactive. In contrast, several B_2-selective antagonists inhibited binding. In view of the saturability, specificity, high affinity and reversibility, Tousignant et al. (1992) proposed a novel binding site recognized by B_2 receptor antagonist ligands only. Whether this site corresponds to a BK receptor is clearly questionable. Burch and Kyle (1992) described binding of [^{3}H]-Hoe 140 to a single site in the guinea-pig ileum (K_d, 0.15 nM). A high-affinity radiolabel, [^{125}I]-p-iodophenyl-Hoe 140 ([^{125}I]-PIP-Hoe 140), binds competitively to a single population of B_2 receptors in ileum (K_d, 15 pM) and to the same site as [^{3}H]-BK, since the B_{max} was similar (Brenner et al., 1993; Table 2.5). Agonists competed better against [^{3}H]-BK, whereas antagonists competed better against [^{125}I]-PIP-Hoe 140, leading to a suggestion that there are different agonist- and antagonist-preferring conformations of the receptor. Leeb-Lundberg et al. (1994) carried out binding studies with [^{3}H]-NPC 17731 (see Table 2.1) in rat and bovine myometrium, and found this ligand to bind to the same site as [^{3}H]-BK, and was displaced by B_2 receptor antagonists.

Photoaffinity radiolabels based on B_2 receptor agonists (Steinmetzer et al., 1995) and antagonists (Schumann et al., 1995) have been designed, and show potential for receptor distribution studies. As an imaging agent, technetium-radiolabeled Hoe 140 can be used to visualize B_2 receptors using scintillography following intravenous injection in rats (Stahl et al., 1995).

2.2.4 Molecular Biology and Isolation of B_2 Receptors

A detailed account of the molecular pharmacology of the kinin receptors is given in Chapter 3 of this volume, and some details relevant to the current discussion are presented here. Early attempts to solubilize and purify BK receptors had fairly low yields (Fredrick and Odya, 1987; Snell et al., 1990). Increased yield was obtained in human fibroblasts using the detergent CHAPS (Faussner et al., 1991), and the solubilized material bound BK with high affinity (1.7 nM), with a displacement potency order for analogs similar to that of intact fibroblasts. The apparent molecular weight (MW) of this receptor material was 250,000 (Faussner et al., 1991). Ransom et al. (1992b) solubilized guinea-pig ileal BK receptors in a stable configuration, and reported a single binding site with identical affinity in longitudinal and circular smooth muscle, and the mucosal layer. The affinity of [^{3}H]-BK binding for the solubilized receptor was about ten-fold lower than for the membrane preparation (Ransom et al., 1992b; see Table 2.5).

Yaqoob and Snell (1994) solubilized and partially purified receptor material from the rat uterus utilizing cross-linking to [^{125}I]-BK with disuccinimidyl suberate, and purification to homogeneity. This yielded a 81 kDa product, in agreement with exclusion chromatography. This protein retained binding properties for BK and [DPhe^7]-BK, and biochemical studies showed it to be a hydrophobic acidic glycoprotein, so the increased mass as compared to the cloned rat uterus receptor can apparently be explained, at least in part, by inclusion of sugar residues.

2.2.4.1 Receptor Isolation and Solubilization Studies

Expression of mammalian B_2 receptors was first described by Mahan and Burch (1990; see also Farmer and Burch, 1992), wherein BK increased $^{45}Ca^{2+}$ efflux and 1,4,5-triphosphate (IP$_3$) accumulation in Xenopus oocytes previously injected with mRNA from murine fibroblasts. Responses were inhibited by B_2 receptor antagonists, but not a B_1 receptor antagonist. The approximate size of the mRNA was 4.5 kb. Phillips et al. (1992) described expression of functional BK receptors in Xenopus oocytes, with mRNA prepared from various sources. The mRNA obtained from rat uterus, NG108-15 cells, and a human fibroblast cell line, WI38, yielded similar responses to BK, as did total mRNA from dorsal root ganglion neurons. However, no responses were obtained in oocytes injected with rat brain mRNA. Hoe 140 blocked responses to BK with all types of tissue mRNA, whereas a B_1 receptor antagonist was inactive, except in WI38 cells. Thus, this cell line was proposed to encode both B_1 and B_2 receptors. (see also Section 2.1.4). An increase in the magnitude of responses from NG108-15 and uterine mRNA was obtained following enrichment of the receptor encoding mRNA, and a size of 4.5 kb was estimated, in agreement with the size estimate in murine fibroblast clones (Mahan and Burch, 1990).

2.2.4.2   Receptor Cloning Studies: Species Characteristics

Cloning of the gene, and expression of the B_2 receptor from rat uterus was described by McEachern *et al.* (1991), wherein it was shown to belong to the seven transmembrane-domained G protein-coupled super-family, with a predicted sequence of 366 amino acids and a MW of 41,696 Da. This sequence contains putative sites for phosphorylation by protein kinases A and C (PKA and PKC). The receptor expressed in *Xenopus* oocytes was not stimulated by desArg^9-BK, although it was activated by BK and Lys-BK. Although [Thi^{5,8},DPhe^7]-BK was a weak partial agonist, it inhibited responses to BK. The partial agonist activity led these authors to propose this receptor was of the putative "smooth muscle" rather than the "neuronal" type (see Section 2.3.2.1). Lower levels of mRNA encoding the B_2 receptor were found in other rat tissues, and the potential for expression of multiple receptor forms was suggested by the occurrence of multiple message species. High stringency analysis demonstrated that the rat, human and guinea-pig genome each contain only a single B_2 receptor gene (McEachern *et al.*, 1991).

Hess *et al.* (1992) cloned a human lung fibroblast BK gene and pharmacologically characterized the expressed receptor. The cDNA encoded a 364 amino-acid protein, having characteristics of a G protein-coupled receptor. The predicted amino-acid sequence had 81% identity to the rat B_2 receptor (McEachern *et al.*, 1991). Transfection of the human clone into COS-7 cells resulted in specific [^{3}H]-BK binding-sites (affinity, 0.13 nM). Importantly, two B_2 receptor antagonists had high affinity, while B_1 receptor-selective ligands were inactive. At high BK concentrations, there was evidence of a lower affinity binding site ($K_d > 10$ nM). Three potential sites of N-glycosylation are present in the rat receptor (McEachern *et al.*, 1991), and are conserved in the human form (Hess *et al.*, 1992). The sequence also contains several consensus sites for PKA- and PKC-dependent phosphorylation, which may be involved in desensitization, a phenomenon often marked in BK receptors (Munoz and Leeb-Lundberg, 1992; Munoz *et al.*, 1993). The B_2 receptor in human tissues showed highest levels in the pancreas, testes, brain and heart, a similar distribution to that seen with the cloned rat receptor.

Eggerickx *et al.* (1992) cloned the gene encoding the human B_2 receptor from genomic DNA by low stringency PCR, and reported a high degree of homology with the rat receptor. Following expression in *Xenopus* oocytes and stable transfection into CHO cells, binding studies showed a high-affinity B_2 receptor. Coupling of this receptor to PI hydrolysis was demonstrated, but the pharmacological profile did not exactly match that described for the guinea-pig or rat B_2 receptor, and this may be attributable to species-dependent subtypes.

Others have used genomic libraries to clone and sequence the B_2 receptor for human (Ma *et al.*, 1994b), rat (Wang *et al.*, 1994) and mouse (Ma *et al.*, 1994a). These genes each consisted of three exons separated by two introns, with the third exon containing a full-length coding region. The human gene encodes a 364 amino-acid, seven transmembrane protein that shares high sequence homology with the rat and mouse B_2 receptors. The human B_2 receptor is localized to chromosome 14q32, and is expressed in most tissues (Ma *et al.*, 1994b). The rat B_2 receptor gene encodes a 366 amino-acid protein (McEachern *et al.*, 1991; Wang *et al.*, 1994). The mouse B_2 receptor gene encodes a 366 amino-acid receptor (41.47 kDa), and shares high identity with the rat and human sequences.

2.2.4.3   Structural Determinants of Functional Properties of Receptors

Nardone and Hogan (1994) attempted to delineate receptor regions essential for high-affinity agonist binding by mutating the sixth transmembrane segment of the rat B_2 receptor expressed in COS-1 cells. Two amino-acid substitutions markedly reduced the affinity of the receptor for BK (Phe-261 with Val by 1,600-fold; Thr-265 with Ala by 700-fold), but had comparatively little effect on the affinity of the B_2 receptor antagonists, NPC 567 and NPC 17731 (see Table 2.1 for structures). Other mutations had only modest effects on affinity. Even the most dramatically affected mutated receptors were still able to couple to PI turnover. These data were interpreted as showing that BK directly contacts the face of the sixth transmembrane helix at five critical points in the 261–269 region, or that this region stabilizes the agonist-binding conformation of the receptor (Nardone and Hogan, 1994).

2.2.4.4   Other Molecular Biology Applications

Distribution studies for the B_2 receptor have made use of knowledge of sequences in generating antibodies against various regions of the receptor, and have been used in conjunction with microscopy. There is limited information as yet, but the distribution of receptors within the kidney has been worked out in detail, and the presence of B_2 receptors on human neutrophils has also been demonstrated (Haasemann *et al.*, 1994; Figueroa *et al.*, 1995; see Chapter 12). The use of antisense oligonucleotides in "knockout" and "knockdown" experiments, involving disruption of the function of receptors, their G proteins, or of the genes themselves, has received considerable attention (for reviews, see Albert and Morris, 1994; Wahlestedt, 1994). Although such studies with BK receptors are in their infancy, targeted disruption of the mouse B_2 receptor gene resulted in loss of responsiveness to BK in the smooth muscle of the ileum or uterus, and the electrical responsiveness in neurons of the superior cervical ganglion. Yet such animals are fertile and visually indistinguishable from litter mates (Borkowski et al., 1995; see Chapter 3). The development of B_1 receptor knockout animals is eagerly awaited.

2.3   BRADYKININ RECEPTOR SUBTYPES

The division of kinin receptors into either B_1 or B_2 subtypes according to their pharmacology and molecular biology has been relatively straightforward. Receptors in some isolated preparations, however, though seemingly more closely aligned to the B_2 receptor subtype, show some quantitative differences. Those of the guinea-pig tracheal smooth muscle have been termed B_3 (Farmer et al., 1989) in view of these differences.

2.3.1   Tracheal B_3 Receptors

Guinea-pig isolated trachea may contract or relax in response to BK depending on experimental conditions but in no published studies are B_1 receptors involved (Farmer et al., 1989; Field et al., 1992a,b; Da Silva et al., 1995). The existence of B_3 receptors was originally proposed based on two grounds: (1) antagonist affinities; and (2) radioligand-binding studies.

(1) Early intimations of atypical properties of BK receptors in guinea-pig trachea came from the low or undetectable antagonist activity of B_2 receptor antagonists, such as NPC 567 and NPC 349 (Rhaleb et al., 1988; Farmer et al., 1989; Field et al., 1992a,b), although they had reasonable affinities in other tissues (Table 2.4). Antagonists such as NPC 16731 and Hoe 140, which have higher affinities, block responses to BK in guinea-pig trachea, although again with a lower affinity than in most other preparations (Table 2.4). The nonpeptide B_2 receptor antagonist, Win 64338 (Table 2.4), was reported to be inactive even at 1 μM in guinea-pig trachea (Farmer and DeSiato, 1994). In contrast, Pruneau et al. (1995b) noted Win 64338 to have a high affinity in the guinea-pig trachea (pA$_2$, 7.36), similar to that seen in other guinea-pig preparations.

(2) In radioligand-binding experiments with first-generation B_2 antagonists in guinea-pig trachea, Farmer et al. (1989) were unable to displace [^{3}H]-BK binding. In functional experiments, the same antagonists were shown to be inactive, thereby supporting their proposal for a novel B_3 receptor. However, the lack of displacement of [^{3}H]-BK binding with B_2 receptor antagonists has been challenged. Field et al. (1992a,b) showed displacement of [^{3}H]-BK binding with NPC 567, NPC 16731 and Hoe 140, and found the activities of these antagonists in displacing the radioligand to correlate well with their affinities in

antagonizing BK-induced contractions in this preparation. In cultured tracheal smooth muscle cells, Farmer et al. (1991b) showed that NPC 567 exhibited high-affinity displacement of [^{3}H]-BK binding. Subsequently, Farmer and colleagues (1991a) reported that NPC 16731, displaced [^{3}H]-BK in freshly isolated tracheal smooth muscle preparations. In binding studies with Hoe 140 in guinea-pig trachea, Trifilieff et al. (1992) identified a single binding site with an affinity of 55 pM for the antagonist.

From the data discussed above it can be seen that, although there is evidence pointing to the existence of a novel receptor in guinea-pig airways, there are puzzling discrepancies in reports by different groups. Possible factors that could contribute to the differences between groups regarding receptor characteristics may be noted. First, epithelial cells can contribute both to relaxant and contractile responses to BK in the trachea through release of mediators such as prostaglandins and nitric oxide, and BK antagonists seem able to attenuate some of these actions (Bramley et al., 1988, 1989, 1990; Farmer et al., 1989; Schlemper and Calixto, 1994; Da Silva et al., 1995). Also, guinea-pig tracheal epithelium contains two BK binding sites associated with prostaglandin production, and BK antagonists displace binding from both sites (Proud et al., 1993) (see Section 2.2.3). Consequently, although most of the studies mentioned above have used epithelium-denuded preparations, this depends on physical removal, the efficiency of which may vary between laboratories.

Second, there is evidence from concentration–responses curves constructed over a wide range of BK concentrations, that there are two phases to the contraction. For instance, Field et al. (1992a,b) measured curve shifts, owing to antagonists, for BK at concentrations up to 100 μM, whereas the original studies by Farmer et al. (1989), studied BK concentrations up to only 1 μM.

Third, depending upon basal tone, guinea-pig tracheal smooth muscle can relax or contract in response to BK. For instance, Da Silva et al. (1995) described concentration-dependent relaxations in response to BK in epithelium-denuded, precontracted trachea. The BK receptor type was not B_1, since desArg^9-[Leu^8]-BK was inactive, but B_2 receptor antagonists such as Hoe 140 and NPC 17761 (Fig. 2.1) potently antagonized the relaxations. The same group showed that a cyclooxygenase inhibitor abolished BK-induced relaxation, and suggested that products of the arachidonic acid cascade can relax or contract the preparation according to tone, and rejected early suggestions of direct activation of PLA$_2$ by a receptor-independent process (Rhaleb et al., 1988).

Fourth, there is evidence for BK receptor activation of PKC in B_2 receptor-mediated responses in tracheal smooth muscle (Farmer and Burch, 1992; Hall, 1992; Ransom et al., 1992a; Field et al., 1994). In guinea-pig trachea, both phospholipases C and D pathways are involved in BK-induced contractions (Pyne and Pyne, 1993a,b). In addition, they reported different sensitivities of the two pathways to antagonism by NPC 567 and NPC 16731, suggesting two receptors, each coupled preferentially to one or other pathway. It might follow that one or the other might predominate according to experimental conditions, and this could contribute to the conflicting results published regarding sensitivity of responses to BK receptor antagonists reported from different laboratories.

Finally, peptidases metabolize BK and its peptide antagonists and, in tracheal preparations, such metabolic degradation may prevent equilibrium with the receptor (Hall *et al.*, 1990; Da Silva *et al.*, 1992; Skidgel, 1992; Kenakin, 1993). However, peptidase inhibitors have only been used in a small number of studies and, because of the slow nature of responses to BK, and the thickness of tissue preparations, it is possible that degradation may significantly influence peptide concentrations in the biophase.

In conclusion, in relation to the proposed existence of a B_3 receptor, it is not clear how some of the discrepancies between groups have arisen. In relation to possible intraspecies and interspecies subtypes, examination of other species for B_3 receptors would be illuminating. In this respect, Farmer and colleagues described a receptor in ferret trachea that has similar properties to the putative guinea-pig tracheal B_3 receptor. Further, in this preparation, the peptidase inhibitors thiorphan and captopril elicited BK receptor-mediated contractions, suggesting that release of endogenous kinins *in vitro* is a further factor that needs to be considered (Farmer *et al.*, 1992, 1994; Farmer and DeSiato, 1994).

It should be noted that the "B_3 receptor" proposed by Saha *et al.* (1990, 1991) to be expressed in opossum esophagus bears no relation whatsoever to that proposed by Farmer and colleagues in guinea-pig airways (Section 2.3.2.1). This is discussed in detail by others (Farmer and Burch, 1992; Regoli *et al.*, 1993).

2.3.2 Homologs of B_2 Receptors

Intraspecies and interspecies cell surface receptor subtypes occur generally for peptides and other mediators, and the challenge that this poses for drug discovery is discussed in detail by Hall *et al.* (1993). It is clear from cloning and expression studies, that there are interspecies differences in B_2 receptors in terms of amino-acid sequence. It is less clear if there are *intra*species receptor protein subtypes, be they distinct subtypes from different genes, or variants derived by alternative splicing or post-translational processing from a single gene. It should be noted that there is no evidence for more than one B_2 receptor gene within any one species.

2.3.2.1 *Within-species Variants or Subtypes*
Evidence from molecular biology
The possibility of alternative splicing of B_2 receptor gene products from a single gene within one species, to yield

receptor subtypes, has been investigated by Park *et al.* (1994) (see also Chapter 3). In a comparison of the sequences of the rat and human genes, it was considered that the most likely initiation of the whole receptor would be by exon-3 only, but that alternative splices were possible, and these putative extensions to the amino-terminal region of the protein could alter the biological properties of each receptor isoform. In the event, all human cDNAs isolated were missing exon-2, whereas exon-1 spliced to exon-3 was very common. No evidence of cDNA for receptor subtypes within the human or rat was evident, even at the low stringency of <60% identical, a level at which the angiotensin AT_1 receptor shows homology.

Pesquero *et al.* (1994) isolated the gene for the B_2 receptor from a rat genomic library, and detected two different B_2 receptor mRNAs containing or lacking an exon, providing evidence for alternative splicing. Yokoyama *et al.* (1994) cloned two distinct BK receptors from NG108-15 cells, a rat/mouse hybrid line. The first was identical with the rat uterus B_2 receptor, whereas the second had 91% amino-acid homology to the rat uterus or 82% homology to the human B_2 receptors. Genomic DNA cloning revealed that message for the second receptor was derived from the mouse genome. The second receptor was expressed in *Xenopus* oocytes and COS-7 cells, and had functional properties considered to be of the "smooth muscle" receptor subtype. The authors concluded that both the rat and the mouse B_2 receptors are expressed in NG108-15 hybrid cells. This finding is relevant to interpretation of the extensive studies performed in NG108-15 hybrid cells and, in particular, the question of whether the same receptor type can couple to two effectors (e.g. see Section 2.3.1 in relation to PLC and PLD).

Nardone *et al.* (1994) investigated whether differences between so-called "neuronal" and "smooth muscle" B_2 receptors, reported by Llona *et al.* (1987) in rat vas deferens (see below), are due to intraspecies subtypes. In PC12 phaeochromocytoma cells, as a model of neuronal receptors, [Thi[5,8],DPhe[7]]-BK was a competitive B_2 receptor antagonist when tested against BK-stimulated PI turnover. This analog was chosen because of reports of differential agonist/antagonist effects in rat vas deferens (see Section 2.3.2.1). Nardone *et al.* (1994) showed the full-length cDNA encoding the BK receptor expressed in PC12 cells to be virtually identical to the sequence determined in rat uterus, and took the view that B_2 receptors are identical in rat nerve and smooth muscle cells (see also Chapter 3). It may be concluded, therefore, that there is potential for alternative splicing of the B_2 receptor gene in the species studied but there is little evidence pointing to this mechanism *in vivo* for generating receptor variant subtypes.

Radioligand-binding
The evidence relating to the existence of single or multiple B_2 receptors, as outlined above, is unresolved.

Where there is more than one site, the identity of each was not established and, for agonist binding, interpretation is complicated by the coexistence of high- and low-affinity conformations of the receptor. For antagonists, theoretical interpretation of binding affinities presents fewer problems, but published values are generally of apparent affinities (e.g. IC_{50} values) rather than rigorous K_d estimates. Thus, comparison of absolute affinities between different systems is difficult. Nevertheless, the presence in a given cell type of two binding sites with different apparent affinities for a given receptor antagonist should be a reliable indicator of intraspecies homologs.

There is a strong argument for two sites in guinea-pig ileum where two B_2 receptor antagonists displaced $[^{125}I]$-BK binding from each site (Liebmann et al., 1994a). These workers proposed the low-affinity site (K_d, 192 pM) to be associated with contraction. In contrast, only a single site was reported by Ransom et al. (1992a,b) in the muscular and mucosal layers of guinea-pig ileum. Seguin et al. (1992) and Seguin and Widdowson (1993) reported three $[^3H]$-BK binding sites in guinea-pig ileum, but suggested only one to be G protein-coupled.

Functional studies
In rat vas deferens, $[Thi^{5,8},DPhe^7]$-BK was a weak partial agonist with respect to "neurogenic" actions, but a competitive antagonist against the "musculotrophic" actions of BK (Llona et al., 1987). This has led several investigators to use the same analog to determine whether a given cloned and expressed receptor is of the "neuronal" or the "smooth muscle" type. However, since $[Thi^{5,8},DPhe^7]$-BK has substantial intrinsic efficacy, a simpler suggestion would be to explain a different effect on the two actions of BK in terms of receptor reserve or efficiency of coupling. Rifo et al. (1987) reported differences against these two aspects of the action of BK in terms of pA_2 for $[Hyp^3,Thi^{5,8},DPhe^7]$-BK and NPC 349 (see Table 2.4). It should be noted, however, that since the actions on prejunctional effects are measured in terms of changed postjunctional (evoked twitch) responses, in practice it is very difficult to separate the two sites of action adequately.

Saha et al. (1990, 1991) proposed the existence of "B_3", "B_4" and "B_5" receptors in opossum esophagus. However, the view has been taken that these putative subtypes were not sufficiently characterized with agonists or antagonists for consideration as novel receptors, and the concentrations of ligands used were very high (see Hall, 1992; Regoli et al., 1993).

Evidence relating to the existence of a B_3 receptor in guinea-pig airways has been discussed (Section 2.3.1). The overall evidence seems to point to a recognition site with properties similar to other B_2 receptors, but where the affinity of antagonists tends to be low. Whether this receptor represents a specific and novel gene product

(e.g. a product of alternative splicing), or of different post-translational processing, or whether it is similar to the receptor in other guinea-pig tissues, is unknown. Data collected in Table 2.4 suggest that peptide antagonists have relatively consistent affinities within guinea-pig preparations, including the trachea. However, as noted below, these values are lower than in some other species. Interestingly, this affinity relationship is reversed with the nonpeptide antagonist Win 64338.

A summary of most of the published affinity estimates for the more extensively studied BK antagonists is given in Table 2.4. A difference of, say, ten-fold in affinity for an antagonist between two preparations in a given species, might constitute evidence of subtypes within that species. However, inspection of Table 2.4 and Fig. 2.2, shows differences as great as this between the estimates for a single preparation, but from different laboratories. Clearly there are considerable methodological differences and difficulties, some of which have been touched on (Section 2.3.1) in relation to the proposed existence of the B_3 receptor. Possibly, evidence from parallel studies, where one group has used similar conditions, can provide more discriminatory evidence and, where there are such data, there is little evidence of intraspecies subtypes. The pA_2 estimates in Table 2.4 show an increase in affinity in guinea-pig tissues in the order: NPC 567 < NPC 349 < DArg-$[Hyp^3,DPhe^7,Leu^8]$-BK < Win 64338 << Hoe 140. Data for the guinea-pig are plotted in Fig. 2.2a.

2.3.2.2   Between-species Variants or Subtypes

The most relevant and conclusive evidence for species-related B_2 receptor subtypes comes from antagonist affinity estimates made in functional studies. Affinity data for preparations from several species are collected in Table 2.4 with respect to antagonists that have been tested extensively. These data are shown in Fig. 2.2. As mentioned above, even for a given preparation from one species, pA_2 estimates vary by up to one log unit. This can be accounted for in part by differences in experimental protocols. Probably the most important variables are the inclusion or not of cyclooxygenase inhibitors and peptidase inhibitors. Also, Hoe 140 presents special problems, since it acts as an insurmountable antagonist and, in most preparations, can decrease maximum responses to BK (Section 2.2.3). Nevertheless, pA_2 estimates have been made in many preparations, and there seems to be no greater variability than with other antagonists. In theory, for a competitive slow-offset antagonist with a high proportion of spare receptors, affinity estimates should be reasonably good and, even if there is depression on the maximum response, and ratios of EC_{50} values should give reasonable estimates (Kenakin, 1993).

Overall, the pA_2 estimates shown in Table 2.4 show a general increase in the affinity order such that: NPC 567 < NPC 349 < DArg-$[Hyp^3,DPhe^7,Leu^8]$-BK << Hoe 140. Win 64338 is always considerably less potent than

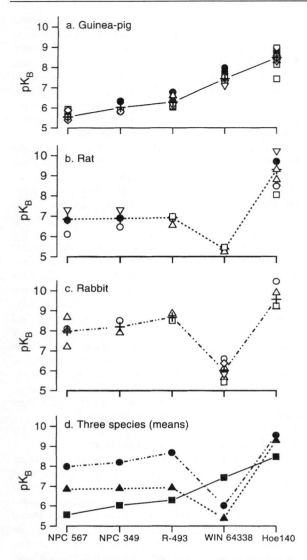

Figure 2.2   A comparison of affinity estimates for five antagonists in preparations from three species: (a) guinea pig; (b) rat; (c) rabbit; and (d) mean values from each species are plotted for comparison. The B_2 receptor antagonists are NPC 567, NPC 349, DArg-[Hyp³,DPhe⁷,Leu⁸]-BK, Win 64338 and Hoe 140. Affinity estimates, plotted as apparent pK_b values, are from Table 2.4. (a) ●, ileum; □, trachea; ◇, taenia; △, colon; ×, urinary bladder; ○, lung parenchyma; ▽, pulmonary artery; –+–, mean. (b) ●, uterus; ◇, urinary bladder; ○, vas deferens; ▽, duodenum; △, portal vein; □, stomach; --+--, mean. (c) ○, iris sphincter; △, jugular vein; □, vena cava; ◇, urinary bladder; ▽, colon; –+–, mean. (d) ■, guinea-pig; ▲, rat; ●, rabbit.

Hoe 140, but its relative and absolute affinity varies with species, as noted.

There are between species differences in receptors (Fig. 2.2). Of the antagonists shown, all peptide analogs show

a greater affinity for rabbit and rat preparations than for those of guinea-pigs. Win 64338, has a potency that varies with species. In rats and rabbits, it is much less active than the [DPhe⁷]-BK series, whereas the reverse is true in guinea pigs. Such differences have been noted in comparative studies within a given laboratory. For instance, on the basis of these and other differences, Regoli and colleagues (Regoli *et al.*, 1993, 1994, 1995) proposed dividing B_2 receptors into B_{2A} (typified by the rabbit, but possibly also including dog and human) and B_{2B} (typified by the guinea-pig, and possibly including the hamster and rat). In an attempt to quantify the proposed close relationship between rabbit and human receptors, Gobeil and Regoli (1994) plotted the apparent binding affinities for agonists and antagonists, in cloned human BK receptors in Chinese hamster ovary (CHO) cells (Eggerickx *et al.*, 1992), against their own functional data in rabbit jugular vein or guinea-pig ileum, and showed the correlation was 0.94 in the former case, as compared with 0.75 in the latter.

Our own studies have also noted species-related differences in B_2 receptors, as well as a marked correlation in affinities of peptide antagonists between species (Field *et al.*, 1988; Everett *et al.*, 1992; Hall, 1992; Hall *et al.*, 1992). Thus, the correlation between the affinities of NPC 567 and Hoe 140, measured in seven preparations from three species, is 0.96 (Hall, 1992). These studies suggested that rabbit BK receptors have a similar, or somewhat higher affinity for peptide antagonists than the rat and these affinities are, in turn, higher than in the guinea-pig. Win 64338 yields data that support this grouping by species, although with reversed affinities, with the rabbit and rat showing a similar trend (see Table 2.4 and Fig. 2.2).

There are insufficient data from most other species, but a species in which peptide antagonists have a low affinity is the pig. In the porcine iris, B_2 receptor antagonists such as Lys-Lys-[Hyp³,Thi⁵,⁸,DPhe⁷]-BK and Hoe 140 have affinities of more than two orders of magnitude lower than in rabbit iris (Everett *et al.*, 1992; Hall, 1992; Fig. 2.3 and Table 2.4). In the mouse, there are few functional data with respect to the native receptor to compare with the data obtained since the cloning of the murine B_2 receptor (see Section 2.2.4.3 and below). However, data for the native receptor shown in Table 2.4 suggest that the mouse BK receptor is very much like that of the rat. Further, the binding affinities of antagonists seen in the cloned murine receptor (Hess *et al.*, 1994a,b) are very similar to those in the rat receptor (Table 2.4 and Fig. 2.2). On the other hand, there are marked differences between mouse and rat receptors as compared to the cloned human receptor (*vide infra*). Thus, functional data suggest there are pronounced species differences in B_2 receptors.

Important evidence of species-related kinin B_2 receptor subtypes comes from molecular biology. Thus, Hess *et al.* (1994a,b) compared cloned human and mouse B_2

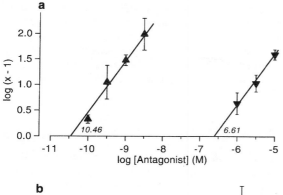

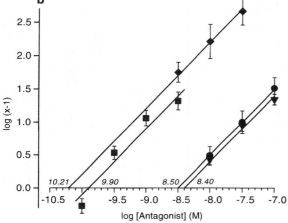

Figure 2.3   Affinity estimates for bradykinin (BK) B$_2$ receptor antagonists in various tissues. Schild plot analyses are plotted with slopes of unity, since all data are compatible with competitive antagonism. (a) Comparison of two antagonists against BK-induced neurogenic contractions in rabbit iris sphincter *in vitro*, where a peptide antagonist, Hoe 140 (▲), has a pK_b estimate of 10.46 ± 0.15 (slope, 1.09 ± 0.27, n = 11), and the nonpeptide antagonist, Win 64338 (▼), has a pK_b of 6.61 ± 0.09 (slope, 0.77 ± 0.23, n = 11). (b) Schild plots for Hoe 140 against functional responses to BK in tissues from four species: rat duodenum (◆), pK_b, 10.21 ± 0.12 (slope, 0.92 ± 0.34, n = 15); rabbit jugular vein (■), pK_b, 9.90 ± 0.06 (slope, 1.11 ± 0.11, n = 19); guinea-pig bladder (●), pK_b, 8.50 ± 0.08 (slope, 1.03 ± 0.21, n = 9); porcine iris sphincter (▼), pK_b, 8.40 ± 0.04 (slope, 0.90 ± 0.10, n = 9). All data are from the authors' laboratory, and are compiled from Everett et al. (1992), Hall et al. (1992, 1995), and previously unpublished.

receptors expressed in cell lines lacking endogenous BK receptors. Human receptors, transfected into CHO cells, activated both PI turnover and arachidonic acid release by independent pathways. This is important in relation to studies where it has been difficult to resolve whether the same receptor could couple to the two different effector pathways. The coding region of the mouse B$_2$ receptor was expressed in COS-7 cells, and this receptor

exhibited classical B$_2$ receptor properties. In a direct comparison of the two species' homologs, both receptors bound BK with a similar affinity (pK_i, 9.0). However, the forms of the receptor from mouse and human differed dramatically in their affinity for peptide antagonists. The murine receptor has 57- and 82-fold higher affinities for NPC 567 and NPC 349, respectively, as compared to the human homolog. On the other hand, [Thi^5,DPhe^7]-BK and Hoe 140 showed only five- and two-fold higher affinities (Hess et al., 1994a,b).

Overall, it is difficult to relate the few quantitative data currently published for transfected clones of human and other B$_2$ receptors to their functional parameters. For instance, taking the data of Hess et al. (1994a,b) for the cloned human receptor gene, NPC 349 and Hoe 140 had relatively high affinity estimates in displacing [^{3}H]-BK binding, giving pK_i values that were of the same order as the few reported values for this species in functional studies (see Table 2.4). In the mouse, the average pA$_2$ for Hoe 140 from functional studies is 9.65, virtually identical to the pK_i reported for binding to the cloned receptor. However, the respective values for the human receptor showed poorer agreement.

In bovine kinin receptors, there may be other evidence of species-related receptor characteristics. With [^{3}H]-BK binding, B$_1$ and B$_2$ receptors have both been proposed to be present on bovine aortic endothelial cells. Thus high-affinity binding is displaced with NPC 349 and [Thi^{5,8},DPhe^7]-BK, whereas binding at the low affinity site is displaced with desArg^9-[Leu^8]-BK (Sung et al., 1988). However, studies measuring antagonist inhibition of kinin-evoked cyclic guanosine monophosphate (cGMP) produced conflicting results regarding the receptor types involved. Thus B$_1$ receptor antagonists inhibit desArg^9-BK-, but not BK-stimulated cGMP production (Wiemer and Wirth, 1992). In contrast, B$_2$ receptor antagonists inhibited both desArg^9-BK- and BK-induced responses (Wirth et al., 1994). The reasons for these discrepancies are not clear.

Returning to the major question posed in this section regarding possible species-related heterology of B$_2$ receptors, there can be little question that the case is proved. The relevance of species-related receptor homologs to drug discovery is a subject for debate (see Hall et al., 1993). However, with B$_2$ receptors it is clear that all known antagonists (but not agonists) distinguish to some degree between extremes in the affinity range, between rat or rabbit receptors on the one hand, and guinea-pig receptors on the other. Whereas it would appear that peptide antagonists favored the former species in terms of affinity, it is of interest that Win 64338, a nonpeptide, reverses this trend. This demonstrates that synthetic antagonists can be designed to show some degree of B$_2$ receptor species-selectivity, a principle already demonstrated at the tachykinin receptors (Hall et al., 1993).

3. *Immunopharmacology of Kinin Receptors*

The kallikrein–kinin system plays an integral role in the acute and chronic inflammatory response. The pro-inflammatory role of the kinin system has been reviewed (Lewis, 1970; Marceau *et al.*, 1983; Roch-Arveiller *et al.*, 1985; Hargreaves *et al.*, 1988; Proud and Kaplan, 1988; Bhoola *et al.*, 1992; Hall, 1992). Results from studies incorporating the use of recently introduced receptor antagonists has led to convincing indications that kinins contribute to the pathophysiology of several inflammatory diseases. The evidence of a role for the kinin system in several specific disease states, is discussed in depth in other chapters of this volume, to which the reader is directed. These include: pancreatitis (Chapter 12), arthritis (Chapters 9 and 13), cardiovascular disease (Chapter 19), urinary tract disorders (Chapter 14), and upper and lower respiratory tract disorders including asthma (Chapter 15) and rhinitis (Chapters 16 and 17). The involvement of kinins in pain, hyperalgesia and in evoking neurogenic inflammation are discussed in Chapter 9 and, in allergic inflammation and hereditary angioedema, in Chapter 16. Also, the roles of the kallikrein–kinin system in sepsis are discussed in Chapter 10.

In the following sections of the present chapter,

therefore, we confine our discussions to an analysis of evidence specifically pointing to a role for BK receptor-mediated actions in inflammation. In particular, we discuss studies where BK receptor antagonists have been influential in establishing a role for endogenous kinins in inflammation (see also Steranka and Burch, 1991; Hall, 1992).

3.1 ACUTE INFLAMMATORY RESPONSES

The acute inflammatory response is characterized by vasodilatation, plasma extravasation, pain and hyperalgesia; together with neutrophil adherence to, and extravasation from, the venule endothelium. Various components of the kinin system have been proposed to contribute to, or modulate, one or other of these aspects of the inflammatory response (Fig. 2.4). Recently it has been shown that both BK and desArg^9-BK induce expression of kininogen binding sites on endothelial cells, a process that has been proposed as a mechanism for increasing the concentration of kinins at the site of inflammation (Zini *et al.*, 1993). Inflammatory effects of kinins are mediated via B$_1$ or B$_2$ receptors by acting directly (on the microvasculature endothelium or smooth muscle) and indirectly (through the antidromic release of tachykinins and calcitonin gene-related peptide (CGRP)

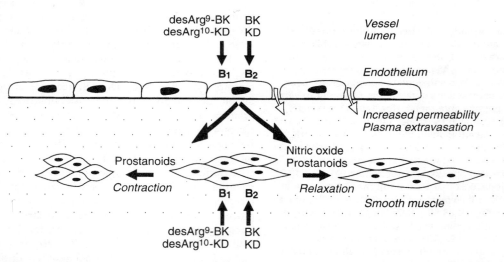

Figure 2.4   Kinin receptors in the vasculature are shown in relation to direct contractile effects on smooth muscle; direct effects on endothelial cells to release nitric oxide, a vasodilator; indirect effects mediated via nitric oxide and/or prostanoids; and multiple effects leading to increased microvascular permeability and plasma extravasation. The predominant effects of bradykinin (BK) or Lys-BK (kallidin, KD) vary widely, and depend upon the type of blood vessel or vascular bed. In large arteries and veins, direct contractile effects may be observed, whereas, in smaller vessels and in the microvasculature, indirect vasodilatation predominates. The latter leads to a powerful overall hypotensive effect. In venules, a prominent effect is an increase in permeability with plasma extravasation. In some vascular beds, kinins induce release from sensory nerves of tachykinins and CGRP, which may contribute to microvascular leak and vasodilatation (not shown; see text). Normally, the effects of BK and Lys-BK are mediated via B$_2$ receptors but, in some pathologic conditions, induction of B$_1$ receptors may allow effects to mediated by the desArg metabolites of BK and Lys-BK, formed from C-terminal deletion by carboxypeptidases such as kininase I.

from primary afferent neurons, or via the release of nitric oxide from the venule endothelium). Moreover, the pro-inflammatory effects resulting from BK receptor stimulation may be amplified through synergy with other mediators, often at the second messenger level. Such interactions have been described with the cytokines (Bathon et al., 1989), and with the eicosanoids (Juan and Lembeck, 1974; Burch et al., 1988; O'Neill and Lewis, 1989; Ljunggren et al., 1991). For example, in 3T3 fibroblasts, priming with cytokines greatly amplifies B_2 receptor BK-stimulated prostaglandin E_2 (PGE_2) synthesis (Burch et al., 1988; Burch and Tiffany, 1989). There was synergy between kinins and IL-1 in human synovial fibroblasts, measured in terms of arachidonic acid release through the phospholipase A_2 (PLA_2) pathway (Cisar et al., 1993). In human gingival fibroblasts there was also a synergistic interaction between BK and desArg[9]-BK and IL-1, as measured by PGE_2 production (Whiteley and Needleman, 1984; Lerner and Modéer, 1991). In the vasculature, the major route of degradation of BK and Lys-BK is by endothelial angiotensin-converting enzyme (ACE; see Chapter 7), particularly in the pulmonary circulation. However, if the function of ACE is impaired, as in shock (see Section 3.6), coagulation, or possibly during therapy with ACE inhibitors, conversion of BK to desArg[9]-BK by other enzymes including plasma carboxypeptidase N may become more important. Given the high selectivity of BK and desArg[9]-BK for B_2 and B_1 receptors, respectively, this may be a common regulatory mechanism. Thus, it is possible that the effects of BK on the vasculature may normally involve both B_2 and B_1 receptors (Guimarães et al., 1986; Sung et al., 1988; D'Orléans-Juste et al., 1989; Nakhostine et al., 1993; DeWitt et al., 1994).

3.1.1   Mast-cell Histamine Release

Kinins release histamine from mast cells and this contributes to their actions. Mast-cell activation by kinins and other cationic amphiphilic peptides has generally been regarded as resulting from a nonreceptor mechanism, and it has been proposed that this involves direct activation of Pertussis toxin-sensitive G proteins with activation of phospholipase C (PLC) (Mousli et al., 1990). Thus, there are no studies demonstrating specific mast-cell binding site, and the potency order of kinins on histamine release relates to the number of positively charged amino-acid residues rather than to the conformational structure of the peptide molecule. Further, both B_1 and B_2 receptor agonists and antagonists can evoke histamine release (Devillier et al., 1988; Bueb et al., 1990, 1993). The mechanism of histamine release is proposed to involve interaction with sialic acid residues on the cell surface, and with G_i-like proteins, with resultant activation of PLC and consequent rise in cytosolic Ca^{2+} (Bueb et al., 1990). Bueb et al. (1993) showed that arginine residues in both positions 1 and 9, as in BK or [DPhe[7]]-BK, is favorable for a histamine-releasing potential, whereas acetylation of the

N-terminal amino-acid causes a dramatic reduction in histamine release.

3.1.2   Blood Flow

The vasodilator and constrictor effects of BK, and the receptors involved in effects of kinins in large blood vessels or perfused organs, are well documented (Regoli and Barabé, 1980; Hall, 1992). B_1 and B_2 receptor binding sites have been demonstrated on the vascular endothelium (Keravis et al., 1991; Sung et al., 1988) and smooth muscle (Hirata et al., 1989; Schneck et al., 1994) of several large vessels. In most cases, vasodilator responses result from both B_2 and/or B_1 receptor-mediated release of nitric oxide and/or prostanoids (Hirata et al., 1989; Schini et al., 1990; Ohde et al., 1991; Toda et al., 1987). Often, the resultant effect and type of receptor involved is dependent on the initial vessel tone (DeWitt et al., 1994). In contrast to larger blood vessels, there is less information regarding the effects and possible control via BK receptors of microvascular blood vessels, although recent studies have demonstrated a vasodilator effect of kinins in the microcirculation of the skin in several species. In monkey skin, intradermal BK evoked a flare-like vasodilatation, although the receptor types, or possible involvement of histamine released from mast cells, was not determined (Treede et al., 1990). In man, intradermal BK also evokes a flare response (Manning et al., 1991) and this probably involves B_2 receptor stimulation since desArg[9]-BK was effective only at very high doses. Further, antihistamine drugs did not inhibit the flare response (Polosa et al., 1993). In both human and rabbit skin, the increase in blood flow elicited by intradermal administration of kinins is mediated by B_2 receptor-mediated release of prostanoids and nitric oxide (Warren and Loi, 1995). A B_2 receptor-induced increase in blood flow has also been demonstrated in the microvasculature of the guinea-pig airways (Yamawaki et al., 1994) and rat gastric mucosa (Pethö et al., 1994). In both cases, the effect of BK was inhibited by Hoe 140.

3.1.3   Microvascular Permeability

Kinins increase microvascular permeability and promote plasma extravasation in many species, including in man, where intradermal BK produces a characteristic weal response (Polosa et al., 1993). Relaxation of arteries and constriction of venules, as described above, leads to an increase in the pressure differential across the capillary bed, thus promoting extravasation. The microvascular leak and edema, evoked by kinins, results from stimulation of B_2 receptors, although under certain conditions, B_1 receptors may also be involved (Campos and Calixto, 1995). Early antagonists of the [DPhe[7]]-BK type, inhibited plasma extravasation evoked by BK in rabbit skin (Schachter et al., 1987; Griesbacher and Lembeck, 1987; Whalley et al., 1987b) and rat skin (Steranka et al., 1989). More recently, Hoe 140 has also been shown to abrogate leakage in various organs, including in the trachea, duodenum and bladder of the rat (Lembeck *et al.*, 1991), and in the airways (Sakamoto *et al.*, 1992, 1993; Campos and Calixto, 1995). A recently developed analog, S 16118 (Félétou *et al.*, 1995a,c), which does not disrupt mast-cells and release histamine, also inhibits extravasation induced by vagal stimulation in guinea-pig airways. In the hamster cheek pouch, NPC 349 (Murray *et al.*, 1991) and S 16118 (Félétou *et al.*, 1995c) inhibited formation of leaky sites in response to BK, indicating an involvement of endothelial B_2 receptors in plasma extravasation. In contrast, in guinea-pig nasal and ocular conjunctiva, BK-induced microvascular leakage is due predominantly to stimulation of trigeminal sensory nerves with release of neuropeptides (Figini *et al.*, 1995). We recently used Win 64338, to demonstrate that BK-evoked plasma extravasation in these two tissues results from B_2 receptor activation (Hall *et al.*, 1995).

Antagonists have been invaluable in establishing a role for endogenous kinins in the microvascular leak produced by inflammatory agents. For example, B_2 receptor antagonists inhibit the microvascular leakage-induced stimuli such as carrageenin (Costello and Hargreaves, 1989; Burch and DeHaas, 1990; Wirth *et al.*, 1991a, 1992a; Damas and Remacle-Volon, 1992), uric acid (Damas and Remacle-Volon, 1992), PLA_2 (Cirino *et al.*, 1991) and endotoxin (Ueno *et al.*, 1995). In contrast, zymosan-induced extravasation was unaffected by Hoe 140 (Damas and Remacle-Volon, 1992). Hoe 140 inhibits plasma extravasation associated with scalding (Wirth *et al.*, 1992a), as well as caerulein-induced edema in rat pancreas (Griesbacher and Lembeck, 1992b) (see Section 3.5 and Chapter 12), and formalin-induced edema in the mouse hind paw (Correa and Calixto, 1993). Also, B_2 receptor stimulation after intrathecal BK administration has been shown to evoke an increase in cutaneous vascular permeability in rats, leading to the proposal that BK has a role as a spinal mediator causing increased peripheral vascular permeability through sensory and cholinergic vagal nerve mechanism involving a spinobulbar pathway (Jacques and Couture, 1990).

3.1.4   Neurogenic Inflammation

Kinins stimulate and sensitize peripheral primary afferent neurons, which results in the release of other pro-inflammatory mediators, including tachykinins and calcitonin gene-related peptide (CGRP) (Geppetti, 1993). This involvement of sensory nerves in inflammation is known as "neurogenic inflammation" (see Chapter 9). Microvascular leakage and edema, resulting from B_2 receptor activation on primary afferent nerves, occurs in several tissues including the mouse paw (Shibata *et al.*, 1986), guinea-pig airways (Nakajima *et al.*, 1994), and guinea-pig nasal mucosa and conjunctiva (Hall *et al.*, 1995) (Fig. 2.5). In guinea-pig trachea, antigen challenge in immunized animals evoked an increase in plasma extravasation, which was inhibited both by a tachykinin NK_1 receptor antagonist and also by the kinin B_2 receptor

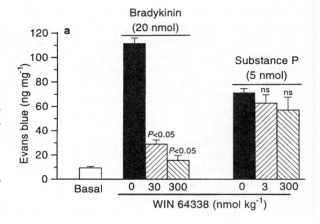

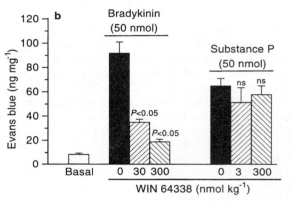

Figure 2.5   The effect of the nonpeptide bradykinin B_2 receptor antagonist, Win 64338, on plasma extravasation in two models of neurogenic inflammation in guinea-pigs. In (a), in the conjunctiva, Win 64338 (30 and 300 nmol/kg) significantly inhibited extravasation in response to bradykinin (20 nmol), but was without effect on substance P (5 nmol). Win 64338, or its vehicle (dimethyl sulfoxide), was administered i.v. 15 min prior to bradykinin of substance P. Mean data ± SEM are shown for at least six experiments. In (b), in the nasal mucosa, Win 64338 (30 and 300 nmol/kg) significantly inhibited extravasation in response to bradykinin (50 nmol), but was without effect on substance P (50 nmol). Data are mean ± SEM of at least five experiments. Extravasation in the absence of treatment (basal) is indicated by open columns. Plasma extravasation was estimated with Evan's blue leakage. For details, see Hall et al. (1995).

antagonist, Hoe 140, suggesting a neurogenic inflammatory mechanism (Bertrand *et al.*, 1993). In anesthetized rats, chemically induced cystitis was inhibited by Hoe 140 via a mechanism involving capsaicin-sensitive sensory nerves (Maggi *et al.*, 1993) (see Chapter 14), and the increase in plasma extravasation evoked by low pH in guinea-pig conjunctiva was also inhibited by this antagonist (Figini *et al.*, 1995) (Fig. 2.5).

3.2 B₁ RECEPTOR INDUCTION AND INFLAMMATION

A role for B_1 receptors has been proposed in certain inflammatory conditions including bone resorption, in a number of chronic inflammatory diseases, including periodontitis, rheumatoid arthritis and osteomyelitis (Ljunggren and Lerner, 1990), chronic cystitis (Marceau et al., 1980), and in persistent inflammatory hyperalgesia (Perkins et al., 1993; Davis et al., 1994) (see Section 3.3, and Chapters 8, 9 and 13). B_1 receptors may modulate enkephalin production, following noxious stimuli to the dental pulp (Inoki and Kudo, 1986), and also where there is neovascularization, which may have relevance to chronic inflammatory disease (Hu and Fan, 1993; see Chapter 18).

The mechanism(s) whereby B_1 receptors contribute to inflammation is unique in as much as they are induced as a response to tissue damage, noxious stimuli and by cytokines. It has been suggested that kinin degradation products, particularly desArg-kinin metabolites, which are inactive at constitutive B_2 receptors, activate newly expressed B_1 receptors, whose synthesis is induced as a result of inflammatory insult, and that this process initiates local inflammatory reactions. Goldberg et al. (1976) first reported an increase in sensitivity with time in vitro of responses to kinins in canine isolated saphenous vein, and this process is now known to result from induction of the expression of B_1 receptors. This phenomenon was subsequently demonstrated in vitro in several vascular tissues (Regoli et al., 1977, 1978; Whalley et al., 1983), in some nonvascular tissues (Marceau et al., 1980; Couture et al., 1982; Boschcov et al., 1984; Marceau and Tremblay, 1986), and in vivo (Regoli et al., 1981). B_1 receptor induction has been demonstrated in almost all B_1 receptor systems studied, with the exceptions of certain cell lines (Goldstein and Wall, 1984; Sung et al., 1988), and in the rat renal vasculature (Guimarães et al., 1986). The characteristics and mechanisms of induction of the B_1 receptor response are thought to involve endogenous cytokines (Bouthillier et al., 1987; DeBlois et al., 1988), and have been reviewed (Marceau, 1983, 1995) (see Chapter 8).

3.3 PAIN AND PERIPHERAL INFLAMMATORY HYPERALGESIA

The algesic and hyperalgesic properties of kinins in many species, including man, have been recognized for some time (Armstrong et al., 1952, 1953; Steranka et al., 1988; Manning et al., 1991; Kindgen-Milles et al., 1994). BK directly stimulates and sensitizes those C- and A-sensory fibers that encode noxious stimuli (e.g. Szolscányi, 1987; Fox et al., 1993). Moreover, BK receptor binding sites have been demonstrated on neuronal pathways associated with pain transmission (Steranka et

al., 1989). There is now substantial evidence that endogenous kinins contribute to pain and hyperalgesia in animal models (Chau et al., 1991; Heapy et al., 1993), and it is clear that locally produced kinins at sites of tissue injury contribute to the ensuing pain and inflammatory hyperalgesia (Armstrong, 1970). The role of kinins in pain and hyperalgesia has been reviewed (Dray and Perkins, 1993), and is detailed in Chapter 9.

In most studies of pain and hyperalgesia, B_1 receptors are not involved since B_1 receptor agonists do not elicit nociception, and B_1 receptor antagonists are ineffective against acute, kinin-induced pain. However, studies in models of persistent hyperalgesia have demonstrated an involvement of B_1 receptors (Perkins et al., 1992, 1993; Davis and Perkins, 1994) and, further, have shown that IL-1 induces expression of B_1 receptor-mediated thermal hyperalgesia in rats (Perkins and Kelly, 1994).

BK-evoked pain and hyperalgesia mediated via B_2 receptors has been demonstrated in the human blister base (Whalley et al., 1987a), reflex, nociceptive hypotension responses to BK perfused through the rabbit ear (Griesbacher and Lembeck, 1987; Lembeck et al., 1991), BK-elicited vascular pain and cutaneous hyperalgesia in rats (Steranka et al., 1988), and mouse abdominal constriction in response to intraperitoneal kaolin or acetic acid (Heapy et al., 1993). There is also evidence from electrophysiological studies that supports a role for B_2 receptors in pain and hyperalgesia. Thus, neurons in the dorsal horn of the spinal cord in rats, and polymodal nociceptors in dog testis spermatic nerve, are excited by BK, effects that are inhibited by B_2 receptor antagonists, but not B_1 receptor antagonists (Haley et al., 1989; Mizumura et al., 1990). Hoe 140, NPC 349 and [DPhe^7]-BK inhibited BK-evoked depolarization of nociceptors in rat isolated spinal cord preparations (Dray et al., 1992; Jeftinija, 1994). Also, B_2 receptors depolarize the central as well as peripheral terminals of primary afferent nerves in rat neonatal spinal cord, indicating a role for locally produced kinins in central nociceptive transmission (Dunn and Rang, 1990; Dray et al., 1992).

B_2 receptor antagonists have proved to be invaluable in establishing a role for endogenous kinins in the pain and hyperalgesia resulting from noxious insults. For example, NPC 349 (Haley et al., 1989) and Hoe 140 (Chapman and Dickenson, 1992) inhibit responses of dorsal horn neurons to subcutaneous formalin, and NPC 567 inhibits urate-induced hyperalgesia in rats (Steranka et al., 1988). A recent study demonstrated inhibition of thermal and mechanical hyperalgesia by the B_1-selective antagonist desArg^9-[Leu^8]-BK, though not by the B_2 receptor antagonist Hoe 140 (Perkins et al., 1992). A recent report, suggests that B_1 and B_2 receptors are involved in the late phase of pain response to formalin in mice (Correa and Calixto, 1993). In rat hind-paws, mechanical hyperalgesia, following carrageenin and endotoxin treatment, was inhibited by Hoe 140 (Ferreira et al., 1993).

3.4 AIRWAYS DISEASE

3.4.1 Lower Airways and Asthma

Evidence supporting a role for kinins in the control of airways function and in asthma, has been reviewed extensively (Collier, 1970; Farmer, 1991; Pongracic et al., 1991; Barnes, 1992) and is the subject of Chapter 15 of this volume. We limit our discussion to studies of BK receptors involved in the effects of kinins in the airways, concentrating on the recent evidence for a role of endogenous kinins in airway inflammation.

Kinins have potent effects in the lower airways of most species, evoking bronchoconstriction, stimulation of sensory nerves (with cough), increased mucus secretion and microvascular leakage with edema (Barnes et al., 1988). In radioligand-binding studies in airways tissue from several species (Field et al., 1992b; Trifilieff et al., 1991), B_2, but not B_1 binding sites have been demonstrated. The autoradiographic distribution of these sites in guinea-pig and human lung correlated well with functional effects of kinins, and the distribution suggests that BK has a direct action on the smooth muscle in peripheral airways, and an indirect effect in the central (proximal) airways (Mak and Barnes, 1991). The presence of B_2 receptors in the airways is also suggested by extensive studies in vitro. In guinea-pig trachea, where kinin effects have been studied extensively, the nature of receptors is discussed in Section 2.3.1. BK, and B_2 receptor-selective agonists such as $[Hyp^3,(TyrMe)^8]$-BK, stimulate human isolated small bronchus, and this response was inhibited by the B_2 receptor antagonist Hoe 140, although not by desArg^9-[Leu^8]-BK (Molimard et al., 1994).

In humans and guinea pigs in vivo, aerosol or instilled BK causes bronchoconstriction indirectly by stimulating peripheral sensory nerve endings, which cause reflex acetylcholine release from postganglionic vagal nerve terminals (Fuller et al., 1987; Ichinose and Barnes, 1990; Ichinose et al., 1990). In guinea-pig airways, i.v. BK-induced bronchoconstriction is unaffected by desArg^9-[Leu^8]-BK and, further, the B_1 receptor agonist, desArg^9-BK, is not a bronchoconstrictor. In contrast, the B_2 receptor antagonists, NPC 349 (Jin et al., 1989; Ichinose and Barnes, 1990; Ichinose et al., 1990) and Hoe 140 (Lembeck et al., 1991; Wirth et al., 1991b, 1993; Sakamoto et al., 1992, 1993), inhibited the bronchoconstrictor response to BK in this species.

In addition to smooth muscle effects in the airways, kinins also activate airway epithelial Cl$^-$ secretion, an effect likely to involve B_2 receptors, which have been localized to epithelial cells (Leikauf et al., 1985; Mak and Barnes, 1991). Further, B_2 receptors may be involved in the control of ciliary beat-frequency induced by BK (Tamaoki et al., 1989). BK, infused intravenously, induced microvascular leakage in the rat trachea, and this was abolished by Hoe 140 (Lembeck et al., 1991). BK (i.v. or inhaled) causes microvascular leakage in guinea-pig

airways and this, too, is inhibited by Hoe 140 (Sakamoto et al., 1992).

In allergic sheep, an experimental model of asthma, inhalation of a peptide B_2 receptor antagonist, NPC 567, significantly inhibited BK-induced neutrophil influx and bronchial hyperreactivity, and the late response to allergen challenge (Abraham, 1991; Abraham et al., 1991a,b).

As mentioned above, aerosol BK causes cough in humans, as well as bronchoconstriction in asthmatic subjects (Fuller et al., 1987). BK-induced cough is not sensitive to aspirin, but is attenuated by a muscarinic antagonist, ipratropium bromide, suggesting a cholinergic reflex mechanism (Fuller et al., 1987). This may have important consequences for future clinical uses of BK antagonists, as the high incidence of ACE inhibitor-induced dry cough during treatment of hypertension in humans, has been speculated to be due to endogenous BK accumulation (Karlberg, 1993; Sunman and Sever; 1993). Therefore, BK antagonists may be useful in treating these adverse effects (Trifilieff et al., 1993).

3.4.2 Upper Airways and Rhinitis

In the upper airways, BK induces symptoms of sore throat and rhinitis, which is independent of atopic state (Proud et al., 1988), and there is now abundant evidence supporting a contribution by kinins to the symptoms of several types of rhinitis in man (Proud et al., 1983; Pongracic et al., 1991; Baraniuk et al., 1994; Rajakulasingam et al., 1992). For example, kinins appear to contribute to both the initial and late phases of allergic rhinitis, and in rhinitis associated with Rhinovirus infections (Proud et al., 1983; Naclerio et al., 1985, 1988). These observations have led to the proposal that BK receptor antagonists may have a role in treating symptoms of the common cold.

In normal human subjects, intranasal administration of BK produces symptoms characteristic of Rhinovirus infection (Proud et al., 1988). B_2 receptors appear to be involved, desArg^9-BK does not produce symptoms (Rajakulasingam et al., 1991). Further, specific binding to B_2 receptor sites has been demonstrated in the nasal turbinate of guinea-pigs (Fugiwara et al., 1989) and humans (Baraniuk et al., 1990). Clinical trials were initiated to investigate the potential of a B_2 receptor antagonist, DArg-[Hyp^3,DPhe^7]-BK (NPC 567) for the treatment of the common cold. Unfortunately, this antagonist failed to inhibit the symptoms, or the increase in vascular permeability in response to exogenous BK (Pongracic et al., 1991), and trials were discontinued. In a further clinical trial, NPC 567 was tested on human volunteers with Rhinovirus-induced colds. Again, however, the antagonist was ineffective (Higgins et al., 1990). The possible reasons for the lack of effect of the antagonists in controlling symptoms are several (see Hall, 1992; Chapter 17). One explanation is the low receptor affinity and susceptibility of this peptide to peptidases, and recent studies with Hoe 140, which is potent and metabolically stable, support this explanation (Austin and Foreman, 1994). Furthermore, Hoe 140 inhibits nasal blockage caused by antigen challenge in subjects with allergic rhinitis, although it had no effect on vascular permeability (Austin et al., 1994). A preliminary report suggests that Hoe 140 does not relieve the symptoms of seasonal allergic rhinitis caused by exposure to pollen (Akbary and Bender, 1993).

3.5 PANCREATITIS

The role of the kallikrein–kinin system in pancreatitis is discussed in detail in Chapter 12. There is evidence that B_2 receptor antagonists may prove beneficial in the treatment of pancreatitis (Yotsumoto et al., 1993). In a model of acute necrotizing pancreatitis induced by sodium taurocholate, Hoe 140 decreased prostaglandin synthesis, plasma lipase levels, and nitric oxide and thromboxane B_2 levels (Closa et al., 1995). Similarly, NPC 349 prevents pancreatic subcellular redistribution of lysosomal enzymes in rats with caerulein-induced pancreatitis (Hirano and Takeuchi, 1994). Similarly, another B_2 receptor antagonist, S 16118, was effective in attenuating caerulein-induced edema in rats (Félétou et al., 1995a).

3.6 ENDOTOXIN AND SEPTIC SHOCK

There is abundant literature pointing to a role for kinins in bacterial endotoxin (lipopolysaccharide, LPS) and septic shock (Colman and Wong, 1979) (see Chapter 10). In rats, infusion of a B_2 receptor antagonist, Lys,Lys-[Hyp^2,$Thi^{5,8}$,$DPhe^7$]-BK, prior to intravenous bolus of endotoxin, substantially reduced the fall in blood pressure (Weipert et al., 1988). In a similar study, NPC 567 inhibited the initial fall in blood pressure and significantly reduced mortality induced by LPS (Wilson et al., 1989). Recently, CP-0127 (see Section 2.2.2.3) was reported to inhibit the BK-induced hypotension in LPS-treated rabbits, and the dual B_1/B_2 receptor antagonist heterodimer, CP-0364 (Section 2.1.2.3), inhibited both BK- and desArg^9-BK-induced hypotension in LPS-treated rabbits (Whalley et al., 1992). This is particularly interesting since it has been known since the 1980s that B_1 receptors are induced by LPS treatment in rabbits (Regoli and Barabé, 1980; Regoli et al., 1981; Marceau, 1995). Thus, in rabbits, a sublethal dose of LPS causes dose-dependent hypotension in response to intra-arterial desArg^9-BK after 5 h (Marceau, 1995).

A potent B_2 receptor antagonist, NPC 17761, administered with a leukocyte recruitment inhibitor, increased survival in a rodent model of sepsis (Otterbein et al., 1993), and NPC 17731 inhibited sepsis-induced acute lung injury (Ridings et al., 1995). The stable and potent antagonist, S 16118, reduced neither mortality in LPS-treated mice, nor the hypotensive responses in LPS-

pretreated rabbits (Félétou et al., 1995a). Also, S 16118 did not increase the survival rate in mice subjected to cecal ligation, a well-known model of Gram-negative sepsis. The reasons for discrepancies between the efficacy of the various BK receptor antagonists in these models, is not clear.

A systematic investigation in a variety of species, of the relative roles of B_1 and B_2 receptors in septic shock would be of value in addressing these questions. Stewart (1995) noted that, in some varieties of sepsis, ACE virtually disappears from the pulmonary circulation (which normally cleaves about 99% of BK into inactive fragments in a single passage). This leaves other enzymes, including carboxypeptidase N (a soluble enzyme in plasma that gives BK a circulating $t_{1/2}$ of around 15 s), whose desArg products of BK and Lys-BK both activate B_1 receptors, including those mediating hypotension. Given that there is evidence that LPS, and other initiators of inflammation can induce B_1 receptor expression (Regoli et al., 1990b; Marceau, 1995; Stewart, 1995), it is perhaps not surprising that antagonists with both B_1 receptor and B_2 receptor blocking activity are more effective in some animal models of hypotensive shock with cardiovascular collapse.

3.7 INFLAMMATORY JOINT DISEASE

There is evidence for the involvement of kinins in some types of inflammatory joint disease (Sharma, 1991; Bhoola et al., 1992; Sharma and Buchanan, 1994) (see Chapter 13). A role for B_1 receptors (and B_2 receptors, though with different time-courses) has been proposed to mediate bone resorption in chronic inflammatory diseases, including periodontitis, rheumatoid arthritis and osteomyelitis (Ljunggren and Lerner, 1990). In inflamed rat knees, B_1 receptor activation evokes an increase in plasma extravasation whereas, in normal animals, the involvement of B_2 receptors predominates (Cruwys et al., 1994).

3.8 CENTRAL NERVOUS SYSTEM

Kinins have been implicated in certain CNS inflammatory conditions, and kinin receptor antagonists have been tested in animal models of brain and spinal cord trauma, and ischemia (Ellis et al., 1987; Francel, 1992; Walker et al., 1995). Since B_2 receptor binding sites have been demonstrated on cerebral blood vessels and on the central endings of primary afferent neurons, and vasodilatation is blocked by B_2 receptor antagonists, it has been suggested that B_2 receptor stimulation may contribute to the vasodilatation and pain of migraine (Ellis et al., 1987; Fujiwara et al., 1988; Homayoun and Harik, 1991; Macfarlane et al., 1991). Finally, cerebral microglia are the main immune effector cells in the CNS (Perry et al., 1993), and B_2 receptor binding sites, and membrane potential changes due to BK, have been demonstrated on these cells (Cholewinski *et al.*, 1991; Hosli and Hosli, 1993a,b; Hosli *et al.*, 1993).

4.   Summary and Future Directions

In the summary section of an earlier review (Hall, 1992), it was noted that the circumstances of the discovery of a mediator tend to have a great influence on future directions of research. The point was made that earlier studies of the roles and actions of kinins were largely vascular, and the hope was expressed that the emphasis might move to a more catholic range of investigations. Looking at the worldwide annual output of *ca* 300 papers on kinins since then, it can be seen that these hopes have been realized. Major recent topics have been the development of novel and effective BK receptor antagonists, and considerable advances in our understanding of the receptors and their genes, largely through the applications of molecular biology. The question that now arises is, what pathophysiological states are likely to be successfully dissected, and hopefully treated, through application of such new-found knowledge and techniques?

In considering the range of putative roles for BK in physiology and pathophysiology, the interested reader may refer to an exhaustive and provocative recent review by Stewart (1995), which catalogs an extensive list ranging through topics in pain, inflammation, edema, shock, rhinitis and asthma. Some indications are given as to types of antagonist being used in various sorts of clinical trial. As expected, peptide antagonists are more likely to be efficacious where they can be applied locally, for instance in rhinitis, the treatment of burn pain and in some types of neurogenic inflammation. More stable and nonpeptide antagonists, which can be administered systemically, are more likely to be used in acute treatment of otherwise fatal conditions such as shock and edema. As yet, there seems less progress in areas where there is defective ion transport such as cholera, Crohn's disease and cystic fibrosis, or in chronic inflammatory conditions such as rheumatoid and osteoarthritis. Application of knowledge regarding the role of B_1 receptors seems mainly in the area of the treatment of septic shock. However, with the increased awareness of the role of B_1 receptors in hyperalgesia, and when active and stable receptor antagonists for this subtype become available, rapid progress may reasonably be expected.

A new aspect of our understanding that may come from molecular biology studies of the BK receptors is the possible identification of factors predisposing to disease in humans. These might include phenotypic variants in receptor genes, or alternative splicing or post-translational processing of the receptor protein. Certainly the future looks bright, and this period may come to be regarded as the renaissance of kinin research.

5.   Acknowledgement

Judith Hall thanks the Wellcome Trust for a Research Fellowship.

6.   References

Abraham, W.M. (1991). Bradykinin antagonists in a sheep model of allergic asthma. In "Bradykinin Antagonists: Basic and Clinical Research" (ed. R.M. Burch), pp. 261–276. Marcel Dekker, New York.

Abraham, W.M., Ahmed, A., Cortes, A., Soler, M., Farmer, S.G., Baugh, L.E. and Harbeson, S.L. (1991a). Airway effects of inhaled bradykinin, substance P, and neurokinin A in sheep. J. Allergy Clin. Immunol. 87, 557–564.

Abraham, W.M., Burch, R.M., Farmer, S.G., Sielczak, M.W., Ahmed, A. and Cortes, A. (1991b). A bradykinin antagonist modifies allergen-induced mediator release and late bronchial responses in sheep. Am. Rev. Respir. Dis. 143, 787–796.

Akbary, M.A. and Bender, N. (1993). Efficacy and safety of HOE 140 (a bradykinin antagonist) in patients with mild-to-moderate seasonal allergic rhinitis. Proceedings of Kinins '93, Brazil, Abstract 22.02.

Albert, P.R. and Morris, S.J. (1994). Antisense knockouts: molecular scalpels for the dissection of signal transduction. Trends Pharmacol. Sci. 15, 250–254.

Armstrong, D. (1970). Pain. In "Handbook of Experimental Pharmacology: Bradykinin, Kallidin and Kallikrein" (ed. E.G. Erdös), pp. 434–481. Springer-Verlag, Berlin.

Armstrong, D., Dry, R.M.L., Keele, C.A. and Markham, J.W. (1952). Pain-producing substances in blister fluid and in serum. J. Physiol. 117, 4P–5P.

Armstrong, D., Dry, R.M.L., Keele, C.A. and Markham, J.W. (1953). Observations on chemical excitants of cutaneous pain in man. J. Physiol. 120, 326–351.

Austin, C.E. and Foreman, J.C. (1994). A study of the action of bradykinin and bradykinin analogues in the human nasal airway. J. Physiol. 478, 351–356.

Austin, C.E., Foreman, J.C. and Scadding, G.K. (1994). Reduction by Hoe 140, the B_2 kinin receptor antagonist, of antigen-induced nasal blockage. Br. J. Pharmacol. 111, 969–971.

Babbedge, R., Dray, A. and Urban, L. (1995). Bradykinin depolarises the rat isolated superior cervical ganglion via B_2 receptor activation. Neurosci. Lett. 193, 161–164.

Babiuk, C., Marceau, F., St-Pierre, S. and Regoli, D. (1982). Kininases and vascular responses to kinins. Eur. J. Pharmacol. 78, 167–174.

Baenziger, N.L., Jong, Y.J., Yocum, S.A., Dalemar, L.R., Wilhelm, B., Vavrek, R. and Stewart, J.M. (1992). Diversity of B_2 bradykinin receptors with nanomolar affinity expressed in passaged IMR90 human lung fibroblasts. Eur. J. Cell Biol. 58, 71–80.

Bao, G., Qadri, F., Stauss, B., Stauss, H., Gohlke, P. and Unger, T. (1991). HOE 140, a new highly potent and long-acting bradykinin antagonist in conscious rats. Eur. J. Pharmacol. 200, 179–182.

Barabé, J., Drouin, J.-N., Regoli, D. and Park, W.K. (1977). Receptors for bradykinin in intestinal and uterine smooth muscle. Can. J. Physiol. Pharmacol. 55, 1270–1285.

Barabé, J., Babiuk, C. and Regoli, D. (1982). Binding of [^{3}H]des-Arg^9-BK to rabbit anterior mesenteric vein. Can. J. Physiol. Pharmacol. 60, 1551–1555.

Baraniuk, J.N., Lundgren, J.D., Mizoguchi, H., Peden, D., Gawin, A., Merida, M., Shelhamer, J.H. and Kaliner, M.A. (1990). Bradykinin and respiratory mucous membranes: analysis of bradykinin binding site distribution and secretory responses in vitro and in vivo. Am. Rev. Respir. Dis. 141, 706–714.

Baraniuk, J.N., Silver, P.B., Kaliner, M.A. and Barnes, P.J. (1994). Perennial rhinitis subjects have altered vascular, glandular, and neural responses to bradykinin nasal provocation. Int. Arch. Allergy Immunol. 103, 202–208.

Barnes, P.J. (1992). Bradykinin and asthma. Thorax 47, 979–983.

Barnes, P.J., Chung, K.F. and Page, C.P. (1988). Inflammatory mediators and asthma. Pharmacol. Rev. 40, 49–84.

Bascands, J.-L., Pecher, C., Rouaud, S., Emond, C., Tack, J.L., Bastie, M.J., Burch, R., Regoli, D. and Girolami, J.-P. (1993). Evidence for existence of two distinct bradykinin receptors on rat mesangial cells. Am. J. Physiol. 264, F548–F556.

Bathon, J.M. and Proud, D. (1991). Bradykinin antagonists. Annu. Rev. Pharmacol. Toxicol. 31, 129–162.

Bathon, J.M., Proud, D., Krackow, K. and Wigley, F.M. (1989). Preincubation of human synovial cells with IL-1 modulates prostaglandin E$_2$ release in response to bradykinin. J. Immunol. 143, 579–586.

Bathon, J.M., Manning, D.C., Goldman, D.W., Towns, M.C. and Proud, D. (1992). Characterization of kinin receptors on human synovial cells and upregulation of receptor number by interleukin-1. J. Pharmacol. Exp. Ther. 260, 384–392.

Benetos, A., Gavras, H., Stewart, J.M., Vavrek, R.J., Hatinoglou, S. and Gavras, I. (1986). Vasodepressor role of endogenous bradykinin assessed by a bradykinin antagonist. Hypertension 8, 971–974.

Bertrand, C., Nadel, J.A., Yamawaki, I. and Geppetti, P. (1993). Role of kinins in the vascular extravasation evoked by antigen and mediated by tachykinins in guinea pig trachea. J. Immunol. 151, 4902–4907.

Bhoola, K.D., Elson, C.J. and Dieppe, P.A. (1992a). Kinins–key mediators in inflammatory arthritis? Br. J. Rheumatol. 31, 509.

Bhoola, K.D., Figueroa, C.D. and Worthy, K. (1992b). Bioregulation of kinins: kallikreins, kininogens and kininases. Pharmacol. Rev. 44, 1–80.

Birch, P.J., Fernandes, L.B., Harrison, S.M. and Wilkinson, A. (1991). Pharmacological characterisation of the receptors mediating the contractile responses to bradykinin in the guinea pig ileum and rat uterus. Br. J. Pharmacol. 102, 170P.

Borkowski, J.A., Ransom, R.W., Seabrook, G.R., Trumbauer, M., Chen, H., Hill, R.G., Strader, C.D. and Hess, J.F. (1995). Targeted disruption of a B$_2$ bradykinin receptor gene in mice eliminates bradykinin action in smooth muscle and neurons. J. Biol. Chem. 270, 13706–13710.

Boschcov, P., Paiva, A.C., Paiva, T.B. and Shimuta, S.I. (1984). Further evidence for the existence of two receptor sites for bradykinin responsible for the diphasic effect in the rat isolated duodenum. Br. J. Pharmacol. 83, 591–600.

Bouthillier, J., DeBlois, D. and Marceau, F. (1987). Studies on the induction of pharmacological responses to des-Arg^9-bradykinin in vitro and in vivo. Br. J. Pharmacol. 92, 257–264.

Braas, K.M., Manning, D.C., Perry, D.C. and Snyder, S.H. (1988). Bradykinin analogues: differential agonist and antagonist activities suggesting multiple receptors. Br. J. Pharmacol. 94, 3–5.

Bramley, A.M., Samhoun, M.M. and Piper, P.J. (1988). Modulation of bradykinin induced responses by indomethacin and the removal of the epithelium of the guinea pig trachea in vitro. Br. J. Pharmacol. 95, 774P.

Bramley, A.M., Samhoun, M.M. and Piper, P.J. (1989). Effect of a bradykinin B$_2$ antagonist on responses of intact and rubbed guinea pig trachea in vitro. Br. J. Pharmacol. 98, 786P.

Bramley, A.M., Samhoun, M.N. and Piper, P.J. (1990). The role of the epithelium in modulating the responses of guinea pig trachea induced by bradykinin in vitro. Br. J. Pharmacol. 99, 762–766.

Brenner, N.J., Stonesifer, G.Y., Schneck, K.A., Burns, H.D. and Ransom, R.W. (1993). [^{125}I]PIP HOE 140, a high affinity radioligand for bradykinin B$_2$ receptors. Life Sci. 53, 1879–1886.

Brown, J.F., Whittle, B.J.R. and Hanson, P.J. (1992). Bradykinin stimulates PGE$_2$ release in cell fractions isolated from the rat gastric mucosa via a B$_1$ receptor. Biochem. Soc. Trans. 20, 276S.

Bueb, J.-L., Mousli, M., Bronner, C., Rouot, B. and Landry, Y. (1990). Activation of G$_i$-like proteins, a receptor-independent effect of kinins in mast cells. Mol. Pharmacol. 38, 816–822.

Bueb, J.-L., Mousli, M., Landry, Y. and Regoli, D. (1993). Structure-activity studies of bradykinin analogues on rat mast cell histamine release. Peptides 14, 685–689.

Burch, R.M. and DeHaas, C.J. (1990). A bradykinin antagonist inhibits carrageenin edema in rats. Naunyn-Schmiedeberg's Arch. Pharmacol. 342, 189–193.

Burch, R.M. and Kyle, D.J. (1992). Recent developments in the understanding of bradykinin receptors. Life Sci. 50, 829–838.

Burch, R.M. and Tiffany, C.W. (1989). Tumor necrosis factor causes amplification of arachidonic acid metabolism in response to interleukin 1, bradykinin, and other agonists. J. Cell Physiol. 141, 85–89.

Burch, R.M., Connor, J.R. and Axelrod, J. (1988). Interleukin 1 amplifies receptor-mediated activation of phospholipase A$_2$ in 3T3 fibroblasts. Proc. Natl. Acad. Sci. USA 85, 6306–6309.

Burch, R.M., Farmer, S.G. and Steranka, L.R. (1990). Bradykinin receptor antagonists. Med. Res. Rev. 10, 237–269.

Butt, S.K., Dawson, L.G. and Hall, J.M. (1995). Bradykinin B$_1$ receptors in the rabbit urinary bladder: Induction of responses, smooth muscle contraction, and phosphatidylinositol hydrolysis. Br. J. Pharmacol. 114, 612–617.

Cahill, M., Fishman, J.B. and Polgar, P. (1988). Effect of des arginine^9-bradykinin and other bradykinin fragments on the synthesis of prostacyclin and the binding of bradykinin by vascular cells in culture. Agents and Actions 24, 224–231.

Calixto, J.B., Yunes, R.A., Rae, G.A. and Medeiros, Y.S. (1991). Nonpeptide bradykinin antagonists. In "Bradykinin Antagonists. Basic and Clinical Research" (ed. R.M. Burch), pp. 97–129. Marcel Dekker, New York.

Campos, A.H. and Calixto, J.B. (1994). Mechanisms involved in the contractile responses of kinins in rat portal vein rings: Mediation by B$_1$ and B$_2$ receptors. J. Pharmacol. Exp. Ther. 268, 902–909.

Campos, M.M. and Calixto, J.B. (1995). Involvement of B_1 and B_2 receptors in bradykinin-induced rat paw oedema. Br. J. Pharmacol. 114, 1005–1013.

Chahine, R., Adam, A., Yamaguchi, N., Gaspo, R., Regoli, D. and Nadeau, R. (1993). Protective effects of bradykinin on the ischaemic heart: Implication of the B_1 receptor. Br. J. Pharmacol. 108, 318–322.

Chakravarty, S., Wilkins, D.E. and Kyle, D.J. (1993). Design of potent, cyclic peptide bradykinin receptor antagonists from conformationally constrained linear peptides. J. Med. Chem. 36, 2569–2571.

Chakravarty, S., Connolly, M. and Kyle, D.J. (1995). Novel pseudopeptides with high affinities for the human bradykinin B_2 receptor. Peptide Res. 8, 16–19.

Chapman, V. and Dickenson, A.H. (1992). The spinal and peripheral roles of bradykinin and prostaglandins in nociceptive processing in the rat. Eur. J. Pharmacol. 219, 427–433.

Chau, T.T., Lewin, A.C., Walter, T.L., Carlson, R.P. and Weichman, B.M. (1991). Evidence for a role of bradykinin in experimental pain models. Agents and Actions 34, 235–238.

Cheronis, J.C., Whalley, E.T. and Blodgett, J.K. (1992a). Bradykinin antagonists: synthesis and in vitro activity of bissuccinimidolalkane peptide dimers. Agents and Actions Suppl. 38/I, 551–558.

Cheronis, J.C., Whalley, E.T., Nguyen, K.T., Eubanks, S.R., Allen, L.G., Duggan, M.J., Loy, S.D., Bonham, K.A. and Blodgett, J.K. (1992b). A new class of bradykinin antagonists: synthesis and in vitro activity of bissuccinimidoalkane peptide dimers. J. Med. Chem. 35, 1563–1572.

Cheronis, J.C., Whalley, E.T., Allen, L.G., Loy, S.D., Elder, M.W., Duggan, M.J., Gross, K.L. and Blodgett, J.K. (1994). Design, synthesis, and in vitro activity of bis(succinimido)-hexane peptide heterodimers with combined B_1 and B_2 antagonist activity. J. Med. Chem. 37, 348–355.

Cholewinski, A.J., Stevens, G., McDermott, A.M. and Wilkin, G.P. (1991). Identification of B_2 bradykinin binding sites on cultured cortical astrocytes. J. Neurochem. 57, 1456–1458.

Christopher, T.A., Ma, X.-L., Gauthier, T.W. and Lefer, A.M. (1994). Beneficial actions of CP-0127, a novel bradykinin receptor antagonist, in murine traumatic shock. Am. J. Physiol. 266, H867–H873.

Churchill, L. and Ward, P.E. (1986). Relaxation of isolated mesenteric arteries by des-Arg^9-bradykinin stimulation of B_1 receptors. Eur. J. Pharmacol. 130, 11–18.

Cirino, G., Cicala, C., Sorrentino, L. and Regoli, D. (1991). Effect of bradykinin antagonists, NG-monomethyl-L-arginine and L-NG-nitro arginine on phospholipase A_2 induced oedema in rat paw. Gen. Pharmacol. 22, 801–804.

Cisar, L.A., Mochan, E. and Schimmel, R. (1993). Interleukin-1 selectively potentiates bradykinin-stimulated arachidonic acid release from human synovial fibroblasts. Cell. Signal. 5, 463–472.

Claeson, G., Fareed, J., Larsson, C., Kindel, G., Arielly, S., Simonsson, R., Messmore, H.L. and Balis, J.U. (1979). Inhibition of the contractile action of bradykinin on isolated smooth muscle preparations by derivatives of low molecular weight peptides. Adv. Exp. Med. Biol. 120B, 691–713.

Closa, D., Hotter, G., Prats, N., Gelpí, E. and Roselló-Catafau, J. (1995). A bradykinin antagonist inhibited nitric oxide generation and thromboxane biosynthesis in acute pancreatitis. Prostaglandins 49, 285–294.

Collier, H.O.J. (1970). Kinins and ventilation of the lungs. In "Handbook of Experimental Pharmacology. Bradykinin, Kallidin and Kallikrein" (ed. E.G. Erdös), pp. 409–420. Springer-Verlag, Berlin.

Colman, R.W. and Wong, P.Y. (1979). Kallikrein–kinin system in pathologic conditions. In "Handbook of Experimental Pharmacology. Bradykinin, Kallidin and Kallikrein" (ed. E.G. Erdös), pp. 569–607. Springer-Verlag, Berlin.

Corrêa, C.R. and Calixto, J.B. (1993). Evidence for participation of B_1 and B_2 kinin receptors in formalin-induced nociceptive response in the mouse. Br. J. Pharmacol. 110, 193–198.

Costello, A.H. and Hargreaves, K.M. (1989). Suppression of carrageenan-induced hyperalgesia, hyperthermia and edema by a bradykinin antagonist. Eur. J. Pharmacol. 171, 259–263.

Couture, R., Mizrahi, J., Regoli, D. and Devroede, G. (1981). Peptides and the human colon: an in vitro pharmacological study. Can. J. Physiol. Pharmacol. 59, 957–964.

Couture, R., Mizrahi, J., Caranikas, S. and Regoli, D. (1982). Acute effects of peptides on the rat colon. Pharmacology 24, 230–242.

Cruwys, S.C., Garrett, N.E., Perkins, M.N., Blake, D.R. and Kidd, B.L. (1994). The role of bradykinin B_1 receptors in the maintenance of intra-articular plasma extravasation in chronic antigen-induced arthritis. Br. J. Pharmacol. 113, 940–944.

Damas, J. and Remacle-Volon, G. (1992). Influence of a long-acting bradykinin antagonist, Hoe 140, on some acute inflammatory reactions in the rat. Eur. J. Pharmacol. 211, 81–86.

Da Silva, A., Dhuy, J., Waedelé, F., Bertrand, C. and Landry, Y. (1992). Endopeptidase 24.15 modulates bradykinin-induced contraction in guinea pig trachea. Eur. J. Pharmacol. 212, 97–99.

Da Silva, A., Amrani, Y., Trifilieff, A. and Landry, Y. (1995). Involvement of B_2 receptors in the bradykinin-induced relaxation of guinea pig isolated trachea. Br. J. Pharmacol. 114, 103–108.

Davis, A.J. and Perkins, M.N. (1994). Induction of B_1 receptors in vivo in a model of persistent inflammatory mechanical hyperalgesia in the rat. Neuropharmacology 33, 127–133.

Davis, A.J., Kelly, D. and Perkins, M.N. (1994). The induction of des-Arg^9-bradykinin-mediated hyperalgesia in the rat by inflammatory stimuli. Braz. J. Med. Biol. Res. 27, 1793–1802.

DeBlois, D., Bouthillier, J. and Marceau, F. (1988). Effect of glucocorticoids, monokines and growth factors on the spontaneously developing responses of the rabbit isolated aorta to des-Arg^9-bradykinin. Br. J. Pharmacol. 93, 969–977.

Denning, G.M. and Welsh, M.J. (1991). Polarized distribution of bradykinin receptors on airway epithelial cells and independent coupling to second messenger pathways. J. Biol. Chem. 266, 12932–12938.

Devillier, P., Renoux, M., Drapeau, G. and Regoli, D. (1988). Histamine release from rat peritoneal mast cells by kinin antagonists. Eur. J. Pharmacol. 149, 137–140.

DeWitt, B.J., Cheng, D.Y. and Kadowitz, P.J. (1994). des-Arg^9-bradykinin produces tone-dependent kinin B_1 receptor-mediated responses in the pulmonary vascular bed. Circ. Res. 75, 1064–1072.

Doctrow, S.R., Abelleira, S.M., Curry, L.A., Heller-Harrison, R., Kozarich, J.W., Malfroy, B., McCarroll, L.A., Morgan, K.G., Morrow, A.R., Musso, G.F., Smart, J.L., Straub, J.A., Turnbull, B. and Gloff, C.A. (1994). The bradykinin analog RMP-7 increases intracellular free calcium levels in rat brain microvascular endothelial cells. J. Pharmacol. Exp. Ther. 271, 229–237.

Donoso, M.V. and Huidobro-Toro, J.P. (1989). Involvement of postjunctional purinergic mechanisms in the facilitatory action of bradykinin in neurotransmission in the rat vas deferens. Eur. J. Pharmacol. 160, 263–273.

D'Orléans-Juste, P., de Nucci, G. and Vane, J.R. (1989). Kinins act on B_1 or B_2 receptors to release conjointly endothelium-derived relaxing factor and prostacyclin from bovine aortic endothelial cells. Br. J. Pharmacol. 96, 920–926.

Drapeau, G., DeBlois, D. and Marceau, F. (1991). Hypotensive effects of Lys-des-Arg^9-bradykinin and metabolically protected agonists of B_1 receptors for kinins. J. Pharmacol. Exp. Ther. 259, 997–1003.

Drapeau, G., Audet, R., Levesque, L., Godin, D. and Marceau, F. (1993). Development and *in vivo* evaluation of metabolically resistant antagonists of B_1 receptors for kinins. J. Pharmacol. Exp. Ther. 266, 192–199.

Dray, A. and Perkins, M. (1993). Bradykinin and inflammatory pain. Trends Neurosci. 16, 99–104.

Dray, A., Patel, I.A., Perkins, M.N. and Rueff, A. (1992). Bradykinin-induced activation of nociceptors: receptor and mechanistic studies on the neonatal rat spinal cord-tail preparation *in vitro*. Br. J. Pharmacol. 107, 1129–1134.

Drouin, J.-N., Gaudreau, P., St-Pierre, S. and Regoli, D. (1979a). Biological activities of kinins modified at the N- or at the C- terminal end. Can. J. Physiol. Pharmacol. 57, 1018–1023.

Drouin, J.-N., Gaudreau, P., St-Pierre, S.A. and Regoli, D. (1979b). Structure–activity studies of [des-Arg^9]-bradykinin on the B_1 receptor of the rabbit aorta. Can. J. Physiol. Pharmacol. 57, 562–566.

Drouin, J.-N., St-Pierre, S.A. and Regoli, D. (1979c). Receptors for bradykinin and kallidin. Can. J. Physiol. Pharmacol. 57, 375–379.

Dunn, F.W. and Stewart, J.M. (1971). Analogs of bradykinin containing -2-thienyl-L-alanine. J. Med. Chem. 14, 779–781.

Dunn, P.M. and Rang, H.P. (1990). Bradykinin-induced depolarization of primary afferent nerve terminals in the neonatal rat spinal cord in vitro. Br. J. Pharmacol. 100, 656–660.

Eggerickx, D., Raspe, E., Bertrand, D., Vassart, G. and Parmentier, M. (1992). Molecular cloning, functional expression and pharmacological characterization of a human bradykinin B_2 receptor gene. Biochem. Biophys. Res. Commun. 187, 1306–1313.

Ellis, E.F., Heizer, M.L., Hambrecht, G.S., Holt, S.A., Stewart, J.M. and Vavrek, R.J. (1987). Inhibition of bradykinin- and kallikrein-induced cerebral arteriolar dilation by a specific bradykinin antagonist. Stroke 18, 792–795.

Emond, C., Bascands, J.-L., Pecher, C., Cabos-Boutot, G., Pradelles, P., Regoli, D. and Girolami, J.-P. (1990). Characterization of a B_2-bradykinin receptor in rat renal mesangial cells. Eur. J. Pharmacol. 190, 381–392.

Emond, C., Pecher, C., Bascands, J.L., Regoli, D. and Girolami, J.P. (1992). Characterization of two different affinity B_2-kinin binding sites in rat glomeruli. Agents Actions Suppl. 38/I, 390–397.

Erdös, E.G. and Sloane, E.M. (1962). An enzyme in human blood that inactivates bradykinin and kallidins. Biochem. Pharmacol. 11, 585–592.

Etscheid, B.G., Ko, P.H. and Villereal, M.L. (1991). Regulation of bradykinin receptor level by cholera toxin, pertussis toxin and forskolin in cultured human fibroblasts. Br. J. Pharmacol. 103, 1347–1350.

Everett, C.M., Hall, J.M., Mitchell, D. and Morton, I.K.M. (1992). Contrasting properties of bradykinin receptor subtypes mediating contractions of the rabbit and pig isolated iris sphincter pupillae preparation. Agents Actions Suppl. 38/II, 378–381.

Falcone, R.C., Hubbs, S.J., Vanderloo, J.D., Prosser, J.C., Little, J., Gomes, B., Aharony, D. and Krell, R.D. (1993). Characterization of bradykinin receptors in guinea pig gall bladder. J. Pharmacol. Exp. Ther. 266, 1291–1299.

Farmer, S.G. (1991). Airways pharmacology of bradykinin. In "Bradykinin Antagonists. Basic and Clinical Research" (ed. R.M. Burch), pp. 213–236. Marcel Dekker, New York.

Farmer, S.G. and Burch, R.M. (1992). Biochemical and molecular pharmacology of kinin receptors. Annu. Rev. Pharmacol. Toxicol. 32, 511–536.

Farmer, S.G. and DeSiato, M.A. (1994). Effects of a novel nonpeptide bradykinin B_2 receptor antagonist on intestinal and airway smooth muscle: further evidence for the tracheal B_3 receptor. Br. J. Pharmacol. 112, 461–464.

Farmer, S.G., Burch, R.M., Meeker, S.A. and Wilkins, D.E. (1989). Evidence for a pulmonary B_3 bradykinin receptor. Mol. Pharmacol. 36, 1–8.

Farmer, S.G., Burch, R.M., Kyle, D.J., Martin, J.A., Meeker, S.N. and Togo, J. (1991a). DArg[Hyp^3-Thi^5-DTic^7-Tic^8]-bradykinin, a potent antagonist of smooth muscle BK_2 receptors and BK_3 receptors. Br. J. Pharmacol. 102, 785–787.

Farmer, S.G., Ensor, J.E. and Burch, R.M. (1991b). Evidence that cultured airway smooth muscle cells contain bradykinin B_2 and B_3 receptors. Am. J. Respir. Cell Mol. Biol. 4, 273–277.

Farmer, S.G., DeSiato, M.A. and Broom, T. (1992). Effect of kinin receptor agonists and antagonists in ferret isolated tracheal smooth muscle: evidence that bradykinin-induced contractions are mediated by B_2 receptors. Br. J. Pharmacol. 107, 32P.

Farmer, S.G., Broom, T. and DeSiato, M.A. (1994). Effects of bradykinin receptor agonists, and captopril and thiorphan in ferret isolated trachea: evidence for bradykinin generation *in vitro*. Eur. J. Pharmacol. 259, 309–313.

Faussner, A., Heinz-Erian, P., Klier, C. and Roscher, A.A. (1991). Solubilization and characterization of B_2 bradykinin receptors from cultured human fibroblasts. J. Biol. Chem. 266, 9442–9446.

Félétou, M., Germain, M., Thurieau, C., Fauchère, J.-L. and Canet, E. (1994). Agonistic and antagonistic properties of the bradykinin B_2 receptor antagonist, Hoe 140, in isolated blood vessels from different species. Br. J. Pharmacol. 112, 683–689.

Félétou, M., Lonchampt, M., Robineau, P., Jamonneau, I., Thurieau, C., Fauchère, J.-L., Villa, P., Ghezzi, P., Prost, J.-F. and Canet, E. (1995a). Effects of the bradykinin B_2 receptor antagonist S 16118 (*p*-guanidobenzoyl-[Hyp^3,Thi^5,DTic^7,-Oic^8]bradykinin) in different *in vivo* animal models of inflammation. J. Pharmacol. Exp. Ther. 273, 1078–1084.

Félétou, M., Martin, C.A.E., Molimard, M., Naline, E., Germain, M., Thurieau, C., Fauchère, J.-L., Canet, E. and Advenier, C. (1995b). *In vitro* effects of HOE 140 in human bronchial and vascular tissue. Eur. J. Pharmacol. 274, 57–64.

Félétou, M., Robineau, P., Lonchampt, M., Bonnardel, E.,

Thurieau, C., Fauchère, J.-L., Widdowson, P., Mahieu, J.-P., Serkiz, B., Volland, J.-P., Martin, C., Naline, E., Advenier, C., Prost, J.-F. and Canet, E. (1995c). S 16118 (p-guanidobenzoyl-[Hyp^3,Thi^5,DTic^7,Oic^8]bradykinin) is a potent and long-acting bradykinin B$_2$ receptor antagonist, *in vitro* and *in vivo*. J. Pharmacol. Exp. Ther. 273, 1071–1077.

Ferreira, S.H., Lorenzetti, B.B. and Poole, S. (1993). Bradykinin initiates cytokine-mediated inflammatory hyperalgesia. Br. J. Pharmacol. 110, 1227–1231.

Field, J.L., Fox, A.J., Hall, J.M., Magbagbeola, A.O. and Morton, I.K.M. (1988). Multiple bradykinin B$_2$ receptor subtypes in smooth muscle preparations? Br. J. Pharmacol. 92, 284P.

Field, J.L., Hall, J.M. and Morton, I.K.M. (1992a). Bradykinin receptors in the guinea pig taenia caeci are similar to proposed BK$_3$ receptors in the guinea pig trachea, and are blocked by HOE140. Br. J. Pharmacol. 105, 293–296.

Field, J.L., Hall, J.M. and Morton, I.K.M. (1992b). Putative novel bradykinin B$_3$ receptors in the smooth muscle of the guinea pig taenia caeci and trachea. Agents Actions Suppl. 38/I, 540–545.

Field, J.L., Butt, S.K., Morton, I.K.M. and Hall, J.M. (1994). Bradykinin B$_2$ receptors and coupling mechanisms in the smooth muscle of the guinea pig taenia caeci. Br. J. Pharmacol. 113, 607–613.

Figini, M., Javdan, P., Cioncolini, F. and Geppetti, P. (1995). Involvement of tachykinins in plasma extravasation induced by bradykinin and low pH medium in the guinea pig conjunctiva. Br. J. Pharmacol. 115, 128–132.

Figueroa, C.D., Gonzalez, C.B., Grigoriev, S., Alla, S.A., Haasemann, M., Jarnagin, K. and Müller-Esterl, W. (1995). Probing for the bradykinin B$_2$ receptor in rat kidney by anti-peptide and anti-ligand antibodies. J. Histochem. Cytochem. 43, 137–148.

Fox, A.J., Barnes, P.J., Urban, L. and Dray, A. (1993). An *in vitro* study of the properties of single vagal afferents innervating guinea pig airways. J. Physiol. 469, 21–35.

Francel, P.C. (1992). Bradykinin and neuronal injury. J. Neurotrauma 9 Suppl. 1, S27–S45.

Frederick, M. and Odya, C.E. (1987). Characterization of soluble bradykinin receptor-like binding sites. Eur. J. Pharmacol. 134, 45–52.

Fujita, H., Usui, H., Kurahashi, K. and Yoshikawa, M. (1995). Isolation and characterization of ovokinin, a bradykinin B$_1$ agonist peptide derived from ovalbumin. Peptides 16, 785–790.

Fujiwara, Y., Mantione, C.R. and Yamamura, H.I. (1988). Identification of B$_2$ bradykinin binding sites in guinea pig brain. Eur. J. Pharmacol. 147, 487–488.

Fujiwara, Y., Bloom, J.W., Mantione, C.R. and Yamamura, H.I. (1989). Identification of B$_2$ bradykinin binding sites in the guinea pig nasal turbinate. Peptides 10, 701–703.

Fuller, R.W., Dixon, C.M., Cuss, F.M. and Barnes, P.J. (1987). Bradykinin-bronchoconstriction in humans. Mode of action. Am. Rev. Respir. Dis. 135, 176–180.

Galizzi, J.P., Bodinier, M.C., Chapelain, B., Ly, S.M., Coussy, L., Giraud, S., Neliat, G. and Jean, T. (1994). Up-regulation of [^{3}H]-des-Arg^{10}-kallidin binding to the bradykinin B$_1$ receptor by interleukin-1 in isolated smooth muscle cells: correlation with B$_1$ agonist-induced PGI$_2$ production. Br. J. Pharmacol. 113, 389–394.

Gaudreau, P., Barabé, J., St-Pierre, S. and Regoli, D. (1981a).

Pharmacological studies of kinins in venous smooth muscles. Can. J. Physiol. Pharmacol. 59, 371–379.

Gaudreau, P., Barabé, J., St-Pierre, S. and Regoli, D. (1981b). Structure–activity study of kinins in vascular smooth muscles. Can. J. Physiol. Pharmacol. 59, 380–389.

Geppetti, P. (1993). Sensory neuropeptide release by bradykinin: mechanisms and pathophysiological implications. Regul. Pept. 47, 1–23.

Gobeil, F. and Regoli, D. (1994). Characterization of kinin receptors by bioassays. Braz. J. Med. Biol. Res. 27, 1781–1791.

Goldberg, M.R., Chapnick, B.M., Joiner, P.D., Hyman, A.L. and Kadowitz, P.J. (1976). Influence of inhibitors of prostaglandin synthesis on venoconstrictor responses to bradykinin. J. Pharmacol. Exp. Ther. 198, 357–365.

Goldstein, R.H. and Wall, M. (1984). Activation of protein formation and cell division by bradykinin and des-Arg^9-bradykinin. J. Biol. Chem. 259, 9263–9268.

Griesbacher, T. and Lembeck, F. (1987). Effect of bradykinin antagonists on bradykinin-induced plasma extravasation, venoconstriction, prostaglandin E$_2$ release, nociceptor stimulation and contraction of the iris sphincter muscle of the rabbit. Br. J. Pharmacol. 92, 333–340.

Griesbacher, T. and Lembeck, F. (1992a). Analysis of the antagonistic actions of HOE 140 and other novel bradykinin analogues on the guinea pig ileum. Eur. J. Pharmacol. 211, 393–398.

Griesbacher, T. and Lembeck, F. (1992b). Effects of the bradykinin antagonist, HOE 140, in experimental acute pancreatitis. Br. J. Pharmacol. 107, 356–360.

Guimarães, J.A., Vieira, M.A., Camargo, M.J. and Maack, T. (1986). Renal vasoconstrictive effect of kinins mediated by B$_1$-kinin receptors. Eur. J. Pharmacol. 130, 177–185.

Haasemann, M., Figueroa, C.D., Henderson, L., Grigoriev, S., Abd Alla, S., Gonzalez, C.B., Dunia, I., Hoebeke, J., Jarnagin, K. and Cartaud, J. (1994). Distribution of bradykinin B$_2$ receptors in target cells of kinin action. Visualization of the receptor protein in A431 cells, neutrophils and kidney sections. Braz. J. Med. Biol. Res. 27, 1739–1756.

Haley, J.E., Dickenson, A.H. and Schachter, M. (1989). Electrophysiological evidence for a role of bradykinin in chemical nociception in the rat. Neurosci Lett. 97, 198–202.

Hall, J.M. (1992). Bradykinin receptors: pharmacological properties and biological roles. Pharmacol. Ther. 56, 131–190.

Hall, J.M. and Geppetti, P. (1995). Kinins and kinin receptors in the nervous system. Neurochem. Int. 26, 17–26.

Hall, J.M. and Morton, I.K.M. (1991). Bradykinin B$_2$ receptor evoked K$^+$ permeability increase mediates relaxation in the rat duodenum. Eur. J. Pharmacol. 193, 231–238.

Hall, J.M., Fox, A.J. and Morton, I.K.M. (1990). Peptidase activity as a determinant of agonist potencies in some smooth muscle preparations. Eur. J. Pharmacol. 176, 127–134.

Hall, J.M., Mitchell, D., Chin, S.Y., Field, J.L., Everett, C.M. and Morton, I.K.M. (1992). Differences in affinities of bradykinin B$_2$ receptor antagonists may be attributable to species. Neuropeptides 22, 28–29.

Hall, J.M., Caulfield, M.P., Watson, S.P. and Guard, S. (1993). Receptor subtypes or species homologues: Relevance to drug discovery. Trends Pharmacol. Sci. 14, 376–383.

Hall, J.M., Figini, M., Butt, S.K. and Geppetti, P. (1995). Inhibition of bradykinin-evoked trigeminal nerve stimulation by the non-peptide bradykinin B$_2$ receptor antagonist Win 64338 *in vivo* and *in vitro*. Br. J. Pharmacol. 116, 3164–3168.

Hargreaves, K.M., Troullos, E.S., Dionne, R.A., Schmidt, E.A., Schafer, S.C. and Joris, J.L. (1988). Bradykinin is increased during acute and chronic inflammation: therapeutic implications. Clin. Pharmacol. Ther. 44, 613–621.

Heapy, C.G., Shaw, J.S. and Farmer, S.G. (1993). Differential sensitivity of antinociceptive assays to the bradykinin antagonist Hoe 140. Br. J. Pharmacol. 108, 209–213.

Hess, J.F., Borkowski, J.A., Young, G.S., Strader, C.D. and Ransom, R.W. (1992). Cloning and pharmacological characterization of a human bradykinin (BK-2) receptor. Biochem. Biophys. Res. Commun. 184, 260–268.

Hess, J.F., Borkowski, J.A., MacNeil, T., Stonesifer, G.Y., Fraher, J., Strader, C.D. and Ransom, R.W. (1994a). Differential pharmacology of cloned human and mouse B_2 bradykinin receptors. Mol. Pharmacol. 45, 1–8.

Hess, J.F., Borkowski, J.A., Stonesifer, G.Y., MacNeil, T., Strader, C.D. and Ransom, R.W. (1994b). Cloning and pharmacological characterization of bradykinin receptors. Braz. J. Med. Biol. Res. 27, 1725–1731.

Higgins, P.G., Barrow, G.I. and Tyrrell, D.A. (1990). A study of the efficacy of the bradykinin antagonist, NPC 567, in rhinovirus infections in human volunteers. Antiviral Res. 14, 339–344.

Hirano, T. and Takeuchi, S. (1994). A bradykinin antagonist, DArg-[Hyp³,Thi⁵,⁸,DPhe⁷]-bradykinin, prevents pancreatic subcellular redistribution of lysosomal enzyme in rats with caerulein-induced acute pancreatitis. Med. Sci. Res. 22, 803–805.

Hirata, Y., Takata, S. and Takaichi, S. (1989). Specific binding sites for bradykinin and its degradation process in cultured rat vascular smooth muscle cells. Adv. Exp. Med. Biol. 247A, 415–420.

Hock, F.J., Wirth, K., Albus, U., Linz, W., Gerhards, H.J., Wiemer, G., Henke, St., Breipohl, G., König, W. and Knolle, J. (1991). Hoe 140, a new potent and long acting bradykinin-antagonist: in vitro studies. Br. J. Pharmacol. 102, 769–773.

Homayoun, P. and Harik, S.I. (1991). Bradykinin receptors of cerebral microvessels stimulate phosphoinositide turnover. J. Cereb. Blood Flow Metab. 11, 557–566.

Hosli, E. and Hosli, L. (1993a). Autoradiographic localization of binding sites for neuropeptide Y and bradykinin on astrocytes. Neuroreports 4, 159–162.

Hosli, E. and Hosli, L. (1993b). Receptors for neurotransmitters on astrocytes in the mammalian central nervous system. Prog. Neurobiol. 40, 477–506.

Hosli, L., Hosli, E. and Kaser, H. (1993). Colocalization of cholinergic, adrenergic and peptidergic receptors on astrocytes. Neuroreport 4, 679–682.

Hu, D.-E. and Fan, T.-P.D. (1993). [Leu⁸]des-Arg⁹-bradykinin inhibits the angiogenic effect of bradykinin and interleukin-1 in rats. Br. J. Pharmacol. 109, 14–17.

Ichinose, M. and Barnes, P.J. (1990). Bradykinin-induced airway microvascular leakage and bronchoconstriction are mediated via a bradykinin B_2 receptor. Am. Rev. Respir. Dis. 142, 1104–1107.

Ichinose, M., Belvisi, M.G. and Barnes, P.J. (1990). Bradykinin-induced bronchoconstriction in guinea pig in vivo: Role of neural mechanisms. J. Pharmacol. Exp. Ther. 253, 594–599.

Innis, R.B., Manning, D.C., Stewart, J.M. and Snyder, S.H. (1981). [³H]Bradykinin receptor binding in mammalian tissue membranes. Proc. Natl Acad. Sci. USA 78, 2630–2634.

Inoki, R. and Kudo, T. (1986). Enkephalins and bradykinin in dental pulp. Trends Pharmacol. Sci. 7, 275–277.

Issandou, M. and Darbon, J.-M. (1991) Des-Arg⁹ bradykinin modulates DNA synthesis, phospholipase C, and protein kinase C in cultured mesangial cells. Distinction from effects of bradykinin. J. Biol. Chem. 266, 21037–21043.

Jacques, L. and Couture, R. (1990). Studies on the vascular permeability induced by intrathecal substance P and bradykinin in the rat. Eur. J. Pharmacol. 184, 9–20.

Jeftinija, S. (1994). Bradykinin excites tetrodotoxin-resistant primary afferent fibers. Brain Res. 665, 69–76.

Jin, L.S., Seeds, E., Page, C.P. and Schachter, M. (1989). Inhibition of bradykinin-induced bronchoconstriction in the guinea-pig by a synthetic B_2 receptor antagonist. Br. J. Pharmacol. 97, 598–602.

Juan, H. and Lembeck, F. (1974). Action of peptides and other algesic agents on paravascular pain receptors of the isolated perfused rabbit ear. Naunyn-Schmiedeberg's Arch. Pharmacol. 283, 151–164.

Karlberg, B.E. (1993). Cough and inhibition of the renin–angiotensin system. J. Hypertens. 11 Suppl. 3, S49–S52.

Kenakin, T.P. (1993). "Pharmacological Analysis of Drug Receptor Interaction", 2nd edn, pp. 1–483. Raven Press, New York.

Keravis, T.M., Nehlig, H., Delacroix, M.-F., Regoli, D., Hiley, C.R. and Stoclet, J.-C. (1991). High-affinity bradykinin B_2 binding sites sensitive to guanine nucleotides in bovine aortic endothelial cells. Eur. J. Pharmacol. Mol. Pharmacol. 207, 149–155.

Kindgen-Milles, D., Klement, W. and Arndt, J.O. (1994). The nociceptive systems of skin, paravascular tissue and hand veins of humans and their sensitivity to bradykinin. Neurosci. Lett. 181, 39–42.

Kyle, D.J. (1994). Structural features of the bradykinin receptor as determined by computer simulations, mutagenesis experiments, and conformationally constrained ligands: establishing the framework for the design of new antagonists. Braz. J. Med. Biol. Res. 27, 1757–1779.

Kyle, D.J., Hicks, R.P., Blake, P.R. and Klimkowski, V.J. (1991a). Confirmational properties of bradykinin and bradykinin antagonists. In "Bradykinin Antagonists. Basic and Clinical Research" (ed. R.M. Burch), pp. 131–146. Marcel Dekker, New York.

Kyle, D.J., Martin, J.A., Farmer, S.G. and Burch, R.M. (1991b). Design and conformational analysis of several highly potent bradykinin receptor antagonists. J. Med. Chem. 34, 1230–1233.

Leeb-Lundberg, L.M.F. and Mathis, S.A. (1990). Guanine nucleotide regulation of B_2 kinin receptors. Time-dependent formation of a guanine nucleotide-sensitive receptor state from which [³H]bradykinin dissociates slowly. J. Biol. Chem. 265, 9621–9627.

Leeb-Lundberg, L.M.F., Mathis, S.A. and Herzig, M.C.S. (1994). Antagonists of bradykinin that stabilize a G-protein-uncoupled state of the B_2 receptor act as inverse agonists in rat myometrial cells. J. Biol. Chem. 269, 25970–25973.

Leikauf, G.D., Ueki, I.F., Nadel, J.A. and Widdicombe, J.H. (1985). Bradykinin stimulates Cl⁻ secretion and prostaglandin E_2 release by canine tracheal epithelium. Am. J. Physiol. 248, F48–F55.

Lembeck, F., Griesbacher, T., Eckhardt, M., Henke, St., Breipohl, G. and Knolle, J. (1991). New, long-acting, potent bradykinin antagonists. Br. J. Pharmacol. 102, 297–304.

Lerner, U.H. and Modéer, T. (1991). Bradykinin B_1 and B_2 receptor agonists synergistically potentiate interleukin-1-induced prostaglandin biosynthesis in human gingival fibroblasts. Inflammation 15, 427–436.

Levesque, L., Harvey, N., Rioux, F., Drapeau, G. and Marceau, F. (1995). Development of a binding assay for the B_1 receptors for kinins. Immunopharmacology 29, 141–147.

Lewis, G.P. (1970). Kinins in inflammation and tissue injury. In "Handbook of Pharmacology. Bradykinin, Kallidin and Kallikrein" (ed. E.G. Erdös), pp. 516–530. Springer-Verlag, Berlin.

Lewis, R.E., Childers, S.R. and Phillips, M.I. (1985). [^{125}I]Tyr-bradykinin binding in primary rat brain cultures. Brain Res. 346, 263–272.

Liebmann, C., Reissmann, S., Robberecht, P. and Arold, H. (1987). Bradykinin action in the rat duodenum: receptor binding and influence on the cyclic AMP system. Biomed. Biochim. Acta 46, 469–478.

Liebmann, C., Offermanns, S., Spicher, K., Hinsch, K.-D., Schnittler, M., Morgat, J.L., Reissmann, S., Schultz, G. and Rosenthal, W. (1990). A high-affinity bradykinin receptor in membranes from rat myometrium is coupled to pertussis toxin-sensitive G-proteins of the G_i family. Biochem. Biophys. Res. Commun. 167, 910–917.

Liebmann, C., Schnittler, M., Stewart, J.M. and Reissmann, S. (1991). Antagonist binding reveals two heterogenous B_2 bradykinin receptors in rat myometrial membranes. Eur. J. Pharmacol. 199, 363–365.

Liebmann, C., Bossé, R. and Escher, E. (1994a). Discrimination between putative bradykinin B_2 receptor subtypes in guinea pig ileum smooth muscle membranes with a selective, iodinatable, bradykinin analogue. Mol. Pharmacol. 46, 949–956.

Liebmann, C., Mammery, K. and Graness, A. (1994b). Bradykinin inhibits adenylate cyclase activity in guinea pig ileum membranes via a separate high-affinity bradykinin B_2 receptor. Eur. J. Pharmacol. Mol. Pharmacol. 288, 35–43.

Linz, W. and Schölkens, B.A. (1992). A specific B_2-bradykinin receptor antagonist HOE 140 abolishes the antihypertrophic effect of ramipril. Br. J. Pharmacol. 105, 771–772.

Ljunggren, Ö. and Lerner, U.H. (1990). Evidence for BK_1 bradykinin receptor-mediated prostaglandin formation in osteoblasts and subsequent enhancement of bone resorption. Br. J. Pharmacol. 101, 382–386.

Ljunggren, Ö., Vavrek, R., Stewart, J.M. and Lerner, U.H. (1991). Bradykinin-induced burst of prostaglandin formation in osteoblasts is mediated via B_2 bradykinin receptors. J. Bone Miner. Res. 6, 807–815.

Llona, I., Vavrek, R., Stewart, J. and Huidobro-Toro, J.P. (1987). Identification of pre- and postsynaptic bradykinin receptor sites in the vas deferens: evidence for different structural prerequisites. J. Pharmacol. Exp. Ther. 241, 608–614.

Lopes, P., Kar, S., Tousignant, C., Regoli, D., Quirion, R. and Couture, R. (1993). Autoradiographic localization of [^{125}I-Tyr^8]-bradykinin receptor binding sites in the guinea pig spinal cord. Synapse 15, 48–57.

Louttit, J.B. and Coleman, R.A. (1993). The action of Hoe 140 on the bradykinin-induced splenic pressor reflex of the anaesthetized cat. Br. J. Pharmacol. 110, 1317–1320.

Ma, J., Wang, D., Chao, L. and Chao, J. (1994a). Cloning, sequence analysis and expression of the gene encoding the mouse bradykinin B_2 receptor. Gene 149, 283–288.

Ma, J., Wang, D., Ward, D.C., Chen, L., Dessai, T., Chao, J. and Chao, L. (1994b). Structure and chromosomal localization of the gene (BDKRB2) encoding human bradykinin B_2 receptor. Genomics 23, 362–369.

Maas, J., Rae, G.A., Huidobro-Toro, J.P. and Calixto, J.B. (1995). Characterization of kinin receptors modulating neurogenic contractions of the mouse isolated vas deferens. Br. J. Pharmacol. 114, 1471–1477.

Macfarlane, R., Moskowitz, M.A., Sakas, D.E., Tasdemiroglu, E., Wei, E.P. and Kontos, H.A. (1991). The role of neuro-effector mechanisms in cerebral hyperperfusion syndromes. J. Neurosurg. 75, 845–855.

MacNeil, T., Bierilo, K.K., Menke, J.G. and Hess, J.F. (1995). Cloning and pharmacological characterization of a rabbit bradykinin B_1 receptor. Biochim. Biophys. Acta Gene Struct. Express. 1264, 223–228.

Madeddu, P., Anania, V., Parpaglia, P.P., Demontis, M.P., Varoni, M.V., Pisanu, G., Troffa, C., Tonolo, G. and Glorioso, N. (1992). Effects of Hoe 140, a bradykinin B_2 receptor antagonist, on renal function in conscious normotensive rats. Br. J. Pharmacol. 106, 380–386.

Maggi, C.A., Patacchini, R., Santicioli, P., Geppetti, P., Cecconi, R., Giuliani, S. and Meli, A. (1989). Multiple mechanisms in the motor responses of the guinea pig isolated urinary bladder to bradykinin. Br. J. Pharmacol. 98, 619–629.

Maggi, C.A., Santicioli, P., Del Bianco, E., Lecci, A. and Guliani, S. (1993). Evidence for the involvement of bradykinin in chemically-evoked cystitis in anaesthetized rats. Naunyn-Schmiedeberg's Arch. Pharmacol. 347, 432–437.

Mahan, L.C. and Burch, R.M. (1990). Functional expression of B_2 bradykinin receptors from Balb/c cell mRNA in Xenopus oocytes. Mol. Pharmacol. 37, 785–789.

Mak, J.C.W. and Barnes, P.J. (1991). Autoradiographic visualization of bradykinin receptors in human and guinea pig lung. Eur. J. Pharmacol. 194, 37–43.

Manning, D.C., Vavrek, R., Stewart, J.M. and Snyder, S.H. (1986). Two bradykinin binding sites with picomolar affinities. J. Pharmacol. Exp. Ther. 237, 504–512.

Manning, D.C., Raja, S.N., Meyer, R.A. and Campbell, J.N. (1991). Pain and hyperalgesia after intradermal injection of bradykinin in humans. Clin. Pharmacol. Ther. 50, 721–729.

Marceau, F. (1995). Kinin B_1 receptors: A review. Immunopharmacology 30, 1–26.

Marceau, F. and Regoli, D. (1991). Kinin antagonists of the B_1 type and their antagonists. In "Bradykinin Antagonists. Basic and Clinical Research" (ed. R.M. Burch), pp. 33–49. Marcel Dekker, New York.

Marceau, F. and Tremblay, B. (1986). Mitogenic effect of bradykinin and of des-Arg^9-bradykinin on cultured fibroblasts. Life Sci. 39, 2351–2358.

Marceau, F., Barabé, J., St-Pierre, S. and Regoli, D. (1980). Kinin receptors in experimental inflammation. Can. J. Physiol. Pharmacol. 58, 536–542.

Marceau, F., Lussier, A., Regoli, D. and Giroud, J.P. (1983). Pharmacology of kinins: their relevance to tissue injury and inflammation. Gen. Pharmacol. 14, 209–229.

Marceau, F., Levesque, L., Drapeau, G., Rioux, F., Salvino, J.M., Wolfe, H.R., Seoane, P.R. and Sawutz, D.G. (1994). Effects of peptide and nonpeptide antagonists of bradykinin B_2 receptors on the venoconstrictor action of bradykinin. J. Pharmacol. Exp. Ther. 269, 1136–1143.

Marsh, K.A. and Hill, S.J. (1994). Des-Arg^9-bradykinin-induced increases in intracellular calcium ion concentration in single bovine tracheal smooth muscle cells. Br. J. Pharmacol. 112, 934–938.

McEachern, A.E., Shelton, E.R., Bhakta, S., Obernolte, R., Bach, C., Zuppan, P., Fujisaki, J., Aldrich, R.W. and Jarnagin, K. (1991). Expression cloning of a rat B_2 bradykinin receptor. Proc. Natl Acad. Sci. USA 88, 7724–7728.

McIntyre, P., Phillips, E., Skidmore, E., Brown, M. and Webb, M. (1993). Cloned murine bradykinin receptor exhibits a mixed B_1 and B_2 pharmacological selectivity. Mol. Pharmacol. 44, 346–355.

Medeiros, Y.S. and Calixto, J.B. (1993). Analysis of the mechanisms underlying the biphasic responses to bradykinin in circular muscle from guinea pig ileum. Eur. J. Pharmacol. 241, 157–163.

Menke, J.G., Borkowski, J.A., Bierilo, K.K., MacNeil, T., Derrick, A.W., Schneck, K.A., Ransom, R.W., Strader, C.D., Linemeyer, D.L. and Hess, J.F. (1994). Expression cloning of a human B_1 bradykinin receptor. J. Biol. Chem. 269, 21583–21586.

Minshall, R.D., Nakamura, F., Becker, R.P. and Rabito, S.F. (1995). Characterization of bradykinin B_2 receptors in adult myocardium and neonatal rat cardiomyocytes. Circ. Res. 76, 773–780.

Mizumura, K., Minagawa, M., Tsujii, Y. and Kumazawa, T. (1990). The effects of bradykinin agonists and antagonists on visceral polymodal receptor activities. Pain 40, 221–227.

Molimard, M., Martin, C.A.E., Naline, E., Hirsch, A. and Advenier, C. (1994). Contractile effects of bradykinin on the isolated human small bronchus. Am. J. Respir. Crit. Care Med. 149, 123–127.

Mousli, M., Bueb, J.-L., Bronner, C., Rouot, B. and Landry, Y. (1990). G protein activation: a receptor-independent mode of action for cationic amphiphilic neuropeptides and venom peptides. Trends Pharmacol. Sci. 11, 358–362.

Munoz, C.M. and Leeb-Lundberg, L.M.F. (1992). Receptor-mediated internalization of bradykinin. DDT1 MF-2 smooth muscle cells process internalized bradykinin via multiple degradative pathways. J. Biol. Chem. 267, 303–309.

Munoz, C.M., Cotecchia, S. and Leeb-Lundberg, L.M.F. (1993). B_2 kinin receptor-mediated internalization of bradykinin in DDT1 MF-2 smooth muscle cells is paralleled by sequestration of the occupied receptors. Arch. Biochem. Biophys. 301, 336–344.

Murray, M.A., Heistad, D.D. and Mayhan, W.G. (1991). Role of protein kinase C in bradykinin-induced increases in microvascular permeability. Circ. Res. 68, 1340–1348.

Naclerio, R.M., Proud, D., Togias, A.G., Adkinson, N.F., Jr, Meyers, D.A., Kagey-Sobotka, A., Plaut, M., Norman, P.S. and Lichtenstein, L.M. (1985). Inflammatory mediators in late antigen-induced rhinitis. N. Engl. J. Med. 313, 65–70.

Naclerio, R.M., Proud, D., Lichtenstein, L.M., Kagey–Sobotka, A., Hendly, J.O., Sorrentino and Gwaltney, J.M. (1988). Kinins are generated during experimental rhinovirus colds. J. Infect. Dis. 157, 133–142.

Nakajima, N., Ichinose, M., Takahashi, T., Yamauchi, H., Igarashi, A., Miura, M., Inoue, H., Takishima, T. and Shirato, K. (1994). Bradykinin-induced airway inflammation: Contribution of sensory neuropeptides differs according to airway site. Am. J. Respir. Crit. Care Med. 149, 694–698.

Nakhostine, N., Ribuot, C., Lamontagne, D., Nadeau, R. and Couture, R. (1993). Mediation by B_1 and B_2 receptors of vasodepressor responses to intravenously administered kinins in anaesthetized dogs. Br. J. Pharmacol. 110, 71–76.

Nardone, J. and Hogan, P.G. (1994). Delineation of a region in the B_2 bradykinin receptor that is essential for high-affinity agonist binding. Proc. Natl Acad. Sci. USA 91, 4417–4421.

Nardone, J., Gerald, C., Rimawi, L., Song, L. and Hogan, P.G. (1994). Identification of a B_2 bradykinin receptor expressed by PC12 phaeochromocytoma cells. Proc. Natl Acad. Sci. USA 91, 4412–4416.

Ohde, H., Morimoto, S. and Ogihara, T. (1991). Bradykinin suppresses endothelin-induced contraction of coronary artery through its B_2 receptor on the endothelium. Biochem. Int. 23, 1127–1132.

O'Neill, L.A. and Lewis, G.P. (1989). Interleukin-1 potentiates bradykinin- and TNFα-induced PGE_2 release. Eur. J. Pharmacol. 166, 131–137.

Osugi, T., Imaizumi, T., Mizushima, A., Uchida, S. and Yoshida, H. (1987). Role of a protein regulating guanine nucleotide binding in phosphoinositide breakdown and calcium mobilization by bradykinin in neuroblastoma X glioma hybrid NG108-15 cells: effects of pertussis toxin and cholera toxin on receptor-mediated signal transduction. Eur. J. Pharmacol. 137, 207–218.

Otterbein, L., Lowe, V.C., Kyle, D.J. and Noronha-Blob, L. (1993). Additive effects of a bradykinin antagonist, NPC 17761, and a leumedin, NPC 15669, on survival in animal models of sepsis. Agents Actions 39 Suppl. C, C125–C127.

Park, J., Freedman, R., Bach, C., Yee, C., Rohrwild, M., Kaminishi, H., Müller-Esterl, W. and Jarnagin, K. (1994). Bradykinin B_2 receptors in humans and rats: cDNA structures, gene structures, possible alternative splicing, and homology searching for subtypes. Braz. J. Med. Biol. Res. 27, 1707–1724.

Perkins, M.N. and Kelly, D. (1994). Interleukin-1-induced-desArg⁹- bradykinin-mediated thermal hyperalgesia in the rat. Neuropharmacology 33, 657–660.

Perkins, M.N., Burgess, G.M., Campbell, E.A., Hallett, A., Murphy, R.J., Naeem, S., Patel, I.A., Patel, S., Rueff, A. and Dray, A. (1991). HOE 140: a novel bradykinin analogue that is a potent antagonist at both B_2 and B_3 receptors in vitro. Br. J. Pharmacol. 102, 171P.

Perkins, M.N., Campbell, E.A., Davis, A. and Dray, A. (1992). Anti-nociceptive activity of bradykinin B_1 and B_2 antagonists in two models of persistent hyperalgesia in the rat. Br. J. Pharmacol. 107, 237P.

Perkins, M.N., Campbell, E. and Dray, A. (1993). Anti-nociceptive activity of the bradykinin B_1 and B_2 receptor antagonists, desArg⁹,[Leu⁸]-BK and HOE 140, in two models of persistent hyperalgesia in the rat. Pain 53, 191–197.

Perry, V.H., Andersson, P.-B. and Gordon, S. (1993). Macrophages and inflammation in the central nervous system. Trends Neurosci. 16, 268–273.

Pesquero, J.B., Lindsey, C.J., Zeh, K., Paiva, A.C.M., Ganten, D. and Bader, M. (1994). Molecular structure and expression of rat bradykinin B_2 receptor gene. Evidence for alternative splicing. J. Biol. Chem. 269, 26920–26925.

Pethö, G., Jocic, M. and Holzer, P. (1994). Role of bradykinin in the hyperaemia following acid challenge of the rat gastric mucosa. Br. J. Pharmacol. 113, 1036–1042.

Phillips, E., Conder, M.J., Bevan, S., McIntyre, P. and Webb, M. (1992). Expression of functional bradykinin receptors in Xenopus oocytes. J. Neurochem. 58, 243–249.

Polosa, R., Djukanovic, R., Rajakulasingam, K., Palermo, F. and

Holgate, S.T. (1993). Skin responses to bradykinin, kallidin, and [desArg⁹]- bradykinin in nonatopic and atopic volunteers. J. Allergy Clin. Immunol. 92, 683–689.

Pongracic, J.A., Churchill, L., and Proud, D. (1991a) Kinins in rhinitis. In "Bradykinin Antagonists. Basic and Clinical Research" (ed. R.M. Burch), pp. 237–259. Marcel Dekker, New York.

Pongracic, J.A., Naclerio, R.M., Reynolds, C.J. and Proud, D. (1991b). A competitive kinin receptor antagonist, [DArg⁰,Hyp³,DPhe⁷]-bradykinin, does not affect the response to nasal provocation with bradykinin. Br. J. Clin. Pharmacol. 31, 287–294.

Proud, D. and Kaplan, A.P. (1988). Kinin formation: mechanisms and role in inflammatory disorders. Annu. Rev. Immunol. 6, 49–83.

Proud, D., Togias, A., Naclerio, R.M., Crush, S.A., Norman, P.S. and Lichtenstein, L.M. (1983). Kinins are generated *in vivo* following nasal airway challenge of allergic individuals with allergen. J. Clin. Invest. 72, 1678–1685.

Proud, D., Reynolds, C.J., Lacapra, S., Kagey-Sobotka, A., Lichtenstein, L.M. and Naclerio, R.M. (1988). Nasal provocation with bradykinin induces symptoms of rhinitis and a sore throat. Am. Rev. Respir. Dis. 137, 613–616.

Proud, D., Reynolds, C.J., Broomfield, J., Goldman, D.W. and Bathon, J.M. (1993). Bradykinin effects in guinea pig tracheal epithelial cells are mediated through a B₂ kinin receptor and can be inhibited by the selective antagonist Hoe 140. J. Pharmacol. Exp. Ther. 264, 112–131.

Pruneau, D., Duvoid, A., Luccarini, J.M., Bélichard, P. and Bonnafous, J.C. (1995a). The kinin B₁ receptor antagonist des-Arg⁹-[Leu⁸]bradykinin: An antagonist of the angiotensin AT₁ receptor which also binds to the AT₂ receptor. Br. J. Pharmacol. 114, 115–118.

Pruneau, D., Luccarini, J.M., Defrène, E., Paquet, J.L. and Bélichard, P. (1995b). Pharmacological evidence for a single bradykinin B₂ receptor in the guinea pig. Br. J. Pharmacol. 116, 2106–2112.

Pyne, S. and Pyne, N.J. (1993a). Bradykinin stimulates phospholipase D in primary cultures of guinea pig tracheal smooth muscle. Biochem. Pharmacol. 45, 593–603.

Pyne, S. and Pyne, N.J. (1993b). Differential effects of B₂ receptor antagonists upon bradykinin-stimulated phospholipase C and D in guinea pig cultured tracheal smooth muscle. Br. J. Pharmacol. 110, 477–481.

Rajakulasingam, K., Polosa, R., Holgate, S.T. and Howarth, P.H. (1991). Comparative nasal effects of bradykinin, kallidin and [Des-Arg⁹]-bradykinin in atopic rhinitic and normal volunteers. J. Physiol. 437, 577–587.

Rajakulasingam, K., Polosa, R., Church, M.K., Holgate, S.T. and Howarth, P.H. (1992). Kinins and rhinitis. Clin. Exp. Allergy 22, 734–740.

Ransom, R.W., Goodman, C.B. and Young, G.S. (1992a). Bradykinin stimulation of phosphoinositide hydrolysis in guinea pig ileum longitudinal muscle. Br. J. Pharmacol. 105, 919–924.

Ransom, R.W., Young, G.S., Schneck, K. and Goodman, C.B. (1992b). Characterization of solubilized bradykinin B₂ receptors from smooth muscle and mucosa of guinea pig ileum. Biochem. Pharmacol. 43, 1823–1827.

Regoli, D. and Barabé, J. (1980). Pharmacology of bradykinin and related kinins. Pharmacol. Rev. 32, 1–46.

Regoli, D., Barabé, J. and Park, W.K. (1977). Receptors for bradykinin in rabbit aortae. Can. J. Physiol. Pharmacol. 55, 855–867.

Regoli, D., Marceau, F. and Barabé, J. (1978). *De novo* formation of vascular receptors for bradykinin. Can. J. Physiol. Pharmacol. 56, 674–677.

Regoli, D., Marceau, F. and Lavigne, J. (1981). Induction of the B₁ receptors for kinins in the rabbit by a bacterial lipopolysaccharide. Eur. J. Pharmacol. 71, 105–115.

Regoli, D., Drapeau, G., Rovero, P., Dion, S., Rhaleb, N.-E., Barabé, J., D'Orléans-Juste, P. and Ward, P. (1986). Conversion of kinins and their antagonists into B₁ receptor activators and blockers in isolated vessels. Eur. J. Pharmacol. 127, 219–224.

Regoli, D., Rhaleb, N.-E., Dion, S. and Drapeau, G. (1990a). New selective bradykinin receptor antagonists and bradykinin B₂ receptor characterization. Trends Pharmacol. Sci. 11, 156–161.

Regoli, D., Rhaleb, N.-E., Drapeau, G. and Dion, S. (1990b). Kinin receptor subtypes. J. Cardiovasc. Pharmacol. 15, Suppl. 6, S30–S38.

Regoli, D., Rhaleb, N.-E., Tousignant, C., Rouissi, N., Nantel, F., Jukic, D. and Drapeau, G. (1991). New highly potent bradykinin B₂ receptor antagonists. Agents Actions 34, 138–141.

Regoli, D., Jukic, D., Gobeil, F. and Rhaleb, N.-E. (1993). Receptors for bradykinin and related kinins: a critical analysis. Can. J. Physiol. Pharmacol. 71, 556–567.

Regoli, D., Gobeil, F., Nguyen, Q.T., Jukic, D., Seoane, P.R., Salvino, J.M. and Sawutz, D.G. (1994). Bradykinin receptor types and B₂ subtypes. Life Sci. 55, 735–749.

Regoli, D., Jukic, D. and Gobeil, F. (1995). Kinin receptor (B₁, B₂) antagonists and their therapeutic potential. Therapie 50, 9–18.

Rhaleb, N.-E. and Carretero, O.A. (1994). Role of B₁ and B₂ receptors and of nitric oxide in bradykinin-induced relaxation and contraction of isolated rat duodenum. Life Sci. 55, 1351–1363.

Rhaleb, N.-E., Dion, S., D'Orléans-Juste, P., Drapeau, G., Regoli, D. and Browne, R.G. (1988). Bradykinin antagonism: differentiation between peptide antagonists and antiinflammatory agents. Eur. J. Pharmacol. 151, 275–279.

Rhaleb, N.-E., Drapeau, G., Dion, S., Jukic, D., Rouissi, N. and Regoli, D. (1990). Structure-activity studies on bradykinin and related peptides: agonists. Br. J. Pharmacol. 99, 445–448.

Rhaleb, N.-E., Gobeil, F. and Regoli, D. (1992a). Non-selectivity of new bradykinin antagonists for B₁ receptors. Life Sci. 51, PL211.

Rhaleb, N.-E., Gobeil, F. and Regoli, D. (1992b). Non-selectivity of new bradykinin antagonists for B₁ receptors. Life Sci. 51, PL125–PL129.

Rhaleb, N.-E., Rouissi, N., Jukic, D., Regoli, D., Henke, S., Breipohl, G. and Knolle, J. (1992c). Pharmacological characterization of a new highly potent B₂ receptor antagonist (HOE 140: DArg-[Hyp³,Thi⁵,DTic⁷,Oic⁸] bradykinin). Eur. J. Pharmacol. 210, 115–120.

Ridings, P.C., Blocher, C.R., Fisher, B.J., Fowler, A.A., III and Sugerman, H.J. (1995). Beneficial effects of a bradykinin antagonist in a model of gram-negative sepsis. J. Trauma Injury Infect. Crit. Care 39, 81–89.

Rifo, J., Pourrat, M., Vavrek, R.J., Stewart, J.M. and Huidobro-Toro, J.P. (1987). Bradykinin receptor antagonists used to characterize the heterogeneity of bradykinin-induced responses in rat vas deferens. Eur. J. Pharmacol. 142, 305–312.

Roch-Arveiller, M., Giroud, J.-P. and Regoli, D. (1985). Kinins and inflammation. Emphasis on interactions of kinins with various cell types. In "Progress in Applied Microcirculation. White Cells Rheology and Inflammation", pp. 81–95. Karger, Basel.

Roscher, A.A., Klier, C., Dengler, R., Faussner, A. and Müller-Esterl, W. (1990). Regulation of bradykinin action at the receptor level. J. Cardiovasc. Pharmacol. 15 (Suppl. 6), S39–S43.

Saha, J.K., Sengupta, J.N. and Goyal, R.K. (1990). Effect of bradykinin on opossum esophageal longitudinal smooth muscle: Evidence for novel bradykinin receptors. J. Pharmacol. Exp. Ther. 252, 1012–1020.

Saha, J.K., Sengupta, J.N. and Goyal, R.K. (1991). Effects of bradykinin and bradykinin analogs on the opossum lower esophageal sphincter: characterization of an inhibitory bradykinin receptor. J. Pharmacol. Exp. Ther. 259, 265–273.

Sakamoto, T., Elwood, W., Barnes, P.J. and Chung, K.F. (1992). Effect of Hoe 140, a new bradykinin receptor antagonist, on bradykinin- and platelet-activating factor-induced bronchoconstriction and airway microvascular leakage in guinea pig. Eur. J. Pharmacol. 213, 367–373.

Sakamoto, T., Tsukagoshi, H., Barnes, P.J. and Chung, K.F. (1993). Role played by NK_2 receptor and cyclooxygenase activation in bradykinin B_2 receptor mediated-airway effects in guinea pigs. Agents Actions 39, 111–117.

Salvino, J.M., Seoane, P.R., Douty, B.D., Awad, M.M.A., Dolle, R.E., Houck, W.T., Faunce, D.M. and Sawutz, D.G. (1993). Design of potent non-peptide competitive antagonists of the human bradykinin B_2 receptor. J. Med. Chem. 36, 2583–2584.

Santiago, J.A., Garrison, E.A. and Kadowitz, P.J. (1994). Analysis of responses to bradykinin: effects of Hoe-140 in the hindquarters vascular bed of the cat. Am. J. Physiol. 267, H828–H836.

Sawutz, D.G., Singh, S.S., Tiberio, L., Koszewski, E., Johnson, C.G. and Johnson, C.L. (1992). The effect of TNF on bradykinin receptor binding, phosphatidylinositol turnover and cell growth in human A431 epidermoid carcinoma cells. Immunopharmacology 24, 1–10.

Sawutz, D.G., Salvino, J.M., Dolle, R.E., Casiano, F., Ward, S.J., Houck, W.T., Faunce, D.M., Douty, B.D., Baizman, E. and Awad, M.M. (1994a). The nonpeptide Win 64338 is a bradykinin B_2 receptor antagonist. Proc. Natl Acad. Sci. USA 91, 4693–4697.

Sawutz, D.G., Salvino, J.M., Seoane, P.R., Douty, B.D., Houck, W.T., Bobko, M.A., Doleman, M.S., Dolle, R.E. and Wolfe, H.R. (1994b). Synthesis, characterization, and conformational analysis of the D/L-Tic^7 stereoisomers of the bradykinin receptor antagonist DArg^0[Hyp^3,Thi^5,DTic^7,Oic^8]bradykinin. Biochemistry 33, 2373–2379.

Sawutz, D.G., Salvino, J.M., Dolle, R.E., Seoane P.R. and Farmer, S.G. (1995). Pharmacology and structure–activity relationships of the non-peptide bradykinin receptor antagonist, Win 64338. Can. J. Physiol. Pharmacol. 73, 805–811.

Schachter, M., Uchida, Y., Longridge, D.J., Labedz, T., Whalley, E.T., Vavrek, R.J. and Stewart, J.M. (1987). New synthetic antagonists of bradykinin. Br. J. Pharmacol. 92, 851–855.

Scherrer, D., Daeffler, L., Trifilieff, A. and Gies, J.-P. (1995). Effects of Win 64338, a nonpeptide bradykinin B_2 receptor antagonist, on guinea pig trachea. Br. J. Pharmacol. 115, 1127–1128.

Schini, V.B., Boulanger, C., Regoli, D. and Vanhoutte, P.M. (1990). Bradykinin stimulates the production of cyclic GMP via activation of B_2 kinin receptors in cultured porcine aortic endothelial cells. J. Pharmacol. Exp. Ther. 252, 581–585.

Schlemper, V. and Calixto, J.B. (1994). Nitric oxide pathway-mediated relaxant effect of bradykinin in the guinea-pig isolated trachea. Br. J. Pharmacol. 111, 83–88.

Schneck, K.A., Hess, J.F., Stonesifer, G.Y. and Ransom, R.W. (1994). Bradykinin B_1 receptors in rabbit aorta smooth muscle cells in culture. Eur. J. Pharmacol. Mol. Pharmacol. 266, 277–282.

Schumann, C., Steinmetzer, T., Gothe, R., Hoppe, A., Paegelow, I., Liebmann, C., Fabry, M., Brandenburg, D. and Reissmann, S. (1995). Potent photoaffinity labelled and iodinated antagonists of bradykinin. Biol. Chem. Hoppe Seyler 376, 33–38.

Seabrook, G.R., Bowery, B.J. and Hill, R.G. (1995). Bradykinin receptors in mouse and rat isolated superior cervical ganglia. Br. J. Pharmacol. 115, 368–372.

Seguin, L. and Widdowson, P.S. (1993). Effects of nucleotides on [^{3}H]bradykinin binding in guinea pig: Further evidence for multiple B_2 receptor subtypes. J. Neurochem. 60, 752–757.

Seguin, L., Widdowson, P.S. and Giesen-Crouse, E. (1992). Existence of three subtypes of bradykinin B_2 receptors in guinea pig. J. Neurochem. 59, 2125–2133.

Sharif, N.A. and Whiting, R.L. (1991). Identification of B_2-bradykinin receptors in guinea pig brain regions, spinal cord and peripheral tissues. Neurochem. Int. 18, 89–96.

Sharma, J.N. (1991). The role of kinin system in joint inflammatory disease. Eur. J. Rheumatol. Inflamm. 11, 30–37.

Sharma, J.N. and Buchanan, W.W. (1994). Pathogenic responses of bradykinin system in chronic inflammatory rheumatoid disease. Exp. Toxicol. Pathol. 46, 421–433.

Shibata, M., Ohkubo, T., Takahashi, H. and Inoki, R. (1986). Interaction of bradykinin with substance P on vascular permeability and pain response. Jpn J. Pharmacol. 41, 427–429.

Skidgel, R.A. (1992). Bradykinin-degrading enzymes: Structure, function, distribution, and potential roles in cardiovascular pharmacology. J. Cardiovasc. Pharmacol. 20 Suppl. 9, S4–S9.

Snell, P.H., Phillips, E., Burgess, G.M., Snell, C.R. and Webb, M. (1990). Characterization of bradykinin receptors solubilized from rat uterus and NG108-15 cells. Biochem. Pharmacol. 39, 1921–1928.

Snider, R.M. and Richelson, E. (1984). Bradykinin receptor-mediated cyclic GMP formation in a nerve cell population (murine neuroblastoma clone N1E-115). J. Neurochem. 43, 1749–1754.

Stahl, W., Breipohl, G., Kuhlmann, L., Steinsträsser, A., Gerhards, H.J. and Schölkens, B.A. (1995). Technetium-99m-labeled HOE 140: a potential bradykinin B_2 receptor imaging agent. J. Med. Chem. 38, 2799–2801.

Steinmetzer, T., Schumann, C., Paegelow, I., Liebmann, C., Glasmacher, D., Brandenburg, D. and Reissmann, S. (1995). New photoaffinity labelled agonists of bradykinin. Biol. Chem. Hoppe Seyler 376, 25–32.

Steranka, L.R. and Burch, R.M. (1991). Bradykinin antagonists in pain and inflammation. In "Bradykinin Antagonists. Basic and Clinical Research" (ed. R.M. Burch), pp. 191–211. Marcel Dekker, New York.

Steranka, L.R., Manning, D.C., DeHaas, C.J., Ferkany, J.W., Borosky, S.A., Connor, J.R., Vavrek, R.J., Stewart, J.M. and Snyder, S.H. (1988). Bradykinin as a pain mediator: receptors are localized to sensory neurons, and antagonists have analgesic actions. Proc. Natl Acad. Sci. USA 85, 3245–3249.

Steranka, L.R., Farmer, S.G. and Burch, R.M. (1989). Antagonists of B_2 bradykinin receptors. FASEB J. 3, 2019–2025.

Stewart, J.M. (1979). Chemistry and biologic activity of peptides related to bradykinin. In "Handbook of Experimental Pharmacology. Bradykinin, Kallidin and Kallikrein" (ed. E.G. Erdös), pp. 227–285. Springer-Verlag, Berlin.

Stewart, J.M. (1995). Bradykinin antagonists: development and applications. Biopolymers 37, 143–155.

Stewart, J.M. and Vavrek, R.J. (1991). Chemistry of peptide B_2 bradykinin antagonists. In "Bradykinin Antagonists. Basic and Clinical Research" (ed. R.M. Burch), pp. 51–96. Marcel Dekker, New York.

Sung, C.-P., Arleth, A.J., Shikano, K. and Berkowitz, B.A. (1988). Characterization and function of bradykinin receptors in vascular endothelial cells. J. Pharmacol. Exp. Ther. 247, 8–13.

Sunman, W. and Sever, P.S. (1993). Non-angiotensin effects of angiotensin-converting enzyme inhibitors. Clin. Sci. 85, 661–670.

Szolcsányi, J. (1987). Selective responsiveness of polymodal nociceptors of the rabbit ear to capsaicin, bradykinin and ultra-violet irradiation. J. Physiol. 388, 9–23.

Tamaoki, J., Kobayashi, K., Sakai, N., Chiyotani, A., Kanemura, T. and Takizawa, T. (1989). Effect of bradykinin on airway ciliary motility and its modulation by neutral endopeptidase. Am. Rev. Respir. Dis. 140, 430–435.

Toda, N., Bian, K., Akiba, T. and Okamura, T. (1987). Heterogeneity in mechanisms of bradykinin action in canine isolated blood vessels. Eur. J. Pharmacol. 135, 321–329.

Tokumasu, T., Ueno, A. and Oh-ishi, S. (1995). A hypotensive response induced by des-Arg^9-bradykinin in young Brown/Norway rats pretreated with endotoxin. Eur. J. Pharmacol. 274, 225–228.

Tousignant, C., Guillemette, G., Barabé, J., Rhaleb, N.-E. and Regoli, D. (1991). Characterization of kinin binding sites: identity of B_2 receptors in the epithelium and the smooth muscle of the guinea pig ileum. Can. J. Physiol. Pharmacol. 69, 818–825.

Tousignant, C., Regoli, D., Rhaleb, N.-E., Jukic, D. and Guillemette, G. (1992). Characterization of a novel binding site for ^{125}I-Tyr-DArg-[Hyp^3,DPhe^7,Leu^8]bradykinin on epithelial membranes of guinea pig ileum. Eur. J. Pharmacol. 225, 35–44.

Treede, R.-D., Meyer, R.A., Davis, K.D. and Campbell, J.N. (1990). Intradermal injections of bradykinin or histamine cause a flare-like vasodilatation in monkey. Evidence from laser Doppler studies. Neurosci. Lett. 115, 201–206.

Trifilieff, A., Haddad, E.B., Landry, Y. and Gies, J.-P. (1991). Evidence for two high-affinity bradykinin binding sites in the guinea pig lung. Eur. J. Pharmacol. 207, 129–134.

Trifilieff, A., Da Silva, A., Landry, Y. and Gies, J.-P. (1992). Effect of Hoe 140, a new B_2 noncompetitive antagonist, on guinea pig tracheal bradykinin receptors. J. Pharmacol. Exp. Ther. 263, 1377–1382.

Trifilieff, A., Da Silva, A. and Gies, J.P. (1993). Kinins and respiratory tract diseases. Eur. Respir. J. 6, 576–587.

Tropea, M.M., Munoz, C.M. and Leeb-Lundberg, L.M.F. (1992). Bradykinin binding to B_2 kinin receptors and stimulation of phosphoinositide turnover and arachidonic acid release in primary cultures of cells from late pregnant rat myometrium. Can. J. Physiol. Pharmacol. 70, 1360–1371.

Tropea, M.M., Gummelt, D., Herzig, M.S. and Leeb-Lundberg, L.M.F. (1993). B_1 and B_2 kinin receptors on cultured rabbit superior mesenteric artery smooth muscle cells: receptor-specific stimulation of inositol phosphate formation and arachidonic acid release by Des-Arg^9-bradykinin and bradykinin. J. Pharmacol. Exp. Ther. 264, 930–937.

Ueno, A., Tokumasu, T., Naraba, H. and Oh-ishi, S. (1995). Involvement of bradykinin in endotoxin-induced vascular permeability increase in the skin of rats. Eur. J. Pharmacol. 284, 211–214.

Vavrek, R. and Stewart, J.M. (1985). Competitive antagonists of bradykinin. Peptides 6, 161–164.

Vavrek, R.J., Gera, L. and Stewart, J.M. (1992). Pseudopeptide analogs of bradykinin and bradykinin antagonists. Agents Actions Suppl. 38/I, 565–571.

Wahlestedt, C. (1994) Antisense oligonucleotide strategies in neuropharmacology. Trends Pharmacol. Sci. 15, 42–46.

Walker, K., Perkins, M. and Dray, A. (1995). Kinins and kinin receptors in the nervous system. Neurochem. Int. 26, 1–16.

Wang, D.Z., Ma, J.X., Chao, L. and Chao, J. (1994). Molecular cloning and sequence analysis of rat bradykinin B_2 receptor gene. Biochim. Biophys. Acta 1219, 171–174.

Ward, P.E. (1991). Metabolism of bradykinin analogues. In "Bradykinin Antagonists. Basic and Clinical Research" (ed. R.M. Burch), pp. 147–170. Marcel Dekker, New York.

Warren, J.B. and Loi, R.K. (1995). Captopril increases skin microvascular blood flow secondary to bradykinin, nitric oxide, and prostaglandins. FASEB J. 9, 411–418.

Webb, M., McIntyre, P. and Phillips, E. (1994). B_1 and B_2 bradykinin receptors encoded by distinct mRNAs. J. Neurochem. 62, 1247–1253.

Weipert, J., Hoffmann, H., Siebeck, M. and Whalley, E.T. (1988). Attenuation of arterial blood pressure fall in endotoxin shock in the rat using the competitive bradykinin antagonist Lys-Lys-[Hyp^2,Thi^{5,8},DPhe^7]-Bk (B4148). Br. J. Pharmacol. 94, 282–284.

Whalley, E.T., Fritz, H. and Geiger, R. (1983). Kinin receptors and angiotensin converting enzyme in rabbits basilar arteries. Naunyn-Schmiedeberg's Arch. Pharmacol. 324, 296–301.

Whalley, E.T., Clegg, S., Stewart, J.M. and Vavrek, R.J. (1987a). The effect of kinin agonists and antagonists on the pain response of the human blister base. Naunyn-Schmiedeberg's Arch. Pharmacol. 336, 652–655.

Whalley, E.T., Nwator, I.A., Stewart, J.M. and Vavrek, R.J. (1987b). Analysis of the receptors mediating vascular actions of bradykinin. Naunyn-Schmiedeberg's Arch. Pharmacol. 336, 430–433.

Whalley, E.T., Loy, S.D., Modiferri, D., Blodgett, J.K. and Cheronis, J.C. (1992). A novel potent bis(succinimido)-hexane peptide heterodimer antagonist (CP-0364) of bradykinin BK$_1$ and BK$_2$ receptors: in vitro and in vivo studies. Br. J. Pharmacol. 107, 257P.

Whiteley, P.J. and Needleman, P. (1984). Mechanism of enhanced fibroblast arachidonic acid metabolism by mononuclear cell factor. J. Clin. Invest. 74, 2249–2253.

Wiemer, G. and Wirth, K. (1992). Production of cyclic GMP via activation of B_1 and B_2 kinin receptors in cultured bovine aortic endothelial cells. J. Pharmacol. Exp. Ther. 262, 729–733.

Wilson, D.D., de Garavilla, L., Kuhn, W., Togo, J., Burch, R.M. and Steranka, L.R. (1989). DArg-[Hyp^3-DPhe^7]-bradykinin, a bradykinin antagonist, reduces mortality in a rat model of endotoxic shock. Circ. Shock, 27, 93–101.

Wirth, K., Breipohl, G., Stechl, J., Knolle, J., Henke, S. and Schölkens, B. (1991a). DesArg^9-DArg[Hyp^3,Thi^5,DTic^7,Oic^8]-bradykinin (desArg^{10}-[Hoe140]) is a potent bradykinin B$_1$ receptor antagonist. Eur. J. Pharmacol. 205, 217–218.

Wirth, K., Hock, F.J., Albus, U., Linz, W., Alpermann, H.G., Anagnostopoulos, H., Henke, St., Breipohl, G., König, W. and Knolle, J. (1991b). Hoe 140 a new potent and long acting bradykinin-antagonist: *in vivo* studies. Br. J. Pharmacol. 102, 774–777.

Wirth, K.J., Alpermann, H.G., Satoh, R. and Inazu, M. (1992a). The bradykinin antagonist Hoe 140 inhibits carrageenan- and thermically induced paw oedema in rats. Agents Actions Suppl. 38/III, 428–431.

Wirth, K.J., Wiemer, G. and Schölkens, B.A. (1992b). DesArg^{10}[Hoe 140] is a potent B$_1$ bradykinin antagonist. Agents Actions Suppl. 38/II, 406–413.

Wirth, K.J., Gehring, D. and Schölkens, B.A. (1993). Effect of Hoe 140 on bradykinin-induced bronchoconstriction in anesthetized guinea pigs. Am. J. Respir. Crit. Care Med. 148, 702–706.

Wirth, K.J., Schölkens, B.A. and Wiemer, G. (1994). The bradykinin B$_2$ receptor antagonist Win 64338 inhibits the effect of des-Arg^9-bradykinin in endothelial cells. Eur. J. Pharmacol. Mol. Pharmacol. 288, R1–R2.

Wolsing, D.H. and Rosenbaum, J.S. (1991). Bradykinin-stimulated inositol phosphate production in NG108-15 cells is mediated by a small population of binding sites which rapidly desensitize. J. Pharmacol. Exp. Ther. 257, 621–633.

Yamawaki, I., Geppetti, P., Bertrand, C., Chan, B. and Nadel, J.A. (1994). Airway vasodilation by bradykinin is mediated via B$_2$ receptors and modulated by peptidase inhibitors. Am. J. Physiol. 266, L156–L162.

Yang, C.M., Luo, S.-F. and Hsia, H.-C. (1995). Pharmacological characterization of bradykinin receptors in canine cultured tracheal smooth muscle cells. Br. J. Pharmacol. 114, 67–72.

Yaqoob, M. and Snell, C.R. (1994), Purification and characterisation of B$_2$ bradykinin receptor from rat uterus. J. Neurochem. 62, 17–26.

Yokoyama, S., Kimura, Y., Taketo, M., Black, J.A., Ransom, B.R. and Higashida, H. (1994). B$_2$ bradykinin receptors in NG108-15 cells: cDNA cloning and functional expression. Biochem. Biophys. Res. Commun. 200, 634–641.

Yotsumoto, F., Manabe, T. and Ohshio, G. (1993). Bradykinin involvement in the aggravation of acute pancreatitis in rabbits. Digestion 54, 224–230.

Zini, J.M., Schmaier, A.H. and Cines, D.B. (1993). Bradykinin regulates the expression of kininogen binding sites on endothelial cells. Blood 81, 2936–2946.

3. Molecular Pharmacology of Kinin Receptors

J. Fred Hess

1. Introduction

The ability of the nonapeptide, bradykinin (BK), to elicit pain, inflammation and smooth muscle contraction was described initially by Rocha e Silva *et al.* (1949). Subsequent pharmacological studies suggested that at least two receptor subtypes, B_1 and B_2, mediate the biological actions of the kinins (Regoli *et al.*, 1977, 1978; Regoli and Barabé, 1980). Several bioactive kinins are generated by the action of a family of serine proteases, the kallikreins, on protein precursors, the kininogens (Fig. 3.1) (Proud and Kaplan, 1988; Bhoola *et al.*, 1992; see also Chapters 4 and 5). The initial products of kallikrein's digestion of kininogen are BK and Lys-BK (kallidin). Removal of the C-terminal arginine of BK and

Lys-BK by carboxypeptidases generates the peptides, desArg^9-BK and Lys-desArg^9-BK. The pharmacological activity of the carboxypeptidase metabolites of the kinins was initially demonstrated in isolated preparations of rabbit vascular smooth muscle (Regoli *et al.*, 1977) and the receptor mediating these responses was classified as a B_1 receptor. Conversely, the receptor mediating the actions of BK and Lys-BK was classified as the B_2 receptor (Regoli and Barabé, 1980; Marceau, 1995; see Chapter 2).

The development of peptides that antagonized the actions of the kinins selectively further defined B_1 and B_2 receptors as distinct pharmacological entities. In addition, a number of studies utilizing B_2 receptor antagonists suggested the existence of several subtypes of

The Kinin System
ISBN 0–12–249340–0

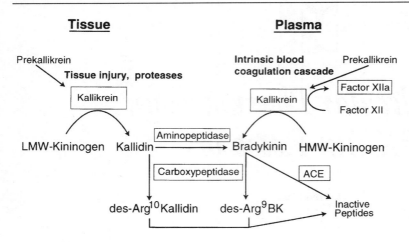

Figure 3.1   Generation of kinin agonists. The two primary pathways for kinin generation via plasma and tissue kallikrein, result in the production of the B_2 receptor agonists, bradykinin (BK) and Lys-BK (kallidin). The action of carboxypeptidases converts these peptides to the B_1 receptor agonists, desArg^9-BK and Lys-desArg^9-BK (desArg^{10}-kallidin).

the B_2 receptor. The recent molecular cloning of cDNAs encoding the kinin receptors has provided direct evidence for the existence of B_1 and B_2 receptors as the products of distinct genes as well as new tools with which to clearly evaluate the question of multiple subtypes of kinin receptors.

2. B_1 Receptors

As noted above, B_1 receptors respond to desArg^9-BK and Lys-desArg^9-BK, produced from BK and Lys-BK, respectively, by the action of carboxypeptidases N or M (Marceau, 1995). Formation of desArg-kinins occurs in conjunction with the proteolytic degradation of BK and Lys-BK to inactive peptides, primarily by angiotensin-converting enzyme (ACE) (Fig. 3.1). It is feasible, therefore, that the local environment of proteases may influence the relative quantities of desArg^9-BK and Lys-desArg^9-BK that are generated *in vivo*, especially at sites of inflammation. In the initial descriptions of B_1 receptors it was noted that this subtype was not apparently expressed constitutively in freshly isolated rabbit blood vessels but, rather, B_1 receptors appeared *de novo* following incubation of tissue preparations for several h *in vitro* (Regoli *et al.*, 1977, 1978). Subsequent work indicated that the expression of B_1 receptors is induced by cytokines, notably interleukin-1β (IL-1β), liberated during the inflammatory response (Regoli *et al.*, 1981; DeBlois *et al.*, 1989; Perkins and Kelly, 1994). These observations led to the proposal that the physiological role of B_1 receptors may be limited primarily to mediating kinin responses in damaged tissues (Marceau, 1995; see Chapter 8).

Although the physiological role of B_1 receptors and their induction is yet to be clearly determined, recent studies *in vivo* indicate that inflammatory hyperalgesia may be mediated in part by B_1 receptors (Dray and Perkins, 1993; Perkins *et al.*, 1993; Perkins and Kelly, 1993) (see Chapter 9). In addition, in a role analogous to

that of B_2 receptors, it has been suggested that B_1 receptors may mediate the protective effects of kinins upon vascular endothelium during myocardial ischemia (Chahine *et al.*, 1993; Lamontagne *et al.*, 1995).

2.1   Pharmacology of B_1 Receptors

The pharmacology of the B_1 receptor was defined primarily in rabbit smooth muscle tissues utilizing kinin receptor agonists and kinin analog antagonists (reviewed by Regoli and Barabé, 1980). Although these landmark studies reported that Lys-desArg^9-BK is more potent than desArg^9-BK, the latter has continued to be the agonist of choice utilized by most investigators to study B_1 receptor-mediated responses. Indeed, in both human cloned B_1 receptors, expressed in *Xenopus* oocytes, and B_1 receptors expressed constitutively in human lung fibroblasts, the affinity for Lys-desArg^9-BK is several orders of magnitude greater than for desArg^9-BK, the "classical" B_1 receptor agonist (Menke *et al.*, 1994).

The first B_1 receptor antagonist was reported following conversion of an agonist (desArg^9-BK) to an antagonist by replacing the phenylalanine at position 8 in BK with leucine, yielding desArg^9-[Leu^8]-BK (Regoli *et al.*, 1977). DesArg^9-[Leu^8]-BK and Lys-desArg^9-[Leu^8]-BK are potent peptide antagonists of B_1 receptors that have been utilized extensively to characterize the pharmacology of this receptor subtype and its physiologic roles.

The recently available radioligands, [^{3}H]-Lys-desArg^9-BK and [^{3}H]-Lys-desArg^9-[Leu^8]-BK, which possess high affinity for the B_1 receptor, have been utilized in binding assays to further characterize the pharmacological properties of B_1 receptors. B_1 receptors present on rabbit aorta smooth muscle cells, maintained in primary culture, have been characterized in both binding experiments and functional studies of signal transduction (Tropea *et al.*, 1993; Schneck *et al.*, 1994). Competition binding studies have indicated that the rabbit aorta B_1

receptor has a 100-fold higher affinity for Lys-desArg^9-BK than for desArg^9-BK (Schneck *et al.*, 1994). The higher affinity for N-terminal lysine was also pronounced when comparing the binding of Lys-BK and BK to the rabbit receptor. Thus, Lys-BK was equipotent with desArg^9-BK, whereas BK was essentially inactive at the rabbit B$_1$ receptor (K_i, 20 μM).

The second-messenger pathways activated by B$_1$ receptors are similar to those coupled to B$_2$ receptors. Studies in rabbit vascular smooth muscle cells have indicated that B$_1$ receptors stimulate phosphotidylinositol (PI) hydrolysis leading to mobilization of intracellular Ca^{2+} [Ca^{2+}]$_i$. In addition, rabbit B$_1$ receptors appear to be coupled to arachidonic acid metabolism, causing biosynthesis and release of prostaglandins (Tropea *et al.*, 1993; Schneck *et al.*, 1994). Earlier studies of B$_2$ receptors indicated that this subtype is coupled to G proteins to stimulate either phospholipase C or phospholipase A$_2$ (Burch and Axelrod, 1987).

2.2 MOLECULAR CLONING OF THE HUMAN B$_1$ RECEPTOR

The first evidence for a mRNA transcript encoding the B$_1$ receptor was obtained using a *Xenopus laevis* oocyte expression system (Phillips *et al.*, 1992). Injection of total mRNA isolated from a human lung fibroblast cell line (WI38), resulted in a robust response to BK and a qualitatively smaller response to desArg^9-BK. Responses to desArg^9-BK, but not BK, were blocked by a B$_1$ receptor antagonist, desArg^9-[Leu^8]-BK. Following the molecular cloning of the B$_2$ receptor, this analysis was extended to provide further evidence for the existence of a distinct B$_1$ receptor transcript (Webb *et al.*, 1994). Co-injection of antisense oligonucleotides directed against the B$_2$ receptor with pools of mRNA, containing either the B$_1$ or the B$_2$ receptor, inhibited responses to BK in *Xenopus* oocytes, but responses to desArg^9-BK were unaffected. Thus, it appeared that the nucleotide sequence encoding the B$_1$ receptor was distinct from that encoding the B$_2$ receptor. Furthermore, size fractionation of the total mRNA indicated that the B$_1$ receptor message was approximately 2 kb in length, compared with the B$_2$ receptor message, which was greater than 4 kb in length.

In our laboratory, we also utilized the *Xenopus* oocyte expression system to isolate a cDNA clone encoding the human B$_1$ receptor (Menke *et al.*, 1994). The prevalent method for detecting ligand-induced mobilization of [Ca^{2+}]$_i$ is to use electrophysiology to monitor the endogenous Ca^{2+}-activated Cl$^-$ channels in the oocytes (Lubbert *et al.*, 1987; Masu *et al.*, 1987). This assay was modified by employing the photoprotein, aequorin, to monitor [Ca^{2+}]$_i$ mobilization (Giladi and Spindel, 1991). Thus, samples were tested for the presence of a transcript encoding the B$_1$ receptor, by measuring aequorin

luminescence following the addition of Lys-desArg^9-BK. This assay provided increased throughput relative to electrophysiological experiments with no apparent loss of sensitivity.

To increase the probability of cloning the B$_1$ receptor, the message encoding the receptor was enriched. This was accomplished (1) by increasing the level of B$_1$ receptor expression prior to preparing total mRNA, and (2) size fractionation of the mRNA. The ability of IL-1β to stimulate B$_1$ receptor expression was established in both *in vitro* and *in vivo* studies (DeBlois *et al.*, 1988, 1989; Perkins and Kelly, 1994; Marceau, 1995). The starting material for cloning the B$_1$ receptor was a human embryonic fibroblast cell line (IMR-90), which has been shown to express B$_1$ receptors (Goldstein and Wall, 1984). Although IMR-90 cells express B$_1$ receptors constitutively, we found that receptor number could be increased by approximately seven-fold following exposure to IL-1β (Menke *et al.*, 1994). Since it was probable that this stimulation of expression was a consequence of increased transcription of the B$_1$ receptor gene, mRNA was prepared from IL-1β-stimulated IMR-90 cells. A sucrose gradient was used to size fractionate total mRNA prepared from IL-1β-induced IMR-90 cells, and fractions collected from the gradient were microinjected into oocytes. These cells were then assayed for increases in [Ca^{2+}]$_i$ in response to either a B$_1$ or a B$_2$ receptor agonist. The results were similar to those observed previously, in that the message encoding the B$_1$ receptor is approximately 1.7 kb, whereas that encoding the B$_2$ receptor is approximately 4.5 kb (Menke *et al.*, 1994; Webb *et al.*, 1994).

A peak fraction of size-fractionated mRNA was used to generate a cDNA library that was divided into pools and assayed using the *Xenopus* oocyte system (Menke *et al.*, 1994), successive rounds of subdividing positive pools resulting in isolation of a single clone. The DNA sequence of this clone encodes a G protein-coupled receptor of 353 amino acids that has the characteristic seven putative transmembrane spanning domains (Fig. 3.2). The sequence of the human B$_1$ receptor is more similar to that of the B$_2$ receptor than to any other known receptor. However, this similarity is only 36% at the amino-acid level. By comparison, the B$_1$ receptor is 30% identical to the angiotensin II type 1 receptor. The relatively low degree of sequence similarity provides a reasonable explanation for the success of the expression cloning approach and the failure of cloning strategies based on sequence homology to the B$_2$ receptor.

2.3 MOLECULAR PHARMACOLOGY OF B$_1$ RECEPTOR SPECIES HOMOLOGS

Expression cloning of the human B$_1$ receptor has facilitated the cloning of B$_1$ receptor homologs from

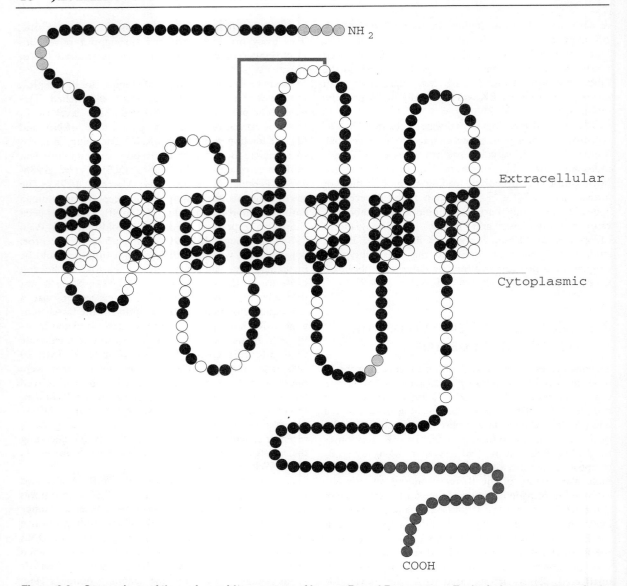

Figure 3.2 Comparison of the amino-acid sequences of human B$_1$ and B$_2$ receptors. Each circle represents a single amino acid, identical amino acids being denoted in white, differences in black and deletions in the B$_1$ receptor in light gray. Deletions in the B$_2$ receptor are shown in dark gray. The seven putative transmembrane-spanning domains are also shown. A putative disulfide bond between two cystine residues in the extracellular loops is shown by the gray line.

other species such as the rabbit and mouse (MacNeil *et al.*, 1995; Hess *et al.*, 1996). Examining the pharmacology of these cloned receptors, which are heterologously expressed in cells lacking endogenous B$_1$ receptors, has revealed significant differences in the interaction of species homologs with kinin agonists. Determination of the DNA sequence encoding the region of kininogen encoding BK may provide an explanation for the divergence in the affinity for kinin agonists (Hess *et al.*, 1996).

2.3.1   Cloned Human B$_1$ Receptors

Competition binding studies were performed in COS cells transiently transfected with human B$_1$ receptor cDNA (Menke *et al.*, 1994), and these experiments revealed that the human B$_1$ receptor exhibits a relatively low affinity for the B$_1$ receptor agonist, desArg^9-BK (K_i, 210 nM). This value is approximately seven times lower than the affinity of the B$_1$ receptor in rabbit aorta smooth muscle cells. In addition, in comparison with the rabbit homolog, the human B_1 receptor has an approximately five-fold lower affinity for desArg^9-[Leu^8]-BK. In contrast, the apparent affinity of cloned human B_1 receptors for [^{3}H]-Lys-desArg^9-BK was 0.4 nM. The differences in affinities for desArg^9-BK in cloned human and rabbit B_1 receptors may be a consequence of pharmacological differences arising from species homologs, or of receptor subtypes encoded by different genes. We addressed this question directly by isolating a cDNA encoding the rabbit B_1 receptor using human B_1 receptor cDNA as a probe for nucleic acid hybridization.

2.3.2   Cloned Rabbit B_1 Receptors

The rabbit B_1 receptor was cloned from a cDNA library from rabbit aorta smooth muscle cells (MacNeil et al., 1995). The cDNA contained an open reading frame encoding a 352-amino-acid protein that has 78% homology to the human B_1 receptor, and the relatively high degree of sequence similarity is consistent with species homologs of the same gene. To support this hypothesis, Southern blot analysis was performed using probes derived either from the rabbit or human B_1 receptor to probe the genomic DNA, from both species, which was digested with restriction endonucleases. These experiments indicated that the rabbit and human probes detected the same restriction fragment from either species' genomic DNA, suggesting that the two sequences are more similar to each other than they are to any others in the rabbit or human genome and, thus, are probably species homologs of the same gene.

The pharmacological profile of the cloned rabbit B_1 receptor was determined in competition binding studies in COS cells transiently transfected with rabbit cDNA (MacNeil et al., 1995), and the binding properties of the cloned receptor were essentially identical to those previously determined for B_1 receptors expressed constitutively in rabbit aorta smooth muscle. Thus, the pharmacological differences that were observed between human and rabbit cloned B_1 receptors are due to variations that are a consequence of the species homologs. That human and rabbit receptors both exhibit relatively high affinity for Lys-desArg^9-BK indicates that this may be the more important endogenous ligand in vivo, particularly in humans. The observation that the affinity for desArg^9-BK varies between the species homologs argues that there is minimal evolutionary selective pressure to maintain a high affinity for this ligand. Thus, the most appropriate B_1 receptor agonist for studies in human cells is Lys-desArg^9-BK.

2.3.3   Cloned Mouse B_1 Receptors

The utility of rodents in studying the role of B_1 receptors in inflammatory hyperalgesia is well established (Perkins et al., 1993; Perkins and Kelly, 1993, 1994; Davis and Perkins, 1994). A mouse B_1 receptor genomic clone was isolated and expressed in COS cells (Hess et al., 1996). The DNA sequence encoding this receptor lies on a single exon, and the deduced amino-acid sequence is 73% identical to the rabbit or human B_1 receptor. Genomic Southern blot analyses of restriction digests of human and mouse genomic DNA indicated that the mouse genomic clone is a species homolog to the previously described human B_1 receptor cDNA. Nevertheless, the mouse B_1 receptor exhibits dramatic differences in affinity for B_1 receptor agonists relative to human and rabbit B_1 receptors.

Human and rabbit B_1 receptors, respectively, exhibit approximately 2,000- and 150-fold higher affinities for Lys-desArg^9-BK relative to desArg^9-BK. In contrast, the mouse B_1 receptor exhibits only a 2–3-fold higher affinity for desArg^9-BK than for Lys-desArg^9-BK. This reversal of selectivity for B_1 receptor agonists in the mouse relative to the human B_1 receptor is a consequence of an approximately 15-fold lower affinity of the mouse receptor for Lys-desArg^9-BK, and a 300-fold higher affinity of the mouse receptor for desArg^9-BK. Considered together, these data suggest that the mouse B_1 receptor is under selective pressure to preserve its affinity for desArg^9-BK, whereas human and rabbit B_1 receptors are under selective pressure to preserve affinity for Lys-desArg^9-BK.

One explanation for the differences in agonist selectivity from human versus mouse B_1 receptors is that the endogenous ligand(s) generated in rodents differ from those in humans. For example, in rats, two kininogen genes, K- and T-kininogen, containing the precursor to BK have been identified (Furuto-Kato et al., 1985). The amino acid preceding BK in rat K-kininogen is arginine, whereas in T-kininogen, it is serine. Recently the sequence of two different cDNAs encoding mouse kininogen were identified, and both are similar to rat K-kininogen and contain an arginine residue preceding BK (Hess et al., 1996). In human and bovine kininogen, on the other hand, the amino-acid residue preceding BK is lysine, and cleavage yields Lys-BK (kallidin) when cleaved by kallikrein (see Chapter 5). Thus, it appears that rodents may lack Lys-BK, although it has not been determined whether rodent kallikrein is capable of liberating Arg-BK. The binding affinity of the mouse B_1 receptor for desArg^9-[Arg-BK] is equivalent to that observed for Lys-desArg^9-BK (approximately 2 nM), and less than that observed for desArg^9-BK. Therefore, it is possible that the higher affinity of mouse B_1 receptors for desArg^9-BK results from an inability of mouse kallikrein to generate Lys-BK or Arg-BK. Further investigation of the substrate specificity of rodent kallikrein and the characterization of the endogenous kinin agonists produced in these species is necessary to evaluate this hypothesis.

3. B_2 Receptors

The initial products of kininogen from kallikrein action, either BK or Lys-BK, are endogenous high-affinity B_2 receptor agonists (Proud and Kaplan, 1988; Bhoola et *al.*, 1992; Hall, 1992). B_2 receptors are expressed in most mammalian tissues where they mediate many of the physiologic actions associated with the kallikrein–kinin system. The acute nociceptive and inflammatory responses elicited by BK are mediated via B_2 receptors present on neurons, endothelial cells, epithelial cells, fibroblasts, and possibly also leukocytes such as macrophages and neutrophils. The vasoactive properties of BK, mediated via B_2 receptors expressed on vascular endothelial cells, are due to release of potent vasodilator substances such as nitric oxide and prostaglandins. Also, the classical studies on the activity of BK on smooth muscle tissues evaluated B_2 receptor activity.

3.1   MOLECULAR CLONING OF THE RAT B_2 RECEPTOR

B_2 receptors probably act through G proteins to stimulate phospholipases C (PLC) and A_2 (PLA$_2$) based on observations that GTP[γS] activates both PLC and PLA$_2$, and that GDP[βS] inhibits BK-mediated phospholipase activation in Swiss 3T3 cells (Burch and Axelrod, 1987). Thus, the biochemical evidence for B_2 receptors coupling to G proteins was fairly well established prior to the molecular cloning of the B_2 receptor. The ability of BK to stimulate Ca^{2+} mobilization via activation of B_2 receptors was utilized to monitor its expression in *Xenopus* oocytes injected with mRNA isolated from either murine 3T3 fibroblasts (Mahan and Burch, 1990) or human WI38 fibroblasts (Phillips *et al.*, 1992). Expression cloning of the B_2 receptor from rat uterus in *Xenopus* oocytes was achieved first by Jarnagin and coworkers who used electrophysiology to monitor Ca^{2+}-activated Cl^- channel activity (McEachern *et al.*, 1991). Rat B_2 receptor cDNA encodes a 366-amino-acid protein that is homologous to other G protein-coupled receptors. Moreover, the rat B_2 receptor contains the seven putative transmembrane spanning domains indicative of a G protein-coupled receptor. The pharmacology of this B_2 receptor was determined, the most potent agonists being BK and Lys-BK, which exhibited EC_{50} (effective concentration to give 50% of maximal activity) values of approximately 2 nM. In contrast, desArg^9-BK, a B_1 receptor agonist, was inactive at 10 μM. Therefore, the rat cloned receptor possessed the pharmacological properties associated with B_2 receptors. The cloned B_2 receptor was classified as a "smooth muscle" subtype (see later, and also Chapter 2) based on its interaction with the B_2 receptor antagonist [Thi^{5,8},DPhe^7]-BK, and its tissue distribution (McEachern *et al.*, 1991). Northern blot analyses indicated receptor expression in rat uterus, vas deferens, kidney, ileum, heart, lung, testis and brain. The presence of several different sized mRNA transcripts that hybridized to B_2 receptor cDNA was proposed to be consistent with multiple subtypes, although genomic

Southern blot analysis indicated only a single gene in rat, guinea pig and human, with a high degree of sequence similarity to the isolated clone.

3.2   MOLECULAR CLONING OF HUMAN AND MOUSE B_2 RECEPTORS

Cloning of the rat B_2 receptor provided a tool with which to isolate other species' homologs. Thus, the human B_2 receptor was cloned and its amino-acid sequence is 81% identical to that of the rat B_2 receptor (Eggerickx *et al.*, 1992; Hess *et al.*, 1992; Powell *et al.*, 1993). The pharmacological properties of the human receptor were consistent with a B_2 subtype. Recently, the human B_2 receptor was mapped to chromosome 14q32 (Powell *et al.*, 1993; Ma *et al.*, 1994b; Kammerer *et al.*, 1995), and polymorphisms in this gene were identified (Braun *et al.*, 1995).

The gene encoding the mouse B_2 receptor was also recently cloned and its pharmacology evaluated (McIntyre *et al.*, 1993; Hess *et al.*, 1994; Ma *et al.*, 1994a; Yokoyama *et al.*, 1994). The mouse B_2 receptor is very similar to that of the rat, in that the nucleic acid sequences of the open reading of these two species exhibit 92% homology. A slightly lower degree of similarity is observed between the human and mouse receptors, being 84% at the nucleic acid level. The mouse B_2 receptor has been mapped to the distal region of chromosome 12, which is homologous to the chromosome region, 14q32, where the human B_2 receptor is located (Taketo *et al.*, 1995). The pharmacology of the mouse cloned B_2 receptor has been examined by several groups, and it exhibits properties consistent with a classical B_2 receptor classification (Hess *et al.*, 1994; Yokoyama *et al.*, 1994). Nevertheless, McIntyre *et al.* (1993) described a mixed B_1/B_2 receptor pharmacology, in that both desArg^9-BK and BK elicited functional responses when the mouse clone was expressed in *Xenopus* oocytes. Furthermore, binding studies with the mouse B_2 receptor heterologously expressed in COS cells revealed high- and low-affinity sites for the B_1 receptor agonist, desArg^9-BK as well as the B_2 receptor antagonist, Hoe 140. The high-affinity site for desArg^9-BK comprised approximately 30% of the binding sites and was subnanomolar. In addition, a high-affinity (37 pM) site was observed for [^{3}H]-BK binding. These data, therefore, argued strongly for a mixed B_1/B_2 pharmacology (McIntyre *et al.*, 1993). The inability to reproduce these data reliably has yet to be explained.

3.3 B_2 RECEPTOR GENE STRUCTURE

It is possible that distinct B_2 receptor subtypes may be generated by alternative splicing. For example, small pharmacological variations arise as a consequence of alternative splicing of the D_2 dopamine receptor gene. In the gene structure of the B_2 receptor of human (Kammerer et al., 1995), rat (Pesquero et al., 1994) and mouse (Ma et al., 1994a), the coding region is assigned to a single large exon and it is, therefore, unlikely that alternative splicing would result in a pharmacologically distinct receptor. However, the B_2 receptor genes contain introns in the 5'-untranslated region and have been demonstrated, in both rats and humans, to span more than 24 kb (Pesquero et al., 1994; Kammerer et al., 1995). The rat and human genes contain three and two introns, respectively, although the inclusion of an additional exon, exon 3, in the rat in some cDNA sequences raises the possibility for alternative splicing. The role for an alternative splice in putative untranslated regions of the cDNA, however, is unknown. Luciferase reporter gene constructs have been generated in order to demonstrate the transcriptional activity of the promoter region from both the human and rat B_2 receptor genes (Pesquero et al., 1994; Kammerer et al., 1995), and these may prove useful in defining the regulatory elements controlling the expression of the B_2 receptor gene.

3.4 B_2 Receptor Subtypes

There are several lines of evidence for multiple B_2 receptor subtypes, as well as a B_3 receptor, which was proposed to be expressed in airway tissues from the guinea pig and sheep. The development of highly potent B_2 receptor antagonists such as Hoe 140, and the isolation of cloned receptors from different species have provided new tools to explore these proposals. Also, the generation of mice in which the gene encoding the B_2 receptor is disrupted has recently provided a unique tool with which to examine the potential role(s) of multiple genes encoding BK receptor subtypes (Borkowski and Hess, 1995).

The development of antagonists was critical to our understanding of the physiological and pathophysiological roles of B_2 receptors. The first antagonists, peptide analogs of BK, were synthesized by Stewart and coworkers, who discovered that replacement of the proline residue at position seven in BK with D-phenylalanine produced a B_2 receptor antagonist, [DPhe^7]-BK (Vavrek and Stewart, 1985). Subsequent modifications of this series of peptides improved their potency, and several studies with these antagonists provided evidence for B_2 receptor subtypes (Llona et al., 1987; Rifo et al., 1987; Braas et al., 1988; Farmer et al., 1989; Liebmann et al., 1991). (See Chapter 2 for detailed discussions on evidence for the existence, or otherwise, of subtypes of B_2 receptors.)

3.4.1 B_3 Receptors

The existence of a pulmonary B_3 receptor was proposed based on the inactivity of a peptide B_2 receptor antagonist, DArg-[Hyp^3,DPhe^7]-BK, to inhibit BK binding to guinea-pig trachea, in addition to the inability of the antagonist to inhibit [^{3}H]-BK binding (Farmer et al., 1989; Farmer and Burch, 1992). Although DArg-[Hyp^3,DPhe^7]-BK was an effective B_2 receptor antagonist in guinea-pig ileum, it was ineffective against BK-induced responses in the trachea from the same species. Furthermore, a two-site curve was observed in binding studies with DArg-[Hyp^3,DPhe^7]-BK in guinea-pig lung membranes, indicating a heterogeneous population of BK receptors in that tissue. The inactivity of B_1 receptor ligands, coupled with weak functional effects of both B_1 and B_2 receptor antagonists in these preparations, indicated that the receptor mediating BK's actions might be a novel receptor, termed "B_3" (Farmer et al., 1989). Additional evidence for a B_3 receptor was presented in experiments examining the first nonpeptide BK receptor antagonist, Win 64338 (Salvino et al., 1993; Sawutz et al., 1994). This compound inhibited BK-induced contractions of guinea-pig ileum but not trachea (Farmer and DeSiato, 1994). The existence of a B_3 receptor is controversial, however, as a potent antagonist, Hoe 140, blocks responses to BK both in ileum and trachea of the guinea-pig (Field et al., 1992a,b; Rhaleb et al., 1992). It may be argued that Hoe 140 is a nonselective antagonist at both B_2 and B_3 receptor subtypes, whereas the other, weaker antagonists may differentiate between the two subtypes. The pharmacological differences in guinea-pig tissues, observed with DArg-[Hyp^3,DPhe^7]-BK may not warrant the classification of a different receptor subtype, and may be due to species differences or the partial agonist properties of some of the [DPhe^7]-BK peptide analogs (Field et al., 1992a,b; Rhaleb et al., 1992). Furthermore, no guinea-pig kinin receptors have yet been cloned. B_2 receptor knockout mice may be useful in addressing this issue (Borkowski and Hess, 1995), although the responsiveness of mouse trachea to kinins, and the effects of antagonists have not been described in the literature.

3.4.2 Smooth Muscle and Neuronal B_2 Receptor Subtypes

Another series of experiments, which suggested the existence of different B_2 receptor subtypes in smooth muscle and nerves, was reported in rat isolated vas deferens (Llona et al., 1987; Rifo et al., 1987). The peptide B_2 receptor antagonist, [Thi^{5,8},DPhe^7]-BK, behaved as a full agonist, stimulating neurogenic contractions, whereas the same compound antagonized the direct, postjunctional effects of BK on vas deferens smooth muscle. In addition, the agonists Met-Lys-BK and BK were 26- and 4-fold more active, respectively, in eliciting neurogenic responses than direct spasmogenic smooth muscle contraction. Moreover, [Hyp^3,Thi^{5,8},Phe^7]-BK, a B_2 receptor antagonist, lacked agonist activity and, yet, was more potent in blocking the smooth muscle responses than neuronal responses. In another study, competition binding studies with the same agents indicated two different binding sites (Liebmann *et al.*, 1991).

Nevertheless, several studies indicate that the proposed existence of different smooth muscle and neuronal B_2 receptor subtypes requires re-evaluation. Thus, heterologous expression of the B_2 receptor isolated from rat uterus indicated that it exhibits similar pharmacology to the so-called "smooth muscle subtype" (McEachern *et al.*, 1991). In contrast, the pharmacology of the human homolog indicates that it may be classified as the "neuronal subtype", with some properties of the so-called "B_3 receptor" (Eggerickx *et al.*, 1992). In addition, cloning of the B_2 receptor from neuronal cell lines such as NG-108 and PC12 revealed that it has properties similar to the smooth muscle B_2 receptor (Nardone *et al.*, 1994; Yokoyama *et al.*, 1994). Furthermore, at least in the mouse, our gene disruption experiments targeting the smooth muscle subtype suggest a single B_2 receptor. Mice in which the B_2 receptor gene was "knocked out" lack responsiveness to BK both in smooth muscle and neuronal preparations (Borkowski *et al.*, 1995).

3.4.3   Species Differences in the Effects of B_2 Receptor Antagonists

Other evidence for different B_2 receptor subtypes arises from studies examining pharmacological differences in effects of some peptide antagonists and in various tissues isolated from different species (Braas *et al.*, 1988; Saha *et al.*, 1990). Finally, the ability of BK to stimulate multiple second-messenger pathways independently, such as PI hydrolysis and arachidonic-acid release, has been suggested to indicate multiple B_2 receptor subtypes (Conklin *et al.*, 1988; Denning and Welsh, 1991).

The molecular cloning of B_2 receptors from different species has permitted an examination of the pharmacological properties of species' homologs (Hess *et al.*, 1994). By expressing these receptors heterologously in a cell line normally lacking BK receptors, the pharmacology of a single gene product can be analyzed. The ability of a single cloned receptor to stimulate both PLC and PLA_2 pathways suggests that one need not propose multiple receptor subtypes to explain the activation of multiple second-messenger pathways. The observation that the mouse B_2 receptor has a 60–80-fold higher affinity for DArg-[Hyp^3,Thi^{5,8},DPhe^7]-BK and a five-fold higher affinity for [Thi^{5,8},DPhe^7]-BK than the human homolog demonstrates that significant pharmacological variability may be a result of species differences. Thus, pharmacological differences observed between receptors from different species should be interpreted with caution before proposing B_2 receptor subtypes.

3.5   TARGETED DISRUPTION OF THE B_2 RECEPTOR

Targeted disruption of the murine B_2 receptor provides a new tool with which to examine its physiological roles (Borkowski *et al.*, 1995). Selective elimination of a single gene should allow the question of whether additional BK receptors exist to be answered unambiguously. Several investigators have initiated studies with the B_2 receptor knockout mice to investigate whether responsiveness to BK is retained *in vitro* and *in vivo*. These mutant animals permit evaluations of the role of B_2 receptors in phenomena such as inflammation and regulation of blood pressure. Genetic ablation of the B_2 receptor eliminates the need to utilize pharmacological antagonists to block its function. Thus, potential problems with pharmacokinetics, nonspecific pharmacology or poor bioavailabilty, utilizing pharmacological agents, particularly peptides, will not be apparent in knockout animals. On the other hand, there may be compensatory changes that occur during the development of the mutant animals that diminish the impact of the disruption.

Mice that are homozygous for the targeted disruption of the B_2 receptor are phenotypically normal by visual inspection (Borkowski *et al.*, 1995). Relative to controls they have normal sensory-motor skills. Clinical pathology indicates that the B_2 receptor knockout mice (BK2r−/−) have normal serum chemistry, hematology and histopathology of the major organs. Breeding of heterozygous (BK2r+/−) animals indicates that the homozygous knockout mice are obtained in the expected 1:2:1 Mendelian ratio. Breeding of male and female BK2r−/− mice indicates that they are fertile and that the litter size does not deviate from controls. Although the kallikrein–kinin system may play a role in reproduction, as suggested (Valdés *et al.*, 1993; Heder *et al.*, 1994), the B_2 receptor does not appear to play a major role in that regard in mice. The lack of a B_2 receptor can readily be demonstrated in binding and functional studies with isolated tissue preparations. In such experiments, BK2r+/+ or BK2r+/− controls exhibit responsiveness to BK whereas BK is inactive in BK2r−/− mouse tissues. Thus, the BK2r−/− mouse lacks B_2 receptors and will be invaluable in studies of the physiological role of this receptor.

4.   Summary

The tools of molecular biology have been responsible for several recent advances in our understanding of the sequence and pharmacology of kinin receptors. The molecular cloning of B_1 and B_2 receptors has revealed them to be members of the superfamily of G protein-coupled receptors. Cloning and pharmacological characterization of B_1 receptors has demonstrated that there are species differences in the affinities of B_1 receptors for different agonists. Similarly, the molecular cloning and characterization of B_2 receptors indicates species differences with respect to antagonist effects and which may account for some of the pharmacological data previously suggestive of receptor subtypes. The targeted disruption of the gene encoding the mouse B_2 receptor provides direct evidence for a single B_2 receptor subtype in smooth muscle and neuronal tissues of this species.

5. Acknowledgements

I thank Dr Catherine Strader for her support and encouragement and my daughters, Kimberly and Laura, for providing joy to each day.

6. References

Bhoola, K.D., Figueroa, C.D. and Worthy, K. (1992). Bioregulation of kinins: kallikreins, kininogens, and kininases. Pharmacol. Rev. 44, 1–80.

Borkowski, J.A. and Hess, J.F. (1995). Targeted disruption of the mouse B_2 bradykinin receptor in embryonic stem cells. Can. J. Physiol. Pharmacol. 73, 773–779.

Borkowski, J.A., Ransom, R.W., Seabrook, E.R., Trumbauer, M., Chen, H., Hill, R.E., Strader, C.D. and Hess, J.F. (1995). Targeted disruption of a B_2 bradykinin receptor gene in mice eliminates bradykinin action in smooth muscle and neurons. J. Biol. Chem. 270, 13706–13710.

Braas, K.M., Manning D.C., Perry, D.C. and Snyder, S.H. (1988). Bradykinin analogues: differential agonist and antagonist activities suggesting multiple receptors. Br. J. Pharmacol. 94, 3–5.

Braun, A., Kammerer, S., Bohme, E., Muller, B. and Roscher, A.A. (1995). Identification of polymorphic sites of the human bradykinin B_2 receptor gene. Biochem. Biophys. Res. Commun. 211, 234–240.

Burch, R.M. and Axelrod, J. (1987). Dissociation of bradykinin-induced prostaglandin formation from phosphatidylinositol turnover in Swiss 3T3 fibroblasts: evidence for G protein regulation of phospholipase A_2. Proc. Natl Acad. Sci. USA 84, 6374–6378.

Chahine, R., Adam, A., Yamaguchi, N., Gaspo, R., Regoli, D. and Nadeau, R. (1993). Protective effects of bradykinin on the ischaemic heart: implication of the B_1 receptor. Br. J. Pharmacol. 108, 318–322.

Conklin, B.R., Burch, R.M., Steranka, L.R. and Axelrod, J. (1988). Distinct bradykinin receptors mediate stimulation of prostaglandin synthesis by endothelial cells and fibroblasts. J. Pharmacol. Exp. Ther. 244, 646–649.

Davis, A.J. and Perkins, M.N. (1994). Induction of B_1 receptors in vivo in a model of persistent inflammatory mechanical hyperalgesia in the rat. Neuropharmacology 33, 127–133.

DeBlois, D., Bouthillier, J. and Marceau, F. (1988). Effects of glucocorticoids, monokines and growth factors on the spontaneously developing responses of the rabbit isolated aorta to des-Arg9-bradykinin. Br. J. Pharmacol. 93, 969–977.

DeBlois, D., Bouthillier, J. and Marceau, F. (1989). Pharmacological modulation of the up-regulated responses to des-Arg9-bradykinin in vivo and in vitro. Immunopharmacology 17, 187–198.

Denning, G.M. and Welsh, M.J. (1991). Polarized distribution of bradykinin receptors on airway epithelial cells and independent coupling to second messenger pathways. J. Biol. Chem. 266, 12932–12938.

Dray, A. and Perkins, M. (1993). Bradykinin and inflammatory pain. Trends Neurosci. 16, 99–104.

Eggerickx, D., Raspe, E., Bertrand, D., Vassart, G. and Parmentier, M. (1992). Molecular cloning, functional expression and pharmacological characterization of a human bradykinin B_2 receptor gene. Biochem. Biophys. Res. Commun. 187, 1306–1313.

Farmer, S.G. and Burch, R.M. (1992). Biochemical and molecular pharmacology of kinin receptors. Annu. Rev. Pharmacol. Toxicol. 32, 511–536.

Farmer, S.G. and DeSiato, M.A. (1994). Effects of a novel nonpeptide bradykinin B_2 receptor antagonist on intestinal and airway smooth muscle. Br. J. Pharmacol. 112, 461–464.

Farmer, S.G., Burch, R.M., Meeker, S.A. and Wilkins, D.E. (1989). Evidence for a pulmonary B_3 bradykinin receptor. Mol. Pharmacol. 36, 1–8.

Field, J.L., Hall, J.M. and Morton, I.K.M. (1992a). Bradykinin receptors in the guinea-pig taenia caeci are similar to proposed BK_3 receptors in guinea-pig trachea, and are blocked by Hoe-140. Br. J. Pharmacol. 105, 293–296.

Field, J.L., Hall, J.M. and Morton, I.K.M. (1992b). Putative novel bradykinin B_3 receptors in the smooth muscle of the guinea-pig taenia caeci and trachea. Agents Actions 38 (Suppl. I), 540–545.

Furuto-Kato, S., Matsumoto, A., Kitamura, N. and Nakanishi, S. (1985). Primary structures of the mRNAs encoding the rat precursors for bradykinin and T-kinin. J. Biol. Chem. 260, 12054–12059.

Giladi, E. and Spindel, E.R. (1991). Simple luminometric assay to detect phosphoinositol-linked receptor expression in Xenopus oocytes. BioTechniques 10, 744–747.

Goldstein, R.H. and Wall, M. (1984). Activation of protein formation and cell division by bradykinin and Des-Arg9-bradykinin. J. Biol. Chem. 259, 9263–9268.

Hall, J.M. (1992). Bradykinin receptors: pharmacological properties and biological roles. Pharmacol. Ther. 56, 131–190.

Heder, G., Bittger, A., Siems, W.-E., Rottman, M. and Kertscher, U. (1994). The enzymatic degradation of bradykinin in semen of various species. Andrologia 26, 295–301.

Hess, J.F., Borkowski, J.A., Young, G.S., Strader, C.D. and Ransom, R.W. (1992). Cloning and pharmacological characterization of a human bradykinin (BK-2) receptor. Biochem. Biophys. Res. Commun. 184, 260–268.

Hess, J.F., Borkowski, J.A., MacNeil, T., Stonesifer, G.Y., Fraher, J., Strader, C.D. and Ransom, R.W. (1994). Differential pharmacology of cloned human and mouse B_2 bradykinin receptors. Mol. Pharmacol. 45, 1–8.

Hess, J.F., Derrick, A.W., MacNeil, T. and Borkowski, J.A. (1996). The agonist selectivity of a mouse B_1 bradykinin receptor differs from human and rabbit B_1 receptors. Immunopharmacology 33, 1–8.

54   J.F. Hess

Kammerer, S., Braun, A., Arnold, N. and Roscher, A.A. (1995). The human bradykinin B2 receptor gene: full length cDNA, genomic organization and identification of the regulatory region. Biochem. Biophys. Res. Commun. 211, 226–233.

Lamontagne, D., Nadeau, R. and Adam, A. (1995). Effect of enalaprilat on bradykinin and des-Arg9-bradykinin release following reperfusion of the ischaemic rat heart. Br. J. Pharmacol. 115, 476–478.

Liebmann, C., Schnittler, M., Stewart, J.M. and Reissmann, S. (1991). Antagonist binding reveals two heterogenous B2 bradykinin receptors in rat myometrial membranes. Eur. J. Pharmacol. 199, 363–365.

Llona, I., Vavrek, R., Stewart, J. and Huidobro-Toro, J.P. (1987). Identification of pre- and postsynaptic bradykinin receptor sites in the vas deferens: evidence for different structural prerequisites. J. Pharmacol. Exp. Ther. 241, 608–614.

Lubbert, H., Hoffman, B.J., Snutch, T.P., Dyke, T., Levine, A.J., Hartig, P.R., Lester, H.A. and Davidson, N. (1987). cDNA cloning of a serotonin-1c receptor by electrophysiological assays of mRNA-injected Xenopus oocytes. Proc. Natl. Acad. Sci. USA 84, 4332–4336.

Ma, J., Wang, D., Chao, L. and Chao, J. (1994a). Cloning, sequence analysis and expression of the gene encoding the mouse bradykinin B2 receptor. Gene, 149, 283–288.

Ma, J., Wang, D., Ward, D.C., Chen, L., Dessai, T., Chao, J. and Chao, L. (1994b). Structure and chromosomal localization of the gene encoding human bradykinin B2 receptor. Genomics 23, 362–369.

MacNeil, T., Bierilo, K.K., Menke, J.G. and Hess, J.F. (1995). Cloning and pharmacological characterization of a rabbit bradykinin B1 receptor. Biochim. Biophys. Acta Gene Struct. Expres. 1264, 223–228.

Mahan, L.C. and Burch, R.M. (1990). Functional expression of B2 bradykinin receptors from Balb/c cell mRNA in Xenopus oocytes. Mol. Pharmacol. 37, 785–789.

Marceau, F. (1995). Kinin B1 receptors: a review. Immunopharmacology 30, 1–26.

Masu, Y., Nakayama, K., Tamaki, H., Harada, Y., Kuno, M. and Nakanishi, S. (1987). cDNA cloning of bovine substance-K receptor through oocyte expression system. Nature (Lond.) 329, 836–838.

McEachern, A.E., Shelton, E.R., Bhakta, S., Obernolte, R., Bach, C., Zuppan, P., Fujisaki, J., Aldrich, R.W. and Jarnagin, K. (1991). Expression cloning of a rat B2 bradykinin receptor. Proc. Natl Acad. Sci. USA 88, 7724–7728.

McIntyre, P., Phillips, E., Skidmore, E., Brown, M. and Webb, M. (1993). Cloned murine bradykinin receptor exhibits a mixed B1 and B2 pharmacological selectivity. Mol. Pharmacol. 44, 346–355.

Menke, J.G., Borkowski, J.A., Bierlo, K.K., MacNeil, T., Derrick, A.W., Schneck, K.A., Ransom, R.W., Strader, C.D., Linemeyer, D.L. and Hess, J.F. (1994). Expression cloning of a human B1 bradykinin receptor. J. Biol. Chem. 269, 21583–21586.

Nardone, J., Gerald, C., Rimawi, L., Song, L. and Hogan, P.G. (1994). Identification of a B2 bradykinin receptor expressed by PC12 phaeochromocytoma cell line. Proc. Natl Acad. Sci. USA 91, 4412–4416.

Perkins, M.N. and Kelly, D. (1993). Induction of bradykinin B1 receptors in vivo in a model of ultra-violet irradiation-induced thermal hyperalgesia in the rat. Br. J. Pharmacol. 110, 1441–1444.

Perkins, M.N. and Kelly, D. (1994). Interleukin-1β induced-desArg9bradykinin-mediated thermal hyperalgesia in the rat. Neuropharmacology 33, 657–660.

Perkins, M.N., Campbell, E. and Dray, A. (1993). Antinociceptive activity of the bradykinin B1 and B2 receptor antagonists, des-Arg9-[Leu8]-BK and HOE 140, in two models of persistent hyperalgesia in the rat. Pain 53, 191–197.

Pesquero, J.B., Lindsey, C.J., Zeh, K., Paiva, A.C.M., Ganten, D. and Bader, M. (1994). Molecular structure and expression of rat bradykinin B2 receptor gene. J. Biol. Chem. 269, 26920–26925.

Phillips, E., Conder, M.J., Bevan, S., McIntyre, P. and Webb, M. (1992). Expression of functional bradykinin receptors in Xenopus oocytes. J. Neurochem. 58, 243–249.

Powell, S.J., Slynn, G., Thomas, C., Hopkins, B., Briggs, I. and Graham, A. (1993). Human bradykinin B2 receptor: nucleotide sequence analysis and assignment to chromosome 14. Genomics 15, 435–438.

Proud, D. and Kaplan, A.P. (1988). Kinin formation: Mechanisms and role in inflammatory disorders. Annu. Rev. Immunol. 6, 49–83.

Regoli, D. and Barabé, J. (1980) Pharmacology of bradykinin and related kinins. Pharmacol. Rev. 32, 1–46.

Regoli, D., Barabé, J. and Park, W.K. (1977). Receptors for bradykinin in rabbit aorta. Can. J. Physiol. Pharmacol. 55, 855–867.

Regoli, D., Marceau, F. and Barabé, J. (1978). De novo formation of vascular receptors for bradykinin. Can. J. Physiol. Pharmacol. 56, 674–677.

Regoli, D., Marceau, F. and Lavigne, J. (1981). Induction of B1-receptors for kinins in the rabbit by a bacterial lipopolysaccharide. Eur. J. Pharmacol. 71, 105–115.

Rhaleb, N.E., Rouissi, N., Jukic, D., Regoli, D., Henke, S., Breipohl, G. and Knolle, J. (1992). Pharmacological characterization of a new highly potent B2 receptor antagonist (HOE 140: D-Arg-[Hyp3,Thi5,D-Tic7,Oic8]bradykinin. Eur. J. Pharmacol. 210, 115–120.

Rifo, J., Pourrat, M., Vavrek, R.J., Stewart, J.M. and Huidobro-Toro, J.P. (1987). Bradykinin receptor antagonists used to characterize the heterogeneity of bradykinin-induced responses in rat vas deferens. Eur. J. Pharmacol. 142, 305–312.

Rocha e Silva, M., Beraldo, W.T. and Rosenfeld, G. (1949). Bradykinin, a hypotensive and smooth muscle stimulating factor released from plasma globulin by snake venoms and by trypsin. Am. J. Physiol. 156, 261–273.

Saha, J.K., Sengupta, J.N. and Goyal, R.K. (1990). Effect of bradykinin on opossum esophageal longitudinal smooth muscle: evidence for novel bradykinin receptors. J. Pharmacol. Exp. Ther. 252, 1012–1020.

Salvino, J.M., Seoane, P.R., Douty, B.D., Awad, M.A., Dolle, R.E., Houck, W.T., Faunce, D.M. and Sawutz, D.G. (1993). Design of potent non-peptide competitive antagonists of the human bradykinin B2 receptor. J. Med. Chem. 36, 2583–2584.

Sawutz, D.G., Salvino, J.M., Dolle, R.E., Casiano, F., Ward, S.J., Houck, W.T., Faunce, D.M., Douty, B.D., Baizman, E., Awad, M.M.A., Marceau, F. and Seoane, P.R. (1994). The nonpeptide Win 64338 is a bradykinin B2 receptor antagonist. Proc. Natl Acad. Sci. USA 91, 4693–4697.

Schneck, K.A., Hess, J.F., Stonesifer, G.Y. and Ransom, R.W. (1994). Bradykinin B1 receptors in rabbit aorta smooth muscle cells in culture. Eur. J. Pharmacol. 266, 277–282.

Taketo, M., Yokoyama, S., Rochelle, J., Kimura, Y., Higashida, H., · Taketo, M. and Seldin, M.F. (1995). Mouse B_2 bradykinin receptor gene maps to distal chromosome 12. Genomics 27, 222–223.

Tropea, M.M., Gummelt, D., Herzig, M.S. and Leeb-Lundberg, L.M.F. (1993). B_1 and B_2 kinin receptors on cultured rabbit superior mesenteric artery smooth muscle cells: receptor-specific stimulation of inositol phosphate formation and arachidonic acid release by des-Arg^9-bradykinin and bradykinin. J. Pharmacol. Exp. Ther. 264, 930–937.

Valdés, G., Corthorn, J., Scicli, A.G., Gaete, V., Soto, J., Ortiz, M.E., Foradori, A. and Saed, G.M. (1993). Uterine kallikrein in the early pregnant rat. Biol. Reprod. 49, 802–808.

Vavrek, R.J. and Stewart, J.M. (1985). Competitive antagonists of bradykinin. Peptides 6, 161–164.

Webb, M., McIntyre, P. and Phillips, E. (1994). B_1 and B_2 bradykinin receptors encoded by distinct mRNAs. J. Neurochem. 62, 1247–1253.

Yokoyama, S., Kimura, Y., Teketo, M., Black, J.A., Ransom, B.R. and Higashida, H. (1994). B_2 bradykinin receptors in NG108-15 cells: cDNA cloning and functional expression. Biochem. Biophys. Res. Commun. 200, 634–641.

4. Gene Expression, Regulation and Cell Surface Presentation of the Kininogens

Alvin H. Schmaier

1. Introduction

This chapter aims to put the field of kininogens in a unique perspective and to indicate key areas for future investigation in a vibrant and evolving field of vascular biology. Although first recognized 21 years ago as a defect associated with a very prolonged activated partial thromboplastin time (APTT) (Colman *et al.*, 1975; Saito *et al.*, 1975; Wuepper *et al.*, 1975), the genetic deficiency of high-molecular-weight kininogen (HMWK) is not associated with bleeding. These features have been both a benefit and a discomfort to investigators in this field. Recognition of HMWK deficiency is greatly facilitated by prolongation of the APTT, although the homozygous HMWK-deficient patient is rare. Alternatively, the prolonged APTT associated with HMWK deficiency has contributed to a masking of the physiological importance of HMWK both as an antithrombin, as well as its role in ordering the assembly and activation of a plasma kallikrein-dependent fibrinolytic pathway on biologic membranes. These activities are paradoxical to any prolongation of a

screening test for bleeding. The weight of current evidence indicates that the kininogens, especially HMWK, are antithrombins and profibrinolytic proteins.

In order to convince the reader of this interpretation, the present chapter discusses the following topics concerning kininogens: gene expression and regulation, protein chemistry, the function of kininogens in relation to their structures, characterization of kininogen expression on biological membranes; and emphasizes the physiological activities of the kininogens based upon function.

2. Kininogen Gene Expression and Regulation

The plasma kininogens, HMWK and low molecular weight kininogen (LMWK), are products of a single gene (Kitamura *et al.*, 1985; Takagaki *et al.*, 1985), which maps to 3q26-qter, the location of α2HS-glycoprotein

Copyright © 1997 Academic Press Limited
All rights of reproduction in any form reserved.

and histidine-rich glycoprotein (Fong *et al.*, 1991; Cheung *et al.*, 1992; Rizzu and Baldini, 1995). A single kininogen gene of 11 exons, consisting of 27 kb, transcribes a unique mRNA for HMWK and LMWK by alternative splicing (Kitamura *et al.*, 1985). HMWK and LMWK share the coding region of the first nine exons, a part of exon 10 containing the bradykinin (BK) sequence, and the first 12 amino acids after the carboxy-terminal BK sequence. Exon 11 codes for the 4 kDa light chain of LMWK. The complete exon 10 contains the full coding sequence for the unique 56 kDa light chain of HMWK. By a process of alternative RNA processing events, HMWK is produced from exon 10. In the rat kininogen gene, a novel mechanism occurs for alternative RNA processing (Kakizuka *et al.*, 1990). Splicing efficiency is controlled by the interaction of U1 small nuclear ribonucleoproteins and the U1 small nuclear RNA (snRNA)-complementary repetitive sequences of the kininogen pre-mRNA. The mRNA for LMWK and HMWK are 1.7 and 3.5 kb, respectively.

Portions of kininogens have been expressed in bacteria with preservation of activity. In particular, Ylinenjarvi *et al.* (1995) expressed kininogen domain 2 in *E. coli* with preservation of the ability to inhibit cathepsin L and papain, but reduced affinity for calpain when compared to LMWK. Likewise, human kininogen domain 3 was expressed in *E. coli* with full preservation of ability to inhibit papain and cathepsin L (Auerswald *et al.*, 1993). Accordingly, portions of the 56 kDa light chain of HMWK have been expressed, which preserve artificial surface-binding abilities (Kunapuli *et al.*, 1993). These studies indicate that recombinant approaches to produce this protein were successful.

Cheung *et al.* (1993) examined the molecular basis for homozygous total kininogen deficiency, "Williams trait", and found no gross DNA deletion or insertion by Southern blot. Rather, a C-to-T transition at nucleotide 587 occurred, changing a CGA (Arg) codon to TGA (Stop) mutation in exon 5, resulting in prevention of synthesis of both HMWK and LMWK (Cheung *et al.*, 1993). This phenotypic defect is similar to that seen in Brown-Norway, Katholiek strain rats, which have the kininogen gene but do not express plasma kininogens (Oh-ishi *et al.*, 1994). The absent plasma kininogen is due to a single point mutation, Ala 163 to Thr, which results in defective secretion from rat liver (Hayashi *et al.*, 1993). Little is known about the regulation of kininogen gene expression. In rats, ovariectomy results in a reduction of kininogen transcripts in the liver, while estrogens increase kininogen mRNA levels (Chen *et al.*, 1992). These results are consistent with the clinical observation that HMWK concentrations increase during pregnancy (Chhibber *et al.*, 1990). In contrast, progesterone treatment reduced kininogen gene expression, resulting in a slight reduction of plasma kininogen levels (Takano *et al.*, 1995). Murine fibroblasts synthesize and secrete kininogens in response to cyclic AMP, forskolin, prostaglandin E_2 and tumor necrosis factor-α (TNF-α) (Takano *et al.*, 1995). Similarly, TNF-α has been recognized to increase kininogen expression in HEP G2 cells (Scott and Colman, 1992). Little else, however, has been shown to influence kininogen levels, primarily because this aspect of kininogen biology has not been studied in detail.

3. *Protein Chemistry and Structure of the Kininogens*

Two mRNAs code for two separate kininogens. LMWK is a 66 kDa β-globulin with a plasma concentration of 160 µg/ml (2.4 µM) and an isoelectric point of 4.7 (Jacobsen and Kriz, 1967; Schmaier *et al.* 1986a). HMWK is a 120 kDa α-globulin with a plasma concentration of 80 µg/ml (0.67 µM) and an isoelectric point of 4.3 (Schmaier *et al.*, 1983, 1986a). Human liver is a source for the cDNA for both kininogens (Kitamura *et al.*, 1985; Takagaki *et al.*, 1985), but human umbilical vein endothelial cells (HUVEC) have also been shown to contain HMWK mRNA and to synthesize the protein (Schmaier *et al.*, 1988a). Kininogen antigen also has been found in platelets, granulocytes, renal tubular cells and skin (Proud *et al.*, 1981; Schmaier *et al.*, 1983; Hallbach *et al.*, 1987; Yamamoto *et al.*, 1987; Gustafson *et al.*, 1989b). Until its cloning, LMWK was also known as α_1-cysteine protease inhibitor (Ohkubo *et al.*, 1984). LMWK gel filters at 66 kDa, and HMWK, although actually 120 kDa, gel filters at 220 kDa. Physical evidence for HMWK being a globular protein also was obtained by electron microscopy studies, wherein it appeared to consist of a linear array of three linked, centralized, globular regions with the two ends thinly connected (Weisel *et al.*, 1994). Cleavage of HMWK by plasma kallikrein leads to a striking change in conformation in HMWK. The central globular region is separated after BK liberation and rearranged with the cysteine protease inhibitory region opposite the prekallikrein-binding region (Weisel *et al.*, 1994). The regions of kininogens are characterized into domains (Fig. 4.1). Separating these domains are serine protease-sensitive regions (Salvesen *et al.*, 1986; Schmaier *et al.*, 1987; Vogel *et al.*, 1988). As discussed below, contiguity of these domains is important for biological functions of kininogens including calpain inhibition, and HMWK and LMWK binding to endothelial cells (Bradford *et al.*, 1993; Hasan *et al.*, 1995a,c). Alternatively, disruption of the contiguity of the HMWK allows for the development of another function, cell antiadhesive activity (Asakura *et al.*, 1992). Nevertheless, the major activity of kininogens, which is to yield kinins, is programmed disruption of the protein as BK is not biologically active unless liberated from its precursor.

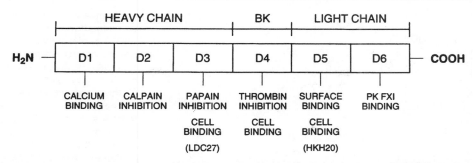

Figure 4.1   The domain structure of high molecular weight kininogen (HMWK). HMWK is divided into three regions: a heavy chain consisting of domains 1–3; the bradykinin (BK) region, domain 4; and a light chain, which consists of domains 5 and 6. Domain 1 binds calcium. Domain 2 is a cysteine protease inhibitory region, particularly of calpain. Domain 3 is also a cysteine protease inhibitory region. It also contains a major cell-binding site. Domain 4 is BK. Its amino-terminal end has α-thrombin inhibitory activity; its carboxy-terminal end is a cell-binding region. Domain 5 is the major cell-binding site and it contains the artificial surface-binding site. Domain 6 is the prekallikrein (PK) and factor XI (FXI) binding sites.

4.   The Domain Structure of the Kininogens

The kininogens are multidomain proteins with activities associated with each domain (Fig. 4.1). Although each domain has one or more specific activities, the protein as a whole participates in several biologic processes. The kininogens are placed on cell-surface receptors such that BK can be liberated at specific sites whereupon it activates kinin receptors and influences the local cellular milieu. Thus, one can view each function of the domains of the kininogens as participating in the whole protein's kinin delivery activity. In general, the kininogens can be divided into three portions: the heavy chain, which is common to both HMWK and LMWK; the kinin moiety; and the light chain, which, as already stated, is unique to HMWK and LMWK (Fig. 4.1). Domains 1–3 comprise kininogen's heavy chain. Little is known about the function of domain 1 except to note that it has a low-affinity Ca^{2+}-binding site, the function of which is unknown (Ishiguro *et al.*, 1987). Although Ca^{2+} is important for phorbol ester-induced upregulation of LMWK and heavy-chain binding to endothelial cells (Zini *et al.*, 1993), the evidence that Ca^{2+} participates in HMWK binding to cells is controversial (Greengard and Griffin, 1984; Gustafson *et al.*, 1986; Van Iwaarden *et al.*, 1988b; Hasan *et al.*, 1995a). Domains 2 and 3 contain the highly conserved amino-acid sequence, Glu-Val-Val-Ala-Gly, found in cysteine protease inhibitors (Salvesen *et al.*, 1986). Both LMWK and HMWK are potent, competitive cysteine protease inhibitors with K_is, respectively, of 2 and 0.5 nM for platelet calpain (Schmaier *et al.*, 1986a; Bradford *et al.*, 1990). Domain 2 is the calpain inhibitory region on kininogens (Schmaier *et al.*, 1987; Auerswald *et al.*, 1993; Bradford *et al.*, 1993), whereas domain 3 inhibits cysteine

proteases except the calpains (Salvesen *et al.*, 1986; Kunapuli *et al.*, 1993). Although the kininogens inhibit cysteine proteases, they are also substrates of this class of enzyme when there is molar excess of enzyme (Schmaier *et al.*, 1986b; Scott *et al.*, 1993). Since kininogens are extracellular proteins, or stored in platelets and granulocytes, it is unclear how they interact with cellular cysteine proteases, which, for the most part, are intracellular (Schmaier *et al.*, 1983; Gustafson *et al.*, 1989b; Schmaier *et al.*, 1990). However, when platelets are activated, calpain translocates to the external membrane where it may be inhibited by plasma or externalized platelet α-granule HMWK (Schmaier *et al.*, 1986a, 1990; Saido *et al.*, 1993).

LMWK and its isolated heavy chain bind to platelets and endothelial cells, indicating that there was a cell-binding region on the kininogen heavy chain (Meloni and Schmaier, 1991; Zini *et al.*, 1993). This was confirmed by direct studies using isolated and recombinant domain 3, which contained the heavy-chain cell-binding region (Jiang *et al.*, 1992; Wachtfogel *et al.*, 1993). Using a computerized model based upon the crystallized structure of cystatin, the sequential amino-acid structure of domain 3 was drawn to reveal three surface-exposed regions, namely a disulfide loop connecting domain 3 to domain 2 and two hairpin loops (Fig. 4.2) (Bode *et al.*, 1988). The cysteine protease inhibitory region of domain 3 consists of these three loops. Using synthetic peptide sequences of the surface-exposed regions, $K_{244}ICVGCPRDIP_{254}$ (KIC11), $N_{276}ATFYFKIDNVKKARVQVVAGKKYFI_{301}$ (NAT26), and $L_{331}DCNAEVYVVPWEKKIYPTVNCQPLGM_{357}$ (LDC27), we sought to determine which of the surface-exposed loops was the domain 3 cell-binding site(s) (Herwald *et al.*, 1995). KIC11, NAT26 and LDC27 each inhibited biotin-HK binding to endothelial cells with IC_{50} values, respectively, of 1,000 μM, 258 μM and

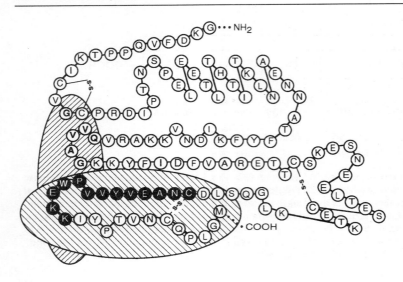

Figure 4.2   A two-dimensional representation of domain 3 (positions 235–357 of plasma kininogen) modeled after the crystal structure of egg white cystatin is shown. The predicted reactive site region (⊘) is formed by three distinct segments of the polypeptide chain: an amino-terminally located segment harboring a critical glycine residue (G); a central segment exposing the consensus sequence of Gln-Val-Val-Ala-Gly (QVVAG) shared by cystatins; and the carboxy-terminal segment holding a critical tryptophan (W) residue. The cell-binding site (◯) of domain 3 overlaps the distal segment of the reactive site. This figure is reproduced from Herwald et al. (1995) with permission from the American Society for Biochemistry and Molecular Biology, Inc.

60 μM. The minimal sequence in LDC27 to inhibit binding was 13 amino acids, starting with C_{333}NAEVYVVPWEKK$_{345}$ (IC$_{50}$, 113 μM) (Herwald et al., 1995). Since papain blocked HMWK binding to endothelial cells, the cysteine protease inhibitory site overlaps with the cell-binding site on domain 3. Thus, the last 27 amino acids of domain 3, which are contiguous with domain 4, the kinin-containing region, comprise its high-affinity cell-binding site.

Another function ascribed to domain 3 was its ability to inhibit α-thrombin-induced platelet activation (Meloni and Schmaier, 1991; Jiang et al., 1992). However, further investigations revealed that the α-thrombin inhibitory region on the kininogens is not on domain 3, but rather, parts of domain 4, the kinin moiety, attached to domain 3 (Hasan et al., 1996). When pure or plasma HMWK is cleaved by plasma kallikrein on an artificial surface, BK is cleaved from the parent protein in three ways (Mori and Nagasawa, 1981; Schmaier et al., 1988b). The first cleavage yields a "nicked" kininogen composed of two disulfide-linked chains (64 and 56 kDa). The second cleavage yields BK (0.9 kDa) and an intermediate kinin-free protein of similar molecular weight to "nicked" HMWK. The third cleavage results in a stable, kinin-free protein composed of two disulfide-linked chains (64 and 46 kDa). However, when kininogens are cleaved in solution, rather than on a surface, this sequence of cleavage does not necessarily occur and BK can remain attached to the heavy or light chain (Tayeh et al., 1994). Since isolated domain 3 was prepared by proteolytic cleavage in solution, we examined both trypsin-cleaved LMWK and domain 3 prepared by tryptic digestion and found that the BK moiety remained attached to LMWK's heavy chain and isolated domain 3 (Hasan et al., 1995c). Investigations were carried out to ascertain whether BK analogs and

breakdown products block α-thrombin. As described below, all of the domain 4 fragments inhibited α-thrombin-induced platelet aggregation (Hasan et al., 1996). These data indicate the thrombin inhibitory activity should be ascribed to domain 4.

In addition to the recently described ability to inhibit α-thrombin, the kinin-containing domain 4 of the kininogens is obviously crucial for the many biological functions associated with free BK. It is beyond the scope of this review to discuss the pharmacology of bradykinin, which is detailed elsewhere in the present volume, except to say that its activity supports the antithrombin, profibrinolytic activity of the kininogens (see below). Following liberation of BK, HMWK is a better substrate for plasma kallikrein and LMWK is a better substrate of tissue kallikrein. However, both kininogens are substrates to both forms of kallikreins. Factor XIIa cleaves HMWK similarly to plasma kallikrein (Wiggins, 1983). Factor XIa initially cleaves HMWK into a 76 and 46 kDa bands. Upon prolonged exposure to Factor XIa, the 46 kDa light chain of HMWK is proteolyzed into smaller, inactive fragments (Scott et al., 1985). Elastase treatment of LMWK facilitates the ability of plasma kallikrein to liberate BK (Sato and Nagasawa, 1988), although procoagulant activity of HMWK is destroyed by elastase. Cathepsin D inactivates kininogen cysteine protease inhibitory activity (Kleniewski and Donaldson, 1988). Another function of domain 4 is to serve as a cell-binding region (Hasan et al., 1994). The carboxy terminal portion of BK and the amino-terminal portion of the light-chain participate as low-affinity (K_d, 1 mM) binding sites to endothelial cells. The importance of the domain 4 cell-binding region is not its isolated affinity to the cell surface but, rather, its ability to hold kininogens in the appropriate conformation for optimal cell binding. For example, intact HMWK binds to endothelial cells with a

K_d of 7 nM and 1×10^7 molecules/cell compared to kinin-free kininogen, which binds to endothelial cells with a K_d of 30 nM and $1-2.6 \times 10^6$ molecules/cell (Hasan et al., 1994, 1995a). These data, obtained using biological surfaces, are consistent with the changes in the shape of HMWK when it is cleaved on an artificial surface (Weisel et al., 1994).

The light chain of LMWK is 4 kDa and consists of one domain ($D5_L$) of unknown function. In contrast, the light chain of HMWK is 56 kDa and consists of domains 5 ($D5_H$) and 6. Initial investigations showed that $D5_H$ served as an additional cell-binding site on platelets, granulocytes and endothelial cells (Meloni et al., 1992; Reddigari et al., 1993a; Wachtfogel et al., 1993; Zini et al., 1993). Sequential and overlapping peptides of $D5_H$ were prepared, and two sequences were reported to participate in cell binding (Hasan et al., 1995b). One was on the amino terminus of the domain and consisted of sequences G_{402}KEQGHTRRHDWGHEKQRK$_{420}$ (GKE19) and H_{421}NLGHGHKHERDQGHGHQRGH$_{441}$ (HNL21), which inhibited biotin-HMWK binding (IC_{50} values, respectively, 792 μM and 215 μM). The second region is on the carboxy terminus of $D5_H$ and comprises two overlapping peptides, H_{479}KHGHGHGKHKNKGKKNGKH$_{498}$ (HKH20) and H_{471}VLDHGHKHKHGHGHGKHKNKGKK$_{494}$ (HVL24), which inhibited HMWK binding with respective IC_{50} values of 0.23 μM and 0.8 μM (Hasan et al., 1995b). Independently of its cell-binding region, $D5_H$ is also the artificial surface-binding region of HMWK (Schmaier et al., 1987). The histidine- and glycine-rich regions have the ability to bind to anionic surfaces, and $D5_H$ is the Zn^{2+}- and heparin-binding region (Retzio et al., 1987; Bjork et al., 1989; DeLa Cadena and Colman, 1992). $D5_H$ contains two histidine- and glycine-rich regions, one on its carboxy terminal and the other on its amino terminal. Using a deletion mutagenesis strategy, it was concluded that the anionic surface-binding region was associated with both histidine- and glycine-rich regions of $D5_H$ (Kunapuli et al., 1993). The peptides HKH20 and HVL24, which are the high-affinity cell-binding regions on $D5_H$, also inhibit the procoagulant activity of HMWK (Hasan et al., 1995b). No other peptides from $D5_H$ exhibited this property. Further, a polyclonal antibody to HKH20 prolonged the procoagulant activity of HMWK in plasma (Hasan et al., 1995b), and these data indicate that the cell-binding and artificial surface-binding region on HMWK are the same, and highly conserved. Peptide HKH20 and HMWK also have the ability to interact with M protein on S. pyogenes (Ben Nasr et al., 1995). It is of interest that the highest affinity cell-binding site for $D5_H$ turns out to be the major artificial surface-binding site. Efforts by many workers over the last two decades to characterize HK binding to artificial surfaces indicated the location of HK's cell-binding site. Finally, when HMWK is cleaved, the residual, kinin-free kininogen has the ability to prevent vitronectin's adhesive interaction with tumor cells, endothelial cells, platelets and monocytes (Asakura et al., 1992), although much more weakly than intact HMWK.

Domain 6 of HMWK has a prekallikrein- and Factor XI-binding site (Tait and Fujikawa, 1986, 1987; Vogel et al., 1990). The affinity of prekallikrein for its binding site on the light chain of HMWK is about 17 nM (Bock and Shore, 1983; Bock et al., 1985). The prekallikrein and Factor XI binding site consists of a 31-residue sequence that contains predominantly β-turn elements (Scarsdale and Harris, 1990). HMWK's procoagulant activity is dependent upon two activities, one being the ability to bind to anionic surfaces via $D5_H$ and the other, the ability to bind prekallikrein and Factor XI to domain 6. Inhibition of either interaction, utilizing monoclonal antibodies directed to these regions, inhibits the procoagulant activity of HMWK (Schmaier et al., 1987; Reddigari and Kaplan, 1989; Kaufmann et al., 1993). Domain 6 serves as the acceptor protein for Factor XI and prekallikrein binding to platelets and endothelial cells (Greengard et al., 1986; Lenich et al., 1995; Motta et al., 1995). As discussed below, prekallikrein binding to bound HMWK initiates a sequence of events that leads to prekallikrein activation on biologic surfaces.

5. Cell Surface Expression of Kininogens

A major impediment to our understanding the importance of contact activation has been the pervasive notion that the system has no biological relevance because it is activated entirely on artificial surfaces. Furthermore, for the past 20 years or so, the physiological negatively charged surface for activation of the contact system is unknown. Nevertheless, for the last 10 years, it has been the hypothesis of my laboratory that the physiological, negatively charged surface for contact activation may, in fact, be the assembly of these proteins on biologic surfaces, namely cell membranes. We have shown that the assembly of contact proteins on endothelial cell membranes leads to a multiprotein complex that results in prekallikrein activation and, therefore, activation of the system (Motta et al., 1995; see Section 5.2).

5.1 MECHANISMS OF KININOGEN CELL-SURFACE PRESENTATION

The pivotal protein for contact system assembly on cell membranes is HMWK. In addition to being contained within platelets, granulocytes and endothelial cells, unoccupied binding sites for HMWK exist on the surfaces of these cell types (Schmaier et al., 1983, 1986b, 1988a; Greengard and Griffin, 1984; Gustafson et al., 1986, 1989b; Van Iwaarden et al., 1988a). Why each of these cells contain kininogens in addition to unoccupied kininogen-binding sites is not known. In platelets, less than 8% of total HMWK is tightly bound to the cell membrane (Schmaier et al., 1983, 1986b). Upon activation, however, 40% of total platelet HMWK is secreted and a further 40% of the total is expressed upon the activated platelet membrane. Although the platelet contribution to plasma HMWK is only 0.23% (Schmaier et al., 1983; Kerbiriou-Nabias et al., 1984), the local concentration of HMWK on or about the activated platelet membrane may exceed 10 times the plasma concentration of this protein, since platelets excrete their granule contents by exocytosis (Schmaier et al., 1983, 1986b).

Most granulocyte-associated HMWK appears to be tightly bound to the cell surface and nonexchangeable (Figueroa et al., 1992; see Chapter 11 of this volume). Granulocytes possess the ability to assemble all of the proteins of the contact system (Figueroa et al., 1992) and neutrophil elastase proteolyzes cell-bound HMWK (Gustafson et al., 1989b). Although our initial investigations suggested that HUVEC can internalize HMWK (Schmaier et al., 1988a; Van Iwaarden et al., 1988b), recent studies indicate that there is no such mechanism (Hasan et al., 1995a). The differences in the apparent amounts of HMWK associated with the endothelial cell membrane, when cells are maintained at 4°C versus 37°C, can be explained in that, at the higher temperature, there is increased expression of kininogen binding sites (Schmaier et al., 1988b; Van Iwaarden et al., 1988b; Berrettini et al., 1992; Zini et al., 1993; Hasan et al., 1995a).

There are characteristic features of kininogens binding to all cells. First, kininogen binding to cells has an absolute requirement for Zn^{2+} (Greengard and Griffin, 1984; Gustafson et al., 1986, 1989b; Schmaier et al., 1988a; Van Iwaarden et al., 1988b). The requirement for Zn^{2+}, however, is not to mediate HMWK binding to the cells by the Zn^{2+}-binding region of domain 5 (Retzio et al., 1987; DeLa Cadena and Colman, 1992). Rather, LK binding to platelets and endothelial cells also has an absolute requirement for Zn^{2+}, indicating the need for this cation is for expression of the kininogen binding site, the putative kininogen receptor (Meloni and Schmaier, 1991; Zini et al., 1993). Although it has been suggested that Ca^{2+} is a cofactor for binding, our investigations reveal that this cation does not influence HMWK binding to unstimulated platelets and endothelial cells (Gustafson et al., 1986; Hasan et al., 1995a). When HMWK or LMWK bind to platelets, granulocytes or endothelial cells, the affinities are similar (Table 4.1). Since the affinity of HK binding to cells in the intravascular compartment is between 7 and 52 nM, and the plasma concentration of HMWK is 670 nM, we postulate that all kininogen-binding sites in the intravascular compartment are saturated in vivo. The number of binding sites for the kininogens on cells in the intravascular compartment

Table 4.1 Kininogen expression on cells in the intravascular compartment

Cell type	K_d (nM)[a]	Number of sites
Platelet		
^{125}I-HMWK[b]	15 ± 4[c]	911 ± 239
^{125}I-LMWK[b]	27 ± 2	647 ± 147
[125]I-D3[b]	39 ± 8	1,227 ± 404
Granulocyte		
^{125}I-HMWK	10 ± 1.3	4.8×10^4
Endothelium		
^{125}I-HMWK, 4°C	52 ± 13	9.3×10^5
^{125}I-LMWK, 4°C	43 ± 8	9.7×10^5
Biotin-HMWK, 4°C	46 ± 8	2.6×10^6
Biotin-HMWK, 4°C	7 ± 3	1.0×10^7

[a] Values determined from direct binding studies.
[b] HMWK and LMWK are high and low molecular weight kininogens, respectively; D3 is domain 3 of HMWK.
[c] Values presented as mean ± s.e. mean.

varies with the cell type, platelets having ~1,000 sites/cell, granulocytes, 50,000 sites/cell, and endothelial cells, ~1,000,000 sites/cell at 4°C and ~10,000,000 sites/cell when maintained at 37°C (Table 4.1) (Gustafson et al., 1986, 1989b; Schmaier et al., 1988a; Jiang et al., 1992; Zini et al., 1993; Hasan et al., 1995a). If the density of distribution of HMWK-binding sites per unit of surface area is calculated, however, each of these cell types has about the same density of distribution per unit of surface area, suggesting that the availability of kininogens for kinin liberation is equally distributed on platelets, granulocytes and endothelial cells in the intravascular compartment.

The expression of kininogens on cell membranes is a complex process. As noted above, there appear to be multiple regions on three domains of kininogen that allow it to interact with a binding site, or putative receptor (Fig. 4.3). As indicated, the interaction sites between HMWK and its putative receptor may be multiple locations, three in domain 3, one domain 4 and two in domain 5 (Hasan et al., 1994, 1995b; Herwald et al., 1995). Clearly the sequence of LDC27 from domain 3 and HKH20 from domain 5 are the high-affinity binding regions on HMWK (Hasan et al., 1995b; Herwald et al., 1995). It is important to appreciate that the binding of even a low-affinity sequence from domain 4, for example, will block whole HMWK from binding to endothelial cells (Hasan et al., 1994). This result suggests that HMWK, and presumably LMWK fit very tightly into their receptors. Indeed, since the K_i and K_d values, calculated from binding studies with HMWK, LMWK and all of their subunits are the same, the two chains of kininogens do not bind to cells in an optimal manner (Jencks, 1981; Meloni et al., 1992). This noncooperative interaction is characterized by a loss of entropy on

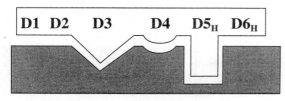

| D1 | D2 | D3 | D4 | D5$_H$ | D6$_H$ |

Kininogen Binding Site

Figure 4.3   High molecular weight kininogen's placement on cell membranes. There are three domains of HMWK that participate in cell binding. Domain 5 has the highest affinity cell-binding site followed by domain 3. The domain 4 cell-binding site has low affinity but is important to hold HMWK in the proper conformation such that it can bind with highest affinity to its binding site, putative receptor.

binding, and suggests that whole HMWK must bend in order to fit into its binding site (Jencks, 1981). In support of this notion, when BK is liberated, the kinin-free HMWK binds to endothelial cells with a lower affinity and number of binding sites (Hasan et al., 1994, 1995a). Likewise, when LMWK is cleaved between domains 1 and 2, such that there is a change in the conformation, there is decreased binding to endothelial cells compared with intact LMWK (Hasan et al., 1995c).

5.2   THE KININOGEN BINDING SITE ON THE ENDOTHELIAL CELL SURFACE

The kininogen binding site on endothelial cells appears to be regulated for several reasons. Thus, inhibitors of anaerobic and aerobic metabolism and the hexose monophosphate shunt abolish the ability of endothelial cells to bind HMWK (Hasan et al., 1995a). Cycloheximide has no effect on HMWK binding to endothelial cells. In addition, temperature or the BK sequence within the kininogens contributes to the level of binding to endothelial cells (Hasan et al., 1994, 1995a). Furthermore, exposure of endothelial cells to BK, via B$_1$ receptors and protein kinase C, results in increased HMWK and LMWK binding (Zini et al., 1993). Heavy-chain of kininogen and LMWK require Ca^{2+} for phorbol-induced upregulation of their endothelial cell-binding site, whereas HMWK does not (Zini et al, 1993). Also, angiotensin-converting enzyme (ACE) inhibitors potentiate the effect of bradykinin on upregulating the HMWK binding site on endothelial cells (Zini et al., 1993). As discussed below, when HMWK binds to endothelial cells, it initiates a series of events that allows for an endothelial cell- or matrix-associated enzyme to activate prekallikrein bound to HMWK (Motta et al., 1995). Thus, BK upregulates

kininogen binding on endothelial cells and, conversely, kininogens influence BK formation. These data indicate that this system is tightly regulated in an autocrine fashion.

In addition to being a presentation receptor, the endothelial HMWK binding site regulates contact activation. Platelet and endothelial cell-bound HMWK is protected from activation by exogenous plasma kallikrein (Meloni et al., 1992; Nishikawa et al., 1992). This function is distinct from the role of HMWK to regulate the activation of prekallikrein bound to HMWK on endothelial cells (Motta et al., 1995). Thus, assembly of contact proteins on endothelial cells allows for a protein complex system that can activate prekallikrein and generate kinins (Motta et al., 1995). Although there is much commonality in the features of HMWK binding to most cells, there are notable differences. For example, HMWK is a noncompetitive inhibitor of fibrinogen binding to granulocytes and activated platelets (Gustafson et al., 1989a). Previous studies indicated that an antibody to the β2-integrin, Mac-1 (CD11b/CD18), did not block HK binding to granulocytes (Gustafson et al., 1989a), although more recent studies indicated otherwise (Wachtfogel et al., 1993). The latter data are difficult to reconcile with our unpublished observation that, in direct binding studies, HMWK binds normally to granulocytes from an individual with a genetic deficiency in CD18, leukocyte adhesion deficiency. In addition, it has been noted that Factor XII and HMWK displace each other from binding to endothelial cells in a non-competitive fashion. These data indicate that the kininogen receptor may be adjacent to the endothelial Factor XII binding site (Reddigari et al., 1993b). Thus, by virtue of binding HMWK, which serves as the cell receptor for prekallikrein and Factor XI, cell surfaces become the loci for assembly of contact proteins (Greengard et al., 1986; Reddigari et al., 1993b; Lenich et al., 1995; Motta et al., 1995). As discussed below, this assembly of proteins allows for a physiological system where these proteins can become activated (Motta et al., 1995). It appears, therefore, that cell membrane assembly through binding is the "physiological" negatively charged surface for activation of the contact system. Furthermore, assembly of the contact system on cell membranes is the major determinant for kinin liberation.

6.   Physiological Activities of the Kininogens

Current evidence indicates that, in addition to being kinin precursor proteins, the kininogens function as antithrombins and profibrinolytic agents in vivo. Although this may seem to contradict the observation that HMWK deficiency is associated with prolonged

Table 4.2   The antithrombin and profibrinolytic activities of the kininogens

Domain	Activity
Bradykinin	Stimulates prostacyclin formation
Bradykinin	Stimulates nitric oxide formation
Bradykinin	Stimulates superoxide formation
Bradykinin	Stimulates tissue plasminogen activator secretion
RPPGF	Prevents α-thrombin from cleaving its receptor
Domain 2	Prevents calpain-related platelet aggregation
Domain 3	Prevents α-thrombin binding to cells
Domain 6	Prekallikrein receptor on endothelial cells

APTT, one must not confuse an assay on an artificial surface *in vitro* from a biologic function. In many ways, kininogens and BK contribute to vessel patency, increased blood flow and antithrombotic/profibrinolytic activities (Table 4.2). BK is a potent agonist of endothelium-derived prostacyclin release (Hong, 1980; Crutchley *et al.*, 1983), superoxide formation (Holland *et al.*, 1990), tissue plasminogen activator release (Smith *et al.*, 1983), nitric oxide formation (Palmer *et al.*, 1987) and endothelium-derived smooth muscle hyper-polarization factor formation (Nakashima *et al.*, 1993). Also, BK, through its ability to stimulate nitric oxide and cGMP formation in endothelial cells, provides a major stimulus to prevent subendothelial smooth muscle proliferation (Boulanger *et al.*, 1990; Schini *et al.*, 1990). In the presence of an intact endothelium, kinins appear to prevent vascular smooth muscle growth and proliferation (Busse and Mulsch, 1990; Imai *et al.*, 1994). Alternatively, when vessels are injured, with subsequent endothelium loss, BK stimulates kinases that can result in vascular smooth muscle growth and proliferation (Dixon *et al.*, 1990, 1994; Khalil and Morgan, 1993). Thus, in an intact vessel, BK maintains blood flow and vessel patency whereas, in the absence of endothelium, BK repairs vessels and may cause vascular smooth muscle proliferation and intimal hypertrophy.

6.1 THROMBIN INHIBITION

In addition to the salutary effects of kinins in maintaining blood vessel patency, the kininogens inhibit α-thrombin-induced platelet and endothelium activation by at least two mechanisms (Meloni and Schmaier, 1991; Jiang *et al.*, 1992; Hasan *et al.*, 1995a) (Table 4.2). The first is an indirect effect whereby kininogen can inhibit platelet calpain. The second mechanism is direct, wherein the kininogens selectively inhibit the activity of α-thrombin to cleave its cell receptors. When α-thrombin activates platelets, cytosolic or internal membrane-associated calpain translocates to be externalized on the activated

cell surface (Schmaier *et al.*, 1990; Saido *et al.*, 1993). The externalized calpain can proteolyze platelet surface glycoproteins such as glycoprotein Ib (Coller, 1982). Calpain also proteolyzes a putative platelet ADP receptor, which exposes the platelet fibrinogen receptor and, thus, allows for aggregation (Puri *et al.*, 1987, 1989). Inhibition of externalized platelet calpain by leupeptin or HMWK results in inhibition of α-thrombin-mediated platelet aggregation by preventing fibrinogen binding (Puri *et al.*, 1987, 1991). These data have been utilized to develop a series of compounds, modeled after kininogen domain 2, which prevent α- and γ-thrombin-induced platelet aggregation without interfering with other platelet agonists or thrombin-induced intracellular platelet activation (Puri *et al.*, 1993; Matsueda *et al.*, 1994).

Although the ability of HMWK to inhibit α-thrombin-induced platelet aggregation is an indirect pathway, some studies suggest that there is more than one mechanism by which kininogens inhibit α-thrombin. Although HMWK and LMWK inhibit both α-thrombin-induced platelet aggregation and secretion (Meloni and Schmaier, 1991), the latter effect is independent of and occurs before aggregation. There must, therefore, be a mechanism by which the kininogens interfere with platelet activation other than simply via inhibition of calpain-related platelet aggregation (Charo *et al.*, 1977). For example, both HMWK and LMWK inhibit α-thrombin in a noncompetitive fashion, but not Phe-Pro-Arg-chloromethylketone-treated thrombin, from binding to platelets and endothelial cells (Meloni and Schmaier, 1991; Puri *et al.*, 1991; Schmaier *et al.*, 1992; Hasan *et al.*, 1995a). This finding provides one explanation of how all platelet activation by α-thrombin can be blocked by large molecular mass proteins like HMWK and LMWK. The ability to inhibit α-thrombin binding to platelets was localized to kininogen domain 3 (Jiang *et al.*, 1992). However, a monoclonal antibody to domain 3, which did not block binding, neutralized HMWK's ability to inhibit α-thrombin-induced platelet activation, suggesting that the cell-binding site and α-thrombin inhibitory regions are different (Jiang *et al.*, 1992). As noted above, the α-thrombin inhibitory region associated with purified domain 3 prepared from proteolytic cleavage was, in fact, attached domain 4 (Hasan *et al.*, 1996).

Although the large molecular mass proteins, HMWK, LMWK and domain 3, each inhibit α-thrombin binding to platelets, domain 4 peptides, BK (Arg-Pro-Pro-Gly-Phe-Ser-Pro-Phe-Arg) and a Met-Lys-BK-Ser-Ser-Arg-Ile-Gly do not (Meloni and Schmaier, 1991; Jiang *et al.*, 1992; Hasan *et al.*, 1996). These data indicate that another mechanism underlies blockade of α-thrombin-induced platelet activation by these peptides. Like the kininogens, the domain 4 peptides do not inhibit α-thrombin's ability to cleave a tripeptide substrate or to clot fibrinogen, suggesting that they may not interact with the α-thrombin active site or anion binding exosite I (Meloni and Schmaier, 1991; Jiang *et al.*, 1992; Hasan *et al.*, 1995a, 1996). Moreover, like HMWK and LMWK, the peptides are neither substrates for nor form complexes with α-thrombin (Meloni and Schmaier, 1991; Hasan *et al.*, 1995a, 1996). Domain 4 peptides do not block ADP-, collagen- or U46619-induced aggregation (Hasan *et al.*, 1996). They do, however, block α-thrombin-induced Ca^{2+} mobilization and γ-thrombin-induced platelet aggregation in plasma (Hasan *et al.*, 1996).

The smallest peptide sequence of domain 4 to date to inhibit α-thrombin-induced platelet activation is Arg-Pro-Pro-Gly-Phe (RPPGF) (Hasan *et al.*, 1996), which is also the major ACE product of BK, with a metabolic degradation rate in plasma of 4.2 h (Shima *et al.*, 1992; Majima *et al.*, 1993). Knowing that this peptide possesses the α-thrombin inhibitory activity indicates that, when this sequence is contained in kininogens, free as BK or degraded into a BK metabolite, its α-thrombin inhibitory activity persists. These data indicate that kininogens contribute to the constitutive anticoagulant environment of the intravascular compartment.

Additional studies were performed to ascertain the mechanism by which RPPGF and related domain 4 peptides inhibit α-thrombin-induced responses. Although RPPGF does not block thrombin receptor agonist-induced platelet activation (Hasan *et al.*, 1996), domain 4 peptides prevent α-thrombin from eliminating an epitope on its receptor wherein α-thrombin cleaves its receptor during activation. This finding suggests that domain 4 peptides prevented α-thrombin from cleaving the cloned thrombin receptor after Arg$_{41}$, a critical step in α-thrombin cell activation (Hasan *et al.*, 1996). Furthermore, RPPGF and HMWK prevent α-thrombin from cleaving a synthetic peptide (NATLDPRSFLLR) spanning the α-thrombin cleavage site between Arg and Ser on the cloned thrombin receptor (Hasan *et al.*, 1996). These combined data indicate that domain 4 peptides and the same sequence in kininogen are selective, proteolytic inhibitors of α-thrombin. RPPGF is a prototype for a new class of very selective inhibitors that interfere with thrombin's ability to cleave the cloned thrombin receptor without interfering with other procoagulant activity.

6.2 PARTICIPATION OF KININOGENS IN FIBRINOLYSIS

In addition to this unique mechanism of α-thrombin inhibition, HMWK participates in cellular fibrinolysis. Since the recognition of HMWK deficiency, this protein has been ascribed to have a role in fibrinolysis, although the specific, physiological mechanism was unclear (Colman *et al.*, 1975; Saito *et al.*, 1975). It has been known for more than 35 years that contact activation can increase total plasma fibrinolysis (Niewiarowski and Prou-Wartelle, 1959). Kallikrein, Factor XIIa and Factor XIa cleave plasminogen directly, albeit much less efficiently than tissue-type plasminogen activator or urokinase plasminogen activator (Colman, 1969; Mandle and Kaplan, 1977, 1979; Goldsmith *et al.*, 1978; Miles *et al.*, 1983). However, plasma kallikrein has been characterized to be a kinetically favorable activator of single-chain urokinase *in vitro* (Ichinose *et al.*, 1986). Recent studies suggest that single-chain urokinase activation by kallikrein is most likely to occur on the platelet and endothelial cell surface (Gurewich *et al.*, 1993; Loza *et al.*, 1994; Lenich *et al.*, 1995).

We were prompted to examine the relationship between prekallikrein assembly on endothelial cells and its participation in single-chain urokinase activation (see Table 4.2) (Motta *et al.*, 1995). When prekallikrein binds to HMWK on endothelial cells, the zymogen becomes activated to kallikrein, as indicated by elaboration of amidolytic activity, changes in the structure of prekallikrein to kallikrein on gel electrophoresis and cleavage of HMWK (Motta *et al.*, 1995). Prekallikrein activation occurs independently of activated forms of Factor XII and its activating enzyme is a Ca^{2+}-requiring metalloprotease rather than a serine protease (Motta *et al.*, 1995). Prekallikrein activation on endothelial cells is kinetically similar to prekallikrein activation by Factor XII on an artificial surface. These data showed, for the first time, that contact protein assembly on endothelial cells results in prekallikrein activation in the absence of Factor XII and an artificial surface (Motta *et al.*, 1995). This assembly of contact proteins allows for a physiological pathway for contact activation. The degree of prekallikrein activation is regulated by HMWK. Increasing HMWK concentrations upregulate the enzyme that activates cell-bound prekallikrein. Thus, HMWK regulates prekallikrein activation, which, in turn, liberates more BK from cell-bound HMWK and removes HMWK from the surface to slow prekallikrein activation (Motta *et al.*, 1995). Also, increased BK increases kininogen binding, which, in turn, decreases soluble kallikrein from cleaving HMWK to liberate more BK (Zini *et al.*, 1993). The pathway for prekallikrein activation and bradykinin liberation is, therefore, tightly regulated.

The prekallikrein activation pathway on endothelial cells results in kinetically favorable conversion of single-chain urokinase into two-chain urokinase in an environment where there is a constitutive molar excess secretion of endothelial cell plasminogen activator inhibitor-1 (Motta *et al.*, 1995). Formation of two-chain urokinase results in a 4.3-fold increase in plasminogen activation. This system for plasminogen activation occurs in an environment where there is no tissue plasminogen activator or fibrin, and the mechanism for single-chain urokinase activation is another pathway for cellular fibrinolysis that is either independent of or conjoined with single-chain urokinase activation associated with its binding to its receptor (Higazi *et al.*, 1995). It is also an explanation of how two-chain urokinase can be formed in plasminogen knockout mice (Carmeliet and Collen, 1995).

7.   Summary

In conclusion, HMWK and LMWK assemble on cell membranes allowing for bradykinin to be liberated in a protected environment, wherein this potent, bioactive peptide can activate its receptors to influence vascular biology. Both kininogens participate in maintaining the constitutive anticoagulant environment of the intravascular compartment by virtue of their α-thrombin inhibitory activities. HMWK also contributes to cellular fibrinolysis by virtue of being a receptor for prekallikrein expression and activation by an endothelial cell-derived enzyme. Both of these activities contribute to the physiology of the vasculature by regulating local blood flow and cellular thrombosis and fibrinolysis.

Note added in proof: Since submission of this manuscript, three proteins have been recognized as the kininogen receptor: gC1qR (J. Biol. Chem. 271:13040, 1996; PNAS 93:8552, 1996), suPAR (Circ. 94:I-42, 1996), and cytokeratin 1 (Thromb Haemost. 76:In Press, 1997). These data are the next frontier in the kininogen field.

8.   Acknowledgement

I appreciate the critical review of this manuscript and the many important discussions I have had with Dr Ahmed A.K. Hasan of the University of Michigan.

9.   References

Asakura, S., Hurley, R.W., Skorstengaard, K., Ohkubo, I. and Mosher, D.F. (1992). Inhibition of cell adhesion by high molecular weight kininogen. J. Cell Biol. 116, 465–476.

Auerswald, E.A., Rossler, D., Mentele, R. and Assfalg-Machleidt, I. (1993). Cloning, expression and characterization of human kininogen domain 3. FEBS Lett. 321, 93–97.

Ben Nasr, A., Herwald, H., Müller-Esterl, W. and Bjork, L. (1995). Human kininogens interact with M protein, a bacterial surface protein and virulence determinant. Biochem. J. 305, 173–180.

Berrettini, M., Schleef, R.R., Heeb, M.J., Hopmeier, P. and Griffin, J.H. (1992). Assembly and expression of intrinsic factor IX activator complex on the surface of cultured human endothelial cells. J. Biol. Chem. 267, 19833–19839.

Bjork, I., Olson, S.T., Sheffer, R.G. and Shore, J.D. (1989). Binding of heparin to human high molecular weight kininogen. Biochemistry 28, 1213–1221.

Bock, P.E. and Shore, J.D. (1983). Protein–protein interactions in contact activation of blood coagulation. Characterization of fluorescein-labeled human high molecular weight kininogen light chain as a probe. J. Biol. Chem. 258, 15079–15068.

Bock, P.E., Shore, J.D., Tans, G. and Griffin, J.H. (1985). Protein–protein interactions in contact activation of blood coagulation. Binding of high molecular weight kininogen and the 5-(iodoacetamido) fluorescein-labeled kininogen light chain to prekallikrein, kallikrein, and the separated kallikrein heavy chain and light chains. J. Biol. Chem. 260, 12434–12443.

Bode, W., Engh, R., Musil, D., Thiele, U., Huber, R., Karshikov, A., Brzin, J., Kos, J. and Turk, V. (1988). The 2.0 Å X-ray crystal structure of chicken egg white cystatin and its possible mode of interaction with cysteine proteinases. EMBO J. 7, 2593–2599.

Boulanger, C., Schini, V.B., Moncada, S. and Vanhoutte, P.M. (1990). Stimulation of cyclic GMP production in cultured endothelial cells of the pig by bradykinin, adenosine diphosphate, calcium ionophore A23187 and nitric oxide. Br. J. Pharmacol. 101, 152–156.

Bradford, H.N., Schmaier, A.H. and Colman, R.W. (1990). Kinetics of inhibition of platelet calpain II by human kininogens. Biochem. J. 270, 83–90.

Bradford, H.N., Jameson, B.A., Adam, A.A., Wassell, R.P. and Colman, R.W. (1993). Contiguous binding and inhibitory sites on kininogen required for the inhibition of platelet calpain. J. Biol. Chem. 268, 26546–26551.

Busse, R. and Mulsch, A. (1990). Induction of nitric oxide synthase by cytokines in vascular smooth muscle cells. FEBS Lett. 275, 87–90.

Carmeliet, P. and Collen, D. (1995). Gene targeting and gene transfer studies of the plasminogen/plasmin system: implications in thrombosis, hemostasis, neointima formation, and atherosclerosis. FASEB J. 9, 934–938.

Charo, I.F., Feinman, R.D. and Detwiler, T.C. (1977). Interrelations of platelet aggregation and secretion. J. Clin. Invest. 60, 866–873.

Chen, L.-M., Chung, P., Chao, L. and Chao, J. (1992). Differential regulation of kininogen gene expression by estrogen and progesterone *in vivo*. Biochim. Biophys. Acta 1131, 145–151.

Cheung, P.P., Cannizzaro, L.A. and Colman, R.W. (1992). Chromosomal mapping of human kininogen gene (KNG) to 3q26-qter. Cytogenet. Cell Genet. 59, 24–26.

Cheung, P.P., Kunapuli, S.P., Scott, C.F., Wachtfogel, Y.T. and Colman, R.W. (1993). Genetic basis of total kininogen deficiency in Williams' trait. J. Biol. Chem. 268, 23361–23365.

Chhibber, G., Cohen, A., Lane, S., Farber, A., Meloni, F. and Schmaier, A.H. (1990). Immunoblotting of plasma in a pregnant patient with hereditary angioedema. J. Lab. Clin. Med. 115, 112–121.

Coller, B.S. (1982). Effects of tertiary amine local anesthetics on von Willebrand factor-dependent platelet function: alterations of membrane reactivity and degradation of GPIb by a calcium-dependent protease(s). Blood 60, 731–743.

Colman, R.W. (1969). Activation of plasminogen by human plasma kallikrein. Biochem. Biophys. Res. Commun. 35, 273–279.

Colman, R.W., Bagdasarian, A., Talamo, R.C., Scott, C.F.,

Seavey, M., Guimaraes, J.A., Pierce, J.V. and Kaplan, A.P. (1975). Williams trait: human kininogen deficiency with diminished levels of plasminogen proactivator and prekallikrein associated with abnormalities of the Hageman factor-dependent pathways. J. Clin. Invest. 56, 1650–1662.

Crutchley, D.J., Ryan, J.W. and Ryan, U.S. (1983). Bradykinin-induced release of prostacyclin and thromboxanes from bovine pulmonary artery endothelial cells. Biochim. Biophys. Acta 751, 99–107.

DeLa Cadena, R.A. and Colman, R.W. (1992). The sequence HGLGHGHEQQHGLGHGH in the light chain of high molecular weight kininogen serves as a primary structural feature for zinc-dependent binding to an anionic surface. Protein Sci. 1, 151–160.

Dixon, B.S., Breckon, R., Fortune, J., Vavrek, R.J., Stuart, R., Marzec-Calvert, K. and Linas, S.L. (1990). Effects of kinins on cultured arterial smooth muscle cells. Am. J. Physiol. 258, C299–C308.

Dixon, B.S., Sharma, R.V., Dickerson, T. and Fortune, J. (1994). Bradykinin and angiotensin II. Activation of protein kinase C in arterial smooth muscle. Am. J. Physiol. 266, C1406–C1420.

Figueroa, C.D., Henderson, L.M., Kaufmann, J., DeLa Cadena, R.A., Colman, R.W., Müller-Esterl, W. and Bhoola, K.D. (1992). Immunovisualization of high (HK) and low (LK) molecular weight kininogen on isolated human neutrophils. Blood 79, 754–760.

Fong, D., Smith, D.I. and Hsieh, W.-T. (1991). The human kininogen gene (KNG) mapped to chromosome 3q26-qter by analysis of somatic cell hybrids using the polymerase chain reaction. Hum. Genet. 87, 189–192.

Goldsmith, G., Saito, H. and Ratnoff, O.D. (1978). The activation of plasminogen by Hageman factor (factor XII) and Hageman factor fragments. J. Clin. Invest. 62, 54–60.

Greengard, J.S. and Griffin, J.H. (1984). Receptors for high molecular weight kininogen on stimulated washed human platelets. Biochemistry 23, 6863–6869.

Greengard, J.S., Heeb, M.J., Ersdal, E., Walsh, P.N. and Griffin, J. H. (1986). Binding of coagulation factor XI to washed human platelets. Biochemistry 25, 3884–3890.

Gurewich, V., Johnstone, M., Loza, J.-P. and Pannell, R. (1993). Pro-urokinase and prekallikrein are both associated with platelets. Implications for the intrinsic pathway of fibrinolysis and for therapeutic thrombolysis. FEBS Lett. 318, 317–321.

Gustafson, E.G., Schutsky, D., Knight, L.D. and Schmaier, A.H. (1986). High molecular weight kininogen binds to unstimulated platelets. J. Clin. Invest. 78, 310–318.

Gustafson, E.G., Lukasiewicz, H., Wachtfogel, Y.T., Norton, K.J., Schmaier, A.H., Niewiarowski S. and Colman R.W. (1989a). B. High molecular weight kininogen inhibits fibrinogen binding to cytoadhesions of neutrophils and platelets. J. Cell Biol. 109, 377–387.

Gustafson, E.G., Schmaier, A.H., Wachtfogel, Y.T., Kaufman, N., Kucich, U. and Colman, R.W. (1989b). Human neutrophils contain and bind high molecular weight kininogen. J. Clin. Invest. 84, 28–35.

Hallbach, J., Adams, G., Wirthensohn, G. and Guder, W.G. (1987). Quantification of kininogen in human renal medulla. Biol. Chem. Hoppe-Seyler 368, 1151–1155.

Hasan, A.A.K., Cines, D.B., Zhang, J. and Schmaier, A.H. (1994). The C-terminus of bradykinin and N-terminus of the light chain of kininogens comprise and endothelial cell binding domain. J. Biol. Chem. 269, 31822–31830.

Hasan, A.A.K., Cines, D.B., Ngaiza, J.R., Jaffe, E.A. and Schmaier, A.H. (1995a). A high molecular weight kininogen is exclusively membrane bound on endothelial cells to influence activation of vascular endothelial cells. Blood, 85, 3134–3143.

Hasan, A.A.K., Cines, D.B., Herwald, H., Schmaier, A.H. and Müller-Esterl, W. (1995b). Mapping the cell binding site on high molecular weight kininogen domain 5. J. Biol. Chem. 270, 19256–19261.

Hasan, A.A.K., Zhang, J., Samuel, M. and Schmaier, A.H. (1995c). Conformational changes in low molecular weight kininogen alters its ability to bind to endothelial cells. Thromb. Haemost. 74, 1088–1095.

Hasan, A.A.K., Amenta, S. and Schmaier, A.H. (1996). Bradykinin and its metabolite, Arg-Pro-Pro-Gly-Phe, are selective inhibitors of α-thrombin-induced platelet activation. Circulation 94, 517–528.

Hayashi, I., Hosniko, S., Makabe, O. and Ohuishi, S. (1993). A point mutation of alanine 163 to threonine is responsible for the defective secretion of high molecular weight kininogen by the liver of brown Norway Katholide rats. J. Biol. Chem. 268, 19219.

Herwald, H., Hasan, A.A.K., Godovac-Zimmermann, J., Schmaier, A.H. and Müller-Esterl, W. (1995). Identification of an endothelial cell binding site on kininogen domain D3. J. Biol. Chem. 270, 14634–14642.

Higazi, A., Cohen, R.L., Henkin, J., Kniss, D., Schwartz, B.S. and Cines, D.B. (1995). Enhancement of the enzymatic activity of single chain urokinase plasminogen activator by soluble urokinase receptor. J. Biol. Chem. 270, 17375–17380.

Holland, J.A., Pritchard, K.A., Pappolla, M.A., Wolin, M.S., Rogers, N.J. and Stemerman, M.B. (1990). Bradykinin induces superoxide anion release from human endothelial cells. J. Cell Physiol. 143, 21–25.

Hong, S.L. (1980). Effect of bradykinin and thrombin on prostacyclin synthesis in endothelial cells from calf and pig aorta and human umbilical cord vein. Thromb. Res. 18, 787–796.

Ichinose, A., Fujikawa, K. and Suyama, T. (1986). The activation of pro-urokinase by plasma kallikrein and its inactivation by thrombin. J. Biol. Chem. 261, 3486–3489.

Imai, T., Hirata, Y., Kanno, K. and Marumo, F. (1994). Induction of nitric oxide synthase by cyclic AMP in rat vascular smooth muscle cells. J. Clin. Invest. 93, 543–549.

Ishiguro, H., Higashiyama, S., Ohkubo, I. and Sasaki, M. (1987). Heavy chain of human high molecular weight and low molecular weight kininogen binds calcium ion. Biochemistry 26, 7021–7029.

Jacobsen, S. and Kriz, M. (1967). Some data on two purified kininogens from human plasma. Br. J. Pharmacol. 29, 25–36.

Jencks, W.P. (1981). On the attribution and additivity of binding energies. Proc. Natl Acad. Sci. USA 78, 4046–4050.

Jiang, Y.P., Müller-Esterl, W. and Schmaier, A.H. (1992). Domain 3 of kininogens contain a cell binding site and a site that modifies thrombin activation of platelets. J. Biol. Chem. 267, 3712–3717.

Kakizuka, A., Ingi, T., Murai, T. and Nakanishi, S. (1990). A set of U1 snRNA-complementary sequences involved in governing alternative RNA splicing of the kininogen genes. J. Biol. Chem. 265, 10102–10108.

Kaufmann, J., Hassemann, M., Modrow, S. and Müller-Esterl, W. (1993). Structural dissection of the multidomain kininogens. Fine mapping of the target epitopes of antibodies interfering with their functional properties. J. Biol. Chem. 268, 9079–9091.

Kerbiriou-Nabias, D.M., Garcia, F.O. and Larrieu, M.-J. (1984). Radioimmunoassays of human high and low molecular weight kininogens in plasmas and platelets. Br. J. Haematol. 56, 273–286.

Khalil, R.A. and Morgan, K.G. (1993). PKC-mediated redistribution of mitogen-activated protein kinase during smooth muscle cell activation. Am. J. Physiol. 265, C406–C411.

Kitamura, N., Kitagawa, H., Fushima, D., Takagaki, Y., Miyata, T. and Nakanishi, S. (1985). Structural organization of the human kininogen gene and a model for its evolution. J. Biol. Chem. 260, 8610–8617.

Kleniewski, J. and Donaldson, V. (1988). Granulocyte elastase cleaves human high molecular weight kininogen and destroys its clot-promoting activity. J. Exp. Med. 167, 1895–1907.

Kunapuli, S.P., DeLa Cadena, R.A. and Colman, R.W. (1993). Deletion mutagenesis of high molecular weight kininogen light chain. Identification of two anionic surface binding subdomains. J. Biol. Chem. 268, 2486–2492.

Lenich, C., Pannell, R. and Gurewich, V. (1995). Assembly and activation of the intrinsic fibrinolytic pathway on the surface of human endothelial cells in culture. Thromb. Haemost. 74, 698–703.

Loza, J.-P., Gurewich, V., Johnstone, M. and Pannell, R. (1994). Platelet-bound prekallikrein promotes prourokinase-induced clot lysis: a mechanism for targeting the factor XII dependent intrinsic pathway of fibrinolysis. Thromb. Haemost. 71, 347–352.

Majima, M., Sunhara, N., Harada, Y. and Katori, M. (1993). Detection of the degradation products of bradykinin by enzyme immunoassays as markers for the release of kinin *in vivo*. Biochem. Pharmacol. 45, 559–567.

Mandle, R.J., Jr and Kaplan, A.P. (1977). Human plasma prekallikrein: mechanism of activation by Hageman factor and participation in Hagemen factor-dependent fibrinolysis. J. Biol. Chem. 252, 6097–6104.

Mandle, R.J., Jr and Kaplan, A.P. (1979). Hageman factor-dependent fibrinolysis: generation of fibrinolytic activity by the interaction of human activated factor XI and plasminogen. Blood 54, 850–862.

Matsueda, R., Umeyama, H., Puri, R.N., Bradford, H.N. and Colman, R.W. (1994). Design and synthesis of a kininogen-based selective inhibitor of thrombin-induced platelet aggregation. Peptide Res. 7, 32–35.

Meloni, F.J. and Schmaier, A.H. (1991). Low molecular weight kininogen binds to platelets to modulate thrombin-induced platelet activation. J. Biol. Chem. 266, 6786–6794.

Meloni, F.J., Gustafson, E.G. and Schmaier, A.H. (1992). High molecular weight kininogen binds to platelet by its heavy and light chains and when bound has altered susceptibility to kallikrein cleavage. Blood 79, 1233–1244.

Miles, L.A., Greengard, J.S. and Griffin, J.H. (1983). A comparison of the abilities of plasma kallikrein, β-factor XIIa, factor XIa and urokinase to activate plasminogen. Thromb. Res. 29, 407–417.

Mori, K. and Nagasawa, S. (1981). Studies on human high molecular weight (HMW) kininogen. II. Structural change in HMW kininogen by the action of human plasma kallikrein. J. Biochem. 89, 1465–1473.

Motta, G., Hasan, A.A.K., Cines, D.B. and Schmaier, A.H. (1995). High molecular weight kininogen and prekallikrein assembly on endothelial cells produce plasminogen activation independent of factor XII. Blood 86 (Suppl. 1), 374a.

Nakashima, M., Mombouli, J.-V., Taylor, A.A. and Vanhoutte, P.M. (1993). Endothelium-dependent hyperpolarization caused by bradykinin in human coronary arteries. J. Clin. Invest. 92, 2867–2871.

Niewiarowski, S. and Prou-Wartelle, O. (1959). Role of the contact factor (Hageman factor) in fibrinolysis. Thromb. Diath. Haemorrh. 3, 593–598.

Nishikawa, K., Shibayama, Y., Calcaterra, E., Kaplan, A.P. and Reddigari, S.R. (1992). Generation of vasoactive peptide bradykinin from human umbilical vein endothelium-bound high molecular weight kininogen by plasma kallikrein. Blood 80, 1980–1988.

Oh-ishi, S., Hayashi, I., Hosiko, S. and Makabe, O. (1994). Molecular mechanism of kininogen deficiency in brown Norway Katholiek rats. Braz. J. Med. Biol. Res. 27, 1803–1815.

Ohkubo, I., Kurachi, K., Takasawa, T., Shiokawa, T. and Sasaki, M. (1984). Isolation of a human cDNA for α_2-thiol proteinase inhibitor and its identity with low molecular weight kininogen. Biochemistry 23, 3891–3899.

Palmer, R.M.J., Ferrige, A.G. and Moncada, S. (1987). Nitric oxide release accounts for the biologic activity of endothelium-derived relaxing factor. Nature (Lond.) 327, 524–526.

Proud, D., Perkins, M., Pierce, J.V., Yates, K.N., Highet, P.F., Herring, P.L., Mangkornkanok/Mark, M., Bahu, R., Carone, F. and Pisano, J.J. (1981). Characterization and localization of human renal kininogen. J. Biol. Chem. 265, 10634–10639.

Puri, R.N., Gustafson, E.J., Zhou, F., Bradford, H.N., Colman, R.F. and Colman, R.W. (1987). Inhibition of thrombin-induced platelet aggregation by high molecular weight kininogen. Trans. Assoc. Amer. Phys. C 232–240.

Puri, R.N., Zhou, F., Colman, R.F. and Colman, R.W. (1989). Cleavage of a 100 kDa membrane protein (Aggregin) during thrombin-induced platelet aggregation is mediated by the high affinity thrombin receptors. Biochem. Biophys. Res. Commun. 162, 1017–1024.

Puri, R.N., Zhou, F., Hu, C.-J., Colman, R.F. and Colman, R.W. (1991). High molecular weight kininogen inhibits thrombin-induced platelet aggregation and cleavage of aggregin by inhibiting binding of thrombin to platelets. Blood 77, 500–507.

Puri, R.N., Matsueda, R., Umeyama, H., Bradford, H.N. and Colman, R.W. (1993). Modulation of thrombin-induced platelet aggregation by inhibition of calpain by a synthetic peptide derived from the thiol-protease inhibitory sequence of kininogens and S-(3-nitro-2-pyridinesulfenyl)-cysteine. Eur. J. Biochem. 214, 233–241.

Reddigari, S. and Kaplan, A.P. (1989). Monoclonal antibody to human high molecular weight kininogen recognizes its prekallikrein binding site and inhibits its coagulant activity. Blood 74, 695–702.

Reddigari, S.R., Kuna, P., Miragliotta, G., Shibayama, Y., Nishikawa, K. and Kaplan, A.P. (1993a). Human high molecular weight kininogen binds to human umbilical vein endothelial cells via its heavy and light chains. Blood 81, 1306–1311.

Reddigari, S.R., Shibayama, Y., Brunnee, T. and Kaplan, A.P. (1993b). Human Hageman factor (Factor XII) and high molecular weight kininogen compete for the same binding site on human umbilical vein endothelial cells. J. Biol. Chem. 268, 11982–11987.

Retzio, A.D., Rosenfeld, R. and Schiffman, S. (1987). Effect of chemical modification on the surface- and protein-binding properties of the light chain of human high molecular weight kininogen. J. Biol. Chem. 262, 3074–3081.

Rizzu, P. and Baldini, A. (1995). Three members of the human cystatin gene superfamily, AHSG, HRG, and KNG, map within one megabase of genomic DNA at 3q27. Cytogenet. Cell Genet. 70, 26–28.

Saito, H., Ratnoff, O.D., Waldmann, R. and Abraham, J.P. (1975). Fitzgerald trait: deficiency of a hitherto unrecognized agent, Fitzgerald factor, participating in surface-mediated reactions of clotting, fibrinolysis, generation of kinins, and the property of diluted plasma enhancing vascular permeability (PF/DIL). J. Clin. Invest. 55, 1082–1089.

Saido, T., Suzuki, H., Yamazaki, H., Tanoue, K. and Suzuki, K. (1993). In situ capture of μ-calpain activation of platelets. J. Biol. Chem. 268, 7422–7426.

Salvesen, G., Parkes, C., Abrahamson, M., Grubb, A. and Barrett, A.J. (1986). Human low-Mr kininogen contains three copies of a cystatin sequence that are divergent in structure and in inhibitory activity for cysteine proteinases. Biochem. J. 234, 429–434.

Sato, F. and Nagasawa, S. (1988). Mechanism of kinin release from human low molecular-mass-kininogen by synergistic action of human plasma kallikrein and leukocyte elastase. Biol. Chem. Hoppe-Seyler 369, 1009–1017.

Scarsdale, J.N. and Harris, R.B. (1990) Solution phase conformation studies of the prekallikrein binding domain of high molecular weight kininogen. J. Prot. Chem. 9, 647–659.

Schini, V.B., Boulanger, C., Regoli, D. and Vanhoutte, P.M. (1990). Bradykinin stimulates the production of cyclic GMP via activation of B_2 kinin receptors in cultured porcine aortic endothelial cells. J. Pharmacol. Exp. Ther. 252, 581–585.

Schmaier, A.H., Zuckerberg, A., Silverman, C., Kuchibhotla, J., Tuszynski, G.P. and Colman, R.W. (1983). High molecular weight kininogen, a secreted platelet protein. J. Clin. Invest. 71, 1477–1489.

Schmaier, A.H., Bradford, H., Silver, L.D., Farber, A., Scott, C.F. and Colman, R.W. (1986a). High molecular weight kininogen is an inhibitor of platelet calpain. J. Clin. Invest. 77, 1565–1573.

Schmaier, A.H., Smith, P.M., Purdon, A.D., White, J.G. and Colman, R.W. (1986b). High molecular weight kininogen. Localization in the unstimulated and activated platelet and activation by a platelet calpain(s). Blood 67, 119–130.

Schmaier, A.H., Schutsky, D., Farber, A., Silver, L.D., Bradford, H.N. and Colman, R.W. (1987). Determination of the bifunctional properties of high molecular weight kininogen by studies with monoclonal antibodies directed to each of its chains. J. Biol. Chem. 262, 1405–1411.

Schmaier, A.H., Kuo, A., Lundberg, D., Murray, S. and Cines, D.B. (1988a). Expression of high molecular weight kininogen on human umbilical vein endothelial cells. J. Biol. Chem. 263, 16327–16333.

Schmaier, A.H., Farber, A., Schein, R. and Sprung, C. (1988b). Structural changes of plasma high molecular weight kininogen after in vitro activation and in sepsis. J. Lab. Clin. Med. 112, 182–192.

Schmaier, A.H., Bradford, H.N., Lundberg, D., Farber, A. and Colman, R.W. (1990). Membrane expression of platelet calpain. Blood 75, 1273–1281.

Schmaier, A.H., Meloni, F.J., Nawarawong, W. and Jiang, Y.P. (1992). PPACK-thrombin is a noncompetitive inhibitor of α-thrombin binding to human platelets. Thromb. Res. 67, 479–489.

Scott, C.F. and Colman, R.W. (1992). Sensitive antigenic determination of high molecular weight kininogen performed by covalent coupling of capture antibody. J. Lab. Clin. Med. 119, 77–86.

Scott, C.F., Purdon, A.D., Silver, L.D. and Colman, R.W. (1985). Cleavage of high molecular weight kininogen (HMWK) by plasma factor XIa. J. Biol. Chem. 260, 10856–10863.

Scott, C.F., Whitaker, E.J., Hammond, B.F. and Colman, R.W. (1993). Purification and characterization of a potent 70-kDa thiol lysyl-proteinase (Lys-gingivain) from Porphyromonas gingivalis that cleaves kininogens and fibrinogen. J. Biol. Chem. 268, 7935–7942.

Shima, C., Majima, M. and Katori, M. (1992). A stable metabolite, Arg-Pro-Pro-Gly-Phe, of bradykinin in the degradation pathway in human plasma. Jpn J. Pharmacol. 60, 111–119.

Smith, D., Gilbert, M. and Owen, W.G. (1983). Tissue plasminogen activator release in vivo in response to vasoactive agents. Blood 66, 835–839.

Tait, J.F. and Fujikawa, K. (1986). Identification of the binding site for plasma prekallikrein in human high molecular weight kininogen to plasma prekallikrein and factor XI. J. Biol. Chem. 61, 15396–15401.

Tait, J.F. and Fujikawa, K. (1987). Primary structural requirements of the binding of human high molecular weight kininogen to plasma prekallikrein and factor XI. J. Biol. Chem. 62, 11651–11656.

Takagaki, Y., Kitamura, N. and Nakanishi, S. (1985). Cloning and sequence analysis of cDNAs for human high molecular weight and low molecular weight prekininogens. J. Biol. Chem. 260, 8601–8609.

Takano, M., Yokoyama, K., Yayama, K. and Okamoto, H. (1995). Murine fibroblasts synthesize and secrete kininogen in response to cyclic-AMP, prostaglandin E_2 and tumor necrosis factor. Biochim. Biophys. Acta 1265, 189–195.

Tayeh, M.A., Olson, S.T. and Shore, J.D. (1994). Surface-induced alterations in the kinetic pathway for cleavage of human high molecular weight kininogen by plasma kallikrein. J. Biol. Chem. 269, 16318–16325.

Van Iwaarden, F., de Groot, P.G., Sixma, J.J., Berrettini, M. and Bouma, B.N. (1988a). High molecular weight kininogen is present in cultured human endothelial cells: localization, isolation, and characterization. Blood 71, 1268–1276.

Van Iwaarden, F., de Groot, P.G. and Bouma, B.N. (1988b). The binding of high molecular weight kininogen to cultured human endothelial cells. J. Biol. Chem. 263, 4698–4703.

Vogel, R., Assfalg-Machleidt, I., Esterl, A., Machleidt, W. and Müller-Esterl, W. (1988). Proteinase-sensitive regions in the heavy chain of low molecular weight kininogen map to the inter-domain junctions. J. Biol. Chem. 263, 12661–12668.

Vogel, R., Kaufmann, J., Chung, D.W., Kellerman, J. and Müller-Esterl, W. (1990). Mapping of the prekallikrein-binding site of human H-kininogen by ligand screening of λgt11 expression libraries. J. Biol. Chem. 265, 12494–12502.

Wachtfogel, Y.T., DeLa Cadena,P R.A., Kunapuli, S.P., Rick, L., Miller, M., Schultze, R.L., Altieri, D.C., Edgington, T.E. and Colman, R.W. (1993). High molecular weight kininogen binds to Mac-1 on neutrophils by its heavy chain (Domain 3) and its light chain (Domain 5). J. Biol. Chem. 269, 19307–19312.

Weisel, J.W., Nagaswami, C., Woodhead, J.L., DeLa Cadena, R.A., Page, J.D. and Colman, R.W. (1994). The shape of high molecular weight kininogen. Organization into structural domains, changes with activation, and interactions with prekallikrein, as determined by electron microscopy. J. Biol. Chem. 269, 10100–10106.

Wiggins, R.C. (1983). Kinin release from high molecular weight kininogen by the action of Hageman factor in the absence of kallikrein. J. Biol. Chem. 260, 10856–10863.

Wuepper, K.D., Miller, D.R. and LaCombe, M.J. (1975). Flaujeac trait: deficiency of human plasma kininogen. J. Clin. Invest. 56, 1662–1672.

Yamamoto, T., Tsuruta, J. and Kambara, T. (1987). Interstitial-tissue localization of high-molecular-weight kininogen in guinea-pig skin. Biochim. Biophys. Acta 916, 332–342.

Ylinenjarvi, K., Prasthofer, T.W., Martin, N.C. and Bjork, I. (1995). Interaction of cysteine proteinase with recombinant kininogen domain 2, expressed in *Escherichia coli*. FEBS Lett. 357, 309–311.

Zini, J.-M., Schmaier, A.H. and Cines, D.B. (1993). Bradykinin regulates the expression of kininogen binding sites on endothelial cells. Blood 81, 2936–2946.

5. The Molecular Biology of the Kallikreins and their Roles in Inflammation

Judith A. Clements

1. Introduction

The kallikreins are a group of serine proteases involved in the post-translational processing of polypeptide precursors. With respect to the kinin system, they consist primarily of plasma kallikrein and tissue or glandular kallikrein. Although these enzymes are similarly named and both generate bioactive peptides, the kinins bradykinin (BK) and Lys-BK (kallidin), from kininogens, they are distinct entities. Plasma kallikrein is encoded by a single gene and bears no sequence or structural relationship to tissue kallikrein. Plasma prekallikrein is expressed solely in the liver, circulating as a complex with its substrate, high molecular weight kininogen (HMWK), before releasing BK at predetermined sites in the vasculature. The primary functions of plasma kallikrein are its crucial involvement in the intrinsic cascade of blood clotting and fibrinolysis (see Chapters 4 and 10 of this volume) and, via the physiological and pathophysiological actions of the kinins, regulation of vascular tone and inflammation.

Tissue kallikrein is a member of a large multigene (*KLK*) family of enzymes, which are highly conserved in sequence and tertiary structure. Tissue kallikrein (or the *KLK1* gene) is almost ubiquitous, being expressed in most tissues at varying levels and acting locally primarily to generate Lys-BK from the liver-derived or locally produced substrate, low molecular weight kininogen (LMWK). Tissue kallikrein, via the action of Lys-BK and its involvement in the regulation of local blood flow and pressure, vascular permeability, Na^+ and glucose transport, inflammation, cellular proliferation and differentiation, contraction of smooth muscle and the induction of pain, has been implicated both in the normal physiology and disorders of the cardiovascular, renal and respiratory systems, brain, gastrointestinal and reproductive tracts. In addition, several studies have shown that tissue kallikrein, at least *in vitro*, acts on numerous other substrates, activating or degrading polypeptide precursors to generate peptides, growth factors, hormones, many of which are also important to physiological processes. A further complication is that other members of the *KLK* gene family of enzymes, in addition to their primary function as activators of specific polypeptide precursors, may also act as kininogenases, albeit more weakly than tissue kallikrein.

In this chapter, the focus will be primarily on the molecular biology of the tissue kallikrein gene family, as many aspects of the regulation of plasma kallikrein expression and its involvement in inflammatory processes are discussed in Chapters 4, 10, 11 and 16 of this volume. Thus, information provided here is primarily to emphasize differences in gene structure and organization between the plasma and tissue kallikrein genes and the latter family of genes (Fig. 5.1). As noted above, greater appreciation has developed of the varied role of the tissue kallikrein–kinin system in human (patho)physiology as the molecular biology and physiology of the tissue kallikrein (*KLK*) gene family of enzymes and other components of this system have been more extensively characterized. Several reviews have covered aspects of this information in the past decade (Fuller and Funder, 1986; Drinkwater *et al.*, 1988a; MacDonald *et al.*, 1988; Proud and Kaplan, 1988; Clements, 1989, 1994; Bhoola *et al.*, 1992; Carbini *et al.*, 1993; Carretero *et al.*, 1993; Scicli *et al.*, 1993; Margolius, 1995). As relatively little research has been performed specifically on the molecular regulation and expression of tissue kallikreins in inflammatory events, this review will describe our current knowledge of the basic molecular biology of this gene family, with both old and new findings, which may lead us to understand better the role of these enzymes in inflammation.

2. The Plasma Kallikrein Gene

Plasma kallikrein or Fletcher factor is encoded by a single gene of about 22 kb in length, which has been characterized in the rat (Beaubien *et al.*, 1991). It is structurally similar in organization to that of the human Factor XI gene (Asakai *et al.*, 1987) and, since both these genes in the human are located on chromosome 4q34–q35, it would appear they have arisen from a

PLASMA KALLIKREIN GENE

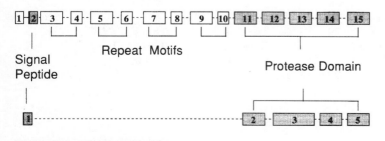

Signal
Peptide

Repeat Motifs

Protease Domain

TISSUE KALLIKREIN GENE FAMILY

Figure 5.1   Schematic diagram of the genomic exon–intron organization of the plasma kallikrein (15 exons) and tissue kallikrein gene (*KLK*) families (5 exons) with signal peptide, repeat motif and protease domains indicated.

common gene duplication. The mouse plasma kallikrein gene is located in the B1–B3 region of chromosome 8 (Beaubien *et al.*, 1991). It is composed of 15 exons, the first 2 exons encoding 5′-untranslated regions and the signal peptide. The next eight exons consist of four repeat domains, which encode the heavy chain of plasma kallikrein, while the last five exons encode the light chain and the protease portion of the molecule, the region which is more structurally and functionally similar to tissue kallikrein (Fig. 5.1). As expected, this C-terminal region contains the three residues, His^{415}, Asp^{464} and Ser^{559}, typical of the active site of serine proteases (Van der Graaf *et al.*, 1982). The promoter of the rat gene has a consensus TATTAA box and CCAAT box elements. Other potential regulatory sequences include a putative estrogen response element at -343 to -330, which may be responsible for the increased hepatic plasma kallikrein mRNA levels in female rats (Seidah *et al.*, 1989; Beaubien *et al.*, 1991). Further putative response elements for the glucocorticoid receptor, thyroid hormone receptor and an Oct-1 element were also described (Beaubien *et al.*, 1991). However, all of these sequences have yet to be functionally examined. The conservation between human, rat and mouse plasma kallikreins, all of which have been now cloned, is approximately 75% (Seidah *et al.*, 1989, 1990; Beaubien *et al.*, 1991).

3. The Tissue Kallikrein Gene Family

3.1 GENERAL OVERVIEW

The exponential rise in new genes and kallikrein-like enzymes, which were identified over the last two decades, led to a haphazard series of gene and enzyme designations. The nomenclature used here, to denote the tissue kallikrein gene families in different species and define individual family members, is the consensus reached by an international working party (Berg *et al.*, 1992a).

The tissue kallikreins are a subgroup of serine proteases, which are encoded by a multigene (*KLK*) family in several species (Evans *et al.*, 1987; Bell and Fahnestock, 1988; Wines *et al.*, 1989; Murray *et al.*, 1990). Generally speaking, they are most closely associated with trypsin in terms of genomic exon–intron arrangement, sequence homology and enzymic action (Neurath, 1989). The structural arrangement of the genes consists of five exons and four introns, and is highly conserved in all species thus far studied, with the splice junctions completely conserved (Evans *et al.*, 1987, 1988; Chen *et al.*, 1988; Schedlich *et al.*, 1988; Lundwall, 1989; Riegman *et al.*, 1989a; Shai *et al.*, 1989; Wines *et al.*, 1989; Chapdelaine *et al.*, 1991; Lin *et al.*, 1993). The coding region of each enzyme extends across

the five exons (Fig. 5.1). The difference in mRNA transcript sizes reported for some of these genes in different species, primarily relates to differences in the length of the 5′- and 3′-untranslated regions also present in exons 1 and 5. There is only one gene in any species, *KLK1*, which encodes (true) tissue kallikrein; that is, the enzyme which generates Lys-BK from the physiological substrate, LMWK (Swift *et al.*, 1982; Ashley and MacDonald, 1985a; Baker and Shine, 1985; Fukushima, *et al.*, 1985; Van Leeuwen *et al.*, 1986; Evans *et al.*, 1988; Takahashi *et al.*, 1988; Inoue *et al.*, 1989; Brady and MacDonald, 1990; Lin *et al.*, 1993; Gauthier *et al.*, 1994). The generic term, tissue kallikreins, is used to describe other enzymes in this highly conserved gene (*KLK*) family, which share considerable sequence and structural homology with *KLK1*/tissue kallikrein. Although many genes have been identified as being members of this family, surprisingly few of the kallikrein-like enzymes have been functionally characterized. Thus, the physiological substrate and thus function of the encoded proteins of the majority of these genes is unknown.

Typically as serine proteases, the tissue kallikreins are characterized by the conserved triad of amino acids (His^{41}-Asp^{90}-Ser^{180}), which, in the correct configuration, is necessary for serine protease catalytic activity (Bode *et al.*, 1983; Neurath, 1989). They are trypsin-like in their actions, usually processing at basic amino acids. Most tissue kallikreins can also cleave synthetic arginyl esters and are often referred to as arginyl esterases (Fiedler, 1979). Thus tissue kallikrein cleaves primarily LMWK at two pairs of amino-acid residues, Met-Lys (Lys-Arg in the rat) and Arg-Ser, to release Lys-BK or BK (rat only) (Maier *et al.*, 1983). The tissue kallikreins do not have as broad a substrate specificity as trypsin, however, as each enzyme appears to have a more distinct substrate specificity, which denotes the physiological function of that enzyme. Other functionally characterized members of this family include rat tonin and T-kininogenase, the mouse renin [γ-renin, prorenin converting enzyme (PRECE)] and growth factor [γ-nerve growth factor (NGF), β-NGF endopeptidase, epidermal growth factor-binding protein (EGF-BP)] processing enzymes and prostate-specific antigen in the human (Greene *et al.*, 1968; Boucher *et al.*, 1974; Frey *et al.*, 1979; Thomas *et al.*, 1981; Poe *et al.*, 1983; Lundgren *et al.*, 1984; Lilja, 1985; Watt *et al.*, 1986; Berg *et al.*, 1987; Blaber *et al.*, 1987; Isackson and Bradshaw, 1984; Drinkwater *et al.*, 1988b; Gutman *et al.*, 1988; Fahnestock *et al.*, 1991; Kim *et al.*, 1991a). From the earlier molecular cloning, biochemical purification and functional studies, it was thought that each *KLK* gene/enzyme had one specific physiological substrate. The increasing number of substrates (see below), which purified preparations of these enzymes can cleave *in vitro*, has led to the suggestion that they may perform a variety of functions in different tissues or physiological circumstances.

3.2 LOCUS, GENOMIC ORGANIZATION, SIZE OF FAMILIES AND SEQUENCE CONSERVATION

3.2.1 The Human Tissue Kallikrein (KLK) Gene Family

The human KLK gene family locus is on chromosome 19q (Evans et al., 1988; Sutherland et al., 1988; Riegman et al., 1989b, 1992; Qin et al, 1991; Richards et al., 1991), at a region analogous to the mouse tissue kallikrein gene (mKlk) family locus on chromosome 7 in the mouse genome (Howles et al., 1984; Evans et al., 1987). The three human characterized genes, KLK1, KLK2 and KLK3, form a cluster spanning just 60 kb in the region 19q.13.3–13.4. KLK2 and KLK3 are separated by 12 kb and are more tightly linked than KLK1, which is located 31 kb from KLK3 (Digby et al., 1989; Riegman et al., 1989b, 1992). The alignment in the genome is KLK1, KLK3 and KLK2 with KLK1 being transcribed in the opposite direction to KLK2 and KLK3.

In marked contrast to the size of the rodent KLK gene families, the size of the human KLK gene family currently stands at just three genes (Table 5.1). From Southern blot analyses, the size of this family has been suggested to vary from just 3–4 (Baker and Shine, 1985; Fukushima et al., 1985; Evans et al., 1988; Riegman et al., 1989b) to as many as 19 genes (Murray et al., 1990). The latter study, however, was complicated by the use of a monkey KLK cDNA probe on human genomic Southerns. As no new genes have been identified since

Table 5.1   Human and other tissue kallikrein (KLK) gene families

Gene	Translated protein
Human	
KLK1	Tissue kallikrein
KLK2	Unknown
KLK3	Prostate-specific antigen (PSA)
Monkey	
cmKLK1	Tissue kallikrein
rmKLK3	PSA
Dog	
dKLK1	Tissue kallikrein
dKLK2	Canine arginine esterase
Pig	
pKLK1	Tissue kallikrein
Guinea pig	
gpKLK1	Tissue kallikrein
gpKLK2	Prostate tissue kallikrein

Lower case characters preceding the gene designation indicate species: cm, cynomolgus monkey; rm, rhesus monkey; d, dog; p, pig; gp, guinea pig.

1988, these three genes may represent the entire human KLK family.

KLK1 encodes human tissue kallikrein and KLK3 encodes prostate-specific antigen or PSA (Baker and Shine, 1985; Fukushima et al., 1985; Watt et al., 1986; Lundwall and Lilja, 1987; Evans et al., 1988; Henttu and Vihko, 1989; Inoue et al., 1989; Lundwall, 1989; Riegman et al., 1989a). The function of the enzyme encoded by KLK2 is unknown (Schedlich et al., 1988). The nucleotide sequence of KLK1 was first provided from human kidney and pancreatic cDNA analysis, and predicted a preprokallikrein of 238 amino acids with a prepro fragment of 24 amino acids (Baker and Shine, 1985; Fukushima et al., 1985). Although just 61–67% homologous with tissue kallikrein from other species (Fiedler and Fritz, 1981; Swift et al., 1982; Van Leeuwen et al., 1986), the residues crucial to tissue kallikrein substrate specificity, such as the Asp^{184} in the bottom of the substrate binding pocket, are conserved (Bode et al., 1983). The KLK2 enzyme also shares this residue suggesting a trypsin/kallikrein-like action, but KLK3/PSA has a serine in this position indicating a more chymotrypsin-like action for this enzyme (Akiyama et al., 1987; Lundwall and Lilja, 1987; Schedlich et al., 1988).

The genomic sequence of the three KLK genes has been elucidated (Evans et al., 1988; Schedlich et al., 1988; Lundwall, 1989; Riegman et al., 1989a). They follow the predicted structural organization of this gene family with five exons and four introns, and the exon/intron splice junctions completely conserved. They are less conserved overall than their rodent counterparts, with their coding regions 52–78% homologous at the amino-acid level. The encoded proteins of the KLK2 and KLK3 genes share greater homology (78%) with both these genes less homologous to KLK1 (66%, KLK2; 52%, KLK3). The 5′- and 3′-flanking regions share similar homology with KLK2 and KLK3 more conserved (>90%) with each other than KLK1. The primary transcripts of KLK2 and KLK3 (~1.5 kb) are ~700 bp longer than that of KLK1 owing to long 3′-untranslated regions. The variant TATA box (TTTAAA), typical of the mouse Klk genes, is present in the KLK1 promoter (Evans et al., 1988). KLK2 and KLK3 share a different variant TTTATA box. All three genes have the variant polyadenylation signal, AGTAAA, although two other alternative signals have been described in variant transcripts of KLK2 and KLK3 (Riegman et al., 1989a, 1991a).

3.2.2 The Monkey Tissue Kallikrein (cmKLK; rmKLK) Gene Families

A cynomolgus monkey KLK1/tissue kallikrein cDNA has been characterized from a monkey renal cDNA library and shown to be 95% homologous with its human counterpart at the nucleotide level and across the exonic sequences (Lin et al., 1993). The cmKLK1 encodes a tissue kallikrein of 257 amino acids, which is 93% conserved with the human protein sequence, although five amino acids shorter, in particular lacking amino-acids 8–10, a feature unique to monkey tissue kallikrein. The key residues important for tissue kallikrein substrate specificity, that is kininogenase activity, are entirely conserved. The monkey tissue kallikrein shares homology with pig, rat and mouse in not having an additional Thr[108], a residue unique to human tissue kallikrein. As with human tissue kallikrein, the conservation drops considerably (60–66%) when the amino-acid sequence of monkey tissue kallikrein is compared with that of the pig, mouse and rat.

A rhesus monkey prostatic KLK3 cDNA encoding the simian counterpart of prostate-specific antigen/PSA has also been recently characterized (Gauthier et al., 1993a). It consists of 1,515 nucleotides, encoding a pre-proenzyme of 261 amino acids, with a long 3′-untranslated region (700 nucleotides) as for KLK3 and the consensus polyadenylation signal typical of all KLK genes. The deduced amino-acid sequence is 89% homologous with PSA compared with 71% for the gene product of KLK2 and 45–55% conservation in comparison with prostate-expressed KLK genes (rKLK8, rKLK9, gpKLK2, dKLK2 or canine arginine esterase) in other species. Other sequence and structural features completely conserved with human PSA include the serine residue at position 183, important in the chymotrypsin-like specificity of these two enzymes. Tyr[93], a residue important for the kininogenase activity of tissue kallikrein, is replaced by serine indicating that rmPSA will lack kininogenase activity as does its human counterpart. That both KLK2 and KLK3 share high conservation with their mammalian homologs was also suggested from recent Southern blot analysis data in which KLK2 and KLK3 positive bands were detected for several nonhuman primate species (macaque, orangutan, chimpanzee, gorilla) but not cow, pig, dog, rabbit, rat or mouse (Karr et al., 1995). Although no monkey genes have yet been cloned, it is probable that the exon–intron arrangement characteristic of this gene family in other species will also be conserved.

3.2.3   The Dog Tissue Kallikrein (dKLK) Gene Family

Two members of the dog KLK family have been characterized to date. dKLK1 encodes dog tissue kallikrein (Deperthes et al., 1995; Gauthier et al., 1994) and dKLK2, canine arginine esterase (Chapdelaine et al., 1984, 1988a, 1991; Lazure et al., 1984). dKLK1 encodes a polypeptide of 261 amino acids with the typical 24-residue prepropeptide and the conserved residues characteristic of the serine protease catalytic triad, and tissue kallikrein substrate binding pocket (Gauthier et al., 1994). In contrast to most other KLK cDNAs, a variant polyadenylation signal (AGTAAA) is found in this cDNA. As for the mammalian KLK gene

products but in contrast to the rodent families, there is greater conservation at the amino-acid level interspecies than between the two dKLK gene products. dKLK1/tissue kallikrein shares just 66% homology with dKLK2 or canine arginine esterase at the amino-acid level but is 74% homologous to human tissue kallikrein. In addition, the predominant tissue-specific pattern of dKLK1 expression in the salivary gland, pancreas and kidney but not prostate, further confirms dKLK1 as encoding tissue kallikrein (Gauthier et al., 1994; Deperthes et al., 1995).

A prostatic cDNA and gene (dKLK2) encoding canine arginine esterase have both been cloned (Chapdelaine et al., 1988a, 1991). As for all other mammalian KLK genes, dKLK2 consists of five exons and four introns with the exon/intron junctions entirely conserved and a variant TATA box (TTTAAA). The dKLK2 mRNA is 873 nucleotides long, encoding a preproenzyme of 260 amino acids. The polyadenylation signal (AGUAAA) is identical to that found in the three human KLK genes but not typical of the rodent families. The deduced amino-acid sequence is identical to the previously published partial N-terminal sequence of the heavy and light chains of canine seminal plasma arginine esterase (Lazure et al., 1984). The dKLK2 gene product or canine arginine esterase is less conserved overall with respect to both other primarily prostate-expressed KLK genes (51–61%: rat/guinea pig/human) and tissue kallikrein (58%: human/pig). This gene is also expressed in the liver, kidney, pancreas and skeletal muscle to some extent (Chapdelaine et al., 1991).

3.2.4   The Mouse Tissue Kallikrein (mKlk) Gene Family

The mKlk gene family was the first KLK gene family to be fully characterized. The mKlk gene family locus is found on chromosome 7, in a 310 kb region which extends over the previously defined Tam-1, Prt-4 and Prt-5 loci, and in a position analogous to the human KLK locus at 19q (Howles et al., 1984; Evans et al., 1987). It consists of 24 genes, the majority closely linked in clusters of at least two genes with the largest cluster consisting of 11 genes (Evans et al., 1987). The intervening regions between the genes are generally 3–8 kb but in two cases extend up to 21 and 24 kb. All appear to be transcribed in the same direction except for mKlk15, which is most closely linked to mKlk14 and transcribed in the opposite direction. The mapping of 24 genes to this locus, which is consistent with the previous reports of 25–30 genes in this family as determined by Southern blot analysis (Mason et al., 1983), presumably reflects the entire mKlk gene family. As for the human family, the structural organization is similar (five exons, four introns) with exon–intron boundaries conserved, although there is some variation in the length of the introns. Ten of the genes are pseudo genes with gross exon–intron rearrangements or in phase termination codons. Of the other genes, 11 are known to be expressed (Mason *et al*, 1983; Ullrich *et al.*, 1984; Evans and Richards, 1985, Van Leeuwen *et al.*, 1986; Blaber *et al.*, 1987; Drinkwater and Richards, 1987, 1988; Drinkwater *et al.*, 1987, 1988b; Evans *et al.*, 1987; Kim *et al.*, 1991a) and the functional enzyme encoded by eight of these has been described (see Table 5.2). All encode a preprokallikrein of 261 amino acids, which includes a hydrophobic signal peptide of 18 amino acids and N-terminal zymogen peptide of 6 amino acids, which when cleaved generates the active enzyme.

mKlk1 clearly encodes tissue kallikrein (Van Leeuwen *et al.*, 1986). Although confirmatory amino-acid sequence data is not available, the deduced sequence from cDNA and genomic nucleotide data shows the key amino-acid residues (Tyr^{93}, $Ser1^{35}$, His^{164}, Asp^{183}, Trp^{205}, Gly^{206}) important for kininogenase activity in other species are clearly conserved. The relatively high levels of *mKlk1* expression, as determined by Northern blot analysis, in the mouse salivary gland, kidney and pancreas is also clearly indicative of the known most abundant sites of tissue kallikrein expression in all species so far studied (Van Leeuwen *et al.*, 1986). Other *mKlk* genes, which encode known functional proteins include *mKlk3* and *mKlk4*, encoding γ-nerve growth factor (NGF) and α-NGF respectively (Ullrich *et al.*, 1984; Evans and Richards, 1985). *mKlk22*, *mKlk13* and *mKlk9* were first described as the epidermal growth factor (EGF)-associated proteins or binding proteins, EGF-BP type A, B and C (Lundgren *et al.*, 1984; Blaber *et al.*, 1987; Drinkwater *et al.*, 1987). More recently, the gene products of *mKlk13* and *mKlk22* have also been identified to act as a prorenin-converting enzyme (PRECE) and a β-NGF endopeptidase (Fahnestock *et al.*, 1991; Kim *et al.*, 1991a). *mKlk16* and *mKlk26* encode other renin processing enzymes, γ-renin and PRECE-2 (Drinkwater *et al.*, 1988b; Kim *et al.*, 1991b).

All of these genes are highly conserved both at the nucleotide (73–95%) and amino-acid level (70–85%). The 5′-flanking regions are similarly conserved (75–96% over some regions) with minor differences of single base deletions or insertions, or point mutations observed for at least the first 500 base pairs for the seven genes for which such comparative sequence data are available (Mason *et al.*, 1983; Evans and Richards, 1985; Van Leeuwen *et al.*, 1986; Drinkwater *et al.*, 1987). All have the variant 5′-TTTAAA-3′ box typical of the *mKlk* promoter regions and a normal polyadenylation signal.

3.2.5 The Rat Tissue Kallikrein (*rKLK*) Gene Family

The *rKLK* gene family currently consists of 13 characterized genes (see Table 5.3) (Ashley and MacDonald, 1985a,b; Chen *et al.*, 1988; Brady *et al.*, 1989; Inoue *et al.*, 1989; Shai *et al.*, 1989; Wines *et al.*, 1989, 1991; Brady and MacDonald, 1990; Ma *et al*, 1992). They are also tightly linked at one locus in the genome, although the chromosomal localization of that locus has not yet been identified (Southard-Smith *et al.*, 1994). Ten of these genes are expressed, while three are pseudogenes (*rKLK5*, *rKLK11*, *rKLK13*). Physical mapping of rat genomic P1 bacteriophage clones has identified two clusters of tightly linked genes encompassing 12 of these 13 genes, separated by less than 75 kb, almost entirely within a 440 kb *Bss*HII fragment of the rat genome (Southard-Smith *et al.*, 1994). One cluster, or the "Odd" contig, spans approximately 176 kb and contains the genes *rKLK1*, *rKLK3*, *rKLK7* and *rKLK9* linked in this specific order. The "Even" contig encompasses the genes *rKLK8*, *rKLK2*, *rKLK6*, *rKLK4*, *rKLK10* and *rKLK12* linked in this order and spans 225 kb. The

---

Table 5.2   The mouse tissue kallikrein (*mKlk*) gene family

Gene	Translated protein
mKlk1	Tissue kallikrein
mKlk3	γ-Nerve growth factor (NGF)
mKlk4	α-NGF
mKlk5	Unknown
mKlk6	Unknown
mKlk8	Unknown
mKlk9	Epidermal growth factor-binding protein (EGF-BP)
mKlk11	Unknown
mKlk13	EGF-BP B/prorenin-converting enzyme (PRECE)
mKlk14	Unknown
mKlk16	γ-Renin
mKlk21	Unknown
mKlk22	EGF-BP A/β-NGF endopeptidase
mKlk24	Unknown
mKlk26	PRECE 2

mKlk2, mKlk7, mKlk10, mKlk12, mKlk15, mKlk17, mKlk18, mKlk19, mKlk23 and *mKlk25* are all pseudogenes.

Table 5.3   The rat kallikrein (*rKLK*) gene family

Gene	Translated protein
rKLK1	Tissue kallikrein
rKLK2	Tonin
rKLK3	Unknown
rKLK4	Unknown
rKLK6	Unknown
rKLK7	Esterase B/proteinase A/K7
rKLK8	K8
rKLK9	Submandibular enzymatic vasoconstrictor (SEV)/S3
rKLK10	T-Kininogenase/antigen γ/proteinase
rKLK12	Unknown
rKLK13	Unknown

rKLK5, rKLK11 and *rKLK13* are all pseudogenes.

pseudogenes, *rKLK5* and *rKLK11*, were mapped to within 3 kb of *rKLK4* and *rKLK6*, respectively. The direction of transcription was determined for four of these genes, two in each of the contigs. *rKLK1* and *rKLK3* in the Odd cluster are transcribed in the opposite direction to *rKLK4* and *rKLK10* in the Even cluster. Whether the entire Odd and Even contigs are transcribed in opposite directions has not been established. Southern blot analysis had predicted 15–20 potential members of the *rKLK* gene family (Ashley and MacDonald, 1985b; Gerald *et al.*, 1986a; Wines *et al.*, 1989), but no new genes have been identified since 1991. Thus, these 13 genes may represent the full *rKLK* gene family locus. In the above study, however, there is a suggestion that a novel *rKLK* family member may have been detected within these two contigs (Southard-Smith *et al.*, 1994). Closer scrutiny of these clones may thus still provide additional members of the *rKLK* gene family in keeping with the Southern data.

In contrast to the mouse, the potential physiological function of just three *rKLK* genes is known. *rKLK1* again encodes tissue kallikrein as determined by the conservation of specific residues necessary for kininogenase activity and the appropriate profile of tissue-specific expression (Swift *et al.*, 1982; Ashley and MacDonald, 1985a,b; Inoue *et al.*, 1989). *rKLK2* encodes tonin, which processes angiotensinogen to angiotensin II *in vitro* and *rKLK10* encodes T-kininogenase, the enzyme which cleaves T-kininogen *in vitro* to release T-kinin (Boucher *et al.*, 1974; Ashley and MacDonald, 1985a; Berg *et al.*, 1987; Gutman *et al.*, 1988; Shai *et al.*, 1989; Wines *et al.*, 1989; Xiong *et al.*, 1990a; Ma *et al.*, 1992). Other *rKLK* gene products (*rKLK7*, *rKLK8*, *rKLK9*) have been biochemically or pharmacologically characterized, but no clear physiological substrate has yet been identified (Chao, 1983; Ashley and MacDonald, 1985a; Khullar *et al.*, 1986; Chao and Chao, 1987; Chen *et al.*, 1988; Brady *et al.*, 1989; Brady and MacDonald, 1990; Elmoujahed *et al.*, 1990; Yamaguchi *et al.*, 1991; Berg *et al.*, 1992b,c). The enzymes encoded by the remaining four expressed genes (*rKLK3*, *rKLK4*, *rKLK6*, *rKLK12*) have yet to be identified.

As for the mouse and human families, the exon–intron arrangement of five exons/four introns and exon–intron splice junctions are entirely conserved in the rat *KLK* gene family. Similarly, the 5′- and 3′-flanking regions are highly conserved with minimal differences being identified as single or double base-pair substitutions, insertions or deletions (Wines *et al.*, 1989, 1991). All of these genes have the variant TATA box (TTTAAA) typical of the rodent *KLK* genes and a putative CAAT box, indicated by the conserved sequence CATCT at approximately position −75 in all characterized rat and mouse *KLK* genes. The coding regions share 85–92% homology amongst these genes and encode prepro-enzymes of 261 amino acids.

3.2.6 Other Tissue Kallikrein Gene Families

In contrast to the rat and mouse families, the size of the *KLK* gene families in all other species appears to be more in keeping with that of the human *KLK* family (Table 5.1). Southern blot analysis of genomic DNA from a range of different species – rabbit, cat, dog, cow, monkey and human – using a monkey kallikrein cDNA probe, suggested that the size of these *KLK* gene families ranged from just 3–4 genes (Murray *et al.*, 1990). This observation is consistent with that previously reported for the human, monkey and dog families (see above), although these findings would need to be verified for other species by the use of species-specific DNA probes and cDNA/genomic cloning of these *KLK* gene families. Similarly, the hamster may only have three *KLK* genes, although this was also a cross-species hybridization performed with a mouse γ-NGF cDNA probe (Howles *et al.*, 1984). However, the use of the same mouse probe on an African rat (*Mastomys natalensis*) genomic Southern blot would suggest that the size of the African rat *KLK* gene family is closer to that of the larger rodent families (Bowcock *et al.*, 1988).

Several other tissue kallikrein family members have been isolated and characterized at the protein level, but not yet at the mRNA or genomic level. These include porcine pancreatic and submandibular tissue kallikrein from which much of the early cloned sequence comparisons were made and guinea-pig salivary gland and prostate tissue kallikreins (Tschesche *et al.*, 1979; Fiedler and Fritz, 1981; Bode *et al.*, 1983; Dunbar and Bradshaw, 1987; Mayer *et al.*, 1989). Although all three enzymes exhibit kininogenase activity, only the pancreatic and salivary gland derived enzymes are considered their species homologues of tissue kallikrein, *per se*.

3.3 TISSUE- AND CELL-SPECIFIC EXPRESSION

3.3.1 *KLK1*/Tissue Kallikrein in the Salivary Gland, Kidney and Pancreas

In all species studied, the *KLK1* gene, which encodes tissue or glandular kallikrein, is expressed predominantly in the salivary gland, pancreas and kidney. Northern blot analysis with *KLK1* cDNA (human, monkey, dog) or gene-specific oligonucleotide probes (rat, mouse) consistently indicate that these three tissues have the highest levels of *KLK1* mRNA (Ashley and MacDonald, 1985b; Fukushima *et al.*, 1985; Van Leeuwen *et al.*, 1986; Clements *et al.*, 1990a,b,c; Lin *et al.*, 1993; Gauthier *et al.*, 1994).

In situ hybridization histochemistry studies localized *KLK1* expression to the granular convoluted tubular and striated duct cells of the submandibular gland and, to a lesser extent, the parotid and sublingual glands of the mouse and rat (Fuller *et al.*, 1986; Van Leeuwen *et al.*, 1987; Penschow *et al.*, 1991a,b; Penschow and Coghlan,

1992). Immunoreactive tissue kallikrein was similarly localized primarily to secretory granules in the apical region of these cell types in the rat, cat, guinea pig, dog and man (Orstavik *et al.*, 1980; Schachter *et al.*, 1980; Simson *et al.*, 1983, 1992; Simson and Chao, 1994). In the mouse, tissue kallikrein has also been immuno-localized to secretory granules at the basolateral surface of these cells, suggesting secretion via fenestrated capillaries underlying the ducts into the general circulation (Penschow and Coghlan, 1993). Tissue kallikrein activity has also been detected in human saliva and the salivary glands of all the above species as well as the pig (Bhoola and Dorey, 1971; Lemon *et al.*, 1979; Mayer *et al.*, 1989).

KLK1 expression has been demonstrated in the luminal cells of the distal convoluted tubule of the rat and mouse kidney and distal tubules, collecting ducts and Henle's loops of human kidney (Fuller *et al.*, 1986; Xiong *et al.*, 1989; Penschow *et al.*, 1991b; Penschow and Coghlan, 1992; Chen *et al.*, 1995b) further confirming the previous immunolocalization data (Orstavik and Ingami, 1982; Figueroa *et al.*, 1984; Barajas *et al.*, 1986; Simson *et al.*, 1987). At the electron microscope level, immunoreactive tissue kallikrein has been localized to both the apical/luminal and basolateral membranes of the cortical distal tubular cells in the rat and human kidney in keeping with both luminal secretion into the urine and involvement, via basolateral secretion, in renal vascular dynamics (Figueroa *et al.*, 1984; Vio and Figueroa, 1985). In another study, *KLK1* expression could not be detected in the normal human kidney, but was detected in the kidneys of nephritis patients indicating both the low level of expression normally and upregulation in an inflammatory condition (Cumming *et al.*, 1994).

Tissue kallikrein has been isolated from pancreatic tissue and secretions, and localized to zymogen granules in the acinar cells of the exocrine pancreas in many species (pig, cat, dog, rat, human) and β-cells of human pancreatic islets (Ole-Moyoi *et al.*, 1979b; Pinkus *et al.*, 1981; Hofmann and Geiger, 1983; Deperthes *et al.*, 1995). It is likely that these cells are also the site of *KLK1* expression and tissue kallikrein biosynthesis, although this has not yet been confirmed by *in situ* hybridization studies. Tissue kallikrein activity has also been detected in a rat pancreatic acinar cell carcinoma cell line (Berg *et al.*, 1985).

3.3.2 *KLK1*/Tissue Kallikrein in Other Tissues

rKLK1 is also expressed at relatively high levels, detectable by Northern blot analysis, in the gastric corpus, duodenum and colon of the rat gastrointestinal tract (Fuller *et al.*, 1989). These findings support the demonstration of kallikrein-like activity in the rat gastrointestinal tract and biosynthetic studies in the rat colon (Uchida *et al.*, 1980; Miller *et al.*, 1984). *KLK1*

expression was demonstrated in a human colonic epithelial cancer cell line (T84) and a *KLK1* cDNA, identical in sequence to that expressed in the pancreas and kidney, was recently cloned from a human colonic cDNA library (Baird *et al.*, 1991; Chen *et al.*, 1995a). Immunoreactive tissue kallikrein has been localized to the mucous cells of the colon of various species, and the intestine and stomach of the cat (Schachter *et al.*, 1986). *In situ* hybridization studies have confirmed *KLK1* expression in the glandular epithelial (goblet) cells of the human colon (Chen *et al.*, 1995a).

rKLK1 expression can also be detected by Northern blot analysis in the rat pituitary and brain (Fuller *et al.*, 1985; Chao *et al.*, 1987; Clements and Fuller, 1990). *rKLK1* mRNA levels are highest in the female anterior pituitary gland and male neurointermediate lobe with lower levels found in the hypothalamus, pineal gland, cerebral cortex, cerebellum and brainstem in keeping with the previously documented levels of immuno-reactive or tissue kallikrein enzyme activity in these tissues and its detection in cerebrospinal fluid (Powers and Nasjletti, 1982, 1983, 1984; Scicli *et al.*, 1984; Kitagawa *et al.*, 1991). The documentation of *KLK* expression in the brain, however, was performed with a *rKLK1* cDNA probe and given the conserved nature of this family thus may not solely indicate *rKLK1* expression (Chao *et al.*, 1987). Immunoreactive tissue kallikrein was localized to the perivascular and vascular walls in the pineal gland, and the ependymal cells lining the third ventricle and cell bodies of the arcuate, supraoptic, paraventricular and ventromedial nuclei of the hypothalamus (Simson *et al.*, 1985; Kitagawa *et al.*, 1991). Tissue kallikrein has also been colocalized with prolactin to the lactotrope in the rat anterior pituitary gland and to prolactin-secreting pituitary adenomas in the human (Jones *et al.*, 1990; Kizuki *et al.*, 1990; Vio *et al.*, 1990). The complementary *in situ* hybridization studies have not yet been performed to confirm these cell types as definitive sites of *KLK1* expression. *KLK1* expression was also recently demonstrated in a series of normal human pituitaries and pituitary tumors, although at a much lower level than observed for the rat pituitary, only being detectable by the RT-PCR technique (Clements *et al.*, 1996).

Other studies using the sensitive detection system of reverse transcriptase–polymerase chain reaction (RT-PCR) have recently identified further sites of *KLK1* expression in the reproductive tract – the mouse mammary gland, the rat ovary, uterus and testis, and the endometrium and myometrium of the human uterus (Clements *et al.*, 1992, 1994, 1995; Valdes *et al.*, 1993; Clements and Mukhtar, 1994; Jahnke *et al.*, 1994). These findings confirm the presence of tissue kallikrein immunoactivity and/or enzyme activity in the rat ovary and uterus, human myometrium and seminal plasma (Malofiejew, 1973; Chao *et al.*, 1981; Geiger and Clausnitzner, 1981; Marin-Grez *et al.*, 1982; Fink *et al.*,

1985; Espey *et al.*, 1986; Gao *et al.*, 1992; Valdes *et al.*, 1993; Brann *et al.*, 1995). Tissue kallikrein has been immunolocalized to the glandular epithelium of the human prostate, epididymal epithelium, Sertoli cells of the human testis, and apical membrane of mouse mammary alveolar cells, and ductal cells of normal human breast and cancer tissue, although these are yet to be verified as sites of local *KLK1* synthesis in these cell types (Saitoh *et al.*, 1987; Rehbock *et al.*, 1995). Tissue kallikrein activity is also present in the human placenta and amniotic fluid, the latter presumably reflecting its placental or perhaps fetal origin (Malofiejew, 1973). These findings overall suggest a more ubiquitous pattern of expression for *KLK1* than previously thought.

Tissue kallikrein immunoreactivity and/or enzyme activity has also been demonstrated in numerous other tissues and biological fluids, further highlighting its ubiquitous expression. It has been detected in human, rat and dog heart tissue extracts, and rat vascular tissue, and localized to the sarcoplasmic reticulum and granules of the atrial myocytes in the rat heart (Nolly *et al.*, 1981, 1985). Expression of a *rKLK* gene family in rat atrial myocytes, although perhaps not solely *rKLK1*, since three kininogenase activities have been demonstrated in this tissue, was confirmed using a *rKLK1* cDNA probe (Xiong *et al.*, 1990b). *rKLK1* is also expressed in the adrenal gland, vascular smooth muscle cells of both large and small blood vessels, and aortic smooth muscle of the rat (Scicli *et al.*, 1989; Oza *et al.*, 1990; Saed *et al.*, 1990). *mKlk1* is expressed in the mouse spleen (Van Leeuwen *et al.*, 1986). Similarly, *KLK* mRNA detected in the rat spleen using a *rKLK1* cDNA probe presumably reflects *rKLK1* expression, since tissue kallikrein immuno- and enzyme activity has been demonstrated in this tissue (Swift *et al.*, 1982; Chao *et al.*, 1984).

Tissue kallikrein has also been demonstrated in nasal secretory glands and pulmonary submucous glands in man, and found in bronchial lavage from asthmatics and nasal secretions from patients with allergic rhinitis (Baumgarten *et al.*, 1986, 1989; Christiansen *et al.*, 1987; see also Chapters 15, 16 and 17 of this volume). Although the definitive *KLK1* expression studies have not been performed for these tissues in man, it is interesting to note that *mKlk1* mRNA was detected in the ducts of the nasal glands of the 19-day-old mouse fetus and neonatal mice, suggesting that this will also be a further site of *KLK1* local expression in man (Penschow and Coghlan, 1992). Tissue kallikrein is present in sweat and was immunolocalized to the acinar dark cells of the secretory fundus of human sweat glands (Hibino *et al.*, 1988a; Poblete *et al.*, 1991). It is also present in synovial fluid from arthritic joints (Worthy *et al.*, 1990; Bhoola and Dieppe, 1991). It is found in human neutrophils, localized to cytoplasmic granules (Figueroa *et al.*, 1988; Figueroa and Bhoola, 1989; see Chapter 11), and particularly prevalent in invasive neutrophils in the inflammatory conditions noted above. However, there are no published reports of *KLK1* gene expression in these latter cells or tissues to date.

3.3.3 Expression of Other *KLK* Genes

In the mouse, in contrast to the documented tissue-specific pattern of expression in the salivary gland, kidney, pancreas, spleen, pituitary and nasal glands for *mKlk1* tissue kallikrein, all other *mKlk* genes are expressed predominantly, if not exclusively, in the submandibular salivary glands as assessed by Northern blot analysis (Evans and Richards, 1985; Drinkwater *et al.*, 1987, 1988b; Evans *et al.*, 1987; Van Leeuwen *et al.*, 1987; Penschow *et al.*, 1991a,b). It is possible that these genes may be expressed in other mouse tissues, but at much lower levels, such that would only be detected by RT-PCR. There are only two other reports of extrasalivary gland *mKlk* expression. A low level of *mKlk21* expression was detected in the Leydig cells of the adult mouse testis by *in situ* hybridization (Penschow *et al.*, 1991b). *mKlk* expression was also detected in mouse mesenteric lymph nodes and Peyer's patches but not in T cells, macrophages or T-cell/macrophage cell lines using a *mKlk6* cDNA probe, but the definitive determination of which *mKlk* gene is expressed in this tissue has not yet been performed (Summers and Hume, 1985).

The *rKLK* gene family has been most well characterized in terms of tissue- and cell-specific expression of its individual members. As in the mouse, all *rKLK* gene family members are expressed in the submandibular salivary glands, however, the majority of these genes are also expressed in a range of other tissues to varying degrees (Swift *et al.*, 1982; Ashley and MacDonald, 1985b; Chen *et al.*, 1988; Shai *et al.*, 1989; Clements *et al.*, 1990c; Ma *et al.*, 1992). In addition to *rKLK1* expression in the rat kidney, *rKLK7* and *rKLK10* mRNA levels can be detected at the Northern blot level, while *rKLK3* and *rKLK9* expression was detected by RT-PCR alone, and *rKLK12* variously reported to be detected by RT-PCR and/or Northern blot analysis (Chen *et al.*, 1988; Brady and MacDonald, 1990; Clements *et al.*, 1990a,b, 1995; Ma *et al.*, 1992; Saed *et al.*, 1992). *rKLK4* mRNA was detected in the rat heart by RT-PCR (Clements *et al.*, 1992). Two *rKLK* mRNAs, which are quite abundant in the prostate and thus easily detected on Northern blots in a similar fashion to the human *KLK2* and *KLK3* mRNAs, are *rKLK8* and *rKLK9* (Chao and Chao, 1987; Clements *et al.*, 1988, 1990a; Brady and MacDonald, 1989). Indeed, both these genes were originally thought to be only expressed in the prostate apart from the salivary glands. The prostatic expression of *rKLK7* and *rKLK12* was also clearly detectable by RT-PCR (Clements *et al.*, 1995). Recent studies, again using RT-PCR, have also demonstrated a low level of *rKLK* gene family expression in the rat ovary and testis. *rKLK3*, *rKLK7*, *rKLK8*, *rKLK9* and *rKLK12*, in addition to *rKLK1*, are all expressed in the rat ovary with *rKLK3*, *rKLK7*, *rKLK9* and *rKLK12* expression localized to the granulosa cell (Clements *et al.*, 1992, 1995). *rKLK3*, *rKLK9* and *rKLK12* mRNA was detected in the adult rat testis with *rKLK12* expression localized to the developing germ cells of the seminiferous tubules (Clements *et al.*, 1992). Although these mRNAs are clearly present at low levels, their localization to specific cell types would suggest a local paracrine/autocrine role in these tissues. In addition, these findings are in keeping with previous reports of kallikrein-like activity, distinct from tissue kallikrein itself, present in these tissues.

In the human, *KLK2* and *KLK3* expression was described as prostate-specific, since this was the only tissue, both normal and cancerous, in which expression was detected by Northern blot analysis (Chapdelaine *et al.*, 1988b; Riegman *et al.*, 1988, 1991a; Henttu *et al.*, 1990; Sharief *et al.*, 1994). Indeed, the enzyme encoded by *KLK3* is named prostate-specific antigen (PSA). More recent studies using RT-PCR have demonstrated that these two genes are also expressed in the human endometrium and pituitary, and that *KLK3* is also expressed in normal breast and breast and ovarian tumor tissue (Clements and Mukhtar, 1994; Monne *et al.*, 1994; Yu *et al.*, 1994, 1995a,b; Clements *et al.*, 1996). PSA is localized to epithelial-derived tumor cells in breast tissue as well as normal breast tissue and is measurable in breast milk (Yu *et al.*, 1994; Yu and Diamandis, 1995a). PSA has been detected in amniotic fluid suggesting that the human placenta and/or fetus also express *KLK3* (Yu and Diamandis, 1995b). PSA has also been detected immunohistochemically in a variety of other tissues and tumors in the female – renal cell, pancreatic, lung and urethral carcinomas, and periurethral gland – but it is not known whether these are also sites of *KLK3* expression (Pollen and Dreilinger, 1984; Pummer *et al.*, 1992; Ebisuno *et al.*, 1995; Kupio *et al.*, 1995; Levesque *et al.*, 1995). The recent findings of a wider pattern of expression for these genes in the human is not surprising given the diversity of tissue-specific expression observed for the rat *KLK* family.

The only two other species for which *KLK* genes other than *KLK1* have been described are the monkey and dog. Both families have a second gene, *rmKLK3* and *dKLK2*, which are expressed predominantly in the prostate (Chapdelaine *et al.*, 1991; Gauthier *et al.*, 1993a). These findings are based solely on Northern blot analysis using tissues derived from the salivary gland, pancreas, kidney and prostate of these animals. Since both these genes are considered the monkey and dog homologs of the human *KLK2* and *KLK3*, and perhaps rat *KLK8* and *KLK9* genes, a wider pattern of expression may also emerge for these genes.

3.3.4   Transcriptional Regulation of Tissue-specific *KLK* Expression

There has been considerable interest in the molecular events, which regulate both the seemingly conserved expression of most *KLK* genes in the salivary gland and the distinct tissue-specific patterns of expression observed for individual genes. Presumably, there are both *cis*- and *trans*-activating tissue-specific elements in the promoters of these genes which regulate such events. At least 500 bp of the 5′-flanking region of all the mouse, rat and human genes have been sequenced and compared. The rat and mouse genes are most conserved with less homology observed between the 5′-flanking regions of the human *KLK* genes as is also the case for coding regions of these genes. Surprisingly, these regions are so highly conserved (70–90%), particularly within species, such that only minor single- or double-base substitution, deletions or insertions are apparent, at least in the rodent genes, suggesting that, if important regulatory elements conferring tissue and/or cell-specific expression are present in this region, minor single- or double-base changes must confer this specificity (Evans *et al.*, 1987, 1988; Schedlich *et al.*, 1988; Riegman *et al.*, 1989a; Wines *et al.*, 1989, 1991). Sequence comparison of these regions for *KLK1* for the mouse, rat and human genes show two conserved sequence elements at −235 to −214 and −177 to −157. The reverse complement of the latter sequence is similar to a *cis* regulatory element in the rat pancreatic serine proteases suggesting that this may be a candidate sequence for directing pancreatic-specific expression (Wines *et al.*, 1989). An alternative explanation might be that the first 500 bp of the 5′-flanking sequences of these genes do not harbor the elements which confer all the tissue-specific expression patterns of individual *KLK* genes and that other important regulatory elements are present further upstream.

3.3.5   Rat *KLK* Transgenics

In comparison with other genes/gene families, relatively few functional studies have been attempted in this area for any of the *KLK* gene family members. To determine the promoter regions conferring tissue specificity of expression for three rat *KLK* genes (*rKLK1*, *rKLK2* and *rKLK8*), transgenic mice and rats were generated using 5′-flanking DNA of varying lengths from each of these genes with an SV40 T-antigen reporter gene (Southard-Smith *et al.*, 1992). Two *rKLK1* constructs of varying length (1.7 kb 5′ *rKLK1*:Tag and a full-length *rKLK1* transgene including 4.5 kb 5′- and 4.7 kb 3′-flanking regions as well as introns) gave essentially similar patterns of expression. Tumors were found in the pancreas, choroid plexus and pituitary, all sites of known or suspected *rKLK1* expression. Of interest was the localization of the pancreatic tumors to the endocrine islets and not the exocrine acinar cells, the site of tissue kallikrein immunolocalization in other studies. On Northern blot analysis, the Tag transgene mRNA was also expressed in the gastrointestinal tract (ileum, duodenum and colon) and sublingual glands, other known sites of *rKLK1* expression, although no tumors were detected at these sites. Surprisingly, a very low level of expression (at the RT-PCR level) was detected in the submandibular glands and kidney, two primary sites of *rKLK1* expression in the normal rat. A similar pattern of expression was observed in the transgenic mice bearing the complete *rKLK1* gene construct with an essentially correct tissue-specific pattern of expression but inappropriate levels of expression compared to the normal animal. Thus it would seem that the regulatory sequences necessary to direct *rKLK1* expression to many of the known sites are present in the first 500 bp of the *rKLK1* promoter. However, those sequences involved in the direction and regulation of salivary gland and renal expression would appear not even to be present in up to 5 kb of 5'- and/or 3'-flanking DNA.

The two other *rKLK*:TAg constructs had 667 bp (*rKLK2*) and 2.5 kb (*rKLK8*) of 5'-flanking sequence directing expression of the transgenes (Southard-Smith *et al.*, 1992). The normal expression pattern of these two genes is that *rKLK2* would appear to be solely expressed in the salivary glands whereas *rKLK8* is also expressed in the prostate in addition to the classical salivary gland expression pattern indicative of most *KLK* genes. Again the transgenic animals did not have any salivary gland tumors, neither did the *rKLK8* transgenics have prostate tumors. Northern blot and RT-PCR analysis also did not detect Tag mRNA in the salivary glands of any of the 15 *rKLK2* transgenics nor expression in 11 other tissues studied. The *rKLK8* transgenics did have a peculiar phenotype. All were stunted in growth compared to their littermates and had an enlarged pancreas with islet tumors, but the exocrine tissue was normal and two animals had benign bone tumors. RT-PCR analysis showed Tag expression in the brain, kidney, gastrointestinal tract, pancreas and parotid salivary gland, all tissues not normally known to express *rKLK8* and a very low level of prostate expression. Thus, as with the *rKLK1* transgenics, the 0.67–2.5 kb of proximal 5'-flanking sequence of these genes has not been sufficient to direct the correct pattern of tissue-specific expression. To overcome a further concern that regulatory sequences of rat transgenes might not be recognized by the host mouse, the latter *rKLK8* transgene had been introduced into a rat germline and, as stated above, no discernible difference in the salivary gland phenotype was observed between the rat and mouse transgenics.

The conclusion of these authors is that the regulatory regions associated with the expression of individual *rKLK* genes may be dominated by a locus control region conferring salivary gland expression on the whole *rKLK* gene family, in a manner akin to that observed for the β-globin gene family where a locus control region, crucial to the transcriptional regulation of this gene family, is located 10–65 kb upstream of the β-globin locus (Southard-Smith *et al.*, 1992). Consistent with this hypothesis, is the mapping of the *rKLK* gene family to two tightly linked clusters and the identification of a potential regulatory element, an unmethylated CpG island, upstream of the Odd contig and the entire *rKLK* locus (Southard-Smith *et al.*, 1994). The specific element(s) which confer salivary gland expression on these genes either individually or the family in its entirety, however, has not yet been identified. It is of interest to also note that the majority of the genes expressed in the rat ovary (*rKLK1*, *rKLK3*, *rKLK7*, *rKLK8*, *rKLK9*, *rKLK12*) (Clements *et al.*, 1995) are derived from the Odd contig (*rKLK1*, *rKLK3*, *rKLK7*, *rKLK9*) suggesting that perhaps a common tissue-specific regulatory element upstream of this gene cluster may direct the ovarian expression of at least these four *rKLK* genes.

3.3.6 Human *KLK* Transgenics

In humans, a full-length *KLK1* gene construct including 800 bp 5'- and 300 bp 3'-flanking DNA was used to generate transgenic mice. In these animals the transgene was predominantly expressed in the pancreas and to a lesser extent in the spleen, but not salivary gland as detected by radioimmunoassay of human tissue kallikrein (Simson *et al.*, 1994). Although *KLK1* is normally expressed at its highest levels in the pancreas compared to the salivary gland and kidney, the complete lack of expression in the salivary gland suggests that the tissue-specific sequences, which direct *KLK1* expression to that organ, must be located upstream from the first 800 bp of 5'-flanking DNA used in this transgene. Since this construct contained the putative pancreatic tissue-specific element at −177 to −157 (Wines *et al.*, 1989), it is perhaps not surprising that pancreatic expression was dominant. These transgenic mice also showed striking histopathological effects in lymphatic tissues including the spleen, thymus and lymph nodes, although no evidence of transgene expression in these tissues was presented. Similar effects were observed in a transgenic line generated from the *KLK1* gene directed by the human metallothionine promoter, thus the pathological effects presumably arise from overexpression of the *KLK1* gene in these tissues (Simson *et al.*, 1994).

Two studies have examined the prostate tissue specificity of the *KLK3*/PSA promoter. A 652 bp 5'-fragment proximal to the *KLK3* gene was used to direct expression of a mutated *ras* gene specifically to the mouse prostate in the expectation that prostate tumors would develop (Schaffner *et al.*, 1995). Tumors were only found in the salivary glands and gastrointestinal tract of these animals and no transgene expression could be detected, even by RT-PCR, in the prostate gland or lung, liver, spleen and kidney. PSA*ras* transgene expression in the salivary gland tumors was confirmed by *in situ* hybridization. These authors attributed this finding to both species specificity of the human *KLK3*/PSA promoter in not being able to direct prostate-specific expression in the mouse, and a conservation of sequence between the *KLK3*/PSA promoter and other mouse *Klk* genes in this region, which could confer salivary gland and perhaps gastrointestinal expression. Given that little difference in phenotype was observed between the mouse and rat *rKLK* transgenics (Southard-Smith *et al.*, 1992), this is probably not a consideration. It is likely, however, that this 652 bp of the *KLK3*/PSA promoter did not contain the appropriate sequence(s) to direct prostatic expression. Similar to that seen for the *rKLK* transgenics (Southard-Smith *et al.*, 1992), it is clear that these 5'-flanking regions have been taken out of their *in vivo* context and can no longer function to direct normal expression of the relevant *KLK* gene. Given the high conservation of these 5'-flanking regions even between species, and the known salivary gland dominant but otherwise tissue-specific pattern of expression for individual *KLK* genes in all species, it is probable that all *KLK* genes will contain similar sequences, which will direct expression to many tissues, but the dominant tissue-specific regulatory sequences acting as enhancers or silencers will be located further upstream.

In contrast, in the second study, a 620 bp fragment of 5'-flanking DNA of the *KLK3*/PSA gene was able to direct prostate-specific expression in the LNCaP human prostate tumor cell line, a cell line which normally expresses PSA (Pang *et al.*, 1995). However, this construct could not direct expression in the non-PSA-producing human prostate cell lines, DU145 and PC3, or a renal (R11) or a breast (MCF-7) cell line. The transfected construct appeared to inhibit the endogenous promoter competitively suggesting that DNA binding proteins critical to the activation of the *KLK3*/PSA gene are both present in these cells and bind to this 620 bp 5'-flanking region. Clearly if the 620 and 652 bp regions used in these two studies are almost identical, as it appears they are, then a silencer sequence, which negates prostate-specific expression, may be present in the larger DNA fragment approximately 32 bp upstream from the 5' end of the 620 bp fragment.

3.4   TRANSCRIPTIONAL AND POST-TRANSCRIPTIONAL HORMONAL REGULATION

3.4.1   Post-transcriptional Regulation of *KLK1* Expression and Tissue Kallikrein Activity

Historically, the *KLK* gene family of kallikrein-like enzymes is regarded as being regulated by androgens, particularly with regard to salivary gland expression and enzyme activity. In keeping with this observation, early studies suggested androgen-dependent changes in tissue kallikrein immunoreactive or enzyme activity, biosynthesis or mRNA levels in the salivary gland, as well as kidney, although not pancreas (Chao and Margolius, 1983; Miller *et al.*, 1984; Gerald *et al.*, 1986b). Recent studies, using more specific DNA probes, have shown

more conclusively that *KLK1* expression is not regulated by androgens in the salivary gland and kidney (Van Leeuwen *et al.*, 1986, 1987; Clements *et al.*, 1988, 1990a; Penschow *et al.*, 1991b). This is reflected in the essentially equivalent levels of *rKLK1/mKlk1* mRNA in both these glands for male and female animals. Thus, *KLK1* or tissue kallikrein is the one *KLK* gene family member/enzyme in the salivary gland which is not androgen dependent in its expression/activity, nor does it appear to be regulated by androgens in any tissue/species so far studied.

Thyroid hormone has also been widely considered to be a primary regulator of *KLK* gene family expression and/or kallikrein enzyme activity, particularly in the rodent salivary gland (Chao and Margolius, 1983). Again, studies using gene-specific probes suggest that *rKLK1/mKlk1* is the one gene in the *KLK* family which is not thyroid hormone dependent in its expression in the salivary gland (Van Leeuwen *et al.*, 1987; Clements *et al.*, 1990a,b; Penschow *et al.*, 1991b). Similarly, the purported thyroid hormone regulation of renal tissue kallikrein activity or expression has not been substantiated when more gene-specific tools were used.

Another widely studied area is the salt and mineralo/glucocorticoid regulation of renal tissue kallikrein or urinary tissue kallikrein activity. Various studies both in man and rats have shown that chronic salt depletion and mineralocorticoid treatment can alter urinary tissue kallikrein and/or renal kallikrein activity, although the precise mechanisms involved in this regulation are still largely unknown (Margolius *et al.*, 1974; Bonner *et al.*, 1981; Rapp *et al.*, 1982; Marin-Grez *et al.*, 1984; Bascands *et al.*, 1987; Weinberg *et al.*, 1987). Some of these studies in the rat have also been extended to examine *rKLK* mRNA levels or synthesis rates, however, none have conclusively demonstrated real specific changes in *rKLK1* expression or synthesis with mineralocorticoid treatment (Miller *et al.*, 1984, 1985; Fuller *et al.*, 1986). This primarily reflects the use of cDNA probes or polyclonal antibodies, which cross-hybridize or cross-react with several gene family members instead of gene-specific *rKLK1* tools. Of interest is the recent observation that furosemide treatment alters the distribution of *mKlk1* expression in the kidney to include cells in the loop of Henle in the renal outer medulla (Penschow and Coghlan, 1995). The significance of this finding is unclear. Similarly, the definitive regulation of salivary gland *KLK1* expression or tissue kallikrein enzyme activity by mineralocorticoids or salt deprivation has not been established (Miller *et al.*, 1984, 1985).

Glucocorticosteroids, which regulate a wide range of inflammatory responses, have been implicated in the regulation of tissue kallikrein activity. Glucocorticosteroids appear to decrease tissue kallikrein biosynthesis in the rat kidney, although this study has not been confirmed at the level of gene expression (Jaffa *et al.*, 1990). Glucocorticosteroid treatment decreased *KLK1* gene expression and tissue kallikrein biosynthesis in a rat pancreatic acinar cell-line (Rosewicz *et al.*, 1991). Cortisol administration to adrenalectomized rats increased rat salivary gland tissue kallikrein activity and altered *rKLK* mRNA levels in general, but this effect on *rKLK1* gene expression, specifically, has not been reported (Chao and Margolius, 1983; Gerald *et al.*, 1986b). The potential effects of glucocorticosteroids on tissue kallikrein activity and/or *KLK1* gene expression in any other tissue or cell type, such as the lung, nasal mucosa or infiltrating neutrophils, which have reported tissue kallikrein-associated inflammatory responses, have not yet been evaluated.

Insulin also may regulate renal kallikrein synthesis and activity. Streptozotocin-induced diabetic rats had lower *rKLK1* mRNA levels and immunoreactive tissue kallikrein levels than the normal controls, a situation which could be reversed by insulin administration (Mayfield *et al.*, 1986; Jaffa *et al.*, 1992). Similarly, tissue kallikrein synthesis and prokallikrein activation was lowered in these rats (Jaffa *et al.*, 1987). The mechanism of this regulatory event is unclear, although it is interesting to note that tissue kallikrein is able to cleave proinsulin *in vitro* perhaps suggesting a more complex interrelationship between these two proteins (Ole-Moiyoi *et al.*, 1979a).

The regulation of rat anterior pituitary *rKLK1* gene expression and tissue kallikrein activity by estrogen has been extensively studied. Clearly, estrogen increases *rKLK1* mRNA levels and tissue kallikrein enzyme and immunoreactivity both in the normal rat anterior pituitary and an estrogen-induced rat pituitary tumor model (Clements *et al.*, 1986, 1989; Powers, 1986, 1987; Chao *et al.*, 1987; Hatala and Powers, 1987, 1988a; Fuller *et al.*, 1988). In keeping with estrogen dependence of expression, anterior pituitary levels of *rKLK1* mRNA and tissue kallikrein activity are substantially higher in females than in males; similarly changes in kallikrein enzyme activity and/or *rKLK1* mRNA levels through puberty and pregnancy also appear to reflect this estrogen dependence (Powers and Nasjletti, 1984; Fuller *et al.*, 1985; Clements *et al.*, 1986, 1989; Powers and Westlin, 1987; Hatala and Powers, 1988b). The pattern of *KLK1* expression in the human endometrium across the menstrual cycle with increased *KLK1* mRNA levels around mid-cycle at a time of high circulating estrogen levels is also suggestive of estrogen regulation (Clements *et al.*, 1994). In contrast, estrogens do not appear to affect levels of *rKLK1* expression in the rat salivary gland (Clements *et al.*, 1990c). *rKLK1* is also expressed in the rat neurointermediate lobe, with higher levels found in the male compared with the female gland (Powers and Nasjletti, 1983; Clements *et al.*, 1986). Rat pituitary kallikrein activity and *rKLK1* expression in both these pituitary compartments is also regulated by dopamine. The administration of the dopaminergic agents, haloperidol and bromocriptine, variously increase

or decrease both female anterior pituitary *rKLK1*/tissue kallikrein and that found in the male neurointermediate lobe (Powers, 1985; Powers and Hatala, 1986; Pritchett and Roberts, 1987; Fuller *et al.*, 1988). The regulation of tissue kallikrein activity or *KLK1* expression by other agents, particularly those implicated in the regulation of inflammatory responses, such as the cytokines, has not been reported.

3.4.2 Hormonal Regulation of *KLK1* at the Transcriptional Level

The hormonal and other regulatory mechanisms, which control *KLK1* expression at the transcriptional level, have not been studied in detail. The 5'-flanking sequence of the human *KLK1* gene contains several sequence motifs similar in sequence to that previously reported for consensus estrogen, progestin, glucocorticoid or cAMP response elements (Murray *et al.*, 1990). These putative elements, however, have not been functionally tested. As stated above, the 5'-flanking regions of the mouse, rat and human *KLK1* genes are quite well conserved with generally minor single- or double-base changes apparent between the rodent genes, but larger differences between the rodent and human genes. If these putative hormonal regulatory sequences in the human *KLK1* gene are compared with the aligned rodent genes, only one of these sequences, the putative cAMP-responsive element, has a motif (at −235 to −214) that is well conserved between these three genes (Van Leeuwen *et al.*, 1986; Evans *et al.*, 1988; Wines *et al.*, 1989). This latter region contains the fully conserved CCCCACCC sequence, a motif which has been identified as an AP-2 binding site, a transcription factor implicated in the regulation of many genes. Further studies are necessary to determine whether the hormonal changes observed above for *rKLK1* expression/tissue kallikrein activity are transcriptionally regulated and mediated by any of these putative elements.

Some of the above hormonal responses recorded at the post-transcriptional level for *rKLK1*, however, are not thought to be a direct result of primary transcriptional hormone activation of *rKLK1* expression but secondary to other cellular changes generated by the hormone action. Thus, the relatively slow induction of rat pituitary *rKLK1* expression and increased tissue kallikrein activity by estrogen over 2–4 days is suggestive of such a secondary event (Clements *et al.*, 1986, 1989; Hatala and Powers, 1987, 1988a; Clements and Fuller, 1990). In contrast, the dopaminergic effects on rat pituitary *rKLK1* expression was clearly shown to be transcriptionally regulated by nuclear transcriptional run-on assay. The induction of *rKLK1* gene transcription was rapid in the neuro-intermediate lobe, within 1 h, in comparison to the anterior pituitary where 16 h elapsed before *rKLK1* transcription was activated (Pritchett and Roberts, 1987). The further characterization of the molecular sequences and transcription factors/DNA binding proteins, which regulate this event, has not been reported. Whether the above putative cAMP regulatory element and/or AP-2 binding site mentioned above will be shown to be involved in this response is yet to be elucidated.

One other study has explored the potential thyroid hormone receptor-binding capacity of 5'-flanking regions of the *mKlk1*/tissue kallikrein and *mKlk3* genes, two mouse *Klk* genes differentially regulated by thyroid hormone at the mRNA level (Barlow *et al.*, 1989). Thyroid hormone receptor binding was observed for a construct containing the first 776 bp of the *mKlk1* 5'-flanking region whereas no discernible binding was seen with a comparable region of the *mKlk3* gene. These results were somewhat surprising, since *mKlk1* mRNA levels are unresponsive to thyroid hormone treatment compared with *mKlk3* expression, which is upregulated by thyroid hormone, albeit with a slow time course of induction (Van Leeuwen *et al.*, 1987). This study, however, did not determine whether transcriptional activation occurred in concert with the preferential DNA binding capacity of the thyroid hormone receptor for the *mKlk1* gene.

3.4.3 Post-transcriptional Regulation of Expression and Enzyme Activity of Other *KLK* Genes

The two hormones which are considered the primary regulators of gene expression and/or enzyme activity for the remainder of the *KLK* genes in all species so far studied are androgen and thyroid hormone. In terms of salivary gland expression, it has been a hallmark of the *KLK* gene families in both the rat and mouse that all gene family members, except for *rKLK1/mKlk1*, are androgen and thyroid-hormone dependent in their expression (Evans and Richards, 1985; Drinkwater *et al.*, 1987, 1988b; Van Leeuwen *et al.*, 1987; Clements *et al.*, 1988, 1990a,b; Penschow *et al.*, 1991a,b). Thus, the salivary gland mRNA levels of all these *KLK* genes are higher in males than in females and castration depletes, while androgen administration to castrate animals restores *rKLK/mKlk* gene expression. Thyroid hormone administration to both mice and rats similarly increases both *KLK* gene expression. The enzymes encoded by some of these genes, α-NGF, EGF-BP and tonin, are similarly regulated by androgens and thyroid hormone (Bhoola *et al.*, 1973; Chao and Margolius, 1983; Isackson and Bradshaw, 1984; Blaber *et al.*, 1987).

The other organ in which androgen dependence of *KLK* gene expression and/or enzyme activity predominates is the prostate. In the rat, human, dog and guinea-pig, one to two *KLK* genes have been described for each species (*rKLK8*, *rKLK9*, *KLK2*, *KLK3*/PSA, *dKLK2*, *gpKLK2*) and all are regulated by androgens both at the level of gene expression and enzyme activity. As noted for the salivary gland, *KLK* mRNA and enzyme

levels clearly decline rapidly following castration but are quickly restored on androgen treatment (Chapdelaine *et al.*, 1984, 1988a,c; Lazure *et al.*, 1984; Clements *et al.*, 1988; Winderickx *et al.*, 1989; Riegman *et al.*, 1991a; Montgomery *et al.*, 1992; Young *et al.*, 1992; Lee *et al.*, 1995). Thyroid hormone and the growth factors, EGF, and transforming growth factors (TGF) α and β, also affect *KLK2* and *KLK3* expression and/or PSA secretion from the human prostate tumor LNCaP cell line, perhaps in part by regulating androgen receptor expression (Esquenet *et al.*, 1995; Henttu and Vihko, 1993). In this cell line, tumor-promoting phorbol ester (TPA) also decreases the androgen induction of PSA (Andrews *et al.*, 1992).

The hormonal regulation of gene expression or enzyme activity of these and other *KLK* genes or enzymes recently described in other tissues, such as the human uterus, breast and pituitary, the rat ovary, testis, kidney, heart and vasculature, and mouse Leydig cells (see above) remains to be established.

3.4.4 Hormonal Regulation of Other *KLK* Genes at the Transcriptional Level

The two *KLK* genes for which transcriptional activation by androgens has been extensively studied are the two human genes, *KLK2* and *KLK3*. Nuclear transcriptional run-on assays with nuclei isolated from LNCaP cells confirmed the rapid transcriptional activation (within 6 h) of *KLK2* and *KLK3* expression by the human androgen receptor (Riegman *et al.*, 1991b; Wolf *et al.*, 1992; Young *et al.*, 1992). Deoxyribonuclease (DNase) footprint analysis of the *KLK3* promoter with LNCaP cell nuclear extracts and chloramphenicol acetyltransferase (CAT) reporter gene functional assays of *KLK3*/PSA promoter:androgen receptor activity have identified a region at −170 to −156 bp of the *KLK3* promoter critical to the androgen receptor activation of this gene (Riegman *et al.*, 1991b). This sequence is similar to the reverse complement of the consensus sequence for binding of the glucocorticosteroid and progesterone receptors. Mutation of this sequence almost completely abolished the transcriptional activation of the *KLK3* gene by androgens in the CPAT reporter gene constructs further demonstrating the importance of this sequence in androgen receptor induction of *KLK3*/PSA gene expression. In addition, a similar sequence, which presumably regulates the androgen induction of *KLK2* gene expression, has been identified in the *KLK2* promoter (Murtha *et al.*, 1993). More recently, the 5'-flanking region of the *dKLK2* gene has been similarly characterized (Gauthier *et al.*, 1993b). Although a putative androgen response element with 73% homology to that found in the *KLK2* and *KLK3* genes was identified in a similar region at −172 to −148 bp of the *dKLK2* gene, this sequence did not appear to confer androgen-inducible gene expression in the transient transfection system used (Dube *et al.*, 1995). These authors have concluded that the transcriptional activation of *dKLK2* gene expression by androgens occurs via a different mechanism than that found for the human *KLK* genes.

The molecular mechanisms by which thyroid hormone and androgens regulate salivary gland *KLK* expression in rodents are yet to be determined, although previous studies in the mouse suggest that such changes may not be the result of a primary transcriptional hormonal activation but be secondary to other hormonally induced cellular events (Van Leeuwen *et al.*, 1987). This observation is further supported by the lack of thyroid hormone receptor binding sites in the proximal 5'-flanking region of the *mKlk3* gene, when expression of this gene in the mouse salivary gland is upregulated by thyroid hormone treatment (Barlow *et al.*, 1989). An alternative explanation might be that a limited region of the *mKlk3* promoter was tested. Similarly, the cellular mechanisms determining growth factor-initiated *KLK2/KLK3* expression in the human prostate tumor LNCaP cell line are also yet to be determined.

3.5 PREDICTED SUBSTRATE SPECIFICITY AND FUNCTION

3.5.1 *KLK1*/Tissue Kallikrein: Function via Kinin Generation

The primary substrate of tissue kallikrein is LMWK from which the multifunctional peptides, Lys-BK or BK are generated. Thus, the predominant role of tissue kallikrein in those tissues where it is expressed, presumably, is via the actions of kinins. Below, the roles of the tissue kallikrein–kinin system in a variety of (patho-)physiological conditions are discussed. (The roles of the kinins themselves in myriad inflammatory diseases are reviewed in detail throughout this volume). As noted above, tissue kallikrein has been localized to many tissues and specific cell types. The kidney, pancreas and heart are clearly sites of *KLK1* expression indicating a local tissue kallikrein–kinin system is active. Other tissues/cell types have still to be verified as sites of *KLK1* expression and not just as harboring sequestered tissue kallikrein.

The involvement of the tissue kallikrein–kinin system in the renal and cardiovascular systems has been studied extensively. These studies have primarily examined the role of tissue kallikrein in the local regulation of blood flow, vascular permeability and Na$^+$/water homeostasis (Roman *et al.*, 1988; Ishiguro *et al.*, 1995; Margolius, 1995; Saitoh *et al.*, 1995). Tissue kallikrein has thus been implicated in hypertension and the inflammatory conditions of cardiovascular ischemia (see Chapter 19), renal nephritis and diabetic renal disease (Margolius *et al.*, 1974; Horwitz *et al.*, 1978; Holland *et al.*, 1980; Mayfield *et al.*, 1984; Ader *et al.*, 1985; Jaffa *et al.*, 1987, 1992; Schölkens *et al.*, 1987; Harvey *et al.*, 1990;

Cumming *et al.*, 1994; Majima and Katori, 1995). Relatively little is known of the molecular mechanisms involved in these events, although tissue kallikrein activity is altered by Na$^+$, K$^+$, glucocorticosteroids, mineralocorticosteroids and insulin administration (see above) or β$_2$-adrenoceptor stimulation (Girolami *et al.*, 1990). No studies have addressed the involvement of these agents or other hormonally induced transcription factors in the activation of *KLK* gene expression. Some, however, have investigated the potential association of the *KLK* locus in genetically acquired forms of hypertension both in man and animal models (Berry *et al.*, 1989; Woodley-Miller *et al.*, 1989). The regulatory regions in the promoter of the *rKLK1* gene, which may lead to hypertension in the rat, have been defined (Wang *et al.*, 1995a), and the therapeutic advantage of *KLK1* gene treatment explored in spontaneous hypertensive rats (Wang *et al.*, 1995b; Xiong *et al.*, 1995). Of interest is the recent mapping of a gene associated with progressive familial heart block to chromosome 19q13, close to the *KLK1*/tissue kallikrein locus (Brink *et al.*, 1995).

The kallikrein–kinin system is clearly involved in pancreatitis (see Chapter 12). One factor involved in this action may be the glucocorticosteroids. The regulation of *KLK* biosynthesis in pancreatic cells by this group of hormones has been clearly established (Rosewicz *et al.*, 1991). Whether this regulation is mediated by the putative glucocorticosteroid response element described in the 5'-flanking region of the *KLK1* gene is yet to be determined (Murray *et al.*, 1990). In other tissues where the kallikrein–kinin system has been implicated in inflammation, elucidation of the molecular events regulating tissue kallikrein activity awaits the definitive demonstration of *KLK1* gene expression at these sites (Curd *et al.*, 1980; Suzuki *et al.*, 1987; Hibino *et al.*, 1988b; Baumgarten *et al.*, 1989; Burch *et al.*, 1989; Selwyn *et al.*, 1989; Christiansen *et al.*, 1992; Rahman *et al.*, 1995). An important aspect in the cascade of events involved in the inflammatory response is neutrophil invasion, and it will be important to determine if *KLK1* is expressed in these cells and the molecular events regulating such expression.

KLK1/tissue kallikrein may also be involved in inflammatory responses or conditions of the urogenital tract. *KLK1* is expressed in the human endometrium and the levels of *KLK1* mRNA change across the menstrual cycle with a profile suggestive of estrogen regulation as does urinary kallikrein activity (Albano *et al.*, 1994; Clements *et al.*, 1994). Similar changes were observed for tissue kallikrein activity in the rat uterus across the estrous cycle and during pregnancy (Valdes *et al.*, 1993; Corthorn and Valdes, 1994; Brann *et al.*, 1995). The cyclic proliferative preparation of the human endometrium for embryo implantation has been described as an inflammatory-like response. Thus, the implication that *KLK1*/tissue kallikrein is involved in these changes suggests a role in physiological inflammation-like responses. Similarly, tissue kallikrein activity is altered in the gonadotropin-stimulated rat ovarian model of ovulation induction (Espey *et al.*, 1986; Gao *et al.*, 1992). These and other authors have suggested an involvement of tissue kallikrein, via BK, in responses such as the hyperemia and edema typical of folliculogenesis, which occurs at ovulation. It will be of interest to determine whether these events require the transcriptional regulation of *KLK1* gene expression by estrogen and/or gonadotropins via the estrogen or cAMP response elements suggested to be present in the proximal 5'-flanking region of the *KLK1* gene. Urinary kallikrein activity is also elevated in women with the painful inflammatory condition of the bladder, interstitial cystitis (Zuraw *et al.*, 1994). Tissue kallikrein also has been implicated in male reproductive function via enhancement of sperm motility and testicular blood flow (Saitoh and Kumamoto, 1988; Schill *et al.*, 1989; Haidl and Schill, 1993), although the mechanism of action is unknown.

An increasing important role is emerging for tissue kallikrein, via the action of BK, in tumor cells. BK may play a role via cell proliferation, vascular permeability and the regulation of tumor blood flow (Matsumara *et al.*, 1988; Roberts and Gullick, 1989). Localization of tissue kallikrein to prolactinomas and a rat pancreatic acinar cell carcinoma cell line and *KLK1* expression to human pituitary tumors suggests an important role in tumor growth and development (Berg *et al.*, 1985; Jones *et al.*, 1989, 1990; Clements *et al.*, 1996). (**Editor's note**: The roles of the kinins as growth factors, and their involvement in angiogenesis are discussed in detail in Chapter 18 of this volume.)

3.5.2 *KLK1*/Tissue Kallikrein: Function via Other Substrates

Tissue kallikrein processes a variety of substrates *in vitro*, including the matrix metalloproteinases, procollagenase and progelatinase, proinsulin, prorenin, angiotensinogen, prolactin, atrial natriuretic factor, vasoactive intestinal peptide, an enkephalin-containing precursor, BAM 22P, and low-density lipoprotein (Mutt and Said, 1974; Eeckhout and Vaes, 1977; Sealey *et al.*, 1978; Derkx *et al.*, 1979; Ole-Moiyoi *et al.*, 1979a; Prado *et al.*, 1983; Briggs *et al.*, 1984; Carden *et al.*, 1984; Currie *et al.*, 1984; Ideishi *et al.*, 1987; Tschesche *et al.*, 1989; Powers and Hatala, 1990). It is not clear whether this reflects a true multisubstrate capability of tissue kallikrein *in vivo* or whether the kallikrein preparations were contaminated with other enzymes. Since most of the *KLK* gene family members have now been cloned, the potential availability of recombinant forms of these enzymes will help.

The potential involvement of tissue kallikrein in the activation of collagenase and gelatinase from their latent pro-forms is especially important in terms of inflammatory processes (Tschesche *et al.*, 1989; Desrivieres *et al.*, 1993; Menashi *et al.*, 1994). These matrix metalloproteinases are clearly critical in facilitating matrix barrier penetration, endothelial cell migration, leukocyte infiltration and tissue remodeling in many tissues and (patho)-physiological circumstances. In one study, tissue kallikrein was immunolocalized to the intima of porcine aortic vessel walls in close proximity to the endothelial cell source of the putative matrix metalloproteinase substrates (Desrivieres *et al.*, 1993). Such colocalization and other studies will undoubtedly provide further confirmation of the role of tissue kallikrein as an important regulator of matrix invasion and tissue remodeling via the activation of matrix metalloproteinases.

3.5.3 Potential Function of Other *KLK* Gene Products

Since many of the *KLK* gene products have not yet been characterized at the protein level, a physiological function is yet to be assigned to a large number of these genes. Those which have been identified include the mouse growth factor processing enzymes, γ-NGF, β-NGF endopeptidase and EGF-BP, and the human PSA, an enzyme which degrades insulin-like growth factor binding protein-3 (IGFBP-3), thus regulating the bioavailability of insulin-like growth factor (IGF-1) (Cohen *et al.*, 1992). Others include enzymes which generate (poly)peptides with a vasoactive response such as γ-renin and PRECE in the mouse, and tonin, submandibular enzymatic vasoconstrictor (SEV) and T-kininogenase in the rat. The role of these enzymes in inflammatory responses clearly must relate to the action of the generated growth factor or vasoactive peptide.

In addition to its role in the regulation of IGF-1 bioavailability (Peehl, 1995), the *KLK3* gene product, PSA, also degrades the abundant seminal vesicle protein, seminogelin, a process which serves to assist in the liquefaction of the seminal clot (Lilja *et al.*, 1987). More recently, PSA has also been shown to degrade the extracellular matrix proteins, laminin and fibronectin (Webber *et al.*, 1995). These authors suggest this action of PSA facilitates invasion by prostate cancer cells in the neoplastic prostate and the later metastatic phases of this disease. Clearly, a role for PSA could also be speculated to extend to matrix penetration/invasion and tissue remodeling, as for the matrix metalloproteinases, in a variety of inflammatory conditions.

The rat *KLK* enzyme, T-kininogenase, releases T-kinin from the substrate, T-kininogen, a protein which has been demonstrated to play a role in the acute-phase response in the rat. In contrast to the clear upregulation of T-kininogen gene expression and protein levels during the acute-phase response, the levels of T-kininogenase, which is encoded by the *rKLK10* gene, are unaltered (Baussant *et al.*, 1988; Chao *et al.*, 1988, 1989). Thus, it is not clear whether the generation of T-kinin is the primary effector in this response or whether

T-kininogen is primarily acting in its other role as a cysteine protease inhibitor (Moreau *et al.*, 1986; Gauthier *et al.*, 1988).

4. The Complex Interactions of KLK Enzymes, Their Inhibitors and Substrates in Inflammation

As will be clear from other chapters on the kinins, kininases and kinin receptors, the regulation of the expression and function of the kallikrein (*KLK*) gene family of enzymes cannot be regarded in isolation as we view the overall role of the kallikrein–kinin system in inflammation. A further aspect of this equation are the endogenous inhibitors of the tissue kallikreins. A specific endogenous inhibitor of tissue kallikrein is kallistatin and it is expressed primarily in the liver (Chao *et al.*, 1986). Rat and human kallistatins have been characterized at both the gene and protein level, and are considered to be members of the serine protease inhibitor (serpin) superfamily (Chao *et al.*, 1986, 1990; Chai *et al.*, 1994; Chao and Chao, 1995). Other members of this family include protein C inhibitor, α_1-antitrypsin and α_1-antichymotrypsin, inhibitors which also bind kallikrein and PSA, another human *KLK* enzyme (Espana *et al.*, 1995; Geiger *et al.*, 1981; Lilja *et al.*, 1991). Kallistatin is clearly involved in the regulation of blood pressure since the somatic delivery of this inhibitor to transgenic mice over-expressing tissue kallikrein reverses the hypotensive phenotype of these animals (Ma *et al.*, 1995). That kallistatin may play a role in the regulation of the kallikrein–kinin system in inflammation is suggested by the finding that liver kallistatin mRNA levels are reduced after acute inflammation, presumably allowing a more active tissue kallikrein–kinin system to proceed (Chao *et al.*, 1990). Another important aspect is the activation of tissue kallikrein from its inactive zymogen form. Although activation can be accomplished by trypsin, thermolysin and other rat kallikreins *in vitro*, it is not clear what enzymes perform this function *in vivo*, or the mechanisms which trigger or regulate this activation (Fiedler, 1979; Noda *et al.*, 1985; Takada *et al.*, 1985; Kamada *et al.*, 1990; Takenobu *et al.*, 1990).

A further magnitude to the complexity of the kallikrein–kinin system is apparent when one considers that the factors which regulate the expression and action of the various components may act at different levels, either singularly to initiate a cascade of regulatory events or simultaneously at several levels. Glucocorticosteroids increase the biosynthesis or activity of *KLK1*, at least in the pancreas and kidney, although whether this is a general effect of glucocorticosteroids on *KLK1*/tissue kallikrein has not been examined in other tissues (Jaffa *et al.*, 1990; Rosewicz *et al.*, 1991). Glucocorticosteroids also increase the synthesis of T-kininogen by rat hepatocytes in culture, although *in vivo* circulating levels of T-kininogen fall with treatment with dexamethasone (Baussant *et al.*, 1988; Howard *et al.*, 1990). The effects of this steroid on kinin receptor, kininase or kallistatin expression or action, however, are less clear. Similarly, there are few studies to date which have examined the effects of cytokines on the action of the kallikrein–kinin system. None have looked specifically at *KLK1*/tissue kallikrein regulation. The cytokines, interferon and interleukin-6, affect hepatic kininogen expression, the latter possibly via interleukin-6 response elements identified in the promotor of that gene (Chen and Liao, 1993; Ito *et al.*, 1988). The effects of these inflammatory agents on kininogen action, however, may serve to activate the cysteine protease inhibitor function of the kininogens rather than to serve as a substrate for tissue kallikrein in the generation of BK. BK can also stimulate tumor necrosis factor and interleukin-1 release from macrophages (Tiffany and Burch, 1989) suggesting a regulatory feedback event may occur.

As noted at the outset of this review, very little is known of the molecular events that regulate the expression of *KLK1*/tissue kallikrein or other *KLK* genes/enzymes in inflammatory responses. The involvement of tissue kallikrein, via the effects of BK, in inflammatory conditions are described in other chapters. In this chapter, an overview of the molecular organization, tissue-specific expression and regulation of the tissue kallikrein (*KLK*) gene families has been provided. Clearly, there is still considerable work to be done in order to understand fully the molecular mechanisms controlling the expression of not only *KLK1*/tissue kallikrein, but other *KLK* genes/enzymes and the kallikrein–kinin system as a whole in inflammation, and indeed other (patho)-physiological conditions where they have been implicated.

5. References

Ader, J.-L., Pollock, D.M., Butterfield, M.I. and Arendshorst, W.J. (1985). Abnormalities in kallikrein excretion in spontaneously hypertensive rats. Am. J. Physiol. 248, F396–F403.

Akiyama, K., Nakamura, T., Iwanaga, S. and Hara, M. (1987). The chymotrypsin-like activity of human prostate-specific antigen, γ-seminoprotein. FEBS Lett. 225, 168–172.

Albano, J.D., Campbell, S.K., Farrer, A. and Millar, J.G. (1994). Gender differences in urinary kallikrein excretion in man: variation throughout the menstrual cycle. Clin. Sci. 86, 227–231.

Andrews, P.E., Young, C.Y.-F., Montgomery, B.T. and Tindall, D.J. (1992). Tumor-promoting phorbol ester down-regulates the androgen induction of prostate-specific antigen in a human prostatic adenocarcinoma cell line. Cancer Res. 52, 1525–1529.

Asakai, R., Davie, E.W. and Chung, D.W. (1987). Organization of the gene for human Factor XI. Biochemistry 26, 7221–7228.

Ashley, P.L. and MacDonald, R.J. (1985a). Kallikrein-related mRNAs of the rat submaxillary gland: nucleotide sequences of four distinct types including tonin. Biochemistry 24, 4512–4519.

Ashley, P.L. and MacDonald, R.J. (1985b). Tissue-specific expression of kallikrein-related genes in the rat. Biochemistry 24, 4520–4525.

Baird, A.W., Miller, D.H., Schwartz, D.A. and Margolius, H.S. (1991). Enhancement of kallikrein production and kinin sensitivity in T84 cells by growth in nude mouse. Am. J. Physiol. 261, C822–C827.

Baker, A.R. and Shine, J. (1985). Human kidney kallikrein: cDNA cloning and sequence analysis. DNA 4, 445–450.

Barajas, L., Powers, K., Carretero, O.A., Scicli, A.G. and Ingami, T. (1986). Immunocytochemical localization of renin and kallikrein in the rat renal cortex. Kidney Int. 29, 965–970.

Barlow, J.W., Raggatt, L.E., Drinkwater, C.C., Lyons, I.G. and Richards, R.I. (1989). Differential binding of thyroid hormone receptors to mouse glandular kallikrein gene promotors: evidence for multiple binding regions in the mGK-6 gene. J. Mol. Endocrinol. 3, 79–84.

Bascands, J.L., Girolami, J.P., Pecher, C., Moatti, J.P., Manuel, Y. and Suc, J.M. (1987). Compared effects of a low and high sodium diet on the renal and urinary concentration of kallikrein in normal rats. J. Hypertens. 5, 311–315.

Baumgarten, C.R., Nichols, R.C., Naclerio, R.M. and Proud, D. (1986). Concentrations of glandular kallikrein in human nasal secretion during experimentally induced allergic rhinitis. J. Immunol. 137, 1323–1328.

Baumgarten, C.R., Scwarting, R. and Kunkel, G. (1989). Localization of glandular kallikrein in nasal mucosa of allergic and nonallergic individuals. Adv. Exp. Med. Biol. 247B, 523–528.

Baussant, T., Michaud, A., Bouhnik, J., Savoie, F., Alhenc-Gelas, F. and Corvol, P. (1988). Effect of dexamethasone on kininogen production by a rat hepatoma cell line. Biochem. Biophys. Res. Commun. 154, 1160–1166.

Beaubien, G., Rosinski-Chupin, I., Mattei, M.G., Mbikay, M., Chretien, M. and Seidah, N.G. (1991). Gene structure and chromosomal localization of plasma kallikrein. Biochemistry 30, 1628–1635.

Bell, R.A. and Fahnestock, M. (1988). Sequence and comparative analysis of three cDNAs from Mastomys natalensis, an African rat. J. Cell Biol. 107, 615a.

Berg, T., Johansen, L., Bergundhaugen, H., Hansen, L.J., Reddy, J.K. and Poulsen, K. (1985). Demonstration of kallikrein in a rat pancreatic acinar cell carcinoma. Cancer Res. 45, 226–234.

Berg, T., Holck, M. and Johansen, L. (1987). Isolation, characterization and localization of antigen γ, a serine proteinase of the "kallikrein-family" in the rat submandibular gland. Biol. Chem. Hoppe-Seyler 368, 1455–1467.

Berg, T., Bradshaw, R.A., Carretero, O.A., Chao, J., Chao, L., Clements, J.A., Fahnestock, M., Fritz, H., Gauthier, F., MacDonald, R.J., Margolius, H.S., Morris, B.J., Richards, R.I. and Scicli, A.G. (1992a). A common nomenclature for members of the tissue (glandular) kallikrein gene families. In "Recent Progress on Kinins" (eds H. Fritz, W. Müller-Esterl, M. Jochum, A. Roscher and K. Luppertz), Agents and Actions, Vol. 38/I, pp. 19–25. Birkhauser Verlag, Basel.

Berg, T., Wassdal, I. and Sletten, K. (1992b). Immuno-histochemical localization of rat submandibular gland esterase B (homologous to the RSKG-7 kallikrein gene) in relation to other serine proteases of the kallikrein family. J. Histochem. Cytochem. 40, 83–92.

Berg, T., Schoyen, H., Wassdal, I., Hull, R., Gerskowitch, V.P. and Toft, K. (1992c). Characterization of a new kallikrein-like enzyme (KLP-S3) of the rat submandibular gland. Biochem. J. 281, 819–828.

Berry, T.D., Hasstedt, S.J., Hunt, S.C., Wu, L.L., Smith, J.B., Ash, K.O., Kuida, H. and Williams, R.R. (1989). A gene for high urinary kallikrein may protect against hypertension in Utah kindreds. Hypertension 13, 3–8.

Bhoola, K.D. and Dieppe, P.A. (1991). Kinins: inflammation-signalling peptides in joint disease. Eur. J. Rheumatol. Inflamm. 11, 66–75.

Bhoola, K.D. and Dorey, G. (1971). Kallikrein trypsin-like proteases and amylase in mammalian submaxillary glands. Br. J. Pharmacol. 43, 784–793.

Bhoola, K.D., Dorey, G. and Jones, C.W. (1973). The influence of androgens on enzymes (chymotrypsin- and trypsin-like proteases, renin, kallikrein and amylase) and on cellular structure of the mouse submaxillary gland. J. Physiol. 235, 503–522.

Bhoola, K.D., Figueroa, C.D. and Worthy, K. (1992). Bioregulation of kinins: kallikreins, kininogens, and kininases. Pharmacol. Rev. 44, 1–80.

Blaber, M.I., Isackson, P.J. and Bradshaw, R.A. (1987). A complete cDNA sequence for the major epidermal growth factor binding protein in the male mouse submandibular gland. Biochemistry 26, 6742–6749.

Bode, W., Chen, Z., Bartels, K., Kutzbach, C., Schmidt-Kastner, G. and Bartunik, H. (1983). Refined 2A X-ray crystal structure of porcine pancreatic kallikrein A, a specific trypsin-like serine protease. J. Mol. Biol. 164, 237–282.

Bonner, G., Autenrieth, R., Marin-Grez, M., Rascher, W. and Gross, F. (1981). Effects of sodium loading, deoxycortico-sterone acetate and corticosterone on urinary kallikrein excretion. Horm. Res. 14, 87–94.

Boucher, R., Asselin, J. and Genest, J. (1974). A new enzyme leading to the direct formation of angiotensin II. Circ. Res. 34/35 (Suppl I), 203–209.

Bowcock, A.M., Fahnestock, M., Goslin, K. and Shooter, E.M. (1988). The NGF and kallikrein genes of mouse, the African rat Mastomys natalensis and man: their distribution and mode of expression in the salivary gland. Mol. Brain Res. 3, 165–172.

Brady, J.M. and MacDonald, R.J. (1990). The expression of two kallikrein gene family members in the rat kidney. Arch. Biochem. Biophys. 278, 342–349.

Brady, J.M., Wines, D.R. and MacDonald, R.J. (1989). Expression of two kallikrein gene family members in the rat prostate. Biochemistry 28, 5203–5210.

Brann, D.W., Greenbaum, L., Mahesh, V.B. and Gao, X. (1995). Changes in kininogens and kallikrein in the plasma, brain and uterus during pregnancy in the rat. Endocrinology 136, 46–51.

Briggs, J., Marin-Grez, M., Steipe, B., Schubert, G. and Schnermann, J. (1984). Inactivation of atrial natriuretic substance by kallikrein. Am. J. Physiol. 247, F480–F484.

Brink, P.A., Ferreira, A., Moolman, J.C., Weymar, H.W., Vandermerwe, P.L. and Corfield, V.A. (1995). Gene for progressive familial heart block type I maps to chromosome 19q13. Circulation 91, 1633–1640.

Burch, R.M., Connor, J.R. and Tiffany, C.W. (1989). The kallikrein–kinin system in chronic inflammation. Agents Actions 27, 258–260.

Carbini, L.A., Scicli, A.G. and Carretero, O.A. (1993). The molecular biology of the kallikrein-kinin system: III. The human kallikrein gene family and kallikrein substrate. J. Hypertens. 11, 893–898.

Carden, A.C., Witt, K.R., Chao, J., Margolius, H.S., Donaldson, V.H. and Jackson, R.L. (1984). Degradation of apolipoprotein B-100 of human plasma low density lipoprotein by tissue and plasma kallikrein. J. Biol. Chem. 259, 8522–8526.

Carretero, O.A., Carbini, L.A. and Scicli, A.G. (1993). The molecular biology of the kallikrein–kinin system: I. General description, nomenclature and the mouse gene family. J. Hypertens. 11, 693–697.

Chai, K.X., Ward, D.C., Chao, J. and Chao, L. (1994). Molecular cloning, sequence analysis, and chromosomal localization of the human protease inhibitor 4 (kallistatin) gene (PI4). Genomics 23, 370–378.

Chao, J. (1983). Purification and characterization of rat esterase A, a plasminogen activator. J. Biol. Chem. 258, 4434–4439.

Chao, J. and Chao, L. (1987). Identification and expression of kallikrein gene family in rat submandibular and prostate glands using monoclonal antibodies as specific probes. Biochim. Biophys. Acta 910, 233–239.

Chao, J. and Chao, L. (1995). Biochemistry, regulation and potential function of kallikistatin. Biol. Chem. Hoppe-Seyler 376, 705–713.

Chao, J. and Margolius, H.S. (1983). Differential effects of testosterone, thyroxine and cortisol on rat submandibular gland versus renal kallikrein. Endocrinology 113, 2221–2225.

Chao, J., Buse, J.B., Shimamoto, K. and Margolius, H.S. (1981). Kallikrein-induced uterine contraction independent of kinin formation. Proc. Natl Acad. Sci. USA 78, 6154–6156.

Chao, J., Chao, L. and Margolius, H.S. (1984). Isolation of tissue kallikrein in rat spleen by monoclonal antibody-affinity chromatography. Biochim. Biophys. Acta 801, 244–249.

Chao, J., Tillman, D.M., Wang, M., Margolius, H.S., Chao, L. and Chapman, I.D. (1986). Identification of a new tissue kallikrein-binding protein. Biochem J. 239, 325–331.

Chao, J., Chao, L., Swain, C.C., Tsai, J. and Margolius, H.S. (1987). Tissue kallikrein in rat brain and pituitary: regional distribution and estrogen induction in the anterior pituitary. Endocrinology 120, 475–482.

Chao, J., Swain, C., Chao, S., Xiong, W. and Chao, L. (1988). Tissue distribution and kininogen gene expression after acute phase inflammation. Biochim. Biophys. Acta 964, 329–339.

Chao, J., Chai, K.X., Chen, L.M., Xiong, W., Chao, S., Woodley-Miller, C., Wang, L.X., Lu, H.S. and Chao, L. (1990). Tissue-kallikrein-binding protein is a serpin. I. Purification, characterization and distribution in normotensive and spontaneously hypertensive rats. J. Biol. Chem. 265, 16394–16401.

Chao, S., Chao, L. and Chao, J. (1989). Sex dimorphism and inflammatory regulation of T-kininogen and T-kininogenase. Biochim. Biophys. Acta 991, 477–483.

Chapdelaine, P., Dube, J.Y., Frenette, G. and Tremblay, R.R. (1984). Identification of arginine esterase as the major androgen-dependent protein secreted by dog prostate and preliminary molecular characterization in seminal plasma. J. Androl. 5, 206–210.

Chapdelaine, P., Ho-Kim, M.-A., Tremblay, R.R. and Dube, J.Y. (1988a). Nucleotide sequence of the androgen-dependent arginine esterase mRNA of canine prostate. FEBS Lett. 232, 187–192.

Chapdelaine, P., Paradis, G., Tremblay, R.R. and Dube, J.Y. (1988b). High level of expression in the prostate of a human glandular kallikrein mRNA related to prostate-specific antigen. FEBS Lett. 236, 205–208.

Chapdelaine, P., Potvin, C., Ho-Kim, M.-A., Larouche, L., Bellemare, G., Tremblay, R.R. and Dube, J.Y. (1988c). Androgen regulation of canine prostatic arginine esterase mRNA using cloned cDNA. Mol. Cell. Endocrinol. 56, 63–70.

Chapdelaine, P., Gauthier, E., Ho-Kim, M.A., Bissonnette, L., Tremblay, R.R. and Dube, J.Y. (1991). Characterization and expression of the prostatic arginine esterase gene, a canine glandular kallikrein. DNA Cell Biol. 10, 49–59.

Chen, H.M. and Liao, W.S.L. (1993). Differential acute-phase response of rat kininogen genes involves type-I and type-II interleukin 6 response elements. J. Biol. Chem. 266, 25311–25319.

Chen, L.-M., Richards, G.P., Chao, L. and Chao, J. (1995a). Molecular cloning, purification and in situ hybridization of human colon kallikrein. Biochem. J. 307, 481–486.

Chen, L.-M., Song, Q., Chao, L. and Chao, J. (1995b). Cellular localization of tissue kallikrein and kallistatin mRNAs in human kidney. Kidney Int. 48, 690–697.

Chen, Y., Chao, J. and Chen L. (1988). Molecular cloning and characterization of two rat renal kallikrein genes. Biochemistry 27, 7189–7196.

Christiansen, S.C., Proud, D. and Cochrane, C.G. (1987). Detection of tissue kallikrein in the bronchoalveolar lavage of asthmatic subjects. J. Clin. Invest. 79, 188–197.

Christiansen, S.C., Proud, D., Sarnoff, R.B., Juergens, U., Cochrane, C.G. and Zuraw, B.L. (1992). Elevation of tissue kallikrein and kinin in the airways of asthmatic subjects after endobronchial allergen challenge. Am. Rev. Respir. Dis. 145, 900–905.

Clements, J.A. (1989). The glandular kallikrein family of enzymes: tissue-specific expression and hormonal regulation. Endocr. Rev. 10, 393–419.

Clements, J.A. (1994). The human kallikrein gene family: a diversity of expression and function. Mol. Cell. Endocrinol. 99, C1–C6.

Clements, J.A. and Fuller, P.J. (1990). Tissue kallikrein in the pituitary and brain. Front. Neuroendocrinol. 1, 38–51.

Clements, J.A. and Mukhtar, A. (1994). Glandular kallikreins and prostate-specific antigen are expressed in the human endometrium. J. Clin. Endocrinol. Metab. 78, 1536–1539.

Clements, J.A., Fuller, P.J., McNally, M., Nikolaidis, I. and Funder, J.W. (1986). Estrogen regulation of kallikrein gene expression in the rat anterior pituitary. Endocrinology 119, 268–273.

Clements, J.A., Matheson, B.A., Wines, D.R., Brady, J.M., MacDonald, R.J. and Funder, J.W. (1988). Androgen dependence of specific kallikrein gene family members expressed in the rat prostate. J. Biol. Chem. 31, 16132–16137.

Clements, J.A., Matheson, B.A., MacDonald, R.J. and Funder, J.W. (1989). The expression of the kallikrein gene family in the rat pituitary: oestrogen effects and the expression of an additional family member in the neurointermediate lobe. J. Neuroendocrinol. 1, 199–203.

Clements, J.A., Matheson, B.A. and Funder, J.W. (1990a). Tissue-specific developmental expression of the kallikrein gene family in the rat. J. Biol. Chem. 265, 1077–1081.

Clements, J.A., Matheson, B.A. and Funder, J.W. (1990b). Tissue-specific regulation of the expression of rat kallikrein gene family members by thyroid hormone. Biochem. J. 3, 745–750.

Clements, J.A., Matheson, B.A., MacDonald, R.J. and Funder, J.W. (1990c). Oestrogen administration and the expression of the kallikrein gene family in the rat submandibular gland. J. Steroid Biochem. 35, 55–60.

Clements, J.A., Mukhtar, A., Ehrlich, A. and Fuller, P. (1992). A re-evaluation of the tissue-specific pattern of expression of the rat kallikrein gene family. In "Recent Progress on Kinins" (eds. H. Fritz, W. Müller-Esterl, M. Jochum, A. Roscher and K. Luppertz), Agents and Actions Suppl., Vol. 38/I, pp. 34–41. Birkhauser Verlag, Basel.

Clements, J.A., Mukhtar, A., Ehrlich, A. and Yap, B. (1994). Kallikrein gene expression in the human uterus. Braz. J. Med. Biol. Res. 27, 1855–1863.

Clements, J.A., Mukhtar, A., Holland, A.M., Ehrlich, A. and Fuller, P.J. (1995). Kallikrein gene expression in the rat ovary: localization to the granulosa cell. Endocrinology 136, 1137–1144.

Clements, J.A., Mukhtar, A., Verity, K., Pullar, M., McNeill, P., Cummins, J. and Fuller, P.J. (1996). Kallikrein gene expression in human pituitary tissues. Clin. Endocrinol. 44, 223–231.

Cohen, P., Graves, H.C.B., Peehl, D.M., Kamarei, M., Guidice, L.C. and Rosenfeld, R.G. (1992). Prostate-specific antigen (PSA) is an insulin-like growth factor binding protein-3 protease found in seminal plasma. J. Clin. Endocrinol. Metab. 75, 1046–1053.

Corthorn, J. and Valdes, G. (1994). Variations in uterine kallikrein during cycle and early pregnancy in the rat. Biol. Reprod. 50, 1261–1264.

Cumming, A.D., Walsh, T., Wojtacha, D., Fleming, S., Thomson, D. and Jenkins, D.A. (1994). Expression of tissue kallikrein in human kidney. Clin. Sci. 87, 5–11.

Curd, J.G., Prograis, L.J., Jr and Cochrane, C.G. (1980). Detection of active kallikrein in induced blister fluids of hereditary angiodema patients. J. Exp. Med. 152, 742–747.

Currie, M.G., Gellen, D.M., Chao, J., Margolius, H.S. and Needleman, P. (1984). Kallikrein activation of a high molecular weight atrial peptide. Biochem. Biophys. Res. Commun. 120, 461–466.

Deperthes, D., Gauthier, E.R., Chapdelaine, P., Lazure, C., Tremblay, R.R. and Dube, J.Y. (1995). Identification of glandular kallikrein in dog pancreas and determination of its distribution. Biochim. Biophys. Acta 1243, 291–294.

Derkx, F.H.M., Tan-Tjiong, H.L., Man In't Veld, A.J., Schalekamp, M.P.A. and Schalekamp, M.A.H. (1979). Activation of inactive plasma renin by tissue kallikreins. J. Clin. Endocrinol. Metab. 49, 765–769.

Desrivieres, S., Lu, H., Peyri, N., Soria, C., Legrand, Y. and Menashi, S. (1993). Activation of the 92 kDa type IV collagenase by tissue kallikrein. J. Cell. Physiol. 157, 587–593.

Digby, M., Zhang, X.Y. and Richards, R.I. (1989). Human prostate-specific antigen (PSA) gene structure and linkage to the kallikrein-like gene, hGK-1. Nucleic Acids Res. 17, 2137.

Drinkwater, C.C. and Richards, R.I. (1987). Sequence of the mouse glandular kallikrein gene, mGK-5. Nucleic Acids Res. 15, 10052.

Drinkwater, C.C. and Richards, R.I. (1988). Sequence of mGK-11, a mouse glandular kallikrein gene. Nucleic Acids Res. 16, 10918.

Drinkwater, C.C., Evans, B.A. and Richards, R.I. (1987). Mouse glandular kallikrein genes: identification and characterization of the genes encoding the epidermal growth factor binding proteins. Biochemistry 26, 6750–6756.

Drinkwater, C.C., Evans, B.A. and Richards, R.I. (1988a). Kallikreins, kinins and growth factor biosynthesis. Trends Biochem. Sci. 13, 169–172.

Drinkwater, C.C., Evans, B.A. and Richards, R.I. (1988b). Sequence and expression of mouse γ-renin. J. Biol. Chem. 263, 8565–8568.

Dube, J.Y., Chapdelaine, P., Guerin, S., Leclerc, S., Rennie, P.S., Matusik, R.J. and Tremblay, R.R. (1995). Search for androgen response elements in the proximal promoter of the canine prostate arginine esterase gene. J. Androl. 16, 304–311.

Dunbar, J.C. and Bradshaw, R.A. (1987). Amino acid sequence of guinea pig prostate kallikrein. Biochemistry 29, 3471–3478.

Ebisuno, S., Miyai, M. and Nagareda, T. (1995). Clear cell adenocarcinoma of the female urethra showing positive staining with antibodies to prostate-specific antigen and prostatic acid phosphatase. Urology 45, 682–685.

Eeckhout, Y. and Vaes, G. (1977). Further studies on the activation of procollagenase, the latent precursor of bone collagenase. Biochem. J. 166, 21–31.

Elmoujahed, A., Gutman, N., Brillard, M. and Gauthier, F. (1990). Substrate specificity of two kallikrein family gene products isolated from the rat submaxillary gland. FEBS Lett. 265, 137–140.

Espana, F., Fink, E., Sanchezcuenca, J., Gilabert, J., Estelles, A. and Witzgall, K. (1995) Complexes of tissue kallikrein with protein C inhibitor in human semen and urine. Eur. J. Biochem. 234, 641–649.

Espey, L.L., Miller, D.H. and Margolius, H.S. (1986). Ovarian increase in kinin generating capacity in PMSG/hCG-primed immature rat. Am. J. Physiol. 251, E362.

Esquenet, M., Swinnen, J.V., Heyns, W. and Verhoeven, G. (1995). Triiodothyronine modulates growth, secretory function and androgen receptor concentration in the prostatic carcinoma cell line LNCaP. Mol. Cell. Endocrinol. 109, 105–111.

Evans, B.A. and Richards, R.I. (1985). Genes for the alpha and gamma subunits of mouse nerve growth factor are contiguous. EMBO J. 4, 133–138.

Evans, B.A., Drinkwater, C.C. and Richards, R.I. (1987). Mouse glandular kallikrein genes. Structure and partial sequence analysis of the kallikrein gene locus. J. Biol. Chem. 262, 8027–8034.

Evans, B.A., Yun, Z.X., Close, J.A., Tregear, G.W., Kitamura, N., Nakanishi, S., Callen, D.F., Baker, E., Hyland, V.J., Sutherland, G.R. and Richards, R.I. (1988). Structure and chromosomal localization of the human renal kallikrein gene. Biochemistry 27, 3124–3129.

Fahnestock, M., Woo, J.E., Lopez, G.A., Snow, J., Walz, D.A., Arici, M.J. and Morley, W.C. (1991). β-NGF-endopeptidase: structure and activity of a kallikrein encoded by the gene mGK-22. Biochemistry 30, 3443–3450.

Fiedler, F. (1979). Enzymology of glandular kallikreins. In "Handbook of Experimental Pharmacology Supplement: Bradykinin, Kallidin and Kallikrein" (ed. E.G. Erdös), Vol. 23, pp. 103–161. Springer-Verlag, New York.

Fiedler, F. and Fritz, H. (1981). Striking similarity of pig pancreatic kallikrein, mouse nerve growth factor gamma-subunit and rat tonin. Hoppe-Seylers Z. Physiol. Chem. 362, 1171–1175.

Figueroa, C.D. and Bhoola, K.D. (1989). Leucocyte tissue kallikrein; an acute phase signal for inflammation. In "The Kallikrein–Kinin System in Health and Disease" (eds. H. Fritz, I. Schmidt and G. Dietze), pp. 311–320. Limbach Verlag, Braunschweig.

Figueroa, C.D., Caorsi, L., Subiabre, J. and Vio, C.P. (1984). Immunoreactive kallikrein localization in the rat kidney: an immunoelectron microscopic study. J. Histochem. Cytochem. 32, 117–121.

Figueroa, C.D., Maciver, A.G., Mackenzie, J.C. and Bhoola, K.D. (1988). Identification of a tissue kallikrein in human polymorphonuclear leucocytes. Br. J. Haematol. 72, 321–328.

Fink, E., Schill, W.-B., Fiedler, F., Krassnig, F., Geiger, R. and Shimamoto, K. (1985). Tissue kallikrein of human seminal plasma is secreted by the prostate gland. Biol. Chem. Hoppe-Seyler 366, 917–924.

Frey, P., Forand, R., Maciag, T. and Shooter, E.M. (1979). The biosynthetic precursor of epidermal growth factor and the mechanism of its processing. Proc. Natl Acad. Sci. USA 76, 6294–6298.

Fukushima, D., Kitamura, N. and Nakanishi, S. (1985). Nucleotide sequence of cloned cDNA for human pancreatic kallikrein. Biochemistry 24, 8037–8043.

Fuller, P.J. and Funder, J.W. (1986). The cellular physiology of glandular kallikrein. Kidney Int. 29, 953–964.

Fuller, P.J., Clements, J.A., Whitfield, P.L. and Funder, J.W. (1985). Kallikrein gene expression in the rat anterior pituitary. Mol. Cell. Endocrinol. 39, 99–105.

Fuller, P.J., Clements, J.A., Nikolaidis, I., Hiwatari, M. and Funder, J.W. (1986). Expression of the renal kallikrein gene in mineralocorticoid-treated and genetically hypertensive rats. J. Hypertens. 4, 427–433.

Fuller, P.J., Matheson, B.A., MacDonald, R.J., Verity, K. and Clements, J.A. (1988). Kallikrein gene expression in estrogen-induced pituitary tumors. Mol. Cell. Endocrinol. 60, 225–232.

Fuller, P.J., Verity, K., Matheson, B.A. and Clements, J.A. (1989). Kallikrein gene expression in the rat gastrointestinal tract. Biochem. J. 264, 133–136.

Gao, X., Greenbaum, L.M., Mahesh, V.B. and Brann, D.W. (1992). Characterization of the kinin system in the ovary during ovulation in the rat. Biol. Reprod. 47, 945–951.

Gauthier, E.R., Chapdelaine, P., Tremblay, R.R. and Dube, J.Y. (1993a). Characterization of rhesus monkey prostate specific antigen cDNA. Biochim. Biophys. Acta 1174, 207–210.

Gauthier, E.R., Chapdelaine, P., Tremblay, R.R. and Dube, J.Y. (1993b). Transcriptional regulation of dog prostate arginine esterase gene by androgens. Mol. Cell. Endocrinol. 94, 155–163.

Gauthier, E.R., Dumas, C., Chapdelaine, P., Tremblay, R.R. and Dube, J.Y. (1994). Characterization of canine pancreas kallikrein cDNA. Biochim. Biophys. Acta 1218, 102–104.

Gauthier, F., Gutman, N., Moreau, T. and Elmoujahed, A. (1988). Possible relationship between the restricted biological function of rat T-kininogen (thiostatin) and its behaviour as an acute phase reactant. Hoppe-Seylers Z. Biol. Chem. 369, 251–255.

Geiger, R. and Clausnitzner, B. (1981). Isolation of enzymatically active tissue kallikrein from human seminal plasma by immunoaffinity chromatography. Hoppe-Seylers Z. Physiol. Chem. 362, 229–231.

Geiger, R., Stuckstedte, U., Clausnitzner, B. and Fritz, H. (1981). Progressive inhibition of human glandular (urinary) kallikrein by human serum and identification of the progressive anti-kallikrein as α1–antitrypsin (α1-proteinase inhibitor). Hoppe-Seylers Z. Physiol. Chem. 362, 317–325.

Gerald, W.L., Chao, J. and Chao, L. (1986a). Immunological identification of rat tissue kallikrein cDNA and characterization of the kallikrein gene family. Biochim. Biophys. Acta 866, 1–14.

Gerald, W.L., Chao, J. and Chao, L. (1986b). Sex dimorphism and hormonal regulation of rat tissue kallikrein mRNA. Biochim. Biophys. Acta 867, 16–23.

Girolami, J.P., Bascands, J.L., Valet, P., Pecher, C. and Cabos, G. (1990). β_2–Adrenergic inhibition of kallikrein release from rat kidney cortical slices. Am. J. Physiol. 258, F1425–F1431.

Greene, L.A., Shooter, E.M. and Varon, S. (1968). Enzymatic activities of mouse nerve growth factor and its subunits. Proc. Natl Acad. Sci. USA 60, 1383–1388.

Gutman, N., Moreau, T., Alhenc-Gelas, F., Baussant, T., Elmoujahed, A., Akpona, S. and Gauthier, F. (1988). T-Kinin release from T-kininogen by rat submaxillary gland endopeptidase K. Eur. J. Biochem. 171, 577–582.

Haidl, G. and Schill, W.B. (1993). Changes of sperm tail morphology after kallikrein treatment. Arch. Androl. 31, 1–8.

Harvey, J.N., Jaffa, A.A., Margolius, H.S. and Mayfield, R.K. (1990). Renal kallikrein and hemodynamic abnormalities of diabetic kidney. Diabetes 39, 299–304.

Hatala, M.A. and Powers, C.A. (1987). Dynamics of estrogen induction of glandular kallikrein in the rat anterior pituitary. Biochim. Biophys. Acta 926, 258–263.

Hatala, M.A. and Powers, C.A. (1988a). Glandular kallikrein in estrogen-induced pituitary tumors: time course of induction and correlation with prolactin. Cancer Res. 48, 4158–4163.

Hatala, M.A. and Powers, C.A. (1988b). Development of the sex difference in glandular kallikrein and prolactin levels in the anterior pituitary of the rat. Biol. Reprod. 38, 846–852.

Henttu, P. and Vihko, P. (1989). A cDNA coding for the entire human prostate specific antigen shows high homologies to the human tissue kallikrein genes. Biochem. Biophys. Res. Commun. 160, 903–910.

Henttu, P. and Vihko, P. (1993). Growth factor regulation of gene expression in the human prostatic carcinoma cell line LNCaP. Cancer Res. 53, 1051–1058.

Henttu, P., Lukkarinen, O. and Vihko, P. (1990). Expression of the gene coding for human prostate-specific antigen and related hGK-1 in benign and malignant tumors of the human prostate. Int. J. Cancer 45, 654–660.

Hibino, T., Takemura, T. and Sato, K. (1988a). Demonstration of glandular kallikrein and angiotensin converting enzyme (kininase II) in human eccrine sweat. J. Invest. Dermatol. 90, 569.

Hibino, T., Isaki, S., Kimura, H., Izaki, M. and Kon, S. (1988b). Partial purification of plasma and tissue kallikreins in psoriatic epidermis. J. Invest. Dermatol. 90, 505–510.

Hofmann, W. and Geiger, R. (1983). Human tissue kallikrein. I. Isolation and characterization of human pancreatic kallikrein from duodenal juice. Hoppe-Seylers Z. Physiol. Chem. 364, 413–423.

Holland, O.B., Chud, J.M. and Braunstein, H. (1980). Urinary kallikrein excretion in essential and mineralocorticoid hypertension. J. Clin Invest. 65, 347–356.

Horwitz, D., Margolius, H.S. and Keiser, H.R. (1978). Effects of dietary potassium and race on urinary excretion of kallikrein and aldosterone in man. J. Clin. Endocrinol. Metab. 47, 296–299.

Howard, E.F., Thompson, Y.G., Lapp, C.A. and Greenbaum, L.M. (1990). Reduction of T-kininogen messenger RNA levels by dexamethasone in the adjuvant-treated rat. Life Sci. 46, 411–417.

Howles, P.N., Dickinson, D.P., DiCaprio, L.L., Woodworth-Gutai, M. and Gross, K.W. (1984). Use of a cDNA recombinant for the gamma-subunit of mouse nerve growth factor to localize members of this multigene family near the TAM-1 locus on chromosome 7. Nucleic Acids Res. 12, 2791–2805.

Ideishi, M., Ikeda, M. and Arakawa, K. (1987). Direct angiotensin II formation by rat submandibular gland kallikrein. J. Biochem. (Tokyo). 102, 859–868.

Inoue, H., Fukui, K. and Miyake, Y. (1989). Identification and structure of the rat true tissue kallikrein gene expressed in the kidney. J. Biochem. (Tokyo). 105, 834–840.

Isackson, P.J. and Bradshaw, R.A. (1984). The α-subunit of mouse 7s nerve growth factor is an inactive serine protease. J. Biol. Chem. 259, 5380–5383.

Ishiguro, T., Shimamoto, K., Ura, N., Nomura, N., Hayashi, M. and Iimura, O. (1995). The pathophysiological role of renal dopamine and kallikrein in deoxycorticosterone acetate (DOCA)-salt treated rats. Clin. Exp. Hypertens. 17, 1287–1299.

Ito, N., Yayama, K. and Okamoto, H. (1988). Stimulation of hepatic T-kininogen production by interferon. FEBS Lett. 229, 247–250.

Jaffa, A.A., Miller, D.H., Bailey, G.S., Chao, J., Margolius, H.S. and Mayfield, R.K. (1987). Abnormal regulation of renal kallikrein in experimental diabetes. Effects of insulin on prokallikrein synthesis and activation. J. Clin. Invest. 80, 1651–1659.

Jaffa, A.A., Miller, D.H., Silva, R.H., Margolius, H.S. and Mayfield, R.K. (1990). Regulation of renal kallikrein synthesis and activation by glucocorticoids. Kidney Int. 38, 212–218.

Jaffa, A.A., Chai, K.X., Chao, J., Chao, L. and Mayfield, R.K. (1992). Effects of diabetes and insulin on expression of kallikrein and renin genes in the kidney. Kidney Int. 41, 789–795.

Jahnke, G.D., Chao, J., Walker, M.P. and Diaugustine, R.P. (1994). Detection of a kallikrein in the mouse lactating mammary gland – a possible processing enzyme for the epidermal growth factor precursor. Endocrinology 135, 2022–2029.

Jones, T.H., Brown, B.L. and Dobson, P.R.M. (1989). Bradykinin stimulates phosphoinositide metabolism and prolactin secretion in rat anterior pituitary cells. J. Mol. Endocrinol. 2, 47–53.

Jones, T.H., Figueroa, C.D., Smith, C., Cullen, D.R. and Bhoola, K.D. (1990). Characterization of a tissue kallikrein in human prolactin-secreting adenomas. J. Endocrinol. 124, 327–331.

Kamada, M., Furuhata, N., Yamaguchi, T., Ikekita, M., Kizuki, K. and Moriya, H. (1990). Observations of tissue prokallikrein activation by some serine proteases, arginine esterases in rat submandibular gland. Biochem. Biophys. Res. Commun. 166, 231–237.

Karr, J.F., Kantor, J.A., Horan Hand, P., Eggensperger, D.L. and Schlom, J. (1995). The presence of prostate-specific antigen-related genes in primates and the expression of recombinant human prostate-specific antigen in a transfected murine cell line. Cancer Res. 55, 2455–2462.

Khullar, M., Scicli, G., Carretero, O.A. and Scicli, A.G. (1986). Purification and characterization of a serine protease (esterase B) from rat submandibular glands. Biochemistry 25, 1851–1857.

Kim, W.-S., Nakayama, K., Nakagawa, T., Kawamura, Y., Haraguchi, K. and Murakami, K. (1991a). Mouse submandibular gland prorenin-converting enzyme is a member of the glandular kallikrein family. J. Biol. Chem. 29, 19283–19287.

Kim, W.-S., Nakayama, K. and Murakami, K. (1991b). The presence of two types of prorenin converting enzymes in mouse submandibular gland. FEBS Lett. 293, 142–144.

Kitagawa, A., Kizuki, K., Moriya, H., Kudo, M. and Noguchi, T. (1991). Localization of kallikrein in rat pineal gland. Endocrinol. Jpn. 38, 109–112.

Kizuki, K., Kitagawa, A., Takahashi, M., Moriya, H., Kudo, M. and Noguchi, T. (1990). Immunohistochemical localization of kallikrein within the prolactin-producing cells of the rat anterior pituitary gland. J. Endocrinol. 127, 317–323.

Kupio, T., Ekfors, T.O., Nikkanen, V. and Nevalainen, T.J. (1995). Acinar cell carcinoma of the pancreas – report of three cases. APMIS 103, 69–78.

Lazure, C., Leduc, R., Seidah, N.G., Chretien, M., Dube, J.Y., Chapdelaine, P., Frenette, G., Paquin, R. and Tremblay, R.R. (1984). The major androgen-dependent protease in dog prostate belongs to the kallikrein family: confirmation by partial amino acid sequencing. FEBS Lett. 175, 1–7.

Lee, C., Sutkowski, D.M., Sensibar, J.A., Zelner, D., Kim, I., Amsel, I., Shaw, N., Prins, G.S. and Kozlowski, J.M. (1995). Regulation of proliferation and production of prostate-specific antigen in androgen-sensitive prostatic cancer cells, LNCaP, by dihydrotestosterone. Endocrinology 136, 796–803.

Lemon, M., Fiedler, F., Forg-Brey B., Hirschauer, C., Leysath, G. and Fritz, H. (1979). The isolation and properties of pig submandibular kallikrein. Biochem. J. 177, 159–168.

Levesque, M., Yu, H., Dcosta, M., Tadross, L. and Diamandis, E.P. (1995). Immunoreactive prostate-specific antigen in lung tumors. J. Clin. Lab. Anal. 9, 375–379.

Lilja, H. (1985). A kallikrein-like serine protease in prostatic fluid cleaves the predominant seminal vesicle protein. J. Clin Invest. 76, 1899–1903.

Lilja, H., Oldbring, J., Rannevik, G. and Laurell, C.B. (1987). Seminal vesicle-secreted proteins and their reactions during gelation and liquefaction of human semen. J. Clin. Invest. 80, 281–285.

Lilja, H., Christensson, A., Dahlen, U., Matikainen, M.-J., Nilssen, O., Pettersson, K. and Lovgen, T. (1991). Prostate-specific antigen in serum occurs predominantly in complex with a-1-antichymotrypsin. Clin. Chem. 37, 1618–1625.

Lin, F.-K., Lin, C.H., Chou, C.C., Chen, K., Lu, H.S., Bacheller, W., Herrera, C., Jones, T., Chao, J. and Chao, L. (1993). Molecular cloning and sequence analysis of the monkey and human tissue kallikrein genes. Biochim. Biophys. Acta 1173, 325–328.

Lundgren, S., Ronne, H., Rask, L. and Peterson, P.A. (1984). Sequence of an epidermal growth factor-binding protein. J. Biol. Chem. 259, 7780–7784.

Lundwall, A. (1989). Characterization of the gene for prostate-specific antigen, a human glandular kallikrein. Biochem. Biophys. Res. Commun. 161, 1151–1159.

Lundwall, A. and Lilja, H. (1987). Molecular cloning of human prostate specific antigen cDNA. FEBS Lett. 214, 317–322.

Ma, J., Chao, J. and Chao, L. (1992). Molecular cloning and characterization of rKLK10, a cDNA encoding T-kininogenase from rat submandibular glands. Biochemistry 31, 10922–10928.

Ma, J.X., Yang, Z.R., Chao, J. and Chao, L. (1995). Intramuscular delivery of rat kallikrein-binding protein gene reverses hypotension in transgenic mice expressing tissue kallikrein. J. Biol. Chem. 270, 451–455.

MacDonald, R.J., Margolius, H.S. and Erdös, E.G. (1988). Molecular biology of tissue kallikrein. Biochem. J. 253, 313–321.

Maier, M., Austen, K.F. and Spragg, J. (1983). Kinetic analysis of the interaction of human tissue kallikrein with single-chain human high and low molecular weight kininogens. Proc. Natl Acad. Sci. USA 80, 3928–3932.

Majima, M. and Katori, M. (1995). Approaches to the development of novel antihypertensive drugs: crucial role of the renal kallikrein–kinin system. Trends Pharmacol. Sci. 16, 239–246.

Malofiejew, M. (1973). Kallikrein-like activity in human myometrium, placenta and amniotic fluid. Biochem. Pharmacol. 22, 123–127.

Margolius, H.S. (1995). Kallikreins and kinins – some unanswered questions about system characteristics and roles in human disease. Hypertension 26, 221–229.

Margolius, H.S., Horwitz, D., Pisano, J.J. and Keiser, H.R. (1974). Urinary kallikrein in hypertension: relationship to sodium intake and sodium-retaining steroids. Circ. Res. 35, 820–825.

Marin-Grez, M., Schaechtelin, G. and Hermann, K. (1982). Kininogenase activity in rat uterus homogenates. Hoppe-Seylers Z. Physiol. Chem. 363, 1359–1364.

Marin-Grez, M., Bonner, G. and Gross, F. (1984). The influence of isotonic saline administration on the urinary excretion of kallikrein in rats. Biochem. Pharmacol. 33, 3585–3590.

Mason, A.J., Evans, B.A., Cox, D.R., Shine, J. and Richards, R.I. (1983). Structure of mouse kallikrein gene family suggests a role in specific processing of biologically active peptides. Nature (Lond.) 303, 300–307.

Matsumara, Y., Kimura, M., Yamamoto, T. and Maeda, H. (1988). Involvement of the kinin generating cascade in enhanced vascular permeability in tumour tissue. Jpn J. Cancer Res. 79, 1327–1334.

Mayer, G., Bhoola, K.D. and Fiedler, F. (1989). Tissue kallikreins of the guinea-pig. Adv. Exp. Med. Biol. 247B, 201–206.

Mayfield, R.K., Margolius, H.S., Levine, J.H., Wohltmann, H.J., Loadholt, C.B. and Colwell, J.A. (1984). Urinary kallikrein excretion in insulin-dependent diabetes mellitus and its relationship to glycemic control. J. Clin. Endocrinol. Metab. 59, 278–286.

Mayfield, R.K., Margolius, H.S., Bailey, G.S., Miller, D.H., Sens, D.A., Squires, J. and Namm, D.H. (1986). Urinary and tissue kallikrein in the streptozotocin-diabetic rat. Diabetes 34, 22–28.

Menashi, S., Fridman, R., Desrivieres, S., Lu, H., Legrand, Y. and Soria, C. (1994). Regulation of 92–kDa gelatinase activity in the extracellular matrix by tissue kallikrein. Ann. N. York Acad. Sci. 732, 466–468.

Miller, D.H., Chao, J. and Margolius, H.S. (1984). Tissue kallikrein synthesis and its modification by testosterone or low dietary sodium. Biochem. J. 218, 237–243.

Miller, D.H., Lindley, J.G. and Margolius, H.S. (1985). Tissue kallikrein levels and synthesis rates are not changed by an acute physiological dose of aldosterone. Proc. Soc. Exp. Biol. Med. 180, 121–125.

Monne, M., Croce, C.M., Yu, H. and Diamandis, E.P. (1994). Molecular characterization of prostate-specific antigen messenger RNA expressed in breast tumors. Cancer Res. 54, 6344–6347.

Montgomery, B.T., Young, C.Y., Bilhartz, D.L., Andrews, P.E., Prescott, J.L., Thompson, N.F. and Tindall, D.J. (1992). Hormonal regulation of prostate-specific antigen (PSA) glycoprotein in the human prostatic adenocarcinoma cell line, LNCaP. The Prostate 21, 63–73.

Moreau, T., Gutman, N., Elmoujahed, A., Esnard, F. and Gauthier, F. (1986). Relationship between the cysteine–protease-inhibitory function of rat T-kininogen and the release of immunoreactive kinin upon trypsin treatment. Eur. J. Biochem. 159, 341–346.

Murray, S.R., Chao, J., Lin, F. and Chao, L. (1990). Kallikrein multigene families and the regulation of their expression. J. Cardiovasc. Pharmacol. 15 (Suppl.), S7–S15.

Murtha, P., Tindall, D.J. and Young, C.Y. (1993). Androgen induction of a human prostate-specific kallikrein, hKLK2: characterization of an androgen response element in the 5′ promoter region of the gene. Biochemistry 32, 6459–6464.

Mutt, V. and Said, S.I. (1974). Structure of porcine vasoactive intestinal octacosapeptide. The amino-acid sequence. Use of kallikrein in its determination. Eur. J. Biochem. 42, 581–589.

Neurath, H. (1989). The diversity of proteolytic enzymes. In "Proteolytic Enzymes – A Practical Approach" (eds. R.J. Beynon and J.S. Bond), pp. 1–13. Oxford University Press, Oxford.

Noda, Y., Takada, Y. and Erdös, E.G. (1985). Activation of human and rabbit prokallikrein by serine and metallo-proteases. Kidney Int. 27, 630–635.

Nolly, H., De Vito, E., Carrerra, R. and Koninckx, A. (1981). Kinin-releasing enzyme in cardiac tissue. In "Hypertension" (ed. H. Villareal), pp. 75–84. John Wiley and Sons, New York.

Nolly, H., Scicli, A., Scicli, G. and Carretero, O.A. (1985). Characterization of a kininogenase from rat vascular tissue resembling tissue kallikrein. Circ. Res. 56, 816–821.

Ole-Moiyoi, O., Seldin, D.C., Spragg, J., Pinkus, G. and Austen, K.F. (1979a). Sequential cleavage of pro-insulin by human pancreatic kallikrein and a human pancreatic kinase. Proc. Natl Acad. Sci. USA 76, 3612–3616.

Ole-Moiyoi, O., Pinkus, G.S., Spragg, J. and Austen, K.F. (1979b). Identification of human glandular kallikrein in the beta cell of the pancreas. N. Engl. J. Med. 300, 1289–1294.

Orstavik, T.B. and Ingami, T. (1982). Localization of kallikrein in the rat kidney and its anatomical relationship to renin. J. Histochem. Cytochem. 30, 385–360.

Orstavik, T.B., Brandtzaeg, P., Nustad, K. and Pierce, J.V. (1980). Immunohistochemical localization of kallikrein in human pancreas and salivary glands. J. Histochem. Cytochem. 28, 557–562.

Oza, N.B., Schwartz, J.H., Goud, M.D. and Levinsky, N.G. (1990). Rat aortic smooth muscle cells in culture express kallikrein, kininogen and bradykinase activity. J. Clin. Invest. 85, 597–600.

Pang, S., Taneja, S., Dardashiti, K., Cohan, P., Kaboo, R., Sokoloff, M., Tso, C.L., Dekernion, J.B. and Belldegrun, A.S. (1995). Prostate tissue specificity of the prostate-specific antigen promoter isolated from a patient with prostate cancer. Human Gene Ther. 6, 1417–1426.

Peehl, D.M. (1995). Prostate specific antigen role and function. Cancer 75 (Suppl. S), 2021–2026.

Penschow, J.D. and Coghlan, J.P. (1992). Sites of glandular kallikrein gene expression in fetal mice. Mol. Cell. Endocrinol. 88, 23–30.

Penschow, J.D. and Coghlan, J.P. (1993). Secretion of glandular kallikrein and renin from the basolateral pole of mouse submandibular duct cells: an immunocytochemical study. J. Histochem. Cytochem. 41, 95–103.

Penschow, J.D. and Coghlan, J.P. (1995). Furosemide treatment alters the distribution of kallikrein gene expression on kidneys of mice. Exp. Nephrol. 3, 280–287.

Penschow, J.D., Haralambidis, J. and Coghlan, J.P. (1991a). Location of glandular kallikrein mRNAs in mouse submandibular gland at the cellular and ultrastructural level by hybridization histochemistry using ^{32}P- and ^{3}H-labeled oligodeoxyribonucletide probes. J. Histochem. Cytochem. 39, 835–842.

Penschow, J.D., Drinkwater, C.C., Haralambidis, J. and Coghlan, J.P. (1991b). Sites of expression and induction of glandular kallikrein gene expression in mice. Mol. Cell. Endocrinol. 81, 135–146.

Pinkus, G.S., Maier, M., Seldin, D.C., Ole-Moiyoi, O., Austen, K.F. and Spragg, J. (1981). Immunohistochemical localization of glandular kallikrein in the endocrine and exocrine pancreas. J. Histochem. Cytochem. 31, 1279–1288.

Poblete, M.T., Reynolds, N.J., Figueroa, C.D., Burton, J.L., Müller-Esterl, W. and Bhoola, K.D. (1991). Tissue kallikrein and kininogen in human sweat glands and psoriatic skin. Br. J. Dermatol. 124, 236–241.

Poe, M., Wu, J.K., Florance, J.R., Rodkey, J.A., Bennett, C.D. and Hoogsteen, K. (1983). Purification and properties of renin and γ-renin from the mouse submaxillary gland. J. Biol. Chem. 258, 2209–2216.

Pollen, J.J. and Dreilinger, A. (1984). Immunohistochemical identification of prostatic acid phosphatase and prostate specific antigen in female periurethral glands. Urology 23, 303–304.

Powers, C.A. (1985). Dopamine receptor blockade increases glandular kallikrein in the neuro-intermediate lobe of the rat pituitary. Biochem. Biophys. Res. Commun. 127, 668–672.

Powers, C.A. (1986). Anterior pituitary glandular kallikrein: trypsin activation and estrogen regulation. Mol. Cell Endocrinol. 46, 163–174.

Powers, C.A. (1987). Elevated glandular kallikrein in estrogen-induced pituitary tumours. Endocrinology 120, 429–431.

Powers, C.A. and Hatala, M.A. (1986). Dopaminergic regulation of the estrogen-induced kallikrein in the rat anterior pituitary. Neuroendocrinology 44, 462–469.

Powers, C.A. and Hatala, M.A. (1990). Prolactin proteolysis by glandular kallikrein: in vitro reaction requirements and cleavage sites and detection of processed prolactin in vivo. Endocrinology 127, 1916–1927.

Powers, C.A. and Nasjletti, A. (1982). A novel kinin-generating protease (kininogenase) in the porcine anterior pituitary. J. Biol. Chem. 257, 5594–5598.

Powers, C.A. and Nasjletti, A. (1983). A kininogenase resembling glandular kallikrein in the rat pituitary pars intermedia. Endocrinology 112, 1194–1200.

Powers, C.A. and Nasjletti, A. (1984). A major sex difference in kallikrein-like activity in the rat anterior pituitary. Endocrinology 114, 1841–1844.

Powers, C.A. and Westlin, W.F. (1987). Glandular kallikrein levels in the rat anterior pituitary during the estrous cycle and pregnancy. Biol. Reprod. 37, 306–310.

Prado, E.S., Prado de Carvalho, L., Araujo-Viel, M.S., Ling, N. and Rossier, J. (1983). A met-enkephalin-containing-peptide, Bam22P, as a novel substrate for glandular kallikreins. Biochem. Biophys. Res. Commun. 112, 366–371.

Pritchett, D.B. and Roberts, J.L. (1987). Dopamine regulates expression of the glandular-type kallikrein gene at the transcriptional level in the pituitary. Proc. Natl Acad. Sci. USA 84, 5545–5549.

Proud, D. and Kaplan, A.P. (1988). Kinin formation: mechanisms and role in inflammatory disorders. Annu. Rev. Immunol. 6, 49–83.

Pummer, K., Wirnsberger, G., Purstner, P., Stettner, H. and Wandschneider, G. (1992). False positive prostate-specific antigen values in the sera of women with renal cell carcinoma. J. Urol. 148, 21–23.

Qin, H., Kemp, J., Yip, M., Lam-Po-Yang, P.R.L. and Morris, B.J. (1991). Localization of human glandular kallikrein-1 gene to chromosome 19q13.3–13.4 by in situ hybridization. Hum. Hered. 41, 222–226.

Rahman, M.M., Lemon, M.J.C., Elson, C.J., Dieppe, P.A. and Bhoola, K.D. (1995). Proinflammatory role of tissue kallikrein in modulating pain in inflamed joints. Br. J. Rheum. 34, 88–90.

Rapp, J.P., McPartland, R.P. and Sustarsic, D.I. (1982). Anomalous response of urinary kallikrein to desoxycorticosterone in Dahl salt-sensitive rats. Hypertension 4, 20–26.

Rehbock, J., Buchinger, P., Hermann, A. and Figueroa, C. (1995). Identification of immunoreactive tissue kallikrein in human ductal breast carcinomas. J. Cancer Res. Clin. Oncol. 121, 64–68.

Richards, R.I., Holman, K., Shen, Y., Kozman, H., Harley, H., Brook, D. and Shaw, D. (1991). Human glandular kallikrein genes: genetic and physical mapping of the KLK1 locus using a highly polymorphic microsatellite PCR marker. Genomics 11, 77–82.

Riegman, P.H.J., Klaassen, P., Van der Korput, J.A.G.M., Romijin, J.C. and Trapman, J. (1988). Molecular cloning and characterization of novel prostate antigen cDNAs. Biochem. Biophys. Res. Commun. 155, 181–188.

Riegman, P.H.J., Vlietstra, R.J., Van Der Korput, J.A.G.M., Romijin, J.C. and Trapman, J. (1989a). Characterization of the prostate-specific antigen gene: a novel human kallikrein-like gene. Biochem. Biophys. Res. Commun. 159, 95–102.

Riegman, P.H.J., Vlietstra, R.J., Klaassen, P., Van der Korput, J.A.G.M., Guerts van Kessel, A., Romijin, J.C. and Trapman, J. (1989b). The prostate-specific antigen gene and the human glandular kallikrein-1 gene are tandemly located on chromosome 19. FEBS Lett. 247, 123–126.

Riegman, P.H.J., Vlietstra, R.J., Van der Korput, J.A.G.M., Romijn, J.C. and Trapman, J. (1991a). Identification and androgen-regulated expression of two major human glandular kallikrein-1 (hGK-1) mRNA species. Mol. Cell. Endocrinol. 76, 181–190.

Riegman, P.H.J., Vlietstra, R.J., Van der Korput, J.A.G.M., Brinkmann, A.O. and Trapman, J. (1991b). The promoter of the prostate-specific antigen gene contains a functional androgen responsive element. Mol. Endocrinol. 5, 1921–1930.

Riegman, P.H.J., Vlietstra, R.J., Suurmeijer, L., Cleutjens, C.B.J.M. and Trapman, J. (1992). Characterization of the human kallikrein locus. Genomics 14, 6–11.

Roberts, R.A. and Gullick, W.J. (1989). Bradykinin receptor number and sensitivity to ligand stimulation of mitogenesis by expression of mutant ras oncogene. J. Cell Sci. 94, 527–535.

Roman, R.J., Kaldunski, M.L., Scicli, A.G. and Carretero, O.A. (1988). Influence of kinins and angiotensin II on the regulation of papillary blood flow. Am. J. Physiol. 255, F690–F698.

Rosewicz, S., Detjen, K., Logsdon, C.D., Chen, L.M., Chao, J. and Riecken, E.O. (1991). Glandular kallikrein gene expression is selectively down-regulated by glucocorticoids in pancreatic AR42J cells. Endocrinology 128, 2216–2222.

Saed, G.M., Carretero, O.A., Macdonald, R.J. and Scicli, A.G. (1990). Kallikrein messenger RNA in rat arteries and veins. Circ. Res. 67, 510–516.

Saed, G.M., Beierwaltes, W.H., Carretero, O.A. and Scicli, A.G. (1992). Submandibular enzymatic vasoconstrictor messenger RNA in rat kidney. Hypertension 19 (Suppl. II), 262–267.

Saitoh, S. and Kumamoto, Y. (1988). Effect of a kallikrein on testicular blood circulation. Arch. Androl. 20, 51–65.

Saitoh, S., Kumamoto, Y., Shimamoto, K. and Iimura, O. (1987). Kallikrein in the male reproductive system. Arch. Androl. 19, 133–147.

Saitoh, S., Scicli, A.G., Peterson, E. and Carretero, O.A. (1995). Effect of inhibiting renal kallikrein on prostaglandin E(2), water, and sodium excretion. Hypertension 25, 1008–1013.

Schachter, M., Peret, M.W., Moriwaki, C. and Rodrigues, J.A.A. (1980). Localization of kallikrein in submandibular gland of cat, guinea-pig, dog and man by the immunoperoxidase method. J. Histochem. Cytochem. 28, 1295–1300.

Schachter, M., Longridge, D.J., Wheeler, G.D., Mehta, J.G. and Uchida, Y. (1986). Immunocytochemical and enzyme histochemical localization of kallikrein-like enzymes in colon, intestine, and stomach of rat and cat. J. Histochem. Cytochem. 34, 927–934.

Schaffner, D.L., Barrios, R., Shaker, M.R., Rajagopalan, S., Huang, S.L., Tindall, D.J., Young, C.Y.F., Overbeek, P.A., Lebovitz, R.M. and Lieberman, M.W. (1995). Transgenic mice carrying a PSARasT24 hybrid gene develop salivary gland and gastrointestinal tract neoplasms. Lab. Invest. 72, 283–290.

Schedlich, L.J., Bennetts, B.H. and Morris, B.J. (1988). Primary structure of a human glandular kallikrein gene. DNA 6, 429–437.

Schill, W.-B., Miska, W., Parsch, E.-M. and Fink, E. (1989). Significance of the kallikrein–kinin system in andrology. In "The Kallikrein–Kinin System in Health and Disease" (eds. H. Fritz, I. Schmidt and G. Dietze), pp. 171–204. Limbach-Verlag, Braunschweig.

Schölkens, B.A., Linz, W., Lindpaintner, K. and Ganten, D. (1987). Angiotensin deteriorates but bradykinin improves cardiac function following ischaemia in isolated rat hearts. J. Hypertens. 5 (Suppl. 5), 7–9.

Scicli, A.G., Forbes, G., Nolly, H., Dujouny, M. and Carretero, O.A. (1984). Kallikreins–kinins in the central nervous system. Clin. Exp. Hypertens. A6, 1731–1738.

Scicli, A.G., Carbini, L.A. and Carretero, O.A. (1993). The molecular biology of the kallikrein–kinin system: II. The rat gene family. J. Hypertens. 11, 775–780.

Scicli, G., Nolly, H., Carretero, O.A. and Scicli, A.G. (1989). Glandular kallikrein-like enzyme in adrenal glands. Adv. Exp. Med. Biol. 247B, 217–222.

Sealey, J.E., Atlas, S.A., Laragh, J.H., Oza, N.B. and Ryan, J.W. (1978). Human urinary kallikrein converts inactive to active renin and is a possible physiological activator of renin. Nature (Lond.) 275, 144–145.

Seidah, N.G., Ladenheim, R., Mbikay, M., Hamelin, J., Lutfalla, G., Rougeon, F., Lazure, C. and Chretien, M. (1989). The cDNA structure of rat plasma kallikrein. DNA 8, 563–574.

Seidah, N.G., Sawyer, N., Hamelin, J., Mion, P., Beaubien, G., Brachpapa, L., Rochemont, J., Mbikay, M. and Chretien, M. (1990). Mouse plasma kallikrein: cDNA structure, enzyme characterization and comparison of protein and mRNA levels among species. DNA Cell. Biol. 9, 737–748.

Selwyn, B., Figueroa, C.D., Fink, E., Swan, A., Dieppe, P.A. and Bhoola, K.D. (1989). A tissue-kallikrein in the synovial fluid of patients with rheumatoid arthritis. Ann. Rheum. Dis. 48, 128–133.

Shai, S., Woodley-Miller, C., Chao, J. and Chao, L. (1989). Characterization of genes encoding rat tonin and a kallikrein-like serine protease. Biochemistry 28, 5334–5343.

Sharief, F.S., Mohler, J.L., Sharief, Y. and Li, S.S.-L. (1994). Expression of human prostatic acid phosphatase and prostate specific antigen genes in neoplastic and benign tissues. Biochem. Mol. Biol. Int. 33, 567–574.

Simson, J.A.V. and Chao, J. (1994). Subcellular distribution of tissue kallikrein and Na, K-ATPase α-subunit in rat parotid striated duct cells. Cell Tiss. Res. 275, 407–417.

Simson, J.A.V., Fenters, R. and Chao, J. (1983). Electron microscopic immunostaining of kallikrein in rat submandibular glands. J. Histochem. Cytochem. 31, 301–306.

Simson, J.A.V., Dom, R., Chao, J., Woodley, C. and Margolius, H.S. (1985). Immunocytochemical localization of tissue kallikrein in brain ventricular epithelium and hypothalamic cell bodies. J. Histochem. Cytochem. 33, 951–953.

Simson, J.A.V., Rowell, C., Barrett, J.M., King, J. and Chao, J. (1987). Rat urinary kallikrein localization in kidney: effects of fixation. Histol. J. 19, 633–642.

Simson, J.A.V., Chao, J. and Chao, L. (1992). Localization of kallikrein gene family proteases in rat tissues. In "Recent Progress on Kinins" (eds. H. Fritz, W. Müller-Esterl, M. Jochum, A. Roscher and K. Luppertz), pp. 595–602. Birkhauser Verlag, Basel.

Simson, J.A.V., Wang, J., Chao, J. and Chao, L. (1994). Histopathology of lymphatic tissues in transgenic mice expressing human tissue kallikrein gene. Lab. Invest. 71, 680–687.

Southard-Smith, M., Lechargo, J., Wines, D.R., MacDonald, R.J. and Hammer, R.E. (1992). Tissue-specific expression of kallikrein family transgenes in mice and rats. DNA Cell Biol. 11, 345–358.

Southard-Smith, M., Pierce, J.C. and MacDonald, R.J. (1994). Physical mapping of the rat tissue kallikrein family in two gene clusters by analysis of P1 bacteriophage clones. Genomics 22, 404–417.

Summers, K.M. and Hume, D.A. (1985). Expression of glandular kallikrein genes in lymphoid and hemopoietic tissues and cell lines. Lymphokine Res. 4, 229–235.

Sutherland, G.R., Baker, E., Hyland, V.J., Callen, D.F., Close, J.A., Tregear, G.W., Evans, B.A. and Richards, R.I. (1988). Human prostate specific antigen is a member of the glandular kallikrein gene family locus at 19q13. Cytogent. Cell Genet. 48, 205–207.

Suzuki, M., Ito, A., Mori, Y., Hayashi, Y. and Matsuta, K. (1987). Kallikrein in synovial fluid with rheumatoid arthritis. Biochem. Med. Metab. Biol. 37, 177–183.

Swift, G.H., Daghorn, J.C., Ashley, P.L., Cummings, S.W. and MacDonald, R.J. (1982). Rat pancreatic kallikrein mRNA: nucleotide sequence and aminoacid sequence of the encoded preproenzyme. Proc. Natl Acad. Sci. USA 79, 7263–7267.

Takada, Y., Skidgel, R.A. and Erdös, E.G. (1985). Purification of human urinary prokallikrein. Identification of the site of activation by the metalloproteinase themolysin. Biochem. J. 232, 851–858.

Takahashi, S., Irie, A. and Miyake, Y. (1988). Primary structure of human urinary prokallikrein. J. Biochem. 104, 22–29.

Takenobu, Y., Takaoka, M., Suyuma, E., Yano, M. and Mormoto, S. (1990). Conversion of rat urinary prokallikrein to its active form. Biochem. Int. 21, 417–423.

Thomas, K.A., Baglan, N.C. and Bradshaw, R.A. (1981). The amino acid sequence of the gamma-subunit of mouse submaxillary gland 7s nerve growth factor. J. Biol. Chem. 256, 9156–9166.

Tiffany, C.W. and Burch, R.M. (1989). Bradykinin stimulates tumour necrosis factor and interleukin-1 release from macrophages. FEBS Lett. 247, 189–192.

Tschesche, H., Mair, G., Godec, G., Fiedler, F., Ehret, W., Hirschauer, C., Lemon, M. and Fritz, H. (1979). The primary structure of porcine glandular kallikreins. In "Kinins II: Biochemistry, Pathophysiology and Clinical Aspects" (eds. S. Fuji, H. Moriya and T. Suzuki), pp. 245–260. Plenum Press, New York.

Tschesche, H., Kohnert, U., Fedrowitz, J. and Oberhoff, R. (1989). Tissue kallikrein effectively activates latent degrading metalloenzymes. Adv. Exp. Med. Biol. 247A, 545–548.

Uchida, K., Niinobe, M., Kato, H. and Fuji, S. (1980). Purification and properties of rat stomach kallikrein. Biochim. Biophys. Acta 614, 501–510.

Ullrich, A., Gray, A., Wood, W.I., Hayflick, J. and Seeburg, P.H. (1984). Isolation of a cDNA clone coding for the γ-subunit of mouse nerve growth factor using a high-stringency selection procedure. DNA 3, 387–392.

Valdes, G., Corthorn, J., Scicli, G.A., Gaete, V., Soto, J., Ortiz, M.E., Foradori, A. and Saed, G. (1993). Uterine kallikrein in the early pregnant rat. Biol. Reprod. 49, 802–808.

Van der Graaf, F., Tans, G., Bouma, B.N. and Griffin, J.H. (1982). Isolation and functional properties of the heavy and light chains of human plasma kallikrein. J. Biol. Chem. 257, 14300–14305.

Van Leeuwen, B.H., Evans, B.A., Tregear, G.W. and Richards, R. (1986). Mouse glandular kallikrein genes. Identification, structure and expression of the renal kallikrein gene. J. Biol. Chem. 261, 5529–5535.

Van Leeuwen, B.H., Penschow, J.D., Coghlan, J.P. and Richards, R.I. (1987). Cellular basis for the differential response of mouse kallikrein genes to hormonal induction. EMBO J. 6, 1705–1713.

Vio, C.P. and Figueroa, C.D. (1985). Subcellular localization of renal kallikrein by ultrastructural immunocytochemistry. Kidney Int. 28, 36–42.

Vio, C.P., Roa, J.P., Silva, R. and Powers, C.A. (1990). Localization of immunoreactive glandular kallikrein lactotrophs of the rat anterior pituitary. Neuroendocrinology 51, 10–14.

Wang, C., Chen, Y.P., Chao, L. and Chao, J. (1995a). Regulatory elements in the promotor region of the renal kallikrein gene in normotensive vs hypertensive rats. Biochem. Biophys. Res. Commun. 217, 113–122.

Wang, C., Chao, L. and Chao, J. (1995b). Direct gene delivery of human tissue kallikrein reduces blood pressure in spontaneously hypertensive rats. J. Clin. Invest. 95, 1710–1716.

Watt, K.W.K., Lee, P.J., M'Timkulu, T., Chan, W.-P. and Loor, R. (1986). Human prostate-specific antigen: structural and functional similarity with serine proteases. Proc. Natl Acad. Sci. USA 83, 3166–3170.

Webber, M.M., Waghray, A. and Bello, D. (1995). Prostate-specific antigen, a serine protease, facilitates human prostate cancer cell invasion. Clin. Cancer. Res. 1, 1089–1094.

Weinberg, M., Belknaps, S., Trebbin, W. and Solomon, R.J. (1987). Effects of changing salt and water balance on renal kallikrein, kininogen and kinin. Kidney Int. 31, 836–841.

Winderickx, J., Swinnen, K., Van Dijck, P., Verhoeven, G. and Keyns, W. (1989). Kallikrein-related protease in the rat ventral prostate: cDNA cloning and androgen regulation. Mol. Cell. Endocrinol. 62, 217–226.

Wines, D.R., Brady, J.M., Pritchett, D.B., Roberts, J.L. and MacDonald, R.J. (1989). Organization and expression of the rat kallikrein gene family. J. Biol. Chem. 264, 7653–7662.

Wines, D.R., Brady, J.M., Southard, E.M. and MacDonald, R.J. (1991). Evolution of the rat kallikrein gene family: gene conversion leads to functional diversity. J. Mol. Evol. 32, 476–492.

Wolf, D.A., Schulz, P. and Fittler, F. (1992). Transcriptional regulation of prostate kallikrein-like genes by androgen. Mol. Endocrinol. 6, 753–762.

Woodley-Miller, C., Chao, J. and Chao, L. (1989). Restriction fragment length polymorphisms mapped in spontaneously hypertensive rats using kallikrein probes. J. Hypertens. 7, 865–871.

Worthy, K., Figueroa, C.D., Dieppe, P.A. and Bhoola, K.D. (1990). Kallikreins and kinins: mediators in inflammatory joint disease? Int. J. Exp. Pathol. 71, 587–601.

Xiong, W., Chao, L. and Chao, J. (1989). Renal kallikrein mRNA localization by in situ hybridization. Kidney Int. 35, 1324–1329.

Xiong, W., Chen, L. and Chao, J. (1990a). Purification and characterization of a kallikrein-like T-kininogenase. J. Biol. Chem. 265, 2822–2827.

Xiong, W., Chen, L., Woodley-Miller, C., Simson, J.V. and Chao J. (1990b). Identification, purification and localization of tissue kallikrein in rat heart. Biochem. J. 267, 639–646.

Xiong, W., Chao, J. and Chao, L. (1995). Muscle delivery of human kallikrein gene reduces blood pressure in hypertensive rats. Hypertension 25, 715–719.

Yamaguchi, Y.T., Carretero, O.A. and Scicli, A.G. (1991). A

novel serine protease with vasoconstrictor activity coded by the kallikrein gene S3. J. Biol. Chem. 266, 5011–5017.

Young, C.Y., Andrews, P.E., Montgomery, B.T. and Tindall, D.J. (1992). Tissue-specific and hormonal regulation of human prostate-specific glandular kallikrein. Biochemistry 31, 818–824.

Yu, H. and Diamandis, E.P. (1995a). Prostate-specific antigen in milk of lacating women. Clin. Chem. 41, 54–58.

Yu, H. and Diamandis, E.P. (1995b). Prostate-specific antigen immunoreactivity in amniotic fluid. Clin. Chem. 41, 204–210.

Yu, H., Diamandis, E.P. and Sutherland, D.J.A. (1994). Immunoreactive prostate-specific antigen levels in female and male breast tumors and its association with steroid hormone receptors and patient age. Clin. Chem. 27, 75–79.

Yu, H., Diamandis, E.P., Monne, M. and Croce, C.M. (1995a). Oral contraceptive-induced expression of prostate-specific antigen in the female breast. J. Biol. Chem. 270, 6615–6618.

Yu, H., Diamandis, E.P., Levesque, M., Asa, S.L., Monne, M. and Croce, C.M. (1995b). Expression of the prostate-specific antigen gene by a primary ovarian carcinoma. Cancer Res. 55, 1603–1606.

Zuraw, B.L., Sugimoto, S., Parsons, C.L., Hugli, T., Lotz, M. and Koziol, J. (1994). Activation of urinary kallikrein in patients with interstitial cystitis. J. Urol. 152, 874–878.

6. Immunological Probes for the Bradykinin B₂ Receptor. A Toolbox

Werner Müller-Esterl

1. Introduction

Receptors for kinins are classified as two major subtypes, B₁ and B₂, although other subtypes may exist (see Chapter 2). B₁ receptors are activated by carboxy-terminally truncated kinins, such as desArg[10]-Lys-BK whereas bradykinin (BK) and kallidin (Lys-BK) are B₂ receptor agonists. Molecular cloning has revealed the primary structures of B₁ (Menke *et al.*, 1994) and B₂ receptors (McEachern *et al.*, 1991), and identified them as members of the G protein-coupled receptor family characterized by seven membrane-spanning α-helices (see Chapter 3). In some tissues, B₁ receptor expression is induced by cytokines such as interleukin-1, whereas the B₂ receptor is thought to be expressed constitutively, and is ubiquitous (Marceau, 1995; see Chapters 2 and 8).

The signaling pathways of B₂ receptors have been explored in detail (Farmer and Burch, 1992). B₂ receptors are preferentially coupled to G proteins of the G_q subtype (Gutowski *et al.*, 1991), which activate the phospholipase (PL) C-mediated cascade. This results in phosphatidylinositol (PI) turnover, and a transient rise in intracellular free Ca^{2+} ($[Ca^{2+}]_i$) (Lee *et al.*, 1993). The initial increase in $[Ca^{2+}]_i$ is followed by Ca^{2+} extrusion, which counteracts the Ca^{2+} influx (Quitterer *et al.*, 1995), and serves to regulate $[Ca^{2+}]_i$. B₂ receptor-mediated release of diacylglycerol, another hydrolysis product of PLC, results in translocation of specific protein kinase C isoforms (Tippmer *et al.*, 1994). B₂ receptors also couple to the PLA_2 pathway, which releases the prostaglandin precursor, arachidonic acid (Burch and Axelrod, 1987).

The Kinin System
ISBN 0–12–249340–0

Copyright © 1997 Academic Press Limited
All rights of reproduction in any form reserved.

Though the amino-acid sequence of the B_2 receptor has been deduced from its cDNA, and its transmembrane topology has been predicted from the corresponding hydropathy plots (see Chapter 3), the structure–function relationships, localization and regulation of B_2 receptors have remained elusive. To address these issues, we have developed an immunological "toolbox" for the B_2 receptor, which contains four sets of gadgets: (1) anti-idiotypic antibodies that are cross-reactive with the receptor; (2) antiligand antibodies for the cross-linked receptor–ligand complex; (3) antipeptide antibodies to the predicted loops which cross-react with the receptor; and (4) antibodies to the purified receptor (still under construction). In the present article, I review the procedures that led to these specific immunological probes, and their applications for the localization and functional analysis of the B_2 receptor.

2. Immunological Tools for the B_2 Receptor – Methodology

2.1 ANTI-IDIOTYPIC ANTIBODIES

The anti-idiotypic antibody (AIA) approach is a powerful strategy that allows construction of receptor-directed probes without prior cloning or isolation of a given receptor. In 1991, when Jarnagin and coworkers first cloned the rat BK B_2 receptor, we had developed and characterized AIAs that cross-react with B_2 receptors from several species, suggesting that the antibodies bore the internal image of an epitope associated with the endogenous "antigen", BK itself (Haasemann et al., 1991). These antibodies have proven useful in the initial isolation and characterization of the B_2 receptor protein (Abd Alla et al., 1993), and in mapping of a BK binding site exposed by the receptor (Abd Alla et al., 1996a).

The strategy to produce AIA to the B_2 receptor was comprised of three steps (Fig. 6.1) as follows. First, monoclonal antibodies (mAbs) against BK were generated by somatic cell fusion. These were isotyped as immunoglobulin G1 (IgG1) kappa type, and their target epitopes mapped with BK, Lys-BK and kinin receptor antagonists, and fragments thereof, revealing three sets of mAb: MBK1, MBK2 and MBK3 (Table 6.1). When the immunological binding affinities of these mAbs were compared with the known pharmacological specificities of BK analogs, striking similarities were noted between MBK3 and the B_2 receptor. AIAs against MBK3 were raised in rabbits and sheep. Inhibition and competition experiments on the level of the antigen (ligand), the idiotype, and the anti-idiotype demonstrated their mutual specificities (Haasemann et al., 1991). Thus, AIAs against MBK3 recognized a particular idiotype, which was conformation dependent and associated with the antigen-binding site of the antibody (MBK3).

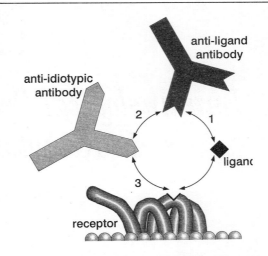

Figure 6.1   Rationale of the anti-idiotypic approach. The strategy is based on the assumption that the structure of a paratope of an idiotypic antiligand antibody resembles the structure of the ligand's receptor binding site. Anti-idiotypic antibodies raised against the paratope of the idiotypic antibodies, which bear the internal image of the ligand, can cross-react with the cognate receptor. The strategy comprises: (1) production, characterization and purification of idiotypic antiligand antibodies; (2) generation, characterization and selection of anti-idiotypic antibodies directed to the paratopes of the idiotypic antibody; (3) analysis of the structural and functional cross-reactivity of anti-idiotypic antibodies with the corresponding receptor.

Table 6.1   Antibodies to ligands for the bradykinin B_2 receptor

Ligand	Linker	Carrier	Antibody
Bradykinin	EDC	BSA	MBK1, MBK2, MBK3
Bradykinin	EDC	KLH	AS348
Hoe 140	EDC	KLH	MHO1, MHO2
Hoe 140	EDC	KLH	AS255

EDC, 1-ethyl-3-(3-dimethylaminopropyl-)carbodiimide–HCl; BSA, bovine serum albumin; KLH, keyhole limpet hemocyanin; Hoe 140, DArg-[Hyp^3,Thi^5,DTic^7,Oic^8]-bradykinin (Hock et al., 1991).

Binding of this AIA of B_2 receptors of human fibroblasts and guinea-pig ileum demonstrated that these AIA cross-react with the corresponding receptor in both species. The AIA against MBK3 stimulated both PI turnover in human fibroblasts and prostaglandin biosynthesis in mouse fibroblasts, the latter being inhibited by a B_2 receptor antagonist, DArg-[Hyp^3,DPhe^7]-BK. Thus, this AIA bears the internal image of a BK epitope, and acts as a B_2 receptor agonist. In the following sections, I have detailed the key procedures that allowed the development of AIA to the B_2 receptor.

2.1.1 Preparation of Peptide Carrier Conjugates (Haasemann *et al.*, 1991)

In order to synthesize conjugates of BK and bovine serum albumin (BSA) by the carbodiimide method (Goodfriend *et al.*, 1964), 1 mg of BK and 0.5 mg BSA, corresponding to a molar ratio of 100:1, are dissolved in 1 ml of distilled water, and coupled by the addition of 50 μl of 1-ethyl-3-(3-dimethylaminopropyl-)carbodiimide–HCl (EDC, 200 mg/ml). The reaction is stopped by extensive dialysis against distilled water. Typically 5–6 moles of BK are coupled/mole BSA using [^{3}H]-BK as tracer. Alternatively, kinins are conjugated to keyhole limpet hemocyanin (KLH) by EDC.

2.1.2 Generation of Idiotypic Antibodies

2.1.2.1 *Immunization Scheme*

For producing mAbs to BK, female BALB/c mice are injected subcutaneously (s.c.) with 100 μg of the BK–BSA conjugate, emulsified in complete Freund's adjuvant. Boosters of 100 μg of conjugate in incomplete Freund's adjuvant are given at intervals of 4 weeks after the initial immunization. Four days before the fusion, 100 μg of the conjugate and 50 μg of the unmodified peptide, dissolved in 250 μl 0.15 M NaCl are injected intraperitoneally (i.p.). The fusion of 2×10^8 immune spleen cells with 5×10^7 X63/Ag8.653 myeloma cells is performed (Galfrè and Milstein, 1981) using 50% polyethylene glycol 4000 as the fusogenic agent. Screening for anti-BK Ig is done by the indirect enzyme-linked immunoassay (ELISA) described below. Colonies secreting specific antibodies are subcloned three times by the limiting dilution method. Multiple myelomas are induced by i.p. injections of 10^7 hybrid cells in BALB/c mice primed with 2,6,4,10-tetramethylpentadecane.

2.1.2.2 *Indirect ELISA (Engvall and Perlmann, 1971)*

To screen for specific antibodies, peptides (2 μg/ml in 16 mM Na$_2$CO$_3$, 35 mM NaHCO$_3$, pH 9.6) or peptide conjugates (0.5 μg/ml in 100 mM NaCl, 100 mM Na acetate, pH 5.5) are immobilized overnight on microtiter plates. The plates are washed with phosphate-buffered saline (PBS), pH 7.4, and serial dilutions (2^n) of the hybridoma supernatants (starting dilution of 1:2, v/v), immune sera (1:50) or purified antibodies (5 μg/ml) are made in PBS containing 0.05% Tween 20, 2% BSA (PBS/Tween/BSA), and applied to the titer plates for 2 h. After washing with PBS, a 0.2–0.5 μg/ml solution of a horseradish peroxidase-conjugated secondary antibody in PBS/Tween/BSA is added for 3 h, followed by the substrate solution of 0.1% (w/v) diammonium 2,2-azino-*bis*(3-ethyl-2,3-dihydrobenzthiazoline-6-sulfonate) (ABTS), 0.012% v/v H$_2$O$_2$ in 100 mM citric acid, 100 mM NaH$_2$PO$_4$, pH 4.5, for 30 min. The change in absorbance is read at 405 nm. All incubations are at 37°C except for the coating step which is at 4°C.

2.1.2.3 *Competitive ELISA*

Competitive ELISA is employed for the determination of the peptide specificity of the idiotypes. Serial dilutions (2^n) of 50 μg/ml of peptides are prepared, and aliquots (150 μl each) are preincubated for 90 min with an equal volume of the MBK anti-idiotypes (0.05–0.5 μg/ml). The mixtures (200 μl each) are transferred to titer plates coated with 2 μg/ml of BK (or desArg^9-BK) and incubated for 90 min. Bound MBK antibodies are detected by a peroxidase-labeled sheep antimouse Ig (0.5 μg/ml). Other conditions are as specified above for the indirect ELISA. To test for the mutual specificity of the various components of the idiotypic network, competitive ELISA tests are performed as detailed in Table 6.2.

2.1.2.4 *Radioimmunoassay*

For testing the specificity of anti-BK mAb, a competitive radioimmunoassay (RIA) is applied. Antibodies are dissolved in 10 mM NaH$_2$PO$_4$, pH 7.2, 140 mM NaCl, 30 mM ethylenediamine tetra-acetic acid (EDTA), 3 mM 1,10-phenanthroline, 0.5 g/l NaN$_3$, 0.2 g/l merthiolate (incubation buffer) containing 1% w/v ovalbumin (final antibody concentration 20 μg/ml). Serial dilutions (2^n) of the kinin peptides are made in the incubation buffer, and aliquots of 50 μl are mixed with equal volumes of the antibody solution. A total of 50 μl of a solution of [^{125}I]-Tyr^0-BK (4×10^5 cpm/ml) in the incubation buffer is added, and the mixture is incubated overnight at 4°C. The idiotypes are precipitated by an antiserum (from sheep) against mouse Ig in the presence of 3% w/v

Table 6.2   Production of anti-idiotypic antibodies (AIA). The components of the competitive ELISAs used for characterization of the antibodies are listed with the concentrations of the coating substances, binding proteins and competitors. The starting concentration of a serial dilution (2^n) is given for each competitor. Peroxidase-labeled secondary antibodies against rabbit Ig (from sheep) or mouse Ig (from rabbit) were used throughout (0.2–0.5 μg/ml)

Experimental objective	Coating antigen	Reporter antibody	Competitor
Peptide specificity	MBK3 0.5 μg/ml	AIA3 0.03 μg/ml	Peptides[a] 50 μg/ml
Idiotype specificity	MBK3 0.5 μg/ml	AIA3 0.03 μg/ml	MBK3 100 μg/ml
Anti-idiotype specificity	Bradykinin 2 μg/ml	MBK3 0.5 μg/ml	AIA3 10 μg/ml
Anti-idiotype specificity	MBK3 0.5 μg/ml	AIA3 0.02 μg/ml	B$_2$ receptor 5–350 fmol/ml

MBK3, monoclonal antibody to BK (idiotype); AIA3, anti-idiotypic antibodies to MBK3.
[a] "Peptides" designates bradykinin receptor agonists and antagonists and other analogs (Haasemann *et al.*, 1991).

polyethyleneglycol 6000. The precipitate is pelleted by centrifugation ($3,800g$, 20 min), washed, and the bound radioactivity measured in a γ-counter. For mAb MBK1, which bound poorly to Tyr^0-BK, the Farr (1958) technique was applied as follows. Serial dilutions (4^n) of the peptides in PBS (starting with 1 $\mu g/ml$) are mixed with 50 μl of a solution of MBK1 (2.5 $\mu g/ml$) in PBS containing 2 mg/ml of bovine nonimmune Ig and 6×10^5 cpm/ml of [^{3}H]-desArg^{10}-Lys-BK (specific activity 80 Ci/mmol), and the mixture is incubated for 90 min at 25°C. Antibodies are precipitated by addition of a saturated $(NH_4)_2SO_4$ solution, and the precipitate is filtered over filters pretreated with polyethyleneimine (2 g/l) for 2 h. The filters are washed with 50% w/v $(NH_4)_2SO_4$, and the bound radioactivity is quantified in a liquid scintillation counter.

2.1.2.5   Antibody Purification and Modification (Hock et al., 1990; Haasemann et al., 1991; Abd Alla et al., 1996a)

mAbs are purified from ascites by 45% w/v $(NH_4)_2SO_4$-precipitation followed by DE-52 cellulose chromatography. To select specific antibodies (Müller-Esterl et al., 1988), affinity chromatography is performed on BK-BSA conjugates (10 mg) covalently attached to Affigel 10 (3 ml). Homogeneity of the purified Ig is assessed by sodium dodecyl sulfate (SDS) electrophoresis and/or isoelectric focusing (pH 3–9). For biotinylation, the affinity-purified antibodies are dialyzed against 0.1 M NaHCO$_3$, pH 8.0, and diluted to 1 mg/ml. A total of 100 μl of biotin-ϵ-aminocaproyl-N-hydroxysuccinimide (biotin-X-NHS) in dimethyl sulfoxide (DMSO) (Subba Rao et al., 1983) is added to 1 ml of the Ig solution (18-fold molar excess of the biotin-X-NHS over protein). After incubation for 4 h at 25°C, the mixture is dialyzed against PBS. The recovery of the biotinylated Ig is 65–80% of the starting material. For radioiodination the affinity-purified antibodies (1 mg) are dissolved in 100 ml of PBS and incubated with 2 mCi carrier-free Na[^{125}I] on a solid phase of Iodogen (100 mg/tube) for 10 min (Fraker and Speck, 1978) to a specific activity of 5–15 $\mu Ci/\mu g$. Unreacted iodine is separated by gel filtration over a Sephadex G50 column or by anion exchange chromatography over Dowex-1.

Affinity-purified antibodies are proteolyzed with papain following established procedures (Mage, 1980). The resultant F(ab) and Fc fragments are separated on Mono Q using a linear gradient of 0–300 mM NaCl in 20 mM Tris, pH 7.5. Elution is followed by the indirect ELISA using the cognate antigen or applying specific antisera against mouse Fab and Fc, respectively.

2.1.3   Production of Anti-idiotypic Antibodies

Affinity-purified MBK antibodies (from mouse) are used as the antigens for production of AIA in rabbits or sheep. To monitor the anti-idiotypic response, sandwich ELISA is performed in a competitive version ("anti-idiotypic detection assay"; AIDA) (Haasemann et al., 1991). Serial dilutions (2^n) of anti-idiotypic antisera (starting dilution of 1:20) are mixed with an equal volume (100 μl each) of 2 $\mu g/ml$ BK-BSA conjugate, and applied to a titer plate precoated with the idiotypic antibody. Bound conjugate is detected by biotinylated anti-BK antibodies (from sheep, 1 $\mu g/ml$) followed by the preformed biotin–avidin peroxidase complex (3 $\mu g/ml$). The presence of AIA, but not of anti-allotypic or anti-isotypic antibodies, results in a reduction of conjugate binding. For affinity purification, the majority of the anti-allotypic and anti-isotypic antibodies are removed by adsorption on immunoaffinity matrices prepared from CNBr-activated Sepharose and nonimmune polyclonal mouse Ig or unrelated monoclonal mouse IgG1-kappa. AIA are positively selected from the depleted immune sera on Sepharose-bound idiotypic Ig.

2.1.4   Applications of Anti-idiotypic Antibodies

2.1.4.1   Binding of [^{125}I]-Labeled Anti-idiotypes to the B$_2$ Receptor (Abd Alla et al., 1993)

Human foreskin fibroblasts ($10^6/100$ μl corresponding to 30–50 fmol of the receptor) (Roscher et al., 1983) are suspended in Dulbecco's PBS containing 0.5 mM EDTA, and washed three times with ice-cold Roswell Park Memorial Institute (RPMI) 1640 medium including 1 mg/ml BSA and 20 mM N-2-hydroxyl-ethylpiperazine-N'-2 ethane sulphonic acid (HEPES), pH 7.2. The cells are resuspended in ice-cold Hanks balanced salt solution (NaCl substituted by KCl) including 2 mM bacitracin, 1 mM phenylmethane-sulfonylfluoride (PMSF), 1 $\mu g/ml$ N-[N(L3-trans-carboxyoxiran-2-carbonyl)-Lleucyl]-agmatin (TCLA), 2 μM enalapril and 1 mg/ml BSA. The cells are incubated under gentle agitation for 4 h at 4°C with 1.0 μCi of [^{125}I]-labeled AIA (from rabbit; final concentration 5×10^{-8} M) in the presence or absence of 100-fold molar excess of the unlabeled ligand or of an nonspecific antibody (preimmune rabbit Ig). Receptor-bound AIA are separated by centrifugation ($9,000g$, 2 min, 4°C) of the suspension through 500 μl of a mixture of dibutylphthalate and 1-bis(2-ethyl-hexyl)phthalate (1.1/1.0, v/v) (Schreurs et al., 1990). Cell-bound radioactivity is measured in a γ-counter.

2.1.4.2   Stimulation of Prostaglandin Synthesis in Mouse Fibroblasts (Haasemann et al., 1991)

SV-T2 murine fibroblasts are cultured in 24-well plates to about 50% confluence in Dulbecco's modified Eagle's medium containing 10% v/v calf serum. For the prostaglandin biosynthesis assay (Burch and Axelrod, 1987), media are aspirated and the wells washed with the serum-free medium containing 20 mM HEPES, pH 7.0. To each well, 200 µl of the medium containing DArg-[Hyp^3,DPhe^7]-BK, a B$_2$ receptor antagonist, or vehicle is added, and the incubation is continued for 10 min. Media are aspirated then replaced with the identical media containing 0–30 µg/ml of the AIA and 0–100 nM DArg-[Hyp^3,DPhe^7]-BK. In controls, 1 µM BK is applied to induce maximum PGE$_2$ synthesis. Incubations are for 5 min throughout. Aliquots of media are taken for RIA of PGE$_2$.

2.1.4.3 Affinity Chromatography of the B$_2$ Receptor (Abd Alla et al., 1993)

Affinity-purified AIA are coupled to Affigel 10 (25 mg/ml). The resultant gel is extensively washed with 100 ml of 0.2 M glycine, pH 2.2, followed by an equal volume of PBS. For receptor solubilization, membrane suspensions of human foreskin fibroblasts (1 ml) are treated for 50 min at 4°C with 10 ml solubilization buffer (20 mM piperazine-N,N′-bis(2-ethanesulfonic acid) (PIPES), 5 mM EDTA, 1 mM PMSF, 1 µg/ml TCLA, pH 6.8, containing 4 mM 3-[(3-cholamidopropyl)dimethyl-ammonio]-l-propane sulfonic acid (CHAPS)). The solubilizate is centrifuged for 40 min at 100,000g to remove particulate material, and the clear supernatant collected. Typical binding activity is 3–5 pmol/mg protein. A total of 50 ml of the solubilizate (from 12 mg of crude fibroblast membranes) is added to the gel (1 ml), and incubated for 4 h at 4°C under gentle agitation. The immunoaffinity matrix is washed three times with 50 ml 5 mM EDTA, 150 mM NaCl, 1% w/v Triton X-100, 20 mM Tris(hydroxymethyl)-aminomethane (Tris)–HCl, pH 7.4, and then with 10 ml 0.2 M NaSCN, pH 5.6, to remove nonspecifically bound proteins. The bound receptor is eluted with 0.2 M glycine, pH 2.2, and immediately neutralized with 1 M KOH. Eluates from 12 cycles are combined and concentrated by centrifugation through a Centricon unit (exclusion limit of 10 kDa) using the CHAPS solubilization buffer. The final solution containing the purified receptor is divided into aliquots and stored at −20°C until use. The overall recovery of functionally active receptor is estimated to be 2–5% of the starting activity suggesting that major losses occur during the isolation procedure due to denaturation and/or incomplete binding of the receptor.

2.2 ANTILIGAND ANTIBODIES

An alternative method to probe for hormone receptors is the ligand conjugate/antiligand (LAL) approach (Fig. 6.2) (Figueroa et al., 1995), wherein ligands, such as BK as an agonist, or Hoe 140 (DArg-[Hyp^3,Thi^5, DTic^7,Oic^8]-BK) as an antagonist, are coupled covalently to a carrier protein. These ligand conjugates are bound to the B$_2$ receptor, and the resultant ligand:conjugate–receptor complex is indirectly detected by antibodies to the respective ligand, i.e., anti-BK antibodies (MBK1,

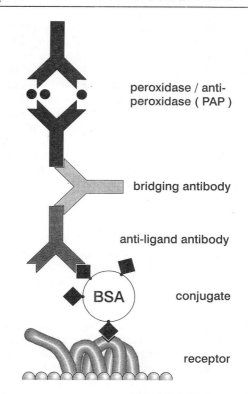

Figure 6.2   Schematic representation of the ligand conjugate/antiligand technique that uses ligand molecules coupled to bovine serum albumin (BSA). After B$_2$ receptor binding, ligands are visualized by the antiligand antibody, a bridging antibody, the peroxidase–antiperoxidase complex and the chromogenic/fluorogenic substrate (from bottom to top).

peroxidase / anti-peroxidase (PAP)

bridging antibody

anti-ligand antibody

BSA   conjugate

receptor

MBK2, MBK3) or anti-Hoe 140 antibodies (MHO1, MHO2) (Haasemann et al., 1991; Abd Alla et al., 1993) (see Table 6.1). Coupling of the B$_2$ ligands to a carrier protein is essential because application of "free" Hoe 140 or BK produces a very weak signal at ligand concentrations ⩽1 mM. This may indicate that the receptors' interaction sites for the ligands overlap their antigenic epitopes, i.e., once bound to the receptor, the ligands are no longer accessible for their antibodies and vice versa. The utilization of BSA conjugates eliminates the problems associated with the antigenic recognition of the free ligands, and provides a sensitive and specific method that rests on the presence of at least two accessible ligands on the surface of the carrier protein. The LAL approach to probe for receptor-binding sites may complement the traditional strategies using radiolabeled ligands or antireceptor antibodies. The LAL strategy is particularly useful in settings where antibodies to a given receptor are not readily available, and agonists or antagonists are. Also, probing for "functional" receptors with intact binding sites may be accomplished by LAL, e.g., in cells transfected with a receptor DNA. One limitation may occur when the chemical coupling of the ligand interferes with its intrinsic affinity for the corresponding receptor. Because the LAL method rests on the preservation of the ligand binding site in the receptor, the fixation and processing conditions of the sample are critical. Furthermore, application of anti-ligand antibodies might yield false-positive results in settings where endogenous ligands and/or their precursors are present. This latter problem is circumvented by utilizing unnatural ligands such as Hoe 140. Below, are summarized the critical methods that allowed the development of antiligand antibodies as probes for ligand-bound B_2 receptors.

2.2.1   Ligand–carrier Conjugates

2.2.1.1   *Preparation of the Conjugates (Figueroa et al., 1995)*

Coupling of B_2 receptor ligands, such as BK and Hoe 140, to BSA or KLH is performed using EDC (see Section 2.1.1). For controls, a conjugate of angiotensin II and BSA is prepared. Typically a molar ratio of 1.39 ± 0.05 of peptide over carrier protein is found by quantitative amino-acid analysis of a Hoe 140-BSA conjugate using β-2-thienylalanine and 4-hydroxyproline as the reference residues (Figueroa *et al.*, 1995). This indicates that ≥30% of the conjugate population has two or more Hoe peptides conjugated per molecule of BSA. To avoid potential problems due to varying coupling efficiency it is recommended to use a single-conjugate preparation throughout a study.

2.2.1.2   *Characterization of Ligand Conjugates (Figueroa et al., 1995)*

A competitive receptor-binding assay is performed (Abd Alla *et al.*, 1993). Membranes prepared from human foreskin fibroblasts are incubated for 120 min at 4°C with 2 nM (0.1 μCi) [2,3-prolyl-3,4-^{3}H]-BK and 2 mM bacitracin in 20 mM PIPES, pH 6.8, in the absence or presence of increasing concentrations (1 pM–10 μM) of competitor. Protein-bound and protein-free ligand are separated by filtration over glass filters precoated with 0.3% w/v polyethyleneimine. Filter-bound radioactivity is measured in a liquid scintillation counter, and specific binding is calculated as the difference between total and nonspecific binding.

2.2.2   Production of Antiligand Antibodies (Abd Alla *et al.*, 1993; Figueroa *et al.*, 1995)

mAb are raised in mice by s.c. injection of 100 μg of the Hoe 140-KLH conjugate emulsified in complete Freund's adjuvant (see Section 2.1.2.1). Polyclonal antibodies are raised according to standard procedures (see Section 2.3.1.2). The specific immune response is followed by indirect ELISA on microtiter plates coated with 2 μg/ml of the peptides or 0.5 μg/ml of the conjugates (see Section 2.1.2.2). Affinity chromatography is performed on Hoe 140 covalently coupled to Affigel 10 (10 mg of peptide per ml gel). In this way specific murine mAb to Hoe 140, MHO1 and MHO2 (IgG subtype), and rabbit polyclonal antiserum to Hoe 140 were produced (Abd Alla *et al.*, 1993).

2.2.3   Application of Antiligand Antibodies

2.2.3.1   *Affinity Cross-linking of B_2 Receptors (Abd Alla et al., 1996a)*

CHAPS-solubilized membrane fractions from human fibroblasts (200–500 μg of total protein corresponding to 1–2 pmol of B_2 receptor protein) are incubated with saturating concentrations of the ^{125}I-labeled ligand in 2 ml of solubilization buffer for 60 min at 4°C (see Section 2.1.4.3). Solutions (100 mM) of the homo-bifunctional cross-linkers, disuccinimidyl tartrate or 1,5-difluoro-2,4-dinitrobenzene in DMSO are added to the membrane suspension to a final concentration of 1 mM. The suspension is shaken for 90 min at 4°C, and 200 μl of 1 M Tris, pH 7.2, is added to terminate the reaction. Following an additional incubation for 10 min, the membranes are pelleted (10,000*g*, 20 min, 4°C), and the resultant cross-linked receptor preparations are stored at −70°C until use.

2.2.3.2   *Affinity Chromatography of the Cross-linked B_2 Receptor (Abd Alla et al., 1993)*

To facilitate isolation of the ligand-bound receptor, Hoe 140 (1 nM) is cross-linked to the receptor solubilized from fibroblast membranes (12 mg of protein, corresponding to 36–60 pmol of receptor) by difluorodinitrobenzene. The reaction is stopped by the addition of 0.1 M Tris, and the unbound ligand removed by gel filtration over Sephadex G50. The resultant protein mixture (50 ml) is applied to an immunoaffinity column prepared of affinity-purified rabbit antibodies to Hoe 140 covalently coupled to Affigel 10, essentially as described in Section 2.1.4.3. Recovery of the receptor protein typically ranges from 1% to 2% of the starting material, as judged from parallel experiments where the [^{125}I]-HPP-Hoe 140-labeled receptor has been applied.

2.2.3.3   *Indirect Immunodetection of the B_2 Receptor (Figueroa et al., 1995)*

Dewaxed and rehydrated renal tissue sections, fixed in periodate–lysine–paraformaldehyde are washed with 0.05 M Tris–HCl, pH 7.6, containing 1 mM EDTA, 1 mM EGTA, 20 μM captopril, 0.1% w/v BSA, and incubated with concentrations of BK-BSA or Hoe 140-BSA (18 h, 4°C). Ligand–conjugate concentrations range from 0.05 to 20 μg/ml. For comparison, 10^{-7}–10^{-3} M of uncoupled BK or Hoe 140 are used to incubate consecutive tissue sections. After incubation, the sections are washed three times for 3 min with ice-cold 0.05 M Tris–HCl, pH 7.6, containing 1 mM EDTA, 1 mM EGTA, 20 μM captopril, and 0.1% w/v BSA, air-dried and fixed with acetone for 5 min. The sections are pretreated with methanol and 1% v/v H$_2$O$_2$ in 0.05 M Tris–HCl buffer, pH 7.6, for 15 min each to block pseudo-peroxidase activity. Bound ligand conjugates are detected by the peroxidase/antiperoxidase method (Sternberger et al., 1970) using specific antiserum to Hoe 140 (AS255, 1:1,000), mAb to Hoe 140 (MHO1, culture supernatant, 1:100; MHO2, ascites fluid, 1:5,000), or mAb to BK (MBK1, MBK2, MBK3; affinity purified mAb; 10 μg/ml) diluted in 0.05 M Tris–HCl, pH 7.6, containing 1 mM EDTA, 1 mM EGTA, 20 μM captopril and 0.1% w/v BSA. Controls are carried out: (1) by omission of the ligand conjugate from the incubation mixture; (2) by omission of the primary antibody; or (3) by replacement of the primary antibody with an isotype-matched mAb (IgG1). Specificity of the conjugate binding is challenged by competition experiments using an excess of uncoupled BK (10^{-3} M) and/or Hoe 140 (10^{-3} M) for conjugate displacement.

2.3   ANTIPEPTIDE ANTIBODIES

In recent years, an increasing number of G protein-coupled peptide hormone receptors have been pharmacologically characterized and defined by cloning of cDNAs (Van Rhee and Jacobson, 1996). In contrast, purification of the corresponding receptor proteins has proven extremely difficult. In this situation, i.e., the availability of the predicted protein structure and the absence of efficient procedures to produce highly purified receptor proteins, synthesis of peptide segments derived from the predicted protein structure of these receptors and the generation of antibodies to these peptides have proven extremely valuable. In subsequent paragraphs, I review the methods that have been successfully used to produce a large variety of antipeptide antibodies to the B$_2$ receptor.

2.3.1   Production, Characterization and Derivatization of Antipeptide Antibodies

2.3.1.1   Peptide Synthesis and Coupling (Figueroa et al., 1995; Abd Alla et al., 1996a)

Peptide segments are selected from the sequences of rat and human B$_2$ receptors (McEachern et al., 1991; Hess et al., 1992) by the criteria of hydrophilicity, relative location in the receptor protein and presence of a cysteine residue, if present, for coupling (Fig. 6.3). Peptides covering the predicted eight intracellular and extracellular domains of the transmembrane-spanning receptor are synthesized using the fluorenylmethyloxycarbonyl (Fmoc) or the t-butyloxycarbonyl (tBoc) chemistry (Table 6.3). The identity of the peptides, purified by high-performance liquid chromatography (HPLC), is routinely established by Edman degradation and electron-spray mass spectrometry. After characterization, each peptide is covalently coupled to KLH via bifunctional cross-linking using N-succinimidyl-6-maleimido capronic acid (EMCS) (Liu et al., 1979) or EDC (Goodfriend et al., 1964).

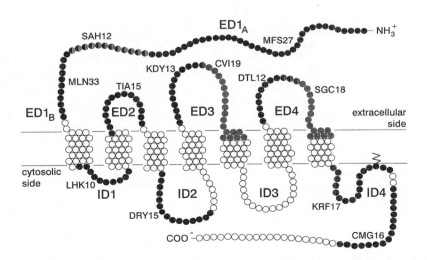

Figure 6.3   Positions of the peptides in the rat B$_2$ receptor. A model of the B$_2$ receptor topology based on hydrophobicity plots and a transmembrane α-helix hypothesis is presented. The positions of the peptides from extracellular domains (ED) and intracellular (ID) domains used to raise antisera are marked by filled or shaded circles. For nomenclature and compositions of the synthetic peptides, see Table 6.3.

Table 6.3 Peptides used for the production of rabbit antisera to the rat B_2 receptor

Peptide (designation)[a]	Linker[b]	Domain[c]	Positions[d]
MFSPWKISMFLSVREDSVPTTASFSAD (MFS27)	—	ED1$_A$	—[e]
MLNVTLQGPTLNGTFAQSKCPQVEWLGWLNTIQ (MLN33)	—	ED1$_B$	1–33
SAHNGTFSEVNC (SAH12)	EMCS	ED1	11–22
TIANNFDWLFGEVLC (TIA15)	EMCS	ED2	91–105
KDYREEGHNVTAC (KDY13)	EMCS	ED3	174–186
CVIVYPSRSWEVFTNMLNN (CVI19)	EMCS	ED3	186–204
DTLLRLGVLSGC (DTL12)	EMCS	ED4	268–279
SGCWNERAVDIVTYQISSY (SGC18)	EMCS	ED4	277–294
LHKTNCTVAE (LHK10)	EMCS	ID1	59–68
DRYLALVKTMSMGRM<u>C</u> [f] (DRY16)	EMCS	ID2	129–143
CMGESVQMENSMGTLR (CMG16)	EMCS	ID4	331–346
KRFRKKSREVYQAI<u>S</u>RK [g] (KRF17)	EDC	ID4	312–328

[a] Peptides are identified by the first three amino-terminal residues using the one-letter code, followed by the total number of residues in that peptide.
[b] EMCS, N-succinimidyl-6-maleimido capronic acid; EDC, 1-ethyl-3-(3-dimethyl-amino-propyl) carbodiimide–HCl; the carrier protein is KLH.
[c] ED, extracellular domain; ID, intracellular domain (Sharif and Hanley, 1992).
[d] The relative positions of the peptides in the sequences of the rat B_2 receptor (McEachern et al., 1991) and of the human B_2 receptor (Hess et al., 1992) are indicated. The listed peptides are derived from the rat sequence except for MFS27 and MLN33 which are from the human sequence.
[e] Note that the sequence of MFS27 precedes the predicted amino terminus of the human B_2 receptor (Hess et al., 1992), and corresponds to the translation product derived from the first in-frame AUG codon of the human B_2 gene (Abd Alla et al., 1996b).
[f] The underlined cysteine residue is not present in the native sequence.
[g] The underlined serine residue substitutes for a cysteine residue in the native sequence.

2.3.1.2 Generation of Antipeptide Antibodies (Figueroa et al., 1995; Abd Alla et al., 1996a)

The peptide conjugates are used to elicit antisera in various species following established immunization procedures (Müller-Esterl et al., 1988). Larger peptides such as MLN33 can be used for immunization without prior coupling to a carrier protein (Table 6.3). Female New Zealand white rabbits or BALB/c mice are injected s.c. at multiple sites with 80 µg of the respective antigen in 1 ml of Freund's complete adjuvant. Intravenous booster injections of 100 µg antigen each are given after 14, 28 and 56 days of the first injection. The titers of specific antipeptide antibodies are followed by indirect ELISA using microtiter plates coated with 2 µg/ml of the unmodified peptide or 0.5 µg/ml of the uncoupled carrier protein (Kaufmann et al., 1993). This procedure allows the assessment of the antibody titers to the peptide and to the carrier protein. The same assay is employed to monitor the specificity of the antibodies using authentic antigen and unrelated peptides. Cross-reactivity of the antipeptide antibodies with the corresponding B_2 receptor is demonstrated by Western blot analysis (see below). When homologous peptides derived from kinin receptors of various species are available, indirect ELISA may help to judge the interspecies cross-reactivity among the antibodies.

2.3.1.3 Affinity Purification of Antibodies (Abd Alla et al., 1996a)

To purify the specific antibodies, peptides are covalently coupled to Affi-Gel 10 (5 mg/ml of gel). The antiserum (5 ml/ml of gel) is applied and incubated under gentle agitation for 12 h at 4°C. The affinity matrix is washed three times with PBS, the bound antibodies eluted with 0.2 M glycine, pH 2.5, and neutralized with 1 M KOH. Antibodies are desalted and concentrated using a 30,000 Da exclusion-limit filtration unit. Purity and specificity of the purified antibodies is analyzed by SDS–polyacrylamide gel electrophoresis (PAGE) and ELISA, respectively (see above). Biotinylation, radiolabeling and fragmentation of the purified antibodies is carried out according as in Section 2.1.2.5.

2.3.2 Application of Antipeptide Antibodies (Abd Alla et al., 1996a)

2.3.2.1 Western Blotting and Immunoprinting

For immunoprint analysis, proteins are resolved by SDS-PAGE (Laemmli, 1970) and transferred to polyvinyl-idenedifluoride (PVDF) sheets using semidry blotting (Kyhse-Anderson, 1984). The sheets are treated with 50 mM Tris, 0.2 M NaCl, pH 7.4, containing 5% w/v nonfat dry milk and 0.1% w/v Tween 20 for 1 h. Antisera are diluted 1:1,000 in 50 mM Tris, 0.2 M NaCl, pH 7.4, containing 2% w/v BSA. After 30 min incubation at 37°C the PVDF sheets are washed five times for 15 min each with 50 mM Tris, 0.2 M NaCl, pH 7.4, and incubated for 30 min with peroxidase-labeled goat antirabbit antibody, diluted 1:5,000 in the same buffer. After extensive washing, bound antibody is visualized by the chemiluminescence.

2.3.2.2 Flow Cytometry

For fluorescence-activated cell sorting (FACS), confluent HF-15 cells containing 0.5–1 pmol B_2 receptor per mg protein are harvested with PBS, 0.5 mM EDTA, pH 7.4. The cells are washed twice with ice-cold RPMI 1640 containing 0.1% w/v BSA, 20 mM Na-HEPES, pH 7.4 (incubation medium). The cells (1×10^6) are suspended in incubation medium containing the antisera, 1:100 v/v, and incubated for 1 h at 4°C. After washing three times, fluorescein isothiocyanate (FITC)-conjugated goat antirabbit Ig, diluted 1:80 v/v, is added. The cells are incubated for 1 h, washed, fixed with 2% formaldehyde, and analyzed by FACS.

2.3.2.3 *Immunoaffinity Purification of the Receptor*

Immunoselected antipeptide antibodies (see above) can be used for affinity purification of the receptor. A total of 15 mg of affinity-purified antibodies are covalently bound to 1 ml of Affi-Gel 10. Transfected cells overexpressing the B₂ receptor are used for receptor purification. Membranes from Sf9 cells infected with baculovirus encoding human B₂ cDNA (100 pmol B₂ receptor per 20 mg of total membrane protein) are solubilized with 2% w/v sodium deoxycholate in PBS, 1 mM PMSF, 1 µg/ml TCLA, and 2 mM leupeptin. After solubilization, the deoxycholate is diluted to 0.1% w/v by the addition of 20 mM HEPES, pH 7.4, containing 150 mM NaCl and 1 mM EDTA. Glycerol (final concentration 10% v/v) and Triton X100 (0.1% w/v) are added and the solution is applied to the immunoaffinity matrix overnight. The affinity matrix is washed extensively with 20 mM HEPES, pH 7.4, containing 150 mM NaCl and 1 mM EDTA, and the bound proteins are eluted with 0.2 M glycine, pH 2.5, supplemented with 10% v/v 1,4-dioxane. The protein fraction is neutralized with 1 M Tris, pH 8.0, and concentrated by filtering at an exclusion limit of 30,000 Da. The purity of the enriched B₂ receptor is assessed by SDS-PAGE and silver staining. Note that the apparent molecular mass of the recombinant B₂ receptor from Sf9 cells is markedly different from that of the native receptor, which is likely to be due to incomplete processing of the carbohydrate side chains (Abd Alla *et al.*, 1996a).

2.3.2.4 *Competition Studies with B₂ Receptor Ligands*

To map the binding sites for agonists and antagonist, domain-directed antipeptide antibodies can be employed in competition studies with radiolabeled ligands. Membranes or confluent HF-15 fibroblasts (Roscher *et al.*, 1983) on 24-well plates in RPMI 1640 buffered with 20 mM HEPES, pH 7.4, are incubated with [^{125}I]HPP-Hoe 140 (0.5 nM, specific activity 1367 Ci/mmol) or with [^{3}H]BK (2 nM, specific activity 98 Ci/mmol) in the presence of affinity-purified antibodies (5×10^{-11}–1×10^{-5} M). After 2 h incubation at 4°C, the cells are washed three times with ice-cold 20 mM HEPES, pH 7.4, dissolved in 1 M NaOH, and the radioactivity measured. Similarly, ligand-induced displacement of radiolabeled antipeptide antibodies against the B₂ receptor (see above)

can be studied. To this end, confluent HF-15 cells on 24-well plates are washed twice with the HEPES. Then 0.5 ml of HEPES containing 1 mM PMSF, 1 µg/ml TCLA, and 2 mM leupeptin is added to each well. The cells are incubated at 4°C with [^{125}I]-labeled immunoselected antibodies (5×10^{-8} M, specific activity 0.2 mCi/mg) in the presence or absence of 1×10^{-5} M BK or Hoe 140. After 2 h at 4°C, the cells are washed three times with ice-cold buffer, dissolved in 1 M NaOH, and the total cellular radioactivity determined.

2.3.3 Immunocytochemistry (Figueroa *et al.*, 1995, 1996)

2.3.3.1 *Tissue Preparation*

Tissues (kidney, striated skeletal muscle) are obtained under ether anesthesia from rats, and fixed in Bouin's fluid, 4% v/v formaline–saline or periodate–lysine–paraformaldehyde for 24 h at room temperature (McLean and Nakane, 1974). The samples are then dehydrated in ethanol and embedded for sectioning. Alternatively, tissues are frozen rapidly in liquid N₂ and sectioned immediately in a cryostat at −30°C. Frozen sections can be stored at −70°C.

2.3.3.2 *Peroxidase Method for Light Microscopy*

For light microscopy, immunostaining is performed according to the peroxidase–antiperoxidase (PAP) method of Sternberger *et al.* (1970) with modifications (Figueroa *et al.*, 1984a,b, 1988). Prior to immunostaining, frozen tissue is fixed for 10 min in periodate–lysine–paraformaldehyde or 3.7% v/v formalin, pH 7.4, whereas sections from embedded tissues are dewaxed and rehydrated. Both frozen and embedded tissue sections are treated with a mixture of absolute methanol and 10% v/v of H₂O₂ for 15 min, and washed with 50 mM Tris–HCl, 150 mM NaCl, pH 7.8. Sections are incubated overnight at 22°C with the antibodies diluted in 50 mM Tris–HCl, 150 mM NaCl, pH 7.8, supplemented with 1% w/v Ig-free BSA. The second antibody (antirabbit or antimouse IgG), at a dilution of 1:80, and the PAP complex (from rabbit or mouse), at 1:100, are applied for 30 min each. Peroxidase activity is visualized by incubating the sections in 0.1% w/v of 3,3′-diaminobenzidine and 0.03% v/v H₂O₂ for 15 min at room temperature in the dark. Counterstaining is with Harris' hematoxylin for 15 s, followed by dehydration and mounting.

2.3.3.3 *Peroxidase Method for Electron Microscopy*

For electron microscopy, tissue slices are fixed in periodate–lysine–paraformaldehyde for 4 h at room temperature as described. Sections (40 µm) are cut, washed with 50 mM Tris–HCl, 150 mM NaCl, pH 7.8, and incubated with the antipeptide antibodies at a dilution of 1:200 to 1:1,000 overnight at 4°C under constant agitation. After washing, the sections are immunostained by the PAP method (Figueroa *et al.*,

1984a,b). Alternatively, the sections are incubated with 10 nM gold-labeled antirabbit Ig. With this method, the reaction can be enhanced with silver methenamine (Figueroa *et al.*, 1984b; Rodriguez *et al.*, 1984). Sections are embedded in an epoxy mixture following standard procedures. Controls with nonimmune rabbit serum are performed simultaneously with incubation lasting for the same period.

2.3.3.4   *Immunofluorescence (Haasemann et al., 1994; Abd Alla et al., 1996a)*

For the immunofluorescence studies, A431 cells are grown on glass coverslips for 48 h. Three hours before the experiment, medium is replaced by RPMI 1640 supplemented with 0.5% v/v fetal calf serum. Cells are washed three times with 60 mM PIPES, 25 mM HEPES, 10 mM EDTA, 2 mM $Mg(CH_3COO)_2$, pH 6.9, and fixed for 30 min with 3% w/v paraformaldehyde in the same buffer adjusted to pH 7.5. Excess paraformaldehyde is quenched by the addition of 50 mM NH_4Cl in PBS, pH 7.4, followed by 30 min incubation with PBS, pH 7.4 containing 0.3% w/v gelatin. For immunostaining, the cells are treated for 1 h at room temperature with antipeptide antisera, 1:100 in 0.3% gelatin/PBS. The first antibody is detected by a rhodamine-coupled donkey antirabbit Ig, 1:100 in 0.3% gelatin/PBS. The coverslips are embedded in Moviol® and viewed under a microscope. With human neutrophils, cells are floated on to circular coverslips, washed with PBS, and fixed and permeabilized with 4% formaldehyde for 10 min followed by 0.2% Triton-X100 in PBS for 2 min. Unfixed cells remain untreated. The coverslips are washed three times for 10 min each with PBS containing 1% human IgG, 1% BSA and 0.2% NaN_3, which blocks nonspecific binding to the Fc receptors, and prevents internalization of B_2 receptors by unfixed cells. The cells are incubated with fluorescein-conjugated antirabbit $F(ab)_2$ for 1 h, washed and viewed.

2.3.3.5   *Specificity Controls*

The controls for the immunostaining procedures are carried out by omission of the specific antibody, by its replacement with nonimmune rabbit serum, or by preabsorption with the cognate antigen. To this end, the antiserum mixture (1:2,000) is preabsorbed overnight at 4°C with an excess (100 μg/ml) of the same peptides used to elicit the antisera to the B_2 receptor. The resultant immunoprecipitates are removed by centrifugation at $13,000g$ for 15 min, and the supernatant is used for staining as above.

3.   *Perspectives*

Anti-idiotypic antibodies, antiligand antibodies and antipeptide antibodies are useful tools in the study of B_2 receptors. This set of immunological tools may be complemented by antibodies to recombinant fusion proteins where portions of the B_2 receptor are expressed in a fused form with a bacterial protein such as maltose-binding protein or glutathione-S-transferase. The ultimate goal, however, is to raise polyclonal and mAbs that recognize the intact, native or recombinant B_2 receptor with high affinity and specificity. The purification of G protein-coupled receptors to apparent homogeneity is a notably extremely difficult task that has not yet been accomplished for the kinin B_2 receptor, at least on a preparative scale. The availability of specific antibodies will undoubtedly promote structural analyses, for example, of post-translational modifications such as phosphorylation (Blaukat *et al.*, 1996), acylation and/or glycosylation, of the B_2 receptor. Hopefully this immunbiological toolbox will spur exploration aspects such as receptor signal transduction, desensitization, internalization, downregulation, and/or recycling. Furthermore, it will boost the efforts of many researchers who localize B_2 receptors in various cells, tissues and organs of the human body. Last, but not least, well-defined antibodies or their fragments may aid in attempts to crystallize a G protein-coupled receptor in complexed form with a corresponding immunoglobulin fragment. *Per aspera ad astra.*

4.   *Acknowledgements*

I wish to express my gratitude to my colleagues and friends who have provided many gadgets for the immunological toolbox: Drs Said Abd Alla, Kanti Bhoola, Andree Blaukat, Jörg Buschko, Carlos Figueroa, Stella Grigoriev, Martina Haasemann, Kurt Jarnagin, Armin Maidhof and Ursula Quitterer. This work was supported in part by grants from the Deutsche Forschungsgemeinschaft (Mu 598/4-2), the Volkswagen Foundation (I-68857) and the Fonds der Chemischen Industrie (#163323).

5.   *References*

Abd Alla, S., Buschko, J., Quitterer, U., Maidhof, A., Haasemann, M., Breipohl, G., Knolle, J. and Müller-Esterl, W. (1993). Structural features of the human BK B_2 receptor probed by agonists, antagonists, and anti-idiotypic antibodies. J. Biol. Chem. 268, 17277–17285.

Abd Alla, S., Quitterer, U., Grigoriev, S., Maidhof, A., Haasemann, M., Jarnagin, K. and Müller-Esterl, W. (1996a). Extracellular domains of the bradykinin B_2 receptor involved in ligand binding and agonist sensing defined by anti-peptide antibodies. J. Biol. Chem. 271, 1748–1755.

Abd Alla, S., Godovac-Zimmermann, J., Braun, A., Roscher, A.A., Müller-Esterl, W. and Quitterer, U. (1996b). The structure of the BK B_2 receptors amino-terminus. Biochemistry 35, 7514–7519.

Blaukat, A., Abd Alla, S., Lohse, M.J. and Müller-Esterl, W. (1996) Ligand-induced phosphphorylation/dephosporylation of the endogenous bradykinin B$_2$ receptor from human fibroblasts. J. Biol. Chem. 271, 32366–32374.

Burch, R.M. and Axelrod, J. (1987). Dissociation of bradykinin-induced prostaglandin formation from phosphatidylinositol turnover in Swiss 3T3 fibroblasts. Evidence for a G protein regulation of phospholipase A$_2$. Proc. Natl Acad. Sci. USA 84, 6374–6381.

Engvall, E. and Perlmann, P. (1971). Enzyme-linked immunosorbent assay (ELISA). Quantitative assay of immunoglobulin G. Immunochemistry 8, 871–874.

Farmer, S.G. and Burch, R.M. (1992). Biochemical and molecular pharmacology of kinin receptors. Annu. Rev. Pharmacol. Toxicol. 32, 511–536.

Farr, R.S. (1958). A quantitative immunochemical measure of the primary interaction of I-BSA and antibody. J. Infect. Dis. 103, 239–243.

Figueroa, C.D., Caorsi, I., Subiabre, J. and Vio, C.P. (1984a). Immunoreactive kallikrein localization in the rat kidney: An immuno electron microscopic study. J. Histochem. Cytochem. 32, 117–121.

Figueroa, C.D., Caorsi, I. and Vio, C.P. (1984b). Visualization of renal kallikrein in luminal and basolateral membranes: effect of the tissue processing method. J. Histochem. Cytochem. 32, 1238–1240.

Figueroa, C.D., MacIver, A.G., Mackenzie, J.C. and Bhoola, K.D. (1988). Localization of immunoreactive kininogen and tissue kallikrein in the human nephron. Histochemistry 89, 437–442.

Figueroa, C.D., Gonzalez, C.B., Grigoriev, S., Abd Alla, S., Haasemann, M., Jarnagin, K. and Müller-Esterl, W. (1995). Probing for the bradykinin B$_2$ receptor in rat kidney by anti-peptide and anti-ligand antibodies. J. Histochem. Cytochem. 43, 137–148.

Figueroa, C.D., Dietze, G. and Müller-Esterl, W. (1996). Immunolocalization of the bradykinin B$_2$ receptors on skeletal muscle cells. Diabetes 45, S1, 24–28.

Fraker, P.J. and Speck, J.C. (1978). Protein and cell membrane iodination with sparingly soluble chloramide, 1,3,4,6-tetrachloro-3a,6-a-diphenyl-glycouril. Biochem. Biophys. Res. Commun. 80, 849–857.

Galfrè, G. and Milstein, C. (1981). Preparation of monoclonal antibodies: strategies and procedures. Meth. Enzymol. 73, 3–46.

Goodfriend, T.L., Levine, L. and Fasman, G.D. (1964). Antibodies to bradykinin and angiotensin: a use of carbodiimides in immunology. Science 144, 1344–1345.

Gutowski, S., Smrcka, A., Nowak, L., Wu, D.G., Simon, M. and Sternweis, P.C. (1991). Antibodies to the α$_q$ subfamily of guanine nucleotide binding regulatory protein α subunits attenuate activation of phosphatidylinositol-4,5-bisphosphate hydrolysis by hormones. J. Biol. Chem. 266, 20519–20524.

Haasemann, M., Buschko, J., Faussner, A., Roscher, A.A., Hoebeke, J., Burch, R.M. and Müller-Esterl, W. (1991). Anti-idiotypic antibodies bearing the internal image of a bradykinin epitope: production, characterization, and interaction with the kinin receptor. J. Immunol. 147, 3882–3892.

Haasemann, M., Figueroa, C.D., Henderson, L.M., Grigoriev, S., Abd Alla, S., Gonzales, C.B., Dunia, I., Hoebeke, J., Jarnagin, K., Cartaud, J., Bhoola, K.D. and Müller-Esterl, W.

(1994). Distribution of bradykinin B$_2$ receptors in target cells of kinin action. Visualization of the receptor protein in A431 cells, neutrophils and kidney sections. J. Med. Biol. Res. 27, 1739–1756.

Hess, J.F., Borkowski, J.A., Young, G.S., Strader, C.D. and Ransom, R.W. (1992). Cloning and pharmacological characterization of a human bradykinin (BK-2) receptor. Biochem. Biophys. Res. Commun. 184, 260–268.

Hock, J., Vogel, R., Linke, R.P. and Müller-Esterl, W. (1990). High molecular weight kininogen-binding site of prekallikrein probed by monoclonal antibodies. J. Biol. Chem. 265, 12005–12011.

Hock, F.J., Wirth, K., Albus, U., Linz, W., Gerhards, H.J., Wiemer, G., Henke, S., Breipohl, G., König, W., Knolle, J. and Schölkens, B.A. (1991). HOE140, a new potent and long acting bradykinin antagonist: in vitro studies. Br. J. Pharmacol. 102, 769–777.

Kaufmann, J., Haasemann, M., Modrow, S. and Müller-Esterl, W. (1993). Structural dissection of the multi-domain kininogens. Fine mapping of the target epitopes of antibodies interfering with their functional properties. J. Biol. Chem. 268, 9079–9091.

Kyhse-Anderson, J. (1984). Electroblotting of multiple gels: a simple apparatus without buffer tank for rapid transfer of proteins from polyacrylamide to nitrocellulose. J. Biochem. Biophys. Meth. 10, 203–209.

Laemmli, U.K. (1970). Cleavage of structural proteins during the assembly of the head of bacteriophage T4. Nature (Lond.) 277, 680–685.

Lee, K.M., Toscas, K., and Villereal, M.L. (1993). Inhibition of bradykinin- and thapsigargin-induced Ca^{2+} entry by tyrosine kinase inhibitors. J. Biol. Chem. 268, 9945–9948.

Liu, F.T., Zinnecker, M., Hamaoka, T., and Katz, D.H. (1979). New procedures for preparation and isolation of conjugates of proteins and a synthetic copolymer of D-amino acids and immunochemical characterization of such conjugates. Biochemistry 15, 690–693.

Mage, M.G. (1980). Preparation of Fab Fragments from IgGs of different animal species. Meth. Enzymol. 70, 142–150.

Marceau, F. (1995). Kinin B$_1$ receptors: a review. Immunopharmacology 30, 1–26.

McEachern, A.E., Shelton, E.R., Bhakta, S., Obernolte, R., Bach, C., Zuppan, P., Fujisaki, J., Aldrich, R.W. and Jarnagin, K. (1991). Expression cloning of a rat B$_2$ bradykinin receptor. Proc. Natl Acad. Sci. USA 88, 7724–7728.

McLean, I.W. and Nakane, P.K. (1974). Periodate–lysine–paraformaldehyde fixative: a new fixative for immuno-electron microscopy. J. Histochem. Cytochem. 22, 1077–1083.

Menke, J.G., Borkowski, J.A., Bierilo, K.K., MacNeil, T., Derrick, A.W., Schneck, K.A., Ransom, R.W., Strader, C.D., Linemeyer, D.L. and Hess, F.J. (1994). Expression cloning of a human B1 bradykinin receptor. J. Biol. Chem. 269, 21583–21586.

Müller-Esterl, W., Johnson, D.A., Salvesen, G. and Barrett, A.J. (1988). Human kininogens. Meth. Enzymol. 163, 240–256.

Quitterer, U., Schröder, C., Müller-Esterl, W. and Rehm, H. (1995). Effects of bradykinin and endothelin-1 on the calcium homeostasis of mammalian cells. J. Biol. Chem. 270, 1992–1999.

Rodriguez, E.M., Yulis, R., Peruzzo, B., Alvial, G. and Andrade,

R. (1984). Standardization of various applications of metacrylate embedding and silver methenamine for light and electron microscopy immunocytochemistry. Histochemistry 81, 253–263.

Roscher, A.A., Manganiello, V.C., Jelsema, C.L. and Moss, J. (1983). Receptors for bradykinin in intact cultured human fibroblasts. Identification and characterization by direct binding study. J. Clin. Invest. 72, 626–635.

Schreurs, J., Arai, K.I. and Miyajima, A. (1990). Evidence for a low-affinity interleukin-3 receptor. Growth Factors 2, 221–233.

Sharif, M. and Hanley, M.R. (1992). Peptide receptors. Stepping up the pressure. Nature (Lond.) 357, 279–280.

Sternberger, L.H., Hardy, P.H., Jr, Cuculis, J.J. and Meyer, H.G. (1970). The unlabeled antibody enzyme method of immunohistochemistry: preparation and properties of soluble antigen–antibody complex (horseradish peroxidase– antihorseradish peroxidase) and its use in identification of spirochetes. J. Histochem. Cytochem. 18, 315–333.

Subba Rao, P.V., McCartney-Francis, N.L. and Metcalfe, D.D. (1983). An avidin–biotin microELISA for rapid measurement of total and allergen specific human IgE. J. Immunol. Meth. 57, 71–85.

Tippmer, S., Quitterer, U., Kolm, V., Faussner, A., Roscher, A., Mosthaf, L., Müller-Esterl, W. and Häring, H. (1994). Bradykinin induces translocation of the protein kinase C isoforms α, ϵ and ζ. Eur. J. Biochem. 225, 297–304.

Van Rhee, A.M. and Jacobson, K.A. (1996). Molecular architectures of G protein-coupled receptors. Drug Devel. Res. 37, 1–38.

7. Metabolism of Bradykinin by Peptidases in Health and Disease

Ervin G. Erdös *and* Randal A. Skidgel

The Kinin System
ISBN 0–12–249340–0

1. Introduction

The enzymatic breakdown of the kinins affects the duration of their biological actions, as the plasma half-life of intravenously injected bradykinin (BK) is in the range of seconds (Erdös, 1966; Kariya *et al.*, 1982; Bhoola *et al.*, 1992). Kinins are cleaved *in vitro* and *in vivo* by enzymes that belong to families such as zinc-metallopeptidases (Skidgel, 1993), serine peptidases and proteases (Erdös, 1966, 1979; Erdös and Yang, 1970; Jackman *et al.*, 1990), astacin-like metallopeptidases (Bond and Beynon, 1995) and catheptic enzymes (Greenbaum and Sherman, 1962). After the discovery by Werle, Rocha e Silva and their associates that kallikrein or trypsin decrease blood pressure by releasing the kinins, bradykinin (BK) and Lys-BK (kallidin), in plasma and that these peptides are then inactivated, the term "kininase" was coined to denote the inactivating enzymes (Frey *et al.*, 1950, 1968; Erdös *et al.*, 1963; Rocha e Silva, 1963; Erdös, 1989; Bhoola *et al.*, 1992; see Chapter 1 of this volume).

Subsequently, a kininase I-like enzyme was shown to release Arg from the C-terminus of BK and Lys-BK (Erdös *et al.*, 1963), whereas kininase II liberates the Phe[8]-Arg[9] dipeptide of BK (Erdös and Yang, 1967; Yang and Erdös, 1967). Moreover, kininase II is identical with angiotensin I-converting enzyme (ACE; Yang *et al.*, 1970, 1971; Skidgel and Erdös, 1993). Early studies noted that sulfhydryl compounds protect kinins against enzymatic inactivation (Rocha e Silva, 1963; Erdös, 1966; Frey *et al.*, 1968; Erdös and Yang, 1970) and that Lys-BK can be converted to BK in plasma (Erdös *et al.*, 1963; Webster and Pierce, 1963). BK is also inactivated by removal of Arg[1] by an enzyme in human erythrocytes (Erdös *et al.*, 1963) or in rat lung by aminopeptidase P, both catalyzing the same reaction (Simmons and Orawski, 1992). Vane (1969) has noted the importance of the pulmonary circulation in the metabolism of vasoactive substances such as BK as well as angiotensin I and 5-hydroxytryptamine. These and other pioneering experiments are reviewed elsewhere (Erdös and Yang, 1970; Erdös, 1979; Skidgel and Erdös, 1993).

Lys-BK has nine, and BK eight peptide bonds that can be hydrolyzed (Bhoola *et al.*, 1992). Lys-BK is converted to BK by the cleavage of the Lys[1]-Arg[2] bond (Erdös *et al.*, 1963; Webster and Pierce, 1963), and BK subsequently, and possibly simultaneously, is inactivated by other enzymes at any of the eight remaining bonds (Erdös *et al.*, 1963) (Fig. 7.1). Recently, many peptidases that cleave BK have been characterized and their primary structures established (Jackman *et al.*, 1990; Skidgel, 1992, 1993; Bond and Beynon, 1995). Although BK has long been utilized as a convenient substrate for enzyme assays *in vitro*, the importance of several of these mechanisms *in vivo* remains to be determined. Furthermore, although the term "kininase" is convenient to describe an enzyme's action on a single type of substrate, it can be misleading as other substrates (e.g., angiotensin I) may

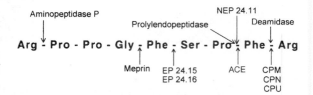

Figure 7.1 Peptide bonds in bradykinin cleaved by peptidases. Arrows show the primary site of cleavage, as established using purified enzymes and bradykinin *in vitro*. NEP 24.11, neutral endopeptidase 24.11; EP 24.15, endopeptidase 24.15; EP 24.16, endopeptidase 24.16; ACE, angiotensin I-converting enzyme; CPM, carboxypeptidase M; CPN, carboxypeptidase N; CPU, carboxypeptidase U.

be cleaved by "kininases". We know now that the kininases cleave many peptides, frequently more avidly than kinins (Skidgel, 1992; Skidgel and Erdös, 1993). Their roles in peptide metabolism depend on their location, access to a particular substrate, and the kinetics of hydrolysis (Skidgel, 1992). A vital question arises in determining which enzymes affect the actions of BK *in vivo*.

With the introduction of therapeutically useful enzyme inhibitors, other questions arise (Gavras *et al.*, 1978; Cushman and Ondetti, 1980; Gavras and Gavras, 1987, 1993; Ondetti, 1994). For example, how much of the effect of a given enzyme inhibitor in the clinic can be attributed to protecting BK, as opposed to other substrates, against enzymatic breakdown? These questions are especially appropriate for ACE inhibitors, which are administered to millions of patients. Inhibitors of neutral endopeptidase 24.11 (NEP or neprilysin, also known as enkephalinase) are also being tested clinically (Elsner *et al.*, 1992). The attractive idea of employing a second generation of inhibitors that can inhibit two enzymes, e.g., both NEP and ACE, will make the issues even more complex (Gros *et al.*, 1991, Gonzalez-Vera *et al.*, 1995, Flynn *et al.*, 1995, French *et al.*, 1994, French *et al.*, 1995). Figure 7.1 illustrates the amino-acid sequence of BK and the sites of actions of peptidases. Of the numerous enzymes that cleave BK, ACE, neutral endopeptidase, aminopeptidase P and carboxypeptidases N and M are probably the most important *in vivo*, although this can vary from species to species.

2. Kininase II (Angiotensin I-converting Enzyme)

Because of the wide distribution of ACE in mammalian species, its presence on the vascular endothelium, and the low K_m of BK with this enzyme (Table 7.1), ACE is a major kininase in most organs. Based upon its physical properties, enzymatic characteristics and immunological

Table 7.1   Kinetic constants for hydrolysis of bradykinin by various peptidases

Enzyme	Species	K_m (µM)	k_{cat} (min^{-1})	k_{cat}/K_m (µM^{-1} min^{-1})	Reference
ACE/kininase II	Human	0.18	660	3,667	Jaspard *et al.* (1993)
ACE N-domain	Human	0.54	300	555	Jaspard *et al.* (1993)
ACE C-domain	Human	0.24	480	2,000	Jaspard *et al.* (1993)
Endopeptidase 24.11	Human	120	4,770	39.8	Gafford *et al.* (1983)
Endopeptidase 24.15	Rat	67	2,028	30.3	Orlowski *et al.* (1989)
Meprin	Mouse	520	1,320	2.5	Wolz *et al.* (1991)
Carboxypeptidase M	Human	16	147	9.2	Skidgel *et al.* (1989)
Carboxypeptidase N	Human	19	58	3.1	Skidgel *et al.* (1984b)
Carboxypeptidase U	Human	10,000	7,260	0.7	Tan and Eaton (1995)
Prolylendopeptidase	Pig	7.5	95	12.7	Ward *et al.* (1987)
Aminopeptidase P	Rat	21	720	34.3	Orawski and Simmons (1995)

ACE, angiotensin I-converting enzyme; ACE N-domain, ACE in which the two zinc-binding histidine residues in the C-domain active site (His[959] and His[963]) have been mutated to Lys, leaving only the N-domain active site functional; ACE C-domain, ACE in which the two zinc-binding histidine residues in the N-domain active site (His[361] and His[365]) have been mutated to Lys, leaving only the C-domain active site functional.

cross-reactivity, ACE enzymes in different species and various tissues are quite similar and likely represent the two-domain somatic ACE (Skidgel and Erdös, 1993). There is also a testicular enzyme, germinal ACE, which contains only the C-terminal half of somatic ACE (Soubrier *et al.*, 1993a,b). From human material, an active, truncated ACE was purified which contains only the N-terminal half (the N-domain) of the somatic ACE (Deddish *et al.*, 1994).

2.1   Localization

ACE is present in vascular beds bound to the plasma membrane of endothelial cells as an ectoenzyme, where it cleaves circulating peptides such as BK. It is also present in subendothelial structures, although in lower concentrations, and in epithelial and neuroepithelial cells. Polyclonal antibodies against either the human kidney or lung enzyme could not distinguish ACE prepared from endothelial, epithelial or neuroepithelial cells, or plasma (Skidgel and Erdös, 1993). ACE activity is higher in arterial than in venous endothelial cells in culture (Johnson *et al.*, 1982). The lungs, because of their heavy vascularization, and the capillary beds in the retina and brain (Igic, 1985) are especially rich in ACE. Monoclonal antibodies also located ACE to the human myocardial vascular endothelium (Falkenhahn *et al.*, 1995). Epithelial cells generally contain more ACE than endothelial cells. The human kidney, for example, contains 5–6 times more ACE per unit wet weight than does the lung. In man and most animals, with the exception of the rat, the kidney proximal tubular brush border is a rich source of ACE (Hall *et al.*, 1976; Schulz *et al.*, 1988). ACE activity is also very high in other microvillar structures of brush border epithelial linings, for example, in the small intestine, choroid plexus and placenta (Skidgel *et al.*, 1987a,b; Skidgel and Erdös, 1993).

In the rat brain, high ACE activity was found in the subfornical organ, area postrema, substantia nigra and locus ceruleus. In the human brain, ACE activity is most concentrated in the caudate nucleus (Skidgel *et al.*, 1987a). Possibly the highest concentration of ACE in any tissue was found in the choroid plexus (Igic *et al.*, 1977) on the ventricular surface of the epithelial cells (Rix *et al.*, 1981; Defendini *et al.*, 1983). The choroid plexus may also be the source of ACE detected in the cerebrospinal fluid.

In rat and rabbit brain, ACE was localized by using labeled inhibitors [125I]-MK351A or [3H]-captopril in autoradiography (Rogerson *et al.*, 1995), and the results are in agreement with immunohistochemical studies (Correa *et al.*, 1985; Skidgel *et al.*, 1987a; Sakaguchi *et al.*, 1988; Chai *et al.*, 1991; Skidgel and Erdös, 1993; Rogerson *et al.*, 1995). ACE is also present in many other human and animal tissues including fish gills, the electric organ of *Torpedo marmorata* (Lipke and Olson, 1988), or the housefly (Cornell *et al.*, 1995), and in human male reproductive tract (Erdös *et al.*, 1985), chorionic membranes, and cultured chorion cells (Alhenc-Gelas *et al.*, 1984). ACE also seems to be a well-conserved enzyme. This was confirmed by molecular cloning and sequencing, showing, for example, that the deduced amino-acid sequences of human and mouse ACE are 83% identical (Soubrier *et al.*, 1988, 1993a,b; Bernstein *et al.*, 1989). Even the *Drosophila* ACE has over 60% sequence similarity (Tatei *et al.*, 1995).

2.2   Physical and Structural Properties

ACE purified from various human tissues is a single-chain glycoprotein. Its estimated molecular weight ranges from 140 to 170 kDa in SDS-PAGE (Skidgel and Erdös, 1993). The apparent discrepancy in values probably arises in part from the heterogeneity in carbohydrate content and experimental variability in electrophoretic conditions. ACE is heavily glycosylated, the human renal enzyme containing approximately 25% carbohydrate (Weare *et al.*, 1982). There are differences, however, in the carbohydrate moieties. For example, human renal ACE has only traces of sialic acid, while the human lung enzyme contains up to 20 sialic acid residues per molecule, in agreement with the hypothesis that the lung is the source of circulating ACE (Weare *et al.*, 1982). Owing to its high sialic acid content, ACE is protected from uptake by liver lectins. The distribution of potential glycosylation sites is uneven, the N-domain containing ten and the C-domain seven (Soubrier *et al.*, 1988). This is consistent with the finding of extensive glycosylation (37% by weight) on the naturally occurring active N-domain in human ileal fluid (Deddish *et al.*, 1994). The molecular weight of ACE deduced from the cDNA sequence (without carbohydrate) is 146 kDa (Skidgel and Erdös, 1993).

2.3 Enzymatic Properties

Similar to many other peptidases, ACE is a zinc metalloenzyme. Thus, chelating agents and sulfhydryl compounds inhibit ACE by complexing the active site Zn^{2+} (Erdös and Yang, 1970; Erdös, 1979). ACE has a pH optimum above neutral, and activity drops steeply at acid pH primarily due to the dissociation of Zn^{2+} from histidine in the active center (Ehlers and Riordan, 1990).

Initially, the idea that BK and angiotensin I are hydrolyzed by the same enzyme protein (Yang *et al.*, 1970b, 1971) was controversial as angiotensin I conversion was almost entirely chloride dependent, while cleavage of BK occurs in the absence of Cl^- at approximately 30–35% of the maximal rate (Igic *et al.*, 1973; Erdös, 1979). Studies with rabbit lung ACE showed that anion activation depends on both the structure of the substrate and the pH of the medium (Ehlers and Riordan, 1990). Accordingly, the substrates were divided into three classes, BK being a Class II substrate with a lower activation constant for Cl^- than angiotensin I, a Class I substrate. Weare (1982) described two anion binding sites in ACE, one of them being the primary activation site.

The concentration of Cl^- ions *in vivo* appears to be high enough for ACE to be fully active in most tissues, although, at some sites, Cl^- concentrations may fluctuate enough to regulate ACE activity with some substrates. It is possible that the Cl^- ion sensitivity of the membrane-bound enzyme *in vivo* differs from that of the solubilized, purified enzyme *in vitro* especially with angiotensin I substrate (Igic *et al.*, 1972). Recent studies with membrane-bound recombinant ACE showed that the Cl^-dependence of the N- and C-domains was the same as that determined for the soluble enzyme (Jaspard and Alhenc-Gelas, 1995).

ACE potentially has many functions in tissues other than the vascular endothelium, although these are not as well understood as its role in cleaving circulating or locally released kinins. For example, on the renal proximal tubular brush border, ACE may inactivate kinins that enter the nephron after glomerular filtration, which would otherwise interfere with renal auto-regulation (Scicli *et al.*, 1978; Skidgel *et al.*, 1987b; Carretero and Scicli, 1995).

2.4 Soluble ACE

Although the enzyme is bound tightly to plasma membranes, ACE also exists in soluble form. In addition to blood, soluble ACE is found in urine, lung edema fluid, amniotic fluid, seminal plasma, cerebrospinal fluid and in homogenates of prostate and epididymis (Skidgel and Erdös, 1993). The level of ACE in human plasma is very low, around 10^{-9} M (Alhenc-Gelas *et al.*, 1983). Among laboratory animals, guinea-pig plasma has the highest activity ACE (Yang *et al.*, 1971). In humans, plasma ACE activity varies significantly from person to person (Soubrier *et al.*, 1993a).

2.4.1 Release of Bound ACE

ACE is membrane bound with the majority of the protein projecting into the extracellular space. ACE contains a potential 17 amino-acid membrane-spanning region near the C-terminus. A mutant ACE lacking this trans-membrane domain was secreted primarily into the medium, in contrast to cells transfected with the full-length cDNA, which synthesized mainly membrane-bound ACE (Wei *et al.*, 1991a,b). Since it is tightly bound to membranes, ACE must be solubilized prior to purification, either with detergent (Erdös and Yang, 1967) or by cleaving it enzymatically from the cell membrane (Nishimura *et al.*, 1976). It was also shown that trypsin-liberated ACE had a lower molecular weight than ACE mobilized from the membrane with detergent (Erdös and Gafford, 1983), providing evidence for the removal of a small anchor peptide by trypsin. The proteolytic release of germinal ACE expressed in cells may involve activation of a protein kinase C (Ehlers *et al.*, 1995). Cleavage sites in the peptide chain for mobilizing the enzyme were established (Beldent *et al.*, 1993; Ramchandran *et al.*, 1994). The so-called α-secretase may be one of the enzymes solubilizing ACE (Oppong and Hooper, 1993). In some lung tissues, a metalloprotease is responsible for the solubilization of membrane-bound ACE (Oppong and Hooper, 1993). The mechanisms underlying release of ACE from membranes *in vivo*, and its subsequent appearance in body fluids, are unclear, but proteolytic cleavage of the membrane anchor peptide on the C-domain is likely. Since the discovery of separate, active single N-domain ACE (Deddish *et al.*, 1994), this issue deserves further scrutiny.

2.5 CLONING AND SEQUENCING

The molecular cloning and sequencing of the cDNA for human (Soubrier *et al.*, 1988) and mouse (Bernstein *et al.*, 1989) enzymes revealed that ACE has two domains, each with a Zn^{2+}-binding site and an active center (Soubrier *et al.*, 1993a). This two-domain ACE was named "somatic ACE", while the enzyme extracted from rabbit testicles and which contains only the C-domain active site was called "germinal ACE" (Ehlers *et al.*, 1989; Kumar *et al.*, 1989; Lattion *et al.*, 1989). Each active site contains two histidines and a glutamic acid to coordinate the zinc atom (Soubrier *et al.*, 1993a), as well as a catalytic glutamic acid, which is presumed to be a base donor. The overall sequence identity of the two domains is 67%, while it is 89% around the active centers (Soubrier *et al.*, 1993a). Although both domains are catalytically active (Wei *et al.*, 1991a), inhibitors react differently with the active sites at the N- and C-domains (Perich *et al.*, 1991, 1994). For example, captopril has a lower K_i for the N-domain active site, while at optimal Cl^- concentrations, the K_i of other inhibitors is lower for the C-domain active site (Wei *et al.*, 1992).

Germinal ACE contains only the C-domain, the C-terminal half of endothelial ACE, and 67 unique amino acids at the N-terminus (Soubrier *et al.*, 1988, 1993a; Ehlers *et al.*, 1989). As is true for the yet unexplained high concentrations of other proteases and peptidases in the genital tract, the function of testicular ACE is unknown. Germinal ACE is probably the most ancient form of the enzyme and a gene-duplication event is likely to have occurred an estimated 600 million years ago to give rise to the two-domain form (Soubrier *et al.*, 1993a,b). Indeed, *Drosophila* ACE is a single-domain enzyme, like testicular ACE, but structurally it represents the soluble form of the enzyme (Cornell *et al.*, 1995).

2.6 VARIATIONS IN ACE ACTIVITY IN DISEASE

After the gene structure of ACE was deduced, a polymorphism was discovered in the gene sequence, due either to the presence or absence of a 287 bp fragment in intron 16. The importance of this insertion/deletion polymorphism (ACE-II/DD) to BK metabolism is not known. In ACE-II homozygotes, ACE activity was lower in T-lymphocytes and in plasma than in ACE-DD subjects (Costerousse *et al.*, 1993; Soubrier *et al.*, 1993a,b; Cambien and Soubrier, 1995).

Changes in the level of ACE have been studied in a variety of conditions. Kinetics of BK hydrolysis show that it has a much lower K_m and a higher specificity constant than angiotensin I (Jaspard *et al.*, 1993; Skidgel and Erdös, 1993), and it follows that changes in plasma ACE levels are more likely to affect kinin inactivation than angiotensin I conversion. Under normal conditions, plasma ACE is very

likely to originate from vascular endothelial cells (Skidgel and Erdös, 1993). In sarcoidosis, the lymph nodes contain a high concentration of ACE, which, when released, raises the enzyme level in the circulation (Lieberman, 1974, 1985; Bunting *et al.*, 1987; Silverstein *et al.*, 1976, 1979; Grönhagen-Riska, 1979). Other granulomatous diseases such as Gaucher's disease or leprosy can also elevate circulating ACE (Silverstein *et al.*, 1978; Dhople *et al.*, 1985; Lieberman, 1985). ACE is very low in monocytes and macrophages, but it can be induced by glucocorticosteroids (Friedland *et al.*, 1978), although its level decreases in plasma of sarcoidosis patients treated with the hormone. Glucocorticosteroids also elevate ACE concentrations in endothelial cells (Mendelsohn *et al.*, 1982). Thyroid hormones markedly affect plasma ACE levels (Reiners *et al.*, 1988), which are increased by hyperthyroidism and reduced by hypothyroidism (Grönhagen-Riska *et al.*, 1985; Brent *et al.*, 1984). In Addison's disease, and in silicosis, asbestosis and berylosis, ACE levels are elevated (Falezza *et al.*, 1985; Bunting *et al.*, 1987). Paradoxically, chronic administration of an ACE inhibitor leads to a higher ACE level in plasma (Fyhrquist *et al.*, 1983). In the brain, where it is localized on cell membranes, ACE concentration is lower in Huntington's chorea (Butterworth 1986; Arregui *et al.*, 1978).

2.6.1 Acute Lung Injury

In the lungs, because most circulating ACE is presumably released from vascular endothelial cells, injury to the lungs affects ACE levels in plasma. In acute lung injury there is an early increase in blood ACE (Heck and Niederle, 1983), and the enzyme also appears in pleural effusions and lavage fluid in perfused lungs (Igic *et al.*, 1972, 1973; Dragovic *et al.*, 1993). In malignant lung tumors and in leukemia, ACE activity in plasma is lower than normal (Heck and Niederle, 1983; Schweisfurth *et al.*, 1985a,b). Blood levels of ACE also tend to be lower in Hodgkin's disease and multiple myeloma (Romer and Emmertsen, 1980). In acute respiratory distress syndrome (ARDS), ACE activity decreases in serum (Bedrossian *et al.*, 1978; Johnson *et al.*, 1985b). Studies with monoclonal antibodies raised against somatic ACE suggested that the N-domain of human ACE is immunodominant (Danilov *et al.*, 1994). Monoclonal antibodies were also raised which inhibited only the N-domain of ACE, also using BK as substrate (Danilov *et al.*, 1994). These antibodies were used in a hypoxic rat lung model of pulmonary hypertension to localize ACE, which was reduced in capillary endothelium of hypoxic animals but increased in the small muscularized pulmonary arteries (Morrell *et al.*, 1995).

2.7 HYDROLYSIS OF KININS BY ACE

2.7.1 Bradykinin

The degree of participation of the two active sites of ACE in BK hydrolysis was established using mutant and wild-type recombinant forms of the enzyme. In the mutants, one active site was eliminated by deletion or by mutation of the two zinc-binding histidine residues (Jaspard *et al.*, 1993). In agreement with previous findings, BK was cleaved by the sequential removal of the C-terminal dipeptides to yield BK_{1-7} (Fig. 7.1) and BK_{1-5}. BK is a preferred substrate of ACE over angiotensin I, since its specificity constant (k_{cat}/K_m) is about 20 times higher (Table 7.1). The N-domain alone contributes 26% of the BK degrading activity, while the C-domain is responsible for 76% of the activity in producing BK_{1-7} (Jaspard *et al.*, 1993). The C-domain active site is more dependent for its activation on Cl^-. Hydrolysis of angiotensin I by the Cl^- sensitive C-domain was activated at optimal Cl^- concentration about 100-fold while, under similar conditions, that of BK was activated only five-fold. The k_{cat}/K_m for BK is 3.6 times higher with the C-domain than with the N-domain (Jaspard *et al.*, 1993; Table 7.1). Inactivation of BK was investigated further using a naturally occurring shorter version of ACE containing only the N-domain. BK was hydrolyzed about twice as fast by the C-domain (germinal ACE) than by the N-domain (Deddish *et al.*, in preparation).

2.7.2 DesArg⁹-Bradykinin

Although ACE is frequently called a peptidyl dipeptidase, it can release a C-terminal tripeptide from desArg⁹-BK (Fig. 7.2), the latter being a product of carboxypeptidase N or M (Inokuchi and Nagamatsu, 1981). The k_{cat} of the tripeptide release is higher than the cleavage of the C-terminal dipeptide from BK, but the K_m of desArg⁹-BK is much higher than that of BK, resulting in a much lower k_{cat}/K_m for desArg⁹-BK (Inokuchi and Nagamatsu, 1981; Oshima *et al.*, 1985). ACE also liberates protected C-terminal di- and tripeptides (e.g., Gly-Leu-Met-NH₂) of substance P and even the protected N-terminal tripeptide (<Glu-His-Trp) of luteinizing hormone releasing-hormone (LHRH; Skidgel and Erdös, 1993). This N-terminal tripeptide is cleaved mainly by the N-domain (Deddish *et al.*, 1994).

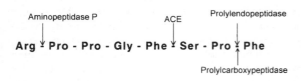

Figure 7.2   Peptide bonds in desArg⁹-bradykinin cleaved by various peptidases. Arrows show the primary site of cleavage. Although it has not been determined, it is likely that neutral endopeptidase 24.11 cleaves the Gly⁴-Phe⁵ bond in desArg⁹-bradykinin, as this bond is hydrolyzed in bradykinin under prolonged incubation conditions (Gafford *et al.*, 1983). Possibly, meprin, endopeptidase 24.15 and endopeptidase 24.16 also cleave desArg⁹-bradykinin at the same sites as in bradykinin (see Fig. 7.1), but this has not been reported.

2.8   INHIBITION OF ACE AND POTENTIATION OF THE EFFECTS OF ENDOGENOUS BRADYKININ

2.8.1   Effects Attributed to Inhibition of Bradykinin Degradation

The potentiation of BK-induced effects *in vitro* and *in vivo* by inhibitors of kininases appeared initially to be due solely to prolongation of the half-life of this peptide, which is so readily degraded (Erdös *et al.*, 1963; Frey *et al.*, 1968; Erdös, 1966). However, the widespread clinical use of ACE inhibitors (Gavras *et al.*, 1978; Gavras and Gavras, 1987, 1993), along with both the development of angiotensin II receptor antagonists (Timmermans and Smith, 1994), and B_1 and B_2 kinin receptor antagonists (Regoli and Barabé, 1980; Kyle and Burch, 1993; Bhoola *et al.*, 1992; Linz *et al.*, 1995), led to renewed interest in the role of kinins as mediators of ACE inhibition. (**Editor's note:** The role of endogenous kinins in the therapeutic effects of ACE inhibitors is reviewed in Chapter 19 of this volume.)

It is unlikely that ACE inhibitors exert their major effects by raising the level of circulating kinins (Bönner, 1995). Rather these drugs likely inhibit breakdown of locally formed kinins (Carretero and Scicli, 1989, 1995). ACE is inhibited by a variety of compounds *in vitro*. Reagents which react with the Zn cofactor of the enzyme, (e.g., EDTA, *o*-phenanthroline, SH compounds), snake venom peptides, and the active form of clinically used inhibitors (e.g., enalaprilat) belong here (Erdös and Yang, 1970; Erdös, 1979). Circulating blood plasma can also inhibit ACE; the inhibition appears to be abolished by dilution of plasma (Erdös, 1979). ACE substrates can also be competitive inhibitors, and even the N-terminal tripeptide (Arg-Pro-Pro) of BK inhibits ACE *in vitro* (Oshima and Erdös, 1974). From studies of the sequence of snake venom inhibitors, synthetic ACE inhibitors were designed to retain the C-terminal proline (Kato and Suzuki, 1970; Erdös, 1979; Cushman and Ondetti, 1980; Patchett and Cordes, 1985; Wyvratt and Patchett, 1985).

Because of their two actions in blocking both angiotensin II release and preventing the inactivation of BK, the effects of ACE inhibitors may be interpreted in different ways. This issue is especially relevant when considering the experimental and therapeutic effects of ACE inhibitors in cardiovascular diseases (CONSENSUS, 1987; Gavras and Gavras, 1987, 1993; Pfeffer; 1993, 1995; Parratt, 1994; Ambrosioni *et al.*, 1995). In the coronary arteries, angiotensin II is a vasoconstrictor and, because of its mitogenic actions, causes cell proliferation in subendothelial tissues. In contrast, BK is a vasodilator and antiproliferative, at least in the rat (Swartz *et al.*, 1980; Farhy *et al.*, 1992; Sunman and Sever, 1993; Scicli, 1994; Hartman, 1995; Margolius, 1995; Auch-Schwelk *et al.*, 1993; De Meyer *et al.*, 1995). Enalapril lowers the blood pressure of hypertensive rats; this effect is partially blocked by the administration of a BK receptor antagonist (Carbonell *et al.*, 1988). Captopril enhances skin microvascular blood flow; this was attributed to NO and prostaglandins released by BK (Warren and Loi, 1995). Thus, ACE inhibitors may improve coronary circulation by two mechanisms. BK receptors are present in heart muscles; B_2-type high affinity BK receptors were discovered in rat cardiomyocytes which are coupled to IP_3 production by a G-protein (Minshall *et al.*, 1995a). In hypertensive rats, enalapril lowers the blood pressure, an effect that is blocked partially by a BK receptor antagonist (Carbonell *et al.*, 1988).

In spontaneously hypertensive rats, high doses of ACE inhibitors prevent the development of hypertension and left ventricular hypertrophy, effects which were abolished by a B_2 receptor antagonist, Hoe 140 (Gohlke *et al.*, 1994a,b). ACE inhibitors diminished neointima formation induced by balloon injury to the rat carotid artery and endogenous kinins contribute to this effect (A.G. Scicli, personal communication). The antihypertrophic effects of ACE inhibitors may be independent of the decrease in blood pressure, and kinin receptor antagonists do not reverse this effect of ACE inhibitors (Scicli, 1994). ACE inhibitors, indeed, may prevent wall thickening in rat heart after restenosis but not the neointimal formation in other species (Shaw *et al.*, 1995).

ACE inhibitors reduced myocardial infarct size after myocardial ischemia and reperfusion injury (Liu *et al.*, 1996), but angiotensin II receptor antagonists are not effective, suggesting that kinins were responsible. ACE inhibitors also ameliorated cardiac arrhythmias induced by digoxin or reperfusion, in part via endogenous BK (Linz *et al.*, 1995). On the other hand, prevention of cardiac myocyte necrosis and coronary vascular damage induced by angiotensin II in the rat, was blocked by lisinopril as well as by an angiotensin II receptor antagonist (Kabour *et al.*, 1995). In one study, administration of an NO synthase inhibitor resulted in the development of hypertension and cardiac and renal insufficiency, which was prevented by ramiprilat. The beneficial effect was attributed to increased BK and prostaglandin production (Hropot *et al.*, 1994).

ACE inhibitors, by enhancing the effects of BK on energy metabolism, may improve cardiac metabolism. Dietze (1982) showed that BK potentiates glucose uptake, in part via the release of prostaglandins. ACE inhibitors reduce lactate release from the heart (Linz *et al.*, 1995) by enhancing the effects of BK. The metabolic functions of the guinea-pig heart were improved by an ACE inhibitor, which decreased lactic acid release and increased intracellular glutathione, and these effects were attributed to endogenous BK (Massoudy *et al.*, 1994). Captopril increased glucose utilization in patients and ACE inhibitors can cause hypoglycemia (Torlone and Bolli, 1991). The beneficial effects of ACE inhibition in reducing hyperfiltration in experimental diabetes of rats was attributed to kinins as mediators (Komers and Cooper, 1995).

The roles of BK potentiation in the clinical benefits, or side effects of acute or chronic administration of ACE inhibitors are still being explored. ACE inhibitors given soon after myocardial infarction reduced mortality and the development of severe heart failure (Gavras and Gavras, 1993; Hall *et al.*, 1994; Pfeffer, 1995). Whether or not coughing, a prominent side effect of ACE inhibitors (Israili and Hall, 1992; Semple, 1995) is caused by prolongation of the half-life of BK or by potentiation of BK's effects is unclear. BK may also be involved in a rare but serious side effect of ACE inhibitors, angioneurotic edema (Slater *et al.*, 1988).

2.8.2 Effects Not Attributed to Inhibition of Bradykinin Degradation

The potentiation of the effects of BK by ACE inhibitors go beyond protecting BK from degradation. Decades of research have yielded several compounds that amplify effects of BK, at least in guinea-pig isolated ileum (Vogel *et al.*, 1970). These agents are unrelated in structure, ranging from snake venom peptides and sulfhydryl compounds, known inhibitors of ACE, to proteases and fibrinopeptides, which do not inhibit ACE (Gladner *et al.*, 1963; Edery, 1964, 1965). This issue recently gained particular importance when ACE inhibitors were demonstrated to improve functions in the failing heart via a mechanism that was attributed to the potentiation of the *action* of BK and *not* inhibition of its metabolism (Scicli, 1994; Linz *et al.*, 1995). A detailed discussion is beyond the scope of this review but a brief summary of the findings follows.

The evidence supporting the hypothesis that potentiation of the actions of BK is not necessarily due solely to inhibition of its enzymatic breakdown has been the subject of several publications (e.g., Paegelow *et al.*, 1976). In guinea-pig ileum or heart, the peptide is inactivated slowly, while the potentiation by ACE inhibitors can be almost instantaneous and is unlikely, therefore, to be due to prolonging the half-life of BK (Ufkes *et al.*, 1977). In isolated guinea-pig atrium, ACE inhibitors potentiated effects of BK even when an ACE-resistant BK analog was used (Auch-Schwelk *et al.*, 1993; Minshall *et al.*, 1995b). ACE inhibitors also potentiated the inotropic effects of BK even when the ACE was inhibited completely by sequestration of its metal cofactor (Minshall *et al.*, 1995b).

An ACE inhibitor also enhanced the release of endothelium-derived hyperpolarizing factor by either BK or a metabolically stable BK analog (Ibrahim *et al.*, 1995; Mombouli and Vanhoutte, 1995). This action was attributed to a possible interaction of ACE inhibitors with BK receptors. ACE inhibitors enhance the relaxation of isolated coronary artery in response to BK and this was endothelium-dependent (Auch-Schwelk *et al.*, 1993). Because responses to a degradation-resistant analog of BK were also potentiated by an ACE inhibitor and no hydrolysis of BK was observed *in vitro*, the phenomenon was attributed to a "potentiation" of the effect of BK rather than inhibition of its degradation. It remains to be established, however, whether ACE inhibitors affect BK receptors directly or via interaction with ACE, which may be in close physical proximity, and thereby enhance the action of BK independent of prolongation of its half-life.

3.   *Neutral Endopeptidase 24.11*

Neutral endopeptidase 24.11 (neprilysin, NEP) is a zinc-metallopeptidase with a single active site containing the canonical HEXXH sequence (Erdös and Skidgel, 1989; Skidgel, 1993; Howell *et al.*, 1994). The primary sequence is well conserved among species (Devault *et al.*, 1987), there being only six nonconserved differences between the human and rat NEP (Malfroy *et al.*, 1987, 1988). The enzyme is a transmembrane, single-chain protein of 742 amino acids but, in contrast to ACE, is bound via an uncleaved N-terminal signal peptide (Roy *et al.*, 1993).

3.1   LOCALIZATION

NEP is distributed widely but, in contrast to ACE, its expression in vascular endothelial cells is low (Johnson *et al.*, 1985a; Llorens-Cortes *et al.*, 1992; Howell *et al.*, 1994), and varies within vascular beds (Graf *et al.*, 1995). Epithelial cells, especially in microvillar structures, are rich in NEP (Johnson *et al.*, 1985a; Turner, 1987; Ronco *et al.*, 1988). As with ACE, these include the proximal tubules (Kerr and Kenny, 1974), placenta, (Johnson *et al.*, 1984), intestine, or the choroid plexus (Turner, 1987; Ronco *et al.*, 1988). Its distribution in the brain has been investigated in detail, possibly because neuropeptides, such as enkephalins and substance P, are among its substrates (Turner, 1987). Owing to its wide distribution, actions which are relevant *in vivo* can be different in each organ with the different substrates. As with ACE, the relevance of the high concentration of NEP in the male genital tract, especially in prostate glands (Erdös *et al.*, 1985) is not known. NEP, under the name, "common acute lymphoblastic leukemia antigen" (CALLA or CD10) (Letarte *et al.*, 1988; LeBien and McCormack, 1989) is present in lymphoblasts but is absent from mature lymphocytes. In contrast, it is found in neutrophils but absent from progenitor cells (Connelly *et al.*, 1985; Painter *et al.*, 1988).

3.2   ENZYMATIC PROPERTIES

NEP cleaves peptides at the N-termini of hydrophobic amino acids, although the molecular mass of substrates does not usually exceed 3 kDa. NEP is a second kininase II that releases the C-terminal Phe^8-Arg^9 of BK (Gafford *et al.*, 1983). Although it was discovered as an endopeptidase that cleaves the B-chain of insulin (Kerr and Kenny, 1974), many active peptide substrates for NEP have been described subsequently, and include enkephalins (Schwartz *et al.*, 1985), endothelin (Vijayaraghavan *et al.*, 1990), atrial natriuretic peptide, substance P and a chemotactic peptide (Connelly *et al.*, 1985; Erdös and Skidgel, 1989).

As noted above, NEP cleaves BK at the Pro^7-Phe^8 bond (Fig. 7.1) (Almenoff *et al.*, 1981), which was first shown qualitatively by high pressure liquid chromatography, although prolonged incubation also results in the hydrolysis of the Gly^4-Phe^5 bond (Gafford *et al.*, 1983). By the use of purified human renal NEP, the kinetics of hydrolysis were established (Gafford *et al.*, 1983; Table 7.1). The k_{cat} for BK is higher with NEP than with ACE but, because of the higher K_m (120 μM vs 0.18 μM), the specificity constant, k_{cat}/K_m, of NEP is lower (40 vs 3,667 $μM^{-1}$ min^{-1}).

Sites where NEP may be an important kininase include the epithelial cells of the respiratory tract (Johnson *et al.*, 1985a; Dusser *et al.*, 1988), skeletal muscles (Dragovic *et al.*, 1996b), neutrophils (Connelly *et al.*, 1985; Painter *et al.*, 1988; Skidgel *et al.*, 1991b), renal proximal tubules (Skidgel *et al.*, 1987b; Ura *et al.*, 1987) and possibly human coronary vessels (Graf *et al.*, 1995). Inhibitors of NEP (e.g., thiorphan, phosphoramidon and candoxatril) are thus useful tools in exploring the functions *in vivo* of this peptidase (Schwartz *et al.*, 1985; Bralet *et al.*, 1991; Elsner *et al.*, 1992; Nadel, 1992, 1994; Bertrand *et al.*, 1993; Schilero *et al.*, 1994).

Because of the antiproliferative effects of BK (Farhy *et al.*, 1992), the high concentration of NEP in solid malignant tumors may indicate that BK is a substrate in those cells. Transplanted malignant tumors of rat liver, SK HEP1 malignant human liver cells (Dragovic *et al.*, 1994), and primary liver tumors (Dragovic *et al.*, 1996a) contain as much as 1,000 times more NEP than noncancerous tissues (Dragovic *et al.*, 1994).

3.2.1   Soluble NEP

NEP levels in circulating blood plasma are normally very low, but increase 60–80-fold in ARDS with septic pneumonia (Johnson *et al.*, 1985b). Chronic cholestasis is another condition that enhances plasma NEP activity (Swan *et al.*, 1993). In contrast to plasma, the level of NEP in amniotic fluid is high (Spillantini *et al.*, 1990). Ura and colleagues (1987) found that most kininase activity in rat urine is attributable to high NEP activity, and injection of an NEP inhibitor enhanced the diuretic and natriuretic effects of kinins. These effects could be mediated in part by an interaction of atrial natriuretic peptide and kinins in the rat kidney (Smits *et al.*, 1990; Bralet *et al.*, 1991). NEP was present in high concentration in the developing edema of the perfused rat lung, while its level in the perfusate was low (Dragovic et al., 1993). NEP activity increased in the urine after proximal tubular injury and in serum in end-stage renal disease (Deschodt-Lanckman et al., 1989; Nortier et al., 1993).

3.3   INHIBITION OF NEP AND ITS ROLE AS A KININASE

Inhibition of NEP in rat bronchial epithelium markedly enhances bronchoconstriction induced by substance P and BK (Bertrand et al., 1993). The BK-induced bronchoconstriction (Barnes, 1987) in asthmatic patients is enhanced by NEP inhibitors (Crimi et al., 1995). Indirect actions of BK are also influenced by NEP inhibition. For example, BK releases substance P from nerve endings, which is then cleaved by NEP (Bhoola et al., 1992; Barnes, 1994; Nadel, 1994; Skidgel, 1994). BK stimulates airway ciliary activity via prostaglandin E_2 release in rabbit and this effect is also regulated by NEP (Tamaoki et al., 1989).

Microvascular leakage in guinea-pig airways was potentiated by both ACE and NEP inhibition (Lötvall, 1990; Lötvall et al., 1990). In the rat lung, combined inhibition of NEP and ACE enhanced the effect of a subthreshold dose of BK in the development of edema (Dragovic et al., 1993). Injecting BK into the left ventricle of rats increased the blood flow in the micro-circulation of airways (Yamawaki et al., 1994) and this, too, was potentiated by phosphoramidon, an NEP inhibitor; thus, besides ACE, NEP participates in the inactivation of BK there. In the rat heart, sensory nerve stimulation enhanced myocardial blood flow that was abolished by a BK receptor blocker but potentiated by phosphoramidon (Piedimonte et al., 1994).

Because of the increased reactivity to the combined administration of inhibitors of two peptidases, several compounds that inhibit both NEP and ACE, or ACE and aminopeptidase are now available (Gros et al., 1991; Bralet et al., 1994; French et al., 1994; Flynn et al., 1995; Gonzalez-Vera et al., 1995). For example, a dual inhibitor of ACE and NEP, administered intravenously (i.v.) to rats, potentiates the hypotensive effect of BK more than the action of an inhibitor specific to either enzyme alone (French et al., 1995). Canine renal responses to BK are enhanced by combined administration of ACE and NEP inhibitors (Seymour et al., 1993, 1994). This combination lowered arterial blood pressure and vascular resistance in nonanesthetized dogs (Seymour et al., 1994). Co-administration of inhibitors of ACE and aminopeptidase P is more effective in reducing the blood pressure in hypertensive rats than inhibition of either enzyme alone (A.G. Scicli et al., personal communication).

4.   Hydrolysis of Bradykinin by Carboxypeptidases

Carboxypeptidases catalyze the hydrolysis of the C-terminal peptide bond in peptides and proteins. Their specificity is largely determined by the C-terminal and/or penultimate residue of the substrate (Skidgel, 1996). Carboxypeptidases are subdivided into two general groups; the serine carboxypeptidases and the metallo-carboxypeptidases. Any carboxypeptidase that removes the C-terminal Arg of BK can be termed "kininase I" (Fig. 7.1). For example, the active subunit of carboxypeptidase N has 41% sequence identity with carboxypeptidase M and both enzymes cleave BK. Although deamidase/cathepsin A cleaves the C-terminal Arg of BK (Fig. 7.1) and, therefore, may be classified as a kininase I, it is a serine carboxypeptidase with no homology to carboxypeptidase M or N (Jackman et al., 1990; Skidgel, 1996).

4.1   METALLOCARBOXYPEPTIDASES

Metallocarboxypeptidases catalyze peptide hydrolysis through a mechanism that requires the participation of a tightly bound zinc atom as an essential cofactor (Skidgel, 1996). The metallocarboxypeptidases can be subdivided further into two categories based on their substrate specificities: carboxypeptidase A-type enzymes prefer hydrophobic C-terminal amino acids, whereas the carboxypeptidase B-type enzymes cleave only C-terminal Arg or Lys (Skidgel, 1988). Since the kinins contain a C-terminal Arg, however, only carboxypeptidase B-type enzymes cleave them.

4.1.1   General Characteristics

Because of the essential function of zinc in catalysis, these carboxypeptidases are all inhibited by metal chelating agents such as 1,10-phenanthroline (Skidgel, 1996). All of the B-type carboxypeptidases can be inhibited relatively specifically by small synthetic arginine analogs such as DL-2-mercaptomethyl-3-guanidinoethylthiopropanoic acid (MGTA) and guanidinoethylmercaptosuccinic acid (GEMSA) (Plummer and Ryan, 1981; Skidgel, 1991, 1996).

Depending upon the enzyme employed, they inhibit in the low micromolar to nanomolar range. The pH of the incubation and the form of the enzyme can also affect the affinity for these inhibitors. For example, the IC_{50} of GEMSA for membrane bound carboxypeptidase M at pH 5.5 is 100-fold lower than for the solubilized, purified enzyme at pH 7.5 (Deddish et al., 1989; Skidgel, 1991).

Replacement of zinc in the active center with other divalent cations, most notably by cobalt or cadmium, alters the activity of these enzymes. For example, Co^{2+} increases carboxypeptidase activity by up to ten-fold, depending on the source of enzyme or the substrate

(Folk and Gladner, 1960, 1961; Marinkovic et al., 1977; Skidgel, 1988; Deddish et al., 1989). The pH is an additional factor that affects activation by Co^{2+} (Deddish et al., 1989; Skidgel, 1991); it activates the enzyme more at a low pH. As with carboxypeptidase N (Skidgel et al., 1984b), stimulation of peptidase activity by Co^{2+} is due to an increase in the V_{max}, as the K_m increases in the presence of this metal ion. In contrast, Cd^{2+} inhibits the peptidase activity of the metallocarboxypeptidases with the exception of carboxypeptidase U (plasma carboxypeptidase B), which is stimulated (Skidgel, 1996).

Based on the primary sequences, mammalian metallocarboxypeptidases fall clearly into two groups; those with a high sequence identity with the pancreatic carboxypeptidases and the so-called regulatory carboxypeptidases. Pancreatic carboxypeptidases A and B, mast cell carboxypeptidase A and carboxypeptidase U (Skidgel, 1996) form the first group and carboxypeptidases N, M and E the second group. Within each group, there is significant sequence similarity (40–58%) whereas between groups, the identity is much lower (14–20%). These data, as well as conservation of active site residues, indicate that the metallocarboxypeptidases arose from the same ancestral gene which duplicated and diverged to evolve into two separate precursor genes, one of which gave rise to the pancreatic carboxypeptidase-like enzymes and the other to regulatory carboxypeptidases M, N and E (Tan et al., 1989a; Avilés et al., 1993).

4.1.2   Carboxypeptidase N

The discovery that a plasma carboxypeptidase inactivates BK was the first demonstration that an endogenous carboxypeptidase could regulate the activity of a peptide hormone (Erdös and Sloane, 1962). Although the enzyme has been given several names (e.g., anaphylatoxin inactivator, creatine kinase conversion factor, plasma carboxypeptidase B, arginine carboxypeptidase, lysine carboxypeptidase, protaminase), the original designation, carboxypeptidase N, is used here.

4.1.2.1   Physical and Structural Properties

Size estimates for purified native human carboxypeptidase N range from about 270 to 330 kDa, depending on the method used but 280 kDa is most frequently cited (Oshima et al., 1975; Plummer and Hurwitz, 1978; Levin et al., 1982). Under denaturing conditions, it dissociates into three major bands of 83 kDa, 55 kDa and 48 kDa (Plummer and Hurwitz, 1978; Levin et al., 1982). The 83 kDa protein is a noncatalytic subunit, whereas the 55 kDa and 48 kDa proteins represent two forms of the same active subunit (Levin et al., 1982; Skidgel, 1995). Carboxypeptidase N, therefore, is a tetrameric enzyme comprised of two heterodimers, each heterodimer containing one catalytic and one non-catalytic 83 kDa subunit. The 83 kDa subunit is heavily glycosylated, (about 28% by weight), the carbohydrate composition being typical of proteins containing Asn-

linked complex carbohydrate chains (Plummer and Hurwitz, 1978; Levin et al., 1982). In contrast, the active subunit lacks carbohydrate.

As revealed by molecular cloning, the 50 kDa subunit has sequence similarities to other metallocarboxypeptidases ranging from 14% to 49% (Gebhard et al., 1989). Cloning and sequencing of the 83 kDa subunit, however, revealed that it encodes a 59 kDa protein with no sequence similarity to the 50 kDa active subunit or other carboxypeptidases (Tan et al., 1990). Consistent with its high carbohydrate content, the sequence contains seven potential Asn-linked glycosylation sites and a serine/threonine rich region that may be a site for attachment of O-linked carbohydrate. The most exceptional feature is a domain comprising over half of the protein that contains 12 leucine-rich tandem repeats of 24 amino acids each (Tan et al., 1990; Skidgel and Tan, 1992). This pattern was initially described in the sequence of the leucine-rich α_2-glycoprotein, although a variety of other mammalian proteins contain it (Tan et al., 1990; Skidgel and Tan, 1992; Kobe and Deisenhofer, 1993). The leucine-rich repeat region is likely to be critical for the binding function of these proteins (Kobe and Deisenhofer, 1993). Probably the leucine-rich repeat region in the 83 kDa subunit mediates its interaction with the 50 kDa active subunit to form a heterodimer, and the N- and C-terminal domains of the 83 kDa subunit link the two heterodimers to form the tetramer (Tan et al., 1990; Skidgel and Tan, 1992).

Because of its lack of carbohydrate, small size and relative instability at 37°C (Levin et al., 1982; Skidgel, 1988), the active subunit of carboxypeptidase N by itself probably would not exist in the circulation for long. In vitro, the 83 kDa subunit stabilizes the active subunit at 37°C and at low pH (Levin et al., 1982). This indicates that the 83 kDa subunit, although lacking enzymatic activity, is important in carrying and stabilizing the active subunit in the blood.

4.1.2.2   Enzymatic Properties

Carboxypeptidase N cleaves a variety of substrates containing C-terminal Arg or Lys, generally cleaving Lys faster than Arg, although the penultimate residue also plays an important role, alanine being preferred in many cases (Skidgel, 1995). The pH optimum of carboxypeptidase N is in the neutral range (7.5) and it retains minimal activity at pH 5.5 (Erdös et al., 1964; Deddish et al., 1989). However, Co^{2+} still activates the enzyme at pH 5.5 to a surprising 156% of the activity at pH 7.5 in the absence of Co^{2+} (Deddish et al., 1989). While most of the enzymatic properties of the isolated catalytic 50 kDa subunit (Levin et al., 1982) agree with those of the intact 280 kDa tetramer, the 83 kDa subunit can affect allosterically the interaction of the enzyme with some substrates including the anaphylatoxin C3a, and inhibitors such as protamine (Skidgel et al., 1986; Tan et al., 1989b).

4.1.2.3 Localization

Carboxypeptidase N is synthesized in the liver and released into the circulation where it is present at a relatively high concentration of approximately 30 μg/ml (10^{-7} M) (Erdös, 1979). Little carboxypeptidase N activity is found in liver (Oshima et al., 1975), probably because the enzyme is not stored, but secreted soon after synthesis. Apparently, no other cells or organs synthesize the enzyme, as Northern blot analysis of various organs gave negative results (Tan and Skidgel, unpublished). Although it was reported that carboxypeptidase N immunoreactivity is present on the cell membrane of cultured bovine pulmonary arterial endothelial cells (Ryan and Ryan, 1983), it was not determined whether the cells synthesize the enzyme or, more likely, take it up from the serum in the medium. In contrast, cell membranes of cultured human pulmonary arterial endothelial cells have high levels of carboxypeptidase M, but no detectable carboxypeptidase N (Nagae et al., 1993).

4.1.2.4 Role of Carboxypeptidase N as a Kininase

Carboxypeptidase N was discovered as a kininase that cleaved the C-terminal Arg^9 from BK (Fig. 7.1). Although the k_{cat} value for BK is lower than other naturally occurring substrates (Table 7.1), BK has the lowest K_m (19 μM) of all the peptide substrates tested (Skidgel, 1995). Even though the ability of carboxypeptidase N to cleave kinins in vitro is well documented (Erdös, 1979), its contribution to the degradation of circulating kinins in vivo is probably of secondary importance as BK is rapidly inactivated by ACE during a single passage through the pulmonary circulation (Vane, 1969; Erdös, 1979). Also, although the rate of degradation of circulating BK would be slower, this pathway becomes more significant in patients treated with ACE inhibitors. Levels of circulating kinins may not be relevant to their physiological actions, which are thought to be localized in tissues (Carretero and Scicli, 1989). Rather, they likely represent "spillover" of kinins generated at local sites. Nevertheless, should kinin concentrations increase significantly in the blood, they can have undesirable effects. Thus, blood-borne enzymes may be important in preventing buildup of endogenous peptides in the circulation. The presence of carboxypeptidase N in blood in a relatively high concentration provides evidence that this enzyme may fulfill this function. Further evidence is the observation that no human subjects tested have been found to lack the enzyme completely and even patients with low enzyme levels are rare (Erdös et al., 1965; Mathews et al., 1980). Indeed, genetically low blood levels of carboxypeptidase N (about 20% of normal), owing to decreased hepatic synthesis, were associated with repeated attacks of angioedema in one patient, possibly due to the increased half-life of kinins and/or anaphylatoxins (Mathews et al., 1980, 1986). Conditions that affect hepatic plasma protein synthesis also alter plasma carboxypeptidase N levels, and include cirrhosis of the liver, which causes a decrease, or pregnancy, which causes an increase (Erdös et al., 1965). In a variety of diseases (e.g., cardiovascular disease, diabetes, allergic conditions), there are no changes in carboxypeptidase N levels (Erdös et al., 1965; Mathews et al., 1980), although elevations in enzyme level have been noted in certain cancers, and in the blood and synovial fluid of arthritic patients (Erdös et al., 1965; Mathews et al., 1980; Schweisfurth et al., 1985b; Chercuitte et al., 1987). The relationship of carboxypeptidase N levels to these disease states, if any, is not known.

The evidence for the protective function of carboxypeptidase N in humans is, by necessity, somewhat indirect. Nevertheless, the protamine-reversal syndrome is one condition where low carboxypeptidase N activity may be involved. Protamine is given routinely to neutralize the antithrombotic effects of heparin after extracorporeal circulation. In some patients, this can trigger a catastrophic reaction consisting of pulmonary vasoconstriction, bronchoconstriction and systemic hypotension (Lowenstein et al., 1983; Morel et al., 1987). This reaction has been attributed to the release of thromboxane, and the generation of anaphylatoxins and kinins subsequent to the activation of the complement cascade and factor XII, which activates plasma kallikrein (Colman, 1987; Morel et al., 1987). As protamine is a potent inhibitor of carboxypeptidase N (Tan et al., 1989b), potentially decreased degradation of anaphylatoxins and kinins may contribute to this syndrome. In addition, the carboxypeptidase N concentration decreases to about 50% of normal after initiation of cardiopulmonary bypass, due primarily to dilution of the blood (Rabito et al., 1992). However, the fact that this syndrome is relatively rare (incidence is approximately 1%) indicates that other factors are involved. Because heparin binds protamine and reverses the inhibition of carboxypeptidase N (Tan et al., 1989b), only when protamine is given in excess would a problem develop. In agreement with this possibility, administration of heparin reversed protamine reactions in two patients and was hypothesized to be due to reactivation of carboxypeptidase N (Lock and Hessel, 1990). In addition, the data of Mathews et al. (1980) indicate that carboxypeptidase N levels of 20% of normal or greater are sufficient for a protective role.

Most of the actions of BK are mediated via binding and activation of B_2 receptors (Bhoola et al., 1992). However, cleavage of BK by carboxypeptidase N yields desArg^9-BK, a specific agonist at B_1 receptors, which stimulates a variety of proinflammatory cellular responses (see Chapters 2, 8, 9 and 13 of this volume). The B_1 receptor system is upregulated in response to injury or inflammation, and may be part of the acute phase reaction (Bhoola et al., 1992; see Chapter 8). For example, many isolated tissues respond to desArg^9-BK only after incubation for several hours (DeBlois et al., 1991; Bhoola et al., 1992). Noxious or pro-inflammatory stimuli, such as Triton X-100, endotoxin, or interleukins-1 and -2, induce the expression of B_1 receptors (Bhoola et al., 1992; Crecelius et al., 1986; DeBlois et al., 1991), and this is inhibited by glucocorticosteroids (DeBlois et al., 1988). Thus, conversion of BK to desArg^9-BK by carboxypeptidase N produces an agonist for B_1 receptors and may play an important role in inflammatory or pathological responses. That this occurs in vivo is supported by the finding that blood levels of desArg^9-BK are over 3-fold higher than native BK in normotensive individuals and in patients with low-renin essential hypertension (Odya, et al., 1983).

4.1.3   Carboxypeptidase M

We discovered significant carboxypeptidase B-like activity in membrane fractions obtained from human and animal tissues and cells (Johnson et al., 1984; Skidgel et al., 1984b). After purification and characterization of this unique enzyme, we named it carboxypeptidase "M" to denote the fact that it is membrane-bound (Skidgel et al., 1989).

4.1.3.1   Physical and Structural Properties

Human carboxypeptidase M yields only a single band of 62 kDa, in sodium dodecyl sulphate–polyacrylamide gel electrophoresis (SDS-PAGE) with or without reduction, showing that it is a single-chain protein (Skidgel et al., 1989). A slightly higher value (73 kDa), obtained in gel filtration in the presence of CHAPS, is likely to be caused by the binding of the enzyme to detergent micelles and also to its glycoprotein nature (Skidgel et al., 1989). Because carboxypeptidase M is a glycoprotein, it binds tightly to concanavalin A–Sepharose and its mass is reduced to 47.6 kDa by chemical deglycosylation (Skidgel et al., 1989). These data indicate a 23% carbohydrate content by weight, in agreement with the presence of six potential Asn-linked glycosylation sites in the deduced protein sequence (Tan et al., 1989a).

In subcellular fractions of cells or tissues, most carboxypeptidase M is firmly membrane bound (Skidgel et al., 1984b, 1989; Skidgel, 1988). However, the primary sequence does not contain a true hydrophobic transmembrane spanning region (Tan et al., 1989a). The extreme C-terminus has a weakly hydrophobic region of 15 amino acids, similar to other proteins that are membrane bound via a glycosylphosphatidylinositol (GPI) anchor (Low, 1987). Indeed, carboxypeptidase M can be released from membrane preparations by bacterial phosphatidylinositol-specific phospholipase C (PI-PLC), a characteristic feature of GPI-anchored proteins (Deddish et al., 1990; Tan et al., 1995). Direct evidence for the presence of a GPI anchor on carboxypeptidase M was obtained by labeling cultured Madin Darby canine kidney (MDCK) cells (which have high carboxypeptidase M activity) with [^{3}H]-ethanolamine. Antiserum specific for carboxypeptidase M immunoprecipitated a single radiolabeled band from the solubilized membrane fraction, corresponding in size to that of carboxypeptidase M (Deddish et al., 1990). Carboxypeptidase M is also found in soluble form in various body fluids such as urine, seminal plasma, amniotic fluid and broncho-alveolar lavage fluid (Skidgel et al., 1984a,b, 1988; Dragovic et al., 1995; McGwire and Skidgel, 1995). The mechanism of release of the enzyme has not been determined, but the fact that the hydrophobic portion of the anchor was removed indicates that either a protease or phospholipase is involved (Deddish et al., 1990).

4.1.3.2   Enzymatic Properties

Carboxypeptidase M is activated by cobalt chloride and inhibited by o-phenanthroline, MGTA, GEMSA and cadmium acetate, as are most B-type carboxypeptidases. It has a neutral pH optimum and cleaves only C-terminal Arg or Lys from a variety of substrates including BK and arginine or lysine-extended opioid peptides (Fig. 7.1). Carboxypeptidase M cleaves C-terminal Arg preferentially over Lys, as demonstrated by faster cleavage of Bz-Gly-Arg than Bz-Gly-Lys, although the penultimate residue prominently affects the rate of hydrolysis (Skidgel et al., 1989). Carboxypeptidase M also hydrolyzed naturally occurring peptide substrates and the kinetic constants were determined (Skidgel et al., 1989). Of the substrates tested, BK (with C-terminal Phe^8-Arg^9) has the lowest K_m (16 μM) and the k_{cat}/K_m is 9.2 μM^{-1}min^{-1} (Table 7.1). Of the synthetic substrates tested, the ester substrate (1 mM Bz-Gly-argininic acid) was cleaved fastest (102 micromoles/min/mg).

4.1.3.3   Localization

Carboxypeptidase M is found in a wide variety of tissues and cells. Northern blot analysis showed high levels of carboxypeptidase M mRNA in human placenta, lung and kidney (Nagae et al., 1992). Significant amounts are present in blood vessels, intestine, brain and in peripheral nerves (Skidgel, et al., 1984b, 1991a; Skidgel, 1988; Nagae et al., 1992, 1993). In the lungs, immuno-histochemical studies located carboxypeptidase M on the surface of type I pneumocytes and in pulmonary macrophages (Nagae et al., 1993). In the brain, oligodendrocytes or astrocytes stain positively and, in peripheral nerves, the enzyme is concentrated on the outer aspects of myelin sheaths and Schwann cell membranes (Nagae et al., 1992). Carboxypeptidase M is also present in soluble form in various body fluids, as noted earlier (Skidgel et al., 1984a,b, 1988; McGwire and Skidgel, 1995). Recently it was discovered that monoclonal antibodies, raised against a differentiation-dependent cell surface antigen on white blood cells, are specific for carboxypeptidase M (de Saint-Vis et al., 1995; Rehli et al., 1995). In one study, carboxypeptidase M was almost undetectable on peripheral blood monocytes, but highly expressed after differentiation into macrophages (Rehli et al., 1995). Similarly, carboxypeptidase M was present on pre-B lymphocytes, downregulated on circulating

B lymphocytes, but re-expressed on activated germinal center B cells (de Saint-Vis *et al.*, 1995).

4.1.3.4 Role of Carboxypeptidase M as a Kininase

Owing to its location on the plasma membranes of a wide variety of cells and tissues, where the BK receptor is also located, carboxypeptidase M is likely to regulate the local actions of BK (Skidgel, 1988, 1992). As with carboxypeptidase N, desArg^9-BK, an agonist for B_1 receptors, is produced by carboxypeptidase M (Fig. 7.1). However, because of its location on plasma membranes, carboxypeptidase M may perform this role in a local environment outside the circulation, for example, at sites of inflammation where B_1 receptors may be upregulated (Skidgel, 1992, 1996; Bhoola *et al.*, 1992).

Epidermal growth factor (EGF; urogastrone) is another substrate of carboxypeptidase M with relevance to the kinin system. EGF is a 53 amino-acid (6 kDa) peptide first isolated from mouse submaxillary gland (Carpenter and Wahl, 1990). Studies in liver and fibroblasts indicated that EGF is first cleaved by a carboxypeptidase to remove the C-terminal Arg, producing EGF$_{1-52}$ at the cell surface or in early endosomes (Schaudies and Savage, 1986; Planck *et al.*, 1984; Renfrew and Hubbard, 1991). Recent studies showed that purified carboxypeptidase M readily converts EGF to desArg^{53}-EGF (McGwire and Skidgel, 1995). Incubation of EGF with MDCK cells, which have high carboxypeptidase M activity, resulted in rapid conversion (61% in 2 h) of EGF to desArg^{53}-EGF as the only metabolite. The hydrolysis was blocked completely by a carboxypeptidase inhibitor, MGTA. Similar results were obtained with urine or amniotic fluid where MGTA or immunoprecipitation with specific antiserum to carboxypeptidase M abolished essentially all EGF hydrolysis (McGwire and Skidgel, 1995). However, there was no difference in the mitogenic potency of EGF and desArg^{53}-EGF on MDCK cells and conversion of EGF to desArg^{53}-EGF was not required for the mitogenic effect (McGwire and Skidgel, 1995). Because carboxypeptidase M cleaves both EGF and BK, it is of interest that EGF stimulates the growth of breast stromal cells, whereas BK decreases growth and causes a concentration-dependent inhibition of EGF-stimulated DNA synthesis (Patel and Schrey, 1992). This effect of BK is mediated through B_1 receptors, implying it is first converted to desArg^9-BK in this system. EGF also potentiates the contractile response to desArg^9-BK in rabbit aortic rings (DeBlois *et al.*, 1992).

Carboxypeptidase M may be involved in inflammatory and pathological processes by virtue of regulating the activity of kinins and anaphylatoxins that mediate many inflammatory effects. The importance of the up- or downregulation of carboxypeptidase M during differentiation of monocytes to macrophages and in B lymphocytes (Rehli *et al.*, 1995; de Saint-Vis *et al.*, 1995)

is not known. In the kidney, carboxypeptidase M may control the activity of kinins which are released by kallikrein liberated from the distal tubules (Scicli *et al.*, 1978). Kinins stimulate prostaglandin production, affect sodium and water excretion (Carretero and Scicli, 1989) and mediate amino-acid-induced hyperperfusion and hyperfiltration (Jaffa *et al.*, 1992). In one study, hypertensive human patients excreted significantly more kininases than normal individuals and the major kininase was a kininase I (Iimura, 1987), probably carboxypeptidase M. Thus, increased release of carboxypeptidase M into the urine could be an early sign of renal damage owing to hypertension or other diseases.

Carboxypeptidase M may also have important functions in the lung as indicated by the high level of activity present in membrane fractions from the lungs of bovines, guinea pigs, baboons, dogs, rats and humans (Chodimella *et al.*, 1991; Nagae *et al.*, 1993). Type I cells, which in immunohistochemistry stain strongly for carboxypeptidase M (Nagae *et al.*, 1993), comprise only 8% of the total cells in the lung, yet they account for 93% of the total surface area. The cells are covered with a layer of surfactant and function primarily as a thin gas-permeable membrane between the air space and capillary. The presence of carboxypeptidase M on this surface indicates a protective role. The enzyme may also be readily mobilized from the cell surface or be soluble in the surfactant layer as bronchoalveolar lavage fluid contains high levels of carboxypeptidase M (Dragovic *et al.*, 1993, 1995). The pulmonary synthesis of carboxypeptidase M or its release from the membrane may be upregulated in disease states as enzyme levels in bronchoalveolar lavage fluid were elevated almost five-fold in patients with pneumocystic or bacterial pneumonia or lung cancer (Dragovic *et al.*, 1995). Carboxypeptidase M was also released into the edema fluid of rat lung in an experimental model of lung injury (Dragovic *et al.*, 1993).

BK can cause pulmonary edema and bronchoconstriction when administered to animals. Although the carboxypeptidase inhibitor MGTA does not enhance either of these responses by itself (Ichinose and Barnes, 1990; Chodimella *et al.*, 1991; Dragovic *et al.*, 1993), it causes further potentiation of these pulmonary responses after inhibition of NEP and ACE (Chodimella *et al.*, 1991; Dragovic *et al.*, 1993), suggesting the involvement of all three enzymes in the metabolism of BK. This might be relevant to the persistent dry cough that is one of the major side effects encountered after the administration of ACE (kininase II) inhibitors to hypertensive patients (Israili and Hall, 1992; Semple, 1995). Because this can result from increased concentration of peptides such as BK in the respiratory tract (Morice *et al.*, 1987) it adds to the potential importance of carboxypeptidase M in the lungs of these patients.

Studies in guinea pigs revealed that a carboxypeptidase M-type enzyme exists in the airways, and that

MGTA enhances the noncholinergic bronchoconstrictor response to capsaicin and vagus nerve stimulation (Desmazes et al., 1992). This probably involves the release of peptides from nerve endings, although BK, presumably, is not involved (Desmazes et al., 1992). While the response to MGTA was attributed to inhibition of the activity in the airways, it could have also been due to inhibition of carboxypeptidase M in the vagus nerve, where it is present in high concentration (Nagae et al., 1992).

Whether carboxypeptidase M in pulmonary type I cells has functions unrelated to its enzymatic activity is unknown. Nevertheless, aminopeptidase N on intestinal epithelial cells acts as a receptor for coronaviruses (Yeager et al., 1992), leaving the possibility that carboxypeptidase M also may be a receptor for infectious agents in the lungs. This hypothesis is supported by the recent discovery that a hepatitis B-virus-binding protein is a B-type carboxypeptidase with significant sequence identity to carboxypeptidases M, N and E (Kuroki et al., 1995). Finally, the functions of carboxypeptidase M in many other locations remain to be explored. For example, in the placenta it may protect the fetus from maternally derived peptides. Its location in central nervous system (CNS) myelin and Schwann cells in peripheral nerves is intriguing, and may indicate a role for the growth or protection of neurons.

4.1.4   Carboxypeptidase U

A recent addition to this class of enzymes is an unstable blood-borne carboxypeptidase that is activated during coagulation. In 1989, several groups published that human serum has a higher (about 2–3-fold) arginine-carboxypeptidase activity than plasma, and that the difference could not be explained by changes in carboxypeptidase N activity (Campbell and Okada, 1989; Hendriks et al., 1989; Sheikh and Kaplan, 1989). Interestingly, Erdös and colleagues (1964) noted that serum has a 9% higher carboxypeptidase activity than plasma, but the reason for the difference was not investigated. In retrospect, the reason for the smaller difference than in recent studies was the result of their use of hippuryl-Lys as the substrate, which is cleaved slower by this enzyme compared with carboxypeptidase N, and the addition of cobalt to the assay, which would inhibit carboxypeptidase U (Hendriks et al., 1989; Tan and Eaton, 1995). Further investigations showed that the enzyme is a unique carboxypeptidase (Campbell and Okada, 1989; Hendriks et al., 1990, 1992). In unrelated studies, a plasminogen-binding protein was fortuitously isolated and its sequence had significant homology to the pancreatic carboxypeptidases (Eaton et al., 1991). Although many of the properties of this enzyme are quite similar to those of carboxypeptidase U, Eaton and coworkers believed the two enzymes were different and named the one they isolated "plasma carboxypeptidase B". However, based on similarities in properties and partial sequence information, investigators now believe they are identical (Shinohara et al., 1994; Wang et al., 1994). For the purposes of this review, they will be considered to be the same and the name, carboxypeptidase U, is used to avoid confusion. The same protein (as revealed by N-terminal sequencing) was recently isolated and named "thrombin-activatable fibrinolysis inhibitor (TAFI)" (Bajzar et al., 1995).

4.1.4.1   Physical and Structural Properties

In the initial report, the size of the partially purified carboxypeptidase U was 435 kDa (Hendriks et al., 1990). Complete purification on a plasminogen affinity column yielded a 60 kDa protein (Eaton et al., 1991). This turned out to be the proenzyme, which, when activated by trypsin, yielded an active protein of 35 kDa (Eaton et al., 1991). The size of the enzyme initially reported by Hendriks et al. (1990) may, therefore, represent carboxypeptidase U bound to plasminogen and possibly other proteins in a multimeric complex (Wang et al., 1994). Indeed, treatment of the crude high molecular weight complex with 3 M guanidine, followed by chromatography, resulted in the isolation of an active enzyme with a major protein band at 53 kDa (Wang et al., 1994). Based on the cDNA sequence, N-terminal sequencing after activation and similarities with the pancreatic carboxypeptidases, the active form has a molecular weight of 35 kDa (Eaton et al., 1991). Further work is required to resolve this discrepancy. Although procarboxypeptidase U is a glycoprotein as shown by a reduction in mass from 60 kDa to 45 kDa after enzymatic deglycosylation (Eaton et al., 1991), the four potential Asn-linked glycosylation sites in the sequence are all in the propeptide segment (Eaton et al., 1991).

The reason why carboxypeptidase U is unstable in serum or after partial purification is not fully understood, although it is likely to involve proteolysis by enzymes activated during coagulation. The fact that the $t_{1/2}$ of carboxypeptidase U in its high molecular weight complex was only 15 min, whereas after guanidine treatment and chromatography to isolate the active enzyme (and presumably to remove any associated proteases), the $t_{1/2}$ increased to 55 min (Wang et al., 1994), supports this hypothesis. In addition, the purified proenzyme can be activated by trypsin, but further incubation leads to an additional cleavage, reducing the size of the activated carboxypeptidase from 35 to 25 kDa, which results in inactivation (Eaton et al., 1991). This is in contrast to the finding with carboxypeptidase B where a similar reduction in size does not inactivate the human pancreatic enzyme (Marinkovic et al., 1977). When the activation with trypsin is carried out in the presence of the competitive inhibitor ε-amino caproic acid, the secondary cleavage is blocked without affecting hydrolysis of the propeptide, leading to production of a stable and fully active enzyme (Tan and Eaton, 1995).

4.1.4.2 *Enzymatic Properties*

Carboxypeptidase U does not have a marked preference for either C-terminal Arg or Lys or a penultimate Ala over Gly in short synthetic substrates (Hendriks *et al.*, 1989, 1990, 1992; Eaton *et al.*, 1991; Wang *et al.*, 1994; Tan and Eaton, 1995). However, with larger naturally occurring peptides, the enzyme cleaves C-terminal Arg faster than Lys (Tan and Eaton, 1995). The very low specificity constants (k_{cat}/K_m) of the substrates with carboxypeptidase U are due to the very high K_m values (63–220 mM; Tan and Eaton, 1995), which are about 1,000-fold higher than with carboxypeptidases M or N (Skidgel *et al.*, 1984b, 1989). Carboxypeptidase U has a pH optimum in the neutral range and in contrast to other metallocarboxypeptidases, Co^{2+} inhibits and Cd^{2+} activates the peptidase activity of the enzyme (Hendriks *et al.*, 1989; Tan and Eaton, 1995).

4.1.4.3 *Localization*

Carboxypeptidase U has only been found in blood, where it exists as an inactive proenzyme, which is activated during coagulation, presumably by a serine protease converted from a proenzyme during clotting (Campbell and Okada, 1989; Hendriks *et al.*, 1989; Eaton *et al.*, 1991). Although thrombin (Bajzar *et al.*, 1995) and plasminogen (Wang *et al.*, 1994) activate procarboxypeptidase U, it remains to be determined whether these are the major activation pathways *in vivo*, or if the propeptide can be cleaved by other serine proteases. The level of procarboxypeptidase U in plasma is high, an estimated 2–5 µg/ml (Eaton *et al.*, 1991) and after complete activation, around 10^{-7} M. As with other plasma proteins, procarboxypeptidase U is probably also synthesized in the liver, consistent with the cloning of its cDNA from a human liver library (Eaton *et al.*, 1991).

4.1.4.4 *Role of Carboxypeptidase U as a Kininase*

Because it was discovered relatively recently, the functions of carboxypeptidase U are still being explored. However, it likely plays an important role in the fibrinolytic pathway by regulating lysine-mediated plasminogen binding to proteins and cells, for example, by cleaving C-terminal Lys residues from α_2-antiplasmin, histidine-rich glycoprotein, fibrin, annexin II or α-enolase (Redlitz *et al.*, 1995).

  With regard to its function as a kininase, it is of interest that one of the first groups to identify the enzyme used BK as a substrate (Sheikh and Kaplan, 1989). They noted that the conversion of BK to desArg[9]-BK was much faster in serum than in plasma, being five-fold higher than could be accounted for by carboxypeptidase N (Sheikh and Kaplan, 1989). More recently, another group showed that, after activation of procarboxypeptidase U with trypsin in plasma, the enzyme removed the C-terminal Arg from BK (Shinohara *et al.*, 1994; Fig. 7.1).

Finally, kinetics of BK hydrolysis were determined for the purified enzyme after trypsin activation (Tan and Eaton, 1995). Of all synthetic and naturally occurring substrates tested, BK had the lowest K_m and highest k_{cat}/K_m. However, the reported K_m is extraordinarily high (10 mM; Table 7.1), even though it is lower than that of all other substrates (K_m = 63–290 mM). For example, the K_m for BK with other peptidases ranges from about 0.18 to 520 µM (Table 7.1). This raises the question of the physiological relevance of carboxypeptidase U in degrading BK, which would normally be present at concentrations many orders of magnitude lower than the K_m. Because serum degrades BK five-fold faster than can be accounted for by the carboxypeptidase N level, carboxypeptidase U may be a significant kininase under certain circumstances. The apparent discrepancy between the observed BK degradation in serum and the calculated kinetic constants (Table 7.1) is not understood. Moreover, carboxypeptidase U would not cleave BK in blood because it is completely inactive as a proenzyme in plasma, but it may be activated if, for example, injury initiates the coagulation cascade.

4.2 SERINE CARBOXYPEPTIDASES

The involvement of serine carboxypeptidases in degrading BK *in vivo* has not been explored in detail. There are two lysosomal serine carboxypeptidases that can potentially affect kinin activity in pathological conditions, and these are prolylcarboxypeptidase and deamidase (cathepsin A/lysosomal protective protein). The active sites of serine carboxypeptidases contain a catalytic triad of amino acids characteristic of serine proteases. The order of the residues in the primary sequence (Ser, Asp, His) is unique and is the same as in the prolylendopeptidase (or prolyloligopeptidase) family of serine proteases (Tan *et al.*, 1993). Whereas the deamidase clearly belongs to the serine carboxypeptidase family with regard to sequence identity, prolylcarboxypeptidase has the sequence motifs characteristic of both the serine carboxypeptidases and the prolylendopeptidases, perhaps an indication that prolylcarboxypeptidase is a link between these two families of enzymes (Tan *et al.*, 1993).

4.2.1 Physical and Structural Properties

Prolylcarboxypeptidase (Yang *et al.*, 1968, 1970a) is a soluble single-chain protein of 58 kDa in sodium dodecyl sulphate – polyacrylamide gel electrophoresis (SDS-PAGE) and elutes as a symmetrical peak of 110 kDa in gel-filtration chromatography, indicating a dimer in its native form (Odya *et al.*, 1978; Tan *et al.*, 1993). It is a glycoprotein that contains 12% carbohydrate by weight, consistent with the presence of six potential Asn-linked glycosylation sites in the mature protein sequence (Tan *et al.*, 1993). The cDNA sequence indicates that the protein contains a 30-residue signal peptide and 15-amino-acid propeptide (Tan *et al.*, 1993).

Deamidase was purified to homogeneity from platelets as a substance P-inactivating peptidase (Jackman *et al.*, 1990). In gel filtration, the enzyme has a molecular mass of 94 kDa whereas, in nonreducing polyacrylamide gel electrophoresis, it is 52 kDa indicating it exists as a homodimer. After reduction on SDS-PAGE, the 52 kDa protein dissociates into two chains of 33 and 21 kDa. The 33 kDa chain was labeled with [³H]-diisopropyl-fluorophosphate, indicating it contains the active site serine residue (Jackman *et al.*, 1990). When the first 25 residues of each chain were sequenced, they were found to be identical with the sequences of the two chains of lysosomal protective protein, so named because it binds and maintains the activity and stability of β-galactosidase and neuraminidase in lysosomes (Galjart *et al.*, 1988). A defect in this protein is the cause of a severe genetic disease called galactosialidosis (d'Azzo *et al.*, 1982). Owing to its binding ability, deamidase can also be isolated from lysosomes in a high molecular weight active complex (500–600 kDa) (van der Horst *et al.*, 1989; Potier *et al.*, 1990).

4.2.2 Enzymatic Properties

Similar to other members of the serine protease family, prolylcarboxypeptidase and deamidase are inhibited by some compounds that react with the active site serine residue. For example, [³H]-diisopropylfluorophosphate covalently labels the active site serine in both enzymes (Jackman *et al.*, 1990; Tan *et al.*, 1993). However, they are not inhibited by naturally occurring serine protease inhibitors such as aprotinin, soybean trypsin inhibitor, α_1-antitrypsin, α_1-antichymotrypsin, etc. (Odya *et al.*, 1978; Jackman *et al.*, 1990; Tan *et al.*, 1993). Because they are exopeptidases, they cannot bind to or cleave the internal peptide bonds in these inhibitors that are recognized by other serine proteases. Deamidase is inhibited by inhibitors of chymotrypsin-type enzymes such as Cbz-Gly-Leu-Phe-CHCl₂ and chymostatin due to its preference to cleave substrates with penultimate hydrophobic residues (Jackman *et al.*, 1990).

Both prolylcarboxypeptidase and deamidase have acidic pH optima (approximately 5.0) when hydrolyzing short synthetic peptide substrates (Jackman *et al.*, 1990; Tan *et al.*, 1993). Interestingly, with many longer substrates, both enzymes retain significant activity in the neutral range. For example, at pH 7.0, prolylcarboxy-peptidase cleaves angiotensin II at 63% of the rate observed at pH 5 and deamidase cleaves BK at 72% of the rate at pH 5.5 (Jackman *et al.*, 1990; Tan *et al.*, 1993). In addition, the deamidation of C-terminally amidated peptides (such as -Met¹¹-NH₂ in substance P) by deami-dase is optimal at neutral pH (Jackman *et al.*, 1990). As its name implies, prolylcarboxypeptidase cleaves peptides only if the penultimate residue is proline (Odya *et al.*, 1978; Tan *et al.*, 1993).

Deamidase/cathepsin A deamidates peptides such as substance P, neurokinin A and enkephalinamides, but also cleaves peptides with free carboxy-termini by carboxypeptidase action. It prefers peptides that contain hydrophobic residues in the P_1' and/or P_1 position. For example, it readily cleaves substance P free acid, converts angiotensin I to II (Jackman *et al.*, 1990), cleaves the chemotactic peptide fMet-Leu-Phe (Jackman *et al.*, 1995) and is the most potent endothelin degrading enzyme that has been identified (Jackman *et al.*, 1992, 1993).

4.2.3 Localization

Deamidase and prolylcarboxypeptidase are localized in lysosomes but, after release from cells by stimulation (e.g., platelets, white blood cells), they appear in the extracellular medium or biological fluids. For example, deamidase and prolylcarboxypeptidase were found in urine (Yang *et al.*, 1970a; Miller *et al.*, 1991) and prolylcarboxypeptidase was released into synovial fluid (Kumamoto *et al.*, 1981). In addition, lysosomal enzymes sometimes appear on the plasma membrane after exocytosis where they may be bound to other trans-membrane or membrane-associated proteins (Skidgel *et al.*, 1991b).

Both enzymes are distributed ubiquitously and mRNA expression is highest in human placenta, lung and liver for prolylcarboxypeptidase (Tan *et al.*, 1993), and mouse kidney and placenta for deamidase/protective protein (Galjart *et al.*, 1990). Deamidase is also highly active in macrophages as well as in platelets, endothelial cells and fibroblasts (Jackman *et al.*, 1990, 1992, 1993, 1995). Prolylcarboxypeptidase is also found in white blood cells and fibroblasts, and is expressed at high levels in endothelial cells (Kumamoto *et al.*, 1981; Skidgel *et al.*, 1981).

4.2.4 Role of Serine Carboxypeptidases as Kininases

Because of their localization in lysosomes, serine carboxy-peptidases would not normally have access to extracellular kinins. Since the kinin system is upregulated in inflamma-tory conditions where lysosomal enzymes can be released from leukocytes and other cells, these enzymes may then gain access to extracellular kinins. In addition, lysosomal serine carboxypeptidases could modulate kinin receptor signal transduction by cleaving kinins after ligand-mediated receptor endocytosis (Erdös *et al.*, 1989). This usually involves fusion of the endosomes with lysosomes and results in peptide degradation and receptor recycling (Yamashiro and Maxfield, 1988).

Because the best substrates of deamidase contain C-terminal hydrophobic residues, (e.g., endothelin 1, fMet-Leu-Phe, furylacryloyl-Phe-Phe, angiotensin I) (Jackman *et al.*, 1990, 1992, 1995), it is surprising that BK, with a C-terminal Arg, is a good substrate (Jackman *et al.*,

1990; Fig. 7.1). It is now clear, however, that the presence of a hydrophobic amino acid in the penultimate position allows deamidase to remove nonhydrophobic amino acids by its carboxypeptidase action. Thus, BK (-Phe-Arg) and angiotensin$_{1-9}$ (-Phe-His) are collectively cleaved by the enzyme (Jackman et al., 1990).

Prolylcarboxypeptidase, on the other hand, cannot cleave native BK because it lacks a penultimate Pro residue. The ligand for the B$_1$ receptor, desArg^9-BK, has a C-terminal -Pro-Phe, a sequence that is readily cleaved by the enzyme (Yang et al., 1968, 1970a; Odya et al., 1978; Fig. 7.2). Thus, the carboxypeptidases potentially may be important regulators of the B$_1$ receptor signaling system. Metallocarboxypeptidases, such as carboxypeptidase M or N, or deamidase can generate desArg^9-BK (Fig. 7.1), which binds to B$_1$ receptors, upregulated in inflammation (see Chapter 8). Prolylcarboxypeptidase released from inflammatory cells would then cleave the C-terminal Phe from desArg^9-BK and inactivate it (Fig. 7.2). Previous studies reported that prolylcarboxypeptidase is released into the blood during endotoxin shock (Sorrells and Erdös, 1972), a potent stimulus for the upregulation of B$_1$ receptors (see above).

5. Hydrolysis of Bradykinin by Aminopeptidases

Aminopeptidases catalyze the removal of one amino acid at a time from the N-terminus of peptides and proteins. Although the N-terminal sequence of BK contains Pro in the second position, rendering the nonapeptide resistant to most aminopeptidases, aminopeptidase P specifically cleaves peptides with a Pro in the second position and, therefore, is capable of inactivating BK (Fig. 7.1). Aminopeptidases can also participate in the release of BK. For example, Lys-BK (kallidin) is liberated by tissue kallikrein, whereas BK is the product of plasma kallikrein (Bhoola et al., 1992). Lys-BK is converted to BK by aminopeptidase action in the blood and similar activity has been detected in a variety of tissues (Erdös, 1979). However, as mentioned elsewhere in this chapter, it is not necessary for Lys-BK to be converted to BK before it is inactivated by peptidases; it can be inactivated directly by enzymes such as ACE and carboxypeptidase N (Erdös, 1979).

Dipeptidyl aminopeptidase IV specifically catalyzes the removal of dipeptides from the N-termini of peptides with a Pro in the second position (McDonald and Barrett, 1986), but it cannot cleave BK (Arg^1-Pro^2-Pro^3-...) because it does not hydrolyze Pro-Pro bonds. Nevertheless, as a secondary metabolic step it would release Pro-Pro from desArg^1-BK. Pro-Pro has been detected as one of the metabolites of BK after passing through the lung (Ryan et al., 1968).

5.1 AMINOPEPTIDASE P

It has long been known that aminopeptidase P contributes to BK metabolism in vitro (Erdös et al., 1963; Erdös and Yang, 1966) and in perfused rat lungs in situ (Ryan et al., 1968). The enzyme was partially purified and characterized from pig kidney (Dehm and Nordwig, 1970), but was purified to homogeneity and its properties described in detail only recently (Hooper et al., 1990; Simmons and Orawski, 1992; Orawski and Simmons, 1995; Vergas Romero et al., 1995).

5.1.1 Physical and Structural Properties

Aminopeptidase P is membrane bound via a GPI anchor and can be solubilized with PI-PLC (Hooper and Turner, 1988; Hooper et al., 1990; Simmons and Orawski, 1992; Orawski and Simmons, 1995). Under denaturing conditions in SDS-PAGE, aminopeptidase P runs as a single-chain protein with an M_r = 90,000–95,000, and is a glycoprotein containing about 17–25% carbohydrate by weight (Hooper et al., 1990; Simmons and Orawski, 1992; Orawski and Simmons, 1995). After gel-filtration chromatography, the purified enzyme has a multimeric structure with a molecular mass of 220–360 kDa (Hooper et al., 1990; Simmons and Orawski, 1992; Orawski and Simmons, 1995), which varies depending on the salt concentration (Orawski and Simmons, 1995). Partial protein sequencing of aminopeptidase P, purified from guinea-pig lung and kidney (Denslow et al., 1994), and the full sequence of the pig kidney enzyme (Vergas Romero et al., 1995) show that aminopeptidase P has some sequence similarity to human and E. coli prolidase as well as E. coli aminopeptidase P. Thus, these enzymes may constitute a newly recognized family of proline peptidases.

Early studies indicated the presence of a soluble (cytosolic) form of aminopeptidase P in kidney extracts (Dehm and Nordwig, 1970) and this form of the enzyme has been purified and characterized from other sources, including human lung, human erythrocytes, rat brain and human platelets (Sidorowicz et al., 1984a,b; Harbeck and Mentlein, 1991; Vanhoof et al., 1992). The soluble enzyme differs from the membrane-bound form both in structure and substrate specificity (see below). For example, the soluble aminopeptidase P purified from rat brain has an M_r = 71,000 in SDS-PAGE and a native M_r = 143,000 in gel filtration (Harbeck and Mentlein, 1991).

5.1.2 Enzymatic Properties

Membrane-bound aminopeptidase P cleaves peptides such as BK and neuropeptide Y (NPY), containing Pro in the second position, with a pH optimum in the neutral range (Orawski and Simmons, 1995; Simmons and Orawski, 1992). The tripeptide Gly-Pro-Hyp has been commonly used as a substrate to measure its activity, although the N-terminal tripeptide of BK (Arg-Pro-Pro) is cleaved much faster (Simmons and Orawski, 1992; Orawski and Simmons, 1995). Aminopeptidase P also cleaves longer substrates, but not dipeptides such as Arg-Pro (Simmons and Orawski, 1992; Orawski and Simmons, 1995). Aminopeptidase P is a zinc metallo-peptidase containing one zinc per mole of enzyme (Hooper et al., 1990). The enzyme can be activated in the presence of Mn^{2+} with some substrates and is inhibited by chelating agents. Other inhibitors include sulfhydryl compounds, such as 2-mercaptoethanol, as well as sulfhydryl-reactive reagents, e.g., p-chloro-mercuriphenylsulfonate (Hooper et al., 1990; Simmons and Orawski, 1992; Orawski and Simmons, 1995). Interestingly, many ACE inhibitors also inhibit aminopeptidase P, although generally with a K_i in the micromolar range (Hooper et al., 1992) and Mn^{2+} can enhance their inhibitory effect (Orawski and Simmons, 1995). The inhibition is likely to be due to the presence of Pro or Pro-like structures in the ACE inhibitors, and the effective zinc binding moieties, which would chelate the active site metal of aminopeptidase P as they do in ACE. Whether some of the effects or side effects of ACE inhibitors could be due to inhibition of aminopeptidase P is not known.

The enzymatic properties of the soluble form of aminopeptidase P are very similar to those of the membrane-bound form with regard to pH optimum, inhibitors and activation by Mn^{2+} (Sidorowicz et al., 1984a,b; Harbeck and Mentlein, 1991; Vanhoof et al., 1992). Unlike the membrane-bound enzyme, however, the soluble form readily cleaves NH_2-X-Pro dipeptides (Sidorowicz et al., 1984a,b; Harbeck and Mentlein, 1991; Simmons and Orawski, 1992; Vanhoof et al., 1992; Orawski and Simmons, 1995). This may indicate that the soluble form is a different gene product and not simply the solubilized membrane-bound aminopeptidase P.

5.1.3 Localization

Aminopeptidase P-type activity has been detected in a variety of tissues and, based on purification studies, it is clear that lung and kidney contain high concentrations of membrane-bound aminopeptidase P, whereas the soluble or cytosolic enzyme is also present in brain, erythrocytes and platelets (Dehm and Nordwig, 1970; Sidorowicz et al., 1984a,b; Hooper et al., 1990; Harbeck and Mentlein, 1991; Simmons and Orawski, 1992; Vanhoof et al., 1992; Orawski and Simmons, 1995). In the lung, the membrane-bound form is likely localized on the surface of pulmonary vascular endothelial cells with access to circulating peptides such as BK (Ryan, 1989).

5.1.4 Role of Aminopeptidase P as a Kininase

In the first studies on BK metabolism, it was discovered that one mechanism of inactivation consists of the removal of the N-terminal Arg residue (Erdös et al., 1963). This type of activity in human erythrocytes (Erdös et al., 1963) and porcine kidney (Erdös and Yang,

1966) was initially attributed to a prolidase, but it is now recognized that prolidase only cleaves X-Pro dipeptides, and that aminopeptidase P is the major enzyme responsible for removing the N-terminal Arg^1 from BK (Fig. 7.1).

The overall role of aminopeptidase P in the degradation of BK in vivo has been investigated in detail only recently. Early studies in perfused rat lungs showed that one pathway for BK degradation involved the removal of the N-terminal Arg residue (Ryan et al., 1968). Two recent studies, also in perfused rat lungs, have carefully assessed the relative contributions of ACE and amino-peptidase P to BK metabolism. In both cases, it was concluded that ACE is the major kininase in rat lung, being responsible for about 70% of the total metabolism, and that essentially all of the BK hydrolysis could be accounted for by the combined actions of ACE and aminopeptidase P (Ryan et al., 1994; Prechel et al., 1995). Nevertheless, even when ACE activity was blocked completely, 75% of the BK was still inactivated during a single passage by aminopeptidase P alone (Ryan et al., 1994; Prechel et al., 1995). Studies in vivo corroborate these data where administration of the specific aminopeptidase P inhibitor, apstatin, doubled the hypotensive action of BK in rats (Scicli et al., personal communication). Furthermore, apstatin was much less effective than lisinopril, a specific ACE inhibitor, confirming the primary role of ACE in kinin metabolism in rat lungs. The relevance of these findings is somewhat questionable because rat lungs contain extremely high levels of aminopeptidase P relative to other species (Ryan, 1989). For example, aminopeptidase P levels in rat lungs are 200-fold higher than in rabbit lungs, 30-fold higher than in pig lungs and 100-fold higher than in cat lungs (Ryan, 1989).

Aminopeptidase P could also play a role in the degradation of $desArg^9$-BK (Fig. 7.2), as the in vitro kinetics are essentially identical with those of BK (Simmons and Orawski, 1992; Orawski and Simmons, 1995). On the other hand, the specificity constant for $desArg^9$-BK with ACE is much less favorable than with BK owing to the much higher K_m (120–240 µM vs 0.18 µM for BK) (Inokuchi and Nagamatsu, 1981; Oshima et al., 1985). Thus, ACE would probably play a much less prominent role in degrading $desArg^9$-BK in vivo, whereas the relative importance of aminopeptidase P would likely increase. However, the pathway for $desArg^9$-BK degradation in vivo has not yet been well characterized.

6. Hydrolysis of Bradykinin by Other Endopeptidases

Endopeptidases are enzymes that cleave a peptide in the interior of the peptide chain, with a specificity that depends mainly on the amino acids on one or both sides of the peptide bond being cleaved. Endopeptidase 24.11 is such an enzyme, and its role as a kininase is discussed in Section 3 of the present review. Several other endopeptidases cleave BK, at least *in vitro*, (Fig. 7.1), but their roles as kininases *in vivo* are not fully understood. They are considered briefly below.

6.1 MEPRIN

Meprin was originally purified from a mouse kidney membrane fraction. It is the mouse homolog of human *N*-benzoyl-L-tyrosyl-*p*-aminobenzoic acid (PABA peptide) hydrolase and rat endopeptidase 2 (Beynon *et al.*, 1981; Butler *et al.*, 1987; Dumermuth *et al.*, 1991; Wolz and Bond, 1995). Purification and sequencing studies revealed meprin to consist of two unique but related subunits, α and β, and showed meprin to be a member of the astacin family of metalloproteases (Dumermuth *et al.*, 1991; Gorbea *et al.*, 1991; Jiang *et al.*, 1992; Johnson and Hersh, 1992; Wolz and Bond, 1995). The enzyme is an oligomeric, cell-surface protein, bound via the transmembrane β subunit to which the α subunits are either disulfide-linked or noncovalently bound (Gorbea *et al.*, 1991; Johnson and Hersh, 1994; Marchand *et al.*, 1994). Meprin has only been detected in kidney and intestine, and its expression varies from species to species, and even within different strains of the same species (Gorbea *et al.*, 1991; Jiang *et al.*, 1992, 1993). In contrast to endopeptidase 24.15, endopeptidase 24.11 and ACE, which hydrolyze only short peptides, meprin cleaves large protein substrates such as azocasein (Butler *et al.*, 1987). The α subunit of meprin hydrolyzes peptide and protein substrates longer than seven amino acids and cleaves the Gly-Phe bond of BK (Butler *et al.*, 1987; Fig. 7.1). Of all peptides tested, BK was hydrolyzed fastest among those cleaved at a single site (Wolz *et al.*, 1991; Table 7.1). This finding led to the synthesis of Phe^5(4-nitro)BK and the development of the first convenient spectrophotometric assay for meprin (Wolz and Bond, 1990, 1995). The β subunit, which is primarily in a latent form, does not cleave BK after activation (Kounnas *et al.*, 1991; Wolz and Bond, 1995). Meprin-α has a very broad substrate specificity and does not have strict requirements for residues adjacent to the cleavage site, but seems to prefer Pro in the P_2' or P_3' position (Wolz *et al.*, 1991). Meprin-α also cleaves α-melanocyte-stimulating hormone, neurotensin and LHRH, and it also hydrolyzes angiotensin I and II rather slowly (Wolz *et al.*, 1991). It is unlikely that meprin plays a dominant role in degrading kinins.

6.2 PROLYLENDOPEPTIDASE

Prolylendopeptidase is a cytoplasmic enzyme found in most tissues, with an especially high concentration in brain and kidney (Wilk, 1983). This serine protease has a molecular mass of about 70–77 kDa and is optimally active at pH 7.5 (Wilk, 1983). Cloning and sequencing of the porcine brain enzyme revealed that it differed from classical serine proteases in its sequence around the active site serine (Rennex *et al.*, 1991), and comparison with other known enzymes identified it as a member of a new family of serine proteases (Rawlings *et al.*, 1991). These enzymes have a catalytic triad similar to other serine proteases, but with a unique order of Ser. . . Asp. . . His in the primary sequence (Polgar, 1992).

Prolylendopeptidase cleaves at the C-terminal side of prolyl residues in peptides of about 30 or less amino acids (Wilk, 1983). It hydrolyzes BK by cleavage of the Pro^7-Phe^8 bond (Fig. 7.1), the same site as ACE and NEP (Wilk, 1983). With the purified porcine kidney enzyme, BK has a K_m of 7.5 μM and a V_{max} of 1.37 μmol min^{-1} mg^{-1} (Ward *et al.*, 1987; Table 7.1). Prolylendopeptidase can also cleave desArg^9-BK, releasing the C-terminal Phe (Fig. 7.2), with kinetics similar to those of BK (Ward *et al.*, 1987). As a consequence, the enzyme can inactivate ligands for B_1 and B_2 kinin receptors. Prolylendopeptidase cleaves a variety of other peptides containing Pro (Wilk, 1983).

The role of prolylendopeptidase is not well understood. Because of its localization in the cytosol, it would not normally have access to peptide hormones synthesized or secreted extracellularly and, even if it were present at significant concentrations in blood, its importance to BK metabolism would be minor. Experiments utilizing a potent prolylendopeptidase inhibitor, Cbz-Pro-prolinal, may help to elucidate its role in peptide metabolism (Wilk and Orlowski, 1983). The interpretation of the results with this inhibitor may be complicated by the fact that it is also a relatively potent inhibitor of the activity of prolylcarboxypeptidase, (Tan *et al.*, 1993), which cleaves some of the same substrates.

6.3 ENDOPEPTIDASE 24.15

Endopeptidase 24.15 was originally purified from the soluble fraction of rat brain homogenates and the enzyme from the extracts cleaved a variety of biologically active peptides (Orlowski *et al.*, 1983). Subsequent studies led to the realization that this enzyme is identical with two other enzymes that had been described and purified earlier: Pz-peptidase, which cleaves a synthetic collagenase substrate; and endo-oligopeptidase A, which was first described as a kininase (for reviews, see Tisljar, 1993; Barrett *et al.*, 1995).

Endopeptidase 24.15 has a pH optimum in the neutral range and is a metalloenzyme, as revealed by biochemical studies as well as its primary sequence, which contains the consensus HEXXH motif of zinc metalloenzymes (Orlowski *et al.*, 1983, 1989; Pierotti *et al.*, 1990; Tisljar, 1993). The enzyme is also sensitive to inhibition by sulfhydryl-reactive agents and is most stable and active in the presence of low concentrations of thiol-containing compounds (Orlowski *et al.*, 1983, 1989; Tisljar and Barrett, 1990). This led one group to classify it as a cysteine peptidase and another to propose renaming it "thimet oligopeptidase" to describe this dual nature (Tisljar, 1993; Barrett *et al.*, 1995). However, the enzyme clearly is a metallopeptidase and its sensitivity to thiols is likely to result from the presence of a free cysteine removed five residues from the catalytic center, but this residue is probably not involved in the catalytic mechanism (Pierotti *et al.*, 1990).

Endopeptidase 24.15 is a single-chain enzyme of 645 amino acids with a molecular mass of 73 kDa (Pierotti *et al.*, 1990; Tisljar, 1993). Although it contains a single potential glycosylation site, there is no evidence that this enzyme is glycosylated (Pierotti *et al.*, 1990). Endopeptidase 24.15 is primarily a cytosolic enzyme, an observation that is consistent with the lack of a signal peptide (Pierotti *et al.*, 1990), although up to 20% of the enzyme may be membrane-associated (Acker *et al.*, 1987). How endopeptidase 24.15 is attached to membrane is not clear, as its primary sequence has no obvious transmembrane domain (Pierotti *et al.*, 1990). The enzyme is ubiquitously distributed, being especially highly concentrated in brain and testes (Tisljar, 1993; Barrett *et al.*, 1995).

Endopeptidase 24.15 has a rather broad substrate specificity, but is most active in cleaving at the carboxyl side of hydrophobic aromatic amino acids, especially those that contain an aromatic residue at the P'_3 position (Orlowski *et al.*, 1983). This explains why it readily cleaves BK at the Phe^5–Ser^6 bond with the aromatic Phe^8 residue in the P'_3 position (Orlowski *et al.*, 1983; Fig. 7.1, Table 7.1). Endopeptidase 24.15 hydrolyzes other biologically active peptides including neurotensin, substance P and LHRH, and converts large opioid peptides to enkephalins (Orlowski *et al.*, 1983; Chu and Orlowski, 1985). As with ACE and endopeptidase 24.11, endopeptidase 24.15 only hydrolyzes peptides shorter than about 20 residues (Chu and Orlowski, 1985; Tisljar, 1993).

The cytosolic localization of endopeptidase 24.15 makes its physiological relevance as a kininase questionable. Although a small portion (around 20%) is membrane bound in subcellular fractions of rat brain (Acker *et al.*, 1987), it is not clear whether this represents enzyme present on the exterior surface of the plasma membrane or bound to intracellular membranes. Nevertheless, some studies using the endopeptidase 24.15 inhibitor *N*-[1(RS)-carboxy-3-phenylpropyl]-Ala-Ala-Phe-pAB (cFP-AAF-pAB) (Orlowski *et al.*, 1988) suggested the possible involvement of the enzyme in inactivating BK *in vivo* or *in situ*. For example, cFP-AAF-pAB enhanced BK-induced contractions of rat uterus (Schriefer and Molineaux, 1993) and blocked most BK degradation by rat hypothalamic slices (McDermott *et al.*, 1987). Intravenous infusion of cFP-AAF-pAB to normotensive rats resulted in an immediate drop in blood pressure of up to 50 mm Hg that was blocked by a B_2 receptor antagonist (Genden and Molineaux, 1991). The inhibitor also increased the potency of an i.v. infusion of BK by ten-fold. However, these data were recently shown to be due primarily to indirect inhibition of ACE by cFP-AAF-pAB after hydrolysis of the inhibitor by endopeptidase 24.11 (Cardozo and Orlowski, 1993) as it had no effect in animals pre-treated with an ACE inhibitor (Yang *et al.*, 1994; Telford *et al.*, 1995).

6.4   ENDOPEPTIDASE 24.16

Endopeptidase 24.16 (neurotensin-degrading enzyme; neurolysin) was first described as an enzyme that degrades neurotensin in rat brain membranes (Checler *et al.*, 1983). Because its properties are similar to those of endopeptidase 24.15, it was not clear that it represented a new enzyme. However, subsequent purification and characterization of the enzyme proved that it is unique (Checler *et al.*, 1986, 1995; Millican *et al.*, 1991). Recent studies have revealed that the enzyme is likely to be identical with the soluble angiotensin II-binding protein and rabbit microsomal endopeptidase, and is at least partly localized in the mitochondrial intermembrane space (Barrett *et al.*, 1995; Serizawa *et al.*, 1995). Its sequence identifies it as a member of the same family as endopeptidase 24.15 (Barrett *et al.*, 1995; Serizawa *et al.*, 1995). Purified endopeptidase 24.16 cleaves BK at the same site as endopeptidase 24.15, (Fig. 7.1) (Millican *et al.*, 1991). The cytosolic and mitochondrial localization of endopeptidase 24.16 appears to preclude it from metabolizing BK under normal circumstances.

7.   *Conclusions*

It is clear after decades of research that ACE on the vascular endothelial cell surface is the most important inactivator of blood-borne BK. However, it has also become evident that BK may act primarily in an autocrine and paracrine fashion (Carretero and Scicli, 1989), establishing the importance of local regulation of its activity by enzymes on cell surfaces (Skidgel, 1992). Thus, the assortment of other enzymes that can inactivate BK may be important in a variety of physiological and pathological situations. In addition, most physiological systems have redundant pathways of metabolism so that the abolishment of one pathway is compensated for by the presence of others. This is vividly demonstrated by the pharmacological inhibition of ACE in hypertension. Although some side effects might be attributed to potentiation of responses to BK (e.g., cough, angioedema), the drugs are surprisingly well tolerated, demonstrating the effectiveness of other peptidases in minimizing the potential deleterious effects of excess kinins. The cleavage of BK by other peptidases at specific sites should lead to the future development of peptidase inhibitors that may be useful in conditions where potentiation of the actions of BK and elevation of its concentration are desirable features.

8. Acknowledgement

Some of the studies on carboxypeptidases M and N described here were supported in part by NIH Grants DK41431, HL36082 and HL36473. We thank Sara Thorburn for her help in preparing the manuscript.

9. References

Acker, G.R., Molineaux, C. and Orlowski, M. (1987). Synaptosomal membrane-bound form of endopeptidase-24.15 generates Leu-enkephalin from dynorphin$_{1-8}$, α- and β-neoendorphin, and Met-enkephalin from Met-enkephalin-Arg^6-Gly^7-Leu^8. J. Neurochem. 48, 284–292.

Alhenc-Gelas, F., Weare, J.A., Johnson, R.L., Jr and Erdös, E.G. (1983). Measurement of human converting enzyme level by direct radioimmunoassay. J. Lab. Clin. Med. 101, 83–96.

Alhenc-Gelas, F., Yasui, T., Allegrini, J., Pinet, F., Acker, G., Corvol, P. and Ménard, J. (1984). Angiotensin I-converting enzyme in foetal membranes and chorionic cells in culture. J. Hypertension, 2 (Suppl. 3), 247–249.

Almenoff, J., Wilk, S. and Orlowski, M. (1981). Membrane bound pituitary metalloendopeptidase: Apparent identity to enkephalinase. Biochem. Biophys. Res. Commun. 102, 206–214.

Ambrosioni, E., Borghi, C. and Magnani, B. (1995). The effect of the angiotensin-converting-enzyme inhibitor zofenopril on mortality and morbidity after anterior myocardial infarction. N. Engl. J. Med. 332, 80–85.

Arregui, A., Emson, P.C. and Spokes, E.G. (1978). Angiotensin-converting enzyme in substantia nigra: reduction of activity in Huntington's disease and after intrastriatal kainic acid in rats. Eur. J. Pharmacol. 52, 121–124.

Auch-Schwelk, W., Bossaller, C., Claus, M., Graf, K., Gräfe, M. and Fleck, E. (1993). ACE inhibitors are endothelium-dependent vasodilators of coronary arteries during submaximal stimulation with bradykinin. Cardiovasc. Res. 27, 312–317.

Avilés, F.X., Vendrell, J., Guasch, A., Coll, M. and Huber, R. (1993). Advances in metallo-procarboxypeptidases. Emerging details on the inhibition mechanism and on the activation process. Eur. J. Biochem. 211, 381–389.

Bajzar, L., Manuel, R. and Nesheim, M.E. (1995). Purification and characterization of TAFI, a thrombin-activatable fibrinolysis inhibitor. J. Biol. Chem. 270, 14477–14484.

Barnes, P.J. (1987). Airway neuropeptides and asthma. TiPS 8, 24–27.

Barnes, P.J. (1994). Neuropeptides and asthma. In "Neuropeptides in Respiratory Medicine" (eds. M.A. Kaliner, P.J. Barnes, G.H.H. Kunkel and J.N. Baraniuk), pp. 501–542. Marcel Dekker, New York.

Barrett, A.J., Brown, M.A., Dando, P.M., Knight, C.G., McKie, N., Rawlings, N.D. and Serizawa, A. (1995). Thimet

oligopeptidase and oligopeptidase M or neurolysin. Meth. Enzymol. 248, 529–556.

Bedrossian, C.W.M., Woo, J., Miller, W.C. and Cannon, D.C. (1978). Decreased angiotensin-converting enzyme in the adult respiratory distress syndrome. Am. J. Clin. Pathol. 70, 244–247.

Beldent, V., Michaud, A., Wei, L., Chauvet, M.-T. and Corvol, P. (1993). Proteolytic release of human angiotensin-converting enzyme. Localization of the cleavage site. J. Biol. Chem. 268, 26428–26434.

Bernstein, K.E., Martin, B.M., Edwards, A.S. and Bernstein, E.A. (1989). Mouse angiotensin-converting enzyme is a protein composed of two homologous domains. J. Biol. Chem. 264, 11945–11951.

Bertrand, C., Geppetti, P., Baker, J., Petersson, G., Piedimonte, G. and Nadel, J.A. (1993). Role of peptidases and NK$_1$ receptors in vascular extravasation induced by bradykinin in rat nasal mucosa. J. Appl. Physiol. 74, 2456–2461.

Beynon, R.J., Shannon, J.D. and Bond, J.S. (1981). Purification and characterization of a metallo-endoproteinase from mouse kidney. Biochem. J. 199, 591–598.

Bhoola, K.D., Figueroa, C.D. and Worthy, K. (1992). Bioregulation of kinins: kallikreins, kininogens, and kininases. Pharmacol. Rev. 44, 1–80.

Bond, J.S. and Beynon, R.J. (1995). The astacin family of metalloendopeptidases. Prot. Sci. 4, 1247–1261.

Bönner, G. (1995). Do kinins play a significant role in the antihypertensive and cardioprotective effects of angiotensin I-converting enzyme inhibitors? In "Hypertension: Pathophysiology, Diagnosis, and Management" (eds. J.H. Laragh and B.M. Brenner), pp. 2877–2893. Raven Press, New York.

Bralet, J., Mossiat, C., Gros, C. and Schwartz, J.C. (1991). Thiorphan-induced natriuresis in volume-expanded rats: roles of endogenous atrial natriuretic factor and kinins. J. Pharmacol. Exp. Ther. 258, 807–811.

Bralet, J., Marie, C., Mossiat, C., Lecomte, J.-M., Gros, C. and Schwartz, J.-C. (1994). Effects of alatriopril, a mixed inhibitor of atriopeptidase and angiotensin I-converting enzyme, on cardiac hypertrophy and hormonal responses in rats with myocardial infarction. Comparison with captopril. J. Pharmacol. Exp. Ther. 270, 8–14.

Brent, G.A., Hershman, J.M., Reed, A.W., Sastre, A., Lieberman, J. (1984). Serum angiotensin-converting enzyme in severe nonthyroidal illnesses associated with low serum thyroxine concentration. Ann. Intern. Med. 100, 680–683.

Bunting, P.S., Szalai, J.P. and Katic, M. (1987). Diagnostic aspects of angiotensin converting enzyme in pulmonary sarcoidosis. Clin. Biochem. 20, 213–219.

Butler, P.E., McKay, M.J. and Bond, J.S. (1987). Characterization of meprin, a membrane-bound metalloendopeptidase from mouse kidney. Biochem. J. 241, 29–235.

Butterworth, J. (1986). Changes in nine enzyme markers for neurons, Glia, and endothelial cells in agonal state and Huntington's disease caudate nucleus. J. Neurochem. 47, 583–587.

Cambien, F. and Soubrier, F. (1995). The angiotensin-converting enzyme: molecular biology and implication of the gene polymorphism in cardiovascular diseases. In "Hypertension: Pathophysiology, Diagnosis, and Management" (eds. J.H. Laragh and B. M. Brenner), pp. 1667–1682. Raven Press, New York.

Campbell, W. and Okada, H. (1989). An arginine specific carboxypeptidase generated in blood during coagulation or inflammation which is unrelated to carboxypeptidase N or its subunits. Biochem. Biophys. Res. Commun. 162, 933–939.

Carbonell, L.F., Carretero, O.A., Stewart, J.M. and Scicli, A.G. (1988). Effect of a kinin antagonist on the acute antihypertensive activity of enalaprilat in severe hypertension. Hypertension 11, 239–243.

Cardozo, C. and Orlowski, M. (1993). Evidence that enzymatic conversion of N-[1(R,S)-carboxy-3–phenylpropyl]-Ala-Ala-Phe-p-aminobenzoate, a specific inhibitor of endopeptidase 24.15, to N-[1(R,S)-carboxy-3-phenylpropyl]-Ala-Ala is necessary for inhibition of angiotensin converting enzyme. Peptides 14, 11259–11262.

Carpenter, G. and Wahl, M.I. (1990). The epidermal growth factor family. In "Peptide Growth Factors and their Receptors I, Handbook of Experimental Pharmacology, vol. 95" (eds. M.B. Sporn and A.B. Roberts), pp. 69–171. Springer-Verlag, New York.

Carretero, O.A. and Scicli, A.G. (1989). Kinins paracrine hormone. In "The Kallikrein–Kinin System in Health and Disease" (eds. H. Fritz, I. Schmidt and G. Dietze), pp. 63–78. Limbach-Verlag Braunschweig, Munich.

Carretero, O.A. and Scicli, A.G. (1995). The kallikrein–kinin system as a regulator of cardiovascular and renal function. In "Hypertension: Pathophysiology, Diagnosis, and Management" (eds. J.H. Laragh and B.M. Brenner), pp. 983–999. Raven Press, New York.

Chai, S.Y., McKinley, M.J., Paxinos, G. and Mendelsohn, F.A.O. (1991). Angiotensin-converting enzyme in the monkey (Macaca fascicularis) brain visualized by in vitro autoradiography. Neuroscience 42, 483–495.

Checler, F., Vincent, J.-P. and Kitabgi, P. (1983). Degradation of neurotensin by rat brain synaptic membranes: involvement of a thermolysin-like metalloendopeptidase (enkephalinase), angiotensin converting enzyme, and other unidentified peptidases. J. Neurochem. 41, 375–384.

Checler, F., Vincent, J.-P. and Kitabgi, P. (1986). Purification and characterization of a novel neurotensin-degrading peptidase from rat brain synaptic membranes. J. Biol. Chem. 261, 11274–11281.

Checler, F., Barelli, H., Dauch, P., Dive, V., Vincent, B. and Vincent, J.P. (1995). Neurolysin: purification and assays. Meth. Enzymol. 248, 593–614.

Chercuitte, F., Beaulieu, A.D., Poubelle, P. and Marceau, F. (1987). Carboxypeptidase N (kininase I) activity in blood and synovial fluid from patients with arthritis. Life Sci. 41, 1225–1232.

Chodimella, V., Skidgel, R.A., Krowiak, E.J. and Murlas, C.G. (1991). Lung peptidases, including carboxypeptidase, modulate airway reactivity to intravenous bradykinin. Am. Rev. Resp. Dis. 144, 869–874.

Chu, T.G. and Orlowski, M. (1985). Soluble metalloendopeptidase from rat brain: action on enkephalin-containing peptides and other bioactive peptides. Endocrinology 116, 1418–1425.

Colman, R.W. (1987). Humoral mediators of catastrophic reactions associated with protamine neutralization. Anesthesiology 66, 595–596.

Connelly, J.C., Skidgel, R.A., Schulz, W.W., Johnson, A.R. and Erdös, E.G. (1985). Neutral endopeptidase 24.11 in human neutrophils: cleavage of chemotactic peptide. Proc. Natl Acad. Sci. USA 82, 8737–8741.

CONSENSUS Trial Study Group (1987). Effects of enalapril on mortality in severe congestive heart failure. N. Engl. J. Med. 316, 1429–1435.

Cornell, M.J., Williams, T.A., Lamango, N.S., Coates, D., Corvol, P., Soubrier, F., Hoheisel, J., Lehrach, H. and Isaac, R. E. (1995). Cloning and expression of an evolutionary conserved single-domain angiotensin-converting enzyme from Drosophila melanogaster. J. Biol. Chem. 270, 13613–13619.

Correa, F.M.A., Plunkett, L.M., Saavedra, J.M. and Hichens, M. (1985). Quantitative autoradiographic determination of angiotensin-converting enzyme (kininase II) binding in individual rat brain nuclei with 125I-351A, a specific enzyme inhibitor. Brain Res. 347, 192–195.

Costerousse, O., Allegrini, J., Lopez, M. and Alhenc-Gelas, F. (1993). Angiotensin I-converting enzyme in human circulating mononuclear cells: genetic polymorphism of expression in T-lymphocytes. Biochem. J. 290, 33–40.

Crecelius, D.M., Stewart, J.M., Vavrek, R.J., Balasubramaniam, T.M. and Baenziger, N.L. (1986). Interaction of bradykinin (BK) with receptors on human lung fibroblasts. Fed. Proc. 45, 454.

Crimi, N., Polosa, R., Pulvirenti, G., Magri, S., Santonocito, G., Prosperini, G., Mastruzzo, C. and Mistretta, A. (1995). Effect of an inhaled neutral endopeptidase inhibitor, phosphoramidon, on baseline airway calibre and bronchial responsiveness to bradykinin in asthma. Thorax 50, 505–510.

Cushman, D.W. and Ondetti, M.A. (1980). Inhibitors of angiotensin converting enzyme. Prog. Med. Chem. 17, 42–104.

Danilov, S., Jaspard, E., Churakova, T., Towbin, H., Savoie, F., Wei, L. and Alhenc-Gelas, F. (1994). Structure–function analysis of angiotensin I-converting enzyme using monoclonal antibodies. Selective inhibition of the amino-terminal active site. J. Biol. Chem. 269, 26806–26814.

D'Azzo, A., Hoogeveen, A., Reuser, A.J.J., Robinson, D. and Galjaard, H. (1982). Molecular defect in combined β-galactosidase and neuraminidase deficiency in man. Proc. Natl Acad. Sci. USA 79, 4535–4539.

DeBlois, D., Bouthillier, J. and Marceau, F. (1991). Pulse exposure to protein synthesis inhibitors enhances vascular responses to des-Arg^9-bradykinin: possible role of interleukin-1. Br. J. Pharmacol. 103, 1057–1066.

DeBlois, D., Drapeau, G., Petitclerc, E. and Marceau, F. (1992). Synergism between the contractile effect of epidermal growth factor and that of des-Arg^9-bradykinin or of α-thrombin in rabbit aortic rings. Br. J. Pharmacol. 105, 959–967.

Deddish, P.A., Skidgel, R.A. and Erdös, E.G. (1989). Enhanced Co^{2+} activation and inhibitor binding of carboxypeptidase M at low pH. Similarity to carboxypeptidase H (enkephalin convertase). Biochem. J. 261, 289–291.

Deddish, P.A., Skidgel, R.A., Kriho, V.B., Li, X.-Y., Becker, R.P. and Erdös, E.G. (1990). Carboxypeptidase M in cultured Madin-Darby canine kidney (MDCK) cells. Evidence that carboxypeptidase M has a phosphatidylinositol glycan anchor. J. Biol. Chem. 265, 15083–15089.

Deddish, P.A., Wang, J., Michel, B., Morris, P.W., Davidson, N.O., Skidgel, R.A. and Erdös, E.G. (1994). Naturally occurring active N-domain of human angiotensin I converting enzyme. Proc. Natl Acad. Sci. USA 91, 7807–7811.

Deddish, P.A., Wang, L.-X., Jackman, H.L., Michel, B., Wang, J., Skidgel, R.A. and Erdös, E.G. (1996). Single-domain angiotensin I converting enzyme (kininase II): characterization and properties. J. Pharmacol. Exper. Ther. 279, 1582–1589.

Defendini, R., Zimmerman, E.A., Weare, J.A., Alhenc-Gelas, F. and Erdös, E.G. (1983). Angiotensin-converting enzyme in epithelial and neuroepithelial cells. Neuroendocrinology 37, 32–40.

Dehm, P. and Nordwig, A. (1970). The cleavage of prolyl peptides by kidney peptidases. Partial purification of a "X-Prolyl-Aminopeptidase" from swine kidney microsomes. Eur. J. Biochem. 17, 364–371.

DeMeyer, G.R.Y., Bult, H., Kockx, M.M. and Herman, A.G. (1995). Effect of angiotensin-converting enzyme inhibition on intimal thickening in rabbit collared carotid artery. J. Cardiovasc. Pharmacol. 26, 614–620.

Denslow, N.D., Ryan, J.W. and Nguyen, H.P. (1994). Guinea pig membrane-bound aminopeptidase P is a member of the proline peptidase family. Biochem. Biophys. Res. Commun. 205, 790–1795.

de Saint-Vis, B., Cupillard, L., Pandrau-Garcia, D., Ho, S., Renard, N., Grouard, G., Duvert, V., Thomas, X., Galizzi, J. P., Banchereau, J. and Saeland, S. (1995). Distribution of carboxypeptidase-M on lymphoid and myeloid cells parallels the other zinc-dependent proteases CD10 and CD13. Blood 86, 1098–1105.

Deschodt-Lanckman, M., Michaux, F., De Prez, E., Abramowicz, D., Vanherweghem, J.-L. and Goldman, M. (1989). Increased serum levels of endopeptidase 24.11 ("enkephalinase") in patients with end-stage renal failure. Life Sci. 45, 133–141.

Desmazes, N.A., Lockhart, A., Lacroix, H. and Dusser, D.J. (1992). Carboxypeptidase M-like enzyme modulates the noncholinergic bronchoconstrictor response in guinea pigs. Am. J. Respir. Cell Mol. Biol. 7, 477–484.

Devault, A., Lazure, C., Nault, C., Le Moual, H., Seidah, N.G., Chretien, M., Kahn, P., Powell, J., Mallet, J., Beaumont, A., Roques, B.P., Crine, P. and Boileau, G. (1987). Amino acid sequence of rabbit kidney neutral endopeptidase 24.11 (enkephalinase) deduced from a complementary DNA. EMBO J. 6, 1317–1322.

Dhople, A.M., Howell, P.C., Williams, S.L., Zeigler, J.A. and Storrs, E.E. (1985). Serum angiotensin-converting enzyme in leprosy. Indian J. Leprosy 57, 282–287.

Dietze, G.J. (1982). Modulation of the action of insulin in relation to the energy state in skeletal muscle tissue: possible involvement of kinins and prostaglandins. Mol. Cell. Endocrinol. 25, 127–149.

Dragovic, T., Igi ́, R., Erdös, E.G. and Rabito, S.F. (1993). Metabolism of bradykinin by peptidases in the lung. Am. Rev. Respir. Dis. 147, 1491–1496.

Dragovic, T., Deddish, P.A., Tan, F., Weber, G. and Erdös, E.G. (1994). Increased expression of neprilysin (neutral endopeptidase 24.11) in rat and human hepatocellular carcinomas. Lab. Invest. 70, 107–113.

Dragovic, T., Schraufnagel, D.E., Becker, R.P., Sekosan, M., Votta-Velis, E.G. and Erdös, E.G. (1995). Carboxypeptidase M activity is increased in bronchoalveolar lavage in human lung disease. Am. J. Respir. Crit. Care Med. 152, 760–764.

Dragovic, T., Sekosan, M., Becker, R.P. and Erdös, E.G. (1996a). Detection of neprilysin (neutral endopeptidase 24.11) in human hepatocellular carcinomas by immunocytochemistry. (Submitted).

Dragovic, T., Minshall, R., Jackman, H.L., Wang, L-X. and Erdös, E.G. (1996b). Kininase II-type enzymes, their putative role in muscle energy metabolism. Diabetes 45, S34–S37.

Dumermuth, E., Sterchi, E.E., Jiang, W., Wolz, R.L., Bond, J.S., Flannery, A.V. and Beynon, R.J. (1991). The astacin family of metallopeptidases. J. Biol. Chem. 266, 21381–21385.

Dusser, D.J., Nadel, J.A., Sekizawa, K., Graf, P.D. and Borson, D.B. (1988). Neutral endopeptidase and angiotensin converting enzyme inhibitors potentiate kinin-induced contraction of ferret trachea. J. Pharmacol. Exp. Ther. 244, 531–536.

Eaton, D.L., Malloy, B.E., Tsai, S.P., Henzel, W. and Drayna, D. (1991). Isolation, molecular cloning, and partial characterization of a novel carboxypeptidase B from human plasma. J. Biol. Chem. 266, 21833–21838.

Edery, H. (1964). Potentiation of the action of bradykinin on smooth muscle by chymotrypsin, chymotrypsinogen and trypsin. Br. J. Pharmacol. 22, 371–379.

Edery, H. (1965). Further studies of the sensitization of smooth muscle to the action of plasma kinins by proteolytic enzymes. Br. J. Pharmacol. 24, 485–496.

Ehlers, M.R.W. and Riordan, J.F. (1990). Angiotensin-converting enzyme. Biochemistry and molecular biology. In "Hypertension: Pathophysiology, Diagnosis, and Management" (eds. J.H. Laragh and B.M. Brenner), pp. 1217–1231. Raven Press, New York.

Ehlers, M.R.W., Fox, E.A., Strydom, D.J. and Riordan, J.F. (1989). Molecular cloning of human testicular angiotensin-converting enzyme: The testis isozyme is identical to the C-terminal half of endothelial angiotensin-converting enzyme. Proc. Natl Acad. Sci. USA 86, 7741–7745.

Ehlers, M.R.W., Scholle, R.R. and Riordan, J.F. (1995). Proteolytic release of human angiotensin-converting enzyme expressed in Chinese hamster ovary cells is enhanced by phorbol ester. Biochem. Biophys. Res. Commun. 206, 541–547.

Elsner, D., Kromer, E.P. and Riegger, G.A.J. (1992). Effectiveness of endopeptidase inhibition (candoxatril) in congestive heart failure. Am. J. Cardiol. 70, 494–498.

Erdös, E.G. (1966). Hypotensive peptides: bradykinin, kallidin, and eledoisin. Adv. Pharmacol. 4, 1–90.

Erdös, E.G. (1979). Kininases. In "Bradykinin, Kallidin and Kallikrein. Handbook of Experimental Pharmacology" (ed. E.G. Erdös), pp. 427–487. Springer-Verlag, Heidelberg.

Erdös, E.G. (1989). From measuring the blood pressure to mapping the gene: the development of ideas of Frey and Werle. In "The Kallikrein and Kinin System in Health and Disease. E.K. Frey and E. Werle Memorial Volume" (eds. H. Fritz, I. Schmidt and G. Dietze), pp. 261–276. Limbach-Verlag, Braunschweig.

Erdös, E.G. and Gafford, J.T. (1983). Human converting enzyme. Clin. Exper. Hypertens. A5, 1251–1262.

Erdös, E.G. and Skidgel, R.A. (1989). Neutral endopeptidase 24.11–enkephalinase and related regulators of peptide hormones. FASEB J. 3, 145–151.

Erdös, E.G. and Sloane, E.M. (1962). An enzyme in human blood plasma that inactivates bradykinin and kallidins. Biochem. Pharmacol. 11, 585–592.

Erdös, E.G. and Yang, H.Y.T. (1966). Inactivation and potentiation of the effects of bradykinin. In "Hypotensive Peptides" (eds. E.G. Erdös, N. Back and F. Sicuteri), pp. 235–251. Springer-Verlag, New York.

Erdös, E.G. and Yang, H.Y.T. (1967). An enzyme in microsomal fraction of kidney that inactivates bradykinin. Life Sci. 6, 569–574.

Erdös, E.G. and Yang, H.Y.T. (1970). Kininases. In "Handbook of Experimental Pharmacology", Vol. XXV, pp. 289–323. Springer-Verlag, Heidelberg.

Erdös, E.G., Renfrew, A.G., Sloane, E.M., and Wohler, J.R. (1963). Enzymatic studies on bradykinin and similar peptides. Ann. N. York Acad. Sci. 104, 222–235.

Erdös, E.G., Sloane, E.M., and Wohler, I.M. (1964). Carboxypeptidase in blood and other fluids – I. Properties, distribution, and partial purification of the enzyme. Biochem. Pharmacol. 13, 893–905.

Erdös, E.G., Wohler, I.M., Levine, M.I. and Westerman, P. (1965). Carboxypeptidase in blood and other fluids. Values in human blood in normal and pathological conditions. Clin. Chim. Acta 11, 39–43.

Erdös, E.G., Schulz, W.W., Gafford, J.T. and Defendini, R. (1985). Neutral metalloendopeptidase in human male genital tract. Comparison to angiotensin I converting enzyme. Lab. Invest. 52, 437–447.

Erdös, E.G., Wagner, B.A., Harbury, C.B., Painter, R.G., Skidgel, R.A. and Fa, X.-G. (1989). Down-regulation and inactivation of neutral endopeptidase 24.11 (enkephalinase) in human neutrophils. J. Biol. Chem. 264, 14519–14523.

Falezza, G., Santonastaso, C.L., Parisi, T. and Muggeo, M. (1985). High serum levels of angiotensin-converting enzyme in untreated Addison's disease. J. Clin. Endocrinol. Metab. 61, 496–498.

Falkenhahn, M., Franke, F., Bohle, R.M., Zhu, Y.-C., Stauss, H.M., Bachmann, S., Danilov, S. and Unger, T. (1995). Cellular distribution of angiotensin-converting enzyme after myocardial infarction. Hypertension 25, 219–226.

Farhy, R.D., Hoo, K.-L., Carretero, O.A. and Scicli, A.G. (1992). Kinins mediate the antiproliferative effect of ramipril in rat carotid artery. Biochem. Biophys. Res. Commun. 182, 283–288.

Flynn, G.A., French, J.F. and Dage, R.C. (1995). Dual inhibitors of angiotensin-converting enzyme and neutral endopeptidase: design and therapeutic rationale. In "Hypertension: Pathophysiology, Diagnosis, and Management" (eds. J.H. Laragh and B.M. Brenner), pp. 3099–3114. Raven Press, New York.

Folk, J.E. and Gladner, J.A. (1960). Cobalt activation of carboxypeptidase A. J. Biol. Chem. 235, 60–63.

Folk, J.E. and Gladner, J.A. (1961). Influence of cobalt and cadmium on the peptidase and esterase activities of carboxypeptidase B. Biochim. Biophys. Acta 48, 139–147.

French, J.F., Flynn, G.A., Giroux, E.L., Mehdi, S., Anderson, B., Beach, D.C., Koehl, J.R. and Dage, R.C. (1994). Characterization of a dual inhibitor of angiotensin I-converting enzyme and neutral endopeptidase. J. Pharmacol. Exp. Ther. 268, 180–186.

French, J.F., Anderson, B.A., Downs, T.R. and Dage, R.C. (1995). Dual inhibition of angiotensin-converting enzyme and neutral endopeptidase in rats with hypertension. J. Cardiovasc. Pharmacol. 26, 107–113.

Frey, E.K., Kraut, H. and Werle, E. (1950). "Kallikrein Padutin". Ferdinand Enke Verlag, Stuttgart.

Frey, E.K., Kraut, H., Werle, E., Vogel, R., Zickgraf-Rüdel, G. and Trautschold, I. (1968). "Das Kallikrein-Kinin-System und seine Inhibitoren". Ferdinand Enke Verlag, Stuttgart.

Friedland, J., Setton, C. and Silverstein, E. (1978). Induction of angiotensin-converting enzyme in human monocytes in culture. Biochem. Biophys. Res. Commun. 83, 843–849.

Fyhrquist, F., Grönhagen-Riska, C., Hortling, L., Forslund, T. and Tikkanen, I. (1983). Regulation of angiotensin-converting enzyme. J. Hypertens. 1(Suppl. 1), 25–30.

Gafford, J.T., Skidgel, R.A., Erdös, E.G. and Hersh, L.B. (1983). Human kidney "enkephalinase", a neutral metalloendopeptidase that cleaves active peptides. Biochemistry 22, 3265–3271.

Galjart, N.J., Gillemans, N., Harris, A., van der Horst, G.T.J., Verheijen, F.W., Galjaard, H. and d'Azzo, A. (1988). Expression of cDNA encoding the human "protective protein" associated with lysosomal-galactosidase and neuraminidase: homology to yeast proteases. Cell 54, 755–764.

Galjart, N.J., Gillemans, N., Meijer, D. and d'Azzo, A. (1990). Mouse "protective protein". cDNA cloning, sequence comparison, and expression. J. Biol. Chem. 265, 4678–4684.

Gavras, I. and Gavras, H. (1987). The use of ACE inhibitors in hypertension. In "Angiotensin Converting Enzyme Inhibitors" (eds. J.B. Kostis and E.A. DeFelice), pp. 93–122. Alan R. Liss, New York.

Gavras, I. and Gavras, H. (1993). ACE inhibitors: a decade of clinical experience. Hosp. Pract. July 15, 61–71.

Gavras, H., Faxon, D.P., Berkoben, J., Brunner, H.R. and Ryan, T.J. (1978). Angiotensin converting enzyme inhibition in patients with congestive heart failure. Circulation 58, 770–776.

Gebhard, W., Schube, M. and Eulitz, M. (1989). cDNA cloning and complete primary structure of the small, active subunit of human carboxypeptidase N (kininase I). Eur. J. Biochem. 178, 603–607.

Genden, E.M. and Molineaux, C.J. (1991). Inhibition of endopeptidase-24.15 decreases blood pressure in normotensive rats. Hypertension 18, 360–365.

Gladner, J.A., Murtaugh, P.A., Folk, J.E. and Laki, K. (1963). Nature of peptides released by thrombin. Ann. N. York Acad. Sci. 104, 47–52.

Gohlke, P., Linz, W., Schölkens, B.A., Kuwer, I., Bartenbach, S., Schnell, A. and Unger, T. (1994a). Angiotensin-converting enzyme inhibition improves cardiac function. Role of bradykinin. Hypertension 23, 411–418.

Gohlke, P., Kuwer, I., Bartenbach, S., Schnell, A. and Unger, T. (1994b). Effect of low-dose treatment with perindopril on cardiac function in stroke-prone spontaneously hypertensive rats: role of bradykinin. J. Cardiovasc. Pharmacol. 24, 462–469.

Gonzalez-Vera, W., Fournie-Zaluski, M.-C., Pham, I., Laboulandine, I., Roques, B.-P. and Michel, J.-B. (1995). Hypotensive and natriuretic effects of RB 105, a new dual inhibitor of angiotensin converting enzyme and neutral endopeptidase in hypertensive rats. J. Pharmacol. Exp. Ther. 272, 343–351.

Gorbea, C.M., Flannery, A.V. and Bond, J.S. (1991). Homo- and heterotetrameric forms of the membrane-bound metalloendopeptidases meprin A and B. Arch. Biochem. Biophys. 290, 549–553.

Graf, K., Koehne, P., Gräfe, M., Zhang, M., Auch-Schwelk, W. and Fleck, E. (1995). Regulation and differential expression of neutral endopeptidase 24.11 in human endothelial cells. Hypertension 26, 230–235.

Greenbaum, L.M. and Sherman, R. (1962). Studies on catheptic carboxypeptidase. J. Biol. Chem. 237, 1082–1085.

Grönhagen-Riska, C. (1979). Angiotensin-converting enzyme. I. Activity and correlation with serum lysozyme in sarcoidosis, other chest or lymph node diseases and healthy persons. Scand. J. Resp. Dis. 60, 83–93.

Grönhagen-Riska, C., Fyhrquist, F., Välimäki, M. and Lambert, B.-A. (1985). Thyroid hormones affect serum angiotensin I converting enzyme levels. Acta Med. Scand. 217, 259–264.

Gros, C., Noël, N., Souque, A., Schwartz, J.-C., Danvy, D., Plaquevent, J.-C., Duhamel, L., Duhamel, P., Lecomte, J.-M. and Bralet, J. (1991). Mixed inhibitors of angiotensin-converting enzyme (EC 3.4.15.1) and enkephalinase (EC 3.4.24.11): rational design, properties, and potential cardiovascular applications of glycopril and alatriopril. Proc. Natl Acad. Sci. USA 88, 4210–4214.

Hall, A.S., Tan, L.-B. and Ball, S.G. (1994). Inhibition of ACE/kininase II, acute myocardial infarction, and survival. Cardiovasc. Res. 28, 190–198.

Hall, E.R., Kato, J., Erdös, E.G., Robinson, C.J.G. and Oshima, G. (1976). Angiotensin I-converting enzyme in the nephron. Life Sci. 18, 1299–1303.

Harbeck, H.T. and Mentlein, R. (1991). Aminopeptidase P from rat brain. Purification and action on bioactive peptides. Eur. J. Biochem. 198, 451–458.

Hartman, J.C. (1995). The role of bradykinin and nitric oxide in the cardioprotective action of ACE inhibitors. Ann. Thorac. Surg. 60, 787–792.

Heck, I. and Niederle, N. (1983). Angiotensin-converting-enzym-aktivität während zytostatischer Therapie bei Patienten mit primär inoperablem Bronchialkarzinom. Klin. Wochenschr. 61, 923–927.

Hendriks, D., Scharpé, S., van Sande, M. and Lommaert, M.P. (1989). Characterisation of a carboxypeptidase in human serum distinct from carboxypeptidase N. J. Clin. Chem. Clin. Biochem. 27, 277–285.

Hendriks, D., Wang, W., Scharpé, S., Lommaert, M.P. and van Sande, M. (1990). Purification and characterization of a new arginine carboxypeptidase in human serum. Biochim. Biophys. Acta 1034, 86–92.

Hendriks, D., Wang, W., van Sande, M. and Scharpé, S. (1992). Human serum carboxypeptidase U: a new kininase? Agents Actions Suppl. 38/I, 407–413.

Hooper, N.M. and Turner, A.J. (1988). Ectoenzymes of the kidney microvillar membrane. Aminopeptidase P is anchored by a glycosyl-phosphatidylinositol moiety. FEBS Lett. 229, 340–344.

Hooper, N.M., Hryszko, J. and Turner, A.J. (1990). Purification and characterization of pig kidney aminopeptidase P. A glycosyl-phosphatidylinositol-anchored ectoenzyme. Biochem. J. 267, 509–515.

Hooper, N.M., Hryszko, J., Oppong, S.Y. and Turner, A.J. (1992). Inhibition by converting enzyme inhibitors of pig kidney aminopeptidase P. Hypertension 19, 281–285.

Howell, S., Boileau, G. and Crine, P. (1994). Neutral endopeptidase (EC 3.4.24.11): constructed molecular forms show new angles of an old enzyme. Biochem. Cell Biol. 72, 67–69.

Hropot, M., Grötsch, H., Klaus, E., Langer, K.H.., Linz, W., Wiemer, G. and Schölkens, B.A. (1994). Ramipril prevents the detrimental sequels of chronic NO synthase inhibition in rats: hypertension, cardiac hypertrophy and renal insufficiency. Naunyn-Schmiedeberg's Arch. Pharmacol. 350, 646–652.

Ibrahim, B., Mombouli, J.V., Ballard, K., Regoli, D. and Vanhoutte, P.M. (1995). Interactions between kinins and ACE-inhibitors in canine arteries. FASEB J. 9 (Suppl I), A561 (Abstract).

Ichinose, M. and Barnes, P.J. (1990). The effect of peptidase inhibitors on bradykinin-induced bronchoconstriction in guinea-pigs in vivo. Br. J. Pharmacol. 101, 77–80.

Igic, R. (1985). Kallikrein and kininases in ocular tissues. Exp. Eye Res. 41, 117–120.

Igic, R., Erdös, E.G., Yeh, H.S.J., Sorrells, K. and Nakajima, T. (1972). Angiotensin I converting enzyme of the lung. Circ. Res. 31(Suppl II), 51–61.

Igic, R., Nakajima, T., Yeh, H.S.J., Sorrells, K. and Erdös, E.G. (1973). Kininases. In "Pharmacology and the Future of Man – Proceedings of the 5th International Congress of Pharmacology" (ed. G. Acheson), pp. 307–319. Karger, Basel.

Igic, R., Robinson, C.J.G. and Erdös, E.G. (1977). Angiotensin I converting enzyme activity in the choroid plexus and in the retina. In "Central Actions of Angiotensin and Related Hormones" (eds. J.P. Buckley and C.M. Ferrario), pp. 23–27. Pergamon Press, New York.

Iimura, O. (1987). Pathophysiological significance of kallikrein–kinin system in essential hypertension. In "Renal Function, Hypertension and Kallikrein–Kinin System" (eds. O. Iimura and H.S. Margolius), pp. 3–18. Hokusen-Sha Publishers, Tokyo, Japan.

Inokuchi, J.-I. and Nagamatsu, A. (1981). Tripeptidyl carboxypeptidase activity of kininase II (angiotensin-converting enzyme). Biochim. Biophys. Acta 662, 300–307.

Israili, Z.H. and Hall, W.D. (1992). Cough and angioneurotic edema associated with angiotensin-converting enzyme inhibitor therapy. A review of the literature and pathophysiology. Ann. Intern. Med. 117, 234–242.

Jackman, H.L., Tan, F., Tamei, H., Beurling-Harbury, C., Li, X.-Y., Skidgel, R.A. and Erdös, E.G. (1990). A peptidase in human platelets that deamidates tachykinins: probable identity with the lysosomal "protective protein". J. Biol. Chem. 265, 11265–11272.

Jackman, H.L., Morris, P.W., Deddish, P.A., Skidgel, R.A. and Erdös, E.G. (1992). Inactivation of endothelin I by deamidase (lysosomal protective protein). J. Biol. Chem. 267, 2872–2875.

Jackman, H.L., Morris, P.W., Rabito, S.F., Johansson, G.B. and Erdös, E.G. (1993). Inactivation of endothelin 1 by an enzyme of the vascular endothelial cells. Hypertension 21, 925–928.

Jackman, H.L., Tan, F., Schraufnagel, D., Dragovic, T., Dezsö, B., Becker, R.P. and Erdös, E.G. (1995). Plasma membrane-bound and lysosomal peptidases in human alveolar macrophages. Am. J. Respir. Cell Mol. Biol. 13, 196–204.

Jaffa, A.A., Vio, C.P., Silva, R.H., Vavrek, R.J., Stewart, J.M., Rust, P.F. and Mayfield, R.K. (1992). Evidence for renal kinins as mediators of amino acid-induced hyperperfusion and hyperfiltration in the rat. J. Clin. Invest. 89, 1460–1468.

Jaspard, E. and Alhenc-Gelas, F. (1995). Catalytic properties of the two active sites of angiotensin I-converting enzyme on the cell surface. Biochem. Biophys. Res. Commun. 211, 528–534.

Jaspard, E., Wei, L. and Alhenc-Gelas, F. (1993). Differences in the properties and enzymatic specificities of the two active sites of angiotensin I-converting enzyme (kininase II). Studies with bradykinin and other natural peptides. J. Biol. Chem. 268, 9496–9503.

Jiang, W., Gorbea, C.M., Flannery, A.V., Beynon, R.J., Grant, G.A. and Bond, J.S. (1992). The α subunit of meprin A. Molecular cloning and sequencing, differential expression in inbred mouse strains, and evidence for divergent evolution of the α and β subunits. J. Biol. Chem. 267, 9185–9193.

Jiang, W., Sadler, P.M., Jenkins, N.A., Gilbert, D.J., Copeland, N.G. and Bond, J.S. (1993). Tissue-specific expression and chromosomal localization of the α subunit of mouse meprin A. J. Biol. Chem. 268, 10380–10385.

Johnson, A.R., John, M. and Erdös, E.G. (1982). Metabolism of vasoactive peptides by membrane-enriched fractions from human lung tissue, pulmonary arteries, and endothelial cells. Ann. N. York Acad. Sci. 384, 72–89.

Johnson, A.R., Skidgel, R.A., Gafford, J.T. and Erdös, E.G. (1984). Enzymes in placental microvilli: angiotensin I converting enzyme, angiotensinase A, carboxypeptidase, and neutral endopeptidase ("enkephalinase"). Peptides 5, 789–796.

Johnson, A.R., Ashton, J., Schulz, W. and Erdös, E.G. (1985a). Neutral metalloendopeptidase in human lung tissue and cultured cells. Am. Rev. Resp. Dis. 132, 564–568.

Johnson, A.R., Coalson, J.J., Ashton, J., Larumbide, M. and Erdös, E.G. (1985b). Neutral metalloendopeptidase in serum samples from patients with adult respiratory distress syndrome: comparison with angiotensin-converting enzyme. Am. Rev. Resp. Dis. 132, 1262–1267.

Johnson, G.D. and Hersh, L.B. (1992). Cloning a rat meprin cDNA reveals the enzyme is a heterodimer. J. Biol. Chem. 267, 13505–13512.

Johnson, G.D. and Hersh, L.B. (1994). Expression of meprin subunit precursors. Membrane anchoring through the subunit and mechanism of zymogen activation. J. Biol. Chem. 269, 7682–7688.

Kabour, A., Henegar, J.R., Devineni, V.R. and Janicki, J.S. (1995). Prevention of angiotensin I induced myocyte necrosis and coronary vascular damage by lisinopril and losartan in the rat. Cardiovasc. Res. 29, 543–548.

Kariya, K., Yamauchi, A., Hattori, S., Tsuda, Y. and Okada, Y. (1982). The disappearance rate of intraventricular bradykinin in the brain of the conscious rat. Biochem. Biophys. Res. Commun. 107, 1461–1466.

Kato, H. and Suzuki, T. (1970). Amino acid sequence of bradykinin-potentiating peptide isolated from the venom of *Agkistrodon halys blomhoffii*. Proc. Jpn. Acad. 46, 176–181.

Kerr, M.A. and Kenny, A.J. (1974). The purification and specificity of a neutral endopeptidase from rabbit kidney brush border. Biochem. J. 137, 477–488.

Kobe, B. and Deisenhofer, J. (1993). Crystal structure of porcine ribonuclease inhibitor, a protein with leucine-rich repeats. Nature (Lond.) 366, 751–756.

Komers, R. and Cooper, M.E. (1995). Acute renal hemodynamic effects of ACE inhibition in diabetic hyperfiltration: role of kinins. Am. J. Physiol. 268, F588–F594.

Kounnas, M.Z., Wolz, R.L., Gorbea, C.M. and Bond, J.S. (1991). Meprin-A and -B. Cell surface endopeptidases of the mouse kidney. J. Biol. Chem. 266, 17350–17357.

Kumamoto, K., Stewart, T.A., Johnson, A.R. and Erdös, E.G. (1981). Prolylcarboxypeptidase (angiotensinase C) in human cultured cells. J. Clin. Invest. 67, 210–215.

Kumar, R.S., Kusari, J., Roy, S.N., Soffer, R.L. and Sen, G.C. (1989). Structure of testicular angiotensin-converting enzyme. A segmental mosaic enzyme. J. Biol. Chem. 264, 16754–16758.

Kuroki, K., Eng, F., Ishikawa, T., Turck, C., Harada, F. and Ganem, D. (1995). gp180, A host cell glycoprotein that binds duck hepatitis B virus particles, is encoded by a member of the carboxypeptidase gene family. J. Biol. Chem. 270, 15022–15028.

Kyle, D.J and Burch, R.M. (1993). A survey of bradykinin receptors and their antagonists. Curr. Opin. Invest. Drugs 2, 5–20.

Lattion, A.-L., Soubrier, F., Allegrini, J., Hubert, C., Corvol, P. and Alhenc-Gelas, F. (1989). The testicular transcript of the angiotensin I-converting enzyme encodes for the ancestral, non-duplicated form of the enzyme. FEBS Lett. 252, 99–104.

LeBien, T.W. and McCormack, R.T. (1989). The common acute lymphoblastic leukemia antigen (CD10) – Emancipation from a functional enigma. Blood 73, 625–635.

Letarte, M., Vera., S., Tran, R., Addis, J.B.L., Onizuka, R.J., Quackenbush, E.J., Jongeneel, C.V. and McInnes, R.R. (1988). Common acute lymphocytic leukemia antigen is identical to neutral endopeptidase. J. Exp. Med. 168, 1247–1253.

Levin, Y., Skidgel, R.A. and Erdös, E.G. (1982). Isolation and characterization of the subunits of human plasma carboxypeptidase N (kininase I). Proc. Natl Acad. Sci. USA 79, 4618–4622.

Lieberman, J. (1974). Elevation of serum angiotensin converting enzyme (ACE) level in sarcoidosis. Am. J. Med. 59, 365–372.

Lieberman, J. (1985). Angiotensin-converting enzyme (ACE) and serum lysozyme in sarcoidosis. In "Sarcoidosis" (ed. J. Lieberman), pp. 145–159. Grune and Stratton, Orlando, FL.

Linz, W., Wiemer, G., Gohlke, P., Unger, T. and Schölkens, B.A. (1995). Contribution of kinins to the cardiovascular actions of angiotensin-converting enzyme inhibitors. Pharmacol. Rev. 47, 25–49.

Lipke, D.W. and Olson, K.R. (1988). Distribution of angiotensin-converting enzyme-like activity in vertebrate tissues. Physiol. Zool. 61, 420–428.

Liu, Y.-H., Yang, X.-P., Sharov, V.G., Sigmon, D.H., Sabbah, H.N. and Carretero, O.A. (1996). Paracrine systems in the cardioprotective effect of angiotensin-converting enzyme inhibitors on myocardial ischemia/reperfusion injury in rats. Hypertension 27, 7–13.

Llorens-Cortes, C., Huang, H., Vicart, P., Gasc, J.-M., Paulin, D. and Corvol, P. (1992). Identification and characterization of neutral endopeptidase in endothelial cells from venous or arterial origins. J. Biol. Chem. 267, 14012–14018.

Lock, R. and Hessel, E.A. (1990). Probable reversal of protamine reactions by heparin administration. J. Cardiothor. Anesth. 4, 605–608.

Lötvall, J. (1990). Tachykinin- and bradykinin-induced airflow obstruction and airway microvascular leakage. Modulation by neutral endopeptidase and angiotensin converting enzyme, pp. 1–76. Ph.D. Thesis. University of Göteborg, Sweden.

Lötvall, J.O., Tokuyama, K., Barnes, P.J. and Chung, K.F. (1990). Bradykinin-induced airway microvascular leakage is potentiated by captopril and phosphoramidon. Eur. J. Pharmacol. 200, 211–217.

Low, M.G. (1987). Biochemistry of the glycosyl-phosphatidylinositol membrane protein anchors. Biochem. J. 244, 1–13.

Lowenstein, E., Johnston, E.W., Lappas, D.G., D'Ambra, M.N., Schneider, R.C., Daggett, W.M., Akins, C.W. and Philbin, D.M. (1983). Catastrophic pulmonary vasoconstriction associated with protamine reversal of heparin. Anesthesiology 59, 470–473.

Malfroy, B., Schofield, P.R., Kuang, W.-J., Seeburg, P.H., Mason, A.J. and Henzel, W.J. (1987). Molecular cloning and amino acid sequence of rat enkephalinase. Biochem. Biophys. Res. Commun. 144, 59–66.

Malfroy, B., Kuang, W.-J., Seeburg, P.H., Mason, A.J. and Schofield, P.R. (1988). Molecular cloning and amino acid sequence of human enkephalinase (neutral endopeptidase). FEBS Lett. 229, 206–210.

Marchand, P., Tang, J. and Bond, J.S. (1994). Membrane association and oligomeric organization of the alpha and beta subunits of mouse meprin A. J. Biol. Chem. 269, 15388–15393.

Margolius, H.S. (1995). Kallikreins and kinins. Some unanswered questions about system characteristics and roles in human disease. Hypertension 26, 221–229.

Marinkovic, D.V., Marinkovic, J.N., Erdös, E.G. and Robinson, C.J.G. (1977). Purification of carboxypeptidase B from human pancreas. Biochem. J. 163, 253–260.

Massoudy, P., Becker, B.F. and Gerlach, E. (1994). Bradykinin accounts for improved postischemic function and decreased glutathione release of guinea pig heart treated with the angiotensin-converting enzyme inhibitor ramiprilat. J. Cardiovasc. Pharmacol. 23, 632–639.

Mathews, K.P., Pan, P.M., Gardner, N.J. and Hugli, T.E. (1980). Familial carboxypeptidase N deficiency. Ann. Intern. Med. 93, 443–445.

Mathews, K.P., Curd, J.G. and Hugli, T.E. (1986). Decreased synthesis of serum carboxypeptidase N (SCPN) in familial SCPN deficiency. J. Clin. Immunol. 6, 87–91.

McDermott, J.R., Gibson, A.M. and Turner, J.D. (1987). Involvement of endopeptidase 24.15 in the inactivation of bradykinin by rat brain slices. Biochem. Biophys. Res. Commun. 146, 154–158.

McDonald, J.K. and Barrett, A.J. (1986). "Mammalian Proteases: A Glossary and Bibliography. Vol. 2, Exopeptidases". Academic Press, London.

McGwire, G.B. and Skidgel, R.A. (1995). Extracellular conversion of epidermal growth factor (EGF) to des-Arg[53]-EGF by carboxypeptidase M. J. Biol. Chem. 270, 17154–17158.

Mendelsohn, F.A.O., Lloyd, C.J., Kachel, C. and Funder, J.W. (1982). Induction by glucocorticoids of angiotensin-converting enzyme production from bovine endothelial cells in culture and rat lung in vivo. J. Clin. Invest. 70, 684–692.

Miller, J.J., Changaris, D.G. and Levy, R.S. (1991). Angiotensin carboxypeptidase activity in urine from normal subjects and patients with kidney damage. Life Sci. 48, 1529–1535.

Millican, P.E., Kenny, A.J. and Turner, A.J. (1991). Purification and properties of a neurotensin-degrading endopeptidase from pig brain. Biochem. J. 276, 583–591.

Minshall, R.D., Nakamura, F., Becker, R.P. and Rabito, S.F. (1995a). Characterization of bradykinin B$_2$ receptors in adult myocardium and neonatal rat cardiomyocytes. Circ. Res. 76, 773–780.

Minshall, R.D., Vogel, S.M., Miletich, D.J. and Erdös, E.G. (1995b). Potentiation of the inotropic actions of bradykinin on the isolated guinea pig left atria by the angiotensin-converting enzyme/kininase II inhibitor, enalaprilat. Circulation 92 (Suppl.), I-221(Abstract).

Mombouli, J.-V. and Vanhoutte, P.M. (1995). Endothelium-derived hyperpolarizing factor(s) and the potentiation of kinins by converting enzyme inhibitors. Am. J. Hypertens. 8, 19S–27S.

Morel, D.R., Zapol, W.M., Thomas, S.J., Kitain, E.M., Robinson, D.R., Moss, J., Chenoweth, D.E. and Lowenstein, E. (1987). C5a and thromboxane generation associated with pulmonary vaso- and bronchoconstriction during protamine reversal of heparin. Anesthesiology 66, 597–604.

Morice, A.H., Brown, M.J., Lowry, R. and Higenbottam, T. (1987). Angiotensin-converting enzyme and the cough reflex. Lancet 2, 1116–1118.

Morrell, N.W., Atochina, E.N., Morris, K.G., Danilov, S.M. and Stenmark, K.R. (1995). Angiotensin-converting enzyme expression is increased in small pulmonary arteries of rats with hypoxia-induced pulmonary hypertension. J. Clin. Invest. 96, 1823–1833.

Nadel, J.A. (1992). Membrane-bound peptidases: endocrine, paracrine, and autocrine effects. Am. J. Respir. Cell Mol. Biol. 7, 469–470.

Nadel, J.A. (1994). Modulation of neurogenic inflammation by peptidases. In "Neuropeptides in Respiratory Medicine" (eds. M.A. Kaliner, P.J. Barnes, G.H.H. Kunkel and J.N. Baraniuk), pp. 351–371. Marcel Dekker, New York.

Nagae, A., Deddish, P.A., Becker, R.P., Anderson, C.H., Abe, M., Skidgel, R.A. and Erdös, E.G. (1992). Carboxypeptidase M in brain and peripheral nerves. J. Neurochem. 59, 2201–2212.

Nagae, A., Abe, M., Becker, R.P., Deddish, P.A., Skidgel, R.A. and Erdös, E.G. (1993). High concentration of carboxypeptidase M in lungs: presence of the enzyme in alveolar type I cells. Am. J. Respir. Cell Mol. Biol. 9, 221–229.

Nishimura, K., Hiwada, K., Ueda, E. and Kokubu, T. (1976). Solubilization of angiotensin-converting enzyme from rabbit lung using trypsin treatment. Biochim. Biophys. Acta 452, 144, 150.

Nortier, J., Abramowicz, D., Najdovski, T., Kinnaert, P., Vanherweghem, J.-L., Goldman, M. and Deschodt-Lanckman, M. (1993). Urinary endopeptidase 24.11 as a new marker of proximal tubular injury. In "Kidney, Proteins and Drugs: An Update". Contrib. Nephrol., Vol. 101 (eds. C. Bianchi, V. Bocci, F.A. Carone and R. Rabkin), pp. 169–176. Karger, Basel.

Odya, C.E., Marinkovic, D. Hammon, K.J., Stewart, T.A. and Erdös, E.G. (1978). Purification and properties of prolylcarboxypeptidase (angiotensinase C) from human kidney. J. Biol. Chem. 253, 5927–5931.

Odya, C.E., Wilgis, F.P., Walker, J.F. and Oparil, S. (1983). Immunoreactive bradykinin and [des-Arg[9]]-bradykinin in low-renin essential hypertension – before and after treatment with enalapril (MK421). J. Lab. Clin. Med. 102, 714–721.

Ondetti, M.A. (1994). From peptides to peptidases: a chronicle of drug discovery. Annu. Rev. Pharmacol. 34, 1–16.

Oppong, S.Y. and Hooper, N.M. (1993). Characterization of a secretase activity which releases angiotensin-converting enzyme from the membrane. Biochem. J. 292, 597–603.

Orawski, A.T. and Simmons, W.H. (1995). Purification and properties of membrane-bound aminopeptidase P from rat lung. Biochemistry 34, 11227–11236.

Orlowski, M., Michaud, C. and Chu, T.G. (1983). A soluble metalloendopeptidase from rat brain. Purification of the enzyme and determination of specificity with synthetic and natural peptides. Eur. J. Biochem. 135, 81–88.

Orlowski, M., Michaud, C. and Molineaux, C.J. (1988). Substrate-related potent inhibitors of brain metalloendopeptidase. Biochemistry 27, 596–602.

Orlowski, M., Reznik, S., Ayala, J. and Pierotti, A.R. (1989). Endopeptidase 24.15 from rat testes. Isolation of the enzyme and its specificity toward synthetic and natural peptides, including enkephalin-containing peptides. Biochem. J. 261, 951–958.

Oshima, E. and Erdös, E.G. (1974). Inhibition of the angiotensin I converting enzyme of the lung by a peptide fragment of bradykinin. Experientia 30, 733.

Oshima, G., Kato, J. and Erdös, E.G. (1975). Plasma carboxypeptidase N, subunits and characteristics. Arch. Biochem. Biophys. 170, 132–138.

Oshima, G., Hiraga, Y., Shirono, K., Oh-ishi, S., Sakakibara, S. and Kinoshita, T. (1985). Cleavage of des-Arg⁹-bradykinin by angiotensin I-converting enzyme from pig kidney cortex. Experientia 41, 325–328.

Paegelow, I., Reissmann, S. and Arold, H. (1976). Der Einfluß potenzierender Faktoren (BPF) auf die Bradykininwirkung in vitro. Acta Biol. Med. Germ. 35, 235–244.

Painter, R.G., Dukes, R., Sullivan, J., Carter, R., Erdös, E.G. and Johnson, A.R. (1988). Function of neutral endopeptidase on the cell membrane of human neutrophils. J. Biol. Chem. 263, 9456–9461.

Parratt, J.R. (1994). Cardioprotection by angiotensin-converting enzyme inhibitors: the experimental evidence. Cardiovasc. Res. 28, 183–189.

Patchett, A.A. and Cordes, E.H. (1985). The design and properties of N-carboxyalkyldipeptide inhibitors of angiotensin converting enzyme. In "Advances in Enzymology and Related Areas of Molecular Biology," Vol. 57 (ed. A. Meister), pp. 1–84. John Wiley and Sons, New York.

Patel, K.V. and Schrey, M.P. (1992). Inhibition of DNA synthesis and growth in human breast stromal cells by bradykinin: evidence for independent roles of B_1 and B_2 receptors in the respective control of cell growth and phospholipid hydrolysis. Cancer Res. 52, 334–340.

Perich, R.B., Jackson, B., Attwood, M.R., Prior, K. and Johnston, C.I. (1991). Angiotensin-converting enzyme inhibitors act at two different binding sites on angiotensin-converting enzyme. Pharm. Pharmacol. Lett. 1, 41–43.

Perich, R.B., Jackson, B. and Johnston, C.I. (1994). Structural constraints of inhibitors for binding at two active sites on somatic angiotensin converting enzyme. Eur. J. Pharmacol. 266, 201–211.

Pfeffer, M.A. (1993). Angiotensin-converting enzyme inhibition in congestive heart failure: Benefit and perspective. Am. Heart J. 126, 789–793.

Pfeffer, M.A. (1995). ACE inhibition in acute myocardial infarction. N. Engl. J. Med. 332, 118–120.

Piedimonte, G., Nadel, J.A., Long, C.S. and Hoffmann, J.I.E. (1994). Neutral endopeptidase in the heart. Neutral endopeptidase inhibition prevents isoproterenol-induced myocardial hypoperfusion in rats by reducing bradykinin degradation. Circ. Res. 75, 770–779.

Pierotti, A., Dong, K.-W., Glucksman, M.J., Orlowski, M. and Roberts, J.L. (1990). Molecular cloning and primary structure of rat testes metalloendopeptidase EC 3.4.24.15. Biochemistry 29, 10323–10329.

Planck, S.R., Finch, J.S., and Magun, B.E. (1984). Intracellular processing of epidermal growth factor. J. Biol. Chem. 259, 3053–3057.

Plummer, T.H., Jr and Hurwitz, M.Y. (1978). Human plasma carboxypeptidase N. Isolation and characterization. J. Biol. Chem. 253, 3907–3912.

Plummer, T.H., Jr and Ryan, T.J. (1981). A potent mercapto bi-

product analogue inhibitor for human carboxypeptidase N. Biochem. Biophys. Res. Commun. 98, 448–454.

Polgar, L. (1992). Structural relationship between lipases and peptidases of the prolyl oligopeptidase family. FEBS Lett. 311, 281–284.

Potier, M., Michaud, L., Tranchemontagne, J. and Thauvette, L. (1990). Structure of the lysosomal neuraminidase-galactosidase-carboxypeptidase multienzymic complex. Biochem. J. 267, 197–202.

Prechel, M.M., Orawski, A.T., Maggiora, L.L. and Simmons, W.H. (1995). Effect of a new aminopeptidase P inhibitor, apstatin, on bradykinin degradation in the rat lung. J. Pharmacol. Exp. Ther. 275, 1136–1142.

Rabito, S.F., Anders, R., Soden, W. and Skidgel, R.A. (1992). Carboxypeptidase N concentration during cardiopulmonary bypass in humans. Can. J. Anaesth. 39, 54–59.

Ramchandran, R., Sen, G.C., Misono, K. and Sen, I. (1994). Regulated cleavage-secretion of the membrane-bound angiotensin converting enzyme. J. Biol. Chem. 269, 2125–2130.

Rawlings, N.D., Polgar, L. and Barrett, A.J. (1991). A new family of serine-type peptidases related to prolyl oligopeptidase. Biochem. J. 279, 907–908.

Redlitz, A., Tan, A.K., Eaton, D.L. and Plow, E.F. (1995). Plasma carboxypeptidases as regulators of the plasminogen system. J. Clin. Invest. 96, 2534–2538.

Regoli, D. and Barabé, J. (1980). Pharmacology of bradykinin and related kinins. Pharmacol. Rev. 32, 1–46.

Rehli, M., Krause, S.W., Kreutz, M. and Andreesen, R. (1995). Carboxypeptidase M is identical to the MAX.1 antigen and its expression is associated with monocyte to macrophage differentiation. J. Biol. Chem. 270, 15644–15649.

Reiners, C., Gramer-Kurz, E., Pickert, E. and Schweisfurth, H. (1988). Changes of serum angiotensin I-converting enzyme in patients with thyroid disorders. Clin. Physiol. Biochem. 6, 44–49.

Renfrew, C.A. and Hubbard, A.L. (1991). Sequential processing of epidermal growth factor in early and late endosomes in rat liver. J. Biol. Chem. 266, 4348–4356.

Rennex, D., Hemmings, B.A., Hofsteenge, J. and Stone, S.R. (1991). cDNA cloning of porcine brain prolyl endopeptidase and identification of the active-site seryl residue. Biochemistry 30, 2195–2203.

Rix, E., Ganten, D., Schüll, Unger, T. and Taugner, R. (1981). Converting-enzyme in the choroid plexus, brain, and kidney: immunocytochemical and biochemical studies in rats. Neurosci. Lett. 22, 125–130.

Rocha e Silva, M. (1963). The physiological significance of bradykinin. Ann. N. York Acad. Sci. 104, 190–211.

Rogerson, F.M., Schlawe, I., Paxinos, G., Chai, S.Y., McKinley, M.J. and Mendelsohn, F.A.O. (1995). Localization of angiotensin-converting enzyme by in vitro autoradiography in the rabbit brain. J. Chem. Neuroanat. 8, 277–243.

Romer, F.K. and Emmertsen, K. (1980). Serum angiotensin converting enzyme in malignant lymphomas, leukaemia and multiple myeloma. Br. J. Cancer 42, 314–318.

Ronco, P., Pollard, H., Galceran, M., Delauche, M., Schwartz, J.C. and Verroust, P. (1988). Distribution of enkephalinase (membrane metalloendopeptidase, E.C. 3.4.24.11) in rat organs. Detection using a monoclonal antibody. Lab. Invest. 58, 210–217.

Roy, P., Chatellard, C., Lemay, G., Crine, P. and Boileau, G. (1993). Transformation of the signal peptide membrane anchor domain of a type II transmembrane protein into a cleavable signal peptide. J. Biol. Chem. 268, 2699–2704.

Ryan, J.W. (1989). Peptidase enzymes of the pulmonary vascular surface. Am. J. Physiol. 257, L53–L60.

Ryan, J.W., Roblero, J. and Stewart, J.M. (1968). Inactivation of bradykinin in the pulmonary circulation. Biochem. J. 110, 795–797.

Ryan, J.W., Berryer, P., Chung, A.Y. and Sheffy, D.H. (1994). Characterization of rat pulmonary vascular aminopeptidase P in vivo: role in the inactivation of bradykinin. J. Pharmacol. Exp. Ther. 269, 941–947.

Ryan, U.S. and Ryan, J.W. (1983). Endothelial cells and inflammation. In "Clinics in Laboratory Medicine", Vol. 3 (ed. P.A. Ward), pp. 577–599. W.B. Saunders, Philadelphia.

Sakaguchi, K., Chai, S.Y., Jackson, B., Johnston, C.I. and Mendelsohn, F.A.O. (1988). Inhibition of tissue angiotensin converting enzyme. Hypertension 11, 230–238.

Schaudies, P.R. and Savage, R.C. (1986). Intracellular modification of ^{125}I-labeled epidermal growth factor by normal human foreskin fibroblasts. Endocrinology 118, 875–882.

Schilero, G.J., Almenoff, P., Cardozo, C. and Lesser, M. (1994). Effects of peptidase inhibitors on bradykinin-induced bronchoconstriction in the rat. Peptides 15, 1445–1449.

Schriefer, J.A. and Molineaux, C.J. (1993). Modulatory effect of endopeptidase inhibitors on bradykinin-induced contraction of rat uterus. J. Pharmacol. Exp. Ther. 266, 700–706.

Schulz, W.W., Hagler, H.K., Buja, L.M. and Erdös, E.G. (1988). Ultrastructural localization of angiotensin I converting enzyme (EC 3.4.15.1) and neutral metallo-endopeptidase (EC 3.4.24.11) in the proximal tubule of the human kidney. Lab. Invest. 59, 789–797.

Schwartz, J.-C., Costentin, J. and Lecomte, J.-M. (1985). Pharmacology of enkephalinase inhibitors. Trends Pharmacol. Sci. 6, 472–485.

Schweisfurth, H., Heinrich, J., Brugger, E., Steinl, C. and Maiwald, L. (1985a). The value of angiotensin-converting enzyme determinations in malignant and other diseases. Clin. Physiol. Biochem. 3, 184–192.

Schweisfurth, H., Schmidt, M., Brugger, E., Maiwald, L. and Thiel, H. (1985b). Alterations of serum carboxypeptidases N and angiotensin I-converting enzyme in malignant diseases. Clin. Biochem. 18, 242–246.

Scicli, A.G. (1994). Increases in cardiac kinins as a new mechanism to protect the heart. Hypertension 23, 419–421.

Scicli, A.G., Gandolfi, R. and Carretero, O.A. (1978). Site of formation of kinins in the dog nephron. Am. J. Physiol. 234, F36–F40.

Semple, P.F. (1995). Putative mechanisms of cough after treatment with angiotensin converting enzyme inhibitors. J. Hypertens. 13 (Suppl. 3), S17–S21.

Serizawa, A., Dando, P.M. and Barrett, A.J. (1995). Characterization of a mitochondrial metallopeptidase reveals neurolysin as a homologue of thimet oligopeptidase. J. Biol. Chem. 270, 2092–2098.

Seymour, A.A., Asaad, M.M., Lanoce, V.M., Langenbacher, K.M., Fennell, S.A. and Rogers, W.L. (1993). Systematic hemodynamics, renal function and hormonal levels during inhibition of neutral endopeptidase 3.4.24.11 and angiotensin-converting enzyme in conscious dogs with pacing-induced heart failure. J. Pharmacol. Exp. Ther. 266, 872–883.

Seymour, A.A., Sheldon, J.H., Smith, P.L., Asaad, M. and Rogers, W.L. (1994). Potentiation of the renal responses to

bradykinin by inhibition of neutral endopeptidase 3.4.24.11 and angiotensin-converting enzyme in anesthetized dogs. J. Pharmacol. Exp. Ther. 269, 263–270.

Shaw, L.A., Rudin, M. and Cook, N.S. (1995). Pharmacological inhibition of restenosis: learning from experience. Trends Pharmacol. Sci. 16, 401–404.

Sheikh, I.A. and Kaplan, A.P. (1989). Mechanism of digestion of bradykinin and lysylbradykinin (kallidin) in human serum. Role of carboxypeptidase, angiotensin-converting enzyme and determination of final degradation products. Biochem. Pharmacol. 38, 993–1000.

Shinohara, T., Sakurada, C., Suzuki, T., Takeuchi, O., Campbell, W., Ikeda, S., Okada, N. and Okada, H. (1994). Pro-carboxypeptidase R cleaves bradykinin following activation. Int. Arch. Allergy Immunol. 103, 400–404.

Sidorowicz, W., Canizaro, P.C. and Behal, F.J. (1984a). Kinin cleavage by human erythrocytes. Am. J. Hematol. 17, 383–391.

Sidorowicz, W., Szechinski, J., Canizaro, P.C. and Behal, F.J. (1984b). Cleavage of the Arg^1-Pro^2 bond of bradykinin by a human lung peptidase: isolation, characterization and inhibition by several β-lactam antibiotics. Proc. Soc. Exp. Biol. Med. 175, 503–509.

Silverstein, E., Friedland, J., Lyons, H.A. and Gourin, A. (1976). Markedly elevated angiotensin converting enzyme in lymph nodes containing non-necrotizing granulomas in sarcoidosis. Proc. Natl. Acad. Sci. USA 73, 2137–2141.

Silverstein, E., Friedland, J. and Vuletin, J.C. (1978). Marked elevation of serum angiotensin-converting enzyme and hepatic fibrosis containing long-spacing collagen fibrils in type 2 acute neuronopathic Gaucher's disease. Am. J. Clin. Pathol. 69, 467–470.

Silverstein, E., Pertschuk, L.P. and Friedland, J. (1979). Immunofluorescent localization of angiotensin converting enzyme in epithelioid and giant cells of sarcoidosis granulomas. Proc. Natl Acad. Sci. USA 76, 6646–6648.

Simmons, W.H. and Orawski, A.T. (1992). Membrane-bound aminopeptidase P from bovine lung. Its purification, properties, and degradation of bradykinin. J. Biol. Chem. 267, 4897–4903.

Skidgel, R.A. (1988). Basic carboxypeptidases: regulators of peptide hormone activity. Trends Pharmacol. Sci. 9, 299–304.

Skidgel, R.A. (1991). Assays for arginine/lysine carboxypeptidases: Carboxypeptidase H (E; Enkephalin convertase), M and N. In "Methods in Neurosciences: Peptide Technology", Vol. 6 (ed. P.M. Conn), pp. 373–385. Academic Press, Orlando, FL.

Skidgel, R.A. (1992). Bradykinin-degrading enzymes: structure, function, distribution, and potential roles in cardiovascular pharmacology. J. Cardiovasc. Pharmacol. 20, S4–S9.

Skidgel, R.A. (1993). Basic science aspects of angiotensin converting enzyme and its inhibitors. In "Heart Failure: Basic Science and Clinical Aspects" (eds. J.K. Gwathmey, G.M. Briggs and P.D. Allen), pp. 399–427. Marcel Dekker, New York.

Skidgel, R.A. (1994). Pulmonary peptidases: general principles of peptide metabolism and molecular biology of angiotensin converting enzyme, neutral endopeptidase 24.11, and carboxypeptidase M. In "Neuropeptides in Respiratory Medicine" (eds. M.A. Kaliner, P.J. Barnes, G.H.H. Kunkel and J.N. Baraniuk), pp. 301–312. Marcel Dekker, New York.

Skidgel, R.A. (1995). Human carboxypeptidase N (lysine carboxypeptidase). In "Methods in Enzymology", Vol. 248, "Proteolytic enzymes", Part E, "Aspartic, Metallo and Other

Peptidases" (ed. A.J. Barrett), pp. 653–663. Academic Press, Orlando, FL.

Skidgel, R.A. (1996). Structure and function of mammalian zinc carboxypeptidases. In "Zinc Metalloproteases in Health and Disease" (ed. N.M. Hooper), pp. 241–243. Taylor and Francis, London (in press).

Skidgel, R.A. and Erdös, E.G. (1993). Biochemistry of angiotensin converting enzyme. In "The Renin Angiotensin System", Vol. 1 (eds. J.I.S. Robertson and M.G. Nicholls), pp. 10.1–10.10. Gower Medical Publishers, London.

Skidgel, R.A. and Tan, F. (1992). Structural features of two kininase I-type enzymes revealed by molecular cloning. Agents Actions Suppl. 38/I, 359–367.

Skidgel, R.A.,Wickstrom, E., Kumamoto, K. and Erdös, E.G. (1981). Rapid radioassay for prolylcarboxypeptidase (angiotensinase C). Anal. Biochem. 118, 113–119.

Skidgel, R. A., Davis, R.M. and Erdös, E.G. (1984a). Purification of a human urinary carboxypeptidase (kininase) distinct from carboxypeptidase A, B, or N. Anal. Biochem. 140, 520–531.

Skidgel, R.A., Johnson, A.R. and Erdös, E.G. (1984b). Hydrolysis of opioid hexapeptides by carboxypeptidase N. Presence of carboxypeptidase in cell membranes. Biochem. Pharmacol. 33, 3471–3478.

Skidgel, R.A., Kawahara, M.S. and Hugli, T.E. (1986). Functional significance of the subunits of carboxypeptidase N (kininase I). Adv. Exp. Med. Biol. 198A, 375–380.

Skidgel, R.A., Defendini, R. and Erdös, E.G. (1987a). Angiotensin I converting enzyme and its role in neuropeptide metabolism. In "Neuropeptides and Their Peptidases" (ed. A.J. Turner), pp. 165–182. Ellis-Horwood, Chichester.

Skidgel, R.A., Schulz, W.W., Tam, L.T. and Erdös, E.G. (1987b). Human renal angiotensin I converting enzyme and neutral endopeptidase. Kidney Int. 31, S45–S48.

Skidgel, R.A., Deddish, P.A. and Davis, R.M. (1988). Isolation and characterization of a basic carboxypeptidase from human seminal plasma. Arch. Biochem. Biophys. 267, 660–667.

Skidgel, R.A., Davis, R.M. and Tan, F. (1989). Human carboxypeptidase M: Purification and characterization of a membrane-bound carboxypeptidase that cleaves peptide hormones. J. Biol. Chem. 264, 2236–2241.

Skidgel, R.A., Anders, R.A., Deddish, P.A. and Erdös, E.G. (1991a). Carboxypeptidase (CP) M and H in small intestine. FASEB J. 5, A1578.

Skidgel, R.A., Jackman, H.L. and Erdös, E.G. (1991b). Metabolism of substance P and bradykinin by human neutrophils. Biochem. Pharmacol. 41, 1335–1344.

Slater, E.E., Merrill, D.D., Guess, H.A., Roylance, P.J., Cooper, W.D., Inman, W.H.W. and Ewan, P.W. (1988). Clinical profile of angioedema associated with angiotensin-converting enzyme inhibition. J. Am. Med. Assoc. 260, 967–970.

Smits, G.J., McGraw, D.E. and Trapani, A.J. (1990). Interaction of ANP and bradykinin during endopeptidase 24.11 inhibition: renal effects. Am. J. Physiol. 258, F1417–F1424.

Sorrells, K. and Erdös, E.G. (1972). Prolylcarboxypeptidase in biological fluids. In "The Fundamental Mechanisms of Shock" (eds. L.B. Hinshaw and B.G. Cox), pp. 393–397. Plenum, New York.

Soubrier, F., Alhenc-Gelas, F., Hubert, C., Allegrini, J., John, M., Gregear, G. and Corvol, P. (1988). Two putative active centers in human angiotensin I-converting enzyme revealed by molecular cloning. Proc. Natl Acad. Sci. USA 85, 9386–9390.

Soubrier, F., Hubert, C., Testut, P., Nadaud, S., Alhenc-Gelas,

F. and Corvol, P. (1993a). Molecular biology of the angiotensin I converting enzyme: I. Biochemistry and structure of the gene. J. Hypertens. 11, 471–476.

Soubrier, F., Wei, L., Hubert, C., Clauser, E., Alhenc-Gelas, F. and Corvol, P. (1993b). Molecular biology of the angiotensin I converting enzyme: II. Structure-function. Gene polymorphism and clinical implications. J. Hypertens. 11, 599–604.

Spillantini, M.G., Sicuteri, F., Salmon, S. and Malfroy, B. (1990). Characterization of endopeptidase 3.4.24.11 (enkephalinase) activity in human plasma and cerebrospinal fluid. Biochem. Pharmacol. 39, 1353–1356.

Sunman, W. and Sever, P.S. (1993). Non-angiotensin effects of angiotensin converting enzyme inhibitors. Clin. Sci. 85, 661–670.

Swan, M.G., Vergalla, J. and Jones, E.A. (1993). Plasma endopeptidase 24.11 (enkephalinase) activity is markedly increased in cholestatic liver disease. Hepatology 18, 556–558.

Swartz, S.L., Williams, G.H., Hollenberg, N.K., Levine, L., Dluhy, R.G. and Moore, T.J. (1980). Captopril-induced changes in prostaglandin production. Relationship to vascular responses in normal man. J. Clin. Invest. 65, 1257–1264.

Tamaoki, J., Kobayashi, K., Sakai, N., Chiyotani, A., Kanemura, T. and Takizawa, T. (1989). Effect of bradykinin on airway ciliary motility and its modulation by neutral endopeptidase. Am. Rev. Respir. Dis. 140, 430–435.

Tan, A.K. and Eaton, D.L. (1995). Activation and characterization of procarboxypeptidase B from human plasma. Biochemistry 34, 5811–5816.

Tan, F., Chan, S.J., Steiner, D.F., Schilling, J.W. and Skidgel, R.A. (1989a). Molecular cloning and sequencing of the cDNA for human membrane-bound carboxypeptidase M: comparison with carboxypeptidases A, B, H and N. J. Biol. Chem. 264, 13165–13170.

Tan, F., Jackman, H., Skidgel, R.A., Zsigmond, E.K. and Erdös, E.G. (1989b). Protamine inhibits plasma carboxypeptidase N, the inactivator of anaphylatoxins and kinins. Anesthesiology 70, 267–275.

Tan, F.,Weerasinghe, D.K., Skidgel, R.A., Tamei, H., Kaul, R.K., Roninson, I.B., Schilling, J.W. and Erdös, E.G. (1990). The deduced protein sequence of the human carboxypeptidase N high molecular weight subunit reveals the presence of leucine-rich tandem repeats. J. Biol. Chem. 265, 13–19.

Tan, F., Morris, P.W., Skidgel, R.A. and Erdös, E.G. (1993). Sequencing and cloning of human prolylcarboxypeptidase (angiotensinase C): similarity to both serine carboxypeptidase and prolylendopeptidase families. J. Biol. Chem. 268, 16631–16638.

Tan, F., Deddish, P.A. and Skidgel, R.A. (1995). Human carboxypeptidase M. In "Methods in Enzymology", Vol. 248, "Proteolytic Enzymes", Part E, "Aspartic, Metallo and Other Peptidases" (ed. A.J. Barrett), pp. 663–675. Academic Press, Orlando, FL.

Tatei, K., Cai, H., Ip, Y.T. and Levine, M. (1995). Race: a drosophila homologue of the angiotensin-converting enzyme. Mech. Develop. 51, 157–168.

Telford, S.E., Smith, A.I., Lew, R.A., Perich, R.B., Madden, A.C. and Evans, R.G. (1995). Role of angiotensin converting enzyme in the vascular effects of an endopeptidase 24.15 inhibitor. Br. J. Pharmacol. 114, 1185–1192.

Timmermans, P.B.M.W.M. and Smith, R.D. (1994). Angiotensin II receptor subtypes: selective antagonists and functional correlates. Eur. Heart J. 15 (Suppl. D), 79–87.

Tisljar, U. (1993). Thimet oligopeptidase – a review of a thiol dependent metallo-endopeptidase also known as Pz-peptidase endopeptidase 24.15 and endo-oligopeptidase. Biol. Chem. Hoppe-Seyler 374, 91–100.

Tisljar, U. and Barrett, A.J. (1990). Thiol-dependent metallo-endopeptidase characteristics of Pz-peptidase in rat and rabbit. Biochem. J. 267, 531–533.

Torlone, E. and Bolli, G.B. (1991). Angiotensin converting enzyme inhibition improves insulin sensitivity in Type 2 diabetes mellitus. Arch. Gerontol. Geriatr. (Suppl.2), 287–290.

Turner, A.J. (1987). Endopeptidase-24.11 and neuropeptide metabolism. In "Neuropeptides and Their Peptidases" (ed. A.J. Turner), pp. 183–201. Ellis-Horwood, Chichester.

Ufkes, J.G.R., Aarsen, P.N. and van der Meer, C. (1977). The mechanism of action of two bradykinin-potentiating peptides on isolated smooth muscle. Eur. J. Pharmacol. 44, 89–97.

Ura, N., Carretero, O.A. and Erdös, E.G. (1987). Role of renal endopeptidase 24.11 in kinin metabolism in vitro and in vivo. Kidney Int. 32, 507–513.

van der Host, G.T.J., Galjart, N.J., d'Azzo, A., Galjaard, H. and Verheijen, F.W. (1989). Identification and in vitro reconstitution of lysosomal neuraminidase from human placenta. J. Biol. Chem. 264, 1317–1322.

Vane, J.R. (1969). The release and fate of vaso-active hormones in the circulation. Br. J. Pharmacol. 35, 209–242.

Vanhoof, G., De Meester, I., Goossens, F., Hendriks, D., Scharpe, S. and Yaron, A. (1992). Kininase activity in human platelets: cleavage of the Arg^1-Pro^2 bond of bradykinin by aminopeptidase P. Biochem. Pharmacol. 44, 479–487.

Vergas Romero, C., Neudorfer, I., Mann, K. and Schäfer, W. (1995). Purification and amino acid sequence of aminopeptidase P from pig kidney. Eur. J. Biochem. 229, 262–269.

Vijayaraghavan, J., Scicli, A.G., Carretero, O.A., Slaughter, C., Moomaw, C. and Hersh, L.B. (1990). The hydrolysis of endothelins by neutral endopeptidase 24.11 (enkephalinase). J. Biol. Chem. 265, 14150–14155.

Vogel, R., Werle, E. and Zickgraf-Rüdel, G. (1970). Neuere aspekte der kininforschung. I. Potenzierung und Blockierung der biologischen Kininwirkung. Z. klin. Chem. klin Biochem. 8, 177–185.

Wang, W., Hendriks, D.F. and Scharpé, S.S. (1994). Carboxypeptidase U, a plasma carboxypeptidase with high affinity for plasminogen. J. Biol. Chem. 269, 15937–15944.

Ward, P.E., Bausback, H.H. and Odya, C.E. (1987). Kinin and angiotensin metabolism by purified renal post-proline cleaving enzyme. Biochem. Pharmacol. 36, 3187–3193.

Warren, J.B. and Loi, R.K. (1995). Captopril increases skin microvascular blood flow secondary to bradykinin, nitric oxide, and prostaglandins. FASEB J. 9, 411–418.

Weare, J.A. (1982). Activation/inactivation of human angiotensin I converting enzyme following chemical modifications of amino groups near the active site. Biochem. Biophys. Res. Commun. 104, 1319–1326.

Weare, J.A., Gafford, J.T., Lu, H.S. and Erdös, E.G. (1982). Purification of human kidney angiotensin I converting enzyme using reverse immunoadsorption chromatography. Anal. Biochem. 123, 310–319.

Webster, M.E. and Pierce, J.V. (1963). The nature of the kallidins released from human plasma by kallikreins and other enzymes. Ann. N. York Acad. Sci. 104, 91–107.

Wei, L., Alhenc-Gelas, F., Corvol, P. and Clauser, E. (1991a). The two homologous domains of human angiotensin I-converting enzyme are both catalytically active. J. Biol. Chem. 266, 9002–9008.

Wei, L., Alhenc-Gelas, F., Soubrier, F., Michaud, A., Corvol, P. and Clauser, E. (1991b). Expression and characterization of recombinant human angiotensin I-converting enzyme. Evidence for a C-terminal transmembrane anchor and for a proteolytic processing of the secreted recombinant and plasma enzymes. J. Biol. Chem. 266, 5540–5546.

Wei, L., Clauser, E., Alhenc-Gelas, F. and Corvol, P. (1992). The two monologous domains of human angiotensin I-converting enzyme interact differently with competitive inhibitors. J. Biol. Chem. 267, 13398–13405.

Wilk, S. (1983). Minireview: prolyl endopeptidase. Life Sci. 33, 2149–2157.

Wilk, S. and Orlowski, M. (1983). Inhibition of rabbit brain prolyl endopeptidase by N-benzyloxycarbonyl-prolyl-prolinal, a transition state aldehyde inhibitor. J. Neurochem. 41, 69–75.

Wolz, R.L. and Bond, J.S. (1990). Phe^5(4-nitro)-bradykinin: a chromogenic substrate for assay and kinetics of the metallo-endopeptidase meprin. Anal. Biochem. 191, 314–320.

Wolz, R.L. and Bond, J.S. (1995). Meprins A and B. Meth. Enzymol. 248, 325–345.

Wolz, R.L., Harris, R.B. and Bond, J.S. (1991). Mapping the active site of meprin-A with peptide substrates and inhibitors. Biochemistry 30, 8488–8493.

Wyvratt, M.J. and Patchett, A.A. (1985). Recent developments in the design of angiotensin-converting enzyme inhibitors. Med. Res. Rev. 5, 483–531.

Xu, Y., Wellner, D. and Scheinberg, D.A. (1995). Substance P and bradykinin are natural inhibitors of CD13/aminopeptidase N. Biochem. Biophys. Res. Commun. 208, 664–674.

Yamashiro, D.J. and Maxfield, F.R. (1988). Regulation of endocytic process by pH. Trends Pharmacol. Sci. 9, 190–194.

Yamawaki, I., Geppetti, P., Bertrand, C., Chan, B. and Nadel, J. A. (1994). Airway vasodilation by bradykinin is mediated via B_2 receptors and modulated by peptidase inhibitors. Am. J. Physiol. 266, L156–L162.

Yang, H.Y.T. and Erdös, E.G. (1967). Second kininase in human blood plasma. Nature (Lond.) 215, 1402–1403.

Yang, H.Y.T., Erdös, E.G. and Chiang, T.S. (1968). New enzymatic route for the inactivation of angiotensin. Nature (Lond.) 218, 1224–1226.

Yang, H.Y.T., Erdös, E.G., Chiang, T.S., Jenssen, T.A. and Rodgers, J.G. (1970a). Characteristics of an enzyme that inactivates angiotensin II (angiotensinase C). Biochem. Pharmacol. 19, 1201–1211.

Yang, H.Y.T., Erdös, E.G. and Levin, Y. (1970b). A dipeptidyl carboxypeptidase that converts angiotensin I and inactivates bradykinin. Biochim. Biophys. Acta 214, 374–376.

Yang, H.Y.T., Erdös, E.G. and Levin, Y. (1971). Characterization of a dipeptide hydrolase (kininase II; angiotensin I converting enzyme). J. Pharmacol. Exp. Ther. 177, 291–300.

Yang, X.-P., Saitoh, S., Scicli, A.G., Mascha, E., Orlowski, M. and Carretero, O.A. (1994). Effects of a metallo-endopeptidase-24.15 inhibitor on renal hemodynamics and function in rats. Hypertension 23, I-235–I-239.

Yeager, C.L., Ashmun, R.A., Williams, R.K., Cardellichio, C.B., Shapiro, L.H., Look, A.T. and Holmes, K.V. (1992). Human aminopeptidase N is a receptor for human coronavirus 229E. Nature (Lond.) 357, 420–422.

8. Kinin B₁ Receptor Induction and Inflammation

François Marceau

1. The Kinin B₁ Receptor: An Inducible, G Protein-coupled Receptor

The kinin B_1 receptor was defined initially as mediating the contractile effect of bradykinin (BK)-related peptides in the rabbit isolated aorta (Regoli et al., 1977). The B_1 receptor is selectively sensitive to kinin metabolites without the C-terminal arginine residue, desArg⁹-BK and Lys-desArg⁹-BK (also referred to as "desArg¹⁰-kallidin"), produced via kininase I pathways of metabolism (see Chapter 7 of this volume). The B_1 receptor seems to be upregulated rapidly in immunopathology under the influence of cytokines, and is further regulated by growth factors. The B_1 receptor, now cloned and sequenced (Menke et al., 1994; see Chapter 3), is a member of the family of G coupled-receptors. Sequence analysis reveals that the most closely related receptors are the BK B_2 receptor and the angiotensin AT_1 and AT_2 receptors

(Menke et al., 1994). The relationship to the B_2 receptor is intellectually satisfying, as both B_1 and B_2 receptors are activated by kinins derived from the kininogens (see Chapters 4 and 5). However, the homology between human B_1 and B_2 receptor sequences is not extensive in absolute terms (36% identity of amino acids), suggesting that both receptor systems diverged early in evolution and that they may be components of significantly different regulatory systems.

The B_1 receptor is particularly interesting because it is upregulated following some types of tissue injury. Many years ago, the time- and protein synthesis-dependent appearance of responsiveness to desArg⁹-BK in rabbit smooth muscle preparations was observed (Regoli et al., 1978). This author is still fascinated by the rapid appearance of a sizeable contractile response to exogenous kinins after 2–3 h of incubation of the rabbit isolated aorta in physiological buffer. A rapid and specific genetic program recruits the expression of what we know now to be a G protein-coupled receptor in smooth muscle cells, fibroblasts and a few other cell types (see below). However, receptor populations are not static and several other examples of G protein-coupled receptors, which are upregulated following tissue injury, are becoming known. For instance, lipopolysaccharide (LPS) injection in the rabbit upregulates in 16 h the number of neutrophil formylated peptide receptors, but without gain of function (Goldman et al., 1986), and viral infection upregulates two orphan G protein-coupled receptors in human B lymphocytes (Birkenbach et al., 1993).

Most pharmacological effects of kinins are mediated by the B_2 receptors, present in most normal tissues and organs, and only responsive to the native kinin sequences that include a C-terminal Arg. However, accumulating evidence suggests that B_1 receptors may amplify, and in some instances, take the place of B_2 receptors in immunopathology. Earlier review papers followed an historical approach to B_1 receptor definition and regulation (Marceau and Regoli, 1991; Marceau et al., 1983; Marceau, 1995). The present text reviews the evidence for B_1 receptor expression and function in selected experimental systems.

2. Place of the B_1 Receptor in Kinin Receptor Classification

2.1 CRITERIA FOR DEFINING THE KININ RECEPTOR SUBTYPES

The kinins influence many tissue and cell functions by stimulating membrane receptors. The present discussion will be limited to receptor types fully defined by a series of six criteria (Table 8.1), although it cannot be excluded that additional subtypes may be identified in the future. The first three criteria are pharmacological as formulated by Schild (1973). That the rabbit aorta responded more to desArg⁹-BK than to BK was the first hint of kinin receptor heterogeneity (Regoli et al., 1977). The potency order of agonists, as established in separate binding assays in a human cell line, IMR-90 lung fibroblasts, shows that the removal of the C-terminal Arg is essential for high affinity for the B_1 receptor and extremely detrimental for affinity for B_2 receptor (Table 8.1). This holds true for the Lys-BK (kallidin) and Lys-desArg⁹-BK, which exhibit much higher affinity for human B_1 receptors than the homolog pair BK and desArg⁹-BK. The only natural kinin sequence with a subnanomolar affinity for B_1 receptors is Lys-desArg⁹-BK, an observation of particular significance that is discussed below. In potency estimates from bioassay experiments, the affinity of BK and of Lys-BK for B_1 receptors is more or less overestimated owing to a partial conversion of these native sequences into their desArg⁹ homologs (discussed previously by Marceau and Regoli, 1991).

The development of kinin receptor antagonists has been intensively pursued for more than two decades (Stewart, 1995). Antagonists are very useful to define kinin receptor subtypes (Table 8.1) but, as more potent and selective agents are produced, the criterion of receptor affinity is applied differently from one text to another. The first described sequence-related, competitive antagonist of kinin receptors was desArg⁹-[Leu⁸]-BK (Regoli et al., 1977). This prototype was selectively effective against BK- and desArg⁹-BK-induced

Table 8.1   Six criteria to define receptor subtypes for kinins

Criterion	B_1 receptor	B_2 receptor
I. Agonists' order of potency[a]	desArg⁹-BK (590) > BK (7800) Lys-desArg⁹-BK (0.5) > Lys-BK (62)	BK (2) > desArg⁹-BK (>30,000) Lys-BK (7) > Lys-desArg⁹-BK (27,000)
II. Affinity of antagonists[a]	desArg⁹-[Leu⁸]-BK (440)[a] Lys-desArg⁹-[Leu⁸]-BK (1.3)	Hoe 140 (0.4) Win 64338 (64)
III. Lack of cross-desensitization	No cross-tachyphylaxis in BK- and desArg⁹-BK-induced increases in [Ca²⁺]ᵢ in cell lines that co-express both receptor types (Bascands et al., 1993; Smith et al., 1995).	
IV. Distinct amino-acid or nucleotide sequences	Human B_1 receptor sequence displays 36% homology with human B_2 receptor (Menke et al., 1994)	Highly homologous (>80%) B_2 receptor sequences available for human, rat, mouse, rabbit (McEachern et al., 1991; Hess et al., 1992; McIntyre et al., 1993; Bachvarov et al., 1995).
V. Second messengers	Phospholipase C-dependent, sustained increase in [Ca²⁺]ᵢ (Bascands et al., 1993; Smith et al., 1995).	Phospholipase C-dependent, transient increase in [Ca²⁺]ᵢ and ligand–receptor complex internalization.
VI. Pattern of expression	Absent from normal tissues, but rapidly inducible following tissue injury.	Preformed in many cell types. Receptor population is not rapidly modulated.

[a] Numerical values between parentheses represent affinity estimates. Values shown are IC_{50} values (nM, B_1 receptors) or K_i (nM, B_2 receptors) in binding competition assays in human IMR-90 cell lines that spontaneously co-expresses both receptor subtypes. The B_1 receptor ligand was 1 nM [³H]-Lys-desArg⁹-BK (Menke et al., 1994), and the B_2 receptor ligand was 1–2 nM ³H-BK (Sawutz et al., 1992a, 1994).

contractile responses of rabbit isolated aorta, and this tissue was obviously different from the vast majority of preparations that had been utilized previously in the study of kinin pharmacology. However, the rabbit aorta, for the first time, allowed the definition of a kinin receptor based on both a potency order of agonists and on the affinity of antagonists (Regoli and Barabé, 1980). The other broad receptor subtype, B$_2$, was not confirmed until over a decade, when the first generation of antagonists based on [DPhe^7]-BK was produced (Vavrek and Stewart, 1985).

Progress has been rapid in the development of many B$_2$ receptor antagonists (Stewart, 1995), but two such agents are of notable importance. These are a degradation-resistant peptide, Hoe 140 (DArg-[Hyp^3,Thi^5,DTic^7,Oic^8]-BK or icatibant), first described by Hock et al. (1991), and the first nonpeptide kinin antagonist, Win 64338 (phosphonium, [[4-[[2-[[bis-(cyclohexylamino)methylene]amino]-3-(2-naphthalenyl) 1-oxopropyl]amino]-phenyl]-methyl]-tributyl, chloride, monohydrochloride) reported by Sawutz et al. (1994). Both agents are highly selective for B$_2$ receptors. Thus, the four antagonists presented in Table 8.1 exhibit a moderate to high affinity, and an almost ideal selectivity for receptor subtypes that can be exploited in pharmacological and physiological studies. In addition, the wide use of Hoe 140 has moderated the claim that there are B$_2$ receptor subtypes, as these claims were based largely on the discrepant affinities and partial agonist activities of earlier prototypes of B$_2$ receptor antagonists. (**Editor's note:** This topic is discussed in some detail in Chapter 2 of this volume.)

In a system that expresses both B$_1$ and B$_2$ receptor types, it is theoretically possible to use the third Schild criterion, namely, the absence of cross-desensitization, to support the existence of distinct pharmacological entities. This approach is useful, because adaptive mechanisms such as tachyphylaxis involve in part specific domains of each receptor molecule (Strader et al., 1994). Although seldom applied to kinins, good examples of such studies are based on cell lines that co-express both B$_1$ and B$_2$ receptors. These include rat mesangial cells (Bascands et al., 1993) and bovine pulmonary artery endothelial cells (BPAEC) (Smith et al., 1995). In both cell lines, it was possible to downregulate a functional response (the acute increase of [Ca^{2+}]$_i$) to BK without depressing that to desArg^9-BK, and vice versa.

Molecular biology studies of kinin receptors essentially exploit the two first Schild criteria to identify a particular cDNA or genomic DNA sequence, in controlled expression studies that involve both binding and functional assays. This has been the case for B$_1$ receptor expression cloning (Menke et al., 1994; see Chapter 3). The ligand-binding capability of a receptor (its "discriminative" function) is determined by its amino-acid sequence. Thus, the corresponding mRNA contains sufficient information to apply these criteria. Distinct B$_1$

and B$_2$ receptors co-exist in the human genome (Menke et al., 1994) and, most probably, in the genome of most, if not all, mammals. B$_2$ receptor homologs have been described in four species (Table 8.1) and exhibit a degree of amino-acid identity far greater than the human B$_1$ and B$_2$ sequences. Yet, significant discrepancies in the pharmacology of B$_2$ receptor antagonists are found between related B$_2$ receptors from different species expressed under controlled conditions (Hess et al., 1994; Bachvarov et al., 1995). Some parsimony should be used under these circumstances and new receptor subtypes based on species variability should not be artificially created. This type of variability probably exists for B$_1$ receptors (see below).

G protein-coupled receptors are linked to several second-messenger systems that are either activated or suppressed when the agonist activates the receptor. A given receptor is believed to bind to one or more G proteins, which are intermediates on a specific enzyme or, perhaps, ion channel. How the G protein selectivity is encoded in specific receptor domains is not clear (Strader et al., 1994), but there is no doubt that the identity of second messenger(s) has an interest for receptor classification (e.g., the broad classes of α- and β-adrenoceptors in different systems). However, B$_1$ and B$_2$ receptors are both primarily coupled via phospholipase C (PLC) activation. Increased phosphatidylinositol (PI) turnover has been documented by a number of investigators following B$_1$ receptor activation (Issandou and Darbon, 1991; Bascands et al., 1993; Levesque et al., 1993; Tropea et al., 1993; Butt et al., 1995; Smith et al., 1995). A transient increase in [Ca^{2+}]$_i$, dependent largely on intracellular stores results from the effect of IP$_3$, one of the direct products of PLC, and has been observed in response to B$_1$ receptor stimulation (Bascands et al., 1993; Marsh and Hill, 1994; Smith et al., 1995). The weak mitogenic effect of B$_1$ receptor agonists on rabbit vascular smooth muscle cells is apparently dependent on another product of PLC, diacylglycerol, which activates a protein kinase C (Levesque et al., 1995b). The smooth muscle contraction mediated by B$_1$ receptors (in rabbit aorta) probably involves the cooperation of protein kinase C and Ca^{2+} (Levesque et al., 1993). In other systems, secondarily released mediators formed by Ca^{2+}-dependent enzymes, and including nitric oxide (Pruneau and Bélichard, 1993) and eicosanoids (deBlois and Marceau, 1987; Levesque et al., 1993), mediate the effects of desArg^9-BK.

Although interaction of B$_1$ receptor with a specific G protein has not been demonstrated, it is not necessary to postulate the recruitment of second messengers or released mediators different from those recruited by B$_2$ receptors. Thus, the identity of second messengers is of limited value in the classification of kinin receptors. However, the subtle differences that have been noted in systems that co-express both receptor subtypes may indicate molecular and physiological differences between B_1 and B_2 receptors. The increase in $[Ca^{2+}]_i$ is more persistent, less subject to tachyphylaxis, and less inhibited by Ni^{2+} when elicited by B_1 receptor stimulation, as opposed to B_2 receptor activation, in BPAEC (Smith et al., 1995). The persistence of B_1 receptor-elicited $[Ca^{2+}]_i$ responses, relative to those produced by B_2 receptor stimulation, also occurs in rat mesangial cells (Bascands et al., 1993). It is also worth mentioning that a B_1 receptor binding assay, conducted at 37°C failed to show internalization of ligand–receptor complexes in rabbit aorta smooth muscle cells (Levesque et al., 1995a). In contrast, B_2 receptors expressed in MF-2 smooth muscle cells mediate internalization, and facilitate the proteolytic degradation of BK (Munoz and Leeb-Lundberg, 1992).

Receptor populations are not static and their expression is likely to be affected by tissue-specific and physiological factors. A general statement about kinin receptors is that the B_1 type is generally absent from healthy tissues and animals, but rapidly induced following some type of injuries. This statement is developed extensively below and important exceptions will be discussed. It should be noted upregulation may reveal a molecular difference not in the receptor structure itself, but in its associated genomic promoter sequence. Conversely, B_2 receptors are preformed in many tissues and cell types, and their functions do not appear to be modified' to any significant degree by physiological conditions. Little is known about B_2 receptor regulation, which is believed to be constitutive, although experimental conditions that increase cellular cyclic AMP can modestly increase the binding capacity in rat mesenteric artery smooth muscle cells (Dixon, 1994).

A sequence analysis of the promoter of the human B_2 receptor gene (Ma et al., 1994) suggests that it contains response elements for cyclic AMP and interleukin-6 (IL-6), but these findings are not yet confirmed. Tumor necrosis-α (TNF-α) and IL-1 are reported to modify B_2 receptor B_{max} values modestly, although in opposing directions in two-cell systems (Bathon et al., 1992; Sawutz et al., 1992b).

2.2 PROBLEMS AND CONTROVERSIES IN THE DEFINITION OF THE B_1 RECEPTOR

The simple classification in Table 8.1 is not strictly applicable to all known experimental systems. Kinin B_1 receptor subtypes may be postulated, but have not yet been substantiated. The first distinct B_1 receptor to be cloned was of human origin (Menke et al., 1994). Competition binding studies, performed on human IMR-90 cells and COS-7 cells transfected with an expression vector containing the cloned sequence, show structure–activity relationships that are generally similar to those observed in rabbit cells and tissues, a notable exception being the low relative affinity of human B_1 receptors for desArg^9-BK. This peptide is about seven times less potent than Lys-BK, and 2,000-fold less potent than Lys-desArg^9-BK to compete for [^{3}H]Lys-desArg^9-BK binding on the human receptor. Thus, a more appropriate ligand for stimulating B_1 receptors, across species, may be Lys-desArg^9-BK and, even in the rabbit, it is effective in subnanomolar concentrations. This has led to speculation that Lys-desArg^9-BK is the physiological B_1 receptor ligand (Drapeau et al., 1991).

McIntyre et al. (1993) reported that COS-7 cells transfected with an expression vector coding for the murine B_2 receptor express a mixed population of B_1 (~30%) and B_2 subtypes (~70%), based on the competition for a single ligand, [^{3}H]-BK, with different peptides specific for each receptor class. Thus, an undetermined translational or post-translational mechanism may generate both types of pharmacological effects. This observation, however, has not been reproduced, as Hess et al. (1994) observed the pharmacological profiles of a "classical" B_2 receptor in this transfection system.

Further species variations were reported, based on the pharmacological studies of cultured bovine aortic endothelial cells (BAEC), which express both B_1 and B_2 receptors. Functional responses to desArg^9-BK in these cells are antagonized by desArg^9-[Leu^8]-BK, a B_1 receptor antagonist, and also by antagonists that are specific for B_2 receptors (Table 8.1), namely Hoe 140 (Wiemer and Wirth, 1992) and Win 64338 (Wirth et al., 1994). However, the antagonism by Hoe 140 against responses elicited by desArg^9-BK was not reproduced by others in BPAEC (Smith et al., 1995). Thus, the bovine B_1 receptor is somewhat unique as far as its susceptibility to antagonists for reasons unknown.

The simple scheme of kinin receptor regulation presented in Table 8.1 may be challenged. Healthy dogs and cats exhibit cardiovascular responses to desArg^9-BK without apparent immunopathology. Hypotension, natriuresis and renal vasodilator responses were observed in normal dogs injected with this peptide (Lortie et al., 1992; Nakhostine et al., 1993). Complex hemodynamic effects are also produced by desArg^9-BK in feline pulmonary circulation (DeWitt et al., 1994). In each case, the B_2 receptors co-exist and the identity of each receptor type was also validated with appropriate antagonists. A strain of Sprague–Dawley rats is reported to respond to desArg^9-BK by small hypertensive responses in vivo (Cunoosamy and Wheeldon, 1995), and a smooth muscle preparation from these animals, in contrast to other studies in rats (see below), was reported to be responsive to the B_1 receptor agonist constitutively (Boxall et al., 1995). The significance of these findings is difficult to determine.

The B_1 receptor may be expressed constitutively in species such as the dog and the cat (related animals that belong to the order Carnivora) owing to a species-dependent difference in the gene promoter region.

Unregulated B$_1$ receptors may represent an unusual subtype, as it has been noted that desArg^9-[Leu^8]-BK is a partial agonist in dogs (Nakhostine *et al.*, 1993; Rangachari *et al.*, 1993). In species such as the rat and the pig (Siebeck *et al.*, 1989), a small basal B$_1$ receptor population may result from strain-specific genetic differences or, perhaps, the health status of the "control" animal population. It is clear that bacterial products can upregulate significantly responses to desArg^9-BK in pigs and rats (Siebeck *et al.*, 1989; Tokumasu *et al.*, 1995), although similar data have not been reported for dogs and cats. As the B$_1$ receptor is now a defined molecular entity, additional approaches (Northern blot, immuno-histochemistry) may be used to analyze these systems.

Several cultured cell lines are reported to exhibit B$_1$ receptor mediated responses and to bind ligands such as [^{3}H]-Lys-desArg^9-BK (Marceau, 1995), and examples are discussed in Section 3.2. How the environment *in vitro* affects B$_1$ receptor regulation is not clear, but the presence of a basal, constitutive receptor population may be a cell culture-induced artifact. This is particularly clear for rabbit vascular smooth muscle cells (see below). However these cells, as well as IMR-90 cells, retain the capacity to upregulate their B$_1$ receptor population in response to IL-1 (see below).

2.3 RELATIONSHIP OF THE B$_1$ RECEPTOR WITH SPECIFIC FORMS OF KININOGEN AND KALLIKREINS: A SEPARATE TISSUE KALLIKREIN–KININ SYSTEM?

B$_1$ receptors may be subject to a selective pressure across species to maintain a subnanomolar affinity to Lys-desArg^9-BK, the most potent B$_1$ receptor agonist in the human and rabbit (see above). It may be significant that this peptide derives from Lys-BK. Compartments in the kallikrein–kinin system may be postulated based on several observations (Fig. 8.1). The protein precursor of BK, high molecular weight kininogen (HMWK), is plasma-borne, and that for Lys-BK, low molecular weight kininogen (LMWK), is distributed in tissues (Bhoola *et al.*, 1992; see Chapter 4).

LMWK is synthesized *de novo* by fibroblasts, notably under the influence of TNF-α (Takano *et al.*, 1995). The corresponding kallikreins (respectively, the plasma and tissue, or glandular, forms) are similarly compartmentalized. Leukocytes are a rich source of tissue kallikrein (Bhoola *et al.*, 1992; see Chapter 11), which, in addition is synthesized *de novo* following trauma or inflammation in tissues such as human kidney (Cumming *et al.*, 1994). The Arg carboxypeptidase (kininase I) activities necessary to convert native kinins into their desArg^9 forms may be provided by the soluble carboxypeptidase N in the plasma system, and by the widely distributed and membrane-bound carboxypeptidase M in tissues (Skidgel, 1988; see Chapter 7). Thus, the right-hand side of Fig. 8.1 is composed of proteins that are carried to tissues by infiltrating leukocytes and/or inducible in inflamed tissues (including the B$_1$ receptor), whereas the left-hand side is the classical plasma kallikrein–kinin system composed of preformed plasma proteins and stimulated by the contact system. Lys-BK is preferentially formed by tissue kallikrein in the nasal secretions of subjects with allergic or viral rhinitis (Naclerio *et al.*, 1988; Proud *et al.*, 1983) and in the bronchoalveolar lavage fluid of asthmatic subjects following allergen challenge (Christiansen *et al.*, 1987). The B$_2$ receptors probably belong to both the tissue and plasma systems, because native Lys-BK has high affinity (Table 8.1). The B$_1$ receptor would clearly belong, therefore, to the postulated extravascular kallikrein–kinin system, and to the realm of inflammation and tissue injury.

The metabolic pathways for kinins indicated in Fig. 8.1 are well known, and notably illustrated by a study of kinin metabolism in the nasal discharge following antigenic

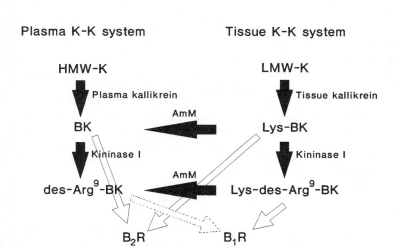

Plasma K-K system

Tissue K-K system

HMW-K → (Plasma kallikrein) → BK → (Kininase I) → des-Arg^9-BK

LMW-K → (Tissue kallikrein) → Lys-BK → (Kininase I) → Lys-des-Arg^9-BK

AmM

AmM

B$_2$R   B$_1$R

Figure 8.1 Postulated compartments of the kallikrein–kinin (K-K) system. HMW-K, high molecular weight kininogen; LMW-K, low molecular weight kininogen; AmM, aminopeptidase M; B$_1$R and B$_2$R, B$_1$ and B$_2$ receptors.

challenge in atopic patients (a form of inflammatory exudate) (Proud *et al.*, 1987; see Chapter 17). Lys-BK is the major kinin in this system, and both aminopeptidase and carboxypeptidase (kininase I) activities react with exogenous Lys-BK to yield BK, Lys-desArg^9-BK and desArg^9-BK. The latter is the final and stable kinin metabolite in this form of inflammatory exudate. The relative importance of the kininase I pathway for BK formed in plasma by the contact system is probably not great, the yield of conversion being small in activated plasma due to BK being degraded efficiently by kininase II (Raymond *et al.*, 1995; see Chapter 7). Kinin concentrations in venous blood may not accurately reflect the tissue kallikrein–kinin system in inflamed tissues. No systematic attempt to measure Lys-desArg^9-BK separately from desArg^9-BK in plasma, however, has been reported.

3. *Evidence for B$_1$ Receptor Upregulation in Immunopathology*

3.1 EVIDENCE *IN VIVO*

Table 8.2 summarizes systems that support the concept of tissue injury regulating B$_1$ receptor expression. In many of these systems, B$_2$ receptors co-exist and mediate similar effects. Thus, B$_1$ receptor upregulation by injury may *extend* the tissue response to desArg^9-kinin metabolites. This idea of extension is not only a matter of intensity, but may relate to duration, as B$_1$ receptor-induced responses may be less subject to tachyphylaxis (see Table 8.1 and related discussion). In some systems, the B$_1$ receptor-mediated response becomes more evident with time as the effect mediated by B$_2$ receptors fades significantly (Kachur *et al.*, 1986; Campos and Calixto, 1995). In that sense, B$_1$ receptor-mediated responses may replace those induced by B$_2$ receptor stimulation during the course of chronic inflammation.

The responses presented in Table 8.2 have been observed either *in vivo* at certain times after noxious stimuli, or *in vitro*. Incubation for long periods of time *in vitro* is itself a powerful stimulus to upregulate B$_1$ receptor-mediated responses in preparations such as rabbit aorta (Fig. 8.2), but it is the early response to kinins in these preparations, recorded 1 h or less after animal sacrifice, that represents the status of the *in vivo* responsiveness to B$_1$ receptor agonists in most studies of smooth muscle contraction. Judicious use of a protein synthesis inhibitor, such as cycloheximide, can "freeze" the population of B$_1$ receptors close to the pre-sacrifice level in LPS-treated or control rabbits, thus minimizing the effect of tissue incubation *in vitro* (Fig. 8.2).

Hypotensive responses to desArg^9-kinins are observed in rabbits and pigs treated with bacterial products (Table 8.2). Only younger Brown–Norway rats have been reported to be sensitized to desArg^9-BK by LPS, and this

may be due to low sensitivity of adult animals to LPS (Tokumasu *et al.*, 1985). In these species, sensitized with LPS, hypotensive effects of BK are not modified, and the cardiovascular system, therefore, responds to stimulation of either B$_1$ or B$_2$ receptors with hypotensive responses that are prevented with receptor subtype-specific antagonists. Rabbits exhibit a high relative potency of Lys-desArg^9-BK, relative to desArg^9-BK (Drapeau *et al.*, 1991), and this species has been used to test metabolically resistant analogs of B$_1$ receptor agonists and antagonists (Drapeau *et al.*, 1991, 1993). The mechanisms underlying hypotension are complex, a fall in peripheral vascular resistance accounting for the early response following a bolus injection of a B$_1$ receptor agonist in LPS-pretreated rabbits, but the prolonged hypotension, associated with persistent receptor stimulation, is due to low cardiac output, high prostaglandin release and sympathetic nervous system activation (Audet *et al.*, 1997). No massive plasma loss occurs in LPS-pretreated rabbits stimulated with B$_1$ receptor agonists by the intravascular route (Audet *et al.*, 1997) or by intradermal injection (Nwator and Whalley, 1989). In contrast, exogenous desArg^9-BK was capable of inducing exudation and edema in rats following inflammatory stimuli (Table 8.2).

The rat has been used in various quantitative models of hyperalgesia (Table 8.2). Depending on the intensity of noxious pretreatments, localized or generalized hyperalgesia was produced in response to exogenous B$_1$ receptor agonists in 4–24 h. Treatments included ultraviolet irradiation of the skin, injection of mycobacterial products, or cytokine injection (Table 8.2). BK, via B$_2$ receptors, is hyperalgesic in control rats and, in general, remained so in these models. Prostaglandins are of particular importance as secondary mediators in these models, at least for the cytokine-induced sensitivity to desArg^9-BK (Davis and Perkins, 1994b). Interestingly, LPS injection in the rat brain is associated with fever and diffuse hyperalgesia, which are reduced by administration of indomethacin or Hoe 140, not B$_1$ receptor antagonists (Walker *et al.*, 1995). Contrary to B$_2$ receptors present on sensory and other neurons, there is no evidence for the expression of B$_1$ receptors by neural tissue at the present time (Davis *et al.*, 1996).

Thus, several physiological effector systems are influenced by B$_1$ receptor agonists *in vivo*, and fairly complex relationships may exist between them. For instance, afferent nerve terminals may be recruited by desArg^9-BK and increase vascular permeability through an axon reflex in turpentine-treated rats (Walker *et al.*, 1994). (**Editor's note:** the subtypes of kinin receptors, and their regulation in pain and hyperalgesia is discussed in detail in Chapter 9.)

3.2 EVIDENCE *IN VITRO*

Several systems *in vitro* are useful to characterize the effects of B$_1$ receptor stimulation (Table 8.3). The rabbit isolated

Table 8.2   Pretreatments that induce responsiveness to B$_1$ receptor agonists *in vivo*

Species/ pretreatment	Response to agonist	B$_1$ receptor antagonist[b]	References
HEMODYNAMIC RESPONSES			
Rabbit			
Intravenous LPS (5–20 h), MDP or IL-1β[a]	Hypotension	Yes	Marceau *et al.* (1983, 1984), Bouthillier *et al.* (1987), deBlois *et al.* (1991), Nwator and Whalley (1989), Drapeau *et al.* (1991, 1993)
Blockade of protein synthesis	Hypotension	No	deBlois *et al.* (1991)
Pig			
E. coli sepsis (4 h)	Hypotension	No	Siebeck *et al.* (1989)
Rat			
I.v. LPS (24 h)	Hypotension	Yes	Tokumasu *et al.* (1995)
Genetic hypertension (SHR[c])	Intracerebroventricular B$_1$ antagonist causes hypotension	Yes	Alvarez *et al.* (1992)
VASCULAR PERMEABILITY RESPONSES			
Rat			
Immune complex arthritis	Local B$_1$ agonist causes MVL	Yes	Cruwys *et al.* (1994)
Turpentine edema	Local B$_1$ agonist causes MVL	Yes	Walker *et al.* (1994)
Daily intraplantar BK for 7 days	DesArg^9-BK causes paw edema	Yes	Campos and Calixto (1995)
CONTRACTILITY			
Rabbit			
I.v. LPS (5–20 h)	Immediate desArg^9-BK-induced response of isolated blood vessels	No	Regoli *et al.* (1981), deBlois *et al.* (1989), Sugihara *et al.* (1991)
Immune complex arthritis	Immediate response of isolated aorta to desArg^9-BK	No	Farmer *et al.* (1991)
Injury *in vivo* to carotid artery	Immediate response of carotid artery to desArg^9-BK *in vitro*	No	Pruneau *et al.* (1994)
Rat			
Septic peritonitis	Immediate response of portal vein artery to desArg^9-BK	No	Mastrangelo *et al.* (1995)
Chemical cystitis	Immediate response of isolated bladder	Yes	Marceau *et al.* (1980), Roslan *et al.* (1995)
Induction (21 d) of granuloma	Immediate contraction of isolated granulation tissue	No	Appleton *et al.* (1994)
HYPERALGESIA			
Rat			
UV skin irradiation	Hyperalgesia	Yes	Perkins and Kelly (1993), Perkins *et al.* (1993)
Intra-articular adjuvant injection	Hyperalgesia	Yes	Davis and Perkins (1994a), Perkins *et al.* (1993)
Intra-articular IL-1, IL-2 or IL-8	Prostaglandin-dependent hyperalgesia	Yes	Davis and Perkins (1994a)
OTHERS			
Rat			
Subcutaneous sponge containing BK or IL-1	Angiogenesis	Yes	Hu and Fan (1993)
Chemical colitis	Increased mucosal ion transport	Yes	Kachur *et al.* (1986)

[a] LPS, lipopolysaccharide or endotoxin; MDP, Muramyl dipeptide. Captopril and enalapril pretreatments have been claimed to induce weak responsiveness to desArg^9-BK in the rabbit (Nwator and Whalley, 1989), but this is controversial (deBlois *et al.*, 1991). In many of the above systems, the B$_2$ receptors are also expressed, and mediate similar effects to B$_1$ receptor stimulation.
[b] "Yes" or "No" indicate whether a B$_1$ receptor antagonist was used to validate the response as being mediated by B$_1$ receptors.
[c] SHR, spontaneously hypertensive rats; MVL, microvascular leakage; UV, ultraviolet.

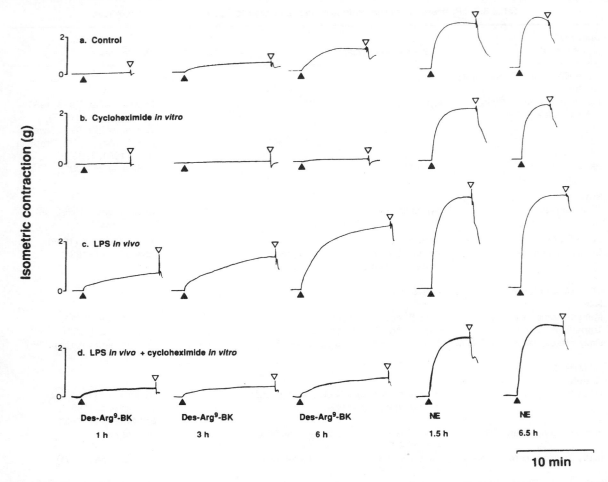

Figure 8.2   Contractile effect of desArg⁹-BK (1.7 μM) and norepinephrine (NE, 100 nM) on the rabbit aortic strip, as a function of incubation time _in vitro_. The two strips of the same aorta isolated from a healthy rabbit were initially insensitive to desArg⁹-BK (tracings a and b), whereas a significant response was recorded in paired strips isolated from a rabbit pretreated with LPS (50 μg i.v., 5 h before sacrifice) (tracings c and d). The subsequent upregulation of responses to the kinin (tracings a and c) was inhibited in the corresponding paired strip continuously exposed to cycloheximide (71 μM) (tracings b and d). Closed symbols refer to the application of agents and open symbols to washout of stimulants. Taken from deBlois _et al._ (1989) with permission.

aorta is contracted by kinins uniquely via B_1 receptors, and was utilized in the initial descriptions of this entity (Regoli _et al._, 1977). Cultured rabbit smooth muscle cells from the aorta or other arteries have also been used to characterize various B_1 receptor-dependent phenomena in vascular pharmacology (Table 8.3). The human line, IMR-90, was used as a source of mRNA for expression cloning of the human B_1 receptor (Menke _et al._, 1994). Some information is also available from rat mesangial cells and bovine endothelial cells (Table 8.3), lines that co-express both B_1 and B_2 receptors, although it is not known whether the B_1 receptor expression is upregulated in these cells. B_1 receptor expression and regulation in many isolated cell and tissue systems are reviewed elsewhere (Marceau, 1995), and not discussed further here.

4.   Mechanisms of B_1 Receptor Induction

Rabbit aorta exhibits selective sensitization to kinins _in vitro_, as responses to norepinephrine (Fig. 8.2) do not change with time of incubation. This tissue was the basis of the initial studies of B_1 receptor induction. Upregulation of contractile responses is abolished by pretreatment _in vitro_ with a protein synthesis inhibitor, cycloheximide, or a RNA synthesis inhibitor, actinomycin D, in rabbit aorta and other tissues (Regoli _et al._, 1978; Marceau _et al._, 1980; Bouthillier _et al._, 1987; deBlois _et al._, 1988, 1991; Altinkurt and Öztürk, 1990; Campos and Calixto, 1994). A protein translocation inhibitor, brefeldin A, also prevents the spontaneous upregulation of B_1 receptor-mediated

Table 8.3   Published experimental systems useful for studying B$_1$ receptors *in vitro*

Model	B$_1$ and B$_2$ receptors present	Effect of IL-1 on B$_1$ receptor expression	Other stimuli of B$_1$ receptor induction	B$_1$ receptor-mediated DNA synthesis?	B$_1$ receptor binding assays	Secondary mediators released	Structure–activity relationship studies	B$_1$ receptor-mediated PI hydrolysis
Rabbit aorta contraction	No	Accelerated induction of contraction (deBlois et al., 1988, 1989, 1991)	IL-2, oncostatin M, EGF, inhibition of protein synthesis, LPS, MDP	Unknown	Unknown	PGI$_2$ (Levesque et al., 1993)	Regoli and Barabé (1980), Wirth et al. (1991), Sawutz et al. (1994)	Unknown
Rabbit vascular smooth muscle cells in culture	Aorta – no Mesenteric artery – yes	Upregulated binding and functional responses (Galizzi et al., 1994; Levesque et al., 1993, 1995a,b)	EGF, LPS, oncostatin M	Yes (Levesque et al., 1995b)	Galizzi et al. (1994), Schneck et al. (1994), Levesque et al. (1995a)	PGI$_2$	Tropea et al. (1993), Schneck et al. (1994), Levesque et al. (1995a,b)	Levesque et al. (1993), Tropea et al. (1993), Schneck et al. (1994)
Human fibroblasts (IMR-90 cells)	Yes	Upregulated binding (Menke et al., 1994)	TNF-α, oncostatin M, blockade of protein synthesis	Yes (Goldstein and Wall, 1984)	Menke et al. (1994)	PGE$_2$, PGF$_{2α}$ and TxA$_2$ (Goldstein and Wall, 1984)	Menke et al. (1994)	Unknown
Rat mesangial cells	Yes	Unknown	Unknown	Yes (Issandou and Darbon, 1991; Bascands et al., 1993)	Bascands et al. (1993)	Unknown	Bascands et al. (1993)	Issandou and Darbon (1991), Bascands et al. (1993)
Bovine endothelial cells	Yes	Unknown	Unknown	Unknown	Unknown	Nitric oxide and PGI$_2$ (D'Orléans-Juste et al., 1989)	Wiemer and Wirth (1992), Wirth et al. (1994), Smith et al. (1995)	Smith et al. (1995)

response (Audet *et al.*, 1994). Brefeldin A inhibits protein transport between the endoplasmic reticulum and the Golgi apparatus, and blocks the secretion or transport to the membrane of proteins possessing a signal peptide, without blocking protein synthesis (see Audet *et al.*, 1994). The glycosylation inhibitor, tunicamycin, also significantly depresses the spontaneous upregulation of responses to desArg^9-BK in rabbit isolated aorta (Audet *et al.*, 1994).

Metabolic inhibitors affect only the contractions in response to B$_1$ receptor stimulation. When B$_2$ receptors are co-expressed in a given preparation, the responses mediated by these receptors remain stable (Campos and Calixto, 1994). However, in the rabbit isolated aorta, if protein synthesis inhibitors are applied as a "pulse", the appearance of the responsiveness proceeds at a greatly increased rate (deBlois *et al.*, 1991).

The basis of these observations is likely to be *de novo* synthesis of B$_1$ receptors induced by tissue injury (in this case, tissue isolation and incubation *in vitro*). This process would require the transcriptional activation of one or more genes, expression of mRNA(s), translation at the level of ribosomes and, at least for the receptor protein, which is predicted to possess a signal peptide and glycosylation sites (Menke *et al.*, 1994), processing through the endoplasmic reticulum–Golgi apparatus pathway. A null level of B$_1$ expression is postulated in normal tissues, which accounts for the observation that kinin B$_1$ receptor agonists are biologically inert in healthy individuals of several species.

Upregulation of B$_1$ receptor-mediated responses from a null initial level occurs in all rabbit, rat and pig smooth muscle preparations tested (Marceau, 1995), and also in human colon (Couture *et al.*, 1981). Preparations, such as rabbit carotid artery and mesenteric artery, which relax in response to desArg^9-BK, also acquire responsiveness to the agonist *in vitro* in a time- and protein synthesis-dependent manner (deBlois and Marceau, 1987; Pruneau and Bélichard, 1993).

All of the cellular components necessary for B$_1$ receptor upregulation are presumably present in rabbit isolated aorta. Upregulation of responses to desArg^9-BK is increased moderately by LPS or muramyl dipeptide (MDP) *in vitro* (Bouthillier *et al.*, 1987; deBlois *et al.*, 1989) and increased vigorously by incubating tissues with IL-1, IL-2 or oncostatin M (deBlois *et al.*, 1988, 1989; Levesque *et al.*, 1995b). Glucocorticosteroids suppress upregulation, either spontaneous or stimulated by LPS, MDP, IL-1, IL-2, or pulse of protein synthesis inhibitors (deBlois *et al.*, 1988, 1989, 1991). Several other cytokines and other agents failed to reveal any effect on B$_1$ receptor induction (deBlois *et al.*, 1989).

All stimulants named above (LPS, MDP, IL-1, IL-2, oncostatin M) are likely to interfere with the autocrine production of cytokines, singularly of IL-1, in cells of mesenchymal origin. The "pulse exposure" to cycloheximide constitutes a "superinduction" condition

for the production of autocrine IL-1 (deBlois *et al.*, 1991). The inhibitory effect of glucocorticosteroids may be explained by the suppression of IL-1 synthesis and perhaps also by an additional action more proximal to B$_1$ receptor expression, as dexamethasone depressed responses in IL-1-treated tissues (deBlois *et al.*, 1988). The postulated role of autocrine cytokine(s) in B$_1$ receptor upregulation is supported by the demonstration of spontaneous IL-1 production by fresh rabbit aortic smooth muscle cells following isolation and incubation *in vitro*, even under aseptic conditions (Clinton *et al.*, 1991). It is also consistent with the model of cardiovascular B$_1$ receptor upregulation in rabbits *in vivo*. The cytokine releaser, MDP, as well as IL-1 itself, mimic the effect of LPS (Bouthillier *et al.*, 1987; deBlois *et al.*, 1991). Autocrine IL-1, which may exert its activity on the producing cell even as a high molecular weight precursor form (Clinton *et al.*, 1991), is difficult to investigate because it is not readily inhibited by probes such as neutralizing antibodies to IL-1 or the natural IL-1 receptor antagonist, which are restricted to the extracellular compartment. However, intra-articular injection of IL-1, IL-2 or IL-8 induced a B$_1$ receptor-mediated hyperalgesia response *in vivo* over several hours in rats, and IL-1 receptor antagonist prevented the effect of all three cytokines (Davis and Perkins, 1994b). This suggests that the effect of IL-2 and of IL-8 were mediated by IL-1 secretion. Thus, several cytokines can influence the expression of functional responses to B$_1$ receptor agonists.

Rabbit vascular smooth muscle cells have been utilized in binding assays to monitor the effect of cytokine treatments on the kinin B$_1$ receptor population (Table 8.3). Correlating binding signals to biochemical responses in the same cells has been particularly helpful. Prolonged pretreatments with epidermal growth factor (EGF), IL-1 or LPS increase significantly the abundance of B$_1$ receptors (B_{max}), without affecting affinity (Galizzi *et al.*, 1994; Schneck *et al.*, 1994; Levesque *et al.*, 1995a). Oncostatin M, on the other hand, has a relatively modest effect on B_{max} (Levesque *et al.*, 1995a). PI turnover in response to a B$_1$ receptor agonist is also increased by cytokine pretreatments, in a manner that correlated to changes in receptor B_{max} (Levesque *et al.*, 1993, 1995a; Schneck *et al.*, 1994). Biochemical responses to desArg^9-BK, which are more distal to receptor activation, such as the prostaglandin I$_2$ (PGI$_2$) synthesis and release (Galizzi *et al.*, 1994) or DNA synthesis (Levesque *et al.*, 1995b), may be amplified to a greater extent by IL-1 or oncostatin M. This suggests additional postreceptor interactions between B$_1$ receptor agonists and cytokines.

As the B$_1$ receptor is a defined molecular entity, direct study of mRNA levels is now possible (Fig. 8.3). The heart has been chosen for a pilot study, as the Langendorff preparation of rabbit coronary artery is unresponsive to desArg^9-BK in normal animals but

A. Rabbit heart

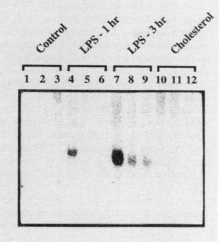

B. Human IMR-90 cells

Figure 8.3   Induction of the mRNA coding for B₁ receptor (Northern blots). (A). LPS treatment of rabbits *in vivo* induced the B₁ receptor mRNA (1.5 kb signal) in heart tissue from three out of three animals after 3 h, and in one out of three in 1 h. The signal was not detectable in control animal hearts or in hearts from rabbits fed a high cholesterol (0.5%) diet for 10 weeks prior to sacrifice. **(B).** The B₁ receptor mRNA (1.5 kb) is constitutively expressed in the human IMR-90 fibroblast line, but considerably upregulated by interleukin-1β (IL-1β; 0.5 ng/ml for 2.5 h). Rabbit and human probes for B₁ receptors were gifts from Dr J.F. Hess, Merck Research Laboratories.

responsive (vasodilation) in LPS-treated rabbits (Regoli *et al.*, 1981). Rabbit B₁ receptor mRNA is not detectable in control heart tissue but increases gradually over 3 h post-LPS injection (Fig. 8.3A), thus mimicking the functional response. It is interesting to note that LPS

itself is relatively ineffective to upregulate B₁ receptors *in vitro* (Table 8.3). In contrast, *in vivo*, LPS may cause cytokine secretion by leukocytes. B₁ receptor induction is not a common response to all vascular pathology. Thus, preliminary experiments on dietary atherosclerosis in the rabbit are negative in this respect (Fig. 8.3A). IL-1 amplifies considerably the B₁ receptor mRNA content of IMR-90 cells (Fig. 8.3B), occurring in as early as 30 min (not shown).

The promoter region analysis of the B₁ receptor gene, not performed yet, should provide useful molecular information on cytokine, growth factor and glucocorticosteroid responsive elements, which could be responsible for the regulation of receptor expression. It is not known whether inflammatory cytokines will influence directly the transcriptional rate or whether other pathophysiological or tissue-specific factors are required for the expression of B₁ receptors. Tissue-specific transcription factors may be suspected, as the origin of cells determines whether the upregulation of responses to B₁ agonists can be observed.

5.   Summary and Conclusion

While many receptor populations are regulated, the B₁ receptor is a peculiar example of a G protein-coupled receptor that is completely and rapidly inducible. Various noxious treatments vigorously upregulate B₁ receptors (tissue isolation and incubation *in vivo*, LPS injection *in vivo*, etc.), but these treatments may ultimately recruit the cytokine network and, singularly, autocrine IL-1. The delayed appearance of B₁ receptors during inflammation has the potential to amplify tissue responses to kinins, because tissues then become responsive to long-lived desArg⁹ metabolites of native kinins. It is also possible that B₁ receptors are less prone to tachyphylaxis than B₂ receptors. The relative selectivity of B₁ receptors for the agonist Lys-desArg⁹-BK suggests that they may belong to a "tissue-type" kallikrein–kinin system, with LMWK and tissue kallikrein being the kinin-generating proteins. The full pathophysiological and therapeutic consequences of the duality of kinin receptors have not been yet fully assessed. Clinical research and medicinal chemistry efforts are needed to bring these findings into medicine.

6.   References

Ahluwalia, A. and Perretti, M. (1995). Involvement of bradykinin B₁ receptors in the polymorphonuclear leukocyte accumulation induced by IL-1β *in vivo* in the mouse. J. Immunol. 156, 269–274.

Altinkurt, O. and Öztürk, Y. (1990). Bradykinin receptors in isolated rat duodenum. Peptides 11, 39–44.

Alvarez, A.L., Delorenzi, A., Santajuliana, D., Finkielman, S., Nahmod, V.E. and Pirola, C.J. (1992). Central bradykininergic system in normotensive and hypertensive rats. Clin. Sci. 82, 513–519.

Appleton, I., Rosenblatt, L. and Willoughby, D.A. (1994). Synergistic effect of recombinant human epidermal growth factor and desArg⁹-bradykinin, on rat croton oil-induced granulation tissue contraction. Br. J. Pharmacol. 111, 290P.

Audet, R., Petitclerc, E., Drapeau, G., Rioux, F. and Marceau, F. (1994). Further analysis of the upregulation of bradykinin B$_1$ receptors in isolated rabbit aorta by using metabolic inhibitors. Eur. J. Pharmacol. 271, 551–555.

Audet, R., Rioux, F., Drapeau, D., Marceau, F. (1997). Cardiovascular effects of Sar-[D-Phe⁸]des-Arg⁹-bradykinin, a metabolically protected agonist of B$_1$ receptor for kinins, in the anesthetized rabbit pretreated with a sublethal dose of bacterial lipopolysaccharide. J. Pharmacol. Exp. Ther. (In press).

Bachvarov, D.R., Saint-Jacques, E., Larrivée, J.-F., Levesque, L., Rioux, F., Drapeau, G., Marceau, F. (1995). Cloning and pharmacological characterization of the rabbit bradykinin B$_2$ receptor. J. Pharmacol. Exp. Ther. 275, 1623–1630.

Bascands, J.L., Pecher, C., Rouaud, S., Emond, C., Leung Tack, J., Bastie, M.J., Burch, R., Regoli, D. and Girolami, J.P. (1993). Evidence for existence of two distinct bradykinin receptors on rat mesangial cells. Am. J. Physiol. 264, F548–F556.

Bathon, J.M., Manning, D.C., Goldman, D.W., Towns, M.C. and Proud, D. (1992). Characterization of kinin receptors on human synovial cells and upregulation of receptor number by interleukin-1. J. Pharmacol. Exp. Ther. 260, 384–392.

Bhoola, K.D., Figueroa, C.D. and Worthy, K. (1992). Bioregulation of kinins: kallikreins, kininogens, and kininases. Pharmacol. Rev. 44, 1–80.

Birkenbach, M., Josefsen, K., Yalamanchi, R., Lenoir, G. and Kieff, E. (1993). Epstein–Barr virus-induced genes: first lymphocyte-specific G protein-coupled peptide receptors. J. Virol. 67, 2209–2220.

Bouthillier, J., deBlois, D. and Marceau, F. (1987). Studies on the induction of pharmacological responses to desArg⁹-bradykinin in vitro and in vivo. Br. J. Pharmacol. 92, 257–264.

Boxall, S.J., Wheeldon, A. and Birch, P.J. (1995). Characterization of the bradykinin receptors that mediate contraction of the rat oesophagus in vitro. Br. J. Pharmacol. 114, 225P.

Butt, S.K., Dawson, L.G. and Hall, J.M. (1995). Bradykinin B$_1$ receptors in the rabbit urinary bladder: induction of responses, smooth muscle contraction, and phosphatidylinositol hydrolysis. Br. J. Pharmacol. 114, 612–617.

Campos, A.H. and Calixto, J.B. (1994). Mechanisms involved in the contractile responses of kinins in rat portal vein rings: mediation by B$_1$ and B$_2$ receptors. J. Pharmacol. Exp. Ther. 268, 902–909.

Campos, M.M., and Calixto, J.B. (1995). Involvement of B$_1$ and B$_2$ receptors in bradykinin-induced rat paw oedema. Br. J. Pharmacol. 114, 1005–1013.

Christiansen, S.C., Proud, D. and Cochrane, C.G. (1987). Detection of tissue kallikrein in bronchoalveolar lavage fluids of asthmatic subjects. J. Clin. Invest. 79, 188–197.

Clinton, S.K., Fleet, J.C., Loppnow, H., Salomon, R.N., Clark, B.D., Cannon, J.G., Shaw, A.R., Dinarello, C.A. and Libby, P. (1991). Interleukin-1 gene expression in rabbit vascular tissue in vivo. Am. J. Pathol. 138, 1005–1014.

Couture, R., Mizrahi, J., Regoli, D. and Devroede, G. (1981). Peptides and the human colon: an in vitro pharmacological study. Can. J. Physiol. Pharmacol. 59, 957–964.

Cruwys, S.C., Garrett, N.E., Perkins, M.N., Blake, D.R. and Kidd, B.L. (1994). The role of bradykinin B$_1$ receptors in the maintenance of intra-articular plasma extravasation in chronic antigen-induced arthritis. Br. J. Pharmacol. 113, 940–944.

Cumming, A.D., Walsh, T., Wojtacha, D., Fleming, S., Thomson, D. and Jenkins, D.A.S. (1994). Expression of tissue kallikrein in human kidney. Clin. Sci. 87, 5–11.

Cunoosamy, M.P. and Wheeldon, A. (1995). Haemodynamic effects of bradykinin B$_1$ and B$_2$ receptor activation in the anaesthetised rat. Br. J. Pharmacol. 114, 227P.

Davis, A.J. and Perkins, M.N. (1994a). Induction of B1 receptors in vivo in a model of persistent inflammatory mechanical hyperalgesia in the rat. Neuropharmacology 33, 127–133.

Davis, A.J. and Perkins, M.N. (1994b). The involvement of bradykinin B$_1$ and B$_2$ receptor mechanisms in cytokine-induced mechanical hyperalgesia in the rat. Br. J. Pharmacol. 113, 63–68.

Davis, C.L., Naeem, S., Phagoo, S.B., Campbell, E.A., Urban, L. and Burgess, G.M. (1996). B$_1$ bradykinin receptors and sensory neurones. Br. J. Pharmacol. 118, 1469–1476.

deBlois, D. and Marceau, F. (1987). The ability of desArg⁹-bradykinin to relax rabbit isolated mesenteric arteries is acquired during in vitro incubation. Eur. J. Pharmacol. 142, 141–144.

deBlois, D., Bouthillier, J. and Marceau, F. (1988). Effect of glucocorticoids, monokines and growth factors on the spontaneously developing responses of the rabbit isolated aorta to desArg⁹-bradykinin. Br. J. Pharmacol. 93, 969–977.

deBlois, D., Bouthillier, J. and Marceau, F. (1989). Pharmacological modulation of the up-regulated responses to desArg⁹-bradykinin in vivo and in vitro. Immunopharmacology 17, 187–198.

deBlois, D., Bouthillier, J. and Marceau, F. (1991). Pulse exposure to protein synthesis inhibitors enhances tissue response to desArg⁹-bradykinin: possible role of interleukin-1. Br. J. Pharmacol. 103, 1057–1066.

DeWitt, B.J., Cheng, D.Y. and Kadowitz, P.J. (1994). DesArg⁹-bradykinin produces tone-dependent kinin B$_1$ receptor-mediated responses in the pulmonary vascular bed. Circ. Res. 75, 1064–1072.

Dixon, B.S. (1994). Cyclic AMP selectively enhances bradykinin receptor synthesis and expression in cultured arterial smooth muscle. Inhibition of angiotensin II and vasopressin response. J. Clin. Invest. 93, 2535–2544.

D'Orléans-Juste, P., De Nucci, G. and Vane, J.R. (1989). Kinins act on B$_1$ or B$_2$ receptors to release conjointly endothelium-derived relaxing factor and prostacyclin from bovine aortic endothelial cells. Br. J. Pharmacol. 96, 920–926.

Drapeau, G., deBlois, D. and Marceau, F. (1991). Hypotensive effects of Lys-desArg⁹-bradykinin and metabolically protected agonists of B$_1$ receptors for kinins. J. Pharmacol. Exp. Ther. 259, 997–1003.

Drapeau, G., Audet, R., Levesque, L., Godin, D. and Marceau, F. (1993). Development and in vivo evaluation of metabolically resistant antagonists of B$_1$ receptors for kinins. J. Pharmacol. Exp. Ther. 266, 192–199.

Farmer, S.G., McMillan, B.A., Meeker, S.N. and Burch, R.M.

(1991). Induction of vascular smooth muscle bradykinin B$_1$ receptors *in vivo* during antigen arthritis. Agents Actions 34, 191–193.

Galizzi, J.P., Bodinier, M.C., Chapelain, B., Ly, S.M., Coussy, L., Gireaud, S., Neliat, G. and Jean, T. (1994). Up-regulation of [^{3}H]-desArg^{10}-kallidin binding to the bradykinin B$_1$ receptor by interleukin-1 in isolated smooth muscle cells: correlation with B$_1$ agonist-induced PGI$_2$ production. Br. J. Pharmacol. 113, 389–394.

Goldman, D.W., Enkel, H., Gifford, L.A., Chenoweth, D.E. and Rosenbaum, J.T. (1986). Lipopolysaccharide modulates receptors for leukotriene B$_4$, C5a, and formyl-methionyl-leucyl-phenylalanine on rabbit polymorphonuclear leukocytes. J. Immunol. 137, 1971–1976.

Goldstein, R.H. and Wall, M. (1984). Activation of protein formation and cell division by bradykinin and desArg^9-bradykinin. J. Biol. Chem. 259, 9263–9268.

Hess, J.F., Borkowski, J.A., Young, G.S., Strader, C.D. and Ransom, R.W. (1992). Cloning and pharmacological characterization of a human bradykinin (BK-2) receptor. Biochem. Biophys. Res. Commun. 184, 260–268.

Hess, J.F., Borkowski, J.A., Macneil, T., Stonesifer, G.Y., Fraher, J. and Strader, C.D. (1994). Differential pharmacology of cloned human and mouse B$_2$ bradykinin receptors. Mol. Pharmacol. 45, 1–8.

Hock, F.J., Wirth, K., Albus, U., Linz, W., Gerhards, H.J., Wiemer, G., Henke, S., Breipohl, G., König, W., Knolle, J. and Schölkens, B.A. (1991). Hoe 140 a new potent and long acting bradykinin-antagonist: *in vitro* studies. Br. J. Pharmacol. 102, 769–773.

Hu, D.E. and Fan, T.P.D. (1993). [Leu^8]desArg^9-bradykinin inhibits the angiogenic effect of bradykinin and interleukin-1 in rats. Br. J. Pharmacol. 109, 14–17.

Issandou, M. and Darbon, J.M. (1991). DesArg^9 bradykinin modulates DNA synthesis, phospholipase C, and protein kinase C in cultured mesangial cells. J. Biol. Chem. 266, 21037–21043.

Kachur, J.F., Allbee, W. and Gaginella, T.S. (1986). Effect of bradykinin and desArg^9-bradykinin on ion transport across normal and inflamed rat colonic mucosa. Gastroenterology 90, 1481.

Levesque, L., Drapeau, G., Grose, J.H., Rioux, F. and Marceau, F. (1993). Vascular mode of action of kinin B$_1$ receptors and development of a cellular model for the investigation of these receptors. Br. J. Pharmacol. 109, 1254–1262.

Levesque, L., Harvey, N., Rioux, F., Drapeau, G. and Marceau, F. (1995a). Development of a binding assay for the B$_1$ receptors for kinins. Immunopharmacology 29, 141–147.

Levesque, L., Larrivée, J.-F., Bachvarov, D.R., Rioux, F., Drapeau, G. and Marceau, F. (1995b). Regulation of kinin-induced contraction and DNA synthesis by inflammatory cytokines in the smooth muscle of the rabbit aorta. Br. J. Pharmacol. 116, 1673–1679.

Lortie, M., Regoli, D., Rhaleb, N.E. and Plante, G.E. (1992). The role of B$_1$- and B$_2$-kinin receptors in the renal tubular and hemodynamic response to bradykinin. Am. J. Physiol. 262, R72–R76.

Ma, J.X., Wang, D.Z., Ward, D.C., Chen, L., Dessai, T., Chao, J. and Chao, L. (1994). Structure and chromosomal localization of the gene (BDKRB2) encoding human bradykinin B$_2$ receptor. Genomics 23, 362–369.

Marceau, F. (1995). Kinin B$_1$ receptors: a review. Immunopharmacology 30, 1–26.

Marceau, F. and Regoli, D. (1991). Kinin receptors of B1 type and their antagonists. In "Bradykinin Antagonists, Basic and Clinical Aspects" (ed. R.M. Burch), pp. 33–49. Marcel Dekker, New York.

Marceau, F., Barabé, J., St-Pierre, S. and Regoli, D. (1980). Kinin receptors in experimental inflammation. Can. J. Physiol. Pharmacol. 58, 536–542.

Marceau, F., Lussier, A., Regoli, D. and Giroud, J.P. (1983). Kinins: their relevance to tissue injury and inflammation. Gen. Pharmacol. 14, 209–229.

Marceau, F., Lussier, A. and St-Pierre, S. (1984). Selective induction of cardiovascular responses to desArg^9-bradykinin by bacterial endotoxin. Pharmacology 29, 70–74.

Marsh, K.A. and Hill, S.J. (1994). DesArg^9-bradykinin-induced increases in intracellular calcium ion concentration in single bovine tracheal smooth muscle cells. Br. J. Pharmacol. 112, 934–938.

Mastrangelo, D., Licker, M. and Morel, D.R. (1995). Differential *in vitro* contractile responses of isolated portal vein to norepinephrine, neurokinin B, angiotensin II and bradykinin in septic rats. FASEB J. 9, A885.

McEachern, A.E., Shelton, E.R., Bhakta, S., Obernolte, R., Bach, C., Zuppan, P., Fujisaki, J., Aldrich, R.W. and Jarnagin, K. (1991). Expression cloning of a rat B$_2$ bradykinin receptor. Proc. Natl. Acad Sci. USA 88, 7724–7728.

McIntyre, P., Phillips, E., Skidmore, E., Brown, M. and Webb, M. (1993). Cloned murine bradykinin receptor exhibits a mixed B$_1$ and B$_2$ pharmacological sensitivity. Mol. Pharmacol. 44, 346–355.

Menke, J.G., Borkowski, J.A., Bierilo, K.K., MacNeil, T., Derrick, A.W., Schneck, K.A., Ransom, R.W., Strader, C.D., Linemeyer, D.L. and Hess, J.F. (1994). Expression cloning of a human B$_1$ bradykinin receptor. J. Biol. Chem. 269, 21583–21586.

Munoz, C.M. and Leeb-Lundberg, L.M.F. (1992). Receptor-mediated internalization of bradykinin. DDT$_1$ MF-2 smooth muscle cells process internalized bradykinin via multiple degradative pathways. J. Biol. Chem. 267, 303–309.

Naclerio, R.M., Proud, D., Lichtenstein, L.M., Kagey-Sobotka, A., Hendley, J.O., Sorrentino, J. and Gwaltney, J.M. (1988). Kinins are generated during experimental rhinovirus colds. J. Infect. Dis. 157, 133–142.

Nakhostine, N., Ribuot, C., Lamontagne, D., Nadeau, R., and Couture, R. (1993). Mediation by B$_1$ and B$_2$ receptors of vasodepressor responses to intravenously administered kinins in anaesthetized dogs. Br. J. Pharmacol. 110, 71–76.

Nwator, I.A.A. and Whalley, E.T. (1989). Angiotensin converting enzyme inhibitors and expression of desArg^9-BK (kinin B$_1$) receptors *in vivo*. Eur. J. Pharmacol. 160, 125–132.

Perkins, M.N. and Kelly, D. (1993). Induction of bradykinin B$_1$ receptors *in vivo* in a model of ultra-violet irradiation-induced thermal hyperalgesia in the rat. Br. J. Pharmacol. 110, 1441–1444.

Perkins, M.N., Campbell, E. and Dray, A. (1993). Antinociceptive activity of the bradykinin B$_1$ and B$_2$ receptor antagonists, desArg^9-[Leu^8]-BK and Hoe 140, in two models of persistent hyperalgesia in the rat. Pain 53, 191–197.

Proud, D., Togias, A., Naclerio, R.M., Crush, S.A., Norman,

P.S. and Lichtenstein, L.M. (1983). Kinins are generated *in vivo* following nasal airway challenge of allergic individuals with allergens. J. Clin. Invest. 72, 1678–1685.

Proud, D., Baumgarten, C.R., Naclerio, R.M. and Ward, P.E. (1987). Kinin metabolism in human nasal secretions during experimentally induced allergic rhinitis. J. Immunol. 138, 428–434.

Pruneau, D. and Bélichard, P. (1993). Induction of bradykinin B_1 receptor-mediated relaxation in the isolated rabbit carotid artery. Eur. J. Pharmacol. 239, 63–67.

Pruneau, D., Luccarini, J.M., Robert, C. and Bélichard, P. (1994). Induction of kinin B_1 receptor-dependent vasoconstriction following balloon catheter injury to the rabbit carotid artery. Br. J. Pharmacol. 111, 1029–1034.

Rangachari, P.K., Berezin, M. and Prior, T. (1993). Effects of bradykinin on the canine proximal colon. Regul. Pept. 46, 511–522.

Raymond, P., Drapeau, G., Raut, R., Audet, R., Marceau, F., Ong, H. and Adam, A. (1995). Quantification of desArg^9-bradykinin using a chemiluminescence enzyme immunoassay: application to its kinetic during plasma activation. J. Immunol. Meth. 180, 247–257.

Regoli, D. and Barabé, J. (1980). Pharmacology of bradykinin and related kinins. Pharmacol. Rev. 32, 1–46.

Regoli, D., Barabé, J. and Park, W.K. (1977). Receptors for bradykinin in rabbit aortae. Can. J. Physiol. Pharmacol. 55, 855–867.

Regoli, D., Marceau, F. and Barabé, J. (1978). *De novo* formation of vascular receptors for kinins. Can. J. Physiol. Pharmacol. 56, 674–677.

Regoli, D., Marceau, F. and Lavigne, J. (1981). Induction of B_1-receptors for kinins in the rabbit by a bacterial lipopolysaccharide. Eur. J. Pharmacol. 71, 105–115.

Roslan, R., Campbell, E.A. and Dray, A. (1995). The induction of bradykinin B_1 receptors in the non-inflamed and inflamed rat urinary bladder. Br. J. Pharmacol. 114, 228P.

Sawutz, D.G., Faunce, D.M., Houck, W.T. and Haycock, D. (1992a). Characterization of bradykinin B_2 receptors on human IMR-90 lung fibroblasts: stimulation of $^{45}Ca^{2+}$ efflux by DPhe^7 substituted bradykinin analogues. Eur. J. Pharmacol. Mol. Pharmacol. 227, 309–315.

Sawutz, D.G., Singh, S.S., Tiberio, L., Koszewski, E., Johnson, C.G. and Johnson, C.L. (1992b). The effect of TNF on bradykinin receptor binding, phosphatidylinositol turnover and cell growth in human A431 epidermoid carcinoma cell. Immunopharmacology 24, 1–10.

Sawutz, D.G., Salvino, J.M., Dolle, R.E., Casiano, F., Ward, S.J., Houck, W.T., Faunce, D.M., Douty, B.D., Baizman, E., Awad, M.M.A., Marceau, F. and Seoane, P.R. (1994). The nonpeptide WIN 64338 is a bradykinin B_2 receptor antagonist. Proc. Natl Acad. Sci. USA 91, 4693–4697.

Schild, H.O. (1973). Receptor classification with special reference to β-adrenergic receptors. In "Drug Receptors" (ed. H.P. Rang), pp. 29–36. University Park Press, Baltimore.

Schneck, K.A., Hess, J.F., Stonesifer, G.Y. and Ransom, R.W. (1994). Bradykinin B_1 receptors in rabbit aorta smooth muscle cells in culture. Eur. J. Pharmacol. Mol. Pharm. 266, 277–282.

Siebeck, M., Whalley, E.T., Hoffmann, H., Weipert, J. and Fritz, H. (1989). The hypotensive response to desArg^9-bradykinin increases during *E. coli* septicemia in the pig. Adv. Exp. Med. Biol. 247B, 389–393.

Skidgel, R.A. (1988). Basic carboxypeptidases: regulators of peptide hormone activity. Trends. Pharmacol. Sci. 9, 299–304.

Smith, J.A.M., Webb, C., Holford, J. and Burgess, G.M. (1995). Signal transduction pathways for B_1 and B_2 bradykinin receptors in bovine pulmonary artery endothelial cells. Mol. Pharmacol. 47, 525–534.

Stewart, J.M. (1995). Bradykinin antagonists: development and applications. Biopolymers 37, 143–155.

Strader, C.D., Fong, T.M., Tota, M.R., Underwood, D. and Dixon, R.A.F. (1994). Structure and function of G protein coupled receptors. Annu. Rev. Biochem. 63, 101–132.

Sugihara, M., Todoki, K. and Okabe, E. (1991). The contractile response of isolated lingual arteries from rabbits treated with bacterial lipopolysaccharide via the stimulation of B_1 receptors for kinins. Folia Pharmacol. Jpn. 98, 63–71.

Takano, M., Yokoyama, K., Yayama, K. and Okamoto, H. (1995). Murine fibroblasts synthesize and secrete kininogen in response to cyclic-AMP, prostaglandin E_2 and tumor necrosis factor. Biochim. Biophys. Acta 1265, 189–195.

Tokumasu, T., Ueno, A. and Oh-Ishi, S. (1995). A hypotensive response induced by desArg^9-bradykinin in young Brown/Norway rats pretreated with endotoxin. Eur. J. Pharmacol. 274, 225–228.

Tropea, M.M., Gummelt, D., Herzig, M.S. and Leeb-Lundberg, L.M.F. (1993). B_1 and B_2 kinin receptors on cultured rabbit superior mesenteric artery smooth muscle cells: receptor-specific stimulation of inositol phosphate formation and arachidonic acid release by desArg^9-bradykinin and bradykinin. J. Pharmacol. Exp. Ther. 264, 930–937.

Vavrek, R.J. and Stewart, J.M. (1985). Competitive antagonists of bradykinin. Peptides 6, 161–164.

Walker, M.J.K., Gentry, C. and Perkins, M.N. (1994). Bradykinin B_1 receptor-induced plasma extravasation following persistent local inflammation in rat: an examination of neurogenic and sympathetic components. Br. J. Pharmacol. 113, 27P.

Walker, K., Dray, A. and Perkins, M.N. (1995). Hyperalgesia and fever in rats following i.c.v. administered lipopolysaccharide: effects of bradykinin B_1 and B_2 receptor antagonists. Br. J. Pharmacol. 115, 95P.

Wiemer, G. and Wirth, K. (1992). Production of cyclic GMP *via* activation of B_1 and B_2 kinin receptors in cultured bovine aortic endothelial cells. J. Pharmacol. Exp. Ther. 262, 729–733.

Wirth, K., Hock, F.J., Albus, U., Linz, W., Alpermann, H.G., Anagnostopoulos, H., Henke, S., Breipohl, G., König, W., Knolle, J. and Schölkens, B.A. (1991). Hoe 140, a new potent and long acting bradykinin-antagonist: *in vivo* studies. Br. J. Pharmacol. 102, 774–777.

Wirth, K.J., Schölkens, B.A. and Wiemer, G. (1994). The bradykinin B_2 receptor antagonist WIN 64338 inhibits the effect of desArg^9-bradykinin in endothelial cells. Eur. J. Pharmacol. Mol. Pharmacol. 288, R1–R2.

9. Kinins and Pain

Andy Dray *and* Martin Perkins

1. Introduction

Under normal circumstances, pain serves to warn animals against stimuli that are potentially tissue damaging. In most acute pain, there is usually negligible tissue damage, and the physiological features of pain signaling are transient and reversible. Pain signals are generated in fine afferent C- and Aδ nerve fibers, which respond to a range of intense physiological stimuli including heat, cold and potentially noxious chemicals. Indeed all tissues, with the exception of the neuropil of the central nervous system (CNS), are innervated by such afferent fibers, although their properties differ markedly depending on whether they are somatic afferents (innervating skin, joints, muscles) or visceral afferents (innervating cardiovascular or respiratory tissues, the gastrointestinal tract, or renal and reproductive systems). A proportion of somatic afferent fibers, the polymodal nociceptors, may be specialized for this purpose, often responding to a range of noxious stimuli.

In the viscera there is less evidence for afferents that serve specific nociceptive functions under normal physiological conditions. However, when significant tissue damage occurs, pain is often more persistent and is associated with inflammation and hyperalgesia (characterized by a decreased pain threshold, enhanced pain to suprathreshold stimuli, and often spontaneous pain) around the inflamed region. Activation and sensitization of peripheral nociceptors by chemical mediators partially accounts for this, with an important contribution from central processes leading to a hypersensitivity, which allows signals from normally innocuous stimuli, such as gentle stroking (mediated by large Aβ fibers), to be perceived in the CNS as pain (Woolf, 1994). Pain produced upon minor tissue injury is also a normal protective response and resolves once the injury has healed. More disturbing, however, is the persistent pain arising from chronic pathological lesions or degenerative processes, although, in many cases, no discernible pathology is evident. The chronic pain states that occur with migraine, rheumatoid arthritis, osteoarthritis, low back pain, cancer pain and neuropathic pain are poorly understood, and, for the most part, difficult to treat with currently available analgesics.

Inflammation is a common and complex feature of clinical pain resulting from injury or tissue degeneration. Mediators produced during inflammation are responsible for the events that occur including hyperalgesia, alterations in cell phenotype, and the expression of new molecules (neurotransmitters, enzymes, ion channels and receptors) in the peripheral and central nervous systems (Devor et al., 1993; Levine et al., 1993; McMahon et al., 1993; Dray 1994; Dray et al., 1994; Woolf and Doubell, 1994). Characterization of these changes has identified a number of new molecular targets that may lead to novel treatments for pain.

1.1 THE KININS

Among the inflammatory mediators, the production and actions of kinins are viewed as critical for the initiation of pain, and the exaggeration of sensory signaling that produce allodynia (painful responses to stimuli that are not normally painful) and hyperalgesia (Levine et al., 1993; Dray, 1995; Rang and Urban, 1995). In this chapter, we review the evidence for the pathophysiological role of kinins in the peripheral and central nervous systems, and indicate some of the advances made towards design of new analgesics which act through blockade of kinin receptors.

2. Kinins in the Periphery

2.1 FORMATION AND DEGRADATION OF KININS

Kinins promote all features of the acute inflammatory response including pain, increased blood flow and edema at the site of injury. Two major biochemical cascades produce kinins, one in plasma and a separate system in tissues (see Chapters 2, 4 and 5 of this volume). In the cardiovascular system, kinin formation is initiated following the binding and subsequent activation of Factor XII (Hageman factor) to negatively charged surfaces such as damaged basement membranes or to endotoxins. Concomitantly, high molecular weight kininogen (HMWK), which circulates bound to the inactive form of the plasma kallikrein enzyme, prekallikrein, binds to the vascular endothelial surface. HMWK is the precursor for bradykinin (BK), the most abundant kinin formed in plasma. The activated Factor XII cleaves off the active enzyme,

plasma kallikrein, which then liberates BK from HMWK (reviewed by Bhoola et al., 1992). In tissues, the precursor for the production of kinins (mainly lysyl-BK or kallidin) is low molecular weight kininogen (LMWK) (Bhoola et al., 1992). Proteolytic enzymes, liberated during trauma, activate tissue prekallikrein to form kallikrein, which then liberates Lys-BK or BK in some species from LMWK. Kinins are also produced during acute inflammation following the release of cellular proteases from immune cells (mast cells, basophils) (Bhoola et al., 1992).

BK and Lys-BK are degraded rapidly by kininases such as carboxypeptidases M and N, which cleave off the carboxy-terminal arginine to generate $desArg^9$-BK and Lys-$desArg^9$-BK, respectively, the major kinin metabolites with significant biological activity (see Chapters 7 and 8). Other peptidases such as kininase II (angiotensin-converting enzyme, ACE) and neutral endopeptidase, remove the carboxy-terminal dipeptide Phe^8-Arg^9, with ACE also hydrolyzing the Ser^6-Pro^7 bond to yield biologically inactive metabolites (see Chapter 7).

2.2 KININ RECEPTORS

BK and Lys-BK mediate their effects via two receptor subtypes, B_1 and B_2 receptors, which have been characterized using a variety of peptide agonists and antagonists (Hall, 1992; see Chapter 2). The genes for B_1 and B_2 receptors have been cloned, showing them to be G protein-coupled, seven transmembrane-domained receptors (McEachern et al., 1991; Hess et al., 1992; Menke et al., 1994; see Chapter 3). The preferred B_2 receptor agonist is BK, which mediates the majority of pharmacological effects attributed to kinins. Lys-BK also acts via B_2 receptors, both directly and after its conversion to BK by aminopeptidases. Several selective B_2 receptor antagonists have been developed and, for the most part, are synthetic peptide analogs including DArg-[Hyp^3,DPhe^7]-BK (NPC 567) and DArg-[Hyp^3,Thi^5,DTic^7,Tic^8]-BK (NPC 16731; Farmer and Burch, 1992), DArg-[Hyp^3,Thi^5,DTic^7,Oic^8]-BK (Hoe 140 or Icatibant; Hock et al., 1991) and the dimeric peptides (e.g., CP-0127 or Bradycor; Cheronis et al., 1992). These antagonists have been critical for evaluating the role of kinins in pain and hyperalgesia. A nonpeptide B_2 receptor antagonist, Win 64338, has also been described (Sawutz et al., 1994; see Chapter 2), but there is little information concerning its anti-inflammatory or analgesic properties.

Less is known about the molecular characteristics of the B_1 receptor, although it has little homology with the B_2 receptor (Menke et al., 1994). DesArg^9-BK and Lys-desArg^9-BK have greater affinity for B_1 receptors than the parent peptides (Hall, 1992). The prototype B_1 receptor antagonist is desArg^9-[Leu^8]-BK, a synthetic analog. More potent B_1 receptor antagonists include Lys-desArg^9-[Leu^8]-BK and desArg^{10}-Hoe 140 (Hall, 1992; Marceau,

1995; see Chapter 2). Although B_1 receptors are expressed constitutively in some tissues, in others B_1 receptor expression appears to be upregulated or possibly synthesized *de novo*, under the influence of inflammatory mediators and growth factors (see Chapter 8). The increased expression of B_1 receptors has been suggested to have a more prominent role in certain types of inflammatory hyperalgesia (Dray and Perkins, 1993), as discussed below.

2.3 KININ-INDUCED ACTIVATION OF NOCICEPTORS

Kinins are amongst the most potent endogenous algogenic (pain-producing) substances and BK, in particular, causes pain by direct stimulation of nociceptors (C and Aδ fibers) innervating many tissues (skin, joint, muscle, tooth pulp, viscera) (Handwerker, 1976; Kumazawa and Mizumura, 1980; Manning *et al.*, 1991; Meller and Gebhart, 1992; Mense 1993; Schaible and Grubb, 1993; Ahlquist and Franzen, 1994; Kingden-Milles *et al.*, 1994). In addition to direct activation of sensory fibers, and possibly of greater significance to the pathophysiology of pain, BK induces sensitization of sensory fibers to physical (heat, mechanical; Stevens *et al.*, 1995) and chemical (Rang *et al.*, 1991, 1994) stimuli. Indeed, there is a strong synergy between the actions of BK and other inflammatory algogens such as prostaglandins, 5-hydroxytryptamine (5-HT) and cytokines. Sensitization may occur through enhanced excitability of sensory fibers, owing to lowering of the activation threshold of exogenous stimuli, or be caused by the prolongation of discharges following fiber activation. BK also induces sensitization indirectly by stimulating the release of other inflammatory mediators from a variety of tissues (Devillier *et al.*, 1989; Dray, 1994). Stimulation by

factors such as these also allows the expression of B_1 receptors (Davis *et al.*, 1994; Dray and Perkins, 1993), and BK itself may be an important source of desArg^9-BK, the most abundant B_1 receptor agonist. BK also excites postganglionic sympathetic nerve fibers during inflammation, and releases prostanoids and other mediators which sensitize nociceptors. This mechanism is important in the mediation of BK-induced hyperalgesia (Taiwo and Levine, 1988), although, in skin, this aspect of BK-induced sensitization is controversial (Meyer *et al.*, 1992). In addition BK-induced activation of sympathetic nerves is important for inducing plasma extravasation via release of secondary mediators or via regulation of the blood flow (Green *et al.*, 1992). Various factors involved in BK-induced pain and hyperalgesia are summarized in Fig. 9.1.

2.3.1 Direct Activation of Afferent Fibers via Stimulation of B_2 Receptors

Direct activation of sensory fibers by BK has been indicated strongly in experiments involving intra-arterial (i.a.) injection of the peptide. This route of drug administration avoids secondary effects due to stimulation of other tissues. Indeed, i.a. administration of BK induces a prompt increase in the activity of afferent fibers and initiates pseudomotor reflexes indicative of pain generation (Meller and Gebhart, 1992; Mense, 1993; Schaible and Grubb 1993; Messlinger *et al.*, 1993). The direct activation of afferent fibers is indicated further from studies *in vitro* where there are minimal tissue barriers to prevent direct access of BK to afferent fibers. Thus, application of BK to quiescent peripheral neurons (C and Aδ fibers) induces prompt depolarization of sensory neuronal membranes and generation of action potentials in fine fibers (Burgess *et al.*, 1989a; Dray *et al.*, 1992; Fox *et al.*, 1993). Indeed, there may be differences in the type of sensory neuron activated by BK, since a

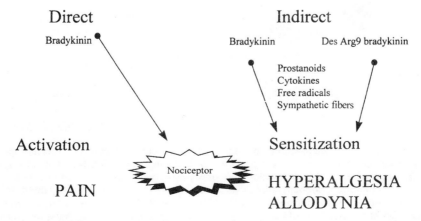

Figure 9.1 Bradykinin activates nociceptors directly to induce pain. There are, however, many indirect mechanisms activated by bradykinin and its metabolite, desArg^9-bradykinin, which induce sensitization of nociceptors leading to hyperalgesia and allodynia.

recent study showed that BK excited and depolarized a subpopulation of neurons that were insensitive to tetrodotoxin (TTX) (Jeftinija, 1994).

Direct activation of nociceptors by BK is via B_2 receptors as indicated from studies with selective antagonists (Hall, 1992; Dray and Perkins, 1993). This receptor subtype is of major importance in producing pain in pathological conditions as several B_2 receptor antagonists are antinociceptive in a variety of experimental models (Steranka et al., 1988; Haley et al., 1992; Perkins et al., 1992; Heapy et al., 1993).

The main intracellular pathway of B_2 receptor signal transduction involves G protein-mediated activation of phospholipase C (PLC), which generates the second messengers, inositol 1,4,5-trisphosphate (IP_3) and diacylglycerol (DAG) by cleavage of membrane phospholipids (Thayer et al., 1988; Burgess et al., 1989a; Gammon et al., 1989). IP_3 stimulates the release of intracellular Ca^{2+} and a rise in cytosolic free Ca^{2+} ($[Ca^{2+}]_i$) (Thayer et al., 1988). DAG activates protein kinase C (PKC), leading to the phosphorylation of various intracellular proteins (Shearman et al., 1989). Indeed, BK-induced cell activation is mimicked by phorbol esters, which activate PKC (Burgess et al., 1989a; Dray et al., 1992; Schepelmann et al., 1993), while PKC inhibitors reduce responses to BK (Dray et al., 1992). In addition, BK-induced nerve depolarization and Ca^{2+} entry are reduced or abolished by inhibition or downregulation of PKC (Burgess et al., 1989a). This may be a mechanism underlying receptor desensitization as discussed below.

While there are a number of Ca^{2+}-dependent effects of BK on sensory neurons, such as neuropeptide release (Geppetti et al., 1990), activation of nitric oxide (NO) synthase, or generation of intracellular cyclic guanosine monophosphate (cGMP), these require extracellular Ca^{2+} (Burgess et al., 1989b). The release of intracellular Ca^{2+} in these cells is a much less effective stimulus than Ca^{2+} entry through voltage-activated channels. Moreover, the mechanism of increased membrane excitability is not likely to involve the release of intracellular Ca^{2+} by IP_3, as this seems to have relatively little effect on sensory neurons. Although BK stimulates voltage-gated Ca^{2+} entry into sensory neurons, this is secondary to membrane depolarization caused, primarily, by an increase in neuronal plasma membrane permeability to Na^+. Indeed, substitution of extracellular Na^+, rather than Ca^{2+}, abolishes the increase in membrane current evoked by BK (Burgess et al., 1989a). The implication of these observations is that the increase in Na^+ conductance can occur through ion channels coupled with the B_2 receptor, and via PKC-induced activation of Na^+ channels. Although there is no direct evidence for this in sensory neurons, other observations have shown that B_2 receptors activation of small sensory neurons was insensitive to TTX, supporting the involvement of receptor-coupled channels, or that activation was due to increased permeability to Na^+ via a specific type of TTX-resistant channel regulated by membrane depolarization (Jeftinija, 1994) (Fig. 9.2).

Interestingly, staurosporine, a PKC inhibitor, does not

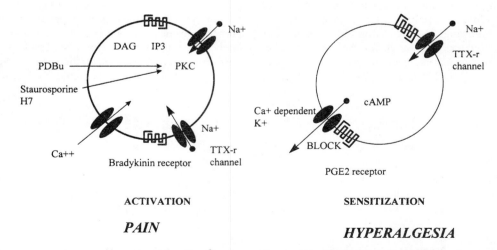

ACTIVATION

PAIN

SENSITIZATION

HYPERALGESIA

Figure 9.2   The left panel shows mechanisms of bradykinin-induced activation of nociceptors. B_2 receptor activation produces a number of second messengers including diacylglycerol (DAG) and inositol 1,4,5-trisphosphate (IP_3). DAG stimulates protein kinase C (PKC), which phosphorylates membrane ion channels. This leads to increased Na^+ conductance and membrane depolarization. Ca^{2+} influx occurs through voltage-activated Ca^{2+} channels. B_2 receptors may also activate tetrodotoxin-resistant (TTX-r) Na^+ channels to induce neuronal depolarization. The right panel shows other indirect mechanisms of increasing neural excitability mediated by receptors for prostaglandin E_2 (PGE_2). The production of cAMP inhibits Ca^{2+}-dependent K^+ conductance, thereby blocking the prolonged spike after-hyperpolarization, normally responsible for regulating neuronal firing rate. In addition, PGE_2 can enhance Na^+ conductance through TTX-resistant channels.

inhibit BK-evoked depolarization of the central terminals of afferent fibers in the spinal cord (Dunn and Rang, 1990) suggesting that PKC activation may not be the only mechanism involved in the effects of BK. Moreover, BK inhibits voltage-gated Ca^{2+} currents in sensory neurons, an action also attributed to be mediated via PKC (Ewald *et al.*, 1988). Although this is hard to reconcile with the ability of BK to activate sensory neurons or evoke neuropeptide release (Andreeva and Rang, 1993), it is possible that regulation of membrane excitability may depend on several mechanisms, which alter the phosphorylation state of several ion channels.

Other possible mechanisms for BK-induced activation of sensory nerves may be through the direct or indirect generation of arachidonic acid and prostanoids via PLC (Gammon *et al.*, 1989; Allen *et al.*, 1992). Thus, activation of nociceptive afferent fibers *in vitro* is attenuated by indomethacin supporting a role for prostanoids in the response to BK (Dray *et al.*, 1992). Indeed, PGE_2 causes a direct increase in membrane permeability to Na^+ ions in sensory neurons (Puttick, 1992), thereby causing neuronal excitation (Schaible and Schmidt, 1988; Birrell *et al.*, 1991) and the release of substance P (Nicol and Cui, 1992). An additional mechanism that may contribute to sensory neuronal activation by prostanoids is the enhancement and prolongation of Na^+ conductance, particularly that mediated via TTX-resistant channels found only in small sensory neurons (Gold *et al.*, 1996).

BK-mediated sensory neuron activation may involve the generation of NO. Small- and medium-sized sensory neurons contain constitutive NO synthase (Aimi *et al.*, 1991; Verge *et al.*, 1992) and biosynthesis of NO occurs during cell activation by increased $[Ca^{2+}]_i$ following stimulation by BK (Bauer *et al.*, 1993). One action of NO is to activate satellite cells in the dorsal root ganglia, since NO donors increase cGMP in these cells (Morris *et al.*, 1992). There is, however, little evidence for direct activation of sensory neurons by NO under normal conditions (McGehee *et al.*, 1992), although intradermal injection of NO induced a delayed burning pain in man (Holthusen and Arndt, 1994). On the other hand, NO donors such as sodium nitroprusside have been postulated to activate cerebral sensory fibers directly, causing release of calcitonin gene-related peptide (CGRP), a potent vasodilator (Wei *et al.*, 1992). Thus, the release of NO may contribute to migraine and other types of head pain (Olesen *et al.*, 1994). Conversely, NO alters the excitability of sensory neurons indirectly by changing their responsiveness to BK. Thus, desensitization to repeated administrations of BK was reduced by NO synthase inhibitors, and increased by activation of guanylate cyclase or cGMP, which mediates the intracellular effects of NO (McGehee *et al.*, 1992; Rueff *et al.*, 1994; Bradley and Burgess, 1993). The molecular mechanisms of these interactions are not yet fully understood.

2.4   DESENSITIZATION TO BRADYKININ

A notable feature of BK-induced activation of sensory neurons or afferent fibers is rapid desensitization, particularly upon repeated administration (Dray *et al.*, 1992). Apart from the rapid degradation of BK *in vivo*, this mechanism may also explain why BK-induced pain is relatively transient. However, this observation is a powerful argument against a significant role for BK in persistent pain. This is unlikely to be the case during injury when BK is produced continuously, and acts with prostanoids and cytokines, or is converted to desArg^9-BK, which acts on B_1 receptors.

B_2 receptor desensitization involves receptor phosphorylation by specific kinases. Thus, B_2 receptor desensitization, measured by changes in membrane conductance or IP_3 production in sensory neurons was enhanced by activation of cGMP-dependent kinase through the production of NO (McGehee *et al.*, 1992; Bradley and Burgess, 1993; Rueff *et al.*, 1994). There is little or no evidence supporting an involvement of cyclic adenosine monophosphate (cAMP) in desensitization to BK (Rueff *et al.*, 1994). On the contrary, increases in intracellular cAMP have been proposed as an important mechanism in the induction of hyperalgesia via sensitization of sensory neurons (Levine *et al.*, 1993; Rang *et al.*, 1994).

Desensitization to BK likely involves uncoupling of the receptor-effector mechanism as well as aggregation and internalization of B_2 receptors (Roscher *et al.*, 1984; Roberts and Gullick, 1990). Concanavalin A, which cross-links membrane proteins, preventing receptor aggregation, inhibits desensitization to BK (Rueff *et al.*, 1994). Moreover, drugs such as phenyl arsine oxide, which prevent internalization of membrane proteins, also inhibit desensitization (Roscher *et al.*, 1984).

2.5   BRADYKININ-INDUCED SENSITIZATION OF SENSORY NEURONS

In many types of cells, BK activates phospholipase A_2 directly to generate prostaglandins, which are powerful and prolonged sensitizing agents in sensory neurons. In keeping with this, cyclooxygenase inhibitors such as indomethacin reduce the excitatory effect of BK on nociceptors. A number of mechanisms have been proposed to underlie this chemical sensitization of afferent neurons. Thus, in visceral sensory neurons (nodose ganglion cells) and, to some extent, in somatic afferent neurons (Undem and Weinreich, 1993) increased excitability is often associated with inhibition of a long-lasting spike after-hyperpolarization (slow AHP). The slow AHP is powerfully regulated by a cAMP-dependent, Ca^{2+}-activated K^+ conductance mechanism, and following a single action potential, normally produces a state of reduced excitability, which limits the number of action potentials that can be evoked upon nerve stimulation. Prostaglandins and BK (through prostanoid formation), inhibit the slow AHP by stimulating cAMP formation, thereby allowing the cell to fire repetitively following initial activation (Weinreich and Wonderlin, 1987). This sensitization mechanism may be common to a number of hyperalgesic substances which generate intracellular cAMP.

As mentioned above, an enhancement and prolongation of membrane excitability, the likely mechanisms for nerve fiber sensitization, can also be produced by alterations in TTX-resistant Na^+ channels. Thus, PGE_2 enhances TTX-resistant but not TTX-sensitive Na^+ currents evoked upon depolarization of small sensory neurons maintained in tissue culture. Both the magnitude and the rate of current inactivation was increased suggesting a mechanism for increased cell firing and the consequent production of hyperalgesia after nociceptor stimulation. In addition, the delay to the increase of the Na^+ current produced by PGE_2 and the prolonged duration of this effect suggested that a secondary diffusable intracellular messenger was involved (Gold et al., 1996) (Fig. 9.2).

2.6   BRADYKININ AND CYTOKINE PRODUCTION

There is accumulating evidence that cytokines are important mediators of hyperalgesia and that their involvement is triggered by kinin receptor activity. Thus, hyperalgesia induced by administration of the interleukins (IL), IL-1β, IL-6 and IL-8, or tumor necrosis factor (TNF-α), as well as that induced by carrageenin, bacterial endotoxin (lipopolysaccharide, LPS) and ultraviolet irradiation, are attenuated by specific antibodies to these cytokines or by an IL-1β receptor antagonist (Cunha et al., 1991, 1992; Davis and Perkins, 1994).

Cytokines, particularly IL-1β, induce and upregulate B_1 receptor-mediated responses (Marceau, 1995; see Chapter 8) in vitro and in vivo. A synergistic action of IL-1β on BK B_2 receptor-mediated responses has also been reported (O'Neil and Lewis, 1989; Vandekerchkove et al., 1991; Bathon et al., 1992; Lerner and Modeer, 1991).

With respect to hyperalgesia and pain there is evidence that the interaction between cytokines and kinins is bidirectional. BK-induced hyperalgesia is blocked to varying extents by antisera to TNF-α, IL-1β, IL-6 and IL-8 (Ferreira et al., 1993). Conversely, cytokine-induced hyperalgesia is reversed by both B_1 and B_2 receptor antagonists (Davis and Perkins, 1994; Davis et al., 1994; Perkins et al., 1994). There is, therefore, the potential for a powerful positive feedback loop involving kinins and cytokines leading to sustained hyperalgesia. This "cytokine axis" may be important in maintaining a hyperalgesic state in chronic inflammation. The complexity of such interactions is further highlighted by a recent report showing that one of the "inhibitory" cytokines, IL-10, limits the hyperalgesia induced by the "excitatory" cytokines such as IL-1β, as well as BK (Poole et al., 1995). This is an area where considerably more work is needed to elucidate the precise nature of the interactions between the kinin and cytokine systems in pathophysiology.

2.7   OTHER INDIRECT MECHANISMS OF BRADYKININ-INDUCED SENSITIZATION

Although BK can cause both constriction and relaxation of blood vessels via B_1 or B_2 receptor stimulation (Hall, 1992), it is the receptor-mediated increase in vascular permeability, which enables the extravasation of blood constituents and the release of several neurogenic factors including 5-HT and histamine. These biogenic amines in turn induce the activation and sensitization of nociceptors (Beck and Handwerker, 1974; Dray, 1994; Rang et al., 1994).

Neuropeptide and prostanoid release occurs through BK-induced stimulation of sensory and sympathetic neurons. The neurokinins, substance P and neurokinin A, cause plasma extravasation by stimulating vascular endothelial cells and induce vasodilatation by the release of NO. CGRP release induces a powerful dilation of arterioles, facilitating plasma extravasation by increasing local blood flow (Brain and Williams, 1985). Activation of sympathetic neurons by BK involves inhibition of an M-type K^+ current. This is mediated via B_2 receptors and is blocked by Hoe 140, a potent B_2 receptor antagonist (Babbedge et al., 1995; Jones et al., 1995; Seabrook et al., 1995). Indeed, sympathetic neurons may be involved in BK-mediated mechanical hyperalgesia (Taiwo et al., 1990). However, sympathetic activation may not account entirely for heat hyperalgesia induced by BK, which is unaffected by sympathectomy (Kolzenburg et al., 1992; Meyer et al., 1992). More recently, BK-induced hyperalgesia was shown to be attenuated by indomethacin or by afferent fiber inactivation with capsaicin, but unaffected by chemical sympathectomy (Schuligoi et al., 1994). These findings indicate that the contribution of sympathetic fibers to BK-induced hyperalgesia may require specific conditions and may be restricted to specific types of afferent neurons.

Kinins cause degranulation of mast cells (Ishizaka et al., 1985; Devillier et al., 1989) to release several mediators of inflammatory hyperalgesia including histamine and 5-HT. This mechanism does not require specific receptor stimulation but, rather, the activation of surface molecules to stimulate PLC. The consequent mobilization of Ca^{2+} induces histamine release (Bueb et al., 1990). Kinins also affect immune cell mobility (Kay and

Kaplan, 1975) and may be involved in the recruitment of immune cells to the site of injury and inflammation. By doing so they provide additional sources of endogenous hyperalgesic agents released from macrophages and neutrophils including free radicals, cytokines and arachidonic acid (see also Chapter 11).

2.8 B₁ RECEPTORS AND HYPERALGESIA

Increasing attention has recently been focused on a role for B_1 receptors in inflammatory hyperalgesia (Dray and Perkins, 1993). Although B_1 receptors are expressed constitutively in some tissues, such as arterial smooth muscle (Bouthillier *et al.*, 1987; Rhaleb *et al.*, 1990; Campos and Calixto, 1993), in general, they are not normally encountered. Instead, B_1 receptor expression appears to be increased during inflammation or infection. This is due to immune cell products such as cytokines, particularly IL-1β (Galizzi *et al.*, 1994; Marceau, 1995) (this is discussed in detail in Chapter 8), and may involve the *de novo* synthesis of new receptors. It is also possible, however, that cytokines unmask existing receptors or facilitate B_1 receptor-effector coupling. The significance of increased receptor expression is not clear, but B_1 receptors are important contributors to hyperalgesia during inflammation in response to various agents. The contribution of B_1 receptors can be measured by the increased efficacy and/or potency of B_1 receptor agonists in exacerbating hyperalgesia, and by the finding that B_1 receptor antagonists produce analgesia (Dray and Perkins, 1993; Cruwys *et al.*, 1994; Davis and Perkins,

1994; Perkins and Kelly, 1993). Of further significance is the observation that desArg⁹-BK, which activates B_1 receptors selectively, is present in greater abundance than BK during inflammation (Burch and DeHaas, 1988). This may be due to accelerated conversion of BK to desArg⁹-BK by the increased activity of proteolytic enzymes (see Chapter 7). Studies have indicated that desArg⁹-BK does not activate sensory neurons directly (Dray *et al.*, 1992; Nagy *et al.*, 1993) and this mechanism is unlikely, therefore, to contribute to B_1 receptor agonist-induced hyperalgesia. B_1 receptor-mediated hyperalgesia is attenuated by indomethacin, indicating that it is mediated indirectly by prostaglandins (Davis and Perkins, 1994), or via TNF-α and IL-1β released from leukocytes such as macrophages (Tiffany and Burch, 1989) (Fig. 9.3).

3. Kinins in the Central Nervous System

3.1 DO KININS CAUSE HYPERALGESIA VIA ACTIONS IN THE CNS?

Clinical reports of pain have been associated with many types of CNS disease including traumatic injury to the brain or spinal cord, stroke, multiple sclerosis, tumors, epilepsy, Parkinson's disease and a variety of CNS infections (Bonica, 1991; Tasker *et al.*, 1991; Boivie, 1994). Indeed, central pain syndromes can result from almost any type of CNS pathology as well as some

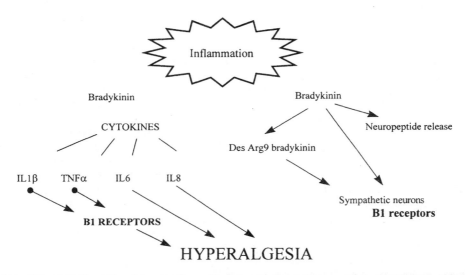

Figure 9.3   **BK production stimulates the formation of several cytokines, which contribute to inflammatory hyperalgesia in different ways. IL-1β and TNFα increase the expression of B₁ receptors, which make a significant contribution to hyperalgesia. Bradykinin also stimulates the release of sensory neuropeptides, which act on sympathetic fibers and the microvasculature. BK is also the important source of desArg⁹-bradykinin, which is a specific agonist at B₁ receptors.**

affective disorders involving anxiety and stress (e.g., fibromyalgia pain, irritable bowel disease). These types of pain appear idiosyncratic, difficult to rationalize and are very difficult to treat (Tasker *et al.*, 1991). Notably, central pain syndromes occur in the absence of any damage to the peripheral nervous system or peripheral tissues. However, while the processing of sensory signals by peripheral nerves appears normal (although this has not been examined in detail), central pain syndromes induce sensations of allodynia and hyperalgesia (Casey, 1991).

3.2   THE OCCURRENCE OF KININS AND KININ BINDING SITES IN THE CNS

Relevant to the present discussion is that the CNS contains all the components of the kallikrein–kinin system necessary for kinin production. Thus, kallikrein and kinins occur in the mammalian brain (Perry and Snyder, 1984; Scicli *et al.*, 1984), and BK-like immunoreactivity has been shown in various regions (Kariya *et al.*, 1985). In addition, a variety of kininases, including aminopeptidase, enkephalinase and kininases have been identified in the brain (Camargo *et al.*, 1973; see Chapter 7).

Although BK binding sites have been localized to cerebral blood vessels, B_2 receptor binding sites have also been identified in other areas of the brain, such as the caudate, subfornical organ and the red nucleus, as well as in the superficial layers of the spinal dorsal horn and in or around ventral horn cells (Correa *et al.*, 1979; Lopes *et al.*, 1995; Murone *et al.*, 1996). Moreover the rapid degradation of BK by brain synaptic membranes (Orawski and Simmons, 1989) is a further indication that kinins are involved in synaptic function. A more specific function with respect to nociception is supported by the finding that BK binding sites were located to the dorsal root and trigeminal ganglia, and on the terminals of primary sensory nerve fibers (Steranka *et al.*, 1988; Fujiwara *et al.*, 1989; Privitera *et al.*, 1992; Lopes *et al.*, 1995).

3.3   THE FUNCTIONS OF KININS IN THE CNS

Although much is known about the actions of BK on peripheral nerve fibers, the functional consequences of its actions on the central terminals of primary afferent nerves are not as well understood. On the one hand, BK does not evoke basal release of the sensory neuropeptide, CGRP, from these afferent terminals. Conversely, BK potentiates CGRP release evoked by dorsal root stimulation, an effect that is prevented by a B_2 receptor antagonist or by indomethacin, suggesting that it is mediated via prostanoid production (Dunn and Rang, 1990; Andreeva and Rang, 1993). Electrophysiological studies

of dorsal-horn nociceptive neurons also support the involvement of B_2 receptors in the responses to noxious peripheral stimulation. For example, following subcutaneous injection of formalin, a powerful irritant, these neurons exhibit an acute phase followed by a more prolonged phase of activation. Only the second phase of the formalin response, which is associated with peripheral inflammation, is reduced by a spinal intrathecal administration of a B_2 receptor antagonist (Chapman and Dickenson, 1992). Interestingly, intrathecal administration of indomethacin also inhibits both phases of the formalin-induced response, suggesting that prostanoid formation is an important factor in the excitation of dorsal-horn nociceptive neurons (Malmberg and Yaksh, 1992). It is possible, therefore, that some peripheral injuries cause B_2 receptor activation of spinal nociceptive afferent terminals and alterations in spinal excitability via prostanoid production. Although the components of the kallikrein–kinin system exist within the spinal cord, the site of kinin production has yet to be identified. Since kinin production has only been demonstrated following tissue insult or infection, it is difficult to account for BK release following the acute activation of peripheral nociceptors.

In behavioral terms, intrathecal BK produces an initial transient nociceptive response, as might be expected from activation of afferent nerve terminals, but this is followed by an apparent *analgesic* response (Laneuville *et al.*, 1989). The latter may result from an initial intense BK-induced depolarization of afferent terminals and spinal dorsal horn neurons, followed by nerve membrane inactivation and block of further afferent nerve activity by exogenous stimuli. It is likely that only high concentrations of exogenous BK produce this effect. However, activation of descending spinal noradrenergic nerve terminals has also been proposed to be involved in BK-mediated antinociception (Laneuville and Couture, 1987; Laneuville *et al.*, 1989). In keeping with this, $[^3H]$-BK binding is reduced after spinal noradrenergic deafferentation (Lopes *et al.*, 1995).

Interestingly cutaneous plasma extravasation was increased following B_2 receptor activation via spinal intrathecal BK administration, and it was suggested that BK and substance P may increase peripheral vascular permeability through a combination of sensory and vagal cholinergic mechanisms (Jacques and Couture, 1990). These observations suggest strongly that kinin-induced effects in the CNS influence the activity of peripheral nerve fibers. This may have important implications with respect to central mechanisms mediating pain and inflammation.

3.4   KININS AND CNS INFLAMMATION

Cerebral inflammatory reactions can follow many different types of insult, including head injury, cerebral ischemia, encephalomyelitis, meningitis and headache (Macfarlane *et al.*, 1991). Increased kininogen and BK formation have been measured in various models of cerebral trauma and ischemia (Maier-Hauff *et al.*, 1984; Germain *et al.*, 1986; Ellis *et al.*, 1987, 1988; Wahl *et al.*, 1993), while kallikrein inhibition with aprotinin-reduced cortical edema produced by cold lesions (Unterberg *et al.*, 1986). Although there is presently relatively little evidence linking kinin receptors with brain inflammation, B_2 receptor activation has powerful effects on brain vasculature. For example, kinins induce endothelium-dependent relaxation in canine brainstem arteries under normal conditions as well as during subarachnoid hemorrhage (D'Orléans-Juste *et al.*, 1985). These effects appear to be mediated both by NO formation and by free radicals (Kontos *et al.*, 1984; Katusic *et al.*, 1993). In addition, superfusion of the cerebral cortex or intra-carotid injections of BK induce fluid leakage from blood vessels in the pial matter, cortical edema, and stimulate the infiltration of macrophages into brain tissue (Unterberg *et al.*, 1984; Gecse *et al.*, 1989). Direct administration of BK into the brain ventricles also produces edema (Unterberg and Baethmann, 1984). The increase in permeability of cerebral blood vessels produced by BK appears to be mainly through cyclo-oxygenase activity and the generation of free radicals (Sarker and Fraser, 1996).

As in the periphery, kinin production may be important for the generation of products of arachidonic acid metabolism. Thus, kinins and arachidonic acid metabolite may activate and sensitize nociceptors, which innervate the dura, pia or blood vessels (Mayberg *et al.*, 1984; Davis and Dostrovsky, 1986, 1988; Andres *et al.*,

1987; Janig and Kolzenburg, 1991), and induce pain and a variety of pro-inflammatory effects including vasodilatation and edema (Chan and Fishman, 1984; Unterberg *et al.*, 1987).

During inflammation, immune-cell recruitment and activation occurs in the CNS. In addition, cerebral microglia play a prominent role as they represent the main immune effector-cell population of the CNS (Perry *et al.*, 1993). Microglia are capable of expressing major histocompatibility complex antigens, they show macrophage-like activity, and release cytokines, NO and free radicals (Rothwell, 1991a,b; Banati *et al.*, 1993). Since microglia and astrocytes express B_2 receptors (Gimple *et al.*, 1992; Hosli and Hosli, 1993), kinins may regulate their activity and the release of inflammatory substances. In addition, activation of microglia or astrocytes releases glutamate, which, through the activation of central excitatory amino-acid receptors, can increase excitability and, in extreme situations, induce cell death (Parpura *et al.*, 1994). Finally, astrocytes and oligodendrocytes are activated by BK to increase $[Ca^{2+}]_i$ and release arachidonic acid (Burch and Tiffany, 1989; Cholewinski *et al.*, 1991; Stephens *et al.*, 1993; He and McCarthy 1994). Figure 9.4 summarizes some of the ways in which kinins can mediate central inflammation and cause hyperalgesia.

3.5   KININS AND CENTRAL PAIN MECHANISMS

Neuropeptide-containing C-fibers, arising mainly from the trigeminal nerve, supply blood vessels, brain parenchyma and the dura mater (Mayberg *et al.*, 1984;

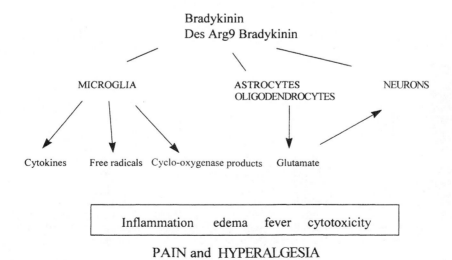

Figure 9.4   Bradykinin stimulates microglia to produce a number of secondary mediators, which contribute to central inflammation, cerebral edema and hyperalgesia. Astrocytes and oligodendrocytes can release glutamate to induce central excitation and possible neurotoxicity.

Arbab *et al.*, 1986). Activation of pial sensory fibers induces the release of substance P and CGRP (Moskowitz *et al.*, 1983; Kurosawa *et al.*, 1995), while electrical stimulation of the trigeminal ganglion or infusion of substance P or neurokinin A, produce vaso-dilatation and plasma extravasation in the dura mater, but not in brain (Markowitz *et al.*, 1987). It is possible that C-fiber activation or sensitization by BK induce similar effects. Indeed, kinin activity has been implicated in headache (Sicuteri, 1970), and B_2 receptors have been identified on cerebral blood vessels and primary afferent nerve terminals (Fujiwara *et al.*, 1989; Homayoun and Harik, 1991; Lopes *et al.*, 1995). More specifically, BK may be important in mediating the pain of migraine, as well as pain in other cerebrovascular disorders and the neurogenic inflammation following various forms of brain injury (Macfarlane *et al.*, 1991). In keeping with this, induction of cerebral inflammation by LPS produces fever (hyperthermia) (Andersson *et al.*, 1992), as well as acute peripheral hyperalgesia to peripheral thermal and mechanical stimuli (Yirmiya *et al.*, 1994; Walker *et al.*, 1996a,b). Both the hyperthermia and hyperalgesia are attenuated or abolished by local but not systemic administration of a B_2 antagonist, Hoe 140, but not by the B_1 receptor antagonists, desArg^9-[Leu^8]-BK or des-Arg^{10}-Hoe 140 (Walker *et al.*, 1996a) (Figs 9.5 and 9.6). In addition, intracerebroventricular as well as i.v.

indomethacin abolishes LPS-induced hyperthermia and hyperalgesia indicating that these acute central kinin-mediated events are secondary to prostanoid formation (Rao and Bhattacharya, 1988; Walker *et al.*, 1996b) (Figs 9.5 and 9.6). To date, although only central B_2 receptors have been implicated in causing pain, B_1 receptor involvement in chronic inflammatory conditions is possible (Germain *et al.*, 1988).

It is notable that antinociceptive effects have also been reported following spinal (Laneuville *et al.*, 1989) or intracerebral injections of BK (Ribeiro *et al.*, 1971; Ribeiro and Rocha e Silva, 1973). The periaqueductal and periventricular gray regions were suggested as possible sites for this effect (Clark, 1979; Couto *et al.*, 1995) and may involve endogenous opioids, as the BK-mediated elevation of nociceptive thresholds was inhibited by naloxone (Burdin *et al.*, 1992). In the spinal cord, however, stimulation of descending noradrenergic nerve terminals has been implicated in the antinociceptive effect.

4.   Conclusions

Kinins are likely to be among the first agents produced at sites of injury or inflammation, and they serve critical roles in signaling tissue distress as well as organizing

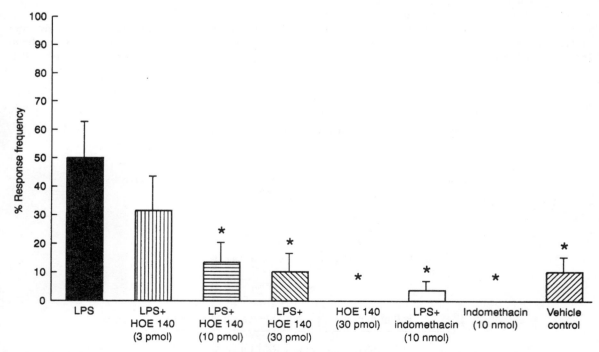

Figure 9.5   The effect of a B_2 receptor antagonist on rat hind paw withdrawal response frequency to von Frey filament stimulation 6 h following i.c.v. administration of vehicle, endotoxin (0.2 µg) alone, co-administration of LPS and Hoe 140, or Hoe 140 alone. Additional groups were treated with LPS + indomethacin or indomethacin alone. * $P < 0.05$ compared to the group treated with LPS alone ($n = 8$/group). Data are presented as mean ± s.e. mean.

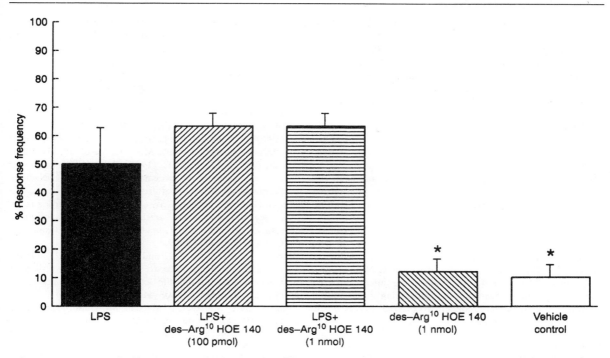

Figure 9.6 The effect of a B_1 receptor antagonist on hind paw withdrawal response frequency to von Frey filament stimulation 6 h following i.c.v. administration of vehicle (open bar), endotoxin (0.2 μg) alone (solid bar), or co-administration of LPS and desArg[10]-Hoe 140 (100 pmol, horizontal-hatched bar; 1 nmol, cross-hatched bar), or desArg[10] -Hoe 140 (1 nmol, alone, vertical-hatched bar). * $P < 0.05$ compared to the group treated with LPS alone (n = 8/group). Data are presented as mean ± s.e. mean.

responses to injury and subsequent repair. Important in this orchestrated series of events is the activation of fine afferents to signal immediate pain followed by prolonged sensitization to exogenous stimuli in order to minimize further tissue disturbance. Kinins interact specifically with B_2 receptors expressed on sensory neurons, and transduce signals that alter membrane excitability and cellular chemistry. B_2 receptor activation also stimulates the release of mediators, such as prostanoids and cytokines from immune cells, vasculature and nerves (afferent and sympathetic), and these substances also contribute to neural sensitization and hyperalgesia.

Little is known, however, about the relative contributions of different kinins and their receptors during different inflammatory conditions. For example, recent studies in mutant, knockout mice, in which the B_2 receptor was deleted, showed an expected lack of responsiveness to BK and reduced thermal hyperalgesia after carrageenin-induced inflammation but, unexpectedly, hyperalgesia induced by Freund's complete adjuvant (FCA) and formalin injection were hardly affected (Boyce *et al.*, 1996). In addition, further detailed studies on the apparent contribution of postganglionic sympathetic fibers to BK-induced hyperalgesia are required. Observations show that inflammation stimulates the expression of B_1 receptors, which also play a prominent

role in hyperalgesia. The substrates and mechanisms associated with this requires further detailed characterization.

The widespread distribution of the kinin system, as well as the localization of kinin receptors in the CNS, begs important questions about their function. Indeed, a role for BK as a central neurotransmitter has been suggested (Snyder, 1980). A role for kinins in central pathophysiology (trauma, ischemia, infection) seems more compelling, but a more detailed examination of the central pharmacology of kinins is necessary. More specifically, kinin activity has powerful effects on pain signaling within the CNS, both within the brain and spinal cord. These data suggest that kinin receptor activation can alter the excitability of afferent terminals in the spinal cord as well as neural excitability in different brain regions. These changes should be investigated with some urgency as they are likely to be important in a variety of centrally mediated pain syndromes, which are difficult to treat with conventional analgesics.

5.   References

Ahlquist, M.L. and Franzen, O.G. (1994). Inflammation and dental pain in man. Endod. Dent. Traumatol. 10, 201–209.

Aimi, Y., Fujimura, M., Vincent, S.R. and Kimura, H. (1991). Localization of NADPH-diaphorase-containing neurons in sensory ganglia of the rat. J. Comp. Neurol. 306, 382–392.

Allen, A.C., Gammon, C.M., Ousley, A.H., McCarthy, K.D. and Morell, P. (1992). Bradykinin stimulates arachidonic acid release through the sequential actions of an sn-1 diacylglycerol lipase and a monoacylglycerol lipase. J. Neurochem. 58, 1130–1139.

Andersson, P.B., Perry, V.H. and Gordon, S. (1992). The acute inflammatory response to lipopolysaccharide in CNS parenchyma differs from that in other body tissues. Neuroscience 48, 169–186.

Andreeva, L. and Rang, H.P. (1993). Effect of bradykinin and prostaglandins on the release of calcitonin gene-related peptide-like immunoreactivity from the rat spinal cord *in vitro*. Br. J. Pharmacol. 108, 185–190.

Andres, K.H., Von During, M., Muszynski, K. and Schmidt, R.F. (1987). Nerve fibres and their terminals of the dura mater encephali of the rat. Anat. Embryol. 175, 289–301.

Arbab, M.A.R., Wiklund, L. and Svendgaard, N.A. (1986). Origin and distribution of cerebral vascular innervation from superior cervical, trigeminal and spinal ganglia investigated with retrograde and anterograde WGA-HRP tracing in the rat. Neuroscience 19, 695–708.

Babbedge, R., Dray, A. and Urban, L (1995). Bradykinin depolarizes the rat isolated superior cervical ganglion via B$_2$ receptor activation. Neurosci. Lett. 193, 161–164.

Banati, R.B., Gehrmann J., Schubert P. and Kreutzberg G.W. (1993). Cytotoxicity of microglia. Glia 7, 111–118.

Bathon, J.M., Croghan, J.E., Goldman, D.W., MacGlashan, D.W. and Proud, D. (1992). Modulation of kinin responses in human synovium by interleukin-1. Agents Actions Suppl. 38, 16–22.

Bauer, M.B., Simons, M.L. Murphy, S. and Gebhart, G.F. (1993). Bradykinin and capsaicin stimulate cyclic GMP production in cultured dorsal root ganglion neurons via a nitrosyl intermediate. J. Neurosci. Res. 36, 280–289.

Beck, P.W. and Handwerker, H.O. (1974). Bradykinin and serotonin effects on various types of cutaneous nerve fibres. Pflügers Arch. 347, 209–222.

Bhoola, K.D., Figueroa, C.D. and Worthy, K. (1992). Bio-regulation of kinins: kallikreins, kininogens, and kininases. Pharmacol. Rev. 44, 1–80.

Birrell, G.J., McQueen, D.S., Iggo, A., Coleman, R.A. and Grubb, B.D. (1991). PGI$_2$-induced activation and sensiti-zation of articular mechanonociceptors. Neurosci. Lett. 124, 5–8.

Boivie, J. (1994). Central pain. In "Textbook of Pain", 3rd edn (eds. P.D. Wall and R. Melzack), pp. 871–902. Churchill Livingstone, London.

Bonica, J.J. (1991). Semantic, epidemiologic, and educational issues. In "Pain and Central Nervous System Disease: The Central Pain Syndromes" (ed. K.L. Casey), pp. 13–29. Raven Press, New York.

Bouthillier, J., Deblois, D. and Marceau, F. (1987). Studies on the induction of pharmacological responses to des-Arg^9-bradykinin *in vitro* and *in vivo*. Br. J. Pharmacol. 92, 257–264.

Boyce, S., Rupniak, N.M.J., Carlson, E.J., Webb, J.K., Borkowski, J.A., Hess, J.F., Strader, C.D. and Hill, R.G. (1996). Nociception and inflammatory hyperalgesia in B$_2$ bradykinin receptor knockout mice. Br. J. Pharmacol. 117, Proc. Suppl. 245P.

Bradley, C. and Burgess, G. (1993). A nitric oxide synthase inhibitor reduces desensitisation of bradykinin-induced activation of phospholipase C in sensory neurones. Trans. Biochem. Soc. 21, 4353.

Brain, S.D. and Williams, T.J. (1985). Inflammatory oedema induced by synergism between calcitonin gene related peptide and mediators of increased vascular permeability. Br. J. Pharmacol. 86, 855–860.

Bueb, J.L., Mousli, M., Bronner, C., Routot, B. and Landry, Y. (1990). Activation of G$_i$-like proteins, a receptor independent effect of kinins in mast cells. Mol. Pharmacol. 38, 816–822.

Burch, R.M. and DeHaas, C. (1988). A bradykinin antagonist inhibits carageenin oedema in the rat. Naunyn-Schmiedeberg's Arch. Pharmacol. 342, 189–193.

Burch, R.M. and Tiffany, C.W. (1989). Tumor necrosis factor causes amplification of arachidonic acid metabolism in response to interleukin 1, bradykinin, and other agonists. J. Cell. Physiol. 141, 85–89.

Burdin, T.A., Graeff, F.G. and Pela, I.R. (1992). Opioid mediation of the antiaversive and hyperalgesic actions of bradykinin injected into the dorsal periaqueductal gray of the rat. Physiol. Behav. 52, 405–410.

Burgess, G.M., Mullaney, J., McNeill, M., Dunn, P. and Rang, H.P. (1989a). Second messengers involved in the action of bradykinin on cultured sensory neurones. J. Neurosci. 9, 3314–3325.

Burgess, G.M., Mullaney, I., McNeill, M., Coote, P.R., Minhas, A. and Wood, J.N. (1989b). Activation of guanylate cyclase by bradykinin in rat sensory neurones is mediated by calcium influx: possible role of the increase in cyclic GMP. J. Neurochem. 53, 1212–1218.

Camargo, A.C., Shapanka, R. and Greene, L.J. (1973). Prepara-tion, assay, and partial characterization of a neutral endo-peptidase from rabbit brain. Biochemistry 12, 1838–1844.

Campos, A.H. and Calixto, J.B. (1993). Mechanisms involved in the contractile responses of kinins in rat portal vein rings: mediation by B$_1$ and B$_2$ receptors. J. Pharmacol. Exp. Ther. 268, 902–909.

Casey, K.L. (1991). Pain and central nervous system disease: a summary and overview. In "Pain and Central Nervous System Disease: The Central Pain Syndromes" (ed. K.L. Casey), pp. 1–11. Raven Press, New York.

Chan, P.H. and Fishman, R.A. (1984). The role of arachidonic acid in vasogenic brain edema. Fed. Proc. 43, 210–213.

Chapman, V. and Dickenson, A.H. (1992). The spinal and peripheral roles of bradykinin and prostaglandins in nociceptive processing in the rat. Eur. J. Pharmacol. 219, 427–433.

Cheronis, J.C., Whalley, E.T., Nguyen, K.T., Eubanks, S.R., Allen, L.G., Duggan, M.J., Loy, S.D., Bonham, K.A. and Blodgett, K. (1992). A new class of bradykinin antagonists: synthesis and *in vitro* activity of bissuccinimidoalkane peptide dimers. J. Med. Chem. 35, 1563–1572.

Cholewinski, A.J., Stevens, G., McDermott, A.M. and Wilkin, G.P. (1991). Identification of B$_2$ bradykinin binding sites on cultured cortical astrocytes. J. Neurochem. 57, 1456–1458.

Clark, W.G. (1979). Kinins and the peripheral and central nervous systems. In "Handbook of Experimental Pharmacology", Vol. XXV (Suppl.) (ed. E.G. Erdös), pp. 312–356. Springer-Verlag, New York.

Correa, F.M., Innis, R.B., Uhl, G.R. and Snyder, S.H. (1979). Bradykinin-like immunoreactive neuronal systems localized histochemically in rat brain. Proc. Natl Acad. Sci. USA 76, 1489-1493.

Couto, L.B., Correa, F.M.A. and Pela, I.R. (1995). Bradykinin microinjection into the locus coeruleus induces antinociception in rats. In "Kinins '95" (eds. J. Stewart, J.C. Cheronis and E.T. Whalley), Abstract B20.

Cruwys, S.C., Garrett, N.E., Perkins, M.N. Blake, D.R. and Kidd, B.L. (1994). The role of bradykinin B_1 receptors in the maintenance of intra-articular plasma extravasation in chronic antigen-induced arthritis. Br. J. Pharmacol. 113, 940–944.

Cunha, F.Q., Lorenzetti, B.B., Poole, S. and Ferreira, S.H. (1991). Interleukin-8 as a mediator of sympathetic pain. Br. J. Pharmacol. 104, 765–767.

Cunha, F.Q., Poole, S., Lorenzetti, B.B. and Ferreira, S.H. (1992). The pivotal role of tumor necrosis factor α in the development of inflammatory hyperalgesia. Br. J. Pharmacol. 107, 660–664.

Davis, A.J. and Perkins, M.N. (1994). The involvement of bradykinin B_1 and B_2 receptor mechanisms in cytokine-induced mechanical hyperalgesia in the rat. Br. J. Pharmacol. 113, 63–68.

Davis, A.J., Kelly, D. and Perkins, M.N. (1994). The induction of des-Arg⁹-bradykinin-mediated hyperalgesia in the rat by inflammatory stimuli. Braz. J. Med. Biol. Res. 27, 1793–1802.

Davis, K.D. and Dostrovsky, J.O. (1986). Activation of trigeminal brain stem nociceptive neurons by dural artery stimulation. Pain 25, 395–401.

Davis, K.D. and Dostrovsky, J.O. (1988). Cerebrovascular application of bradykinin excites central sensory neurons. Brain Res. 446, 401–406.

Devillier, P., Drapeau, G., Renoux, M. and Regoli, D. (1989). Role of the N-terminal arginine in the histamine-releasing activity of substance P, bradykinin and related peptides. Eur. J. Pharmacol. 168, 53–60.

Devor, M., Govrin-Lippman, R. and Angelides, K. (1993). Na⁺ channel immunolocalization in peripheral mammalian axons and changes following nerve injury and neuroma formation. J. Neurosci. 13, 1976–1992.

D'Orléans-Juste, P., Dion, S., Mizrahi, J. and Regoli, D. (1985). Effects of peptides and non-peptides on isolated arterial smooth muscles: role of endothelium. Eur. J. Pharmacol. 114, 9–21.

Dray, A. (1994). Tasting the inflammatory soup: role of peripheral neurones. Pain Rev. 1, 153–171.

Dray, A. (1995). Inflammatory mediators of pain. Br. J. Anaesth. 75, 125–131.

Dray, A. and Perkins, M. (1993). Bradykinin and inflammatory pain. Trends Neurosci. 16, 99–104.

Dray, A., Patel, I.A., Perkins, M.N. and Rueff A. (1992). Bradykinin-induced activation of nociceptors: receptor and mechanistic studies on the neonatal rat spinal cord-tail preparation in vitro. Br. J. Pharmacol. 107, 1129–1134.

Dray, A., Urban, L. and Dickenson, A. (1994). Pharmacology of chronic pain. Trends Pharmacol. Sci. 15, 190–197.

Dunn, P.M. and Rang, H.P. (1990). Bradykinin-induced depolarization of primary afferent nerve terminals in the neonatal rat spinal cord in vitro. Br. J. Pharmacol. 100, 656–660.

Ellis, E.F., Heizer, M.L., Hambrecht, G.S., Holt, S.A., Stewart, J.M. and Vavrek, R.J. (1987). Inhibition of bradykinin- and kallikrein-induced cerebral arteriolar dilation by a specific bradykinin antagonist. Stroke 18, 792–795.

Ellis, E.F., Holt, S.A., Wei, E.P. and Kontos, H.A. (1988). Kinins induce abnormal vascular reactivity. Am. J. Physiol. 255, H397–400.

Ewald, D.A., Matthies, J.G., Perney, T.M., Walker, M.W. and Miller, R.J. (1988). The effect of down regulation of protein kinase C on the inhibition of dorsal root ganglion neuron Ca²⁺ currents by neuropeptide Y. J. Neuroscience 8, 2447–2451.

Farmer, S.G. and Burch, R.M. (1992). Biochemical and molecular pharmacology of kinin receptors. Annu. Rev. Pharmacol. Toxicol. 32, 511–536.

Ferreira, S.H., Lorenzetti, B.B. and Poole, S. (1993). Bradykinin initiates cytokine-mediated inflammatory hyperalgesia. Br. J. Pharmacol. 110, 1227–1231.

Fox, A.J., Barnes, P.J., Urban, L. and Dray, A. (1993). An in vitro study of the properties of single vagal afferents innervating guinea-pig airways. J. Physiol. 469, 21–35.

Fujiwara, Y., Mantione, C.R., Vavrek, R.J., Stewart, J.M. and Yamamura, H.I. (1989). Characterization of (³H)bradykinin binding sites in guinea-pig central nervous system: possible existence of B_2 subtypes. Life Sci. 44, 1645–1653.

Galizzi, J.P., Bodinier, M.C., Chapelain, B., Ly, S.M., Coussy, L., Giraoud, S., Neliat, G. and Jean, T. (1994). Up-regulation of [³H]des-Arg¹⁰-kallidin binding to the bradykinin B_1 receptor by interleukin-1β in isolated smooth muscle cells: correlation with B_1 agonist induced PGI_2 production. Br. J. Pharmacol. 113, 389–394.

Gammon, C.M., Allen, A.C. and Morell, P, (1989). Bradykinin stimulates phosphoinositide hydrolysis and mobilisation of arachidonic acid in dorsal root ganglion neurons. J. Neurochem. 53, 95–101.

Gecse, A., Mezei, Z. and Telegdy, G. (1989). The effect of bradykinin and its fragments on the arachidonate cascade of brain microvessels. Adv. Exp. Med. Biol. 247A, 249–254.

Geppetti, P., Tramontana, M., Santicioli, P., Bianco, E., Giuliani, S. and Maggi, C.A., (1990). Bradykinin-induced release of calcitonin gene-related peptide from capsaicin-sensitive nerves in guinea-pig atria: mechanism of action and calcium requirements. Neuroscience 38, 687–692.

Germain, L., Barabé, J. and Galeano, C. (1986). Blood levels of kinins in experimental allergic encephalomyelitis. J. Neuroimmunol. 13, 135–142.

Germain, L., Barabé, J. and Galeano, C. (1988). Increased blood concentration of des-Arg⁹-bradykinin in experimental allergic encephalomyelitis. J. Neurol. Sci. 83, 211–217.

Gimpl, G., Walz, W., Ohlemeyer, C. and Kettenmann, H. (1992). Bradykinin receptors in cultured astrocytes from neonatal rat brain are linked to physiological responses. Neurosci. Lett. 144, 139–142.

Gold, M.S., Reichling, D.B., Shuster, M.J. and Levine, J.D. (1996). Hyperalgesic agents increase a tetrodotoxin-resistant Na⁺ current in nociceptors. Proc. Natl Acad. Sci. USA 93, 1108–1112.

Green, P.G., Luo, J., Heller, P. and Levine, J.D. (1992). Modulation of bradykinin induced plasma extravasation in the knee joint by sympathetic co-transmitters. Neuroscience 52, 451–458.

Haley, J.E., Dickenson, A.H. and Schachter, M. (1992). Electrophysiological evidence for a role of nitric oxide in prolonged chemical nociception in the rat. Neuropharmacology 31, 251–258.

Hall, J.M. (1992). Bradykinin receptors: pharmacological properties and biological roles. Pharmacol. Ther. 56, 131–190.

Handwerker, H.O. (1975). Influence of prostaglandin E_2 on the discharge of cutaneous nociceptive C-fibres induced by radiant heat. Pflüger's Arch. 355, 116–121.

Handwerker, H.O. (1976). Influence of algogenic substances and prostaglandins on discharges of unmyelinated cutaneous nerve fibres identified as nociceptors. In "Advances in Pain Research and Therapy" (eds. J.J. Bonica and D.G. Albe-Fessard), pp. 41–51. Raven Press, New York.

He, M. and McCarthy, K.D. (1994). Oligodendroglial signal transduction systems are developmentally regulated. J. Neurochem. 63, 501–508.

Heapy, C.G., Shaw, J.S. and Farmer, S.C. (1993). Differential sensitivity of antinociceptive assays to the bradykinin antagonist Hoe 140. Br. J. Pharmacol. 108, 209–213.

Hess, J.F., Borkowski, J.A., Young, G.S., Strader, C.D. and Ransom, R.W. (1992). Cloning and pharmacological characterization of a human bradykinin (BK-2) receptor. Biochem. Biophys. Res. Commun. 184, 260–268.

Hock, F.J., Wirth, K., Albus, U., Linz, W., Gerhards, H.J., Wiemer, G., St. Henke, G., Breipohl, G., Konig, W., Knolle, J. and Schölkens, B.A. (1991). Hoe-140 a new potent long acting bradykinin antagonist: in vitro studies. Br. J. Pharmacol. 102, 769–774.

Holthusen, H. and Arndt, J.O. (1994). Nitric oxide evokes pain in humans on intracutaneous injection. Neurosci. Lett. 165, 71–74.

Homayoun, P. and Harik, S.I. (1991). Bradykinin receptors of cerebral microvessels stimulate phosphoinositide turnover. J. Cereb. Blood Flow Metab. 11, 557–566.

Hosli, E. and Hosli, L. (1993). Receptors for neurotransmitters on astrocytes in the mammalian central nervous system. Prog. Neurobiol. 40, 477–506.

Ishizaka, T., Iwata, M. and Ishizaka, K. (1985). Release of histamine and arachidonate from mouse mast cells induced by glycosylation-enhancing factor and bradykinin. J. Immunol. 134, 1880–1887.

Jacques, L. and Couture, R. (1990). Studies on the vascular permeability induced by intrathecal substance P and bradykinin in the rat. Eur. J. Pharmacol. 184, 9–20.

Janig, W. and Kolzenburg, M. (1991). Receptive properties of pial afferents. Pain 45, 77-85.

Jeftinija, S. (1994). Bradykinin excites tetrodotoxin-resistant primary afferent fibres. Brain Res. 665, 69–76.

Jones, S., Brown, D.A., Milligan, G., Willer, E., Buckley, N.J. and Caulfield, M.P. (1995). Bradykinin excites rat sympathetic neurons by inhibition of M current through a mechanism involving B_2 receptors and G alpha q/11. Neuron 14, 399–405.

Kariya, K., Yamauchi, A. and Sasaki, T. (1985). Regional distribution and characterization of kinin in the CNS of the rat. J. Neurochem. 44, 1892–1897.

Katusic, Z.S., Milde, J.H., Cosentino, F. and Mitrovic, B.S. (1993). Subarachnoid hemorrhage and endothelial L-arginine pathway in small brain stem arteries in dogs. Stroke 24, 392–399.

Kay, A.B. and Kaplan, A.P. (1975). Chemotaxis and haemostasis. Br. J. Haematol. 31, 417–422.

Kingden-Milles, D., Klement, W. and Arndt, J.O. (1994). The nociceptive systems of skin, perivascular tissue and hand veins of humans and their sensitivity to bradykinin. Neurosci. Lett. 181, 39–42.

Kolzenburg, M., Kress, M. and Reeh, P.W. (1992). The nociceptor sensitization by bradykinin does not depend on sympathetic neurons. Neuroscience 46, 465–473.

Kontos, H.A., Wei, E.P., Povlishock, J.T. and Christman, C.W. (1984). Oxygen radicals mediate the cerebral arteriolar dilation from arachidonate and bradykinin in cats. Circ Res. 55, 295–303.

Kumazawa, T. and Mizumura, K. (1980). Chemical responses of polymodal receptors of scrotal contents in dogs. J. Physiol. 299, 219–231.

Kurosawa, M., Messlinger, K., Pawlak, M and Schmidt, R.F. (1995). Increases of meningeal blood flow after electrical stimulation of rat dura mater encephali: mediation by calcitonin-gene related peptide. Br. J. Pharmacol. 114, 1397–1402.

Laneuville, O. and Couture, R. (1987). Bradykinin analogue blocks bradykinin-induced inhibition of a spinal nociceptive reflex in the rat. Eur. J. Pharmacol. 137, 281–285.

Laneuville, O., Reader, T.A. and Couture, R. (1989). Intrathecal bradykinin acts presynaptically on spinal noradrenergic terminals to produce antinociception in the rat. Eur. J. Pharmacol. 159, 273–283.

Lerner, U.H. and Modeer, T. (1991). Bradykinin B_1 and B_2 receptor agonists synergistically potentiate interleukin-1 induced prostaglandin biosynthesis in human gingival fibroblasts. Inflammation 15, 427–436

Levine, J.D., Fields, H.L. and Basbaum, A.I. (1993). Peptides and the primary afferent nociceptor. J. Neurosci. 13, 2273–2286.

Lopes, P., Kar, S., Chretien, L., Regoli, D., Quirion, R. and Couture, R. (1995). Quantitative autoradiographic localization of $[^{125}I\text{-}Tyr^8]$-bradykinin receptor binding sites in the rat spinal cord: effects of neonatal capsaicin, noradrenergic deafferentation, dorsal rhizotomy and peripheral axotomy. Neuroscience 68, 867–881.

Macfarlane, R., Moskowitz, M.A., Sakas, D.E., Tasdemiroglu, E., Wei, E.P. and Kontos, H.A. (1991). The role of neuro-effector mechanisms in cerebral hyperfusion syndromes. J. Neurosurg. 75, 845–855.

Maier-Hauff, K., Baethmann, A.J., Lange, M., Schurer, L. and Unterberg, A. (1984). The kallikrein–kinin system as mediator in vasogenic brain edema, Part 2: Studies on kinin formation in focal and perifocal brain tissue. J. Neurosurg. 61, 97–106.

Malmberg, A.B. and Yaksh, T.L. (1992). Antinociceptive actions of spinal nonsteroidal anti-inflammatory agents on the formalin test in the rat. J. Pharmacol. Exp. Ther. 263, 136–146.

Manning, D.C., Raja, S.N., Meyer, R.A. and Campbell, J.N. (1991). Pain and hyperalgesia after intradermal injection of bradykinin in humans. Clin. Pharmacol. Ther. 50, 721–729.

Marceau, F. (1995). Kinin B_1 receptors: a review. Immunology 30, 1–26.

Markowitz, S., Saito, K. and Moskowitz, M.A. (1987). Neurogenically mediated leakage of plasma protein occurs from blood vessels in dura mater but not brain. J. Neurosci. 7, 4129–4136.

Mayberg, MR., Zervas, N.T. and Moskowitz, M.A. (1984). Trigeminal projections to supratentorial pial and dural blood vessels in cats demonstrated by horseradish peroxidase histochemistry. J. Comp. Neurol. 223, 46–56.

McEachern, A.E., Shelton, E.R., Bhakta, S., Obernolte, R., Bach, C., Zuppan, P., Fujisaki, J., Aldrich, R.W. and Jarnagin, K. (1991). Expression cloning of a rat B_2 bradykinin receptor. Proc. Natl Acad. Sci. USA 88, 7724-7728.

McFadden, R.G. and Vickers, K.E. (1989). Bradykinin augments the *in vitro* migration of nonsensitised lymphocytes. Clin. Invest. Med. 12, 247–253.

McGehee, D.S., Goy, M.F. and Oxford, G.S. (1992). Involvement of the nitric oxide-cyclic GMP pathway in the desensitization of bradykinin responses of cultured rat sensory neurons. Neuron 9, 315–324.

McMahon, S.B., Lewin, G.R. and Wall, P.D. (1993). Central hyperexcitability triggered by noxious inputs. Curr. Opinion in Neurobiology 3, 602–610.

Meller, S.T. and Gebhart, G.F. (1992). A critical review of the afferent pathways and the potential chemical mediators involved in cardiac pain. Neuroscience 48, 501–524.

Menke, J.G., Borkowski, J.A., Bierilo, K., MacNeil, T., Derrick, A.W., Schneck, K.A., Ransom, R.W., Strader, K.D., Linemeyer, D.L. and Hess, J.F. (1994). Expression cloning of a human B_1 bradykinin receptor. J. Biol. Chem. 269, 21583–21586.

Mense, S. (1993). Nociception from skeletal muscle in relation to clinical muscle pain. Pain 54, 241–289.

Messlinger, K., Pawlak, M. Schepelmann, K. and Schmidt, R.F. (1993). Responsiveness of slowly conducting articular afferents to bradykinin: effects of an experimental arthritis. Pain 59, 335–343.

Meyer, R.A., Davis, K.D., Raja, S.N. and Campbell, J.N. (1992). Sympathectomy does not abolish bradykinin-induced cutaneous hyperalgesia in man. Pain 51, 323–327.

Morris, R., Southam, E., Braid, D.J. and Garthwaite, J. (1992). Nitric oxide may act as a messenger between dorsal root ganglion neurones and their satellite cells. Neurosci. Lett. 137, 29–32.

Moskowitz, M.A., Brody, M. and Liu-Chen, L.Y. (1983). *In vitro* release of immunoreactive substance P from putative afferent nerve endings in bovine pia arachnoid. Neuroscience 9, 809–814.

Murone, C., Perick, R.B., Schlawe, I., Chai, S.Y., Casley, D., MacGregor, D.P., Müller-Esterl, W. and Mendelsohn, F.A.O. (1996). Characterization and localization of bradykinin B_2 receptors in the guinea-pig using a radioiodinated Hoe 140 analogue. Eur. J. Pharmacol. 306, 237–247.

Nagy, I., Pabla, R., Matesz, K., Woolf, C.J. and Urban, L. (1993). Cobalt uptake enables identification of capsaicin and bradykinin sensitive sub-populations of rat dorsal root ganglion cells *in vitro*. Neuroscience 56, 167–172.

Nicol, G.D. and Cui, M. (1992). Enhancement by prostaglandin E_2 of bradykinin activation of embryonic rat sensory neurones. J. Physiol. 480, 485–492.

O'Neil, L.A.J. and Lewis, G.P. (1989). Interleukin-1 potentiates bradykinin- and TNFα-induced PGE_2 release. Eur. J. Pharmacol. 166, 131–137.

Olesen, J., Thomsen, L.L. and Iversen, H. (1994). Nitric oxide is a key molecule in migraine and other vascular headaches. TIPS 15, 149–153.

Orawski, A.T. and Simmons, W.H. (1989). Degradation of bradykinin and its metabolites by rat brain synaptic membranes. Peptides 10, 1063–1073.

Parpura, V., Basarsky, T.A., Liu, F., Jeftinija, K., Jeftinija, S. and Haydon, P.G. (1994). Nature (Lond.) 369, 744–747.

Perkins, M.N. and Kelly, D. (1993). Induction of bradykinin B_1 receptors *in vivo* in a model of ultra-violet irradiation-induced thermal hyperalgesia in the rat. Br. J. Pharmacol. 110, 1441–1444.

Perkins, M.N., Campbell, E. and Dray, A. (1992). Anti-nociceptive activity of the bradykinin B_1 and B_2 receptor antagonists, des-Arg^9[Leu^8]-BK and HOE 140 in two models of persistent hyperalgesia in the rat. Pain 53, 191–197.

Perkins, M.N., Kelly, D. and Davis, A.J. (1994). Bradykinin B_1 and B_2 receptor mechanisms and cytokine-induced hyperalgesia in the rat. Can. J. Physiol. Pharmacol. 73, 832–836.

Perry, D.C. and Snyder, S.H. (1984). Identification of bradykinin in mammalian brain. J. Neurochem. 43, 1072–1080.

Perry, V.H., Andersson, P.-B. and Gordon, S. (1993). Macrophages and inflammation in the central nervous system. Trends Neurosci. 16, 268–273.

Poole, S., Cunha, F.Q., Selkirk, S., Lorenzetti, B.B. and Ferreira, S.H. (1995). Cytokine-mediated inflammatory hyperalgesia limited by interleukin-10. Br. J. Pharmacol. 115, 684–688.

Privitera, P.J., Daum, P.R., Hill, D.R. and Hiley, C.R. (1992). Autoradiographic visualization and characteristics of (^{125}I) bradykinin binding sites in guinea pig brain. Brain Res. 577, 73–79.

Puttick, R.M. (1992). Excitatory action of prostaglandin E_2 on rat neonatal cultured dorsal root ganglion cells. Br. J. Pharmacol. 105, 133P.

Rang, H.P. and Urban, L. (1995). New molecules in analgesia. Br. J. Anaesth. 75, 145–156.

Rang, H.P., Bevan, S.J. and Dray, A. (1991). Chemical activation of nociceptive peripheral neurons. Br. Med. Bull. 47, 534–548.

Rang, H.P., Bevan, S.J. and Dray, A. (1994). Nociceptive peripheral neurones: cellular properties. In "Textbook of Pain" (eds. P.D. Wall and R. Melzack), pp. 57–78. Churchill Livingstone, Edinburgh.

Rao, P.J. and Bhattacharya, S.K. (1988). Hyperthermic effect of centrally administered bradykinin in the rat: role of prostaglandins and serotonin. Int. J. Hypertherm. 4, 183–189.

Rhaleb, N.-E., Drapeau, G., Dion, S., Jukic, D., Rouissi, N. and Regoli, D. (1990). Structure–activity studies on bradykinin and related peptides: agonists. Br. J. Pharmacol. 99, 445–448.

Ribeiro, S.A. and Rocha e Silva, M. (1973). Antinociceptive action of bradykinin and related kinins of larger molecular weights by the intraventricular route. Br. J. Pharmacol. 47, 517–528.

Ribeiro, S.A., Corrado, A.P. and Graeff, F.G. (1971). Antinociceptive action of intraventricular bradykinin. Neuropharmacology 10, 725–731.

Roberts, R.A. and Gullick, W.J. (1990). Bradykinin receptors undergo ligand-induced desensitization. Biochemistry 29, 1975–1979.

Roscher, A.A., Manganiello, V., Jesselma, C.L. and Moss, J. (1984). Autoregulation of bradykinin receptors and bradykinin-induced prostacyclin formation in human fibroblasts. J. Clin. Invest. 74, 552–558.

Rothwell, N.J. (1991a). Functions and mechanisms of interleukin 1 in the brain. Trends Pharmacol. Sci. 12, 430–436.

Rothwell, N.J. (1991b). CNS regulation of thermogenesis. Crit. Rev. Neurobiol. 8, 1–10.

Rueff, A. and Dray, A. (1993). Sensitization of peripheral afferent fibres in the in vitro neonatal rat spinal cord-tail by bradykinin and prostaglandins. Neuroscience 54, 527–535.

Rueff, A., Patel, I.A., Urban, L. and Dray, A. (1994). Regulation of bradykinin sensitivity in peripheral sensory fibres of the neonatal rat by nitric oxide and cyclic GMP. Neuropharmacology 33, 1139–1145.

Sarker, M.H. and Fraser, P.A. (1996). Evidence that bradykinin increases permeability of single cerebral microvessels via free radicals. J. Physiol. 479, 368P.

Sawutz, D.G., Salvino, J.M., Dolle, R.E., Casiano, F., Ward, S.J., Houck, W.T., Faunce, D.M., Douty, B.D., Baizman, E., Awad, M.M.A., Marceau, F. and Seoane, P.R. (1994). The nonpeptide WIN 64388 is a bradykinin B$_2$ receptor antagonist. Proc. Natl Acad. Sci. USA 91, 4693–4697.

Schaible, H.-G. and Grubb, B.D. (1993). Afferent and spinal mechanisms of joint pain. Pain 55, 5–54.

Schaible, H.-G. and Schmidt, R.F. (1988). Excitation and sensitization of fine articular afferents from cat's knee joint by prostaglandin E$_2$. J. Physiol. 403, 91–104.

Schepelmann, K., Messlinger, K. and Schmidt, R.F. (1993). The effect of phorbol ester on slowly conducting afferents of the cat's knee joint. Exp. Brain Res. 92, 391–398.

Schuligoi, R., Donnerer, J. and Amann, R. (1994). Bradykinin-induced sensitization of afferent neurons in the rat paw. Neuroscience 59, 211–215.

Scicli, A.G., Forbes, G., Nolly, H., Dujouny, M. and Carretero, O.A. (1984). Kallikrein–kinins in the central nervous system. Clin. Exp. Hypertens. A6, 1731–1738.

Seabrook, G.R., Bowery, B.J. and Hill, R.G. (1995). Bradykinin receptors in mouse and rat isolated superior cervical ganglion. Br J. Pharmacol. 115, 368–372.

Shearman, M.S., Sekiguchi, K. and Nishizuka, Y. (1989). Modulation of ion channel activity: a key function of the protein kinase C enzyme family. Pharmacol. Rev. 41, 211–237.

Sicuteri, F. (1970). Bradykinin and intracranial circulation in man. Handbook Exp. Pharmacol. 25, 482–515.

Snyder, S.H. (1980). Brain peptides as neurotransmitters. Science 209, 976–983.

Stephens, G.J., Marriott, D.R., Djamgoz, M.B.A. and Wilkin, G.P. (1993). Electrophysiological and biochemical evidence for bradykinin receptors on cultured rat cortical oligodendrocytes. Neurosci. Lett. 153, 223–226.

Steranka, L.R., Manning, D.C., DeHaas, C.J., Ferkany, J.W., Borosky, S.A., Connor, J.R., Vavrek, R.J., Stewart, J.M. and Snyder, S.H. (1988). Bradykinin as a pain mediator: receptors are localized to sensory neurons, and antagonists have analgesic actions. Proc. Natl Acad. Sci. USA 85, 3245–3249.

Stevens Negus, S., Butelman, E.R., Gatch, M.B. and Woods, J.H. (1995). Effects of morphine and ketorolac on thermal allodynia induced by prostaglandin E$_2$ and bradykinin in rhesus monkeys. J. Pharmacol. Exp. Ther. 274, 805–814.

Taiwo, Y. and Levine, J.D. (1988). Characterization of the arachidonic acid metabolites mediating bradykinin and noradrenaline hyperalgesia. Brain Res. 458, 402–406.

Taiwo, Y.O., Heller, P.H. and Levine, J.D. (1990). Characterization of distinct phospholipases mediating bradykinin and noradrenaline hyperalgesia. Neuroscience 39, 523–531.

Tasker, R.R., de Carvalho, G. and Dostrovsky, J.O. (1991). The history of central pain syndromes, with observations concerning pathophysiology and treatment. In "Pain and Central Nervous System Disease: The Central Pain Syndromes" (ed. K.L. Casey), pp. 31–58. Raven Press, New York.

Thayer, S.A., Perney, T.M. and Miller, R.J. (1988). Regulation of calcium homeostasis in sensory neurons by bradykinin. J. Neurosci. 8, 4089–4097.

Tiffany, C.W. and Burch, R.M. (1989). Bradykinin stimulates tumor necrosis factor and interleukin-1 release from macrophages. FEBS Lett. 247, 189–192.

Undem, B.J. and Weinreich, D. (1993). Electrophysiological properties and chemosensitivity of guinea pig nodose ganglion neurons in vitro. J. Autonom. Nerv. Syst. 44, 17–34.

Unterberg, A. and Baethmann, A.J. (1984). The kallikrein–kinin system as mediator in vasogenic brain edema. Part 1: Cerebral exposure to bradykinin and plasma. J. Neurosurg. 61, 87–96.

Unterberg, A., Wahl, M. and Baethmann, A. (1984). Effects of bradykinin on permeability and diameter of pial vessels in vivo. J. Cereb. Blood Flow Metab. 4, 574–585.

Unterberg, A., Dautermann, C., Baethmann, A. and Müller-Esterl, W. (1986). The kallikrein–kinin system as mediator in vasogenic brain edema, Part 3: inhibition of the kallikrein–kinin system in traumatic brain swelling. J. Neurosurg. 64, 269–276.

Unterberg, A., Wahl, M., Hammersen, F. and Baethmann, A. (1987). Permeability and vasomotor response of cerebral vessels during exposure to arachidonic acid. Acta Neuropathol. 73, 209–219.

Vandekerchkove, F., Opdenakker, G., Van Ranst, M., Lenaerts, J.-P., Put, W., Billiau, A. and Van Damme, J. (1991). Bradykinin induces interleukin-6 and synergizes with interleukin-1. Lymphokine Cytokine Res. 10, 285–289.

Verge, V.M.K., Xu, Z., Xu, X.-J., Wiesenfelt-Hallin, Z. and Hokfelt, T. (1992). Marked increase in nitric oxide synthase mRNA in rat dorsal root ganglia after peripheral axotomy: in situ hybridization and functional studies. Proc. Natl Acad. Sci. USA 89, 11617–11621.

Wahl, M., Schilling, L., Unterberg, A. and Baethmann, A. (1993). Mediators of vascular and parenchymal mechanisms in secondary brain damage. Acta Neurochir. Suppl. 57, 64–72.

Walker, K., Dray, A. and Perkins, M. (1996a). Hyperalgesia in rats following intracerebroventricular administration of endotoxin: effects of bradykinin B$_1$ and B$_2$ receptor antagonist treatment. Pain 65, 211–219.

Walker, K., Dray, A. and Perkins, M. (1996b). Development of hyperthermia following intracerebroventricular administration of endotoxin in the rat: effects of kinin B$_1$ and B$_2$ receptor antagonists. Br. J. Pharmacol. 117, 684–688.

Wei, P., Moskowitz, M.A., Boccalini, P. and Kontos, H.A. (1992). Calcitonin gene-related peptide mediates nitroglycerin and sodium nitroprusside-induced vasodilatation in feline cerebral arterioles. Circ. Res. 70, 1313–1319.

Weinreich, D. and Wonderlin, W.F. (1987). Inhibition of calcium-dependent spike after-hyperpolarization increases excitability of rabbit visceral sensory neurones. J. Physiol. 394, 415–427.

Woolf, C.J. (1994). The dorsal horn: state-dependent sensory processing and the generation of pain. In "Textbook of Pain", 2nd edn (eds. P.D. Wall and R. Melzack), pp. 101–112. Churchill Livingstone, Edinburgh.

Woolf, C.J. and Doubell, T.P. (1994). The pathophysiology of chronic pain – increased sensitivity to low threshold Aβ-fibre inputs. Curr. Biol. 4, 525–534.

Yirmiya, R., Rosen, H., Donchin, O. and Ovadia, H. (1994). Behavioral effects of lipopolysaccharide in rats: involvement of endogenous opioids. Brain Res. 648, 80–86.

10. The Kallikrein–Kinin System in Sepsis Syndrome

Robin A. Pixley *and* Robert W. Colman

1. Sepsis Syndrome

The Systemic Inflammatory Response Syndrome (SIRS) is the pathophysiological response to severe clinical insults (Bone, 1993) including infection, pancreatitis, hemorrhagic shock, ischemia, multiple trauma and immune-mediated organ injury. The symptoms of SIRS include tachycardia, hypothermia, tachypnea, reduced organ perfusion or dysfunction, manifested by altered cerebral function, hypoxemia, elevated plasma lactate and low urine output. Sepsis or sepsis syndrome, is the SIRS resulting from infection and, when it is accompanied by hypotension that is unresponsive to fluid therapy, it is referred to as "septic shock".

Diagnosis of the sepsis syndrome does not rest upon a positive blood culture (bacteremia), but it requires a suspected initial site of infection. Exposure to Gram-negative or Gram-positive bacteria, fungi or bacterial endotoxins or proteoglycan polysaccharides can trigger the release of mediators from cells such as monocytes, macrophages, neutrophils, platelets and endothelial cells (Bone, 1991; Parillo, 1993). Some of the more important mediators in SIRS include tumor necrosis factor-α (TNF-α), platelet-activating factor (PAF), interleukins 1, 6 and 8, and proteases such as cathepsin G. Collectively, these substances of pathogen or host origin can activate the plasma kallikrein–kinin system, coagulation, and the complement system.

1.1   THE KALLIKREIN–KININ SYSTEM

The plasma kallikrein–kinin system (also known as the "contact system") is a cascade of proteinases, which, when activated, generates the nonapeptide, bradykinin (BK), from high molecular weight kininogen (HMWK). There is also cross-activation of C1 (the initial protein of the classical complement pathway), hydrolysis of pro-urokinase to yield urokinase (which activates cell-associated fibrinolysis), and activation of Factor XI of the intrinsic blood coagulation cascade. The plasma kallikrein–kinin system is triggered following activation of Factor XII (Hageman Factor) by endotoxin (Kalter *et al.*, 1983) and microbial proteases (Molla *et al.*, 1989) to yield Factor XIIa, or indirectly by autoactivation through injury to endothelium.

Factor XIIa cleaves plasma prekallikrein, to its active form, kallikrein (Fig. 10.1), which, in turn, can further activate more Factor XII to Factor XIIa. HMWK associated with prekallikrein forms a noncovalent complex and aids prekallikrein in the feedback activation by carrying it to a common activating surface where Factor XII or Factor XIIa are present. When the prekallikrein-HMWK is activated to kallikrein-HMWK, kallikrein can cleave HMWK at two sites, causing release of BK. BK, via activation of cell surface B_2 receptors on the vascular endothelium, releases nitric oxide (NO) and prostaglandin I_2 (PGI_2), both potent vasodilators. Thus, activation of the plasma kallikrein– kinin system can contribute significantly to the hypotension that occurs in sepsis. Factor XIIa also contributes to coagulation by activating Factor XI to XIa, thereby initiating the intrinsic cascade of coagulation.

Plasma kallikrein can also stimulate neutrophil chemotaxis, aggregation, elastase release and oxygen consumption *in vitro* (Kaplan *et al.*, 1972; Goetzl and Austen, 1974; Schapira *et al.*, 1982; Wachtfogel *et al.*, 1983). Furthermore, neutrophil activation *in vivo*, assessed by increased complexes in plasma between human neutrophil elastase (HNE) and α_1-proteinase inhibitor (α_1-PI), occurs in sepsis and is associated with a poor prognosis (Egbring *et al.*, 1977; Duswald *et al.*, 1985; Nuijens *et al.*, 1992). Thus, the kallikrein–kinin system may also contribute to HNE-mediated tissue damage seen in sepsis syndrome.

1.2   THE COMPLEMENT SYSTEM

The classical complement pathway is activated by immune complexes, whereas the alternative pathway is activated on the surfaces of damaged host tissues and invading microorganisms. A precursor protein, C1 (800 kDa) is composed of a 460 kDa subunit (C1q) associated with two molecules each of the 83 kDa polypeptides, C1s and C1r (C1q(r,s)$_2$; Fig. 10.2). Upon activation, C1q dissociates and activates the protease form of C1s. Activated C1s initiates a cascade resulting in formation of the membrane attack complex (MAC), as well as release of C3a, C4a and C5a (Fig. 10.2). These anaphylatoxins, acting on cell-surface receptors on mast cells and basophils, can release histamine and serotonin, thereby contributing to microvascular leakage. C3a and C5a are also chemotactic for monocytes and neutrophils.

Inactivation of C4b2a occurs by dissociation of C2a from the complex and cleavage of bound C4b (Fig. 10.2). Factor I cuts C4b at two sites releasing a large fragment, C4c, and leaves the C4d attached to the cell membrane. Inactivation of C3b occurs in a similar manner.

Direct activation of complement by bacteria and their products occurs, since endotoxin and intact bacteria can activate complement via the alternative pathway (Morrison and Kline, 1977; Grossman and Leive, 1984). However, studies *in vitro* have demonstrated a mechanism for activation of the classical pathway in that another cleavage product of Factor XIIa, Factor XIIf (30 kDa), can activate C1 through cleavage of C1r (Ghebrehiwet *et al.*, 1981, 1983). Alternatively, kallikrein can cleave the C1 components, resulting in destruction of C1 (Cooper *et al.*, 1980). Kallikrein has been demonstrated to replace Factor D in the alternative pathway, generating a C3 convertase by cleaving Factor B (DiScipio, 1982). Kallikrein can generate C5a from C5 (Wiggins *et al.*, 1981), suggesting that the kallikrein–kinin system may also activate complement.

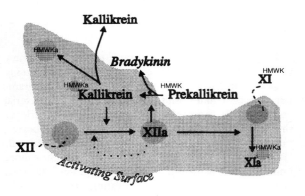

Kallikrein-Kinin Cascade

Figure 10.1   The components of the kallikrein–kinin system. HMWK, high molecular weight kininogen; HMWKa, kinin-free HMWK; XI and XII, Factors XI and XII respectively.

1.3   C1-INHIBITOR

Factor XIIa, kallikrein and C1 are highly regulated by a serine protease inhibitor (SERPIN), C1-inhibitor (also known as C1 esterase inhibitor or C1-inactivator). Table 10.1 illustrates the relative enzyme inhibitory activity and

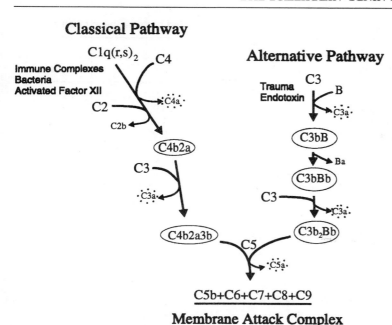

Figure 10.2   Classical and alternative pathways of complement activation.

Table 10.1   Comparisons of the relative enzyme inhibitory activities and predicted relative effectiveness of C1-inhibitor, α_2-antiplasmin (α_2-AP), α_2-macroglobulin (α_2-M), α_1-proteinase inhibitor (α_1-PI) and antithrombin III (ATIII)

Enzyme	Inhibitor	k ($M^{-1} min^{-1}$ $\times 10^3$)	[Plasma] (μM)	k_{obs} ($min^{-1} \times 10^{-3}$)	Relative effectiveness Ratio	Relative effectiveness %
Factor XIIa	C1-inhibitor	220	1.7	374	110	91.9[a]
	α_2-AP	11	1.1	12.1	4	3.0
	α_2-M	5.0	3.6	17.5	5	4.3
	ATIII	1.3	2.6	3.4	1	0.8
Kallikrein	C1-inhibitor	102	1.7	1,734	37	40.7[a]
	α_2-M	690	3.6	2,484	53	58.2
	ATIII	18	2.6	46.8	1	1.1
Factor XIa	α_1-PI	13	45.5	591.5	9	73.6
	C1-inhibitor	40	1.7	68	1	8.5[a]
	α_2-AP	60	1.1	66	1	8.2
	ATIII	30	2.6	78	1.2	9.7
C1s	C1-inhibitor	78	1.7	132.6	—	100[a]
C1r	C1-inhibitor	90	1.7	153	—	100[a]

[a] Denotes C1-inhibitor contribution values. Data are from Sim *et al.* (1980), Schapira *et al.* (1981), Scott *et al.* (1982) and Pixley *et al.* (1985).

the predicted relative effectiveness of inhibitors of the kallikrein–kinin system and complement. C1-inhibitor is the only SERPIN known to inactivate C1s and C1r, and also contributes over 90% of the plasma's capacity to inhibit Factor XIIa (De Agostini *et al.*, 1984; Pixley *et al.*, 1985). In addition, C1-inhibitor is a major kallikrein inhibitor, contributing 40.7% of the plasma capacity to inhibit this enzyme (Schapira *et al.*, 1981). A second major kallikrein inhibitor, α_2-macroglobulin (α_2-M) contributes 58.2% of the regulatory activity. We capitalized on the formation of kallikrein:α_2-M (Kal:α_2-

M) complexes by developing an assay for contact activation in blood samples obtained from SIRS patients and primates (Kaufman *et al.*, 1991; Pixley *et al.*, 1992, 1993, 1995). C1-inhibitor regulates Factor XIa, with 8.5% of the latter forming a complex with the inhibitor (Scott *et al.*, 1982). The primary inhibitor in regulating Factor XIa is α_1-PI, with α_2-antiplasmin (α_2-AP), and antithrombin III (AT-III) contributing the remainder (Table 10.1).

Table 10.2 illustrates the relative effectiveness of C1-inhibitor to inhibit various plasma proteinases. Plasma

Table 10.2   Comparison of the effectiveness of C1-inhibitor to inhibit activated proteinases in plasma

Zymogen/ Inhibitor	[Plasma] (μM)	Mass (kDa)	Enzyme	Rate of inactivation by C1-inhibitor[a]		
				k_2 ($M^{-1} s^{-1} \times 10^3$)	k_{obs} ($s^{-1} \times 10^{-3}$)	Relative rate
Factor XII	0.37	80	Factor XIIa[b]	3.7	6.3	5.7
Prekallikrein	0.4	88	Kallikrein[c]	17	28.9	26.3
Factor XI[d]	0.04	(80)$_2$[d]	Factor XIa[d]	0.7	1.1	1
C1s[e]	0.3	83	Active C1s[be]	1.3	2.2	2
C1r[e]	0.3	83	Active C1r[be]	1.5	2.6	2.4
C1-inhibitor	1.7	105	—	—	—	—

[a] k_2 is the second-order rate constant under physiological conditions. k_{obs} is the observed rate constant determined from the plasma concentration of C1-inhibitor and k_2 for each enzyme.
[b] Major or sole SERPIN is active C1-inhibitor.
[c] α_2-Macrogobulin is also a significant inhibitor of kallikrein.
[d] Factor XI and Factor XIa are dimers of 80 kDa each.
[e] Present as C1q(C1s)$_2$(C1r)$_2$. C1-inhibitor binds active C1s and C1r, and dissociates them from the C1 complex leaving C1q attached to the activating surface and free C1r/s–C1-inhibitor complexes.
Data from Sim *et al.* (1980), Schapira *et al.* (1981), Scott *et al.* (1982), Pixley *et al.* (1985).

concentrations of C1-inhibitor are most effective in inactivating kallikrein. It is clear that sustained activation of the contact and complement systems, which can lead to depletion of C1-inhibitor, will result in reduced inhibitory regulation of kallikrein and Factor XIIa. This decrease in the regulatory molecule allows the reciprocal activation to proceed between these two enzymes, resulting in sustained release of BK and neutrophil activation. Continued depletion of C1-inhibitor can also stimulate the complement system by leaving C1r and C1s unchecked, thereby allowing formation of the MAC with ensuing cell lysis. Further depletion of C1-inhibitor would also decrease concentrations to a critical value where Factor XIa is regulated, allowing fibrin generation, via Factor IX from the intrinsic coagulation cascade, although α_1-PI is a major regulator of this enzyme. It is interesting that C1-inhibitor has not been demonstrated to regulate proteases other than Factor XIIa, kallikrein and Factor XIa, generated in the intrinsic or extrinsic coagulation cascades. This suggests that low levels of C1-inhibitor may favor activation of the kallikrein–kinin system in the absence of appreciable coagulation. In shock, where organ perfusion is reduced, a local environment exists in which C1-inhibitor is depleted and not further supplied from blood. Thus, both the kinin and complement systems may be fully activated, and uncontrolled, further exacerbating inflammation.

1.4   EXPERIMENTAL DETECTION OF CONTACT ACTIVATION

One can predict some of the abnormalities expected in human diseases where the contact system may be activated (Table 10.3). If Factor XII is activated by exposure to, e.g., cell membranes, extracellular matrix, lipid micelles, etc., or cleaved by enzymes such as HNE or bacterial proteases, Factor XII and prekallikrein are

converted to active enzymes that react rapidly with C1-inhibitor or α_2-M to form enzyme:inhibitor complexes. Thus, during contact activation, one might expect decreases in plasma concentrations of functional Factor XII and prekallikrein, and concomitant increases in antigenic Factor XIIa:inhibitor and kallikrein:inhibitor complexes. However, the plasma concentration of zymogens may increase owing to increased synthesis during an acute-phase attack, and levels of complexes may decrease owing to clearance by macrophages or hepatocytes. The antigenic values for Factor XII and

Table 10.3   Predicted changes in kallikrein–kinin system proteins in sepsis

Functional assay	Value	Immunological assay	Value
Factor XII	⇓	Factor XII	⇔
		Factor XIIa:C1-inhibitor[a]	⇑
Prekallikrein	⇓	Prekallikrein	⇔
		Kal:C1-inhibitor[a]	⇑
		Kal:α_2-M[a]	⇑
Factor XI	⇓ ⇔ ⇑	Factor XI	⇑ ⇔
		Factor XIa:C1-inhibitor[a]	⇑
		Factor XIa:α_1-PI[a]	⇑
HMWK	⇓	HMWK	⇓
		Bradykinin	⇑
C1s	⇓	C1s:C1-inhibitor[a]	⇑
C1r	⇓	C1r:C1-inhibitor[a]	⇑
C1-inhibitor	⇓	C1-inhibitor	⇑ ⇔
		i-C1-inhibitor[b]	⇑
α_2-M	⇓	α_2-M	⇓ ⇔

⇑ Elevated above normal; ⇓ decreased; ⇔ unchanged.
[a] Complexes.
[b] Inactive inhibitor cleaved by bacterial protease and elastase.
HMWK, high molecular weight kininogen; Kal, kallikrein; α_2-M, α_2-macroglobulin; α_1-PI, α_1 proteinase inhibitor; i-C1-inhibitor, inactive C1-inhibitor.

prekallikrein would be expected to remain similar since most immunological assays detect both zymogen and enzyme-complexed forms of these molecules. Since normal concentrations of Factor XI in blood are relatively low (0.04 μM), increases may be observed owing to acute-phase protein synthesis or cellular release.

Levels of HMWK drop during contact activation, since HMWK is further cleaved and inactivated by Factor XIa (Scott et al., 1985). Measurable BK degradation products may increase, but only transiently since 95% of BK is cleared during one passage through the lungs. C1s and C1r functional activity decrease owing to complement activation and consumption, accompanied by increases in levels of C1r:C1-inhibitor and C1s:C1-inhibitor. Functional levels of C1-inhibitor decrease, being consumed during its inactivation of Factor XIIa, kallikrein, C1s and C1r, and by cleavage by HNE and bacterial proteases producing inactive i-C1-inhibitor, which would increase. α_2-M levels decrease owing to its inactivation of kallikrein as well as many other enzymes that this inhibitor regulates, while levels of Kal:α_2-M complexes increase.

2. Clinical Studies of the Kallikrein–Kinin System

Earlier studies examined functional zymogens and inhibitors in patients with sepsis or septic shock, with single points to assess levels of the proteins at time of diagnosis. Either coagulant or amidolytic assays were used for Factor XII, prekallikrein, HMWK and Factor XI, and which depended on biological activity. Functional inhibition of C1-inhibitor and α_2-M was determined using exogenous kallikrein as a substrate.

2.1 CLINICAL SEPSIS

In an initial investigation, Mason et al. (1970) obtained single plasma samples from 54 individuals who were divided into 4 groups: normal; hypotensive owing to blood loss; normotensive with bacteremia; and hypotensive with bacteremia. Those patients who were hypotensive owing to blood loss did not differ from normal subjects in functional levels of Factor XII, prekallikrein and C1-inhibitor. Those with septicemia alone showed no significant decrease in these proteins. Only the hypotensive septicemic group showed significantly decreased levels of contact factors, indicating activation of the kinin system. In another study (Smith-Erichsen et al., 1982), plasma was examined for 21 days in 14 patients with septic shock, 7 of whom died. All patients had significantly lower than normal levels of Factor XII (<60%), prekallikrein (<70%) and HMWK (<50%). Patients who subsequently died in this study showed little improvement in these values. The patients

who recovered had gradual increases in their Factor XII levels, and were normal within 2–4 weeks after the shock period. Martínez-Brotóns et al. (1987) examined the initial levels of Factor XII, prekallikrein, HMWK, α_2-M, and C1-inhibitor in uncomplicated sepsis and septic shock. This study involved 24 cases of uncomplicated sepsis (Gram-positive and Gram-negative), all of whom survived, as well as 12 cases of fatal septic shock. They reported significantly lowered levels of Factor XII, prekallikrein, HMWK and α_2-M in fatal septic shock, whereas, in cardiogenic shock, Factor XII, prekallikrein and α_2-M levels were lower than controls, but HMWK levels were normal. Functional C1-inhibitor levels were increased in uncomplicated sepsis but reduced in fatal sepsis. No significant differences occurred in the levels of contact system proteins between Gram-positive and Gram-negative bacteremia in patients with fatal outcome with the exception that patients with Gram-negative organisms had lower C1-inhibitor levels. An alteration in the kinin system in septic shock was postulated to represent a specific response to sepsis instead of being a nonspecific consequence of hemodynamic instability (Martínez-Brotóns et al., 1987).

Carvalho et al. (1988) noted activation of the contact system in the acute respiratory distress syndrome (ARDS) due to trauma or sepsis (Table 10.4). In patients with sepsis, Factor XII, prekallikrein and HMWK levels were significantly reduced, compared to normal controls. C1-inhibitor functional activity was reduced but antigen levels were elevated compared to control. α_2-M levels were also significantly reduced. These observations indicated consumption of inhibitor, presumably due to enzyme:inhibitor complex formation and/or proteolysis to the inactive i-C1-inhibitor, in addition to an acute-phase response of increased C1-inhibitor production, possibly with reduced clearance of complexes.

In patients with severe sepsis, Nuijens et al. (1988,

Table 10.4 Contact activation in sepsis, and assay value changes in contact system proteins. Data are presented as percent changes in patient plasma values (n = 38) relative to normal subjects (n = 10–24)

Plasma protein	Assay type	% Decrease
Factor XII	Functional	56
Factor XI	Functional	(−1)
Prekallikrein	Functional	77
Prekallikrein	Antigenic	48
HMWK	Functional	49
HMWK	Antigenic	20
C1-Inhibitor	Functional	55
C1-Inhibitor	Antigenic	(−75)
α_2-Macroglobulin	Antigenic	38
Antithrombin III	Antigenic	42

HMWK, high molecular weight kininogen. Parentheses indicate a net increase. Data from Carvalho et al. (1988).

Table 10.5 Plasma levels of kallikrein-kinin system proteins in healthy subjects, and in patients with sepsis at time of hospital admission

Plasma protein	Units	Sepsis (48)	Normal (31)
Factor XII	%PNP	35[a]	98
Prekallikrein	%PNP	34[b]	91
C1-inhibitor	%PNP	103	103
i-C1-inhibitor	%PNP	322[a]	99
Factor XIIa:C1-inhibitor[b]	U/ml[a]	< 0.005	< 0.005
Kallikrein:C1-inhibitor[b]	U/ml	0.007	0.006
C1:C1-Inhibitor[b]	U/ml	0.04	0.04
α_2-Macroglobulin	μM	2.33[b]	3.71

Median values: %PNP, % protein in pooled normal plasma; U/ml, 1 Unit/ml of maximum amounts detected after full activation of PNP by dextran sulfate (Factor XIIa or kallikrein) or of serum by aggregated IgG (C1).
[a] Significantly different from control.
[b] Complexes. Numbers in parentheses indicate numbers of subjects. i-C1-inhibitor, inactive C1-inhibitor.
Data from Nuijens et al. (1988, 1989) and Abbink et al. (1991).

Table 10.6 Plasma levels of kallikrein–kinin system proteins in healthy subjects, and in septic patients with SIRS at time of hospital admission

Plasma protein	Units	SIRS (23)	Normal (20–22)
Factor XII:Fn[a]	U/ml	1.00	1.16
Prekallikrein:Fn	U/ml	0.67**	1.04
Prekallikrein:Ag[a]	U/ml	0.52**	1.00
HMWK:Fn	U/ml	0.72**	1.11
HMWK:Ag	U/ml	0.86*	1.04
Kal:α_2-M:Ag[b]	nM	0.97**	0.00
Factor XI:Fn	U/ml	0.71	0.88

All values are means, and numbers in parentheses are numbers of subjects.
[a] Fn denotes functional protein, and Ag denotes antigenic.
[b] Antigenic complex between kallikrein and α_2-macroglobulin.
** $P < 0.01$, * $P < 0.05$, level of statistically significant difference from control. SIRS, systemic inflammatory response syndrome; HMWK, high molecular weight kininogen. Data are from Pixley et al. (1995).

1989) and Abbink et al. (1991) reported that functional levels of C1-inhibitor were not different from normal individuals in their total patient population, but were significantly lower in patients in shock than in those who remained normotensive. Data indicated that complement or contact-system activation are associated with shock (Table 10.5). Factor XII and prekallikrein concentrations were decreased, and no significant differences in enzyme:C1-inhibitor complexes were found. The most plausible explanation for this finding is that clearance rates of enzyme:inhibitor complexes from the blood are too rapid to enable detection of complexes. Levels of i-C1-inhibitor markedly increased, indicating the presence of dysfunctional inhibitor with a slower clearance rate, whereas α_2-M was lower than in controls (Table 10.5). Patients in shock had lower levels than patients without shock (not shown), and lower levels of functional α_2-M correlated with those of Factor XII and prekallikrein.

2.2 SYSTEMIC INFLAMMATORY RESPONSE SYNDROME

We measured plasma kallikrein–kinin system proteins in SIRS patients admitted to an intensive care unit with a view to determining whether admission levels have prognostic value in clinical sepsis (Pixley et al., 1995). Since patients manifesting SIRS are in different phases of the syndrome, we analyzed maximal or minimal levels of kinin system proteins in blood samples taken upon admission to the study, at 2, 12, 24, 48 and 72 h, and at time of discharge or death.

2.2.1 Admission values

All patients had significantly lower values than normal individuals for antigenic prekallikrein (PK:Ag), functional

prekallikrein (PK:Fn) and HMWK:Fn (Table 10.6). Elevated levels of kallikrein (Kal):α_2-M complexes were observed in all groups. These results indicate that the contact system is activated in SIRS, with consumption of prekallikrein and HMWK, and formation of enzyme:inhibitor products upon admission. Levels of Factor XII and Factor XI were not different from normal, supporting earlier studies (Pixley et al., 1992; DeLa Cadena et al., 1993).

Patient groups were compared according to survival, to whether initially they were hypotensive or normotensive, or had positive or negative blood cultures. All patients were clinically septic by the SIRS criteria. When stratified in this manner there was little difference in the outcomes between the groups suggesting that, although admission values indicated kinin system activation, they did not appear to predict survival.

2.2.2 Serial values

The dynamics of SIRS make it difficult to view assay values of patients serially. To examine kinin system activation, we selected maximal or minimal values of serial samples taken within 48 h in order to determine any prognostic value. Hypotension and the presence of a positive blood cultures were not compared since treatment intervention during the syndrome altered these values. These results are shown in Table 10.7. Minimal kallikrein–kinin protein levels were significantly lower than normal. When the data were compared between the patient groups, only the Factor XII minimal levels in nonsurvivors were significantly lower than survivors, indicating that greater consumption of Factor XII may distinguish nonsurvivors (Table 10.7).

When *maximal* values were compared to normal (Table 10.7), Kal:α_2-M complex values were

Table 10.7   Plasma protein levels in patients with SIRS. Data show minimum and maximum concentrations, and levels at 48 h[a] after admission

Plasma protein	Units	Survivors (n = 14)			Died (n = 9)		
		Min.	Max.	48 h[a]	Min.	Max.	48 h
Factor XII:Fn[b]	U/ml	0.76**	1.43*	1.03	0.57**	0.97	0.69**
Prekallikrein:Fn	U/ml	0.50**	0.87**	0.65**	0.49**	0.72**	0.46**
Prekallikrein:Ag[b]	U/ml	0.30**	0.67*	0.47**	0.39**	0.62**	0.48**
HMWK:Fn	U/ml	0.56**	1.04	0.82**	0.47**	0.80**	0.54**
HMWK:Ag	U/ml	0.69**	1.12	0.88	0.66**	0.83	0.66**
Kal:α_2-M:Ag[c]	nM	0.00	1.20**	0.48	0.00	1.14**	0.68
Factor XI:Fn	U/ml	0.44**	0.86	0.62**	0.50**	0.92	0.63*

[a] Data obtained at 48 h or last value prior to 48 h. All values are means.
[b] Fn denotes functional protein and Ag denotes antigenic.
[c] Antigenic complex between kallikrein and α_2-macroglobulin.
** $P < 0.01$, * $P < 0.05$, level of statistically significant difference from control. SIRS, systemic inflammatory response syndrome; HMWK, high molecular weight kininogen. Data from Pixley et al. (1995).

significantly higher than normal for both groups, although no differences were detected between the patient groups. The Factor XII and Factor XI maximal level comparison shown no significant differences. Prekallikrein maximal levels for antigenic or functional activity were significantly lower than normal, but no differences were evident between the groups. HMWK maximal levels for patients were not different from normal subjects except for functional levels of HMWK (HMWK:Fn) in nonsurvivors, whose maximal levels were significantly different compared to controls (Table 10.7).

2.2.3   48 h values

We also examined a hypothesis that the outcome of disease is predetermined by 48 h. Thus, survivors may closely resemble or approach normal levels of kallikrein–kinin system components as they recover, while nonsurvivors would be at the most severe stage of the syndrome and, therefore, exhibit the most abnormal levels. When compared to healthy normal subjects, 48 h values for Factor XII in survivors remained within the normal range, whereas nonsurvivors had significantly reduced values (Table 10.7). Similarly, when 48 h Factor XII values for survivors were compared with nonsurvivors, a significant difference was found. When comparisons were made between admission and 48 h values within each group, no significant differences were found for Factor XII. This higher level of Factor XII in the survivors versus the 48 h levels in nonsurvivors also indicates that normal or high levels of Factor XII correlate with favorable patient outcome. 48 h prekallikrein levels in both groups were significantly lower than normal (Table 10.7), similar to that observed in their admission values (Table 10.6). These findings indicate that prekallikrein levels fail to distinguish survivors from nonsurvivors.

Although both SIRS groups exhibited dramatic differences in HMWK levels at 48 h, compared to normal, their levels failed to distinguish between survivors and nonsurvivors. When 48 h values were compared to healthy levels, no difference for HMWK:Ag was found for survivors, while a significant difference was noted for HMWK:Fn. The nonsurvivor 48 h HMWK:Ag and HMWK:Fn concentrations were significantly lower than normal.

2.2.4   Post-48 h values

The data in the nonsurvival group after 48 h were not sufficient to justify a valid analysis for prognosis, but may be useful in following the kallikrein–kinin system as a prognostic indicator. When such an analysis is carried out, the last values for nonsurvivors being compared with the last values obtained from survivors (discharge or 72 h), similar results to the 48 h analysis were observed. The first and last value analysis for HMWK:Fn for survivors was significantly higher, with the average value rising from low to near normal levels (0.071 ± 0.063 to 0.956 ± 0.064 U/ml). HMWK:Ag levels in survivors were not significantly different from initial values. Similarly, nonsurvivors levels of HMWK:Fn (0.741 ± 0.119 to 0.516 ± 0.071 U/ml) and HMWK:Ag (0.802 ± 0.104 to 0.721 ± 0.100 U/ml) were not significantly different.

In summary, changes observed in proteins on diagnosis of SIRS at time of admission to the intensive care unit suggested activation of the kallikrein–kinin system. This finding is supported by the lowered HMWK and prekallikrein levels (by 20–40%), and higher Kal:α_2-M complex levels. Since BK may contribute to hypotension, this parameter was used as one of the selection criteria of SIRS. When the initial mean systemic arterial pressure (MSAP) of all patients was examined, hypotension was not prognostic since blood pressures were normal in five out of nine patients who died after the onset of SIRS.

3.   Controlled Studies

Clinical studies differ from experimental models in that the causative microorganisms or microbial toxins may vary greatly, and the septic challenge is less uniform with respect to time course and intensity. Animal models of septicemia and low-dose endotoxinemia in human volunteers allows control of the initiating agent, disease severity and alignment with the pathophysiological symptoms of inflammation. With blood sampling at fixed intervals, correlation of laboratory determinations with hemodynamic changes and organ dysfunction can be performed to formulate hypotheses of pathophysiological mechanisms. Additionally, by observing changes in biochemical assays, the sequence of mediator systems activated can be elucidated.

3.1.   SEPSIS AND THE KALLIKREIN–KININ SYSTEM IN BABOONS

We investigated the role of the kinin system in a model of bacteremia in baboons (Pixley et al., 1992, 1993) to determine if there is an association with changes in blood pressure. Lethal concentrations of E. coli, infused over 2 h, produced hypotension and death, and the data suggested that irreversible hypotension correlated temporally with the prolonged activation of the kinin system. In this model, lethal doses of E. coli cause an initial decline in MSAP after 60 min, followed by a gradual, secondary hypotension (Fig. 10.3). A nonlethal dose also results in a primary decline in MSAP without a secondary decline but, rather, a return to normal blood pressure by 300 min. In the nonlethal model, levels of functional HMWK declined to 80% within 360 min and returned to normal thereafter. A small, but significant increase in Kal:α_2-M occurred at 180 min. No significant correlation were found between the changes occurring with HMWK, Kal:α_2-M and MSAP (not shown).

In the lethal model, no significant changes were observed for Factor XII, Factor XI and prekallikrein (Fig. 10.3). In contrast, there was a marked decline in the levels of HMWK within 30 min, and this was significantly different at 4 and 6 h. The secondary declines in MSAP and HMWK were related temporally, suggesting a relationship. As a reflection of kallikrein activation, Kal:α_2-M formation increased dramatically at 60 min, correlating inversely with the decline in blood pressure. From these studies we determined that significant changes in the kinin system, including the rise in Kal:α_2-M and the decrease in functional HMWK, occurred in the lethal but not in the nonlethal model. Blood levels of HMWK and Kal:α_2-M were the most sensitive indicators of contact activation.

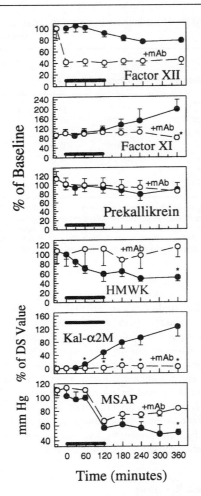

Figure 10.3   Serial measurements of kallikrein–kinin system proteins and mean systemic arterial pressure (MSAP) in a lethal sepsis baboon model. Closed circles represent the response of animals to lethal injections of E. coli. Open circles represent the response of E. coli-infected baboons treated with a monoclonal antibody (mAb) to Factor XII. The solid bar denotes the time of bacterial infusion. HMWK, high molecular weight kininogen; Kal-α2M, kallikrein–α_2-macroglobulin; DS, dextran sulfate.

3.1.1   Use of an Anti-Factor XII Antibody to Establish the Role of the Kallikrein–Kinin System in Sepsis

We utilized a monoclonal, neutralizing antibody (C6B7) to the light chain of Factor XII to investigate the hypothesis that there is a causal relationship between the kallikrein–kinin system and pathophysiological responses to lethal sepsis. This antibody inhibits the coagulant activity of Factor XII and cleavage of HMWK in vitro. All animals had disseminated intravascular coagulation (DIC), which was unaltered by C6B7 and, although antibody significantly prolonged survival times, four of the five animals died from DIC. Despite incomplete inhibition of Factor XII activity in monoclonal antibody (mAb)-treated animals, activation of the kinin system was decreased as evidenced by the minimal formation of Kal:α_2-M, and the preservation of HMWK activity (Fig. 10.3). Moreover, since activation of the kinin system in lethal sepsis seemed to increase Factor XI (Fig. 10.3), this coagulant activity may be due to increased synthesis or release of Factor XI from the liver or to circulating Factor XIa. Factor XI is normally present at 40 nM in humans and, based on our assay, is about the same in baboons. In one normal baboon, where only TNF was injected, a progressive increase in the coagulant activity of Factor XI was observed (not shown). We also found increased Factor XI in humans administered TNF during cancer therapy. Thus, we tentatively ascribe the increase in Factor XI to an effect of TNF.

Antibody treatment did not modulate the initial hypotension associated with sepsis. In both groups of animals (lethal and mAb-treated lethal), initial MSAP declined during the first 2 h of the experiment (Fig. 10.3). In the untreated group only, contact activation occurred as shown by the fall in HMWK and the rise in Kal:α_2-M complexes (Fig. 10.3). We cannot exclude the possibility that some Factor XII activation occurred at the antibody concentrations used, although formation of Kal:α_2-M complexes were not detectable, indicating marked inhibition of kinin system by the mAb.

These observations suggest that, although the contact system was activated in the untreated group, initially there is a more potent mediator affecting the MSAP. In another study in baboons Creasy et al. (1991) reported that TNF and endotoxin levels peak, then decline in the first 2 h. Interleukin (IL)-1 and IL-6 were detected starting at 1 h but peaked at 2–3 h. In another study, anti-TNF antibody, infused 30 min after initiating the infusion of the E. coli, largely eliminated the initial decline observed during the 1 and 2 h, indicating that TNF and the subsequent cytokine release may be a major contributor to the dramatic decline of MSAP during the second hour (Hinshaw et al., 1990). In our study, a more important effect of the antibody against Factor XII was inhibition of the secondary, irreversible hypotension (Fig. 10.3). We conclude that the sustained decline in MSAP after the initial fall may be attributed, at least in part, to BK.

3.1.2 Effects on Complement Activation by the Kallikrein–Kinin System

De Boer et al. (1993) examined activation of the complement system in the lethal sepsis baboon model, by measuring C3b/c, C4b/c (proteolytic products of C3 and C4 released into plasma), and C5b–9 over 6 h (see Fig. 10.2). Release of C4b and C4c reflects the classical activation pathway initiated by C1 activation, and release of C3b and C3c, both the classical and alternative pathway of activation. C5b–9 is the MAC. Selected results of that study are shown in Table 10.8. Almost immediately after the start of E. coli infusion, levels of C3b/c increased steeply to eight times the baseline levels and remained high throughout the experiment. This indicates that complement activation occurs upon introduction of bacteria. C4b/c also increased up to five-fold at 3 h, indicating that part of C3 was activated via the classical pathway. Levels of the C5b–9 (MAC) also increased rapidly, peaking at 3 h. Circulating i-C1-inhibitor in these baboons increased immediately after the start of bacterial administration, gradually increasing 2–4-fold over baseline after 2 h. A rise in C1:C1-inhibitor levels paralleled the increase in i-C1-inhibitor, again reflecting activation of the classical pathway. It is of note that the 1–2 h period in this model is the critical time where a rise in Kal:α_2-M and a decline in HMWK, indicative of kinin production, are evident (Fig. 10.3). Thus, bacterial activation of complement leads to rapid consumption of C1-inhibitor, the major regulatory inhibitor of the kallikrein–kinin system.

We examined treated baboons with lethal sepsis with anti-Factor XII to test the hypothesis that the kinin system may contribute to complement activation (Jansen et al., 1994). Complement activation was confirmed in the lethal control group by the observed increase in C4b/c (classical pathway activation) and C3b/c (alternative pathway), as observed in the study described in (Table 10.8; De Boer et al., 1993). In the antibody-treated

Table 10.8 Activation of the complement system in baboons inoculated with lethal infections of *E. coli* (sepsis). Data shown percent increases in complement proteins above baseline

| Protein | Baseline | Time (min) | | | | | |
		30	60	120	180	240	360
i-C1-inhibitor	0	30	65	170	200	400	272
C1:C1-inhibitor	0	—	—	53	—	—	64
C3b/c	0	837	923	673	1,289	545	716
C4b/c	0	175	284	369	448	80	325
C5b-9	0	893	1,103	1,103	1,137	1,029	733

i-C1-inhibitor, inactive C1-inhibitor.
Data are adapted from De Boer et al. (1993).

animals, decreases in C4b/c and C3b/c were observed at and beyond the critical kinin period of 1–2 h, suggesting a mechanistic connection between complement activation and the kinin system. A possible explanation may be the maintenance of C1-inhibitor levels, protecting against both complement and kallikrein–kinin system activation. Alternatively, direct complement activation may occur due to Factor XII acting on C1.

3.2   The Kallikrein–Kinin System in Human Endotoxinemia

DeLa Cadena *et al.* (1993) reported on contact system proteins obtained from human volunteers administered low doses of endotoxin. These individuals developed a flu-like illness which subsided in 24 h. In 18 volunteers, blood samples were taken at 0, 1, 2, 3, 5 and 24 h after endotoxin injection, and examined for evidence of contact activation (Fig. 10.4). Factor XII and C1-inhibitor levels did not change, whereas HMWK, Factor XI and prekallikrein decreased, and, by 5 h, reached a nadir that was accompanied by a five-fold increase in Kal:α_2-M. Prekallikrein levels were apparently decreased and Kal:α_2-M levels increased – two indicators that the kinin system was activated. As observed in septic patients a decrease in functional C1-inhibitor was not detected (Nuijens *et al.*, 1988, 1989).

4.   *General Findings of the Kallikrein–Kinin System in Sepsis*

Functional and immunological measurements during the sepsis syndrome indicate that contact activation occurs in

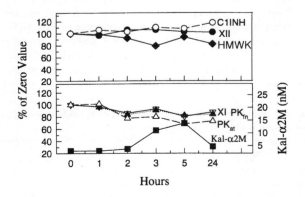

Figure 10.4   Serial measurements of kallikrein–kinin system proteins in a human volunteer nonlethal model of sepsis. C1 INH, C1-inhibitor; XI and XII, Factors XI and XII, respectively; HMWK, high molecular weight kininogen; PK$_{fn}$ and PK$_{at}$, functional and antigenic prekallikrein, respectively; Kal-α2M, kallikrein–α_2 macroglobulin.

the clinical and animal model studies discussed. When activation of the kallikrein–kinin system occurs, only a small percentage of the total concentrations (1–5%) of zymogens need be converted to active enzymes, which effectively cleave their substrates (Scott *et al.*, 1985). Thus, a modest decline in Factor XII and prekallikrein functional levels can indicate a greater release of BK. A large decline would suggest profound activation with large releases of BK. In septic baboons, only modest changes could be measured for Factor XII or prekallikrein, but more dramatic increases in Kal:α_2-M and impressive decreases in functional HMWK concentrations indicated activation of the kinin system (Fig. 10.3). In clinical sepsis or sepsis syndrome, there are major declines in Factor XII and prekallikrein levels (Carvalho *et al.*, 1988; Pixley *et al.*, 1995). Low consumption of Factor XII and prekallikrein, without observed changes in C1-inhibitor, were observed in the minimal activation of the kinin system induced by a very low dose of endotoxin in humans (Fig. 10.4). Activation of Factor XII and prekallikrein, and effects on their substrates is limited by the presence of their respective plasma protease inhibitors. C1-inhibitor is the most critical inhibitor of both of these active enzymes, and can normally limit the feedback activation between the two proteins. Therefore, with mild activation, there may not be a dramatic drop in these zymogens. A drop can occur if there were a severe, continuous activation of the contact system along with consumption of C1-inhibitor to below levels necessary for effective regulation. Additional evidence is illustrated in C1-inhibitor deficiency of hereditary angioedema (HAE), wherein levels of functional C1-inhibitor are <40% of normal. In the acute-attack phase of HAE, levels of Factor XII and prekallikrein decline at most to 50% of normal (Schapira *et al.*, 1983; Cugno *et al.*, 1990). These findings indicate that measuring the functional levels of the Factor XII or prekallikrein may not always be a sensitive indicator of activation of the kallikrein–kinin system in the clinic.

Functional levels of HMWK are more sensitive and the changes are easier to predict, since HMWK is a substrate for kallikrein, which acts in catalytic amounts, i.e., low levels of kallikrein can cleave and release high concentrations of BK and kinin-free HMWK (HMWKa) with coagulant activity. HMWK circulates as a complex with prekallikrein or Factor XI, and in the free form (under normal conditions, the concentration of HMWK in plasma is 0.66 μM, and the concentrations of prekallikrein and Factor XI are 0.40 and 0.04 μM, respectively). As kallikrein cleaves HMWK, BK is released. This cleaved HMWKa has unaltered coagulant activity, but binds more readily to negatively charged surfaces *in vitro*. The HMWKa light chain is then cleaved by proteases, such as Factor XIa (Scott *et al.*, 1985), plasmin (Kleniewski and Donaldson, 1987) or elastase (Kleniewski and Donaldson, 1988) to inactivate its cofactor function. Loss of HMWK activity requires the initial cleavage by kallikrein, which facilitates the inactivation of the HMWKa. Therefore, as observed in clinical sepsis syndrome, as well as in septic baboons, there is apparently a major decline in functional HMWK upon activation of the kinin system.

Blood Kal:α_2-M levels, normally undetectable (Kaufman et al., 1991), increase with the generation of kallikrein. Although Kal:α_2-M is rapidly cleared in vivo, clearance is modulated by cytokines (LaMarre et al., 1993). Another major inhibitor, C1-inhibitor, also inactivates kallikrein. If functional C1-inhibitor is consumed or inactivated, α_2-M becomes a significant regulator of the contact system. This was observed, with the complex being found in patients with sepsis syndrome (Pixley et al., 1995), and in septic baboons (Pixley et al., 1992, 1993).

Although the pathophysiology of hypotension, frequently occurring in patients with bacteremia, is not fully understood, increased vascular permeability and arteriolar vasodilatation are important. Vasodilatation may be initiated by release of bacterial cell wall components such as endotoxin, from Gram-negative organisms and peptidoglycan-polysaccharides from Gram-positive pathogens. These molecules activate the kinin and complement systems with release, respectively, of kinins and C3a and C5a, which induce hypotension. Endotoxin also activates monocytes to produce cytokines, including IL-1 and TNF, which contribute indirectly to vasodilatation and capillary hyperpermeability.

Bacterial proteases activate Factor XII or prekallikrein directly in vitro (Kalter et al., 1983; Molla et al., 1989). Alternatively, the presence of bacteria in the blood can stimulate neutrophils and/or monocytes, causing release of proteolytic enzymes or active oxygen radicals, which in turn perturb endothelial cells, and expose and alter components of the subendothelial layer. Oxygen radicals can oxidase methionines at critical cleavage sites of SERPINs, such as α_1-PI and α_1-AP. Additionally, complement activation consumes C1-inhibitor. These changes can create a reactive environment by providing an activating surface and/or an area protected from SERPINs favoring activation of the kinin system, beginning with Factor XII, and allow sustained activation of the system.

4.1 ACTIVATION SEQUENCE IN BABOON SEPSIS

Evidence suggests a hypothesis to explain the findings with septic baboons for activation of the kinin system upon exposure to bacteria or endotoxin, and aids in interpreting clinical studies. Upon exposure of monocytes to endotoxin, TNF and IL-1 are released. These cytokines act on the vascular endothelium and smooth muscle, causing vasodilatation and a decline in MSAP, as observed during the 1–2 h interval (Fig. 10.3).

During the first hour, the cytokines also stimulate neutrophils, causing margination and degranulation, resulting in endothelial cell damage, thereby exposing contact-activating surfaces, such as elastin, collagen and basement membrane. During the first 60 min, sustained complement activation is occurring in addition to limited but sustained kallikrein–kinin system activation, with BK contributing to vasodilatation. The contribution of BK to the decline in MSAP may be masked by the more dramatic effects of TNF on endothelial cells to stimulate formation of the vasodilators PGI$_2$ and nitric oxide. During the first hour, the contact and complement systems are highly regulated by C1-inhibitor. Formation of Kal:α_2-M complexes during this interval is not significant, since C1-inhibitor predominates. However, C1-inhibitor is consumed by human plasma proteases or by bacterial protease cleavage, resulting in lower concentrations, subsequently allowing an increase in α_2-M regulation of kallikrein after an hour. At 2 h, or the end of bacterial infusion, endotoxin levels decline along with the levels of TNF and IL-1. At this time, C1-inhibitor would be low as the damaged endothelium continues to expose activating surfaces, sustaining contact activation and BK release. Continued release of BK probably sustains and enhances the hypotension after 2 h, allowing for irreversible organ damage and eventual death. The role of the kallikrein–kinin system is clear from the results obtained from modulation of hypotension, in the septic baboon model, with a monoclonal antibody to Factor XII.

This mechanism is supported by the nonlethal sepsis baboon model (Pixley et al., 1992) and the human endotoxin model (DeLa Cadena et al., 1993), wherein the contact system is less active. In the nonlethal baboon model, MSAP declines as a result of cytokine or other factors. However, owing to a lack of sustained contact activation and BK release, MSAP was able to return to baseline after 2 h. In humans, nonlethal doses of endotoxin resulted in a modest decline in MSAP at 3–5 h (Suffredini et al., 1989), during the time that Kal:α_2-M is generated (Fig. 10.4) indicting kinin system activation. This is an indication that BK may be contributing to the secondary decline in blood pressure.

4.2 CONCLUDING REMARKS

Many of the reactions described in this chapter are illustrated in Fig. 10.5. Therapy directed against specific mediators, such as BK, either receptor antagonists or inhibitors of the kallikrein–kinin system enzymes, or antibodies or C1-inhibitor, which can abort a continuing decline in MSAP after the initial bacteremic attack, may allow continued tissue perfusion and aid in preventing extensive organ damage. However, the uncorrected DIC may also be a major contributor to clinical mortality. Inhibition of kinin formation or effects, in combination

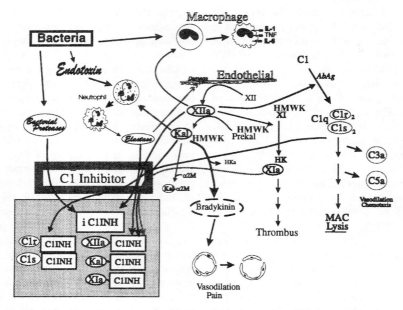

Figure 10.5   Summary of kallikrein–kinin system activation and responses in sepsis syndrome. The presence of bacteria activates many host responses, which lower the normally protective amounts of active C1-inhibitor (C1-INH) owing to its consumption by the cascade, its forming complexes with activated contact and complement factors, or by its cleavage to an inactive form by bacterial proteases. These and other actions provide an environment which allows the prekallikrein–Factor XII feedback activation mechanism to proceed unchecked, giving rise to increased release of bradykinin and related kinins. IL-1 and IL-6, interleukin-1 and -6, respectively; TNF, tumor necrosis factor; XI and XII, Factors XI and XII, respectively; HMWK, high molecular weight kininogen; HMWKa, kinin-free HMWK; Kal, kallikrein; Prekal, prekallikrein; Kal-α2M, kallikrein–α_2 macroglobulin; MAC, membrane attack complex; i-C1-INH, inactive C1-INH.

with a drug to control DIC, may allow the administered antibiotics time to contain the bacterial proliferation, thus decreasing the high mortality rate for sepsis and sepsis syndrome.

5.   References

Abbink, J.J., Nuijens, J.H., Eerenberg, A.J.M., Huijbregts, C.C.M., Strac van Schijndel, R.J.M., Thijs, L.G. and Hack, C.E. (1991). Quantification of functional and inactivated α_2-macroglobulin in sepsis. Thromb. Haemostas. 65, 32–39.

Bone, R.C. (1991). Pathogenesis of sepsis. Ann. Intern. Med. 115, 457–469.

Bone, R. (1992). Sepsis and multiple organ failure: consensus and controversy. In "Mediators of Sepsis" (eds. M. Lamy and L.G. Thijs), pp. 3–12. Springer-Verlag, New York.

Bone, R.C. (1993). Why new definitions of sepsis and organ failure are needed. Amer. J. Med. 95, 348–350.

Carvalho, A.C., DeMarinis, S., Scott, C.F., Silver, L., Schmaier, A.H. and Colman, R.W. (1988). Activation of the contact system of plasma proteolysis in the adult respiratory distress syndrome. J. Lab. Clin. Med. 112, 270–277.

Cooper, N.R., Miles, L.A. and Griffin, J.H. (1980). Effects of

plasma kallikrein and plasmin on the first complement component. J. Immunol. 124, 1517 (abstract).

Creasey, A.A., Stevens, P., Kenney, J., Allison, A.C., Warren, K., Catlett, R., Hinshaw, L. and Taylor, F.B., Jr (1991). Endotoxin and cytokine profile in plasma of baboons challenged with lethal and sublethal Escherichia coli. Circ. Shock 33, 84–91.

Cugno, M., Nuijens, J., Hack, E., Eerenberg, A., Frangi, D., Agostoni, A. and Circardi, M. (1990). Plasma levels of C1-inhibitor complexes and cleaved C1-inhibitor in patients with hereditary angioneurotic edema. J. Clin. Invest. 85, 1215–1220.

De Agostini, A., Lijnen, H.R., Pixley, R.A., Colman, R.W. and Schapira, M. (1984). Inactivation Of Factor XII active fragment in normal plasma. Predominant role of C1 inhibitor. J. Clin. Invest. 73, 1542–1549.

De Boer, J.P., Creasey, A.A., Chang, A., Roem, D., Eerenberg, J.M., Hack, C.E. and Taylor, F.B., Jr (1993). Activation of the complement system in baboons challenged with live Escherichia coli: correlation with mortality and evidence for a biphasic activation pattern. Infect. Immun. 61, 4293–4301.

DeLa Cadena, R.A., Suffredini, A.F., Page, J.D., Pixley, R.A., Kaufman, N., Parrillo, J. and Colman, R.W. (1993). Activation of the kallikrein–kinin system after endotoxin administration to normal human volunteers. Blood 81, 3313–3317.

DiScipio, R.G. (1982). The activation of the alternative pathway C3 convertase by human plasma kallikrein. J. Immunol. 45, 587–595.

Duswald, K.H., Jochem, M., Schramm, W. and Fritz, H. (1985). Released granulocytic elastase: an indicator of pathobiochemical alterations in septicaemia after abdominal surgery. Surgery 98, 892–899.

Egbring, R., Schmidt, W., Fuchs, G. and Havermann, K. (1977). Demonstration of granulocytic proteases in plasma of patients with acute leukemia and septicemia with coagulation defects. Blood 49, 219–231.

Ghebrehiwet, B., Silverberg, M. and Kaplan, A.P. (1981). Activation of the classical pathway of complement by Hageman factor fragment. J. Exp. Med. 153, 665–676.

Ghebrehiwet, B., Randazzo, B.P., Dunn, J.T., Silverberg, M. and Kaplan, A.P. (1983). Mechanisms of activation of the classical pathway of complement by Hageman Factor fragment. J. Clin. Invest. 71, 1450–1455.

Goetzl, E.J. and Austen, K.F. (1974). Stimulation of human neutrophil aerobic glucose metabolism by purified chemotactic factors. J. Clin. Invest. 53, 591–599.

Grossman, N. and Leive, L. (1984). Complement activation via the alternative pathway by purified salmonella lipopolysaccharide is affected by its structure but not its O-antigen length. J. Immunol. 132, 376–385.

Hinshaw, L.B., Tekamp-Olson, P., Chang, A.C.K., Lee, P.A., Taylor, F.B., Jr, Murray, C.K., Peer, G.T., Emerson T.E., Jr, Passey, R.B. and Kuo, G.C. (1990). Survival of primates in LD_{100} septic shock following therapy with antibody to tumor necrosis factor (TNF). Circ. Shock 30, 279–292.

Jansen, P.M., Pixley, R.A., Brouwer, M., de Jong, I.W., Chang, A.C.K., Hack, C.E., Taylor, F.B. Jr., and Colman, R.W. (1996) "Inhibition of factor XII in septic baboons attenuates activation of complement and fibrinolytic systems and reduces the release of interleukin-6 and neutrophil elastase." Blood 87, 2337–2344.

Kalter, E.S., Dijk, W.C., Timmermen, A., Verhoef, J. and Bouma, B.N. (1983). Activation of purified human plasma prekallikrein triggered by cell wall fractions of Escherichia coli and Staphylococcus aureus. J. Infect. Dis. 4, 682–691.

Kaplan, A.P., Kay, A.B. and Austen, K.F. (1972). A prealbumin activator of prekallikrein. III. Appearance of chemotactic activity for human neutrophils by the conversion of prekallikrein to kallikrein. J. Exp. Med. 135, 81–97.

Kaufman, N., Page, J.D., Pixley, R.A., Schein, R., Schmaier, A.H. and Colman, R.W. (1991). α_2-Macroglobulin-kallikrein complexes detect contact system activation in hereditary angioedema and human sepsis. Blood 77, 2660–2667.

Kleniewski, J. and Donaldson, V.H. (1987). Comparison of human high molecular weight kininogen digestion by plasma kallikrein and by plasmin. A revised method of purification of high molecular weight kininogen. J. Lab. Clin. Med. 109, 469–479.

Kleniewski, J. and Donaldson, V.H. (1988). Granulocyte elastase cleaves human high molecular weight kininogen and destroys its clot promoting activity. J. Exp. Med. 167, 1895–1907.

LaMarre, J., Wolf, B.B., Kittler, E.L.W., Queensberry, P.J. and Gonias, S.L. (1993). Regulation of macrophage α_2-macroglobulin receptor/low density lipoprotein receptor-related protein by lipopolysaccharide and interferon-γ. J. Clin. Invest. 91, 1219–1224.

Martínez-Brotóns, F., Oncins, J.R., Mestres, J., Amargó, V. and Reynaldo, C. (1987). Plasma kallikrein–kinin system in patients with uncomplicated sepsis and septic shock – comparison with cardiogenic shock. Thromb. Haemostas. 58, 709–713.

Mason, J.W., Kleebe, U., Dolan, P. and Colman, R.W. (1970). Plasma kallikrein and Hageman factor in Gram-negative bacteremia. Ann. Intern. Med. 73, 545–551.

Molla, A., Yamamoto, T., Akaike, T., Akaike, T., Miyoshi, S. and Maeda, H. (1989). Activation of Hageman factor and prekallikrein and generation of kinin by various microbial proteases. J. Biol. Chem. 264, 10589–10594.

Morrison, D.C. and Kline, L.F. (1977). Activation of the classical and properdin pathways of complement by bacterial lipopolysaccharides (LPS). J. Immunol. 118, 362–368.

Nuijens, J.H., Huijbregts, C.C.M., Eerenberg-Belmer, A.J.M., Abbink, J.J., Strac van Schijndel, R.J.M., Felt-Bersma, R.J.F., Thijs, L.G. and Hack, C.E. (1988). Quantification of plasma factor XIIa-C1-inhibitor complexes in sepsis. Blood 72, 1841–1848.

Nuijens, J.H., Eerenber-Belmer, A.J.M., Huijbregts, C.C.M., Schreuder, W.O., Felt-Bersma, R.J.F., Abbink, J.J., Thijs, L.G. and Hack, C.E. (1989). Proteolytic inactivation of plasma C1 inhibitor in sepsis. J. Clin. Invest. 84, 443–450.

Nuijens, J.H., Abbink, J.J., Wachtfogel, Y.T., Colman, R.W., Eerenberg, A.J.M., Dors, D., Kamp, A.J.M., Strack van Schijndel, R.J.M., Thijs, L.G. and Hack, C.E. (1992). Plasma elastase-α_1-antitrypsin and lactoferrin in sepsis: evidence for neutrophils as mediators in fatal sepsis. J. Lab. Clin. Med. 119, 159–168.

Parillo, J.E. (1993). Pathogenic mechanism of septic shock. N. Engl. J. Med. 328, 1471–1477.

Pixley, R.A., Schapira, M. and Colman, R.W. (1985). The regulation of factor XIIa by plasma proteinase inhibitors. J. Biol. Chem. 260, 1723–1729.

Pixley, R.A., DeLa Cadena, R.A., Page, J.D., Kaufman, N., Wyshock, E.G., Colman, R.W., Chang, A. and Taylor, F.B., Jr (1992). Activation of the contact system in lethal hypotensive bacteremia in a baboon model. Am. J. Pathol. 140, 897–906.

Pixley, R.A., DeLa Cadena, R., Page, J.D., Kaufman, N., Wyshock, E.G., Change, A., Taylor, F.B. and Colman, R.W. (1993). The contact system contributes to hypotension but not disseminated intravascular coagulation in lethal bacteremia. J. Clin. Invest. 91, 61–68.

Pixley, R.A., Zellis, S., Bankes, P., DeLa Cadena, R.A., Page, J.D., Scott, C.F., Kappelmayer, J., Wyshock, E.G., Kelly, J.J. and Colman, R.W. (1995). Prognostic value of assessing contact system activation and factor V in systemic inflammatory response syndrome. Crit. Care Med. 23, 41–51.

Schapira, M., Scott, C.F. and Colman, R.W. (1981). Protection of human plasma kallikrein from inactivation by C1 inhibitor and other protease inhibitors. The role of high molecular weight kininogen. Biochemistry 20, 2738–2743.

Schapira, M., Despand, E., Scott, C.F., Boxer, L.A. and Colman, R.W. (1982). Purified human plasma kallikrein aggregates human blood neutrophils. J. Clin. Invest. 69, 1199–1202.

Schapira, M., Silver, L.D., Scott, C.F., Schmaier, A.H., Prograis, L.J., Curd, J.G. and Colman, R.W. (1983). Prekallikrein activation and high molecular weight kininogen consumption in hereditary angioedema. N. Engl. J. Med. 308, 1050–1053.

Scott, C.F., Schapira, M., James, H.L., Cohen, A.B. and Colman, R.W. (1982). Inactivation of factor XIa by plasma protease inhibitors. Predominant role of α_1-protease inhibitor and protective effect of high molecular weight kininogen. J. Clin. Invest. 69, 844–852.

Scott, C.F., Silver, L.D., Purdon, A.D. and Colman R.W. (1985). Cleavage of high molecular weight kininogen by plasma factor XIa in vitro. J. Biol. Chem. 260, 10856–10863.

Sim, R.B., Arlaud, G.J. and Colomb, M.G. (1980). Kinetics of reaction of human C1-inhibitor with the human complement system proteases C1r and C1s. Biochim. Biophys. Acta 612, 433–449.

Smith-Erichsen, N., Aasen, A.O., Gallimore, M.J. and Amundsen, E. (1982). Studies of components of the coagulation systems in normal individuals and septic shock patients. Circ. Shock 9, 491–497.

Suffredini, A.F., Fromm, R.E., Parker, M.M., Brenner, M., Kovacs, J.A., Wesley, R.A. and Parillo, J.E. (1989). The cardiovascular response of normal humans to the administration of endotoxin. N. Engl. J. Med. 321, 280–287.

Wachtfogel, Y.T., Kuicich, U., James, H.L., Scott, C.F., Schapira, M., Zimmerman, M., Cohen, A.B. and Colman, R.W. (1983). Human plasma kallikrein releases neutrophil elastase during blood coagulation. J. Clin. Invest. 72, 1672–1677.

Wiggins, R.C., Giglas, P.C. and Henson, P.M. (1981). Chemotactic activity generated from the fifth component of complement by plasma kallikrein of the rabbit. J. Exp. Med. 153, 1391–1404.

11. The Kinin System and Neutrophils

Yugen Naidoo *and* Kanti Bhoola

1. Historical Overview

Initial observations on the kallikrein–kinin system were made by a surgeon, E.K. Frey (1926), when he injected human urine into dogs. The observed hypotensive response was attributed to a substance, thought to be a hormone, and named "kallikrein" after a Greek synonym for the pancreas, where it is present in high concentrations (Kraut *et al.*, 1928, 1930) and originally thought to be its source (reviewed by Bhoola *et al.*, 1992). The protein responsible for the hypotensive effect was isolated and subsequently shown to be a nondialysable, thermolabile substance of high molecular weight (Frey and Kraut, 1928; see Chapter 1 of this volume for a historical perspective on the kinin system). These observations were followed by studies by Werle and coworkers (1937), who demonstrated that kallikrein was present in blood, pancreas and salivary glands. Werle and Berek (1948) went on to study the biochemical properties of kallikrein and subsequently discovered it to be a proteolytic enzyme, which released a smooth muscle spasmogen from an inactive precursor (kininogen). Further, it was observed that the biological activity of this substance increased and then rapidly disappeared. The new biologically active molecule was called "kallidin" (Werle and Berek, 1948), which, unlike kallikrein, was dialyzable and thermostable, but similarly caused contractions of guinea-pig isolated ileum and showed marked hypotensive activity (Werle and Grunz, 1939). Kallikrein and kallidin (referred to in this volume as *Lysyl-bradykinin*, or *Lys-BK*) showed different susceptibilities to inactivation, as aprotinin inhibited only the former. Furthermore, Lys-BK was inactivated by plasma and tissue proteases, and this was attributed to degradation by peptidases, the kininases. Kallidin was the first of the kinins to be discovered.

At around the same time, Brazilian pharmacologists discovered another kinin (Rocha e Silva *et al.*, 1949), which was also released from a plasma protein by trypsin or snake venom proteases. They called this new substance "bradykinin" (BK), because it produced a slow contraction of guinea-pig ileum (see Chapter 1). Werle *et al.* (1950) showed that both Lys-BK and BK were released from the same precursor proteins, the high and low molecular weight kininogens (Fritz *et al.*, 1988).

2. Production and Metabolism of Kinins

2.1 KALLIKREINS

The kallikreins are a group of serine proteases present in glandular cells, neutrophils and biological fluids (Bhoola *et al.*, 1979). On the basis of their molecular weights, isoelectric points, substrate specificities, immunological characteristics and type of kinin they release, they are designated as plasma and tissue kallikreins. The molecular biology and roles of kallikreins in inflammation are reviewed in Chapter 5 of the present volume.

2.1.1   Tissue Kallikrein

Although tissue kallikrein was originally discovered in the pancreas, salivary glands and kidney (Frey and Kraut, 1928), it was only recently described in neutrophils (Bhoola *et al.*, 1992). The discovery, isolation and purification of this enzyme in a large variety of tissues (reviewed in Bhoola *et al.*, 1992) has enabled an extensive study of the biochemical, immunological and enzymatic properties of tissue kallikrein. As noted, tissue kallikrein was recently localized in human circulating (Figueroa and Bhoola, 1990) and synovial (Rahman *et al.*, 1994) neutrophils, and with recent immunocyto-chemical and *in situ* hybridization studies in our laboratory, described in Section 3, we hope to advance knowledge with regard to the storage and synthesis sites of tissue kallikrein in these inflammatory cells.

Tissue kallikrein is an acidic glycoprotein with molecular weight ranging from 30 to 45 kDa, and isoelectric point ranging from 3.5 to 4.4. The enzymes from different tissues show similar enzymatic, immunological and chemical properties. Like other serine proteases of the trypsin, chymotrypsin and elastase group, the serine residue in tissue kallikrein forms a triad with histidine and aspartic acid, and this spatial arrangement forms the catalytic triad necessary for its enzymatic activity and kinin generation. Glycosylation sites vary in number and position in different species and cells in which the enzyme is synthesized. One mole of kallikrein has one glycosylation site in the rat and mouse, whereas porcine pancreatic kallikrein has two or three sites (Fritz *et al.*, 1967), and human urinary kallikrein has three asparagine-linked sites (Lu *et al.*, 1989) and three additional oxygen glycosylation sites linked to two serine and one threonine residues (Kellerman *et al.*, 1988). Human urinary kallikrein is a 238-amino-acid poly-peptide with Ile at the amino terminus and Ser at the carboxyl terminus (Geiger *et al.*, 1979), and the carbohydrate segments contain fucose, mannose, galactose, *N*-acetylglucosamine and sialic acid (Moriya *et al.*, 1983).

Tissue kallikrein is synthesized bound to a 17-amino-acid signal peptide, which is cleaved to produce an inactive precursor, prekallikrein, which is found in most tissue and body fluids. This proenzyme can be activated *in vitro* by proteolytic enzymes, such as trypsin (Kamada *et al.*, 1988) or thermolysin, a bacterial metalloprotease. The cleavage site for the conversion to the active form is at the amino terminus of the hydrophobic amino acids, Peptide CAP-II (Fig. 11.1). The amino terminus of the

Figure 11.1   Peptide CAP-II.

proenzyme is Ala, and after removal of a short, seven-amino-acid link peptide, Ile forms the new N-terminus (Fig. 11.1; Takada *et al.*, 1985). The proenzyme can be activated *in vitro* by either trypsin or thermolysin, which hydrolyzes the Arg-Ile bond in the sequence of the activation peptide (Girolami *et al.*, 1986). The endogenous enzyme that performs this function has not been identified.

Tissue kallikrein forms Lys-BK from the preferred substrate low molecular weight kininogen (LMWK), but the enzyme also releases kinins from both LMWK and high molecular weight kininogen (HMWK) *in vitro* (Iwanaga *et al.*, 1977). Formation of Lys-BK by tissue kallikrein involves the cleavage of the Met-Lys bond at the N-terminus in the kinin sequence of the kininogen molecule (Pisano *et al.*, 1974). Tissue kallikrein, in most mammals, does not produce BK because of its inability to accommodate the Lys-Arg-Pro sequence for hydrolysis Lys-Arg on the amino terminus of the peptide.

Kallistatin is a recently discovered, endogenous human tissue kallikrein inhibitor that was purified and cloned by Chao *et al.* (1996). The reactive site of kallistatin has a unique sequence at P2-P1-P1 of Phe-Phe-Ser. Chao *et al.* (1996) showed that P1-Phe is a crucial specificity determinant, and P2-Phe is important for hydrophobic environment of the kallikrein–kallistatin interaction. Enzyme-linked immunosorbent assays (ELISA) and *in situ* hybridization studies have shown that there is a wide distribution of kallistatin in tissues of healthy and diseased individuals. Plasma kallistatin levels are altered in patients with liver disease, sepsis and pre-eclampsia. The expression of kallistatin in the rat is downregulated during acute phase inflammation and is upregulated by sex hormones.

Techniques for determining tissue kallikrein are diverse, and include utilizing kininogen as a substrate for kinin release, with subsequent bioassay (Bhoola *et al.*, 1962), radioimmunoassay (Miwa *et al.*, 1968) or ELISA (Geiger and Miska, 1986), and high-performance liquid chromatography (HPLC) (Kato *et al.*, 1985). In addition, specific antibodies to tissue kallikrein are used to inhibit its activity. Using these techniques, it has recently been established that circulating neutrophils contain a tissue kallikrein (Figueroa and Bhoola, 1989). *In vivo*, it has been suggested that the liver is the main organ that clears tissue kallikrein from the circulation. Although the classical physiological function of tissue kallikrein is Lys-BK generation, other enzymatic functions have been proposed, and these include conversion of inert precursors of enzymes and hormones into biologically active molecules (Bhoola and Dorey, 1971; Bothwell et al., 1979; Mason et al., 1983).

2.1.2 Plasma Kallikrein

Plasma kallikrein differs from tissue kallikrein in its biochemical, immunological and functional characteristics. As a zymogen, plasma kallikrein is synthesized in hepatocytes and secreted into the blood where it circulates as a heterodimer complex bound to its substrate, HMWK. The plasma kallikrein–HMWK complex and Hageman factor (HF) are involved in activating the complement system (see also Chapter 10). Furthermore, the complex plays an important role in the surface-dependent activation of HF that results in blood coagulation, formation of kinins and fibrinolysis. The proenzymes involved in clotting include Factor XII (HF), Factor XI and plasma prekallikrein. Once the cascade is triggered, clotting occurs to initiate thrombus formation together with the formation of active plasma kallikrein, and the release of kinins on endothelial and subendothelial surfaces. The clotting cascade is initiated by the conversion of inactive Factor XII into active HF (HFa) either by exposure to macromolecular anionic surfaces or by enzymatic action of plasma kallikrein (Cochrane et al., 1973). HFa comprises a heavy chain of 50 kDa, which has the binding site for the attachment to anionic surfaces during activation and a light chain of 28 kDa linked together by a disulfide bond (Revak et al., 1977, 1978). The light chain contains the active site (Revak and Cochran, 1976; Kaplan and Silverberg, 1987) for converting the plasma prekallikrein into an active form (Miller et al., 1980; Silver et al., 1980; Silverberg et al., 1980). When sufficient amounts of plasma kallikrein are formed, the cascade progresses rapidly. Plasma kallikrein forms more HFa and drives the reaction forward to activate Factor XI, thereby enhancing clotting, and plasminogen to initiate fibrinolysis. With rising levels, plasma kallikrein cleaves HMWK from which BK is released.

A single gene codes for plasma prekallikrein (see Chapter 5), which is synthesized in the liver, localized in hepatocytes and secreted as an inactive molecule. HMWK is also synthesized in the liver and, when both HMWK and plasma kallikrein are secreted into the circulation, they form a complex with Factor XI. This complex circulates bound to the outer surface of human neutrophils (Henderson et al., 1992; Naidoo et al., 1994a). Plasma prekallikrein exists as a single-chain glycoprotein in two forms of 85 and 88 kDa in plasma (Talamo et al., 1969). The mature human enzyme is composed of 619 amino acids with 371 amino acids at the amino terminus linked to a catalytic chain of 248 residues.

HMWK is the preferred substrate for plasma kallikrein, which releases BK. Also, and in spite of the fact that LMWK is a poor substrate, plasma kallikrein can release BK from this kininogen in the presence of neutrophil elastase. This occurs because the elastase cleaves a fragment from LMWK from which plasma kallikrein can then release BK (Kitamura et al., 1985). It is possible that this mechanism of BK generation occurs in vivo, and this activity has important implications as LMWK has been localized on the external membrane of the neutrophil (Figueroa et al., 1992). Plasma kallikrein has a significant effect on neutrophils and may, therefore, play an important role in inflammation (Henderson et al., 1994).

2.2 KININOGENS

The endogenous substrates for tissue and plasma kallikrein are the kininogens. Cysteine proteinase inhibitor domains also exist in kininogen molecules, and may act as cofactors for inhibition of lysosomal enzymes released during inflammation. Hepatocytes play a role in the synthesis and release of these proteins into circulation (Kitamura et al., 1985; Müller-Esterl, 1988). HMWK and LMWK are found not only in blood and other biological fluids, but also in several types of cells (see also Chapter 4). Kininogens have been identified in the collecting ducts of the human kidney (Proud et al., 1981; Figueroa et al., 1988), platelets (Schmaier et al., 1983, 1986b), endothelial cells (Schmaier et al., 1988; Chapter 4) and human neutrophils (Gustafson et al., 1989). Figueroa et al. (1990) identified HMWK on the external surface human neutrophils, and described for the first time the presence of LMWK on these cells. In contrast, when platelets are activated, HMWK is translocated from the granules to the external surface of the cell membrane where the binding of the HMWK to the membrane probably involves specific acceptor proteins exposed on the outer surfaces. Binding sites for HMWK, which are specific, saturable and reversible, have been identified on the cell membranes of platelets (Greengard and Griffin, 1984; Gustafson et al., 1986; Meloni and Schmaier, 1991), human endothelial cells (Gustafson et al., 1986; van Iwaarden et al., 1989) and neutrophils (Gustafson et al., 1989).

The kininogens are single-chain glycoproteins, which include the kinin sequence interleafed between the two polypeptides that are bridged by single disulfide loops (Kellerman et al., 1986). HMWK and LMWK differ with respect to structure, size and enzymatic susceptibility. HMWK is composed of 626 amino acids and, depending on species of origin, has a molecular mass of 88–120 kDa (Jacobsen, 1966; Komiya et al., 1974; Kerbiriou et al., 1980). LMWK consists of 409 amino acids and varies in mass from 50 to 68 kDa (Kato et al., 1976; Müller-Esterl et al., 1985). Limited proteolytic analyses have shown multiple domains, HMWK having six and LMWK five (Kato et al., 1981; Müller-Esterl et al., 1986) (Fig. 11.2).

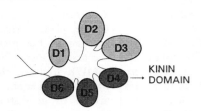

Figure 11.2   High molecular weight kininogen.

The heavy-chain basic structures of HMWK and LMWK are the same, and form the three amino-terminus domains that make up 50–60 kDa. The fourth domain contains the kinin segment followed by a light chain, which varies in the two kininogens (Lottspeich *et al.*, 1985; Kellerman *et al.*, 1986). For HMWK, the light chain has two segments and a molecular weight of 45–58 kDa. The single light chain in LK is smaller, with a sequence of Met-Lys-BK at its amino terminus (Kato *et al.*, 1985). Cysteine proteinases, namely cathepsins B, papain and platelet calpain, are inhibited by domains 2 and 3 of the heavy chain (Ohkubo *et al.*, 1984; Müller-Esterl *et al.*, 1985; Schmaier *et al.*, 1986a). The histidine-rich region of domain 5 in the light chain of HMWK anchors on to anionic surfaces during initiation of the intrinsic clotting cascade. Domain 6 provides a binding site for PPK and Factor XI. Since plasma prekallikrein is coupled to domain 6 of HMWK (Gustafson *et al.*, 1989), conversion of plasma prekallikrein to active kallikrein on the neutrophil membrane may bring about the release of the kinin moiety, stimulate elastase secretion and induce neutrophil aggregation.

Colman (1995) has probed the kininogen molecule with monoclonal antibodies, peptides and deletion mutagenesis to elucidate its interactions with other proteins, surfaces and cells. Domain 2 inhibits platelet calpain, while domain 3 blocks thrombin-induced activation of platelets. Domain 3 also contains a binding site for neutrophils and this was postulated to modulate

neutrophil adhesion. Domain 5 of HMWK contains two regions that react with negatively charged surfaces and cells, including neutrophils and endothelial cells. Domain 6 plays an important role in the cofactor function of kininogen, as this site houses both Factor XII and prekallikrein, the latter liberating BK from HMWK-domain 4 following activation. In order to determine how kininogens place themselves in their binding sites on endothelial cells, Schmaier *et al.* (1995) sought to determine the sequences on kininogen that participate in cell binding and reported that domains 3, 4 and 5 each contain endothelial cell binding sites.

2.3   METABOLISM AND CELLULAR ACTIONS OF KININS

Kinins produce many cellular effects associated with inflammation and these are discussed in detail in other chapters of this volume. These include pain, and arterial dilatation, venoconstriction and endothelial cell retraction, which result in vascular leakage. Kinins also release neuropeptides from sensory nerves, stimulate the synthesis of cytokines, such as interleukin 1 (IL-1) and tumor necrosis factor (TNF), and lipids mediators such as prostaglandins and leukotrienes. Kinins also cause biosynthesis and release of endothelium-derived nitric oxide. Injected intravenously, kinins produce a rapid, transient fall in blood pressure, and high doses cause pulmonary and systemic vascular leak, resulting in cyanosis and edema. In addition, kinins induce bronchoconstriction (see Chapter 15), reflecting direct and indirect effects on bronchial smooth muscle.

Lys-BK is released from LMWK by tissue kallikrein, whereas BK is generated from HMWK by the action of plasma kallikrein (Fig. 11.3). The conversion of Lys-BK to BK may occur in the circulation and tissue fluids through removal of the amino-terminal Lys by aminopeptidases (Rodell *et al.*, 1995). Once produced,

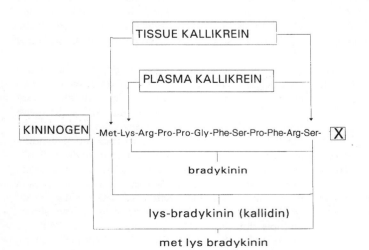

Figure 11.3   Cleavage sites on kininogen molecule for the release of kinins.

both BK and Lys-BK are rapidly inactivated by several distinct kininases and their half-lives in biological fluids are, therefore, usually very short. (The metabolic degradation of kinins is reviewed by Erdös and Skidgel in Chapter 7 of this volume). Circulating carboxypeptidase N and membrane-bound carboxypeptidase M (kininase I) cleave the C-terminal Arg of BK to produce $desArg^9$-BK. Angiotensin-converting enzyme (ACE), also known as "kininase II", cleaves two internal bonds, producing inactive metabolites.

In addition, a neutral endopeptidase, expressed on the surface of endothelial and epithelial cells, and the outer surface neutrophils, degrades kinins.

The cellular effects of kinins are mediated by two major classes of receptors, B_1 and B_2. The cellular distributions, regulation and immunopharmacology of kinin receptors are reviewed in detail in Chapters 2, 3 and 8 of this volume. Expression of the B_2 receptor is ubiquitous and this subtype mediates most of the physiological actions of kinins. The B_1 receptor has attracted interest owing to its apparent upregulation following some types of tissue damage (Marceau, 1995; see Chapter 8).

3. Neutrophils and the Kallikrein–Kinin System

Normal human blood contains leukocytes of different morphologies, and their classification is based broadly on the presence or absence of cytoplasmic granules, as well as the shape of their nuclei. Neutrophils, eosinophils and basophils are the granular leukocytes, and are classified according to specific staining affinities of their intracellular granules. The nongranular leukocytes are divided into monocytes and lymphocytes. Neutrophils, as with all granulocytes, originate from bone-marrow stem cells and the stem cells in turn originate from pluripotential stem cell colony-forming units. The pluripotent cells divide to form progenitor stem cells, some of which become committed to maturing into granulocytes. Maturation of neutrophils from stem cells takes about 10 days, after which the neutrophils remain in bone marrow for 5 days before being released into the circulation, extracellular spaces and biological fluids.

During immunocytochemical studies on human pyelonephrotic kidneys, we observed that the infiltrating neutrophils contained tissue kallikrein, as shown by an intense granular pattern of immunostaining (Figueroa et al., 1988). Immunoreactive tissue kallikrein was identified in neutrophils of normal human blood and bone marrow, specifically in the mature neutrophils as well as in immature forms, such as metamyelocytes and myelocytes (Figueroa and Bhoola, 1989; Bhoola et al., 1992). Moreover, no tissue kallikrein was detected in eosinophils, lymphocytes, macrophages, megakaryocytes and platelets. Additionally, large numbers of neutrophils were observed in the synovial fluid and membranes of patients with rheumatoid arthritis. It is likely, therefore, that neutrophil-derived tissue kallikrein could generate and release Lys-BK in inflamed joints (Melmon et al., 1967). This finding is important since a major role has been postulated for tissue kallikrein in acute inflammation, in which granulocytes are considered to participate in regulating vascular permeability. Although immunoreactive intracellular tissue kallikrein was localized in the neutrophil granules, it has not been determined whether the enzyme is contained in azurophilic or specific granules.

If tissue kallikrein is indeed synthesized in the neutrophil, its expression would be controlled by a single gene that encodes a pre-proenzyme with a 17-amino-acid signal, which is subsequently cleaved during protein translocation within the cells (see Section 2.1.1 and Fig. 11.1). Anders and Kemme (1994) first demonstrated that tissue prekallikrein specifically binds to intact human neutrophils and that structural features of the zymogen are required for interaction with sites on the neutrophil surface. The question as to whether tissue kallikrein is synthesized in situ, or internalized via endocytosis after binding to the neutrophil cell plasma membrane is, as yet, unequivocal.

In myeloid leukemia cells, both tissue kallikrein and prekallikrein have been observed in the cytoplasmic granules of precursor cells (Naidoo et al., 1995). Myelocytes showed significant immunoreactivity while no labeling was observed in other types of precursor cells. Furthermore, monocytes, lymphocytes and eosinophils showed no immunoreactivity. This finding suggests, therefore, that tissue kallikrein is probably synthesized in myeloid stem cells.

Neutrophils participate actively in acute inflammation, initially within blood vessels to where they marginate and then adhere to vascular endothelial cells and, later, when they migrate through the vessel wall to reach the site of inflammation or injury. Macrophages and lymphocytes follow after the initial infiltration by neutrophils, the latter occurring usually within a few minutes (see reviews by Schall and Bacon, 1994; Springer, 1994, 1995). During the early stages of inflammation, BK, produced by the action of plasma kallikrein on HMWK, may be an important mediator of changes in vascular permeability or caliber. This property has led to the conclusion that kinins are likely also to be formed in the protein-rich exudate present in acutely inflamed areas (Bhoola et al., 1992). It has been suggested that, during acute inflammatory reactions, locally generation of chemotactic factors attracts neutrophils to vessel walls, where they play a role in the control of vasodilatation and microvascular permeability (Issekutz, 1984) prior to migrating to the site of inflammation.

Specific, reversible and saturable binding sites for HMWK were previously demonstrated in human neutrophils (Gustafson *et al.*, 1989). Historically, both LMWK and HMWK were thought to be restricted to extracellular fluids, but more recent studies have shown that these molecules are present on or in several cell types (see Chapter 4). For example, when platelets are activated, HMWK is translocated from the α-granules to the external surface of the cell membrane. In contrast, in neutrophils, Figueroa *et al.* (1992) showed that HMWK- and LMWK-immunoreactivity were restricted solely to the cell membrane, being absent from intracellular granules and other organelles. They concluded that clusters of kininogen molecules present on the neutrophil surface represent kininogen bound to specific receptor proteins. Kininogens associated with the neutrophil plasma membrane were thought to have originated from plasma, as it was thought unlikely they were synthesized in the neutrophil and translocated to the cell membrane. With the availability of specific antibodies to plasma kallikrein, it was possible to demonstrate the presence of this protein on the outer surface of the neutrophil membrane (Henderson *et al.*, 1992).

The binding of kininogen to appropriate receptors on the surface of blood cells may serve several functions: (1) surface-bound kininogens on platelets, together with prekallikrein and Factor XI, could locally trigger the endogenous coagulation cascade; or (2) in the case of neutrophils, proteolytic processing of surface-bound LMWK and HMWK by the kallikreins could form kinins, which, by opening endothelial cell junctions (Gabbiani *et al.*, 1970; Oyvin *et al.*, 1970), may promote the local diapedesis of neutrophils and the extravasation of plasma constituents.

4.   Mechanisms Involved in Inflammatory Disorders

The inflammatory response is made up of several stages, the first being changes in vascular tone and increased microvascular permeability, resulting in increased local blood flow and formation of edema. Neutrophils are attracted by chemotactic factors formed in the early stage of inflammation. Chemoattractants initiate migration by causing neutrophils to marginate and adhere to the endothelium of capillaries (Ryan and Majno, 1977; Schall and Bacon, 1994). Two mechanisms have been proposed for this interaction, one depending on neutrophils and, the other, on endothelial cells. Neutrophils migrate into the interstitial spaces via the endothelial gaps. However, the precise mechanisms by which neutrophils form endothelial gaps and migrate through them are unclear at present. However, the immuno-localization of kallikreins and kininogens on the human neutrophil may be significant for diapedesis between vascular endothelial cells (Fig. 11.4). Neutrophils express receptors for HMWK (Lottspeich *et al.*, 1985) and, since plasma prekallikrein colocalizes with HMWK on the neutrophil membrane (Henderson *et al.*, 1992), activation of this enzyme may result in the formation of BK.

Henderson *et al.* (1994) demonstrated that all of the components of the contact system assemble on the surface of the neutrophil cell membrane. HMWK and Factor XII bind to sites exposed on the external surface of the neutrophil. On the other hand, plasma kallikrein is held on the cell surface by virtue of its binding to HMWK. This represents a potentially novel mechanism for kinin formation. The sequence of events involves the formation of kinins from HMWK or LMWK on the surface of the neutrophil either by Factor XII-mediated activation of plasma kallikrein, or by the release of tissue kallikrein in its active form (Fig. 11.5). The locally released kinin may enhance the passage of the neutrophils into the extracellular space by causing the endothelial cells to retract. This mechanism permits the transudation of plasma content by controlling vascular permeability and the passage of circulating neutrophils into the interstitial tissue space surrounding the site of injury or inflammation (Wright and Gallin, 1979).

After being attracted to an inflammatory site, by substances such as C5a and IL-1, and tight adhesion to endothelial cells, neutrophils undergo diapedesis through the endothelial cell gaps. *In vivo* studies suggest

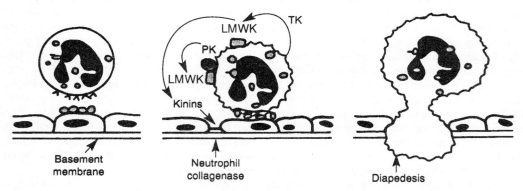

Figure 11.4   Generation of kinins by kallikreins.

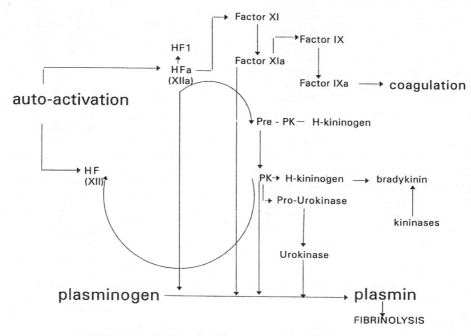

Figure 11.5   Generation of bradykinin by activation of Factor XII.

that the endothelial cells undergo changes during injury and become "sticky" for circulating neutrophils. Although much knowledge has accumulated on the role of contact activation in inflammation, the nature of the physiological contact surface has remained elusive. Henderson and colleagues (1994) suggested a new and novel mechanism whereby formation of kinins from kininogen on the surface of the neutrophil occurs either by the activation of plasma prekallikrein, and/or the release of active tissue kallikrein. It was also proposed that released kinin initiates margination and aids the passage of the neutrophil into the extracellular space by causing the endothelial cells to retract (Henderson *et al.*, 1994).

5.   Clinical Indications

There is a need to discern the functional importance of each of the components of the kallikrein–kinin system. The significance of this system in inflammatory disorders continues to unfold with molecular mapping and by gene characterization and cloning of kinin receptors (see Chapter 3). Kinin-receptor antagonists represent powerful tools for elucidating the pathophysiological actions of kinins, and these agents continue to provide valuable information on the role of the kinins in the clinical manifestations of the systemic inflammatory response syndrome associated with infection, severe trauma, burns and pancreatitis (see Chapters 10 and 12 of this volume).

Enzyme systems identified in the synovial fluids of patients with rheumatoid arthritis include those of plasma transudates and degranulating neutrophils, and cellular damage may occur owing to the enzymatic properties of the kallikreins. Large numbers of neutrophils invade rheumatoid synovial vessels, migrate to the luminal surface of the synovial membrane and accumulate in the synovial fluid (Dularey *et al.*, 1990). Therefore, tissue kallikrein probably reaches the synovial fluid from both neutrophils and plasma exudates. Secretion from even a small number of neutrophils in rheumatoid arthritis synovial fluid could result in the activation of the enzyme and the subsequent formation of kinins during episodes of acute inflammation within the joints.

Most of the plasma kallikrein in synovial fluid exists as a proenzyme complexed to HMWK. Conversion of plasma prekallikrein to its active form may be triggered through activation of HMWK by tissue matrix components, such as proteoglycans, urate crystals or pyrophosphates (Cassim *et al.*, 1996). Plasma kallikrein is also believed to cause significant conversion of latent collagenase to its active form *in vitro* (Dularey *et al.*, 1990), and this could be an important property of plasma kallikrein in the joint spaces. Activated collagenase could be responsible for the extensive cartilage and bone destruction seen in rheumatoid arthritis. In particular, the synovial membrane and the joint space are infiltrated with numerous neutrophils, which carry on their surface the kinin- and fibrin-generating proteolytic enzymes and cascade proteins (Kaplan *et al.*, 1989).

As noted earlier, most of the characteristic actions of kinins appear to be mediated by activation of B_2 receptors. Studies in our laboratory, on neutrophils isolated from whole blood and synovial fluid of patients with rheumatoid arthritis, indicate that the density of the neutrophil B_2 receptors is increased in cells isolated from the circulation, while neutrophils isolated from synovial fluid show a decrease in B_2 receptor density (Cassim et al., 1995).

The inflammatory response to a number of stimuli, including infection, severe trauma and burns, appears to be similar. Acute inflammation is characterized by fever, increased rapid heart and respiratory rates, and leukocytosis. In more serious forms, such as that occurring in sepsis, these symptoms may progress to hypotension, shock, organ dysfunction and failure, and death. Loss in the amount of the kinin moiety residing in the kininogen molecules on the surface of the neutrophil has been demonstrated in sepsis patients and may, therefore, be a manifestation of sepsis (Naidoo et al., 1994b). Patients presenting with pneumonia also show a distinct loss of neutrophil cell surface kinin (Naidoo et al., 1995a), as was the case with the synovial fluid neutrophils (Cassim et al., 1996).

6. References

Anders, J. and Kemme, M. (1994). The binding of tissue prokallikrein to isolated human neutrophils. FEBS Lett. 348, 166–168.

Bhoola, K.D. and Dorey, G. (1971). Kallikrein trypsin-like proteases and amylase in mammalian submaxillary glands. Br. J. Pharmacol. 43, 784–793.

Bhoola, K.D., May May Yi, R., Morley, J. and Shachter, M. (1962). Release of kinin by an enzyme in the accessory sex glands of the guinea pig. J. Physiol. 163, 269–280.

Bhoola, K.D., Lemon, M. and Matthews, M. (1979). Kallikrein in exocrine glands. In "Bradykinin, Kallidin and Kallikrein" (ed. E.G. Erdös), pp. 489–523. Springer-Verlag, Berlin.

Bhoola, K.D., Figueroa, C.D. and Worthy, K. (1992). Bioregulation of kinins: kallikreins, kininogens and kininases. Pharmacol. Rev. 44, 1–80.

Bothwell, M.A., Wilson, W.H. and Shooter, E.M. (1979). The relationship between glandular kallikrein and growth factor-processing proteases of mouse submaxillary gland. J. Biol. Chem. 254, 7287–7294.

Cassim, B., Naidoo, S., Naidoo, Y., Williams, R. and Bhoola, K.D. (1996). Immunolocalisation of the kinin moiety and bradykinin (B_2) receptors on synovial fluid neutrophils in rheumatoid arthritis. Immunopharmacology 33, 321–324.

Chao, J., Karl, X., Chai, and Chao, L. (1996). Tissue kallikrein inhibitors in mammals. Immunopharmacology 32, 67–72.

Cochrane, C.G., Revak, S.D. and Wuepper, K.D. (1973). Activation of Hageman factor in solid and fluid phases: a critical role of kallikrein. J. Exp. Med. 138, 1564–1583.

Colman, R.W. (1995). Kininogens: relationship of structure to physiopathological roles. Immunopharmacology 32, 9–18.

Dularey, B., Dieppe, P.A. and Elson, C.J. (1990). Depressed degranulation response of synovial fluid polymorphonuclear leukocytes from patients with rheumatoid arthritis to IgG aggregates. Clin. Exp. Immunol. 79, 195–201.

Figueroa, C.D. and Bhoola, K.D. (1989). Leukocyte tissue kallikrein: an acute phase signal for inflammation. In "The Kallikrein–Kinin System in Health and Disease" (eds. H. Fritz, I. Schmidt and G. Dietze), pp. 311–320. Limbach-Verlag Braunschweig, Munich.

Figueroa, C.D., MacIver, A.G., Mackenzie, J.C. and Bhoola, K.D. (1988). Localization of immunoreactive kininogen and tissue kallikrein in the human nephron. Histochemistry 89, 437–442.

Figueroa, C.D., Henderson, L.M., Colman, R.W., DeLa Cadena, R.A. and Müller-Esterl, W. (1990). Immunoreactive H- and L-kininogens in human neutrophils. J. Physiol. 425, 65P.

Figueroa, C.D., Henderson, L.M., Kaufman, J., DeLa Cadena, R.A., Colman, R.W., Müller-Esterl, W. and Bhoola, K.D. (1992). Immunovisualisation of high (HK) and low (LK) molecular weight kininogens on isolated human neutrophils. Blood 79, 754–759.

Figueiredo, A.F.S., Salgado, A.H., Siqueira, G.R., Velloso, C.R. and Beraldo, W.T. (1990). Rat uterine contraction by kallikrein and its dependence on uterine kininogen. Biochem. Pharmacol. 39, 763–767.

Frey, E.K. (1926). Zussammenhange zwischen Herzarbeit und Nierentatigkeit. Arch. Klin. Chir. 142, 663.

Frey, E.K. and Kraut, H. (1928). Ein neues Kreislaufhormon und seine Wirkung. Arch. Exp. Pathol. Pharmakol. 133, 1–56.

Fritz, H., Eckert, I. and Werle, E. (1967). Isoliering und charakterisierung von sialinsaurehaltigemund sialinsaurefreiem kallikreinaus schweinepankreas. Hoppe-Seylers Z. Physiol. Chem. 348, 1120–1132.

Fritz, H., Schmidt, I. and Dietze, G. (1988). "The Kallikrein–Kinin System in Health and Disease." International Symposium, Munich, 1988.

Gabbiani, G., Badonnel, M.C. and Majno, G. (1970). Intra-arterial injections of histamine, serotonin or bradykinin: a topographic study of vascular leakage. Proc. Soc. Exp. Biol. Med. 135, 447–452.

Geiger, R. and Miska, W. (1986). Determination of bradykinin by enzyme immunoassay. Adv. Exp. Med. Biol. 198A, 531–536.

Geiger, R., Stuckstedte, U., Forg-Brey, B. and Fink, E. (1979). Human urinary kallikrein-biochemical and physiological aspects. Adv. Exp. Med. Biol. 120A, 235–244.

Girolami, J.P., Alhenc-Gelas, F., Dos Reis, M.L., Bascands, J.L., Suc, J.M., Corvol, P. and Menard, J. (1986). Hydrolysis of high molecular weight kininogen by purified rat urinary kallikrein: identification of bradykinin as the kininogen formed. Adv. Exp. Med. Biol. 198A, 137–145.

Greengard, J.S. and Griffin, J.H. (1984). Receptors for high molecular weight kininogen on stimulated washed human platelets. Biochemistry 23, 6863–6869.

Gustafson, E.J., Schutsky, D., Knight, L.C. and Schmaier, A.H. (1986). High molecular weight kininogen binds to unstimulated platelets. J. Clin. Invest. 78, 310–318.

Gustafson, E.J., Schmaier, A.H., Wachtfogel, Y.T., Kaufman, N., Kulich, U. and Colman, R.W. (1989). Human neutrophils contain and bind high molecular weight kininogen. J. Clin. Invest. 84, 28–35.

Henderson, L.M., Figueroa, C.D., Müller-Esterl, W., Strain, A. and Bhoola, K.D. (1992). Immunovisualisation of plasma prekallikrein in hepatocytes and on neutrophils. Agents Actions Suppl. 38(I), 590–594.

Henderson, L.M., Figueroa, C.D., Müller-Esterl, W. and Bhoola, K.D. (1994). Assembly of contact-phase factors on the surface of the human neutrophil membrane. Blood 84, 474–482.

Issekutz, A.C. (1984). Role of polymorphonuclear leukocytes in the vascular responses of acute inflammation. Lab. Invest. 50, 605–607.

Iwanaga, S., Han, Y.N., Kato, H. and Suzuki, T. (1977). Actions of various kallikreins on HMW kininogen and its derivatives. In "Kininogenases and Kallikreins, IV" (eds. G.L. Haberland, J.W. Rowen and T. Suzuki), pp. 79–90. Schattaeur-Verlag, Stuttgart, New York.

Jacobsen, S. (1966). Separation of two different substrates for plasma kinin-forming enzymes. Nature (Lond.) 210, 98–99.

Kamada, M., Aoki, K., Ikekita, M., Kizuki, K., Moriya, H., Kamo, M. and Tsugita, A. (1988). Generation of α- and β-kallikreins from porcine pancreatic prokallikrein by the action of trypsin. Chem. Pharm. Bull. (Tokyo) 36, 4891–4898.

Kaplan, A.P. and Silverberg, M. (1987). The coagulation-kinin pathway of human plasma. Blood 70, 1–16.

Kaplan, A.P., Reddigari, S. and Silverberg, M. (1989). Assessment of the plasma kinin-forming pathways in allergic diseases. In "The Kallikrein–Kinin System in Health and Disease" (eds. H. Fritz, I. Schmidt and G. Dietze), pp. 143–153. Limback-Verlag, Braunschweig.

Kato, H., Han, Y.N., Iwanaga, S., Suzuki, T. and Komiya, M. (1976). Bovine plasma HMW and LMW kininogens. Structural differences between heavy and light chains derived from kinin-free proteins. J. Biochem. (Tokyo) 80, 1299–1311.

Kato, H., Nagasawa, S. and Iwanaga, S. (1981). HMW and LMW kininogens. Meth. Enzymol. 80, 172–198.

Kato, H., Enjyoji, K., Miyata, T., Hayashi, I., Oh-Ishi, S. and Iwanaga, S. (1985). Demonstration of arginyl-bradykinin moiety in rat HMK kininogen: direct evidence for liberation of bradykinin by rat glandular kallikreins. Biochem. Biophys. Res. Commun. 127, 289–295.

Kellerman, J., Lottspeich, F., Henschen, A. and Müller-Esterl, W. (1986). Completion of the primary structure of human high-molecular-mass kininogen. The amino acid sequence of the entire heavy chain and evidence for its evolution by gene triplication. Eur. J. Biochem. 154, 471–478.

Kellerman, J., Lottspeich, F., Geiger, R. and Deutzmann, R. (1988). Human urinary kallikrein–amino acid sequence and carbohydrate attachment sites. Protein Seq. Data Anal. 1, 177–182.

Kerbiriou, D.M., Bouma, B.N. and Griffin, J.H. (1980). Immunochemical studies of human high molecular weight kininogen and of its complexes with plasma prekallikrein or kallikrein. J. Biol. Chem. 255, 3952–3958.

Kitamura, N., Kitagawa, H., Fukushima, D., Takagaki, Y., Miyata, T. and Nakanishi, S. (1985). Structural organization of the human kininogen gene and a model for its evolution. J. Biol. Chem. 260, 8610–8617.

Komiya, M., Kato, H. and Suzuki, T. (1974). Bovine plasma kininogens. III. Structural comparison of high molecular weight kininogens. J. Biochem. (Tokyo) 76, 833–845.

Kraut, H., Frey, E.K. and Werle, E. (1928). Uber ein neues kreislaufhormon. II. Mittleilung. Hoppe-Seylers Z. Physiol. Chem. 175, 97–114.

Kraut, H., Frey, E.K. and Werle, E. (1930). Der Nachweis eines Kreislaufhormon in der pankreasdruse. Hoppe-Seylers Z. Physiol. Chem. 189, 97–106.

Lottspeich, F., Kellerman, J., Henschen, J., Foertsch, B. and Müller-Esterl, W. (1985). The amino acid sequence of the light chain of human high-molecular-mass kininogen. Eur. J. Biochem. 152, 307–314.

Lu, H.S., Lin, F., Chao, L. and Chao, J. (1989). Human urinary kallikrein. Complete amino acid sequence and sites of glycosylation. Int. J. Pept. Protein Res. 33, 237–249.

Marceau, F. (1995). Kinin B_1 receptors: a review. Immunopharmacology 30, 1–26.

Mason, A.J., Evans, B.A., Cox, D.R., Shine, J. and Richards, R.I. (1983). Structure of mouse kallikrein gene family suggests a role in specific processing of biological active peptides. Nature (Lond.) 303, 300–307.

Melmon, K.L., Webster, K.E., Goldfinger, S.E. and Seegmiller, J.E. (1967). The presence of a kinin in inflammatory synovial effusions from arthritides of varying etiologies. Arthritis Rheum. 10, 13–20.

Meloni, F.J. and Schmaier, A.H. (1991). Low molecular weight kininogen binds to platelets to modulate thrombin-induced platelet activation. J. Biol. Chem. 266, 6786–6794.

Miller, G., Silverberg, M. and Kaplan, A.P. (1980). Auto-activatibility of human hageman factor (factor XII). Biochem. Biophys. Res. Commun. 92, 803–810.

Miwa, I., Erdös, E.G. and Seki, T. (1968). Presence of three peptides in urinary kinin (substance) preparations. Life Sci. 7 (Suppl.), 1339–1343.

Moriya, H., Ikekita, M. and Kizuki, K. (1983). Some aspects of carbohydrate contents in glandular kallikrein. Adv. Exp. Med. Biol. 156A, 309–315.

Müller-Esterl, W. (1988). Kininogens, kinins and kinships. Thromb. Haemost. 61, 2–6.

Müller-Esterl, W., Fritz, H., Machleidt, W., Ritonja, A., Brzin, J., Kotnik, M., Turk, V., Kellerman, J. and Lottspeich, F. (1985). Human plasma kininogens are identical with α$_2$-cysteine proteinase inhibitors. Evidence from immunological, enzymological and sequence data. FEBS Lett. 182, 310–314.

Müller-Esterl, W., Iwanaga, S. and Nakanishi, S. (1986). Kininogens revisited. Trends Biochem. Sci. 11, 336–339.

Naidoo, Y., Snyman, C. and Bhoola, K.D. (1994a). Kinin antagonists – a new family of therapeutic agents. Contin. Med. Educ. 12, 1591–1595.

Naidoo, Y., Snyman, C., Narotam, P.K., Müller-Esterl, W. and Bhoola, K.D. (1994b). Release of the kinin moiety from kininogen on the outer surface of the circulating neutrophil in patients with sepsis. Can. J. Physiol. Pharmacol. 72, 40 (P2.3).

Naidoo, Y., Naidoo, S., Nadar, R.N. and Bhoola, K.D. (1995). Role of prokallikrein and tissue kallikrein in myeloid leukaemic patients. Electron Microscopy Society of South Africa, Proceedings 25, 60.

Naidoo, Y., Nadar, R.N. and Bhoola, K.D. (1996). Role of neutrophil kinin in infection. Immunopharmacology 33, 387–390.

Ohkubo, I., Kurachi, K., Takasawa, C., Shiokawa, H. and Sasaki, M. (1984). Isolation of a human cDNA for α$_2$-thiol proteinase inhibitor and its identity with low molecular weight kininogen. Biochemistry 23, 5691–5697.

Oyvin, I.A., Gaponyuk, P.Y., Oyvin, V.I. and Tokarev, O.Y. (1970). The mechanism of blood vessel permeability derangement under the influence of histamine, serotonin and bradykinin. Experientia 26, 843–844.

Pisano, J.J., Geller, R., Margolius, H.S. and Keiser, H.R. (1974). Urinary kallikrein in hypertensive rats. Acta Physiol. Latin Am. 24, 453–458.

Proud, D., Perkins, M., Pierce, J.V., Yates, K., Highet, P., Herrig, P., Mark, M.M. Bahu, R., Larone, F. and Pisano, J.J. (1981). Characterisation and localisation of human renal kininogen. J. Biol. Chem. 256, 10634–10639.

Rahman, M., Worthy, K., Elson, C.J., Fink, E., Dieppe, P.A. and Bhoola, K.D. (1994). Inhibitor regulation of tissue kallikrein activity in the synovial fluid of patients with rheumatoid arthritis. Br. J. Rheumatol. 33, 215–223.

Revak, S.D. and Cochrane, C.G. (1976). The relationship of structure and function in human Hageman factor. The association of enzymatic and binding activities with separate regions of the molecule. J. Clin. Invest. 57, 852–860.

Revak, S.D., Cochrane, C.G. and Griffin, J.H. (1977). The binding and cleavage characteristics of human Hageman factor during contact activation. A comparison of normal plasma with plasmas deficient in factor XI, prekallikrein or high molecular weight kininogen. J. Clin. Invest. 59, 1159–1167.

Revak, S.D., Cochrane, C.G., Bouma, B.N. and Griffin, J.H. (1978). Surface and fluid phase activities of two forms of activated Hageman factor produced during activation of plasma. J. Exp. Med. 147, 719–729.

Rocha e Silva, M., Beraldo, W.T. and Rosenfeld, G. (1949). Bradykinin hypotensive and smooth muscle stimulating factor released from plasma globulins by snake venoms and by trypsin. Am. J. Physiol. 156, 261–273.

Rodell, T.C., Naidoo, Y. and Bhoola, K.D. (1995). Role of kinins in inflammatory responses. Prospects for drug therapy. Clin. Immunother. 3, 352–361.

Ryan, G.B. and Majno, G. (1977). Acute inflammation: a review. Am. J. Pathol. 86, 185–276.

Schall, T.J. and Bacon, K.B. (1994). Chemokines, leukocyte trafficking, and inflammation. Curr. Opin. Immunol. 6, 865–873.

Schmaier, A.H., Zuckerberg, A., Silverman, C., Kuchibhotla, J., Tuszynski, G.P. and Colman, R.W. (1983). High molecular weight kininogen: a secreted platelet protein. J. Clin. Invest. 71, 1477–1489.

Schmaier, A.H., Bradford, H., Farber, A., Silver, L.D., Schutsky, D., Scott, C.F. and Colman, R.W. (1986a). High molecular weight kininogen inhibits platelet calpain. J. Clin. Invest. 77, 1565–1573.

Schmaier, A.H., Smith, P.M., Purdon, A.D., White, J.G. and Colman, R.W. (1986b). High molecular weight kininogen: localization in the unstimulated and activated platelet and activation by platelet calpain(s). Blood 67, 119.

Schmaier, A.H., Kuo, A., Lundberg, D., Murray, S. and Cines, D.B. (1988). The expression of high molecular weight kininogen on human umbilical vein endothelial cells. J. Biol. Chem. 263, 16327–16333.

Schmaier, A.H., Hasan, D.B., Cines, H., Herwald, J., Godovac-Zimmerman, W. and Müller-Esterl, W. (1995). High molecular weight kininogen assembly on endothelial cells is a multidomain interaction. Communication, Kinin 1995, Denver.

Silver, M.R., Ole-Moiyoi, O., Austen, K.F. and Spragg, J. (1980). Active site radioimmunoassay for human urokallikrein and demonstration by radioimmunoassay of a latent form of the enzyme. J. Immunol. 124, 1551–1555.

Silverberg, M., Dunn, J.T., Garen, L. and Kaplan, A.P. (1980). Autoactivation of human Hageman factor. Demonstration utilizing a synthetic substrate. J. Biol. Chem. 255, 7281–7286.

Springer, T.A. (1994). Traffic signals for lymphocyte recirculation and leukocyte emigration: the multistep paradigm. Cell 76, 301–314.

Springer, T.A. (1995). Traffic signals on endothelium for lymphocyte recirculation and leukocyte emigration. Annu. Rev. Physiol. 57, 827–872.

Takada, Y., Skidgel, R.A. and Erdös, E.G. (1985). Purification of human prokallikrein. Identification of the site of activation by the metalloproteinase thermolysin. Biochem. J. 232, 851–858.

Talamo, R.C., Haber, E. and Austen, K.F. (1969). A radioimmunoassay for bradykinin in plasma and synovial fluid. J. Lab. Clin. Med. 74, 816–827.

van Iwaarden, A.F., de Groot, P.G. and Bouma, B.N. (1989). The binding of high molecular weight kininogen to cultured human endothelial cells. J. Biol. Chem. 263, 4698–4703.

Werle, E. and Berek, U. (1948). Zur Kenntnis des Kallikreins. Z. angew. Chem. 60A, 53.

Werle, E. and Grunz, M. (1939). Zur Kenntnis der darmkontraheirenden uteruser-regenden und blutdrucksenkenden Substanz DK. Biochem. Z. 301, 429–436.

Werle, E., Gotze, W. and Keppler, A. (1937). Über die Wirkung des Kallikreins auf den isolierten Darm und über eine neue darmkontraheirende Substanz. Biochem. Z. 289, 217–233.

Werle, E., Kehl, R. and Koebke, K. (1950). Über bradykinin, kallidin und hypertension. Biochem. Z. 320, 327–383.

Wright, D.G. and Gallin, J.I. (1979). Secretory response of human neutrophils: exocytosis of specific (secondary) granules by human neutrophils during adherence in vitro and during exudation in vivo. J. Immunol. 123, 285–294.

Yung, L.L., Lim, F., Khan, M.M.H., Kunapali, S.P., Rick, L., Colman, R.W. and Cooper, S.L. (1996). Neutrophil adhesion on surfaces preadsorbed with high molecular weight kininogen under well-defined flow conditions. Immunopharmacology 32, 19–23.

12. Putative Roles of Bradykinin and the Kinin System in Pancreatitis

Thomas Griesbacher *and* Fred Lembeck

1. Discovery of the Kallikrein–Kinin System in the Pancreas

The history of the discovery of the kallikrein–kinin system is intimately related with the pancreas and was reviewed in detail previously (Werle, 1970; see also Chapter 1). In the present essay, we describe the studies that led to the identification and characterization of the kallikrein–kinin system in the pancreas and its implication in acute pancreatitis.

1.1 KALLIKREIN

At the end of the 19th century, Bouchard, a pathologist, noted that injection of urine caused cutaneous vasodilatation. Abelous and Bardier (1909a,b) found that a fraction of urine caused peripheral vasodilatation and hypotension, the substance responsible being named "urohypotensine" and, later, *"F-Stoff"* (F-substance, from Frey, 1926). *F-Stoff* was assumed to be a circulating hormone, and was thermolabile but resistant to dialysis (Frey and Kraut, 1928; Kraut *et al.*, 1928). Since high concentrations of *F-Stoff* were found in the pancreas of practically all species studied, this organ was thought to be its primary source (Kraut *et al.*, 1930). Hence, the name "kallikrein" was coined from a Greek synonym for the pancreas (see Chapter 1).

Further investigations showed that kallikrein recovery was increased when tissue extracts were prepared to avoid autolytic processes, indicating that some of the kallikrein in the pancreas was stored for secretion (Werle, 1937). Indeed, kallikrein had been found in the pancreatic exocrine secretions in a form that was largely inactive, but activation could be accomplished by contact with the gut mucosa (Werle and von Roden, 1936). It was determined that trypsin activation by enterokinase preceded activation of kallikrein (Werle and Urhahn, 1940). These results explained previous observations that pancreatic juices reduced blood pressure upon intravenous injection, an effect that was greatly augmented by prior exposure to and activation by the gut contents (Migay and Petroff, 1923). The inactive precursor of kallikrein is known today as "prekallikrein".

There are two groups of kallikreins (EC 3.4.21.8), plasma or serum kallikrein, and glandular or tissue kallikrein. Pancreatic kallikrein is a typical example of glandular kallikrein, as is urinary kallikrein. It should be remembered that the enzymology of these two groups of kallikreins exhibits important differences (Fiedler, 1979). However, kinin-liberating properties are common to all of them. Chapter 5 provides a review on the molecular biology of the kallikreins and their roles in inflammation.

1.2 KININS

Kallikrein-induced hypotension was noted not to be due directly to kallikrein itself but to another substance released from plasma by kallikrein. This newly described mediator had the same actions as kallikrein but was a heat-stable polypeptide. Since it caused contraction of isolated intestinal smooth muscle, this peptide was initially called *"Substanz DK"* (*"darmkontrahierende Substanz"*), or gut-contracting substance; Werle *et al.*, 1937) and, later, "kallidin". The precursor peptide in the plasma was named "kallidinogen" (Werle, 1948). Another peptide with actions similar to kallidin and almost identical chemical properties, bradykinin (BK), was purified from plasma, following its release by snake venom, by Rocha e Silva *et al.* (1949). The discovery of BK is reviewed in Chapter 1 of this volume.

For pathophysiological considerations, the smooth muscle-contracting properties of the kinins may be less important than actions such as vasodilatation, increased microvascular permeability, and induction of pain (Holdstock *et al.*, 1957; Elliott *et al.*, 1960). Since these actions mimic the classical symptoms of inflammation, it has long been considered that the kinins may have extremely important roles as mediators of inflammatory disease (Holdstock *et al.*, 1957).

1.3 KININOGENS

The kinin precursors, the kininogens, are acidic glycoproteins, which are highly specific substrates for the plasma and glandular kallikreins. Kininogens also serve as nonspecific substrates for trypsin, especially when denatured. Pancreatic kallikrein, like that of other tissues, releases kallidin (Lys-BK), whereas plasma kallikrein releases BK (Webster and Pierce, 1963; Pierce, 1968). Two forms of kininogen are distinguished by their molecular weights, with the high molecular weight form (HMWK) being an equally good substrate for both kallikreins, whereas low molecular weight kininogen (LMWK) is the preferred substrate for tissue kallikrein (Jacobsen, 1966). Although the kininogens occur predominantly in the plasma, kininogen has also been found in a variety of tissues (Werle and Zach, 1970) including the pancreas.

1.4 KININ SYSTEM OF THE RAT

A considerable number of studies on the roles of kinins in the pathophysiology of the pancreas have been carried out in rats. In this species, however, another kinin, Ile-Ser-BK or T-kinin, is released from substrates, the T-kininogens (Greenbaum *et al.*, 1992). The name "T-kinin" was coined from the observation that this peptide was cleaved from T-kininogen, not by kallikrein, but by large quantities of trypsin (Okamoto and Greenbaum, 1983a,b).

While the two kininogens of other species including man are generated from a single gene by alternate RNA splicing (Kitamura *et al.*, 1983), the rat possesses at least two separate genes coding for two closely related T-kininogens in addition to a gene coding for a kininogen, which is very similar to that of other species (Cole and Schreiber, 1992). (**Editor's note:** The regulation of genes encoding kinin precursor proteins is detailed in Chapter 3). A specific enzyme, T-kininogenase, for the release of T-kinin also is part of the rat kinin system (Barlas *et al.*, 1983). Nevertheless, T-kinin has pharmacological characteristics identical to those of BK and Lys-BK (Gao *et al.*, 1989).

1.5   IMPLICATIONS FOR PANCREAS PATHOPHYSIOLOGY

Since both trypsin, which can activate kallikrein, and kallikrein itself are present in the pancreas in high quantities, this initiated proposals that kinins may be responsible for the inflammatory symptoms and the severe hypotension observed during acute pancreatitis (Forrell, 1955; Werle *et al.*, 1955). This concept also initiated a quest for inhibitors of trypsin or kallikrein for the therapy of this disease (Frey, 1953; Werle *et al.*, 1958; Forrell, 1961; Forrell and Dobovicnik, 1961).

2.   *Clinical Pancreatitis*

2.1   CLASSIFICATION

The systematic description of the inflammatory and pathological changes occurring in acute pancreatitis has been known for over a century (see Fitz, 1889a,b,c). Investigations at that time focused on those cases exhibiting the most severe clinical symptoms, including death, and it was not until the 1930s that assays for determining amylase activity in blood and urine began to provide major progress in diagnosis. However, no criteria to distinguish different forms of pancreatitis were available until the Symposium of Marseilles (Sarles, 1965) produced the first clinical classification system. The system of classification was initially based primarily on clinical symptoms and anatomical findings. The distinction between acute and relapsing acute pancreatitis on the one hand, and chronic relapsing and chronic pancreatitis on the other, rested also on the assertion that in the first two forms a complete cure would ensue once the cause was removed, whereas the two latter forms would damage the gland functionally and anatomically. Later classification systems included etiologic and histological findings as well as prognostic considerations (reviewed in Sarner, 1993). Acute pancreatitis, by current definition, is an acute condition with typical abdominal pain and pronounced serum elevations of pancreatic enzymes, such as amylase, lipase or elastase. Chronic pancreatitis is characterized by ongoing inflammatory changes, pain and permanent impairment of function. Modern imaging techniques can be used to further characterize the severity of morphological changes in the pancreas.

2.2   EDEMATOUS AND NECROTIZING PANCREATITIS

Investigations on the possible involvement of kinins in pancreatitis have focused on the acute form, since the effects of these peptides (vasodilatation, increased microvascular permeability, and pain) closely mimic acute inflammatory symptoms. Acute pancreatitis has three main etiologies, and these are diseases of the biliary tract, especially tract stones, ethanol abuse and cases of unknown origin. Further factors that are recognized to cause acute pancreatitis include trauma, hyperparathyroidism, hyperlipidemia, some drugs, tumors and infection (Creutzfeldt and Schmidt, 1970; Steer, 1993). Over the past decades, the incidence of acute pancreatitis has increased 2–11-fold (Corfield *et al.*, 1985; Wilson and Imrie, 1990), and mortality has been estimated at about 20% (Corfield *et al.*, 1985; Amurawaiye and Brown, 1991), emphasizing the importance of the disease.

The acute disease can be divided into two major forms, interstitial-edematous pancreatitis and necrotizing–hemorrhagic pancreatitis, which also reflect the severity of the disease. Whereas mortality in the latter is almost 30%, edematous pancreatitis is associated with less than 10% mortality (Gauthier *et al.*, 1981; Banks, 1991; Fernández-Cruz *et al.*, 1994). Although the majority of patients suffer from the milder form, about 15–20% of cases develop necrotizing–hemorrhagic pancreatitis (Büchler, 1991). No particular etiologic factor has been demonstrated to be specific for any given form of acute pancreatitis, in part due to the failure to distinguish between etiologic and pathogenic factors such as the cause(s) of the inflammatory response (McCutcheon, 1968). This is in contrast to the experimental models for the study of acute pancreatitis described below.

For the interpretation of the results from animal models of acute or necrotizing pancreatitis, two cautionary notes are necessary. First, with acute pancreatitis, the pathogenic factors responsible for the induction of the disease and factors involved in its further course are two separate entities. Secondly, the edematous and the necrotizing–hemorrhagic forms of acute pancreatitis are, as far as their pathophysiological mechanisms are concerned, not simply different grades of severity. Necrotizing pancreatitis certainly involves additional pathological mechanisms. Even if one mechanism is present in both, its pathophysiological role might be different in one form as compared to the other.

3.   *Experimental Models of Acute Pancreatitis*

In contrast to the clinical situation, where acute pancreatitis can take either the mild interstitial–edematous course or the severe necrotizing–hemorrhagic course, experimental induction of pancreatitis in laboratory animals selectively generates only one form or the other. Animal models clearly have time-courses for the development of pathological effects and symptoms of pancreatitis that are different from the clinical time-course for symptoms. This has raised some concern about the validity of animal models, especially in evaluating novel drugs (Steinberg and Schlesselman, 1987). However, in the absence of alternatives, animal models are widely used and have yielded a large body of information on the involvement of the kallikrein–kinin system and other pathophysiological mechanisms. In the following we give a brief description of models which are most widely used. Readers searching for more detailed methodological information are referred to a recent excellent review (Lerch and Adler, 1994).

3.1   INTERSTITIAL–EDEMATOUS PANCREATITIS

Intravenous infusion or repeated intraperitoneal injections of high doses of cholecystokinin (CCK) agonists is the model most frequently used for investigations of pathological mechanisms underlying edematous pancreatitis. Caerulein, a CCK analog extracted from frog skin, is usually employed instead of CCK because of its greater metabolic stability *in vivo*. In contrast to physiological exocrine stimulation of the pancreas, this hyperstimulation leads to edematous pancreatitis (Lampel and Kern, 1977; Adler *et al.*, 1979). The underlying mechanisms include intracellular activation of digestive zymogens (Steer and Meldolesi, 1987; Leach *et al.*, 1991), disturbances of intracellular membrane transport mechanisms, leading to colocalization of digestive and lysosomal enzymes (Watanabe *et al.*, 1984; Saluja *et al.*, 1987; Willemer *et al.*, 1990), and discharge of activated enzymes into the interstitial space of the pancreas. The intrapancreatic changes occur rapidly with maximal effects observed within 2–6 h, followed by complete recovery of the animals.

The exocrine function of the pancreas can also be stimulated by cholinoceptor agonists. Experimentally, acute pancreatitis can indeed be induced *in vivo* by cholinergic hyperstimulation (Adler *et al.*, 1983; Bilchik *et al.*, 1990b). Although this model mimics the induction of an acute pancreatitis following anticholinesterase intoxication (Weizman and Sofer, 1992), this procedure is not used routinely because of systemic effects of cholinoceptor agonists.

The ligation of the pancreatic and hepatic ducts, together with stimulation of the secretory function of the pancreas in several species, has been suggested to mimic clinical phenomena associated with acute pancreatitis, particularly biliary tract stones or obstructive diseases of the papilla of Vater. Within hours of ligation, elevation of digestive enzymes in the serum and mild pancreatic edema occur (Dumont and Martelli, 1968; Ohshio *et al.*, 1991). Ultrastructurally, the intracellular colocalization of digestive and lysosomal enzymes can also be demonstrated in this model (Saluja *et al.*, 1989). However, other acute inflammatory changes are absent, especially when no secretory stimulus is applied (Walker, 1987; Walker *et al.*, 1992) and this model, therefore, is used rather for investigations on pancreatic atrophy and regeneration.

3.2   NECROTIZING–HEMORRHAGIC PANCREATITIS

The retrograde injection of bile acids is widely employed to induce necrotizing pancreatitis in experimental animals. In fact, the combined injection of bile and olive oil into the pancreatic duct of a dog was the first experiment reported to induce acute pancreatitis in an animal (Bernard, 1856). Although the effect of this procedure varies with the type of bile acid used, the most frequently used substance is taurocholic acid, which can induce dose-dependent pancreatic damage (Sum *et al.*, 1970; Aho and Nevalainen, 1980). Retrograde intraductal infusion of taurocholate into the *in situ* isolated pancreas of the pig also leads to a severe necrotizing form of pancreatitis (Vollmar *et al.*, 1989). In the same preparation, intra-arterial injection of oleic acid produces a pancreatitis that is predominantly edematous, but includes small areas of necroses spread uniformly over the entire organ (Vollmar *et al.*, 1989, 1991).

Other models of pancreatitis include ligation of the duodenum proximal and distal to the Sphincter of Oddi, with concomitant diversion of the bile flow, resulting in acute necrotizing–hemorrhagic pancreatitis by the action of intestinal contents reaching the pancreas by reflux (McCutcheon, 1964; Johnson and Doppman, 1967). The exact pathogenic mechanism of this model is unclear. Also, omission of choline from the diet of young female mice with concomitant supplement with ethionine leads to acute necrotizing pancreatitis (Lombardi *et al.*, 1975).

Ischemia is a well-known factor involved in the progression from edematous to necrotizing pancreatitis (Klar *et al.*, 1990). Ischemia of the pancreas alone has also been used to induce pathological changes mimicking necrotizing pancreatitis. Thus, the obstruction of terminal arterioles by injection of 8–20 μm microspheres causes necrotizing hemorrhagic pancreatitis in dogs (Pfeffer *et al.*, 1962).

Contrary to the effects of ligating the biliopancreatic duct in most of the common laboratory animals described above, the same procedure in the opossum leads to severe hemorrhagic pancreatitis with a mortality of 100% within 2 weeks (Senninger *et al.*, 1984; Cavuoti *et al.*, 1988), although the early histological changes are very similar to those observed in rats (Samuel *et al.*, 1994). In this model, the obstruction of pancreatic outflow is the critical event triggering acute pancreatitis without a contribution of reflux of bile into the pancreas (Lerch *et al.*, 1993).

3.3 MODELS OF PANCREATITIS *IN VITRO*

The isolated perfused pancreas is used in experiments aimed at inducing changes that resemble those of either edematous or necrotizing pancreatitis. Thus, the intra-arterial infusion of caerulein into the isolated pancreas of dogs perfused with autologous blood causes pancreatic edema, increase in amylase activity in the venous outflow and, contrary to the situation *in vivo*, mild hemorrhage (Nordback *et al.*, 1991). Similar changes including mild hemorrhage can be induced in this model by partial duct obstruction with secretin stimulation, or by a 2 h period of ischemia. Perfusion with oleic acid causes morphologic changes resembling severe hemorrhagic pancreatitis (Nordback *et al.*, 1991). Intra-arterial infusion of bile acids or trypsin into the isolated pancreas of cats also induces massive or focal parenchymal necrosis, respectively (Hong *et al.*, 1988). Elevations of pancreatic enzymes in the venous outflow of the preparation have been described following intraductal injection of taurocholate into the isolated pancreas of rats (Kimura *et al.*, 1992).

These models are not used routinely, since they involve laborious surgical procedures to remove the pancreas. Moreover, results from isolated pancreas may not be extrapolated to the clinical disease, since pancreatitis also involves systemic hemodynamic factors which cannot be mimicked *in vitro*.

3.4 CLINICAL RELEVANCE OF EXPERIMENTAL MODELS

The major objection put forward, regarding the "relevance" of experimental pancreatitis in animals to clinical pancreatitis, is generally that almost all of the stimuli used to induce experimental pancreatitis are not present in the human disease. Nevertheless, the close resemblance of some experimental models with clinical cases of either edematous or necrotizing pancreatitis in morphological, histological, ultrastructural or biochemical respects have led to the view that these models, indeed, have a high relevance to investigation of the pathophysiological mechanisms (Bilchik *et al.*, 1990a; Büchler *et al.*, 1992). Caution is necessary, however, when conclusions from therapeutic measures in experimental models are to be applied to the clinical disease.

4. Research in the Kallikrein–Kinin System in Pancreatitis

Early investigations on the involvement of the kallikrein–kinin system in acute pancreatitis relied on the isolation of factors from patients or animals, and comparisons with known components of the kallikrein–kinin system. Thus, these factors were characterized according to their apparent chemical nature, molecular weights, or according to their effects in bioassay preparations or animals. Since all such results rely on the comparison of an unknown substance(s) with known substances, the conclusions reached were debatable. Nevertheless, the early evidence in favor of a role of kinins in acute pancreatitis was convincing.

4.1 KININ AND KALLIKREIN LEVELS

More recently developed immunoassays for measuring the different forms of the kallikreins and the kininogens have provided much insight. The results from immunoassay measurements of kallikreins, however, must be supplemented by measuring kallikrein activity, since kallikreins can be present as inactive proenzymes or complexed to inhibitors. Several different methods have been described (Fiedler, 1979). The first successful production of kinin antibodies made it possible to develop specific immunoassays for kinins (reviewed by Talamo and Goodfriend, 1979). However, the short plasma half-life of kinins led to substantial differences in the data reported by different groups, although comparisons between the published results at least demonstrate kinin formation.

4.2 BRADYKININ RECEPTOR ANTAGONISTS

Kinin receptors were classified as two subtypes, B_1 and B_2, by Regoli and Barabé (1980), and desArg^9-[Leu^8]-BK is still the most widely used B_1 receptor antagonist. B_2 receptor antagonists have been available since the mid-1980s (Vavrek and Stewart, 1985). Until recently, all known kinin receptor antagonists were peptide analogs of [DPhe^7]-BK and, as such, metabolically labile, thereby limiting their utility for studies *in vivo* (e.g., Griesbacher and Lembeck, 1987; Griesbacher *et al.*, 1989). Recently, more stable antagonists containing modified amino-acid residues have become available. The most popular of these agents has been DArg-[Hyp^3,Thi^5,DTic^7,Oic^8]-BK (more commonly known as "Hoe 140" or, recently, "Icatibant"; Hock *et al.*, 1991; Lembeck *et al.*, 1991; Wirth *et al.*, 1991a). To the same group of peptides belongs S 16118, in which the N-terminal DArginine of Hoe 140 is substituted with *p*-guanidobenzoyl (Thurieau *et al.*, 1994; Félétou *et al.*, 1995a). Another B_2 receptor antagonist is CP 0597, which has a shorter duration of action and seems to have B_1 receptor antagonist properties *in vivo* (Goodfellow *et al.*, 1996). The desArg[10]-derivative of Hoe 140 is a B_1 receptor antagonist (Wirth *et al.*, 1991b). A detailed review of kinin receptors and their antagonists is given in Chapter 2 of this volume.

5. Clinical and Experimental Studies

5.1 KININ, KALLIKREIN AND KININOGEN LEVELS IN CLINICAL PANCREATITIS

Using bovine isolated carotid artery as a bioassay preparation, Thal *et al.* (1963) were the first to estimate the liberation of a kinin-like material in clinical pancreatitis. BK was the only known vasoactive substance that produced contractions of this preparation and such responses induced by biological fluids were, therefore, attributed to BK or similar peptides. In blood samples of four patients with acute or chronic pancreatitis with severe abdominal pain, and in one case of pancreatic contusion, such kinin-like activity was found. One patient with recurrent episodes of pancreatitis showed an increase in the blood content of this factor in parallel with each exacerbation. No kinin-like material could be found in normal human controls. Similarly, increased blood levels of kinin-like peptides were found in chronic alcoholic pancreatitis, although the increase was lower than in acute pancreatitis (Katz *et al.*, 1964).

Significant increases in the plasma content of kallikrein, together with a decrease in kininogen, were described by Orlov and Belyakov (1978) in plasma from patients with acute pancreatitis. These changes were accompanied by an increase in the activity of kininase enzymes. The most severe changes were found in those patients in whom the disease was accompanied by hemodynamic changes. Pancreatic kallikrein was recovered and purified from the plasma of patients with acute pancreatitis (Sumi *et al.*, 1978; Takasugi *et al.*, 1980; Toki *et al.*, 1981). Although almost all of the kallikrein was combined with α_2-macroglobulin (α_2-M), these complexes retained their kinin-forming activity (Sumi *et al.*, 1978). Using antiserum raised against human urinary kallikrein, Shimamoto *et al.* (1984) reported that serum kallikrein levels were low in normal volunteers, but elevated in plasma from patients with acute, acute relapsing or chronic relapsing pancreatitis. No increases in the plasma levels of tissue kallikrein were found in patients with chronic pancreatitis or pancreatic carcinoma. Similarly, the determination of glandular kallikrein in human plasma by radioimmunoassay following partial purification showed that glandular kallikrein was 5–6-fold higher in the plasma of patients with acute pancreatitis as compared to healthy controls (Nishimura *et al.*, 1987).

A reduction in blood levels of prekallikrein in patients with acute pancreatitis was reported by Singh and Howard (1966), and Lasson and Ohlsson (1984, 1986) noted that prekallikrein and kininogen were lower in blood from patients with severe acute pancreatitis as compared to samples from patients with moderate or mild attacks. The effects were even more pronounced in the peritoneal fluid. Since both LMWK and HMWK were decreased, it was concluded that, in addition to plasma kallikrein, other kininogenases must be activated. Prekallikrein and HMWK were detected in the peritoneal fluid of patients with severe acute pancreatitis (Aasen *et al.*, 1989).

Seven patients with acute fulminant pancreatitis investigated by Hiltunen *et al.* (1985) showed a reduction in plasma prekallikrein and high kallikrein activity in the urine. While, in the five survivors, the changes in the activity of these components of the kallikrein–kinin system was restored following recovery, no normalization of the pathological changes was found in the two non-survivors. Aasen *et al.* (1986) defined a proenzyme functional inhibition (PFI) index as the sum of deviations from the normal plasma pool values of plasma prekallikrein, functional kallikrein inhibition, plasminogen, antiplasmin, prothrombin and antithrombin III, with decreased values of the index counting as negative. The PFI index was negative in 37 patients with acute pancreatitis on admission, a strong correlation being made with disease severity. Consumption of plasma prekallikrein in conjunction with reduced plasma levels of glandular kallikrein was also described by Uehara *et al.* (1989). In parallel to these changes in kallikrein and prekallikrein, both HMWK and LMWK in plasma were decreased simultaneously with the onset of circulatory shock. The results were explained tentatively by the action of glandular kallikrein and trypsin of pancreatic origin, which, in turn, might activate plasma kallikrein.

Components of the kallikrein–kinin system were also found in the contents of pancreatic pseudocysts. Thus, Lasson *et al.* (1989) reported that pseudocysts contained a high level of proteolytic activity. Besides having trypsin- and plasmin-like activity, part of the proteolytic activity was found to be kallikrein-like. The protease activity was present both in free form and bound to α_2-M. No kininogens could be determined in the cyst fluid, presumably as a result of the proteolytic activity. A low concentration of immunoreactive glandular kallikrein was found in the lymphatic fluids of patients with acute pancreatitis (Girolami *et al.*, 1989). The immunoreactive kallikrein showed only a weak kininogenase activity, suggesting partial inhibition by antiproteases in the lymph. The excretion of glandular kallikrein in the urine was elevated in a substantial number of patients with pancreatitis (Fabris *et al.*, 1991).

5.1.1 Other Clinical Observations

Hereditary angioedema is caused by a hereditary defect of the C1q esterase inhibitor, leading to attacks of severe generalized edemas caused by the endogenous kinins owing to the reduced inhibition of kinin-liberating enzymes (Schapira *et al.*, 1983; Kaplan *et al.*, 1989). One case report describes a recurrent acute pancreatitis in a woman with this disease (Cutler *et al.*, 1992). (**Editor's note:** The role of the kallikrein–kinin system in hereditary angioedema is described in Chapter 16).

5.2 KININ, KALLIKREIN AND KININOGEN LEVELS IN EXPERIMENTAL STUDIES

The experimental models employed for the investigation of necrotizing–hemorrhagic pancreatitis are much older than those of interstitial–edematous pancreatitis, especially that induced by caerulein (see Section 3). Attempts to measure components of the kallikrein–kinin system in experimental models of acute pancreatitis were, therefore, for a long time only carried out in necrotizing, bile-induced pancreatitis.

5.2.1 Necrotizing–Hemorrhagic Pancreatitis

In parallel to the progression of hypotension during bile-induced pancreatitis in dogs, Thal *et al.* (1963) reported that blood from the splenic and femoral veins contained substances that caused contraction of bovine isolated carotid artery (see Section 5.1). It was concluded, therefore, that BK or a similar peptide must have been liberated. Similar increases in kinin-like material were found in the blood from the femoral and portal veins (Katz *et al.*, 1964).

The isolated rat uterus was used as a bioassay preparation for measuring kinin-like material in portal blood and ascites fluid during bile-induced pancreatitis in dogs (Nugent and Atendido, 1966; Satake *et al.*, 1973b). Kinin-like activity in plasma and lymph from the thoracic duct of these animals showed a progressive increase during an observation period of 3 h (Satake *et al.*, 1973b). High concentrations of kinin-like material were determined in the ascites fluid 5 h after the induction of necrotizing–hemorrhagic pancreatitis in dogs (Satake *et al.*, 1985a,b). Pretreatment with a synthetic protease inhibitor, nafamstat mesilate, decreased the amount of kinins in the ascites fluid below the limit of detection.

During bile-induced pancreatitis in dogs, blood kininogen levels were reduced significantly and correlated with the hemoconcentration, hypotension and the accumulation of ascites (Ryan *et al.*, 1964, 1965; Nugent *et al.*, 1969). Both the symptoms and the kininogen loss were attenuated by pretreatment with aprotinin, a naturally occurring serine protease inhibitor (Ryan *et al.*, 1964, 1965). Nugent *et al.* (1969) reported that a substantial fraction of kininogen was found in the peritoneal fluid such that kininogen levels in the ascites sometimes exceeded that lost from the plasma. Popieraitis and Thompson (1969) showed that, in bile-

induced pancreatitis, the reduction of the kinin-forming material in the blood is more pronounced in the pancreatic venous blood than in arterial blood. The arteriovenous difference was also visible in aprotinin-treated dogs, although it seemed to be less pronounced. Ofstad (1970) measured kininogen and kinins in the blood and ascites from dogs in which hemorrhagic pancreatitis had been induced by intraductal injection of taurocholate and trypsin without prior splenectomy. Kinin-like material was found in most samples drawn from the pancreaticoduodenal vein, but in only less than half of the samples of the femoral vein or from samples of the pancreatic exudate.

In the pig, taurocholate-induced pancreatitis is also accompanied by a reduction in plasma kininogen (Kortmann *et al.*, 1984). Although kininogen consumption seemed to be HMWK rather than LMWK, most of the kinin liberated seemed to be derived from the latter kininogen. The loss in kininogen could also be measured in the blood from the portal vein and in the pancreatic lymph of *in situ*, isolated porcine pancreas during intraductal infusion of taurocholate or intra-arterial infusion of oleic acid, producing, respectively, severe or mild necrotizing pancreatitis (Waldner *et al.*, 1993).

Satake *et al.* (1973a) reported that the ascitic fluid formed during bile-induced pancreatitis in dogs possessed the ability to release kinins from a substrate prepared from plasma according to the method of Jacobson (1966). Jacobson described two substrates, one being a good substrate for plasma kallikrein and the other being a good substrate for both kallikrein types. Although Satake and colleagues did not report which of these substrates was used, they concluded that the kinin-forming enzyme in the peritoneal exudate may have been glandular kallikrein.

During acute necrotizing pancreatitis induced by intraductal injection of taurodeoxycholic acid in rats, kallikrein-like activity in the blood plasma was elevated 6 and 12 h later (Seung and Feldman, 1985). Kallikrein activity was increased in the hemorrhagic ascites at 1 h, and remained elevated for at least 18 h, while the prekallikrein concentrations in the exudate were reduced. The kallikrein inhibitor concentrations were markedly decreased in the peritoneal exudate throughout the experiment.

During taurocholate-induced pancreatitis in pigs, increased plasma kallikrein activity together with a reduction in plasma and functional kallikrein inhibition were found in the peritoneal exudate while no changes in the kallikrein activity was found in the plasma (Ruud *et al.*, 1982, 1984, 1985, 1986a,b). Glandular kallikrein in the plasma, however, was increased almost immediately after the onset of taurocholate-induced pancreatitis in piglets (Kortmann *et al.*, 1984), and bile- or bile/trypsin-induced pancreatitis in pigs (Bläckberg and Ohlsson, 1994). Tissue kallikrein increases further throughout the bile-induced pancreatitis. Increased tissue kallikrein activity was also determined in the portal venous blood and pancreatic lymph of the *in situ,* isolated pancreas of the pig (Vollmar *et al.,* 1991; Waldner *et al.,* 1993).

Even higher values of tissue kallikrein were found in ascites (Kortmann *et al.,* 1984; Waldner *et al.,* 1993; Bläckberg and Ohlsson, 1994). The tissue kallikrein in the exudate was partly free kallikrein and partly kallikrein complexed to α_1-M, α_2-M or to α_1-proteinase inhibitor. The complexes with α_1-M and α_2-M retained their enzymatic activity (Bläckberg and Ohlsson, 1994). The ascites also exhibited increased plasma kallikrein activity together with reduced plasma and functional kallikrein inhibition (Ruud *et al.,* 1982, 1984, 1985, 1986a).

Forrell (1955) had already noted that prekallikrein was reduced in the blood of rats and cats when symptoms of hypotension were visible during bile-induced pancreatitis. Seung and Feldman (1985) reported that the reduction of prekallikrein during taurodeoxycholate-induced pancreatitis in the rat lasted longer than the increases in kallikrein-like activity in the blood plasma.

5.2.2 Interstitial–Edematous Pancreatitis

Shimizu *et al.* (1993) reported that immunoreactive plasma kinin levels increased during caerulein-induced pancreatitis in rats, peaking at 6 h after the induction of pancreatitis, and remaining elevated for at least 36 h. At this time, however, histological signs of pancreatic inflammation such as interstitial edema, vacuolization of acinar cells or infiltration of inflammatory cells had already been resolved. During a 2 h infusion of caerulein to induce acute pancreatitis, we found that immunoreactive kinin levels in pancreatic tissue also increase, peaking at 2 h (Griesbacher and Lembeck, 1993).

5.3 STUDIES WITH BRADYKININ ANTAGONISTS

Since the early bradykinin antagonists suffered from lack of potency and especially exhibited only a very short duration of action *in vivo*, studies using such compounds in experimental models of acute pancreatitis had not been undertaken until a few years ago. However, with the development of bradykinin antagonists with sufficient stability *in vivo* (see above), pharmacological tools are available to explore the role of kinins in experimental models of pancreatitis.

5.3.1 Interstitial–Edematous Pancreatitis

In caerulein-induced edematous pancreatitis in rats, pretreatment with a potent and long-acting B_2 receptor antagonist, Hoe 140, prevents both hypotension and pancreatic edema (Griesbacher and Lembeck, 1992). Elevated blood levels of amylase and lipase were, however, not reduced but, rather, were augmented by Hoe 140. The same effects on pancreatic edema formation, hypotension, hypovolemia and serum enzyme activities

were reported by Félétou *et al.* (1995b) using another B_2 receptor antagonist, S 16118. The augmentation of enzyme activities in the blood is accompanied by a reduction of enzyme accumulation in pancreatic tissue (Griesbacher *et al.,* 1993), indicating that the edema of acute pancreatitis traps a proportion of the enzymes discharged into the interstitial spaces, whereas prevention of edema with a kinin antagonist may improve the removal of activated digestive enzymes from the tissue.

The augmentation by Hoe 140 of the serum enzyme activities was only observed in the early stages of pancreatitis, i.e., up to 2 h, whereas the later enzyme activities were not different from controls (Griesbacher *et al.,* 1995; Lerch *et al.,* 1995). The reduction of pancreatic enzyme accumulation in the pancreas, however, persists for a longer period (Griesbacher *et al.,* 1995). This reduction of the accumulation of activated digestive enzymes in the pancreatic tissue by kinin antagonists could be beneficial in the clinic, since such enzymes are likely to be important in promoting the progression from edematous towards necrotizing pancreatitis (Creutzfeld and Schmidt, 1970; Fernández-del Castillo *et al.,* 1994).

During the first 20 min of caerulein-induced pancreatitis, amylase and lipase activities in pancreatic juice are almost as high as during the initial periods of "physiological" stimulation of the exocrine function with low doses of caerulein (Griesbacher *et al.,* 1995). This might lead to speculation that, during the very early stages of edematous pancreatitis, a residual, direct secretion of amylase by the acinar cells in response to caerulein persists (San Román *et al.,* 1990), whereas the later development of edema would prevent this secretion by compressing the pancreatic ducts. In this case, prevention of the development of the edema with a BK antagonist should restore a more sustained secretion of digestive enzymes into the pancreatic duct. Since, however, pretreatment with Hoe 140 almost abolished the enzyme output via the biliopancreatic duct, even during the first stages of edematous pancreatitis, and also did not inhibit the normal, exocrine pancreatic function (Griesbacher *et al.,* 1995), it is more likely that the initial secretion of pancreatic enzymes into the pancreatic duct is due to "overflow" of enzymes, which have been released into the tissue and trapped by the edema. When the egress of enzymes from the pancreas into the general circulation via absorption into the pancreatic veins and/or via drainage into the lymphatic system (Howard *et al.,* 1949; Egdahl, 1958) is improved, as with Hoe 140 (Griesbacher and Lembeck, 1992; Griesbacher *et al.,* 1995), no stagnation of enzymes in the tissue will occur and no "overflow" into the pancreatic ducts will take place.

The effects of Hoe 140 on serum activities of pancreatic enzymes in caerulein-induced pancreatitis indicate that BK is not involved in the initiation of pancreatitis, i.e., in the mechanisms leading to the disturbances of acinar cell function, but may be an important contributor to the inflammatory response in the pancreas. A role in the initiation of pancreatitis can also be excluded because Hoe 140 does not have an effect on acinar cell vacuole formation typical of acute pancreatitis (Lerch *et al.*, 1992, 1995).

That loss of plasma into the pancreatic edema is the cause of the hypotension was confirmed by the observation that Hoe 140 inhibits hemoconcentration and this, along with inhibition of pancreatic edema during pancreatitis by this antagonist are strictly parallel (unpublished observations). Hoe 140 was also effective when given 10 or 25 min after the induction of pancreatitis, i.e., at times when the pancreatic edema had not reached its maximum extent (Griesbacher *et al.*, 1993).

Pretreatment with Hoe 140 also reduced the levels of immunoreactive kinins in the pancreas during caerulein-induced pancreatitis (Griesbacher and Lembeck, 1993). Although Hoe 140 can inhibit kallikrein in micromolar concentrations (Damas *et al.*, 1995), it is unlikely that the tissue concentrations of the antagonist, following subcutaneous injection of 100 nmol/kg, reach such high values. A possible explanation of the reduced tissue kinin levels could be that the initial release of small amounts of kinins in the pancreas cause increases in the permeability of pancreatic blood vessels and, consequently, the extravasation of plasma kininogen into the tissue. The blockade by a BK antagonist of this supply with further substrate could also prevent a further release of kinins.

The possibility has been raised that kinin receptor antagonism may have deleterious effects in edematous pancreatitis. Thus, Weidenbach *et al.* (1995) noted that, in caerulein-induced pancreatitis, damage to pancreatic blood vessels with hemorrhagic lesions occurred following treatment with Hoe 140, an effect which was not observed in untreated pancreatitis. The results were interpreted as a progression from edematous to hemorrhagic pancreatitis. Hemorrhagic lesions in the pancreas would be an untoward effect that could adversely affect recovery from the condition. Nevertheless, histological evaluations in our laboratory could not reveal conclusive evidence for hemorrhagic lesions as a consequence of kinin receptor blockade (unpublished observations).

Measuring blood glucose levels during caerulein-induced pancreatitis and during a glucose tolerance test in the recovery period on the day after the experiment as an indication for endocrine function of the pancreas, we could not find adverse effects resulting from treatment with Hoe 140 (Griesbacher *et al.*, 1995). Similarly, no adverse effects on resting and stimulated exocrine pancreatic function during and after pancreatitis could be demonstrated, and blood levels of liver enzyme activities were not changed by this drug. Although this investigation failed to show an adverse effect of Hoe 140 in functional respects, the results of Weidenbach *et al.* (1995) underline the need of further studies to clarify the exact role of kinins in edematous pancreatitis.

Maintaining adequate perfusion is essential for organ function during pathological situations. Whether or not microcirculation in the pancreas is decreased in experimental edematous pancreatitis can, however, not be answered conclusively at present. Morphological studies suggest that one of the earliest changes in this model are marked distortions of the pancreatic microvasculature (Kelly *et al.*, 1993), and a drastic reduction in the number of perfused capillaries (Gress *et al.*, 1990). However, other reports suggest that, in edematous pancreatitis, capillary blood cell velocity is increased in the presence of homogeneous capillary perfusion (Klar *et al.*, 1994) or that capillary perfusion itself is increased (Schmidt *et al.*, 1994). Intravital videomicroscopy observations also suggest that pancreatic capillary perfusion is increased in the early stages of caerulein-induced pancreatitis, but returns to baseline values within 3 h (Knoefel *et al.*, 1994). Bloechle *et al.* (1995) found arterial vasodilatation together with a reduced number of perfused capillaries using epiluminescent microscopy *in vivo*. At present, this latter, preliminary study is the only report on effects of a kinin antagonist on microcirculatory perfusion in edematous pancreatitis. Bloechle *et al.* (1995) reported that arterial vasodilatation was unaffected by Hoe 140, injected 15 min after the induction of pancreatitis, whereas the number of perfused capillaries was further reduced. Since impairment of pancreatic microcirculation is thought to be a critical factor in the progression from edematous to necrotizing pancreatitis (Klar *et al.*, 1994), findings like those of Bloechle *et al.* may add to the concern raised by Weidenbach *et al.* (1995) discussed above. The issue of whether BK antagonists are positive or adverse cannot be settled yet, since results of at least some studies with Hoe 140 in models of necrotizing pancreatitis, described below, indicate favorable effects of kinin inhibition.

The mild pancreatic edema, the hyperamylasemia, and the small increase in the number of neutrophils in the pancreas, following ligation of the pancreatic duct together with secretin stimulation in rats, was not affected by the blockade of BK receptors with Hoe 140 (Lerch *et al.*, 1995). However, this experimental model did not induce any changes in the histological appearance of the acinar cells and may not necessarily be related to mechanisms of acute edematous pancreatitis (see Section 3.1).

5.3.2 Necrotizing–Hemorrhagic Pancreatitis

In contrast to studies on experimental edematous pancreatitis, where in general, only one model is used, comparisons of studies in necrotizing experimental pancreatitis are more difficult because of the greater variety of the experimental models (see Section 3). The first report of a BK antagonist in experimental pancreatitis was published by Berg *et al.* (1989) who noted that DArg-[Hyp³,Thi⁵,⁸,DPhe⁷]-BK, a B_2 receptor antagonist, was without effect upon the hypotension associated with taurocholate-induced pancreatitis in rats. However, the antagonist was administered only at a time when the hypotension had fully developed, so that the lack of effect only excludes kinin-mediated vasodilatation but not hypovolemia owing to loss of plasma volume into a kinin-mediated pancreatic edema.

The effects of pretreatment, rather than post-treatment with Hoe 140 were first reported by Kanbe *et al.* (1994) who employed intraductal injection of sodium taurocholate supplemented with trypsin. This protocol induced necrotizing pancreatitis in rats with a mortality of 25–35% within 24 h. Hoe 140 [0.1–1 mg/kg sub-cutaneously (s.c.)], administered immediately before and 3 h after induction of pancreatitis, increased the 24 h survival rate, whereas a higher dose (3 mg/kg) was without effect. The dose is important in this context because the maximally effective dose of Hoe 140, following systemic administration in rats, has been reported to be about 0.13 mg/kg (Lembeck *et al.*, 1991; Wirth *et al.*, 1991a). Although nonspecific or partial agonist effects are not apparent with this antagonist, unless much higher doses are used (Lembeck *et al.*, 1992; Wirth *et al.*, 1991a), the effects of very high doses *in vivo* should be interpreted carefully. The Hoe 140-induced reduction of mortality described by Kanbe *et al.* (1994) at doses of 0.1 or 0.3 mg, however, is unlikely to be such a nonspecific effect. The lowest dose of Hoe 140 also reduced the magnitude of hypotension observed during taurocholate-induced pancreatitis, although the activity of serum amylase or pancreatic edema formation were, however, not affected.

Taurocholate-induced pancreatitis was also employed by Lerch *et al.* (1995) who reported no effect of Hoe 140 on the mortality observed within 30 h. The morpho-logical and histological appearance of the pancreas was not improved by Hoe 140 but, contrary to the seemingly adverse effects of the antagonist in caerulein-induced interstitial pancreatitis described by the same group (Weidenbach *et al.*, 1995, see above), the histology was apparently not worsened. Similarly, the accumulation of neutrophils, the intracellular redistribution of cathepsin B from the lysosome-enriched to the zymogen granule-enriched fraction, or the increase in serum amylase activity remained unaffected by Hoe 140.

Contrary to the lack of effect of Hoe 140 on the increase in serum amylase reported by Lerch *et al.* (1995), Closa and colleagues (1995) showed that increased levels of lipase activity in the blood during taurocholate-induced pancreatitis are reduced by pre-treatment with the kinin antagonist. Hoe 140 also prevented the increase in the production of pancreatic nitric oxide, and greatly attenuated thromboxane B_2 levels in the tissue. The effects were seen as potentially favorable, since the authors had previously implicated both nitric oxide and prostanoids in the pathogenesis of hemorrhagic pancreatitis (Closa *et al.*, 1994).

Bloechle *et al.* (1995) showed a significant improve-ment by Hoe 140 of histopathological alterations by preventing necrosis during taurocholate-induced pan-creatitis in rats. Using epiluminescent microscopy *in vivo*, vasoconstriction in small arteries was noted and the capillaries ceased to be perfused. Administration of Hoe 140 15 min after the intraductal injection of taurocholate had no effect on the arterial vasoconstriction, but preserved capillary perfusion. Leukocyte adherence to the wall of pancreatic blood vessels was prominent during taurocholate-induced pancreatitis, but was greatly reduced by the treatment with Hoe 140 (Bloechle *et al.*, 1995).

Satake *et al.* (1996) recently reported continuous infusion of Hoe 140, starting soon after the induction of a necrotizing pancreatitis by intraductal instillation of bile and trypsin in dogs, significantly improved the survival rate, hypotension, myocardial depression and plasma lactate, and inhibited the increase in plasma levels of β-endorphin. Moreover, increased serum levels of BK were significantly lower in the antagonist-treated group in the later stages of pancreatitis, which is in agreement with our findings of reduced kinin levels in the pancreas during edematous pancreatitis in rats treated with Hoe 140 (Griesbacher and Lembeck, 1993.) (see Section 5.3.1).

Lerch *et al.* (1995) also examined the effects of Hoe 140 in necrotizing pancreatitis induced by choline-deficient ethionine-supplemented diet in mice. No effect of the antagonist on the morphology of the pancreas, inflammatory cell infiltration, subcellular distribution of cathepsin B, or the increased levels of plasma amylase activity was found.

Watanabe *et al.* (1993) investigated the effects of Hoe 140 in necrotizing–hemorrhagic pancreatitis induced by a closed duodenal loop model in rats. Hoe 140 (6 mg/kg s.c.) at 0 and 3 h, significantly attenuated the increased levels of amylase and lipase in the blood serum, reduced the increases of pancreatic weight, and strongly inhibited the accumulation of fluid in the peritoneal cavity.

Another form of necrotizing pancreatitis, complete ischemia of the pancreas for 2 h with subsequent reper-fusion for 5 days, was induced by Hoffmann *et al.* (1995, 1996a,b). The B_2 receptor antagonist, CP 0597, was administered via osmotic minipumps starting from 15 min prior to the end of ischemia, and lasting throughout the following observation period of 5 days. While histological signs of severe necrotizing pancreatitis were present in all untreated animals, no signs of postischemic necrotizing tissue damage were observed in those animals that received the BK antagonist. Furthermore, the high mortality of about 50% associated with the procedure was completely prevented by the BK antagonist.

5.4   SIGNIFICANCE OF KININS DURING ACUTE PANCREATITIS

In the preceding sections we have summarized the work carried out over the past decades on the involvement of the kallikrein–kinin system in acute pancreatitis, and it is clear that the kallikrein–kinin system is activated during the disease. From the available data it is likely that the primary activation takes place in the pancreas itself. However, activation of the kallikrein–kinin system in the blood is also possible. The development of kinin receptor antagonists made it possible to gain more information on the function of the kinins released during acute pancreatitis. Kinins certainly are not involved in the processes initiating the disease. However, it is certain that once the initial damage to the acinar cell function has occurred, kinin release and actions are important in the inflammatory mechanisms that follow.

5.4.1 Vascular Functions

The primary pathophysiological significance of pancreatic kinins will be the induction of pancreatic edema with its systemic consequences of hemoconcentration, hypovolemia and hypotension. The significance of the kinin-mediated pancreatic edema for the pancreas itself is, at present, not clear. While the accumulation of activated digestive enzymes in the pancreas can be prevented by BK receptor antagonists, indicating a pathological role of kinins, the possible aggravation of morphological parameters by pretreatment with Hoe 140 could also indicate that kinin release during the early stages of edematous pancreatitis might be a protective mechanism. In contrast to this effect, the prevention or reduction by kinin antagonists of necrosis in models of necrotizing pancreatitis makes it possible that antagonizing the action of kinins could be beneficial. The apparent differences of the effects of BK antagonists in models of edematous or necrotizing pancreatitis emphasize that these forms of acute pancreatitis should not be seen merely as two different grades of severity of one disease. It is certain that necrotizing–hemorrhagic pancreatitis involves even more pathophysiological mechanisms than the interstitial–edematous form, and it is possible that a given mechanism, such as kinin release, could have different consequences in each form. Data available at present confirm the importance of the kinins in acute pancreatitis, and should be investigated further. Kinin receptor antagonists provide valuable tools for this purpose.

5.4.2 Pain

Acute pancreatitis is associated with severe pain. This clinical sign is also the primary symptom that makes the patient seek medical advice. Nevertheless, investigations on the role of kinins in acute pancreatitis using BK antagonists have, until now, focused on vascular functions of kinins. However, kinins also have potent actions on afferent nerve fibers and are the most potent endogenous nociceptive agents (see Chapter 9 of this volume). This makes them key candidates for the mediation of pain in any pathophysiological situation associated with kinin release.

In terms of a potential of BK antagonists as drugs for the therapy of acute pancreatitis, there clearly is a fundamental difference in the efficacy of such compounds against vascular or neural effects of kinins. In animal studies, the edema of the pancreas can only be prevented by a pretreatment with a BK antagonist. Once the edema has completely developed, a BK antagonist would not be able to exert this effect. In contrast to this, the nociceptive effects of kinins rely on a continuous activation of nociceptors. If the pain during pancreatitis proves to be due to endogenous kinins, then an administration of an antagonist, even in the later stages of acute pancreatitis, might be of great therapeutic interest. This question will only be solved by clinical studies to be performed in future.

Another type of pancreatic pain should also be mentioned in this context. Obstructive diseases of the biliary and pancreatic duct system associated with elevation of intraductal pressure are frequently accompanied by severe pain (Geenen et al., 1989; Moody et al., 1990; Prinz and Greenlee, 1990). Although nociceptive information arising from increases in the intraductal pressure in the biliopancreatic duct is clearly propagated to the central nervous system by capsaicin-sensitive afferent nerves (Griesbacher, 1994), this may not necessarily involve BK as the initiating factor. However, some patients undergoing diagnostic or therapeutic procedures, which cause an increase in intraductal pressure, such as endoscopic retrograde cholangiopancreatography (Lasson et al., 1988) or balloon dilatation (Jaschke et al., 1992), report symptoms of severe pain for several hours. Similarly, pancreatic carcinoma causes severe pain in a high proportion of patients (Bakkevold et al., 1992).

5.4.3 Drug-induced Pancreatitis

A number of drugs including azathioprine, thiazides, estrogens, furosemide, sulfonamides and tetracycline are associated with the development of acute pancreatitis in a small proportion of patients (Mallory and Kern, 1980; Banerjee et al., 1989; Steer, 1993). The mechanisms of induction of pancreatitis by these drugs are difficult to determine. In some cases, an interaction with the kallikrein–kinin system may, at least in part, contribute to effects damaging the pancreas. In most cases a causative role of a drug was assumed if abdominal pain developed when the drug was taken and serum amylase levels were elevated, and the symptoms resolved when the drug was withdrawn, no other cause of pancreatitis being found. Although a causative role of drugs is, therefore, likely for the majority of cases, also another possibility has to be considered. If a very mild form of inflammation develops in the pancreas from a cause unrelated to the intake of a drug, this will not be noticed by the patient if pain is absent. If, however, a drug is given which can increase intrapancreatic factors leading to stimulation of nociceptive nerve fibers, overt pain will result and the pancreatic inflammatory process will be diagnosed.

Of the drugs mentioned, case reports linking episodes of acute pancreatitis to the use of furosemide (Wilson *et al.*, 1967; Jones and Oelbaum, 1975; Buchanan and Cane, 1977; Call *et al.*, 1977) seem to be interesting in this context. Furosemide has the ability to release kinins from endothelial cells, an effect associated with the subsequent release of nitric oxide and prostacyclin via bradykinin B_2 receptors (Wiemer *et al.*, 1994). Kinins released by furosemide during a subclinical and, thus, undiagnosed form of pancreatitis could induce pain, while the autacoids released by the kinins contribute to pancreatic damage. It is also noteworthy that Call *et al.* (1977) described a case of furosemide-induced pancreatitis associated with hyperlipidemia, a factor thought possibly to be able to cause pancreatitis by itself (Steer, 1993).

Angiotensin-converting enzyme (ACE) inhibitors have recently been connected to acute pancreatitis. Published case reports to date seem to be limited to enalapril (Tilkemeier and Thompson, 1988; Adverse Drug Reactions Advisory Committee, 1989; Pedro-Botet *et al.*, 1990; Roush *et al.*, 1991) and lisinopril (Dabaghi, 1991; Maliekal and Drake, 1993; Standridge, 1994). A possible mechanism may involve prolongation of the biological half-life of endogenous kinins owing to inhibition of the major degrading enzyme should be considered. This could also, in cases of otherwise subclinical pancreatic damage, lead to pain.

6.   Possible Physiological Roles of Kinins in the Pancreas

6.1   FROM EARLY DISCOVERIES TO RECENT FINDINGS

The discovery of kallikrein, at first assumed to be a hormone regulating blood pressure, but later defined as an enzyme, in the pancreas was the key to the characterization of the kallikrein–kinin system. In the following years kallikrein was found in many other organs, such as salivary glands, skin, gut, endothelial cells and in plasma. The history of the discovery of the components of the kallikrein–kinin system in the pancreas is given briefly above (see Chapter 1).

The most important tools for the study of the role of endogenous kinins are receptor antagonists. However, it is also possible to assess a possible involvement of kinins when observed symptoms are enhanced following the administration of inhibitors of kininase II (ACE), the major kinin-degrading enzyme (Erdös, 1979; see also Chapter 7 of this volume). Augmentation of effects by such inhibitors can be attributed to endogenous kinins if the effects are blocked by kinin receptor antagonists.

BK antagonists and kininase inhibitors can also be used to explore possible physiological actions of kinins.

The presence of all components of the kallikrein–kinin system in the pancreas makes it conceivable that such physiological actions of kinins exist in this organ. In the following section we wish to present an overview of investigations that support such a physiological role and also summarize recent results from our own laboratory.

6.2   PANCREATIC BLOOD FLOW AND MICROCIRCULATION

Papp *et al.* (1967) demonstrated that the injection of kinins into arteries supplying the pancreas of dogs caused a pronounced increase in pancreatic blood flow. In the isolated perfused pancreas of rats, intra-arterial injection of BK reduced the perfusion pressure via the action of prostacyclin (Sweiry *et al.*, 1994). In anesthetized dogs, BK increased the blood flow in the superior pancreatico-duodenal artery and in the pancreatic microcirculation, and also increased pancreatic oxygen consumption (Pawlik *et al.*, 1994). In contrast to its effects in rat pancreas (Sweiry *et al.*, 1994), the hyperemic and metabolic effects of BK in dogs were significantly reduced by a nitric oxide synthase inhibitor, suggesting a contribution of nitric oxide release (Sweiry *et al.*, 1994).

In cat pancreas, CCK was the most potent vasodilator, increasing the flow by up to five-fold (Hilton and Jones, 1968). The vasodilator effect of CCK was almost certainly localized in the pancreas, since the response was observed within the organ in the absence of a fall of systemic blood pressure. When the effluent from the perfused pancreas was examined for kinin-forming activity, acetylcholine and CCK led to a considerable release of kininogenase. This effect was not observed with secretin. Stimulation of the vagus nerve also increased the blood flow in the pancreas by 60% (Hilton and Jones, 1968). Intravenous infusion of caerulein, in a dose that does not induce pancreatitis but leads to a sustained secretion of digestive enzymes into the pancreatic duct of rats also increased the secretion of active kallikrein into the pancreatic juice during the infusion (Lembeck and Griesbacher, unpublished observations). The observation that stimuli leading to increased secretion of digestive enzymes also increased kallikrein excretion from the pancreas was a strong indication for release of kinins in the pancreas (Hilton and Jones, 1968). Current experiments suggest that Hoe 140 blocks increases in pancreatic blood flow induced by the intravenous injection of caerulein (Lembeck, unpublished observation).

Changes in the pancreatic microcirculation are closely linked to influences on microvascular permeability by opening of gaps between capillary endothelial cells. Plasma extravasation might be compensated to a degree by simultaneous removal of the fluids from the tissue via the lymph (Stürmer, 1966). When the plasma extravasation exceeds this level, it will cause edema in the tissue.

This effect may be enhanced by simultaneous contraction of the postcapillary veins by kinins (Guth *et al.*, 1966; Goldberg *et al.*, 1976; see also Bhoola *et al.*, 1992).

Intravenous infusion of high doses of BK in the rat leads to pancreatic edema, whereas pretreatment with a kininase II inhibitor, captopril, increases sensitivity to this effect of BK. While the infusion of secretin does not induce the formation of edema in the pancreas, it reduces the edema induced by BK or caerulein (Lembeck, unpublished observation) suggesting that fluids extravasated into the tissue are partly removed by fluid secretion into the pancreatic duct induced by secretin.

Opening of endothelial gaps in the microcirculation of the pancreas can be visualized by the injection of Monastral blue B particles (Joris *et al.*, 1982). Pretreatment of rats with captopril induces a pronounced increase in the number of dye particles in the wall of pancreatic capillaries and small veins (Griesbacher and Lembeck, 1995; Lembeck and Griesbacher, 1996). The effect of captopril is abolished by prior pretreatment with Hoe 140. These results indicate that there is continuous formation of kinins in the microcirculation of the pancreas modulating the state of the endothelial cells.

The assumed physiological effects of kinins in the pancreas seem to be changed gradually to pathological effects, i.e., edema formation, depending on the magnitude of stimulation. When "mild pancreatitis" was induced in rats by high doses of caerulein, the capillary perfusion significantly increased during the first 30 min and declined within 3 h, whereas in "moderate and severe pancreatitis", evoked by additional intraductal infusion of glycodeoxycholic acid either alone or together with enterokinase, the capillary perfusion and the hemoglobin oxygen saturation were decreased and progressive tissue ischemia developed (Knoefel *et al.*, 1994).

6.3   PANCREATIC SECRETION

Intravenous infusion of BK, in doses that do not produce edema, induced an increase in the volume of the pancreatic juice collected from the biliopancreatic duct of anesthetized rats following ligation of the hepatic ducts to prevent the dilution of the pancreatic juice with bile fluid. The effect of kinin was dose-dependent and lasted for 20–40 min after the end of the infusion (Lembeck and Griesbacher, 1996).

Although BK by itself increased the volume of the pancreatic juice, it had no effect on the secretory effect of secretin following their co-administration. Similarly, the pro-secretory action of secretin was unaffected by a pretreatment with Hoe 140 (Lembeck, unpublished observations). It is concluded, therefore, that although BK can increase the volume of the pancreatic juice, it does not participate in the secretory actions of secretin. No data are available on effects of kinins on bicarbonate secretion from the pancreas.

6.4   FROM PHYSIOLOGICAL FUNCTION TO PATHOLOGICAL EVENTS

Kinins are known to be released in many different types of inflammation, such as arthritis (see Chapter 13 of this volume), colitis, allergy (see Chapter 16) and nephritis (Hutchinson *et al.*, 1995). The release of kinins in these conditions may be initiated by activation of kinin-releasing enzymes within the body itself or by kinin-liberating enzymes brought into the body by microorganisms (Maeda and Molla, 1989; Kaminishi *et al.*, 1993; Maruo *et al.*, 1993). We have reviewed the literature on the involvement of the kallikrein–kinin system in acute pancreatitis and also its possible physiological functions in the pancreas.

Although kinins clearly are released in this and other pathological situations, it has to be remembered that kinins may also serve physiological or protective functions. Inflammation itself has basically to be regarded as a repair or defense mechanism of the body, or at least an attempt of the body to that effect. During the course of inflammation, some of the mechanisms involved will also have damaging effects on the organism. This is well expressed by the term "pathophysiology", which encompasses pathological and physiological effects. This view is elegantly expressed by Grawitz who has defined inflammation as "the reaction of the irradiated or damaged tissues which still retain vitality" or by Burdon-Sanderson who characterized the process of inflammation as "the succession of changes which occur in a living tissue when it is injured, provided that the injury is not of such a degree as at once to destroy its structure or vitality" (see Florey, 1962). These views are also reflected in the classification of inflammatory stages of Ehrich (1961).

A dual, pathological and protective role of the kallikrein–kinin system is also possible in acute pancreatitis. As reviewed in detail above, treatment of experimental animals with B_2 kinin receptor antagonists has resulted in beneficial effects, pointing towards a pathological function of endogenous kinins, and also in potentially adverse effects, which could be an indication for a protective role of kinin release. It is likely that the exact function of the activation of the kallikrein–kinin system is different in interstitial–edematous and necrotizing–hemorrhagic pancreatitis.

7.   *Perspectives in the Treatment of Pancreatitis*

"Acute pancreatitis remains to represent a vexing clinical entity. The pathobiology of the disease process is ill understood, few effective therapies exist, and several entities plague investigation of this disease" and "No specific medical or surgical therapy is capable of directly limiting pancreatic autodigestion and inflammation" are introductory words of a review about new perspectives on acute pancreatitis (Leach *et al.*, 1992). Small wonder it is then that the surgeon looks upon the pancreas as the "powder keg of the abdomen" (Thal *et al.*, 1963). Medical treatment remains largely custodial, maintaining an adequate circulatory volume, sufficient renal perfusion, supporting respiration and correcting electrolyte or plasma protein abnormalities. The therapeutic efficacy of "putting the pancreas at rest" was uniformly disappointing. The activation of the digestive enzymes within the acinar cells pointed to the essential site of the pathology and this process was investigated in detail. The premature activation of digestive zymogens within the acinar cell itself seems to be the initial event. Although a number of studies have demonstrated the involvement of several proteases, the systemic administration of the protease inhibitor, aprotinin (Trasylol®), so hopefully performed for several decades, turned out to be ineffective or unconvincing in clinical practice. Every kind of drug treatment for acute pancreatitis is hampered by the fact that it is usually delayed for up to 24 h after the onset of the symptoms. Progress, however, was made in the treatment of gallstone pancreatitis, where papillotomy and stone extraction have been demonstrated to be safe and effective.

All the models of experimental pancreatitis developed during recent years have to be regarded under the limitation that they are not perfect copies of acute clinical pancreatitis. They allow us to obtain information about factors involved in the pathobiology of the disease and permit the investigation of possible new therapeutic agents.

The development of kinin receptor antagonists, like Hoe 140 (Icatibant), with a long duration of action *in vivo* (Hock *et al.*, 1991; Lembeck *et al.*, 1991; Wirth *et al.*, 1991a) have made it possible to investigate the role of endogenous kinins in acute pancreatitis (Griesbacher and Lembeck, 1992; Griesbacher *et al.*, 1993, 1995). These and other investigations showed that endogenous kinins play an important role in the pathobiology of acute pancreatitis. However, they also showed that some of the functions of these kinins can only be prevented when the antagonist is administered at a very early stage, e.g., before the pancreatic edema has fully developed.

Although there is certainly a clear difference in the time-courses of experimental pancreatitis, which develops within less than one hour, and the clinical disease, which needs a much longer period of time to develop fully, a pharmacological intervention that is effective only in the very early stages of pancreatitis is a substantial limitation for its potential clinical value. However, there are three points which have to be raised in support of a therapeutic potential even at later stages.

First, a very careful clinical trial with a platelet-activating factor (PAF) antagonist, lexipafant, showed convincing therapeutic effects in a randomized, double-blind, placebo-controlled study (Kingsnorth *et al.*, 1995). PAF can be synthesized by pancreatic acinar cells in response to CCK or caerulein (Soling and Fest, 1986) and, by itself, can stimulate the secretion of enzymes from acinar cells (Soling *et al.*, 1984). Acute pancreatitis can even be induced experimentally by injection of PAF into the superior pancreaticoduodenal artery (Emanuelli *et al.*, 1989). Hence, the positive results of the clinical trial with lexipafant demonstrate that inhibition of a pathophysiological mechanism, which clearly comes into play at an earlier stage as kinin release can have a therapeutic effect.

Secondly, the prevention of the pancreatic edema following a very early treatment might turn out to be associated with adverse consequences as indicated by some investigators who raised concerns about the effects of Hoe 140 in experimental models of edematous pancreatitis (see Section 5.3).

Third, there is a possibility that kinin antagonists may be used as therapeutic agents for the treatment of the pain symptoms during the course of the disease (see Section 5.4.2). If it could be demonstrated that kinins play a significant role in pancreatic inflammatory pain, then antagonists of the BK receptors could be administered even after the initial symptoms have developed.

While early treatment with a BK antagonist might improve the removal of activated enzymes from the tissue via the bloodstream or the lymph system (Griesbacher *et al.*, 1993, 1995), another potential treatment might be effective by improving the elimination of such factors via the pancreatic duct. Conflicting reports exist from experimental studies on whether this is reduced (Keim *et al.*, 1985; Renner *et al.*, 1986) or preserved (Manso *et al.*, 1989, 1992) during early edematous pancreatitis. Nevertheless, the stimulation of juice production by secretin, either released endogenously (Manso *et al.*, 1992) or administered exogenously (Niederau *et al.*, 1985; Renner *et al.*, 1986; Manso *et al.*, 1989), was reported to have beneficial effects both in edematous and in necrotizing experimental pancreatitis. Recent findings in our laboratory agree with these findings, since the infusion of secretin reduces pancreatic edema during caerulein-induced pancreatitis (Lembeck, unpublished observations).

More than a century of investigations into the pathobiology of acute pancreatitis and several decades of research on the possible involvement of the kallikrein–kinin system have accumulated a large body of information about the possible roles of kinins in this disease. The development of selective, long-acting BK receptor antagonists only a few years ago has given this line of research powerful tools for further inquiries into the exact pathophysiological functions of kinins in the pancreas. Whether or not these or other antagonists will eventually be used as therapeutic agents remain to be determined.

8. References

Aasen, A.O., Ruud, T.E., Kaaresen, R. and Stadaas, J.O. (1986). Evaluation of patients with acute pancreatitis by means of chromogenic peptide substrate assays and the proenzyme functional inhibition index. Scand. J. Gastroenterol. Suppl. 126, 40–45.

Aasen, A.O., Ruud, T.E., Roeise, O., Bouma, B.N. and Stadaas, J.O. (1989). Peritoneal lavage efficiently eliminates protease-alpha-2-macroglobulin complexes and components of the contact system from the peritoneal cavity in patients with severe acute pancreatitis. Eur. Surg. Res. 21, 1–10.

Abelous, J.-E. and Bardier, E. (1909a). Les substances hypotensives de l'urine humaine normale. C. R. Soc. Biol. 66, 511–512.

Abelous, J.-E. and Bardier, E. (1909b). L'urohypotensine. J. Physiol. Pathol. Génér. 11, 777–786.

Adler, G., Hupp, T. and Kern, H.F. (1979). Course and spontaneous regression of acute pancreatitis in the rat. Virchows Arch. Pathol. Anat. Histol. 382, 31–47.

Adler, G., Gerhards, G., Schick, J., Rohr, G. and Kern, H.F. (1983). Effects of in vivo cholinergic stimulation of rat exocrine pancreas. Am. J. Physiol. 244, G623–G629.

Adverse Drug Reactions Advisory Committee (1989). Pancreatitis and ACE inhibitors. Austr. Adv. Drug React. Bull., August.

Aho, H.S. and Nevalainen, T.J. (1980). Experimental pancreatitis in the rat. Ultrastructure of sodium taurocholate-induced pancreatic lesions. Scand. J. Gastroenterol. 15, 417–424.

Amurawaiye, E.O. and Brown, R.A. (1991). Acute pancreatitis – 30 years' experience at a teaching hospital. Can. J. Surg. 34, 137–143.

Bakkevold, K.E., Arnesjø, B. and Kambestad, B. (1992). Carcinoma of the pancreas and papilla of Vater: presenting symptoms, signs, and diagnosis related to stage and tumour site. A prospective multicentre trial in 472 patients. Scand. J. Gastroenterol. 27, 317–325.

Banerjee, A.K., Patel, K.J. and Grainger, S.L. (1989). Drug-induced acute pancreatitis. A critical review. Med. Toxicol. Adverse Drug Exp. 4, 186–198.

Banks, P.A. (1991). Infected necrosis: morbidity and therapeutic consequences. Hepatogastroenterology 38, 116–119.

Barlas, A., Okamoto, H. and Greenbaum, L.M. (1983). T-Kininogen – the major plasma kininogen in rat adjuvant arthritis. Biochem. Biophys. Res. Commun. 129, 280–286.

Berg, T., Schlichting, E., Ishida, H. and Carretero, O.A. (1989). Kinin antagonist does not protect against the hypotensive response to endotoxin, anaphylaxis or acute pancreatitis. J. Pharmacol. Exp. Ther. 251, 731–734.

Bernard, C. (1856). "Leçons de physiologie expérimentale". Vol. 2, p. 278. Baillière, Paris.

Bhoola, K.D., Figueroa, C.D. and Worthy, K. (1992). Bioregulation of kinins: kallikreins, kininogenases, and kininases. Pharmacol. Rev. 44, 1–80.

Bilchik, A.J., Leach, S.D., Zucker, K.A. and Modlin, I.M. (1990a). Experimental models of acute pancreatitis. J. Surg. Res. 48, 639–647.

Bilchik, A.J., Zucker, K.A., Adrian, T.E. and Modlin, I.M. (1990b). Amelioration of cholinergic-induced pancreatitis with a selective cholecystokinin receptor antagonist. Arch. Surg. 125, 1546–1549.

Bläckberg, M. and Ohlsson, K. (1994). Studies on the release of tissue kallikrein in experimental pancreatitis in the pig. Eur. J. Surg. 26, 116–124.

Bloechle, C., Betge, S., Athar, H., Abdollahi, F., Kuehn, R., Kusterer, K. and Izbicki, J.R. (1995). Bradykinin-antagonist icatibant preserves pancreatic microcirculation in S-taurocholate, but not in cerulein-induced pancreatitis of the rat. Pancreas 11, 421.

Buchanan, N. and Cane, R.D. (1977). Frusemide-induced pancreatitis. Br. Med. J. 2, 1417.

Büchler, M. (1991). Objectification of the severity of acute pancreatitis. Hepatogastroenterology 38, 101–108.

Büchler, M., Friess, H., Uhl, W. and Beger, H.G. (1992). Clinical relevance of experimental acute pancreatitis. Eur. Surg. Res. 24 (Suppl. 1), 85–88.

Call, T., Malarkey, W.B. and Thomas, F.B. (1977). Acute pancreatitis secondary to furosemide with associated hyperlipidemia. Am. J. Dig. Dis. 22, 835–838.

Cavuoti, O.P., Moody, F.G. and Martinez, G. (1988). Role of pancreatic duct occlusion with prolamine (Ethibloc) in necrotizing pancreatitis. Surgery 103, 261–366.

Closa, D., Hotter, G., Prats, N., Bulbena, O., Roselló-Catafau, J., Fernández-Cruz, L. and Gelpí, E. (1994). Prostanoid generation in the early stages of acute pancreatitis: a role for nitric oxide. Inflammation 18, 469–480.

Closa, D., Hotter, G., Prats, N., Gelpí, E. and Roselló-Catafau, J. (1995). A bradykinin antagonist inhibited nitric oxide generation and thromboxane biosynthesis in acute pancreatitis. Prostaglandins 49, 285–294.

Cole, T.J. and Schreiber, G. (1992). The structure and expression of the genes for T-kininogen in the rat. In "Recent Progress on Kinins: Biochemistry and Molecular Biology of the Kallikrein–Kinin System", Agents Actions Suppl. 38/I (eds. H. Fritz, W. Müller-Esterl, M. Jochum, A. Roscher and K. Luppertz), pp. 292–299. Birkhäuser, Basel.

Corfield, A.P., Cooper, M.J. and Williamson, R.C.N. (1985). Acute pancreatitis: a lethal disease of increasing incidence. Gut 26, 724–729.

Creutzfeldt, W. and Schmidt, H. (1970). Aetiology and pathogenesis of pancreatitis (current concepts). Scand. J. Gastroenterol. 5 (Suppl. 6), 47–62.

Cutler, A.F., Yousif, E.A. and Blumenkehl, M.L. (1992). Hereditary angioedema associated with pancreatitis. South. Med. J. 85, 1149–1150.

Dabaghi, S. (1991). ACE inhibitors and pancreatitis. Ann. Intern. Med. 115, 330–331.

Damas, J., Bourdon, V. and Souza Pinto, J.-C. (1995). The myostimulating effect of tissue kallikrein on rat uterus. Naunyn-Schmiedebergs Arch. Pharmacol. 351, 535–541.

Dumont, R.E. and Martelli, A.B. (1968). Pathogenesis of pancreatic edema following exocrine duct obstruction. Ann. Surg. 168, 302–309.

Egdahl, R.H. (1958). Mechanism of blood enzyme changes following the production of experimental pancreatitis. Ann. Surg. 148, 389–399.

Ehrich, W.E. (1961). Inflammation. Prog. Surg. 1, 1–70.

Elliott, D.F., Horton, E.W. and Lewis, G.P. (1960). Actions of pure bradykinin. J. Physiol. 153, 473–480.

Emanuelli, G., Montrucchio, G., Gaia, E., Dughera, L., Corvetti, G. and Gubetta, L. (1989). Experimental acute pancreatitis induced by platelet-activating factor in rabbits. Am. J. Pathol. 134, 315–326.

Erdös, E.G. (1979). Kininases. In "Bradykinin, Kallidin and Kallikrein", Handbook of Experimental Pharmacology, Vol. 25, Suppl. (ed. E.G. Erdös), pp. 428–487. Springer-Verlag, Heidelberg.

Fabris, C., Panozzo, M.P., Basso, D., Del Favero, G., Plebani, M., Zaninotto, M., Fogar, P., Meggiato, T., Scalon, P., Ferrara, C. and Naccarato, R. (1991). Urinary kallikrein excretion in chronic pancreatic diseases. Am. J. Nephrol. 11, 386–390.

Félétou, M., Robineau, P., Lonchampt, M., Bonnardel, E., Thurieau, C., Fauchère, J.-L., Widdowson, P., Mahieu, J.-P., Serkiz, B., Volland, J.-P., Martin, C., Naline, E., Advenier, C., Prost, J.-F. and Canet, E. (1995a). S 16618 (p-guanidobenzoyl-[Hyp^3,Thi^5,DTic^7,Oic^8]bradykinin) is a potent and long-acting bradykinin B$_2$ receptor antagonist, in vitro and in vivo. J. Pharmacol. Exp. Ther. 273, 1071–1077.

Félétou, M., Lonchampt, M., Robineau, P., Jamonneaut, I., Thurieau, C., Fauchère, J.-L., Villa, P., Ghezzi, P., Prost, J.-F. and Canet, E. (1995b). Effects of the bradykinin B$_2$ receptor antagonist S 16618 (p-guanidobenzoyl-[Hyp^3,Thi^5,DTic^7, Oic^8]bradykinin) in different in vivo animal models of inflammation. J. Pharmacol. Exp. Ther. 273, 1078–1084.

Fernández-Cruz, L., Navarro, S., Valderrama, R., Sáenz, A., Guarner, L., Aparisi, L., Espi, A., Jaurietta, E., Marruecos, L., Gener, J., De Las Heras, G., Pérez-Mateo, M., Garcia Sabrido, J.L., Roig, J. and Carballo Bardají, F. (1994). Acute necrotizing pancreatitis: a multicenter study. Hepatogastroenterology 41, 185–189.

Fernández-del Castillo, C., Schmidt, J., Warshaw, A.L. and Rattner, D.W. (1994). Interstitial protease activation is the central event in progression to necrotizing pancreatitis. Surgery 116, 497–507.

Fiedler, F. (1979). Enzymology of glandular kallikreins. In "Bradykinin, Kallidin and Kallikrein", (Handbook of Experimental Pharmacology, Vol. 25, Suppl. (ed. E.G. Erdös), pp. 103–161. Springer, Heidelberg, New York.

Fitz, R.H. (1889a). Acute pancreatitis. A consideration of pancreatic hemorrhage, hemorrhagic, suppurative, and gangrenous pancreatitis, and of disseminated fat necrosis (Part I). Boston Med. Surg. J. 120, 181–187.

Fitz, R.H. (1889b). Acute pancreatitis. A consideration of pancreatic hemorrhage, hemorrhagic, suppurative, and gangrenous pancreatitis, and of disseminated fat necrosis (Part II). Boston Med. Surg. J. 120, 205–207.

Fitz, R.H. (1889c). Acute pancreatitis. A consideration of pancreatic hemorrhage, hemorrhagic, suppurative, and gangrenous pancreatitis, and of disseminated fat necrosis (Part III). Boston Med. Surg. J. 120, 229–234.

Florey, H.W. (1962). In "General Pathology" (ed. H.W. Florey), pp. 21–39. Lloyd-Luke, London.

Forrell, M.M. (1955). Zur Frage des Entstehungsmechanismus des Kreislaufkollapses bei der akuten Pankreasnekrose. Gastroenterologia 84, 225–251.

Forrell, M.M. (1961). Aktivierung und Inaktivierung proteolytischer Fermente des Pankreas und ihre klinische Bedeutung. Deutsch. Med. Wochenschr. 86, 981–984.

Forrell, M.M. and Dobovicnik, W. (1961). Über die Möglichkeit durch Inaktivierung des Trypsins die akute Pankreatitis kausal zu beeinflussen. Klin. Wochenschr. 39, 47–51.

Frey, E.K. (1926). Zusammenhänge zwischen Herzarbeit und Nierentätigkeit. Langenbecks Arch. Klin. Chir. 142, 663–668.

Frey, E.K. (1953). Zur Therapie der Pankreatitis. Therapie-woche 4, 323.

Frey, E.K. and Kraut, H. (1928). Ein neues Kreislaufhormon und seine Wirkung. Naunyn-Schmiedebergs Arch. Exp. Pharmakol. 133, 1–56.

Gao, X., Regoli, D., Stewart, J.M., Vavrel, R.J. and Greenbaum, L.M. (1989). Studies on T-kinin receptors. Pharmacologist 31, 3.

Gauthier, A., Escoffier, J.M., Camatte, R. and Sarles, H. (1981). Severe acute pancreatitis. Clin. Gastroenterol. 10, 209–224.

Geenen, J.E., Hogan, W.J., Dodds, W.J., Toouli, J. and Venu, R.P. (1989). The efficacy of endoscopic sphincterotomy after cholecystectomy in patients with Sphincter-of-Oddi dysfunction. N. Engl. J. Med. 320, 82–87.

Girolami, J.P., Pecher, C., Bascands, J.L., Cabos-Boutot, G., Vega-Vidalle, C., Colle, A., Adam, A. and Suc, J.M. (1989). Direct radioimmunoassay of active and inactive human glandular kallikrein: some physiological and pathological variabilities. J. Immunoassay 10, 221–236.

Goldberg, M.R., Chapnick, B.M., Joiner, P.D., Hyman, A.L. and Kadowitz, P.J. (1976). Influence of inhibitors of prosta-glandin synthesis on venoconstrictor responses to bradykinin. J. Pharmacol. Exp. Ther. 198, 357–365.

Goodfellow, V.S., Marathe, M.V., Kuhlmann, K.S., Fitzpatrick, T.D., Cuadrado, D., Hanson, W., Zuzack, J.S., Ross, S.E., Wieczorek, M., Burkhard, M. and Whalley, E.T. (1996). Bradykinin receptor antagonists containing N-substituted amino acids: in vitro and in vivo B$_2$ and B$_1$ receptor antagonist activity. J. Med. Chem. 39, 1472–1484.

Greenbaum, L.M., Howard, E., Albus, U., Lapp, C. and Gao, X.X. (1992). T-Kininogen, processing and functions. In "Recent Progress on Kinins: Biochemistry and Molecular Biology of the Kallikrein–Kinin System", Agents Actions Suppl. 38/I (eds. H. Fritz, W. Müller-Esterl, M. Jochum, A. Roscher and K. Luppertz), pp. 300–306. Birkhäuser, Basel.

Gress, T.M., Arnold, R. and Adler, G. (1990). Structural alterations of pancreatic microvasculature in cerulein-induced pancreatitis in the rat. Res. Exp. Med. 190, 401–412.

Griesbacher, T. (1994). Blood pressure reflexes following activation of capsaicin-sensitive afferent neurones in the biliopancreatic duct of rats. Br. J. Pharmacol. 111, 547–554.

Griesbacher, T. and Lembeck, F. (1987). Effect of bradykinin antagonists on bradykinin-induced plasma extravasation, venoconstriction, prostaglandin E$_2$ release, nociceptor stimulation and contraction of the iris sphincter muscle in the rabbit. Br. J. Pharmacol. 92, 333–340.

Griesbacher, T. and Lembeck, F. (1992). Effects of the bradykinin antagonist, HOE 140, in experimental acute pancreatitis. Br. J. Pharmacol. 107, 356–360.

Griesbacher, T. and Lembeck, F. (1993). Endogenous kinins in experimental acute pancreatitis. International Congress on Inflammation, Vienna, October 1993, P88.

Griesbacher, T. and Lembeck, F. (1995). Investigations into pathological and possible physiological roles of kinins in the pancreas. Pancreas 11, 430.

Griesbacher, T., Lembeck, F. and Saria, A. (1989). Effects of the bradykinin antagonist B4310 on smooth muscles and blood pressure in the rat, and its enzymatic degradation. Br. J. Pharmacol. 96, 531–538.

Griesbacher, T., Tiran, B. and Lembeck, F. (1993). Pathological events in experimental acute pancreatitis prevented by the bradykinin antagonist, Hoe 140. Br. J. Pharmacol. 108, 405–411.

Griesbacher, T., Kolbitsch, C., Tiran, B. and Lembeck, F. (1995). Effects of the bradykinin antagonist, icatibant (Hoe 140), on pancreas and liver functions during and after caerulein-induced pancreatitis in rats. Naunyn-Schmiedeberg's Arch. Pharmacol. 352, 557–564.

Guth, P.S., Bobbin, R., Cano, G. and Amaro, J. (1966). Venoconstriction induced by bradykinin in the rabbit ear. In "Hypotensive Peptides" (eds. E.G. Erdös, N. Back and F. Sicuteri), pp. 396–506. Springer, New York.

Hilton, S.M. and Jones, M. (1968). The role of plasma kinin in functional vasodilatation in the pancreas. J. Physiol. 195, 521–533.

Hiltunen, J., Kaukinen, S., Pessi, T. and Ylitalo, P. (1985). The kallikrein–kinin system in operated patients with acute pancreatitis. Acta Chir. Scand. 151, 371–375.

Hock, F.J., Wirth, K., Albus, U., Linz, W., Gerhards, H.J., Wiemer, G., Henke, S., Breipohl, G., König, W., Knolle, J. and Schölkens, B.A. (1991). Hoe 140 a new potent and long acting bradykinin-antagonist: in vitro studies. Br. J. Pharmacol. 102, 769–773.

Hoffmann, T.F., Waldner, H. and Messmer, K. (1995). Bradykinin-antagonist CP-0597 can limit the progression of postischemic pancreatitis. Kinin' 95, International Symposium on Kinins, Denver, September, L52.

Hoffmann, T.F., Kübler, J. and Messmer, K. (1996a). Bradykinin-Antagonisierung bei Ischämie und Reperfusion des Pankreas. Zentralbl. Chir., 121, 412–422, discussion 422–423.

Hoffmann, T.F., Leiderer, R., Waldner, H. and Messmer, K. (1996b). Bradykinin-antagonists HOE140 and CP-0597 diminish microcirculatory injury after ischemia-reperfusion of the pancreas in rats. Br. J. Surg. 83, 189–195.

Holdstock, D.J., Mathias, A.P. and Schachter, M. (1957). A comparative study of kinin, kallidin, and bradykinin. Br. J. Pharmacol. Chemother. 12, 149–158.

Hong, S.S., Case, R.M. and Kim, K.H. (1988). Analysis in the isolated perfused cat pancreas of factors implicated in the pathogenesis of pancreatitis. Pancreas 3, 450–458.

Howard, J.M., Smith, A.K. and Peters, J.J. (1949). Acute pancreatitis: pathways of enzymes into the blood stream. Surgery 26, 161–166.

Hutchinson, F.N., Cui, X.Y. and Webster, S.K. (1995). The antiproteinuric action of angiotensin-converting enzyme is dependent on kinin. J. Am. Soc. Nephrol. 6, 1216–1222.

Jacobsen, S. (1966). Substrates for plasma kinin-forming enzymes in human, dog, and rabbit plasmas. Br. J. Pharmacol. 26, 403–411.

Jaschke, W., Klose, K.J. and Strecker, E.P. (1992). A new balloon-expandable tantalum stent (Strecker-Stent) for the biliary system: preliminary experience. Cardiovasc. Intervent. Radiol. 15, 356–359.

Johnson, R.H. and Doppman, J. (1967). Duodenal reflux and the tiology of pancreatitis. Surgery 62, 462–467.

Jones, P.E. and Oelbaum, M.H. (1975). Frusemide-induced pancreatitis. Br. Med. J. 1, 133–134.

Joris, I., DeGirolami, U., Wortham, K. and Majno, G. (1982). Vascular labelling with Monastral blue B. Stain Technol. 57, 177–183.

Kaminishi, H., Cho, T., Itoh, T., Iwata, A., Kawasaki, K., Hagiharara, Y. and Maeda, H. (1993). Vascular permeability enhancing activity of Porphyromonas gingivalis protease in guinea-pigs. FEMS Microbiol. Lett. 114, 109–114.

Kanbe, T., Naruse, S., Kitagawa, M., Nakae, Y., Kondo, T. and Hayakawa, T. (1994). Effect of a bradykinin receptor antagonist on taurocholate-induced acute pancreatitis in rats. Dig. Dis. Sci. 55, 307.

Kaplan, A.P., Silverberg, M., Ghebrehiwet, B., Atkins, P. and Zweiman, B. (1989). Pathways of kinin formation and role in allergic diseases. Clin. Immunol. Immunopathol. 50, S41–S51.

Katz, W., Silverstein, M., Kobold, E.E. and Thal, A.P. (1964). Trypsin release, kinin production, and shock. Arch. Surg. 89, 322–328.

Keim, V., Adler, G., Haberich, F.J. and Kern, H.F. (1985). Failure of secretin to prevent or ameliorate cerulein-induced pancreatitis in the rat. Hepatogastroenterology 32, 91–96.

Kelly, D.M., McEntee, G.P., McGeeney, K.F. and Fitzpatrick, J.M. (1993). Microvasculature of the pancreas, liver, and kidney in cerulein-induced pancreatitis. Arch. Surg. 128, 293–295.

Kimura, W., Meyer, F., Hess, D., Kirchner, T., Fischbach, W. and Mössner, J. (1992). Comparison of different treatment modalities in experimental pancreatitis in rats. Gastroenterology 103, 1916–1924.

Kingsnorth, A.N., Galloway, S.W. and Formela, L.J. (1995). Randomized, double-blind phase II trial of Lexipafant, a platelet-activating factor antagonist, in human acute pancreatitis. Br. J. Surg. 82, 1414–1420.

Kitamura, N., Takagaki, Y., Furuto, S., Tanaka, T., Nawa, H. and Nakanishi, S. (1983). A single gene for bovine high molecular weight and low molecular weight kininogen. Nature (Lond.) 305, 545–549.

Klar, E., Messmer, K., Warshaw, A.L. and Herfarth, C. (1990). Pancreatic ischaemia in experimental pancreatitis: mechanism, significance and therapy. Br. J. Surg. 77, 1205–1210.

Klar, E., Schratt, W., Foitzik, T., Buhr, H., Herfahrt, C. and Messmer, K. (1994). Impact of microcirculatory flow pattern changes on the development of acute edematous and necrotizing pancreatitis in rabbit pancreas. Dig. Dis. Sci. 39, 2639–2344.

Knoefel, W.T., Kollias, N., Warshaw, A.L., Waldner, H., Nishioka, N.S. and Rattner, D.W. (1994). Pancreatic microcirculatory changes in experimental pancreatitis of graded severity in the rat. Surgery 116, 904–913.

Kortmann, H., Fink, E. and Bönner, G. (1984). The influence of the kallikrein–kinin system in the development of pancreatic shock. Adv. Exp. Med. Biol. 167, 495–503.

Kraut, H., Frey, E.K. and Bauer, E. (1928). Über eine neues Kreislaufhormon. Hoppe-Seylers Z. Physiol. Chem. 175, 97–114.

Kraut, H., Frey, E.K. and Werle, E. (1930). Der Nachweis eines Kreislaufhormons in der Pankreasdrüse. Hoppe-Seylers Z. Physiol. Chem. 189, 97–102.

Lampel, M. and Kern, H.F. (1977). Acute interstitial pancreatitis in the rat induced by excessive doses of a pancreatic secretagogue. Virchows Arch. (Pathol. Anat.) 373, 97–117.

Lasson, Å. and Ohlsson, K. (1984). Changes in the kallikrein–kinin system during acute pancreatitis in man. Thromb. Res. 35, 27–41.

Lasson, Å. and Ohlsson, K. (1986). Kinin activation and protease inhibition in acute pancreatitis in man. Adv. Exp. Med. Biol. 198B, 18–20.

Lasson, Å., Fork, F.-T., Trägårdh, B. and Zederfeldt, B. (1988). The postcholecystectomy syndrome: bile ducts as pain trigger zone. Scand. J. Gastroenterol. 23, 265–271.

Lasson, Å., Göransson, J. and Ohlsson, K. (1989). Pancreatic pseudocyst fluid – a mixture of plasma proteins and pancreatic juice possessing high proteolytic activity. Scand. J. Clin. Lab. Invest. 49, 403–412.

Leach, S.D., Modlin, I.M., Scheele, G.A. and Gorelick, F.S. (1991). Intracellular activation of digestive zymogens in rat pancreatic acini. Stimulation by high doses of cholecystokinin. J. Clin. Invest. 87, 362–366.

Leach, S.D., Gorelick, F.S. and Modlin, I.M. (1992). New perspectives on acute pancreatitis. Scand. J. Gastroenterol. 27 (Suppl. 192), 29–38.

Lembeck, F. and Griesbacher, T. (1996). Pathophysiological and possible physiological roles of kinins in the pancreas. Immunopharmacology 33, 336–338.

Lembeck, F., Griesbacher, T., Eckhardt, M., Henke, S., Breipohl, G. and Knolle, J. (1991). New, long-acting, potent bradykinin antagonists. Br. J. Pharmacol. 102, 297–304.

Lembeck, F., Griesbacher, T. and Legat, F.J. (1992). Lack of significant unspecific effects of HOE 140 and other novel bradykinin antagonists in vitro and in vivo. In "Recent Progress on Kinins: Pharmacological and Clinical Aspects of the Kallikrein–Kinin System", Part I, Agents Actions Suppl. 38/II (eds. G. Bönner, H. Fritz, T. Unger, A. Roscher and K. Luppertz), pp. 414–422. Birkhäuser, Basel.

Lerch, M.M. and Adler, G. (1994). Experimental animal models of acute pancreatitis. Int. J. Pancreatol. 15, 159–170.

Lerch, M.M., Weidenbach, H., Gress, T. and Adler, G. (1992). Effect of the potent bradykinin antagonist HOE140 in acute pancreatitis in the rat. Pancreas 7, 745.

Lerch, M.M., Saluja, A.K., Rünzi, M., Dawra, R., Saluja, M. and Steer, M.L. (1993). Pancreatic duct obstruction triggers acute necrotizing pancreatitis in the opossum. Gastroenterology 104, 853–861.

Lerch, M.M., Weidenbach, H., Gress, T.M. and Adler, G. (1995). Effect of kinin inhibition in experimental pancreatitis. Amer. J. Physiol. 269, G490–G499.

Lombardi, B., Estes, L.W. and Longnecker, D.S. (1975). Acute hemorrhagic pancreatitis (massive necrosis) with fat necrosis induced in mice by DL-ethionine fed with a choline-deficient diet. Am. J. Pathol. 79, 465–480.

Maeda, H. and Molla, A. (1989). Pathogenic potentials of bacterial proteases. Clin. Chim. Acta 185, 357–368.

Maliekal, J. and Drake, C.F. (1993). Acute pancreatitis associated with the use of lisinopril. Ann. Pharmacother. 27, 1465–1466.

Mallory, A. and Kern, F. (1980). Drug-induced pancreatitis: a critical review. Gastroenterology 78, 813–820.

Manso, M.A., de Dios, I., San Román, J.I., Calvo, J.J. and López, M.A. (1989). Effect of secretin on pancreatic juice production in cerulein-induced acute pancreatitis in the rat. Peptides 10, 255–260.

Manso, M.A., San Román, J.I., de Dios, I., Garcia, L.J. and López, M.A. (1992). Cerulein-induced acute pancreatitis in the rat. Study of pancreatic secretion and plasma VIP and secretin levels. Dig. Dis. Sci. 37, 364–368.

Maruo, K., Akaike, T., Inada, Y., Ohkubo, I., Ono, T. and Maeda, H. (1993). Effect of microbial and mite proteases on low and high molecular weight kininogens. J. Biol. Chem. 268, 17711–17715.

McCutcheon, A.D. (1964). Reflux of duodenal contents in the pathogenesis of pancreatitis. Gut 5, 260–265.

McCutcheon, A.D. (1968). A fresh approach to pathogenesis of pancreatitis. Gut 9, 296–310.

Migay, T.J. and Petroff, J.R. (1923). Untersuchungen über die Wirkung des Pankreassaftes auf den Organismus bei parenteraler Einführung. Z. Ges. Exp. Med. 36, 457–473.

Moody, F.G., Calabuig, R., Vecchio, R. and Runkel, N. (1990). Stenosis of the sphincter of Oddi. Surg. Clin. North Am. 70, 1341–1354.

Niederau, C., Ferrell, L.D. and Grendell, J.H. (1985). Caerulein-induced acute necrotizing pancreatitis in mice: protective effects of proglumide, benzotript, and secretin. Gastroenterology 88, 1192–1204.

Nishimura, K., Inoue, Y. and Kokubu, T. (1987). Radio-immunoassay of glandular kallikrein in human plasma after partial purification by immunoaffinity column. Clin. Chim. Acta 162, 341–347.

Nordback, I.H., Clemens, J.A. and Cameron, J.L. (1991). The role of cholecystokinin in the pathogenesis of acute pancreatitis in the isolated pancreas preparation. Surgery 109, 301–306.

Nugent, F.W. and Atendido, W. (1966). Hemorrhagic pancreatitis. Aggressive treatment. Postgrad. Med. 40, 87–94.

Nugent, F.W., Zuberi, S. and Bulan, M.B. (1969). Kinin precursor in experimental pancreatitis. Proc. Soc. Exp. Biol. Med. 130, 566–567.

Ofstad, E. (1970). Formation and destruction of plasma kinins during experimental acute pancreatitis in dogs. Scand. J. Gastroenterol. 5 (Suppl. 5), 1–44.

Ohshio, G., Saluja, A. and Steer, M.L. (1991). Effects of short-term pancreatic duct obstruction in rats. Gastroenterology 100, 196–202.

Okamoto, H. and Grennbaum, L.M. (1983a). Kininogen substrates for trypsin and cathepsin D in human, rabbit and rat plasmas. Life Sci. 32, 2007–2013.

Okamoto, H. and Grennbaum, L.M. (1983b). Isolation and structure of T-kinin. Biochem. Biophys. Res. Commun. 112, 701–708.

Orlov, V. and Belyakov, N. (1978). Blood kallikrein. Kinin system in acute pancreatitis. Am. J. Gastroenterol. 70, 645–648.

Papp, M., Makara, G.B. and Varga, B. (1967). Effets de la bradykinine, de la kallidine, de la sérotonine et de l'histamine sur le flux sanguin du pancréas. Arch. Int. Pharmacodyn. Thér. 165, 31–36.

Pawlik, W.W., Gustaw, P., Pawlik, T., Sendur, R. and Czarnobilski, K. (1994). Nitric oxide as mediator of bradykinin-induced pancreatic vasodilation. Digestion 55, 321.

Pedro-Botet, J., Miralles, R., Coll, J. and Rudies-Prat, J. (1990). Captopril versus enalapril: cough and ACE inhibitors. DICP Ann. Pharmacother. 24, 438–439.

Pfeffer, R.B., Lazzarini-Robertson, A., Jr, Safadi, D., Mixter, G., Secoy, C.F. and Hinton, J.W. (1962). Gradation of pancreatitis, edematous, through hemorrhagic, experimentally produced by controlled injection of microspheres into blood vessels in dogs. Surgery 51, 764–769.

Pierce, J.V. (1968). Structural features of plasma kinins and kininogens. Fed. Proc. 27, 52–57.

Popieraitis, A.S. and Thompson, A.G. (1969). The site of bradykinin release in acute experimental pancreatitis. Arch. Surg. 98, 73–76.

Prinz, R.A. and Greenlee, H.B. (1990). Pancreatic duct drainage in chronic pancreatitis. Hepatogastroenterology 37, 295–300.

Regoli, D. and Barabé, J. (1980). Pharmacology of bradykinin and related kinins. Pharmacol. Rev. 32, 1–46.

Renner, I.G., Wisner, J.R. and Lavigne, B.C. (1986). Partial restoration of pancreatic function by exogenous secretin in rats with ceruletide-induced acute pancreatitis. Dig. Dis. Sci. 31, 305–313.

Rocha e Silva, M., Beraldo, W.T. and Rosenfeld, G. (1949). Bradykinin, a hypotensive and smooth muscle stimulating factor released from plasma globulin by snake venoms and by trypsin. Am. J. Physiol. 156, 261–273.

Roush, M.K., McNutt, R.A. and Gray, T.F. (1991). The adverse effect dilemma: quest for accessible information. Ann. Intern. Med. 114, 298–299.

Ruud, T.E., Aasen, A.O., Kierulf, P., Stadaas, J. and Aune, S. (1982). Studies on components of the plasma kallikrein–kinin system in peritoneal fluid and plasma before and during experimental acute pancreatitis in pigs. Acta Chir. Scand. Suppl. 509, 89–93.

Ruud, T.E., Aasen, A.O., Kierulf, P., Stadaas, J. and Aune, S. (1984). Studies on the kallikrein–kinin system in plasma and peritoneal fluid during experimental pancreatitis. Adv. Exp. Med. Biol. 167, 489–494.

Ruud, T.E., Aasen, A.O., Kierulf, P., Stadaas, J. and Aune, S. (1985). Studies on the plasma kallikrein–kinin system in peritoneal exudate and plasma during experimental acute pancreatitis in pigs. Scand. J. Gastroenterol. 20, 877–882.

Ruud, T.E., Aasen, A.O., Røise, O., Lium, B., Pillgram-Larsen, J. and Stadaas, J. (1986a). Studies on the plasma kallikrein–kinin system in peritoneal exudate and plasma during experimental acute pancreatitis in pigs. Scand. J. Gastroenterol. Suppl. 126, 25–31.

Ruud, T.E., Aasen, A.O., Pillgram-Larsen, J., Stadaas, J. and Aune, S. (1986b). Effects of protease inhibitor pretreatment on hemodynamic performances and survival rate in experimental, acute pancreatitis. Adv. Exp. Med. Biol. 198B, 413–421.

Ryan, J.W., Moffat, J.G. and Thompson, A.G. (1964). Role of bradykinin in the development of acute pancreatitis. Nature (Lond.) 204, 1212–1213.

Ryan, J.W., Moffat, J.G. and Thompson, A.G. (1965). Role of bradykinin system in acute hemorrhagic pancreatitis. Arch. Surg. 91, 14–24.

Saluja, A., Hashimoto, S., Saluja, M., Powers, R.E., Meldolesi, J. and Steer, M.L. (1987). Subcellular redistribution of lysosomal enzymes during caerulein-induced pancreatitis. Am. J. Physiol. 253, G508–G516.

Saluja, A.K., Saluja, M., Villa, A., Leli, U., Rutledge, P., Meldolesi, J. and Steer, M.L. (1989). Pancreatic duct obstruction in rabbits causes digestive zymogen and lysosomal enzyme colocalization. J. Clin. Invest. 84, 1260–1266.

Samuel, I., Toriumi, Y., Yokoo, H., Wilcockson, D.P., Trout, J.J. and Joehl, R.J. (1994). Ligation-induced acute pancreatitis in rats and opossums: a comparative morphologic study of the early phase. J. Surg. Res. 57, 299–311.

San Román, J.I., de Dios, I., Manso, M.A., Calvo, J.J. and López, M.A. (1990). Caerulein-induced acute pancreatitis in the rat. Pancreatic secretory response to cholecystokinin. Arch. Int. Physiol. Biochim. Biophys. 98, 237–243.

Sarles, H. (ed.) (1965). "Pancreatitis: Symposium of Marseilles, 1963". Karger, Basel.

Sarner, M. (1993). Pancreatitis: definitions and classifications. In "The Pancreas. Biology, Pathobiology, and Disease" (eds. V.L.W. Go, E.P. DiMagno, J.D. Gardner, E. Lebenthal,

H.A. Reber and G.A. Scheele), pp. 575–580. Raven Press, New York.

Satake, K., Rozmanith, J.S., Appert, H.E., Carballo, J. and Howard, J.M. (1973a). Hypotension and release of kinin-forming enzyme into ascitic fluid exudate during experimental pancreatitis in dogs. Ann. Surg. 177, 497–502.

Satake, K., Rozmanith, J.S., Appert, H. and Howard, J.M. (1973b). Hemodynamic change and bradykinin levels in plasma and lymph during experimental acute pancreatitis in dogs. Ann. Surg. 178, 659–662.

Satake, K., Koh, I., Nishiwaki, H. and Umeyama, K. (1985a). The management of acute hemorrhagic pancreatitis: toxicity of ascitic fluid from experimental hemorrhagic pancreatitis in dogs. In "Pancreatitis: Its Pathophysiology and Clinical Aspects" (eds. T. Sato and H. Yamauchi), pp. 239–251. Tokyo University Press, Tokyo.

Satake, K., Koh, I., Nishiwaki, H. and Umeyama, K. (1985b). Toxic products in hemorrhagic ascitic fluid generated during experimental acute pancreatitis in dogs and a treatment which reduces their effect. Digestion 32, 99–105.

Satake, K., Ha, H.H. and Hiura, A. (1996). Effects of bradykinin receptor antagonist on the release of beta-endorphin and bradykinin and on hemodynamic changes in a canine model of experimental acute pancreatitis. Pancreas 12, 92–97.

Schapira, M., Silver, L.D., Scott, C.F., Schmaier, A.H., Prograis, L.J., Jr, Curd, J.G. and Colman, R.W. (1983). Prekallikrein activation and high-molecular-weight kininogen consumption in hereditary angioedema. N. Engl. J. Med. 308, 1050–1053.

Schmidt, J., Riedel, D., Mithöfer, K., Ryschich, E., Buhr, H.J., Herfarth, C. and Klar, E. (1994). Quantitative evaluation of pancreatic microcirculation in edematous and necrotizing pancreatitis of the rat. Digestion 55, 333–334.

Senninger, N., Moody, F.G., Van Buren, D.H., Coelho, J.C.U. and Li, Y.F. (1984). Effect of biliary obstruction on pancreatic exocrine secretion in conscious opossums. Surg. Forum 35, 226–228.

Seung, W.P. and Feldman, B.F. (1985). Early phase components of the kallikrein–kinin system in hemorrhagic ascitic fluid and plasma in the rat with induced acute pancreatitis. Am. J. Vet. Res. 46, 1961–1966.

Shimamoto, K., Mayfield, R.K., Margolius, H.S., Chao, J., Stroud, W. and Kaplan, A.P. (1984). Immunoreactive tissue kallikrein in human serum. J. Lab. Clin. Med. 103, 731–738.

Shimizu, I., Wada, S., Okahisa, T., Kanamura, M., Yano, M., Kodaira, T., Nishino, T. Shima, K. and Ito, S. (1993). Radio-immunoreactive plasma bradykinin levels and histological changes during the course of cerulein-induced pancreatitis in rats. Pancreas 8, 220–225.

Singh, L.M. and Howard, J.M. (1966). Serum kallikreinogen levels in postoperative patients and in patients with pancreatitis. Surgery 59, 837–841.

Soling, H.D. and Fest, W. (1986). Synthesis of 1-O-alkyl-2-acetyl-sn-glycero-3-phosphocholine (platelet activating factor) in exocrine glands and its control by secretagogues. J. Biol. Chem. 261, 13916–13922.

Soling, H.D., Eibl, H. and Fest, W. (1984). Acetylcholine-like effects of 1-O-alkyl-2-acetyl-sn-glycero-3-phosphocholine ("platelet activating factor") and its analogues in exocrine secretory glands. Eur. J. Biochem. 144, 65–72.

Standridge, J.B. (1994). Fulminant pancreatitis associated with lisinopril therapy. South. Med. J. 87, 179–181.

Steer, M.L. (1993). Etiology and pathophysiology of acute pancreatitis. In "Pancreas. Biology, Pathobiology, and Disease" (eds. V.L.W. Go, E.P. DiMagno, J.D. Gardner, E. Lebenthal, H.A. Reber and G.A. Scheele), pp. 581–591. Raven Press, New York.

Steer, M.L. and Meldolesi, J. (1987). The cell biology of experimental pancreatitis. N. Engl. J. Med. 316, 144–150.

Steinberg, W.A. and Schlesselman, S.E. (1987). Treatment of acute pancreatitis. Comparison of animal and human studies. Gastroenterology 93, 1420–1427.

Stürmer, E. (1966). The influence of intra-arterial infusions of synthetic bradykinin on flow and composition in dogs. In "Hypotensive Peptides" (eds. E.G. Erdös, N. Back and F. Sicuteri), pp. 368–374. Springer, New York.

Sum, P.T., Bencomse, S.A. and Beck, I.T. (1970). Pathogenesis of bile-induced acute pancreatitis in the dog: Experiments with detergents. Am. J. Dig. Dis. 15, 637–646.

Sumi, H., Takasugi, S. and Toki, N. (1978). Studies on kallikrein–kinin system in plasma of patients with acute pancreatitis. Clin. Chim. Acta 87, 113–118.

Sweiry, J.H., Rühlmann, D.O. and Mann, G.E. (1994). Bradykinin induces rapid endothelium-dependent vasodilation in the perfused rat pancreas: involvement of nitric oxide or prostacyclin? Digestion 55, 339.

Takasugi, S., Toki, N. and Sumi, H. (1980). Studies on a kallikrein–kinin system in plasma of patients with acute pancreatitis: the preparation and characterization of a kallikrein-like enzyme in patient's plasma. Clin. Biochem. 13, 156–159.

Talamo, R.C. and Goodfriend, T.L. (1979). Bradykinin radioimmunoassay. In "Bradykinin, Kallidin and Kallikrein", (Handbook of Experimental Pharmacology, Vol. 25, Suppl. (ed.' E.G. Erdös)), pp. 301–309. Springer, Heidelberg, New York.

Thal, A.P., Kobold, E.E. and Hollenberg, M.J. (1963). The release of vasoactive substances in acute pancreatitis. Am. J. Surg. 105, 708–713.

Thurieau, C., Félétou, M., Canet, E. and Fauchère, J.-L. (1994). p-Guanidobenzoyl-[Hyp^3,Thi^5,$_D$Tic^7,Oic^8]bradykinin is almost completely devoid of the agonistic effect of Hoe 140 on the endothelium-fee femoral artery of the sheep. Biorg. Med. Chem. Lett. 4, 781–784.

Tilkemeier, P. and Thompson, P.D. (1988). Acute pancreatitis possibly related to enalapril. N. Engl. J. Med. 318, 1275–1276.

Toki, N., Sumi, H. and Takasugi, S. (1981). Purification and characterization of kallikrein from plasma of patients with acute pancreatitis. Clin. Sci. 60, 199–205.

Uehara, S., Honjyo, K., Furukawa, S., Hirayama, A. and Sakamoto, W. (1989). Role of the kallikrein–kinin system in human pancreatitis. Adv. Exp. Med. Biol. 247B, 157–162.

Vavrek, R.J. and Stewart, J.M. (1985). Competitive antagonists of bradykinin. Peptides 6, 161–164.

Vollmar, B., Waldner, H., Schmand, J., Conzen, P.F., Goetz, A.E., Habazettl, H., Schweiberer, L. and Brendel, W. (1989). Release of arachidonic acid metabolites during acute pancreatitis in pigs. Scand. J. Gastroenterol. 24, 1253–1264.

Vollmar, B., Waldner, H., Schmand, J., Conzen, P.F., Goetz, A.E., Habazettl, H., Schweiberer, L. and Brendel, W. (1991). Oleic acid induced pancreatitis in pigs. J. Surg. Res. 50, 196–204.

Waldner, H., Vollmar, B., Conzen, P., Götz, A., Lehnert, P., Fink, E., Brendel, W. and Schweiberer, L. (1993). Enzymfreisetzung und Aktivierung der Kallikrein–Kinin-

Systeme bei experimenteller Pankreatitis. Untersuchungen in Pfortaderblut, Pankreaslymphe und Peritonealexudat. Langenbecks Arch. Chir. 378, 154–159.

Walker, N.I. (1987). Ultrastructure of the rat pancreas after experimental duct ligation. The role of apoptosis and intra-epithelial macrophages in acinar cell deletion. Am. J. Pathol. 126, 439–451.

Walker, N.I., Winterford, C.M. and Kerr, J.F.R. (1992). Ultrastructure of the rat pancreas after experimental duct ligation. II. Duct and stromal cell proliferation, differentiation, and deletion. Pancreas 7, 420–434.

Watanabe, O., Baccino, F.M., Steer, M.L. and Meldolesi, J. (1984). Supramaximal caerulein stimulation and ultrastructure of rat pancreatic acinar cell: early morphological changes during development of experimental pancreatitis. Am. J. Physiol. 246, G457–G467.

Watanabe, S., Nishino, T., Chang, J.-H., Shiratori, K., Moriyoshi, Y. and Takeuchi, T. (1993). Effect of Hoe 140, a new potent bradykinin antagonist, on experimental acute pancreatitis in rats. Gastroenterology 104, A342.

Webster, M.E. and Pierce, V.J. (1963). The nature of the kallidins released from human plasma by kallikreins and other enzymes. Ann. N. York Acad. Sci. 104, 91–107.

Weidenbach, H., Lerch, M.M., Gress, T.M., Pfaff, D., Turi, S. and Adler, G. (1995). Vasoactive mediators and the progression from oedematous to necrotizing experimental acute pancreatitis. Gut 37, 434–440.

Weizman, Z. and Sofer, S. (1992). Acute pancreatitis in children with anticholinesterase insecticide intoxication. Pediatrics 90, 204–206.

Werle, E. (1937). Über den Aktivitätszustand des Kallikreins der Bauchspeicheldrüse und ihres äußeren Sekretes beim Hund. Biochem. Z. 290, 129–134.

Werle, E. (1948). Zur Kenntnis des Kallikreins. Angew. Chem. A 60, 53.

Werle, E. (1970). Discovery of the most important kallikreins and kallikrein inhibitors. In "Bradykinin, Kallidin and Kallikrein", Handbook of Experimental Pharmacology, Vol. 25 (eds. E.G. Erdös and A.F. Wilde), pp. 1–6. Springer, Heidelberg, New York.

Werle, E. and Urhahn, K. (1940). Über den Aktivitätszustand des Kallikreins in der Bauchspeicheldrüse. Biochem. Z. 304, 387–396.

Werle, E. and von Roden, P. (1936). Über das Vorkommen von Kallikrein in den Speicheldrüsen und im Mundspeichel. Biochem. Z. 286, 213–219.

Werle, E. and Zach, P. (1970). Verteilung von Kininogen in Serum und Geweben bei Ratten und anderen Säugetieren. Z. Klin. Chem. Klin. Biochem. 8, 186–189.

Werle, E., Götze, W. and Keppler, A. (1937). Über die Wirkung des Kallikreins auf den isolierten Darm und über eine neue darmkontrahierende Substanz. Biochem. Z. 289, 217–233.

Werle, E., Forrell, M.M. and Maier, L. (1955). Zur Kenntnis der blutdrucksenkenden Wirkung des Trypsins. Naunyn-Schmiedebergs Arch. Exp. Pathol. Pharmakol. 225, 369–380.

Werle, E., Tauber, K., Hartenbach, W. and Forrell, M.M. (1958). Zur Frage der Therapie der Pankreatitis. Münch. Med. Wochenschr. 100, 1265–1267.

Wiemer, G., Fink, E., Linz, W., Hropot, M. and Schölkens, B.A. (1994). Furosemide enhances the release of endothelial kinins, nitric oxide and prostacyclin. J. Pharmacol. Exp. Ther. 271, 1611–1615.

Willemer, S., Bialek, R. and Adler, G. (1990). Localization of lysosomal and digestive enzymes in cytoplasmic vacuoles in caerulein-pancreatitis. Histochemistry 94, 161–170.

Wilson, C. and Imrie, C.W. (1990). Changing patterns of incidence and mortality from acute pancreatitis in Scotland, 1961–1985. Br. J. Surg. 77, 731–734.

Wilson, A.E., Mehra, S.K., Gomersall, C.R. and Davies, D.M. (1967). Acute pancreatitis associated with frusemide therapy. Lancet I, 105.

Wirth, K., Hock, F.J., Albus, U., Linz, W., Alpermann, H.H., Anagnostopoulos, H., Henke, S., Breipohl, G., König, W., Knolle, J. and Schölkens, B.A. (1991a). Hoe 140 a new potent and long acting bradykinin-antagonist: *in vivo* studies. Br. J. Pharmacol. 102, 774–777.

Wirth, K., Breipohl, G., Stechl, J., Knolle, J., Henke, S. and Schölkens, B.A. (1991b). Des-Arg^9-DArg[Hyp^3,Thi^5,DTic^7,-Oic^8]bradykinin (desArg^{10}[Hoe 140]) is a potent bradykinin B$_1$ receptor antagonist. Eur. J. Pharmacol. 205, 217–218.

13. The Role of the Kallikrein–Kinin System in Inflammation-induced Bone Metabolism

Ulf H. Lerner

1. Introduction

The integrity of organs within the body is dependent on the balance between formation of new tissue and the breakdown of old. During normal physiology or homeostasis, these processes are in equilibrium but, in pathological conditions or during tissue repair, the normal tissue formation and degradation functions become uncoupled, resulting either in a net loss or increase in the number of cells and the amount of extracellular matrix. In the skeleton, the cells react to local inflammation or to tumors with both anabolic and catabolic reactions leading to osteopenic loss of bone or to sclerotic reactions in the skeleton. In most patients suffering from inflammatory diseases, such as rheumatoid arthritis or periodontitis, a common finding is local disappearance of bone, giving rise to disabled joints or loss of teeth, respectively. Also, patients with

osteomyelitis often show loss of bone tissue, a process called "rarefying osteitis" but, in some cases, new bone formation occurs, a condition referred to as "sclerosing osteitis". Similarly, most malignant tumors that metastasize within or in the vicinity of the skeleton cause loss of bone tissue. However, some tumors, the most common being prostatic cancers, stimulate new bone formation.

In patients with inflammation within or near the skeleton, the common view is that loss of bone results from enhanced breakdown of the mineralized tissue by bone resorbing osteoclasts. It should not be overlooked, however, that decreased bone formation also may contribute to reductions in bone tissue volume, although the relative importance of these two processes in the pathogenesis of inflammation-induced bone loss is not understood. As regards the anabolic reactions in the skeletons of patients with inflammatory diseases (or malignant tumors), these may hypothetically be due to decreased bone resorption and/or enhanced bone formation. The skeletal morphology in such patients, however, suggests that enhanced osteoblastic new bone formation is likely to be the main cause.

Much effort has been made over the last 20 years to discover the important factors that cause the changes of bone-cell activities during inflammation. Observations that the skeletal reactions mostly are local has generated the idea that inflammation-induced bone loss is caused by locally produced factors, although there is evidence indicating that patients with inflammatory diseases also may have a systemic loss of bone. The present paper briefly summarizes knowledge of the pathogenesis of inflammation-induced bone loss, and then reviews in more detail the effects of kinins on bone metabolism. In order to facilitate understanding for nonbone specialists, a short overview of bone-cell biology is provided.

2. BONE-CELL BIOLOGY

In addition to providing a framework of mineralized tissue that helps us to stay upright, bone is a living tissue containing a variety of cells with different functions.

Three different cell types are distinguished, and these are osteoblasts, which form new bone, osteoclasts, which are cells responsible for bone resorption, and osteocytes, which are cells entrapped in the bone tissue.

2.1 OSTEOBLASTS

Osteoblasts are terminally differentiated mesenchymal cells with the capacity to produce mineralized bone (reviewed in Caplan and Pechak, 1987). These cells are not widely spread out in bone but exist as a one-cell layer syncytium that covers all bone tissues. This is true for compact bone in the periphery of an individual bone in the skeleton as well as the trabecular bone network in the inner part of the bones (Fig. 13.1). Osteoblasts produce bone in a two-step process. Initially the unmineralized bone matrix (osteoid) is synthesized, consisting mainly (90%) of collagen fibers type I, but also several glycoproteins and proteoglycans, collectively called "noncollagen bone-matrix proteins". Although our knowledge of bone-matrix proteins has increased during the last decade, the functions and organization of these proteins are largely unknown. The noncollagen proteins include growth factors, such as transforming growth

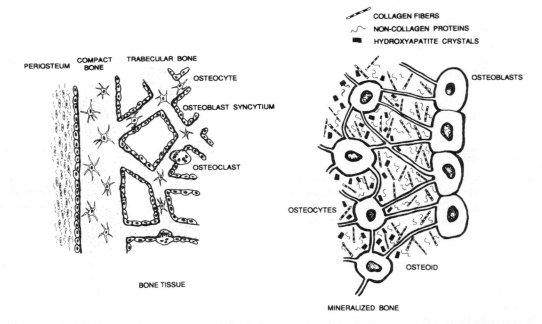

Figure 13.1   (Left) Bone tissues are surrounded by compact bone, which extends into the marrow cavity as a network of trabecular bone. Bone-forming osteoblasts cover all bone surfaces, both at compact and trabecular bone, as a one cell-layer thick syncytium. During bone formation, some of the osteoblasts are entrapped in bone tissue and these cells will form the osteocytes. Occasionally, bone-resorbing multinucleated osteoclasts can be seen at the bone surfaces. (Right) Osteoblasts produce bone initially by releasing collagen fibers and noncollagen proteins into a nonmineralized bone matrix (osteoid). Subsequently, the osteoblasts mineralize the osteoid by deposition of hydroxyapatite crystals in the organic network of proteins. The osteoid closest to the osteoblasts never will be mineralized and persists as a nonmineralized zone. Osteoblasts and osteocytes communicate with each other by a huge amount of cellular extensions.

factor-β (TGF-β) and insulin-like growth factors (IGF-I and IGF-II), as well as the TGF-β-related peptides, or bone morphogenetic proteins (BMP), which are entrapped together with structural proteins in bone tissue (reviewed in Heinegård and Oldberg, 1993; Centrella *et al.*, 1994). The biosynthesis of bone-matrix proteins is followed by hydroxyapatite crystal deposition in the extracellular matrix and the mineralized bone is thus formed (Fig. 13.1). The biochemical mechanisms used by osteoblasts during the mineralization process are essentially unknown. An important point, which is discussed later, is that mineralization is never complete. That is, between the mineralized bone tissue and the osteoblastic syncytium, a zone of unmineralized bone matrix (osteoid) is always present. Consequently, the mineralized part of bone tissue is covered by a zone of osteoid, which in turn is covered by the syncytium of osteoblasts (Fig. 13.1).

2.2 OSTEOCYTES

During bone formation, some of the osteoblasts are entrapped in the osteoid and eventually these cells will be located in lacunae within the mineralized bone (Fig. 13.1). These cells, the osteocytes, have several cell processes which communicate with similar processes from other osteocytes and with cell processes from osteoblasts in the syncytium. The function of osteocytes is largely unknown, but one hypothesis is that these cells, together with the osteoblastic and osteocytic cell processes, form a three-dimensional network, which can sensitize loading on the skeleton and transform it into regulation of osteoblastic formation in weight-bearing bones (Lanyon, 1992).

2.3 OSTEOCLASTS

The mineralized bone can be resorbed only by osteoclasts. This is true both for physiological remodeling of bone as well as for bone breakdown in pathological conditions. Osteoclasts are easily recognized multinuclear giant cells located on bone surfaces or in the Haversian canals within the compact bone tissue. The multinucleated, terminally differentiated osteoclasts are generated from stem cells in hematopoietic tissues. Available data indicate that the osteoclastic cell lineage is closely related to the monocytic cell lineage. The proliferation and differentiation of these precursor cells into mononuclear preosteoclasts, and the further differentiation of these cells into cells that can fuse to multinuclear osteoclasts, are regulated by systemic hormones, as well as by locally produced cytokines and growth factors (reviewed in Mundy and Roodman, 1987; Goldring and Goldring, 1990; Gowen, 1992, 1994; Suda *et al.*, 1992). As shown in Fig. 13.2, interleukin-1 (IL-1), IL-3, IL-6, IL-11, tumor necrosis factors (TNF-α and -β), leukemia inhibitory factor (LIF), macrophage colony-stimulating factor (M-CSF), TGF-β and IGF-I have all been shown to increase osteoclast generation, whereas IL-4, IL-13 and interferon-γ (IFNγ) inhibit osteoclast formation. Systemic hormones known to enhance the number of osteoclasts are parathyroid hormone (PTH), $1,25(OH)_2$-vitamin D3, thyroid hormones and glucocorticosteroids, whereas estrogen and calcitonin negatively influence the generation of osteoclasts. The level at which all these hormones, cytokines and growth factors act has not been outlined in detail. It seems clear, however, that systemic hormones, at least to some extent, act via stimulation of

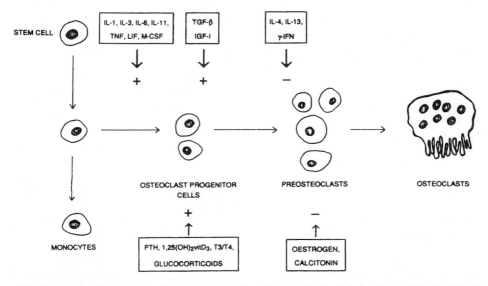

Figure 13.2 Schematic drawing showing the regulation of multinucleated osteoclast formation by cytokines, growth factors and systemic hormones.

local cytokines and, in addition, that some cytokines affect the biosynthesis of others. Furthermore, evidence has been presented indicating that stromal cells in the hematopoietic tissues and osteoblasts in bone tissue exert a paracrine control of osteoclast generation, probably via production of cytokines. In addition, the effects of systemic hormones on osteoclast generation seems to be mediated by a paracrine action of stromal cells and/or osteoblasts.

Lack of functionally active M-CSF results in a skeleton with large amounts of bone owing to a decrease of functional osteoclasts, a condition referred to as "osteopetrosis" (Yoshida et al., 1990). In ovariectomized mice the number of osteoclasts increases rapidly, resulting in massive loss of bone. In IL-6 gene knockout mice, no increase of osteoclasts is seen after ovariectomy (Poli et al., 1994), suggesting an important role for this cytokine in osteoclastogenesis, at least in postmenopausal osteoporosis.

The breakdown of bone starts with attachment of multinucleated osteoclasts to the mineralized bone surface (osteoclasts can not attach to the unmineralized osteoid), probably via an interaction between vitronectin receptors located on the cell membrane in the attachment area and osteopontin (one of the members in the noncollagen protein family in bone extracellular matrix) (reviewed by Vaes, 1988; Hultenby, 1994). In the central part of the attached area, a ruffled border is developed in the cell membrane and, between this part and bone, the resorption process takes place, resulting in a loss of bone and formation of the Howships resorption lacuna. The biochemical mechanisms utilized by the osteoclasts in the resorption process are not fully understood, but recent data indicate that hydroxyapatite crystals are dissolved by a low pH in the lacunae generated by a vacuolar type H^+-ATPase proton pump in the ruffled border. Subsequently, collagen fibers and other noncollagen proteins are degraded by proteases. The proteases responsible for osteoclastic breakdown of bone matrix are not known. Morphological and biochemical studies suggest that release of lysosomal enzymes into the resorption lacunae are involved, and proteolytic enzymes belonging to the cysteine proteinase family seem to play a crucial role (reviewed in Eeckhout et al., 1988). Recently, evidence has been presented suggesting that matrix metalloproteinases also may be involved (Delaissé et al., 1993).

2.4 OSTEOBLAST–OSTEOCLAST INTERACTIONS

It has been demonstrated that, unexpectedly, receptors for hormones and cytokines known to stimulate osteoclastic bone resorption are located on osteoblasts and not on osteoclasts and, in addition, that isolated osteoclasts are not stimulated such agonists unless osteoblasts are present (reviewed in Chambers, 1985; Vaes, 1988). These findings, together with studies indicating that osteoclasts are unable to attach to and degrade bone covered with osteoid, have generated the view (Fig. 13.3), that the resorption process starts by ligand–receptor interactions in osteoblasts resulting in that these cells: (1) degrade the osteoid zone between the osteoblastic syncytium and mineralized bone; (2) secrete one or several factors that activate latent, multinucleated osteoclasts, and enhance the generation of new osteoclasts; and (3) retract from bone, making it possible for the activated osteoclasts to attach to mineralized bone and initiate resorption. One implication of this hypothesis is that osteoblasts are the key cells both in anabolic and catabolic events in bone. Although this view was first presented by Rodan and Martin in 1981, the identity of the paracrine factor(s) from the osteoblasts involved in the activation and generation of osteoclasts is still unknown.

When a certain amount of bone is resorbed in the lacunae, the osteoclasts leave the area. The mechanisms regulating the termination of the resorption process are unknown, but one possibility may be that a calcium sensor in the ruffled border is involved. The fate of postresorption osteoclasts is unknown, but some of them may be involved in a resorption process in another area. During physiological remodeling of bone, it has been known for many years that, after the resorption has ceased and the osteoclasts have disappeared, osteoblasts will appear at the bottom of the lacunae and start to produce new bone (Fig. 13.3; reviewed in Vaes, 1988). During remodeling of bone, when there is equilibrium between resorption and formation, the lacunae will be completely filled up with new bone, i.e., the processes are coupled. However, when osteoclastic resorption is a consequence of a pathologic process leading to a net loss of bone tissue, the amount of new bone formation is less than the volume of the resorbed area, i.e., the processes are uncoupled. The molecular mechanisms involved in physiological coupling between resorption and formation remains elusive. However, one possibility may be that growth factors such as TGF-β and IGF-I, released from the bone matrix during the resorption process, function as coupling factors (Mundy, 1991).

It appears, therefore, that there are tightly regulated paracrine interactions between osteoblasts and osteoclasts during catabolic processes in bone, and between osteoclasts and osteoblasts during anabolic processes, although the biochemical and molecular natures of these intercellular interactions are essentially unknown. It is quite clear that osteoblasts play a key role in osteoclast activation, but whether or not osteoclastic resorption has to precede bone formation in all situations is less clear. It is debatable whether new bone formation can be stimulated by direct activation of osteoblasts.

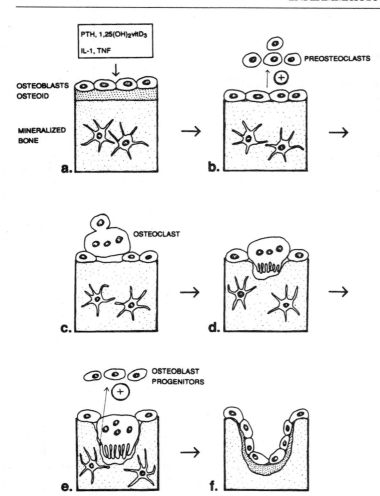

Figure 13.3 Osteoblasts are target cells for hormones and cytokines stimulating bone resorption and the resorptive events start by ligand receptor interactions on osteoblasts followed by catabolic activation of these cells (a). This leads to degradation of the unmineralized osteoid by the osteoblasts and release of factors activating nearby mononuclear preosteoclasts (b). These cells fuse into multinuclear osteoclasts. The osteoblasts retract and thereby give access for the osteoclasts to the mineralized bone surface to which they attach (c). Then the resorption process starts and a Howships lacuna is formed (d). During bone resorption, coupling factors are released which activate preosteoblasts (e). After osteoclasts have disappeared from the lacuna, the activated osteoblasts invade the lacuna and start to produce new bone (f).

3. Inflammation-induced Bone Loss

In rheumatoid arthritis, periodontitis and osteomyelitis, the common view has been that the pathological conditions in the skeleton are a consequence of activities in bone cells directed by paracrine factors released from infiltrating leukocytes. We have presented evidence that bone-cell activities in inflammatory conditions may also be affected by activation of the coagulation cascade and the kallikrein–kinin system (reviewed in Lerner, 1992, 1994). Based upon histopathological studies on periodontitis lesions, demonstrating a predominance of mononuclear leukocytes, and the increasing number of reports that such cells release paracrine factors, once collectively called lymphokines, Horton et al. (1972) first reported the release of a factor from leukocytes with capacity to stimulate bone resorption. It was demonstrated that human peripheral blood leukocytes, in response to activation by antigens, secrete to culture supernatants an activity which stimulates the number of osteoclasts and the rate of bone resorption in cultured fetal rat long bones. This activity, presumed to be a

lymphokine, was called osteoclast-activating factor. Since then, several laboratories have reported that lymphocytes, monocytes and macrophages release peptides which can affect the activity of osteoblasts and osteoclasts, and the osteoclast-activating factor initially reported is now known to be the net effect of several lymphokines and monokines. Thus, IL-1, IL-3, IL-6, IL-11, TNF, LIF and M-CSF all have been shown to stimulate osteoclast generation and bone resorption, whereas IL-4, IL-13 and IFNγ inhibit these processes (Goldring and Goldring, 1990; Gowen, 1992, 1994; Riancho et al., 1993; Girasole et al., 1994; Miyaura et al., 1995). The role of TGF-β is less clear, since this cytokine has been reported both to stimulate and inhibit osteoclast generation and activity (Mundy, 1991). Several of these peptides also change the activity of osteoblasts. Thus, IL-1, IL-4 and TNF inhibit osteoblastic bone formation suggesting that, at least IL-1 and TNF, may decrease the amount of bone in inflammatory conditions both by enhancing bone resorption and decreasing bone formation. The inhibitory effect of IL-4 on bone formation is also suggested by morphological studies on mice overexpressing IL-4, in which the osteopenia developed in the transgenic mice seems to be due to decreased osteoblastic bone formation (Lewis et al., 1993).

It is a well-known feature of chronic inflammatory processes that fibrin deposition can be seen, a phenomenon called extravascular coagulation. This observation indicates that the coagulation cascade has been activated, without being part of hemostasis or thrombosis. It is also well known that vasodilatation, increased vascular permeability and pain are characteristic features of inflammatory processes, and that these reactions, at least in part, are caused by kinin formation. Since both the coagulation cascade and the kallikrein–kinin system is activated by the Hageman factor (coagulation Factor XII), the activation of Hageman factor is likely to play a key role in inflammation. Recent reports have demonstrated that thrombin, the end product of the coagulation cascade responsible for fibrin deposition, and bradykinin (BK), the peptide responsible for the effects on vessel walls, and which is released by activation of the kallikrein–kinin system, both have pleiotropic effects and can exert nonenzymatic, hormone-like (or cytokine-like) effects mediated via specific receptors on a variety of cells. Based upon this information, we have directed our main interests in inflammation-induced bone loss, not to studies on cytokines produced by immunocompetent cells, but to studies on the effects of products involved in nonspecific inflammatory reactions.

4.   Kinins in Inflammation-induced Bone Lesions

The presence of kinins in diseases with inflammation-induced bone loss has been demonstrated by assessments of kininogens, kininogenase activity and kinins in inflammatory exudates (reviewed in Sharma, 1991). Increased kinin levels have been shown in synovial fluids from patients with rheumatoid arthritis, gout or the arthritis associated with psoriasis (Kellermeyer and Graham, 1968; Eisen, 1970). In gout, the kinin levels increase both in response to a spontaneous acute attack or to an intra-articular injection of urate crystals (Melmon et al., 1967). Tissue kallikrein activity has been demonstrated in synovial fluid from rheumatoid arthritics (Worthy et al., 1990a,b) and the level of prekallikrein is significantly higher in synovial fluid from rheumatoid arthritis as compared to those with osteoarthritis (Sharma et al., 1983; Suzuki et al., 1987). Similarly, in arthritic dogs and adjuvant arthritic rats, kallikrein activity in the synovium is increased (Al-Haboubi et al., 1986). In addition, Jasani et al. (1969) detected both high (HMWK) and low (LMWK) molecular weight kininogens in synovial fluids, and Sawai et al. (1979)

reported reduced HMWK in synovial fluid from patients with rheumatoid arthritis. In plasma, levels of kininogen are increased in untreated rheumatoid arthritis patients (Brooks et al., 1974; Sharma et al., 1976), and the high plasma level is decreased by treatment with aspirin or indomethacin (Sharma et al., 1976, 1980). Reports showing that immunoreactive tissue kallikrein, as well as HMWK and LMWK, are detected on the surface of neutrophils (Figueroa et al., 1989, 1990) (see Chapter 11 of this volume) lend further support to the view that kinins are involved in inflammatory diseases. In addition, increased kininogenase activity and kinin levels have been demonstrated in inflammatory exudates collected in gingival pockets between the teeth and inflamed gingiva in dogs with experimental gingivitis (Montgomery and White, 1986).

5.   Effects of Kinins on Bone Metabolism

The effects of kinins on bone metabolism have been studied in bone organ cultures. BK stimulates bone resorption in neonatal mouse calvarial bones cultured for 72–96 h, as assessed by the release of ^{45}Ca from bones dissected from mice, the skeletons of which were labeled in vivo by injection of ^{45}Ca (Gustafson and Lerner, 1984; Lerner et al., 1987a). Also in cultured fetal rat long bones, ^{45}Ca release is augmented by BK, although the response is smaller in magnitude than seen in mouse calvariae (Ljunggren and Lerner, unpublished). BK also stimulates the mobilization of stable calcium and inorganic phosphate from unlabeled mouse calvariae (Gustafson and Lerner, 1984), showing that BK not only causes increased leakage of isotope, but stimulates bone mineral mobilization. The effect on ^{45}Ca release in calvariae is above BK concentrations of 3 nM, with half-maximal effects at 100 nM. Not only are bone inorganic materials released after stimulation with BK, but bone matrix proteins are also released, as assessed by the release of $[^{3}H]$-proline from labeled bones (Lerner et al., 1987a). The release of $[^{3}H]$-proline has been shown to parallel closely that of $[^{3}H]$-hydroxyproline and, since the latter amino acid is a specific indicator of collagen breakdown, these observations indicate that BK stimulates breakdown of both inorganic and organic constituents of bone. Since osteoclasts are the only cells capable of resorbing bone tissue, these findings suggest that BK stimulates bone resorption by enhancing the activity and/or generation of osteoclasts. The observation that calcitonin, a specific physiological osteoclast inhibitor generated in thyroid C-cells, blocks the stimulatory effect of BK on ^{45}Ca release (Lerner et al., 1987a), indicates further that BK-induced bone resorption is mediated by osteoclasts. The inhibition by calcitonin is transient, in agreement with the phenomenon called "escape from calcitonin-induced inhibition", which can be also seen when bone resorption is stimulated with PTH and 1,25(OH)$_2$-vitamin D3 (Wener *et al.*, 1972). Since, however, calcitonin can both inhibit differentiated, multinucleated, actively resorbing osteoclasts and inhibit proliferation, differentiation and fusion of late pre-osteoclasts, it is not possible from such experiments to distinguish between effects on osteoclast activity and osteoclast generation. In subsequent experiments, we found that calcitonin inhibits the actions of BK, both when added concomitantly with the kinin, and when added after BK has been allowed to initiate ^{45}Ca release. The latter finding is likely due to an action of calcitonin on osteoclasts already stimulated by BK. These experiments in neonatal mouse calvariae have been performed in bones preincubated for 24 h, and then stimulated by BK or other agonists. Since morphological studies have shown that the osteoclasts present in the calvariae directly after dissection have disappeared after the preculture period (Lerner *et al.*, unpublished observations), substances including BK that stimulate osteoclastic bone resorption in this model must be able to enhance the generation of new osteoclasts. The action of BK on ^{45}Ca release is delayed, with no effect observed until after 24 h incubation with BK, a finding compatible with such a mechanism, although final proof must await morphological assessment of osteoclast number. It should be added, that the osteoclast precursor cells in the mouse calvariae probably are late precursor cells in the osteoclastic cell lineage, since no hematopoietic tissue is present in these bone explants. Whether or not BK can stimulate osteoclast formation in the presence of the early precursors in bone-marrow culture is not known at present.

BK not only stimulates mineral mobilization and bone-matrix breakdown, but also enhances the release of β-glucuronidase (Gustafson and Lerner, 1984). Although the role of this enzyme in bone resorption is not known, it is a marker of lysosomal enzyme release. In agreement with the view that release of lysosomal enzymes from osteoclasts to the resorption lacuna plays an important role in bone resorption (Eeckhout *et al.*, 1988), there is a significant correlation between the release of β-glucuronidase and mineral mobilization in bones stimulated by BK. The preferential release of lysosomal enzymes is indicated by the lack of an effect of BK on the cytosolic enzyme lactate dehydrogenase.

These observations suggest that in inflammatory processes, activation of plasma prekallikrein (by the Hageman factor) and subsequent formation of BK from HMWK may lead to loss of bone owing to stimulation of osteoclastic bone resorption by the kinin. In addition to circulating kallikrein, several cells in different tissues synthesize tissue prekallikrein (MacDonald *et al.*, 1988; Margolius, 1989). Once activated, extra- or intracellularly, this enzyme generates the decapeptide kallidin (Lys-BK) from both HMWK and LMWK. This peptide can exert biological effects by itself or, alternatively, can exert effects after conversion to BK by arginine aminopeptidase. In mouse calvarial bones, Lys-BK stimulates mineral mobilization and matrix breakdown, by a process inhibited by calcitonin (Gustafson *et al.*, 1986). The relative potencies of BK and Lys-BK have not been reported, but preliminary studies indicate that they are equipotent.

The magnitudes of the bone resorptive effects of BK and Lys-BK are less than those induced by PTH and 1,25(OH)$_2$-vitamin D3. This is likely to be due to degradation of the kinins in long-term organ cultures used for the bone-resorption bioassay, a hypothesis supported by the fact that a kininase II (angiotensin-converting enzyme, or ACE) inhibitor BPP5a potentiates ^{45}Ca release induced by BK, but not those of PTH and PGE$_2$ (Lerner *et al.*, 1987a). Nevertheless, another more potent ACE inhibitor, captopril, did not have such a potentiating effect. Met-Lys-BK, which may be released from kininogens owing to cleavage by neutrophil acid proteases, also has been shown to stimulate bone resorption in mouse calvariae by a mechanism sensitive to calcitonin inhibition (Ljunggren and Lerner, 1988).

The bone-resorptive effects of cytokines such as IL-1 and TNF *in vitro* are inhibited by IFNγ, which has little if any inhibitory effect on bone resorption stimulated by the hormones PTH and 1,25(OH)$_2$-vitamin D3 (Gowen *et al.*, 1986). Such a preferential inhibitory action of IFNγ generates the speculation that IFNγ may exert a negative-feedback control in inflammation-induced bone resorption. Interestingly it has been shown that IFNγ also inhibits BK-induced bone resorption in mouse calvariae *in vitro* in a concentration-dependent manner (Lerner *et al.*, 1991a). The mechanism by which IFNγ acts on bone resorption is not known, but there are indications that it may act at the level of osteoclast generation (Takahashi *et al.*, 1986).

As discussed above, inflammatory mechanisms may cause loss of bone both by stimulating bone resorption and by inhibiting bone formation. Very little is known, however, about the effects of kinins on osteoblast proliferation or bone formation. In some cell types, including embryonic human lung fibroblasts, Balb/c 3T3 cells and NIL8 hamster cells (Goldstein and Wall, 1984; Straus and Pang, 1984; Godin *et al.*, 1991), BK exerts a stimulatory effect on the rate of cell division, whereas such an effect is not evident in freshly isolated fibroblasts from human amniotic fluid, rabbit dermis or human dental-pulp (Marceau and Tremblay, 1986; Sundqvist *et al.*, 1995). In contrast, BK inhibits retinoic acid- and TGF-β-stimulated proliferation of normal rat kidney cells (Lahaye *et al.*, 1994). In the human osteoblastic osteosarcoma cell line MG-63, BK does not stimulate cell proliferation, nor does BK affect biosynthesis of type I collagen or osteocalcin (Rosenquist *et al.*, 1996).

6. Prostaglandins in Kinin-induced Bone Resorption

Most systemic hormones and cytokines known to stimulate bone resorption cause parallel stimulation of prostanoid biosynthesis in bone cells. In line with the view that osteoblasts are the key target cells for such stimulants, the bone cell showing a prostanoid response is the osteoblast. It has been demonstrated *in vivo* and *in vitro* that the bone-resorptive response to some agonists is independent of the prostanoid response, whereas that of others is partially due to the enhancement of prostaglandin biosynthesis. As regards BK, indomethacin abolishes BK-induced prostaglandin biosynthesis, and completely inhibits BK-induced mineral mobilization and bone matrix breakdown. Furthermore, structurally unrelated nonsteroidal anti-inflammatory drugs (NSAID) such as meclofenamic acid, naproxen and flurbiprofen also abolish the bone-resorptive effects of BK (Lerner et al., 1987a). Similarly, bone resorption in response to Lys-BK and Met-Lys-BK is abolished by NSAIDs (Gustafson et al., 1986; Ljunggren and Lerner, 1988). The observation that cyclooxygenase inhibitors completely inhibit the effects of BK in bone indicates that only arachidonic-acid metabolites via the cyclooxygenase pathway are important in the actions of BK on bone resorption. The observation that glucocorticosteroids are inhibitors of prostaglandin biosynthesis in bone, together with reports that hydrocortisone and dexamethasone are potent inhibitors of bone resorption stimulated by BK (Lerner et al., 1987a), further support the view that stimulation of endogenous prostaglandin biosynthesis is a prerequisite for the bone-resorptive effect of BK. Such a view is also compatible with observations in different bone organ culture systems showing that exogenous addition of prostaglandin E_2 (PGE$_2$), and to some degree also prostaglandin I_2 (PGI$_2$), stimulates osteoclastic bone resorption (Raisz et al., 1977; Tashjian et al., 1977). Similar to the kinetics of BK-induced ^{45}Ca release, exogenous PGE$_2$ induces bone resorption that can be seen at 12 h (Lerner et al., 1987b).

Different subpopulations of bone cells from mouse calvariae have been isolated utilizing time sequential enzyme-digestion techniques. One population of these cells expresses receptors for PTH and 1,25(OH)$_2$-vitamin D3, as well as high levels of bone-specific alkaline phosphatase. These cells also synthesize type I collagen and osteocalcin (a specific noncollagen bone matrix protein), and make mineralized bone nodules. Based on these criteria, such cells are regarded as osteoblasts. Addition of BK to isolated osteoblasts results in a burst of PGE$_2$ and 6-keto-PGF$_{1\alpha}$ (the stable breakdown product of PGI$_2$) biosynthesis, which is maximal after 5–10 min (Lerner et al., 1989). The potency of BK is similar to that seen for its effect on different cells, with the EC$_{50}$ being around 10 nM. The observation that its EC$_{50}$ for

prostaglandin biosynthesis is slightly less than that for bone resorption (100 nM) is, again, probably due to differences in degradation in the short-term cell incubations versus the long-term bone organ cultures. The stimulatory effect of BK on PGE$_2$ biosynthesis in mouse calvarial osteoblasts is reversible after removal of BK. However, cells prestimulated with BK cannot respond to a second challenge with the peptide. This is probably not due to changes in arachidonic-acid metabolism, since BK-prestimulated cells respond fully to a subsequent challenge with exogenous arachidonic acid but, rather, is due to homologous desensitization at the BK receptor level, as shown in a mouse fibrosarcoma cell line, HSDM1C1 (Becherer et al., 1982).

BK-induced prostaglandin biosynthesis in osteoblasts is dependent on cell density. Seeding calvarial osteoblasts at a low cell density and then assaying the amounts of PGE$_2$ and 6-keto-PGF$_{1\alpha}$ produced per cell, at different times and with increasing number of cells per dish, has revealed that the prostanoid release is less at high than at low cell densities (Lerner et al., 1989). This effect is not due to differences in culture time, since osteoblasts seeded at increasing cell densities and then attached overnight also produce the largest amounts of PGE$_2$ in response to BK at low cell density. The increasing response to BK at decreasing cell densities may be due to differences in receptor expression. However, the observation that the amounts of prostaglandins produced in response to exogenous arachidonic acid also are highest at low cell density suggest that the different responses are due to alterations in cyclooxygenase activity.

A burst of prostaglandin biosynthesis in response to BK can also be obtained in the cloned nontransformed mouse calvarial osteoblastic line, MC3T3-E1 (Lerner et al., 1989), in a human osteoblastic osteosarcoma cell line (MG-63; Lerner and Bernhold, in preparation) and in nonenzymatically isolated human bone cells (Ljunggren et al., 1990; Rahman et al., 1992), but not in the rat osteosarcoma osteoblastic cells, ROS 17/2.8 and UMR 106-01 (Lerner et al., 1989). The latter finding is unlikely to be due to species differences since BK stimulates formation of PGE$_2$ and 6-keto-PGF$_{1\alpha}$ in primary rat calvarial osteoblasts (Partridge et al., 1985). The differences between the rat osteosarcoma cell lines and the other BK-responsive cells may be related to phenotypic alterations in rat (but not in human) transformed cell lines. Thus, UMR 106-01 cells spontaneously produce only minute amounts of PGE$_2$ and, in addition, respond very weakly to exogenous arachidonic acid, whereas ROS 17/2.8 cells, spontaneously produce high levels of PGE$_2$ and respond to exogenous addition of arachidonic acid with an abundant release of PGE$_2$ (Lerner et al., 1989). These observations indicate that UMR 106-01 expresses very low levels of cyclooxygenase(s) whereas the activities of this enzyme(s) are very high in ROS 17/2.8 cells.

Indomethacin, meclofenamic acid, naproxen,

flurbiprofen, hydrocortisone and dexamethasone are all potent inhibitors of BK-induced PGE_2 release in osteoblasts (Lerner et al., 1989). The action of the different NSAIDs is seen only after a very short preincubation before exposure to BK, whereas the effect of the glucocorticosteroids requires that cells are pretreated for several hours. This may be explained by differences in mechanism of action, the NSAIDs acting directly at the level of cyclooxygenase activity and the steroids exerting their effects via a genomic action. These observations raise the question why simultaneous addition of hydrocortisone or dexamethasone with BK is sufficient to abolish the bone-resorptive effect (Lerner et al., 1987a). It could indicate that the inhibitory effect of the glucocorticosteroids on bone resorption is unrelated to inhibition of prostaglandin biosynthesis, but exerted at a step distal to this process. Speaking against such a view is the recent observation that, in mouse calvariae, glucocorticosteroids potentiate, in an additive manner, the bone-resorptive effect of PGE_2 (Conaway and Lerner, 1995). The inhibitory effect of glucocorticosteroids may, however, be influenced by the possibility that the kininase I metabolite of BK, desArg^9-BK, contributes to the effect of BK. As discussed later, desArg^9-BK also stimulates bone resorption and prostaglandin production (Lerner et al., 1987a; Ljunggren and Lerner, 1990), but the kinetics of the prostaglandin response are delayed substantially as compared with BK, making it possible for glucocorticosteroids to exert an inhibitory action.

7. Kinin Receptors in Bone

The presence of BK receptor subtypes in bone cells has been studied by pharmacological methods using a variety of BK analogs and comparing the rank-order potency (Regoli and Barabé, 1980; Farmer and Burch, 1991; Hall, 1992; Marceau, 1995) (see Chapter 2). BK, Lys-BK, Met-Lys-BK and desArg^9-BK all stimulate bone resorption indicating the presence of both B_1 and B_2 receptors in mouse calvariae (Gustafson et al., 1986; Lerner et al., 1987a; Ljunggren and Lerner, 1988). However, effects on ^{45}Ca release could exclusively, or at least partially, be mediated by B_1 receptors if BK is converted completely to the B_1 receptor agonist desArg^9-BK (and Lys-BK to desArg^9-Lys-BK) by kininase I in calvarial bones. Such a possibility is also indicated by the observation that the bone-resorptive effect of BK persists after pretreatment with kininase I, whereas that of PTH is significantly reduced (Lerner et al., 1987a). That a B_1 receptor antagonist, desArg^9-[Leu^8]-BK, does not inhibit BK-induced bone resorption demonstrates, however, that B_2 and not B_1 receptors mediate the effects of BK (Lerner et al., 1987a). Attempts to demonstrate the presence of B_2 receptors by using a specific antagonist DArg-[Hyp^3,Thi^{5,8},DPhe^7]-BK (NPC 349) in BK-stimu-

lated bone resorption failed, probably due to the metabolic lability of this peptide in the long-term bioassays (Ljunggren and Lerner, unpublished). However, the fact that NPC 349 inhibits BK-induced PGE_2 formation in mouse calvariae incubated for 30 min with the ligands (Ljunggren et al., 1991a), suggests strongly the existence of B_2 receptors in mouse calvarial bones. The involvement of B_1 receptors is shown by the observation that desArg^9-[Leu^8]-BK inhibits the effects of desArg^9-BK, but not those of BK (Ljunggren and Lerner, 1990). These data indicate clearly the presence of both B_1 and B_2 receptors in mouse calvariae linked to stimulation of bone resorption, although the cell types expressing these receptors in the multicellular bone organ is not known at present.

The bone-resorptive effect of desArg^9-BK is abolished by indomethacin, flurbiprofen and hydrocortisone, similar to the observations on B_2 receptor agonist-induced stimulation of bone resorption. In addition, desArg^9-BK and BK both stimulate PGE_2 biosynthesis in calvariae (Ljunggren and Lerner, 1990), suggesting that B_1 and B_2 receptor-induced resorption is linked to enhanced prostanoid formation as an obligatory step in the mechanisms linking kinin receptors to osteoclast activation. There are, however, differences in the kinetics of B_1 and B_2 receptor-induced prostaglandin formation in mouse calvariae, with desArg^9-BK and BK significantly stimulating PGE_2 formation in 72 h cultures, with BK showing such a response within 30 min.

Studies on the relative potencies kinins causing a burst of PGE_2 and 6-keto-$PGF_{1\alpha}$ formation in primary cultures of mouse calvarial osteoblasts demonstrated the following rank order potency:

$$BK = Lys\text{-}BK > Met\text{-}Lys\text{-}BK \gg desArg^9\text{-}BK$$

These data, and observations that NPC 349, but not desArg^9-[Leu^8]-BK, inhibits the acute rise in prostaglandins induced by BK, demonstrate the presence of B_2 receptors on osteoblasts (Ljunggren et al., 1991a). Similar observations were made in the cloned mouse calvarial osteoblastic cell line, MC3T3-E1 where expression of B_2 receptors was also indicated by the observation that BK and NPC 349, but not desArg^9-BK, competed with the binding of [^{3}H]-BK (Ljunggren et al., 1991a). These data suggest that osteoblasts are target cells for B_2 receptor-mediated actions on prostaglandin biosynthesis and bone resorption, similar to other stimulants of bone resorption (Chambers, 1985; Vaes, 1988). The data, however, are more difficult to reconcile with the observation that a B_1 receptor agonist also stimulated bone resorption via a prostaglandin-dependent mechanism. However, when mouse calvarial osteoblasts are exposed to desArg^9-BK in long-term culture, PGE_2 release is seen only after 24 h (Ljunggren and Lerner, 1990). The response to desArg^9-BK was antagonized by desArg^9-[Leu^8]-BK, whereas responses to BK and Lys-BK were unaffected.

It is clear, therefore, that osteoblasts express both B_1 and B_2 receptors, and that their signal transduction mechanisms may differ. The burst of prostaglandin formation linked to B_2 receptors is probably due to BK-induced activation of phospholipase A_2 (PLA_2), a hypothesis awaiting confirmation by direct assay of this enzyme in cells exposed to BK. The delayed enhancement of prostaglandin formation coupled to B_1 receptors may be due to induction of cyclooxygenase(s), similar to observations showing that IL-1 causes delayed-onset prostaglandin biosynthesis in mouse calvarial osteoblasts (Kawaguchi et al., 1994). Circumstantial evidence with guinea-pig aorta cells indicates the possibility that high (μM) concentrations of BK enhance cyclooxygenase activity, concentrations at which the kinin also has affinity for B_1 receptors (Zhang et al., 1991). The result, that pretreatment with glucocortico-steroids inhibits BK-induced PGE_2 biosynthesis in osteoblasts (Lerner et al., 1989), may be indirect evidence that B_2 receptors are coupled to PLA_2, since these steroids induce lipocortin, an endogenous PLA_2 inhibitor (Hirata, 1981). However, glucocorticosteroids also inhibit desArg^9-BK-stimulated bone resorption and prostaglandin formation in mouse calvariae (Ljunggren and Lerner, 1990), which may indicate that gluco-corticosteroid-induced inhibition is either unrelated to lipocortin inhibiting PLA_2 or that PLA_2 activation is not linked exclusively to B_2 receptors.

The delayed onset of B_1 receptor-induced prosta-glandin formation may also be due to B_1 receptors not being expressed constitutively in bone organs or in osteoblasts, but, rather, induced in culture (Marceau, 1995; and Chapter 8 for discussion). This possibility could be analyzed by using radioligand binding of B_1 and B_2 receptors. Since human B_1 (Menke et al., 1994) and rat and human B_2 (McEachern et al., 1991; Hess et al., 1992) receptors are cloned, it will also be possible to study their expression under different conditions, and to investigate whether osteoblasts express both subtypes. As regards the osteoblastic line, ROS 17/2.8, Santora (1989) reported specific binding of a B_1 receptor radioligand, [^{125}I]-[Tyr]-desArg^9-BK, indicating B_1 receptor expression in these cells.

Recently, we demonstrated that the human osteoblast osteosarcoma line, MG-63, responds to BK with a time- and concentration-dependent burst of PGE_2 and 6-keto-$PGF_{1\alpha}$ release (Lerner and Bernhold, in preparation). The acute rise of prostanoid formation in MG-63 cells is seen after challenge with BK, Lys-BK and Met-Lys-BK, but not with desArg^9-BK or desArg^9-Lys-BK. This finding and the observation that NPC 349 and DArg-[Hyp^3,Thi^5,DTic^7,Oic^8]-BK (Hoe 140), B_2 receptor antagonists but not desArg^9[Leu^8]-BK inhibit BK-induced PGE_2 formation, indicate the presence of B_2, but not B_1 receptors linked to the burst of prostanoid formation, similar to the observations in primary mouse calvarial osteoblasts. In MG-63 cells, prolonged incubations result in an increased responsiveness to desArg^9-BK and desArg^9-Lys-BK, indicating induction of B_1 receptor expression.

8.   Kinin Signal Transduction in Bone

BK enhances basal and guanine nucleotide-stimulated adenylate cyclase activity as well as increasing intracellular cyclic adenosine monophosphate (cAMP) accumulation in guinea-pig tracheal smooth muscle cells (Stevens et al., 1994). In contrast, BK indirectly inhibits cAMP formation in rat glomeruli and mesangial cells (Bascands et al., 1993). In mouse calvarial osteoblasts, cAMP levels are rapidly increased by BK (Lerner et al., 1989). Since the effect is abolished by indomethacin, the rise of cAMP is probably secondary to the burst of PGE_2 (known to stimulate cAMP formation in a variety of cell types including osteoblasts). This observation suggests that cAMP may not serve directly as a signal transduction system for BK in osteoblasts, although cAMP analogs, and activators of adenylate cyclase (forskolin and cholera toxin) enhance PGE_2 biosynthesis in osteoblasts (Lerner et al., unpublished).

In the osteoblastic cell line MC3T3-E1, labeled with [^{3}H]-myoinositol, an acute, time-dependent release of polyphosphoinositides is seen in response to BK. Since desArg^9-BK has no effect on inositol 1,4,5-trisphosphate (IP_3) formation, the finding indicates that the effect is linked to B_2 receptors (Ljunggren et al., 1991b). The increased level of IP_3 suggests that B_2 receptors are coupled to phospholipase C (PLC) and turnover of phosphatidylinositol 4,5-bisphosphate (PIP_2) as a signal transduction system, and that IP_3 and diacylglycerol (DAG), the two principal breakdown products of PIP_2, act as second messengers. It is generally accepted that IP_3 causes a rapid, transient rise of intracellular calcium ($[Ca^{2+}]_i$) by mobilizing Ca^{2+} from intracellular stores, and that DAG acts by translocation and activation of protein kinase C (PKC; reviewed in Berridge, 1987). In line with the view that B_2 receptors are linked to phos-phatidylinositol turnover, it has been shown, using microfluorometric cell analysis, that BK causes a rapid transient rise in $[Ca^{2+}]_i$ in MC3T3-E1 cells, an effect that is independent of extracellular Ca^{2+} (Ljunggren et al., 1991b). Ljunggren et al. (1993) also presented indirect evidence for translocation of PKC in response to BK, but not desArg^9-BK, in MC3T3-E1 cells. The effects of BK on IP_3 formation, increase in $[Ca^{2+}]_i$ and PKC translocation are seen at kinin concentrations similar to those causing the burst of PGE_2 formation (EC_{50}, 3 nM) but are unaffected by indomethacin.

The observations suggest that BK-induced PGE_2 formation in osteoblasts (and bone resorption) may be due to sequential activation of PLC, DAG and mono-acylglycerol lipase, and subsequent release of arachidonic acid. An alternative explanation could be that activation of PLC, with formation of IP_3 and DAG, subsequently resulting in a rise of $[Ca^{2+}]_i$ and activation of PKC, may cause activation of PLA_2 and release of arachidonic acid. Stimulation of PKC with phorbol esters and increasing $[Ca^{2+}]_i$ with ionophores (A23187 and ionomycin) result in a rapid stimulation of PGE_2 formation in MC3T3-E1 cells, similar to that induced by BK. These observations and the finding that simultaneous treatment with phorbol esters and Ca^{2+} ionophores causes a synergistic interaction on PGE_2 formation (Ljunggren et al., 1991b, 1993) are compatible with the hypothesis that IP_3 and PKC are intracellular mediators in the action of BK on prostaglandin formation in osteoblasts. The observation that Ca^{2+} ionophores and phorbol esters stimulate bone resorption in mouse calvariae by a mechanism sensitive to indomethacin (Lerner et al., unpublished) argues further for a role for PLC-mediated formation of IP_3 and PKC in the actions of BK on bone resorption and prostaglandin formation.

In contrast to this hypothesis is the view presented by Burch and Axelrod (1987) suggesting that the effects of BK on IP_3 and PGE_2 formation are not linked causally. The latter view is based upon observations in Swiss 3T3 fibroblasts that dexamethasone inhibited PGE_2 biosynthesis, but not the formation of IP_3. The inhibitory effect on PGE_2 release by dexamethasone is likely to be due to induction of the PLA_2 inhibitor, lipocortin, at a step distal to IP_3 formation in our hypothesis. Thus, we interpret the data by Burch and Axelrod (1987) as being compatible with our hypothesis. However, against this view is the finding by Yanaga et al. (1991) that pretreatment with Pertussis toxin inhibited BK-induced PGE_2 formation in MC3T3-E1 cells without affecting $[Ca^{2+}]_i$, indicating that the effect of BK on prostanoid biosynthesis, but not on Ca^{2+}, is mediated via G protein-linked receptors. This observation indicates that the PGE_2 response and elevated $[Ca^{2+}]_i$ can be disassociated.

In summary, it is suggested that BK binds to B_2 receptors on osteoblasts, resulting in a G protein-dependent activation of PLC with the subsequent formation of IP_3 and DAG. These two second messengers increase $[Ca^{2+}]_i$ and activate PKC, resulting in activation of PLA_2, which releases arachidonic acid from phospholipids in the cell membrane (Fig. 13.4). Arachidonic acid is then converted to PGE_2 and PGI_2 by the osteoblasts, and these prostanoids change, by autocrine or paracrine mechanisms, the osteoblastic phenotype into a catabolic state leading to release of paracrine factors. These factors enhance the formation of multinucleated osteoclasts from the precursor pool of mononucleated osteoclast precursors present in the periosteum and endosteum of the bones, as previously described for other stimulants of bone resorption (Fig. 13.5).

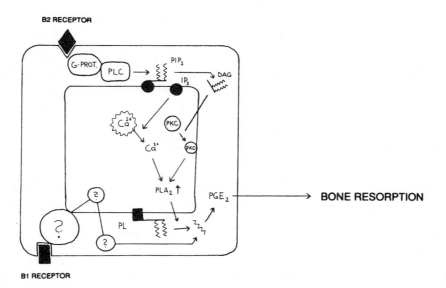

Figure 13.4   Schematic drawing summarizing our hypothesis linking prostaglandin formation to bradykinin (BK) receptors and stimulation of bone resorption. Occupancy of BK B_2 receptors leads to a G protein-dependent activation of phospholipase C (PLC) and subsequent formation of inositol 1,4,5-trisphosphate (IP_3) and diacylglycerol (DAG) from phosphatidylinositol 4,5-bisphosphate (PIP_2). Subsequently, the concentration of intracellular Ca^{2+} is raised and protein kinase C (PKC) is activated, leading to stimulation of phospholipase A_2 (PLA_2). This enzyme liberates free arachidonic acid from phospholipids in the cell membrane and this unsaturated fatty acid is converted to prostaglandins (PGE_2) stimulating bone resorption. Activation of kinin B_1 receptors also leads to prostaglandin formation and bone resorption, but the signal transducing mechanism, as well as the mechanism involved in prostaglandin formation is unknown.

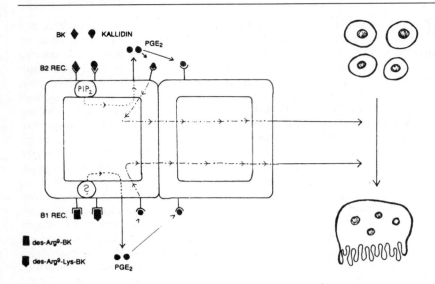

Figure 13.5 Schematic drawing showing the importance of osteoblasts in the mechanism by which bradykinin receptors, via a prostaglandin-dependent mechanism, activate bone resorbing osteoclasts.

9. *Interactions Between Kinins and Cytokines*

It is apparent that several cytokines, as well as kinins, may be responsible for activation of osteoclastic bone resorption in inflammatory diseases. The relative importance of these agonists is not known at present, but the probability that several or all of them are present concomitantly in inflammatory tissues in the vicinity of bone, strongly suggests that positive and negative interactions between these mediators may control the numbers and activity of osteoclasts. As mentioned previously, IL-1 and TNF stimulate bone resorption, whereas IFNγ and IL-4 inhibit this process (Gowen *et al.*, 1986; Watanabe *et al.*, 1990). Such an interplay may also exist between kinins and cytokines.

One possible interaction is suggested by the finding that BK can stimulate IL-1 and TNF production in murine macrophages, and IL-6 formation in human bone cells (Tiffany and Burch, 1989; Rahman *et al.*, 1992). In addition, we demonstrated in human gingival fibroblasts that BK potentiates the stimulatory effect of TNF on the biosynthesis of IL-1 (Yucel-Lindberg *et al.*, 1995). Since IL-1 and TNF stimulate bone resorption *in vitro* and *in vivo*, these observations raise the possibility that BK may exert its effect on bone resorption via induction or potentiation of IL-1 and/or TNF formation. Although such a mechanism may operate *in vivo*, the stimulatory effect of BK on bone resorption *in vitro* is not likely to be due to IL-1 or TNF, since bone resorption in mouse calvariae induced by these cytokines is only slightly reduced by indomethacin (Lerner *et al.*, 1991b, Lerner and Ohlin, 1993), whereas that of BK is abolished (Lerner *et al.*, 1987a). It is not possible, however, to exclude the possibility of cytokines playing an intermediary role, since indomethacin may act at a

level proximal to cytokine induction. The finding that IL-6 production can be induced by BK, together with the finding that IL-6, in some systems *in vitro*, stimulates osteoclast generation and bone resorption (Tamura *et al.*, 1993), suggest the possibility that IL-6 may be involved in the action of BK. This may not be the case in mouse calvariae, since IL-6 does not stimulate bone resorption in this culture system (Lerner, unpublished observation).

Although cytokines may not be involved in BK-induced bone resorption at a step distal to the stimulatory action of BK, the possibility exists that BK and cytokines may act in concert. In line with this view, it has been shown that BK synergistically potentiates the stimulatory effects of IL-1α and IL-1β on bone resorption and prostanoid formation in mouse calvariae (Lerner, 1991a). Similarly, BK synergistically potentiates TNF-α-induced bone resorption and prostaglandin formation in calvarial bones (Lerner, unpublished observations). The synergy between BK and IL-1, and TNF on prostaglandin biosynthesis can also be seen in human gingival and dental-pulp fibroblasts, as well as in human periodontal ligament cells (Lerner and Modéer, 1991; Marklund *et al.*, 1994; Sundqvist and Lerner, 1995). In ongoing studies, we have demonstrated synergy by BK on biosynthesis of PGE_2 and PGI_2 in the human osteoblastic cell line MG-63 in response to IL-1 and TNF (Lerner and Bernhold, unpublished observations). In these cells Lys-BK, desArg[9]-BK and desArg[9]-Lys-BK each synergize the prostanoid formation induced by IL-1β, indicating that both B_1 and B_2 receptors are expressed, and have the capacity to potentiate the actions of IL-1. Similar findings have been made in human dental-pulp fibroblasts (Sundqvist and Lerner, 1995). The molecular mechanisms involved in these interactions are not known, but may be at the level of receptor number or affinity, signal transduction or arachidonic-acid release and/or metabolism.

Studies in gingival fibroblasts show that the synergistic

interaction between BK and IL-1 is not associated with increased intracellular cAMP or a rise in $[Ca^{2+}]_i$ (Lerner *et al.*, 1992), nor can synergistic interactions on arachidonic-acid release be detected (Lerner and Modéer, 1991). BK causes a burst of prostanoid formation, whereas IL-1 and TNF cause a delayed (onset 1–2 h) stimulation of prostanoid formation in both osteoblasts and fibroblasts. The synergy between BK and IL-1 in PGE_2 formation is also delayed (Sundqvist and Lerner, 1995), indicating that the interactions between BK and IL-1 may be due to BK upregulating the mechanism by which IL-1 stimulates prostaglandin formation and not vice versa.

10. Systemic Effects on Bone by Local Inflammation

The skeleton may not only be affected by local inflammatory changes. That bone tissue may be systemically affected is indicated by findings showing that rats with experimental "nonosseus" inflammation develop generalized bone loss (Minne *et al.*, 1984), and that patients with inflammatory bowel disease have an increased risk of osteopenia (Motley *et al.*, 1993). The pathogenesis of such systemic effects is not known, but may be related to the observation that one of the acute-phase reactants, haptoglobin, stimulates bone resorption *in vitro* at concentrations found in sera from patients with inflammatory conditions (Lerner and Fröhlander, 1992). Moreover, the stimulatory effect of BK on PGE_2 formation in osteoblasts is synergistically potentiated by haptoglobin (phenotypes 1-1 as well as 2-1; Fröhlander *et al.*, 1991).

It is apparent that the skeleton can be affected by several local and systemic factors produced in, or as a consequence of, inflammatory processes and that such factors can act in negative-feedback systems and in synergistic cooperation in patients with inflammation-induced bone loss. In this context, it is interesting to note that infusion of PTH causes significantly larger increases of ionized serum calcium in patients with rheumatoid arthritis compared with matched controls without inflammatory disease (af Ekenstam *et al.*, 1990). These findings suggests the possibility of an interplay between endocrine factors and inflammatory mediators, a hypothesis supported by observations *in vitro* of synergistic interactions between PTH and IL-1 on bone resorption in fetal rat long bones (Stashenko *et al.*, 1987), and between PTH and TGF-β on prostanoid formation in mouse calvariae (Lerner, 1991b).

11. Acknowledgements

Studies performed in the author's laboratory have been supported by The Swedish Medical Research Council, The Swedish Association Against Rheumatic Diseases, The Royal 80 Year Fund of King Gustav V and A.-G. Craaford Foundation. The artistic work performed by Mr Lars Mattson is gratefully acknowledged.

12. References

af Ekenstam, E., Hällgren, R., Joborn, C. and Ljunghall, S. (1990). In "Influence of inflammation and corticosteroid treatment on indices of bone turnover and parathyroid function" (ed. E. af Ekenstam). Thesis, Uppsala University.

Al-Haboubi, H.A., Bennett, D., Sharma, J.N., Thomas, G.R. and Zeitlin, I.J. (1986). A synovial amidase acting on tissue kallikrein-selective substrate in clinical and experimental arthritis. Adv. Exp. Med. Biol. 198B, 405–411.

Bascands, J.-L., Pecher, C. and Girolami, J.-P. (1993). Indirect inhibition by bradykinin of cyclic AMP generation in isolated rat glomeruli and mesangial cells. Mol. Pharmacol. 44, 818–826.

Becherer, P.R., Mertz, L.F. and Baenziger, N.L. (1982). Regulation of prostaglandin synthesis mediated by thrombin and B_2 bradykinin receptors in a fibrosarcoma cell line. Cell 30, 243–251.

Berridge, M.J. (1987). Inositol trisphosphate and diacylglycerol, two interacting second messengers. Annu. Rev. Biochem. 56, 159–193.

Brooks, P.M., Dick, W.C., Sharma, J.N. and Zeitlin, I.J. (1974). Changes in plasma kininogen levels associated with rheumatoid activity. Br. J. Clin. Pharmacol. 1, 351P.

Burch, R.M. and Axelrod, J. (1987). Dissociation of bradykinin-induced prostaglandin formation from phosphatidylinositol turnover in Swiss 3T3 fibroblasts: evidence for G protein regulation of phospholipase A_2. Proc. Natl Acad. Sci. USA 84, 6374–6378.

Caplan, A.I. and Pechak, D.G. (1987). The cellular and molecular embryology of bone formation. In "Bone and Mineral Research" (ed. W.A. Peck), pp. 117–183. Elsevier, Amsterdam.

Centrella, M., Horowitz, M.C., Wozney, J.M. and McCarthy, T.L. (1994). Transforming growth factor-β gene family members and bone. Endocr. Rev. 15, 27–39.

Chambers, T.J. (1985). The pathobiology of the osteoclast. J. Clin. Pathol. 38, 241–252.

Conaway, H.H. and Lerner, U.H. (1995). Glucocorticoids stimulate bone resorption *in vitro*. J. Bone Miner. Res. 16, Suppl. 1, abstr. 350.

Delaissé, J.-M., Eeckhout, Y., Neff, L., François-Gillet, C., Henriet, P., Su, Y., Vaes, G. and Baron, R. (1993). (Pro)collagenase (matrix metalloproteinase-1) is present in rodent osteoclasts and in the underlying bone-resorbing compartment. J. Cell Sci. 106, 1071–1082.

Eeckhout, Y., Delaissé, J.-M., Ledent, P. and Vaes, G. (1988). In "The Control of Tissue Damage" (ed. E. Glauert), pp. 297–313. Elsevier, Amsterdam.

Eisen, V. (1970). Formation and function of kinins. Rheumatology 3, 103–168.

Farmer, S.G. and Burch, R.M. (1991). The pharmacology of bradykinin receptors. In "Bradykinin Antagonists in Basic and Clinical Research" (ed. R.M. Burch), pp. 1–31. Marcel Dekker, New York.

Figueroa, C.D., Maciver, A.G. and Bhoola, K.D. (1989).

Identification of a tissue kallikrein in human polymorphonuclear leucocytes. Br. J. Haematol. 72, 321–328.

Figueroa, C.D., Henderson, L.M., Colman, R.W., DeLa Cadena, R.A., Müller-Esterl, W. and Bhoola, K.D. (1990). Immunoreactive L- and H-kininogen in human neutrophils. J. Physiol. 425, 65P.

Fröhlander, N., Ljunggren, Ö. and Lerner, U.H. (1991). Haptoglobin synergistically potentiates bradykinin and thrombin induced prostaglandin biosynthesis in isolated osteoblasts. Biochem. Biophys. Res. Commun. 178, 343–351.

Girasole, G., Passeri, G., Jilka, R.L. and Manolagas, S.C. (1994). Interleukin-11: a new cytokine critical for osteoclast development. J. Clin. Invest. 93, 1516–1524.

Godin, C., Smith, A.D. and Riley, P.A. (1991). Bradykinin stimulates DNA synthesis in competent Balb/c 3T3 cells and enhances inositol phosphate formation induced by platelet-derived growth factor. Biochem. Pharmacol. 42, 117–122.

Goldring, M.B. and Goldring, S.R. (1990). Skeletal tissue response to cytokines. Clin. Orthop. 258, 245–278.

Goldstein, R.H. and Wall, M. (1984). Activation of protein formation and cell division by bradykinin and des-Arg⁹-bradykinin. J. Biol. Chem. 259, 9263–9268.

Gowen, M. (ed.) (1992). "Cytokines and Bone Metabolism". CRC Press, Boca Raton.

Gowen, M. (1994). Cytokines and cellular interactions in the control of bone remodelling. In "Bone and Mineral Research" (eds J.N.M. Heersche and J.A. Kanis), pp. 77–114. Elsevier, Amsterdam.

Gowen, M., Nedwin, G.E. and Mundy, G.R. (1986). Preferential inhibition of cytokine-stimulated bone resorption by recombinant interferon-γ. J. Bone Min. Res. 1, 469–474.

Gustafson, G.T. and Lerner, U.H. (1984). Bradykinin stimulates bone resorption and lysosomal-enzyme release in cultured mouse calvaria. Biochem. J. 219, 329–332.

Gustafson, G.T., Ljunggren, Ö., Boonekamp, P. and Lerner, U. (1986). Stimulation of bone resorption in cultured mouse calvaria by Lys-bradykinin (kallidin), a potential mediator of bone resorption linking anaphylaxis processes to rarefying osteitis. Bone Miner. 1, 267–277.

Hall, J.M. (1992). Bradykinin receptors: pharmacological properties and biological roles. Pharmacol. Ther. 56, 131–190.

Heinegård, D. and Oldberg, Å. (1993). Glycosylated matrix proteins. In "Connective Tissue and Its Heritable Disorders" (eds. P.M. Royce and B. Steinmann), pp. 189–209. Wiley-Liss, New York.

Hess, J.F., Borkowski, J.A., Young, G.S., Strader, C.D. and Ransom, R.W. (1992). Cloning and pharmacological characterization of a human bradykinin (BK-2) receptor. Biochem. Biophys. Res. Commun. 184, 260–268.

Hirata, F. (1981). The regulation of lipomodulin, a phospholipase inhibitory protein, in rabbit neutrophils by phosphorylation. J. Biol. Chem. 256, 7730–7733.

Horton, J.E., Raisz, L.G., Simmons, H.A., Oppenheim, J.J. and Mergenhagen, S.E. (1972). Bone resorbing activity in supernatant fluid from cultured human peripheral blood leukocytes. Science, 177, 793–796.

Hultenby, K. (1994). Ultrastructural studies on sialoproteins and integrins in metaphyseal rat bone. Possible roles in cell–matrix interaction. Karolinska Institutet, thesis.

Jasani, M.K., Katori, M. and Lewis, G.P. (1969). Intracellular enzymes and kinin enzymes in synovial fluid in joint diseases. Ann. Rheum. Dis. 28, 497–512.

Kawaguchi, H., Raisz, L.G., Voznesensky, O.S., Alander, C.B., Hakeda, Y. and Pilbeam, C.C. (1994). Regulation of the two prostaglandin G/H synthases by parathyroid hormone, interleukin-1, cortisol and prostaglandin E₂ in cultured neonatal mouse calvariae. Endocrinology 135, 1157–1164.

Kellermeyer, R.W. and Graham, R.C., Jr (1968). Kinins – possible physiologic and pathologic roles in man. N. Engl. J. Med. 279, 859–866.

Lahaye, D.H.T.P., Afink, G.B., Bleijs, D.A., van Alewijk, D.C.G.J. and van Zoelen, E.J.J. (1994). Effect of bradykinin on loss of density-dependent growth inhibition of normal rat kidney cells. Cell. Mol. Biol. 40, 717–721.

Lanyon, L.E. (1992). Control of bone architecture by functional load bearing. J. Bone Miner. Res. 7, 369–375.

Lerner, U.H. (1991a). Bradykinin synergistically potentiates interleukin-1 induced bone resorption and prostanoid biosynthesis in neonatal mouse calvarial bones. Biochem. Biophys. Res. Commun. 175, 775–783.

Lerner, U.H. (1991b). Parathyroid hormone and transforming growth factor β synergistically stimulate formation of prostaglandin E₂ in neonatal mouse calvarial bones. Acta Physiol. Scand. 143, 133–134.

Lerner, U.H. (1992). Effects of kinins, thrombin and neuropeptides on bone. In "Cytokines and Bone Metabolism" (ed. M. Gowen), pp. 267–298. CRC Press, Boca Raton.

Lerner, U.H. (1994). Regulation of bone metabolism by the kallikrein–kinin system, the coagulation cascade, and the acute-phase reactants. Oral Surg. Oral Med. Oral Pathol. 78, 481–493.

Lerner, U.H. and Fröhlander, N. (1992). Haptoglobin-stimulated bone resorption in neonatal mouse calvarial bones in vitro. Arthr. Rheum. 35, 587–591.

Lerner, U.H. and Modéer, T. (1991). Bradykinin B₁ and B₂ receptor agonists synergistically potentiate interleukin-1-induced prostaglandin biosynthesis in human gingival fibroblasts. Inflammation 15, 427–436.

Lerner, U.H. and Ohlin, A. (1993). Tumor necrosis factors α and β can stimulate bone resorption in cultured mouse calvariae by a prostaglandin-independent mechanism. J. Bone Min. Res. 8, 147–155.

Lerner, U.H., Jones, I.L. and Gustafson, G.T. (1987a). Bradykinin, a new potential mediator of inflammatory-induced bone resorption. Arthr. Rheum. 30, 530–540.

Lerner, U.H., Ransjö, M. and Ljunggren, Ö. (1987b). Prostaglandin E₂ causes a transient inhibition of mineral mobilization, matrix degradation, and lysosomal enzyme release from mouse calvarial bones in vitro. Calcif. Tissue Int. 40, 323–331.

Lerner, U.H., Ransjö, M. and Ljunggren, Ö. (1989). Bradykinin stimulates production of prostaglandin E₂ and prostacyclin in murine osteoblasts. Bone Miner. 5, 139–154.

Lerner, U.H., Ljunggren, Ö., Ransjö, M., Klaushofer, K. and Peterlik, M. (1991a). Inhibitory effects of γ-interferon on bradykinin-induced bone resorption and prostaglandin formation in cultured mouse calvarial bones. Agents Actions 32, 305–311.

Lerner, U.H., Ljunggren, Ö., Dewhirst, F. and Boraschi, D. (1991b). Comparison of human interleukin-1β and its 163–171 peptide in bone resorption and the immune response. Cytokine 3, 141–148.

Lerner, U.H., Brunius, G. and Modéer, T. (1992). On the

signal transducing mechanisms involved in the synergistic interaction between interleukin-1 and bradykinin on prostaglandin biosynthesis in human gingival fibroblasts. Biosci. Rep. 12, 263–271.

Lewis, D.B., Liggitt, H.D., Effmann, E.L., Motley, S.T., Teitelbaum, S.L., Jepsen, K.J., Goldstein, S.A., Bonadio, J., Carpenter, J. and Perlmutter, R.M. (1993). Osteoporosis induced in mice by overproduction of interleukin 4. Proc. Natl Acad. Sci. USA 90, 11618–11622.

Ljunggren, Ö. and Lerner, U.H. (1988). Stimulation of bone resorption by Met-Lys-bradykinin. J. Periodont. Res. 23, 75–77.

Ljunggren, Ö. and Lerner, U.H. (1990). Evidence for BK_1 bradykinin-receptor-mediated prostaglandin formation in osteoblasts and subsequent enhancement of bone resorption. Br. J. Pharmacol. 101, 382–386.

Ljunggren, Ö., Rosenquist, J., Ransjö, M. and Lerner, U.H. (1990). Bradykinin stimulates prostaglandin E_2 formation in isolated human osteoblast-like cells. Biosci. Rep. 10, 121–126.

Ljunggren, Ö., Vavrek, R., Stewart, J.M. and Lerner, U.H. (1991a). Bradykinin induced burst of prostaglandin formation in osteoblasts is mediated via B_2 bradykinin receptors. J. Bone Min. Res. 6, 807–815.

Ljunggren, Ö., Johansson, H., Ljunghall, S., Fredholm, B.B. and Lerner, U.H. (1991b). Bradykinin induces formation of inositol phosphates and causes an increase in cytoplasmic Ca^{2+} in the osteoblastic cell lineage MC3T3-E1. J. Bone Min. Res. 6, 443–452.

Ljunggren, Ö., Fredholm, B.B., Nordstedt, C., Ljunghall, S. and Lerner, U.H. (1993). Role of protein kinase C in bradykinin-induced prostaglandin formation in osteoblasts. Eur. J. Pharmacol. 244, 111–117.

MacDonald, R.J., Margolius, H.S. and Erdös, E.G. (1988). Molecular biology of tissue kallikrein. Biochem. J. 253, 313–321.

Marceau, F. (1995). Kinin B_1 receptors: a review. Immunopharmacology 30, 1–26.

Marceau, F. and Tremblay, B. (1986). Mitogenic effect of bradykinin and of des-Arg^9-bradykinin on cultured fibroblasts. Life Sci. 39, 2351–2358.

Margolius, H.S. (1989). Tissue kallikreins and kinins, regulation and roles in hypertensive and diabetic diseases. Annu. Rev. Pharmacol. Toxicol. 29, 343–364.

Marklund, M., Lerner, U.H., Persson, M. and Ransjö, M. (1994). Bradykinin and thrombin stimulate release of arachidonic acid and formation of prostanoids in human periodontal ligament cells. Eur. J. Orthod. 16, 213–221.

McEachern, A., Shelton, E., Bhakta, S., Obernolte, R., Bach, C., Zuppan, P., Fujisaki, J., Aldrich, R. and Jarnagin, K. (1991). Expression cloning of a rat B_2 bradykinin receptor. Proc. Natl Acad. Sci. USA 88, 7724–7728.

Melmon, K.L., Webster, M.E., Goldfinger, S.E. and Seegmiller, J.E. (1967). Presence of a kinin in inflammatory synovial effusion from arthritides of varying aetiology. Arthr. Rheum. 10, 13–20.

Menke, J.G., Borkowski, J.A., Kierilo, K.K., MacNeil, T., Derrick, A.W., Schneck, K.A., Ransom, R.W., Strader, C.D., Linemeyer, D.L. and Hess, J.F. (1994). Expression cloning of a human B_1 bradykinin receptor. J. Biol. Chem. 269, 21583–21586.

Minne, H.W., Pfeilschifter, J., Scharla, S., Mutschelknauss, S., Schwartz, A., Krempien, B. and Ziegler, R. (1984).

Inflammation-mediated osteopenia in the rat, a new animal model for pathological loss of bone mass. Endocrinology 115, 50–54.

Miyaura, C., Onoe, Y., Ohta, H., Nozawa, S., Nagai, Y., Kaminakayashiki, T., Kudo, I. and Suda, T. (1995). Interleukin-13 inhibits bone resorption by suppressing cyclooxygenase-2 (COX2) mRNA expression and prostaglandin production in osteoblasts. J. Bone Min. Res. 10 (Suppl. 1), Abstr. 80.

Montgomery, E.H. and White, R.R. (1986). Kinin generation in the gingival inflammatory response to topically applied bacterial lipopolysaccharides. J. Dent. Res. 65, 113–117.

Motley, R.J., Clements, D., Evans, W.D., Crawley, E.O, Evans, C., Rhodes, J. and Compston, J.E. (1993). A four-year longitudinal study of bone loss in patients with inflammatory bowel disease. Bone Min. 23, 95–104.

Mundy, G.R. (1991). The effects of TGF-β on bone. In "Clinical Applications of TGF-β" (eds. G.R. Bock and J. Marsh), pp. 137–151. CIBA Foundation Symposium 157. John Wiley & Sons, London.

Mundy, G.R. and Roodman, G.D. (1987). Osteoclast ontogeny and function. In "Bone and Mineral Research" (ed. W.A. Peck), pp. 207–279. Elsevier, Amsterdam.

Partridge, N.C., Hillyard, C.J., Nolan, R.D. and Martin, T.J. (1985). Regulation of prostaglandin production by osteoblast-rich calvarial cells. Prostaglandins 30, 527–539.

Poli, V., Balena, R., Fattori, E., Markatos, A., Yamamoto, M., Tanaka, H., Ciliberto, G., Rodan, G.A. and Costantini, F. (1994). Interleukin-6 deficient mice are protected from bone loss caused by estrogen depletion. EMBO J. 13, 1189–1196.

Rahman, S., Bunning, R.A.D., Dobson, P.R.M., Evans, D.B., Chapman, K., Jones, T.H., Brown, B.L. and Russell, R.G.G. (1992). Bradykinin stimulates the production of prostaglandin E2 and interleukin-6 in human osteoblast-like cells. Biochim. Biophys. Acta 1135, 97–102.

Raisz, L.G., Dietrich, J.W., Simmons. H.A., Seyberth, H.W., Hubbard, W., and Oates, J.A. (1977). Effect of prostaglandin endoperoxides and metabolites on bone resorption in vitro. Nature (Lond.) 267, 532–534.

Regoli, D. and Barabé, J. (1980). Pharmacology of bradykinin and related kinins. Pharmacol. Rev. 32, 1–46.

Riancho, J.A., Zarrabeitia, M.T., Mundy, G.R., Yoneda, T. and González-Marcías, J. (1993). Effects of interleukin-4 on the formation of macrophages and osteoclast-like cells. J. Bone Min. Res. 8, 1337–1344.

Rodan, G.A. and Martin, T.J. (1981). Role of osteoblasts in hormonal control of bone resorption – a hypothesis. Calcif. Tissue Int. 33, 349–351.

Rosenquist, J.B., Ohlin, A. and Lerner, U.H. (1996). Studies on the role of prostaglandins and non-steroidal antiinflammatory drugs on cytokine-induced inhibition of the biosynthesis of bone matrix proteins. Inflamm. Res. 45, 457–463.

Santora, A.C., II (1989). Des-Arg^9-bradykinin stimulation of ROS 17/2.8 rat osteosarcoma adenosine 3′,5′-cyclic monophosphate content, receptor specificity. J. Bone Min. Res. 4 (Suppl. 1), S332 (abstr.).

Sawai, K., Niwa, S. and Katori, M. (1979). The significant reduction of high molecular weight kininogen in synovial fluid of patients with active rheumatoid arthritis. Adv. Exp. Med. Biol. 120B, 195–202.

Sharma, J.N. (1991). The role of the kallikrein–kinin system in joint inflammatory disease. Pharmacol. Res. 23, 105–112.

Sharma, J.N., Zeitlin, I.J., Brooks, P.M. and Dick, W.C. (1976). A novel relationship between plasma kininogen and rheumatoid disease. Agents Actions 6, 148–153.

Sharma, J.N., Zeitlin, I.J., Brooks, P.M., Buchanan, W.W. and Dick, W.C. (1980). The action of aspirin on plasma kininogen and other proteins in rheumatoid patients; relationship to disease activity. Clin. Exp. Pharmacol. Physiol. 7, 347–353.

Sharma, J.N., Zeitlin, I.J., Deodhar, S.D. and Buchanan, W.W. (1983). Detection of kallikrein-like activity in inflamed synovial tissue. Arch. Int. Pharmacodyn. Ther. 262, 279–286.

Stashenko, P., Dewhirst, F.E., Peros, W.J., Kent, R.L. and Ago, J.M. (1987). Synergistic interactions between interleukin-1, tumor necrosis factor and lymphotoxin in bone resorption. J. Immunol. 138, 1464–1468.

Stevens, P.A., Pyne, S., Grady, M. and Pyne, N.J. (1994). Bradykinin-dependent activation of adenylate cyclase activity and cyclic AMP accumulation in tracheal smooth muscle occurs via protein kinase C-dependent and -independent pathways. Biochem. J. 297, 233–239.

Straus, D.S. and Pang, K.J. (1984). Effects of bradykinin on DNA synthesis in resting NIL8 hamster cells and human fibroblasts. Exp. Cell Res. 151, 87–95.

Suda, T., Takahashi, N. and Martin, T.J. (1992). Modulation of osteoclast differentiation. Endocr. Rev. 13, 66–80.

Sundqvist, G. and Lerner, U.H. (1995). Bradykinin and thrombin synergistically potentiate interleukin-1 and tumor necrosis factor induced prostanoid biosynthesis in human dental pulp fibroblasts. Cytokine 8, 168–177.

Sundqvist, G., Rosenquist, J.B. and Lerner, U.H. (1995). Effects of bradykinin and thrombin on prostaglandin formation, cell proliferation and collagen biosynthesis in human dental-pulp fibroblasts. Arch. Oral Biol. 40, 247–256.

Suzuki, M., Ito, A., Mori, Y., Hayash, Y. and Matsuta, K. (1987). Kallikrein in synovial fluid with rheumatoid arthritis. Biochem. Med. Metab. Biol. 37, 177–183.

Takahashi, N., Mundy, G.R. and Roodman, G.D. (1986). Recombinant human interferon-γ inhibits formation of human osteoclast-like cells. J. Immunol. 137, 3544–3549.

Tamura, T., Udagawa, N., Takahashi, N., Miyaura, C., Tanaka, S., Yamada, Y., Koishihara, Y., Ohsugi, Y., Kumaki, K., Taga, T., Kishimoto, T. and Suda, T. (1993). Soluble interleukin-6 receptor triggers osteoclast formation by interleukin 6. Proc. Natl Acad. Sci. USA 90, 11924–11928.

Tashjian, A.H., Jr, Tice, J.E. and Sides, K. (1977). Biological activities of prostaglandin analogues and metabolites on bone in organ culture. Nature (Lond.) 266, 645–647.

Tiffany, C.W. and Burch, R.M. (1989). Bradykinin stimulates tumor necrosis factor and interleukin-1 release from macrophages. FEBS Lett. 247, 189–192.

Vaes, G. (1988). Cellular biology and biochemical mechanism of bone resorption. Clin. Orthopaed. 231, 239–271.

Watanabe, K., Tanaka, Y., Morimoto, I., Yahata, K., Zeki, K., Fujihira, T., Yamashita, U. and Eto, S. (1990). Interleukin-4 as a potent inhibitor of bone resorption. Biochem. Biophys. Res. Commun. 172, 1035–1041.

Wener, J.A., Gorton, S.J. and Raisz, L.G. (1972). Escape from inhibition of resorption in cultures of fetal bone treated with calcitonin and parathyroid hormone. Endocrinology 90, 752–759.

Worthy, K., Bond, P.A., Elson, C.J., Dieppe, P.A. and Bhoola, K.D. (1990a). The effect of α_2-macroglobulin-bound proteases on the measurement of tissue kallikrein in the synovial fluid of patients with RA and other arthritides. J. Physiol. 425, 66P.

Worthy, K., Figueroa, C.D., Dieppe, P.A. and Bhoola, K.D. (1990b). Kallikreins and kinins: mediators in inflammatory joint disease? Int. J. Exp. Pathol. 71, 587–601.

Yanaga, F., Hirata, M. and Koga, T. (1991). Evidence for coupling of bradykinin receptors to a guanine nucleotide binding protein to stimulate arachidonate liberation in the osteoblast-like cell line, MC3T3-E1. Biochem. Biophys. Acta 1094, 139–146.

Yoshida, H., Hayash, S.I., Kunisada, T., Ogawa, M., Nishaws, S., Okamura, H., Sudo, T., Chultz, L.D. and Nishikawa, L. (1990). The murine mutation osteopetrosis in the coding region of the macrophage colony stimulating factor gene. Nature (Lond.) 345, 442–444.

Yucel-Lindberg, T., Lerner, U.H. and Modéer, T. (1995). Effects and interactions of tumour necrosis factor α and bradykinin on interleukin-1 production in gingival fibroblasts. J. Periodont. Res. 30, 186–191.

Zhang, H., Gaginella, T.S., Chen, X. and Cornwell, D.G. (1991). Action of bradykinin at the cyclooxygenase step in prostanoid synthesis through the arachidonic acid cascade. Agents Actions 34, 397–404.

14. Bradykinin as an Inflammatory Mediator in the Urinary Tract

Carlo Alberto Maggi

1. Introduction

Bradykinin (BK) has multiple pro-inflammatory actions suggesting a pathophysiological role in disease. Distributed ubiquitously throughout the body, the cascade leading to production of BK and related kinins occurs in the interstitial spaces following activation of kinin-forming enzymes, the kallikreins (see Chapters 5 and 17 of this volume). When considering BK as a mediator of inflammation in the urinary tract, however, kinins that are present or formed in the urine may also be important in pathophysiology. As discussed below, the presence of kallikrein, kininogen and kinins in the urine raises the question, are conditions of altered permeability of the blood–urine barrier, formed by the epithelial layer lining the urinary tract (urothelium), relevant to the pathogenesis of inflammatory diseases? This chapter reviews the actions of BK as an inflammatory mediator in the urinary tract, the preclinical evidence suggesting the involvement of kinins as inflammatory mediators in the tract, and provides an outline of the possible importance of this mediator in diseases related to the urinary tract.

2. Bradykinin in the Urinary Tract and the Barrier Function of the Urothelium

The urine contains measurable levels of kallikrein, kininogen and BK. Renal kallikrein is produced by the connecting tubule cells in the distal nephrons, and its release is regulated by a number of factors including arginine vasopressin, activity of sympathetic nerves and K^+ ions (reviewed by Bhoola et al., 1992). Low molecular weight kininogen is expressed by principal cells in the distal nephron (Bhoola et al., 1992). The close proximity of the cells containing immunoreactive tissue kallikrein and kininogen, together with a concomitant release of the enzyme and its substrate, makes the formation of kinins in the lumen of the collecting duct inescapable, with the highest concentrations of kinins occurring in final portion of the distal nephron (Scicli et al., 1978). It is believed that renal kinins play a humoral role in the regulation of renal papillary blood flow (e.g., Carretero and Scicli, 1990), and in the control of electrolyte and water transport (reviewed by Coyne and Morrison, 1991). Moreover, kinins present in the urine may regulate renal function indirectly by activating reno-renal reflexes (see Section 3).

The concentrations of kinins in rat ureter urine are in the range of 20–30 ng/ml, and are influenced by the presence of kinin-degrading activity (Yamasu et al., 1989). In the human urine, Lys-BK, BK and Met-Lys-BK can be detected (7.6, 3.4 and 9.5 µg/24 h, respectively) (Hial et al., 1976). Shimomato et al. (1978) reported a concentration of 38 ng/ml of BK-like immunoreactive material in human urine.

Kinins present in the urine may be considered as waste products, which have subserved their physiological role(s) in regulating kidney function. Whether such powerful mediators of pain and inflammation may play another role in the urinary tract is yet to be ascertained. The access of kinins (and other solutes present into the urine) to the wall of the urinary tract (renal pelvis, ureter, urinary bladder and urethra) is prevented by the barrier function of the urothelium. The barrier is ensured by the transitional epithelium lining the urinary tract and by the mucus (glycosaminoglycan) layer covering the apical cells of the urothelium. The glycosaminoglycan layer also inhibits bacterial adherence (Parsons et al., 1980). The integrity of the glycosaminoglycan layer is thought to be crucial for preventing backward diffusion of solutes and its alteration may have a pathophysiological role in interstitial cystitis (see Section 5). Although the apical cells of the mammalian urothelium can actively transport sodium in vitro, it is doubtful whether this occurs in vivo, as the normal urothelium is regarded as being impermeable to the substance(s), which are physi-ologically present in the urine (reviewed by Lewis, 1986; Steers, 1992). Owing to their relatively large size, it appears unlikely that kinins can back diffuse across the urothelium. Furthermore, the mucosal layer is endowed with powerful kinin-degrading activity (Maggi et al., 1989). The impermeability of the urothelium is not absolute, however, and some drugs administered intra-vesically have a sizeable, albeit limited, systemic absorption (Massad et al., 1992). In addition, the permeability of the urothelium is increased and/or modulated by a number of factors, particularly over-distension and inflammation (Lunglmayr and Czech, 1971; Fellows and Marshall, 1972; Au et al., 1991; Tammela et al., 1993). In addition, mechanical trauma, such as that induced by the passage of stones through the ureter lumen, or by catheterization of the urinary bladder are obvious conditions for inducing damage and increasing the permeability of the urothelium. These considerations provide a ground to speculate that, under certain circumstances, kinins present in the urine may gain access to target structures of the urinary tract which express BK receptors. Thus, a possible role of urinary kinins in the pathogenesis of urinary tract diseases can be anticipated (see Section 5), in addition to the action of

kinins generated in the interstitial spaces during inflammation/tissue injury.

3. Kinin Receptors and Actions in the Urinary Tract

Both B_1 and B_2 receptors, as originally defined by Regoli and Barabé (1980), have been identified in the mammalian urinary tract. B_2 receptors are expressed constitutively on a number of cell types, whereas B_1 receptor expression appears to be inducible following prolonged inflammation or tissue injury (Regoli and Barabé, 1980) (see Chapter 8).

Manning and Snyder (1989) reported a detailed autoradiographic study on the distribution of BK receptors in guinea-pig ureter and urinary bladder. In both of these organs, the density of BK receptors is maximal in the lamina propria, especially in the immediate subepithelial layer, while little if any labeling of smooth muscle was reported. The reported localization is consistent with a direct action of BK on primary afferent nerves (see Section 3.3), which form a dense sub- and intra-epithelial plexus in the urinary tract, while the absence of smooth muscle labeling is puzzling, since part of the contractile action of BK on the guinea-pig isolated detrusor muscle appears to be direct (Maggi et al., 1989).

A number of studies have examined the actions of kinins in the urinary tract but, in this chapter, I will concentrate on those which can be considered as specific to this tissue. Nevertheless, it must be remembered that the actions of BK which involve the cellular components of inflammation (reviewed in Marceau et al., 1983) are also relevant for its proposed role as mediator of inflammation in the urinary tract.

3.1 KININS AND SMOOTH MUSCLE TONE

Kinins are potent regulators of smooth muscle tone throughout the urinary tract. Their actions are excitatory or inhibitory depending on experimental factors, and involve both an apparently direct action on smooth muscle and indirect components through the release or production of other mediators.

3.1.1 Kinins and Local Regulation of Pyeloureteral Motility

A few studies have addressed the local motor responses produced by BK on ureter motility. In the guinea-pig isolated renal pelvis, which contracts spontaneously owing to the generation of pacemaker potentials in its proximal region (Zawalinski et al., 1975), BK induces concentration-dependent potentiation of spontaneous activity via binding to B_2 receptors (Maggi et al., 1992a).

BK also induces the release of neuropeptides from capsaicin-sensitive primary afferent nerves, although this response does not apparently contribute at a major extent to its local motor effects (Maggi et al., 1992a).

In dogs and rats, systemic administration of BK stimulates ureter peristalsis and can induce retrograde pressure waves (antiperistalsis) (Catacutan-Labay and Boyarsky, 1966). The ability of the urinary kallikrein–kinin system to stimulate latent pacemakers in the ureter was suggested indirectly by observations that application of dialyzed rat urine, BK or purified kallikrein all induced phasic contractions of the rat isolated ureter (Marin-Grez et al., 1983). A kallikrein antiserum prevented the response to dialyzed rat urine, implying that urinary kallikrein was responsible for this effect, presumably mediated by newly formed kinins.

In the guinea-pig isolated ureter, BK transiently stimulates phasic rhythmic contractions via B_2 receptors. This response was unaffected by indomethacin or capsaicin pretreatment in vitro, suggesting a direct action on smooth muscle cells (Maggi et al., 1992a). In sucrose gap experiments, superfusion of the guinea-pig ureter with BK triggers action potentials and accompanying phasic contractions, apparently by stimulating latent pacemakers in the smooth muscle, a response that is similar to that produced by direct electrical stimulation of the smooth muscle (C.A. Maggi and P. Santicioli, unpublished observations). Latent pacemakers in the smooth muscle may be important in giving rise to antiperistaltic waves by inducing a backward movement of urine toward the renal pelvis/kidney, which is held as a pathogenic mechanism in ascending infections of the urinary tract/pyelonephritis. When using a three-chambered organ bath, which permits a separate perfusion and restricted drug administration to the renal, middle and vesical regions of the guinea-pig ureter (Meini et al., 1995), the application of BK to the middle compartment produced propagated contractions of both the renal and vesical ends of the ureter, thus demonstrating the ability of BK to activate both ortho- and antidromic peristaltic waves (S. Meini and C.A. Maggi, unpublished observations).

3.1.2   Kinins and Local Regulation of Urinary Bladder Motility

BK produces contraction of the isolated urinary bladder from various species including rats (Marceau et al., 1980; Acevedo et al., 1990; Watts and Cohen, 1991), guinea pigs (Falconieri-Erspamer et al., 1973; Maggi et al., 1989), rabbits (Downie and Rouffignac, 1981; Nakahata et al., 1987; Butt et al., 1995), hamsters (Rhaleb et al., 1991; Pinna et al., 1992), dogs (Steidle et al., 1990) and humans (Andersson et al., 1992). B_2 receptors mediate responses of the rat, guinea-pig, hamster and rabbit bladder to kinins. B_1 receptors have been reported to mediate contraction of rat and rabbit bladder smooth muscle (Marceau et al., 1980; Butt et al., 1995; Roslan et al., 1995), with these

responses exhibiting time-dependent upregulation after tissue setup, and magnification of the responses after induction of cystitis. For example, the inflamed rat isolated urinary bladder (48 h after local instillation of 25% turpentine) exhibits immediate and robust contractions in response to the B_1 receptor agonists, desArg⁹-BK and Lys-desArg⁹-BK, which are blocked by a B_1 receptor antagonist (Roslan et al., 1995). In the noninflamed rat bladder, however, the response to B_1 receptor agonists is weak at the beginning of experiments, but increases in magnitude progressively with time for > 7 h after setup of tissues. The latter phenomenon has been demonstrated in many smooth muscle bioassays for B_1 receptors and may be indicative of de novo synthesis of B_1 receptors after tissue manipulation or injury (see Chapter 8). Roslan et al. (1995) also reported that the time-dependent increase in responsiveness of the rat isolated bladder to B_1 receptor stimulation is reduced by cycloheximide and potentiated by interleukin-1β, suggesting the involvement of this cytokine in the synthesis and/or expression of B_1 receptors.

In general, the contractile response of the mammalian isolated bladder smooth muscle to BK is strongly (hamster, rabbit, rat) or partially (guinea pig, human) inhibited by cyclooxygenase inhibitors, and is potentiated by inhibitors of angiotensin-converting enzyme or neutral endopeptidase. Removal of the mucosa greatly enhances the contraction to BK in the guinea-pig bladder, probably because of the epithelial peptidase activity (Maggi et al., 1989). On the other hand, the mucosal layer contributes to the contractile activity of BK in the diabetic rat isolated bladder, apparently being a source of contractile prostanoids (Pinna et al., 1992). The complexity of mechanisms initiated by B_2 receptor occupation was evidenced in the guinea-pig isolated bladder wherein, when the tone was elevated pharmacologically, application of BK produced a transient relaxation, as opposed to the sustained contraction produced in tissues at resting tone. The relaxant response to BK was blocked by apamin, indicating that Ca^{2+} mobilization and activation of small conductance Ca^{2+}-dependent K^+ channels are involved (Maggi et al., 1989). In rabbit bladder smooth muscle, contractile responses to B_1 receptor agonists are linked to stimulation of phosphatidylinositol turnover (Butt et al., 1995).

BK-induced bladder contractions are also evident in vivo. For example, in anesthetized rats intravenous or close intraarterial injection of BK cause bladder contraction. Similarly, topical application of BK to the serosal surface of the bladder induces smooth muscle contractions (Marceau et al., 1980; Maggi et al., 1993; Lecci et al., 1995). The noninflamed rat bladder in vivo responds to intravenous BK with a single phasic contraction, which is blocked selectively by a B_2 receptor antagonist, Hoe 140 (Maggi et al., 1993) (Fig. 14.1). Following induction of inflammation, a contractile response to a B_1 receptor agonist is apparent (Marceau et

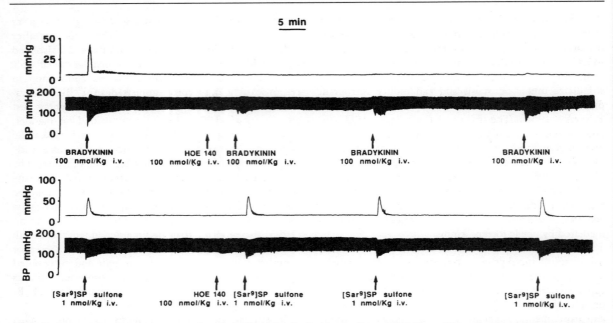

Figure 14.1   A B₂ receptor antagonist, Hoe 140, selectively blocks the contraction of the rat urinary bladder *in vivo* induced by intravenous administration of bradykinin (BK). Experiments were performed in urethane-anesthetized rats, simultaneously recording intravesical pressure (upper tracing in each panel) and blood pressure (BP, lower tracing in each panel). Either BK (100 nmol/kg) or a tachykinin NK₁ receptor agonist, (Sar⁹)Substance P (SP) sulfone (1 nmol/kg) were injected intravenously at regular intervals until reproducible responses were obtained. Either agonist produced a transient phasic increase of intravesical pressure and a transient hypotension. Hoe 140 (100 nmol/kg) was administered intravenously 5 min before the next challenge with the agonist. The upper panel shows that Hoe 140 induced a complete and long-lasting blockade of the bladder contraction in response to BK without affecting the response to the NK₁ receptor agonist. Also the hypotensive response to BK was inhibited by Hoe 140, although at a lower extent than the bladder response. Reproduced with permission from Maggi *et al.* (1993).

al., 1980). When applied topically to noninflamed urinary bladder of anesthetized rats, BK evokes concentration-dependent, complex motor responses consisting of a tonic, low-amplitude contraction on which a series of phasic high-amplitude pressure waves are superimposed (Lecci *et al.*, 1995). The second response is produced through the stimulation of a chemoceptive supraspinal vesico-vesical micturition reflex, and is dramatically reduced by systemic capsaicin pretreatment, indomethacin, bilateral removal of pelvic ganglia or intrathecal administration of tachykinin NK₁ receptor antagonists (Lecci *et al.*, 1995). The first response is likely produced via a direct action of BK on smooth muscle kinin receptor, corresponding to the BK-induced contraction observed in the rat isolated detrusor muscle. Both responses were abolished by Hoe 140, indicating involvement of B₂ receptors (Lecci *et al.*, 1995). From these experiments, one may speculate that locally generated kinins can produce motor disturbances in the regulation of the micturition reflex and, in particular, that they may activate chemoceptive vesical afferent fibers to produce the pain or discomfort experienced during cystitis.

3.2   KININS AND MICROVASCULAR PERMEABILITY

BK also increases microvascular permeability in the urinary bladder and ureter, leading to extravasation of plasma proteins (Saria *et al.*, 1983; Giuliani *et al.*, 1993). BK-induced plasma protein extravasation in the rat bladder is selectively blocked by Hoe 140 (Giuliani *et al.*, 1993) (Fig. 14.2). As discussed below, Hoe 140 inhibits the plasma-protein extravasation occurring during experimental cystitis in rats. At present, there is no information as to whether B₁ receptors are involved in kinin-induced microvascular permeability in the urinary tract.

3.3   KININS AND AFFERENT NERVES

Kinins are amongst the most potent endogenous pain-producing substances (Armstrong *et al.*, 1957; Juan and Lembeck, 1974) (see Chapter 9), and their ability to stimulate afferent nerve discharge *in vivo* is well-known (e.g., Beck and Handwerker, 1974; Mizumura *et al.*,

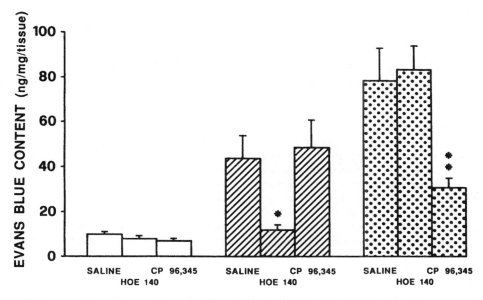

Figure 14.2   Plasma-protein extravasation, quantified by Evans-blue leakage, induced by intravenous bradykinin (BK) (1 μmol/kg i.v., ▨) or Substance P (SP) (3 nmol/kg i.v., ▧) in the urinary bladder of urethane-anesthetized rats. The increased amounts of extracted dye are assumed to reflect the amounts of plasma proteins extravasated following challenge with the agonist. Hoe 140 (100 nmol/kg i.v., 5 min before) abolished the extravasation induced by BK without affecting that induced by SP. An NK$_1$-receptor antagonist (±)-CP 96,345 (10 μmol/kg i.v., 5 min before) inhibited by about 70% the response to SP without affecting the response to BK. ☐, vehicle. Reproduced with permission from Giuliani *et al.* (1993).

1990). BK receptors are expressed by primary afferent neurons and have been localized on both their central and peripheral projections by autoradiography (Steranka *et al.*, 1988). This, coupled with the observations that dorsal root ganglion neurons in culture are excited by BK (Baccaglini and Hogan, 1983; Burgess *et al.*, 1989), provides the basis to speculate that peripherally generated kinins produce a direct stimulatory action on afferent nerves causing pain, activation of visceral reflexes and initiation of neurogenic inflammation (see below). There is evidence that stimulated synthesis of cyclooxygenase products is an important sensitizing mechanism for inducing afferent nerve discharge by kinins. Cyclooxygenase inhibitors greatly reduce this response to kinins, even in the urinary tract (Maggi *et al.*, 1993; Lecci *et al.*, 1995).

Szolcsanyi (1984) showed that BK excites two subpopulations of afferent nerves, R1 and R2, in the rat renal pelvis (Recordati *et al.*, 1980; Moss, 1982). R1 receptors have no resting discharge and do not respond to acute changes in renal arterial, venous or intrapelvic pressures. Rather, they are activated in a bursting pattern of discharge by renal ischemia such as that induced by occlusion of the renal artery. On the other hand, renal pelvis R2 receptors have a basal discharge and respond to backflow of urine, as well as by renal ischemia (Recordati *et al.*, 1980; Moss, 1982). Renal pelvis R2 receptors, but

not R1, are powerfully activated by BK, either injected intraarterially or within the renal pelvis (Szolcsanyi, 1984). Interestingly, an almost complete overlap was found in rat renal pelvis between the population of afferent nerves, which are stimulated by BK, and those which are excited by capsaicin (Szolcsanyi, 1984). In keeping with these observations, superfusion of guinea-pig renal pelvis with BK induces the release of calcitonin gene-related peptide (CGRP) from the peripheral endings of capsaicin-sensitive afferent neurons, and this effect is inhibited by indomethacin (Maggi *et al.*, 1992b).

The physiological significance of renal pelvis chemo-receptor stimulation by BK can be inferred from *in vivo* perfusion of the rat renal pelvis with BK causing increased ipsilateral afferent nerve discharge, and activation of a contralateral reflex increase in urine flow rate and urinary sodium excretion (Kopp and Smith, 1993). Again, these responses to BK were inhibited by indomethacin and mimicked by renal pelvis perfusion with capsaicin (Kopp and Smith, 1993). Interestingly, BK infused into the renal pelvis did not affect systemic cardiovascular parameters, unless administered at high doses. Conversely, BK administered via the renal artery elicits a pressor reflex response (Smits and Brody, 1984). Thus, it appears that different sets of afferent nerves (or different reflex responses encoded by afferent nerves distributing at different peripheral locations) are elicited by BK in the kidney. Intrarenal BK excites the firing of spinoreticular cells in the spinal cord (Ammons, 1988), and which may be involved in relaying renal pain to supraspinal centers and mediate the above mentioned pressor reflex response. The same pathway could also activate the firing of supraoptic hypothalamic nerves (Day and Ciriello, 1987), causing increased vasopressin secretion (Caverson and Ciriello, 1987).

BK-induced, capsaicin-sensitive afferent nerve stimulation has been demonstrated in the ureter and urinary bladder, with simultaneous production of afferent nerve discharge and local release of sensory neuropeptides. Cervero and Sann (1989) showed that a minority of afferent nerves (U1 units, 9% of all mechanosensitive units) in guinea-pig ureter respond to increases in intraluminal pressure, which are within the range of

pressures attained during physiological ureter peristalsis. On the other hand, the great majority of mechanosensitive units (U2 units, 89% of all mechanosensitive afferents) are only excited by mechanical stimuli, which largely exceed the values of intraureteral pressure occurring during peristalsis. U2 units are chemosensitive and excited by both BK and capsaicin, and it was proposed that U2 units are involved in signaling ureter pain (Cervero and Sann, 1989).

Close intra-arterial injection or topical application of BK activates the discharge of bladder sensory nerves (Janig and Morrison, 1986). As discussed above, topical application of capsaicin to the serosal surface of the rat urinary bladder *in vivo* activates a chemoceptive micturition reflex (Lecci *et al.*, 1995). Although the concomitant direct contraction of bladder muscle may be involved in reinforcing afferent discharge, by stimulating

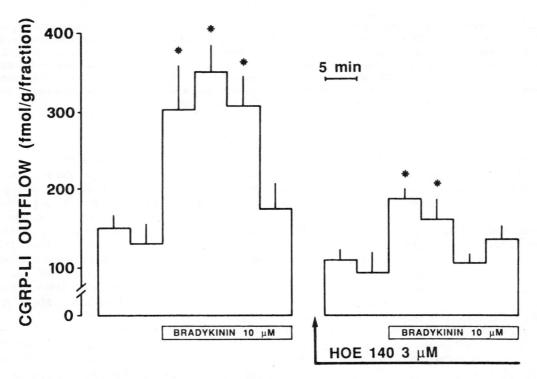

Figure 14.3   Release of calcitonin gene-related peptide-like immunoreactivity (CGRP-LI) from the superfused rat isolated urinary bladder by bradykinin (BK) in the absence and presence of Hoe 140. The urinary bladder was excised from male rats, trimmed and superfused with Krebs' solution containing thiorphan (10 μM). After equilibration, two basal fractions (5 min each) were collected and the perfusion medium was then changed to one containing BK (10 μM), which was left in contact for 20 min. In parallel experiments (right panel) Hoe 140 (3 μM) was present in the perfusion medium from the beginning of the experiment. CGRP-LI in the superfusate was determined by radioimmunoassay. Total evoked release was calculated as the difference between CGRP-LI outflow detected in the presence of BK and that detected in basal fractions. All values are mean ± s.e. mean of four experiments. The asterisks indicate significant increases of CGRP-LI outflow above baseline. Total evoked release of CGRP-LI by BK averaged 436 ± 85 fmol/g of wet weight in controls (calculated from values in left panel), which was reduced by about 50% in the presence of Hoe 140 (221 ± 67 fmol/g of wet weight, $n = 4$, $P < 0.05$ vs controls). Reproduced with permission from Maggi *et al.* (1993).

mechanoreceptors in series with muscle fibers, the chemoceptive micturition reflex activated by BK is most likely linked to a direct stimulatory effect on the capsaicin-sensitive primary afferents. Indeed, capsaicin treatment abolished the chemoceptive reflex response to BK without affecting the local contractile response, and activation of the micturition reflex by distention of the urinary bladder is only partially affected by capsaicin pretreatment. Neurochemical evidence for sensory nerve stimulation by BK has been obtained in rat urinary bladder (Maggi et al., 1993). Application of BK induces marked release of CGRP-like immunoreactivity from this tissue (Fig. 14.3), the neuropeptide originating entirely from capsaicin-sensitive afferent nerve endings. The effect of BK is antagonized by Hoe 140, indicating the involvement of B_2 receptors (Fig. 14.3). Therefore, B_2 receptors are clearly capable of activating both the sensory and "efferent" functions of capsaicin-sensitive afferents in the urinary bladder, thus mimicking the main features of cystitis.

In summary, kinins exert a spectrum of actions in the urinary tract, spanning from local changes in muscle tone, increased vascular permeability, and stimulation of afferent nerves to produce pain and activating reflexes. Constitutively expressed B_2 receptors appear to be involved in all these effects, as judged by the inhibitory actions of Hoe 140. De novo synthesis and expression of B_1 receptors is involved in local changes in muscle tone after inflammation or injury, although the roles of B_1 receptors in microvascular permeability changes and afferent nerve discharge in the urinary tract are unknown. Many effects of BK, acting via B_2 receptor stimulation, are modulated by endogenous prostanoids. Taken together, these findings provide grounds to speculate that kinins, generated in the interstitial spaces during inflammation, or gaining access to the urinary tract wall owing to back diffusion from the urine, can play a role as inflammatory mediators in the urinary tract.

4. Evidence that Kinins are Inflammatory Mediators in the Urinary Tract

The evidence for an active involvement of kinins in inflammatory conditions of the lower urinary tract originates from studies investigating the effect a B_2 receptor antagonist in rat models of cystitis (Giuliani et al., 1993; Maggi et al., 1993; Ahluwalia et al., 1994). In these studies two models of chemically evoked cystitis were employed, acute cystitis induced by instillation of 30% xylene in silicone oil into the urinary bladder of female rats (15–60 min before study), or a subacute cystitis, induced by the intraperitoneal administration of cyclophosphamide in male rats (150 mg/kg, 2–48 h

prior to study). The first model is assumed to originate from the irritant and tissue-damaging action of xylene, following its back diffusion from the lumen into the bladder wall (Maggi et al., 1988). In the second model, which mimics the hemorrhagic cystitis in patients receiving cyclophosphamide, the biotransformation of cyclophosphamide into its active metabolite(s) and the irritant, acrolein, is excreted into the urine, producing cystitis via back diffusion into the bladder wall (Phillips et al., 1961; Cox, 1979; Grinberg-Funes et al., 1990; Maggi et al., 1992b).

In both models, the marked increase in vascular permeability is conveniently quantified by measuring the tissue content of Evans blue bound to extravasated plasma proteins. In parallel to the local inflammatory response, a detrusor hyperreflexia develops, which can be quantified by cystometry as a reduced volume threshold for eliciting the micturition reflex (reduced bladder capacity). The latter effect can be quantified further during nonstop cystometric recordings, as an increased frequency of micturition, thus mimicking the symptoms of urinary frequency experienced by patients with cystitis (see below).

With regard to plasma protein extravasation, Hoe 140 reduced by about 50% the inflammatory response produced by intravesical xylene instillation in female rats (Giuliani et al., 1993) (Fig. 14.4) and that induced by cyclophosphamide administration in male rats (Ahluwalia et al., 1994). At doses effective in reducing cystitis-induced plasma-protein extravasation, Hoe 140 selectively blocked the increase in microvascular permeability induced by BK without affecting the response to other agents, providing evidence for the involvement of endogenous kinins in experimental cystitis. In addition, in both models, techykinin NK_1 receptor antagonists are effective. Thus, in the xylene model, the inhibitory effect of Hoe 140 and that of an NK_1-receptor antagonist on plasma-protein extravasation are additive, the combined blockade of B_2 and NK_1 receptors yielding a total abolition of the plasma extravasation (Giuliani et al., 1993) (Fig. 14.4). In the cyclophosphamide model of cystitis, blockade of NK_1 receptors inhibited plasma protein extravasation to the same degree as Hoe 140, but a combination of the two antagonists was not additive, indicating a common target of action (Ahluwalia et al., 1994). Moreover, the catheterization required for xylene instillation into the female rat bladder produced a moderate degree of plasma-protein extravasation of its own (Giuliani et al., 1993). In these animals, either Hoe 140 or NK_1 receptor blockade abolished the evoked extravasation (Fig. 14.5).

Overall, it appears that kinins are generated during mechanical or chemical irritation of the bladder to produce a local inflammatory response via B_2 receptor stimulation. In different models of cystitis and/or for different degrees of evoked plasma protein extravasation, the effects of Hoe 140 can either be additive with

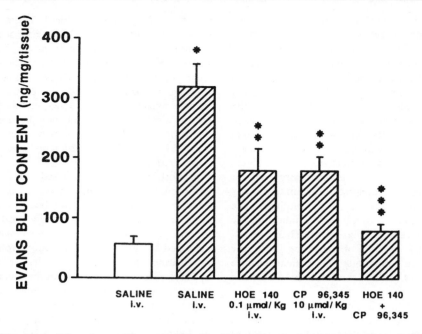

Figure 14.4   Plasma-protein extravasation, quantified by Evans-blue leakage, induced by intravesical instillation of xylene (30%, ▨), in silicone oil (▫), in the urinary bladder of urethane-anesthetized rats. The increased amounts of extracted dye are assumed to reflect the amounts of plasma proteins extravasated following instillation of the irritant. Either Hoe 140 (100 nmol/kg intravenously, 5 min before) or an NK₁-receptor antagonist (±)-CP 96,345 (10 μmol/kg intravenously, 5 min before) inhibited by about 50% the response to xylene and their combined administration eliminated all the response to the irritant. *Significantly different from xylene vehicle (silicone oil), $P < 0.05$. **Significantly different from xylene, $P < 0.05$. ***Significantly different from Hoe 140 and (±)-CP96,345 alone, $P < 0.05$. Reproduced with permission from Giuliani *et al.* (1993).

antagonism at tachykinin NK_1 receptors, or the two systems operate in series to produce an inflammatory response. These results are consistent with a model whereby the production of moderate amounts of kinins causes extravasation indirectly, by activating tachykinin release from sensory nerve endings. Depending upon the stimulus, and possibly upon different amounts or time-courses of kinin production, a direct action of kinins on microvascular permeability also becomes evident. Although this experimental model requires further evaluation, it appears that stimuli producing a local inflammatory response in the bladder wall cause production of kinins, which act on B_2 receptors to produce their effects.

In both the xylene and the cyclophosphamide models of cystitis, a detrusor hyperreflexia develops, characterized by a reduction in bladder capacity and a markedly increased frequency of urination during continuous infusion of saline into the urinary bladder (Maggi *et al.*, 1988, 1992b, 1993) (Fig. 14.6). At doses effective in blocking the bladder contractions in response to intravenous BK in anesthetized rats, Hoe 140 slightly increased the bladder capacity and decreased the micturition frequency in cyclophosphamide-treated rats (Fig. 14.6), without significantly affecting these same

urodynamic parameters in xylene-treated rats (Maggi *et al.*, 1993). It is interesting to note that detrusor hyperreflexia induced by cyclophosphamide is abolished by systemic capsaicin pretreatment of adult rats, while that induced by xylene is unaffected (Maggi *et al.*, 1988, 1992b). As BK produces a B_2 receptor-mediated release of CGRP from capsaicin-sensitive afferents in the rat urinary bladder (Maggi *et al.*, 1993), it appears that different sets of afferent nerves may be involved in the hyperreflexia induced by different irritants, and that the effects of Hoe 140 in cyclophosphamide-induced detrusor hyperreflexia are due to a stimulant action of endogenous kinins on a subset of urinary bladder afferents which are sensitive to capsaicin desensitization during adulthood [see Maggi *et al.* (1993) for further discussion on this point].

In summarizing this section, the available evidence indicates that endogenous kinins, acting via B_2 receptors, are involved in the local inflammatory response and detrusor hyperreflexia in rat models of irritant-induced acute or subacute cystitis. In both cases, the stimulation of afferent discharge and a local neuropeptide release from capsaicin-sensitive afferents seems central to the action of kinins. As mentioned earlier, there is evidence that induction of cystitis produces an upregulation of B_1 receptors mediating bladder contraction.

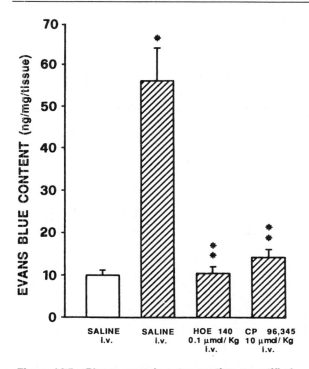

Figure 14.5   Plasma-protein extravasation, quantified by Evans-blue leakage, induced by catheterization [intravesical instillation of the biologically inert vehicle (▢), silicone oil (▨)] in the urinary bladder of urethane-anesthetized rats. The increased amounts of extracted dye are assumed to reflect the amounts of plasma proteins extravasated following catheterization. Either Hoe 140 (100 nmol/kg intravenously, 5 min before) or an NK_1-receptor antagonist ($\pm$)-CP 96,345 (10 μmol/kg intravenously, 5 min before) almost completely eliminated the response to catheterization. *Significantly different from noncatheterized animals, $P < 0.05$. **Significantly different from catheterized animals, $P < 0.05$. Reproduced with permission from Giuliani et al. (1993).

5.   Clinical Implications

The evidence reviewed in this chapter provides a background to speculate on the possible role of kinins as mediators of inflammation in the urinary tract. No clinical trials with BK receptor antagonists in this field have been reported yet. With regard to the upper urinary tract, the powerful stimulant action of kinins on afferent nerves suggests their possible contribution in producing pain during renal colics. In these cases, damage to the urothelium produced during passage of stones may enable urinary kinins to penetrate into the ureter wall to excite sensory nerves, producing pain and a local inflammatory response. As the afferent nerve-stimulating action of kinins is largely prostaglandin dependent (see

Section 3.3), it is interesting to note that cyclooxygenase inhibitors are effective in relieving pain in renal colic (Lundstrom et al., 1982; Oosterlink et al., 1990). Various mechanisms can be proposed for the powerful analgesic effect of these drugs in renal colics, including reduction in renal blood flow and intraureter pressure, by eliminating the facilitating role of endogenous prostanoids on pyeloureter motility (Perlmutter et al., 1993; Santicioli et al., 1995). On the other hand, since the stimulating action of kinins on primary afferent nerves is largely blocked by cyclooxygenase inhibitors, it is possible that blockade of kinin actions on ureter afferents is involved.

In the urinary bladder, it is evident that kinins, either generated in the bladder wall because of an ongoing inflammatory process, or penetrating the bladder wall because of increased permeability of the urothelium, may be involved in the pathogenesis of cystitis. By contracting the detrusor muscle and stimulating afferent nerves, kinins may be involved in both pain or discomfort accompanying cystitis, and in generating an increased frequency of urination. Thus, BK receptor antagonists could prove effective in relieving symptoms of bacterial cystitis, for example, while simultaneously addressing the specific causative agent by appropriate antimicrobial therapy. Apart from cystitis of bacterial origin, at least two other types of cystitis can be considered for clinical evaluation of BK receptor antagonists. Based on the preclinical data described in Section 4, cystitis developing in patients receiving cyclophosphamide therapy could be considered as potentially amenable to amelioration by treatment with kinin receptor antagonists. This side effect of the cytotoxic drug are observed in up to 70% of patients receiving cyclophosphamide and can be severe enough to require interruption of therapy (Foad and Hess, 1976; Stillwell and Benson, 1988).

A second condition to be considered is interstitial cystitis (reviewed by Koziol et al., 1993; Parsons, 1990). This condition is a chronic, aseptic inflammation producing pain in the region of the bladder and pelvis musculature and variable sensory-motor dysfunctions of micturition, and for which no effective therapy is known. The pathogenesis of interstitial cystitis is poorly understood, although there is evidence for a defect of the barrier function of the urothelium in these patients (Parsons et al., 1991; Moskowitz et al., 1994). Because of the defective barrier function, it is proposed that small molecules present into the urine could leak into the bladder interstitium to produce inflammation and irritant voiding symptoms (Parsons, 1990). Although this pathogenic mechanism is not universally accepted as a causative factor of interstitial cystitis (e.g., Chelsky et al., 1994), a role for kinins present in the urine may be postulated, and the hypothesis could be verified by assessing the effects of kinin receptor antagonists in these patients.

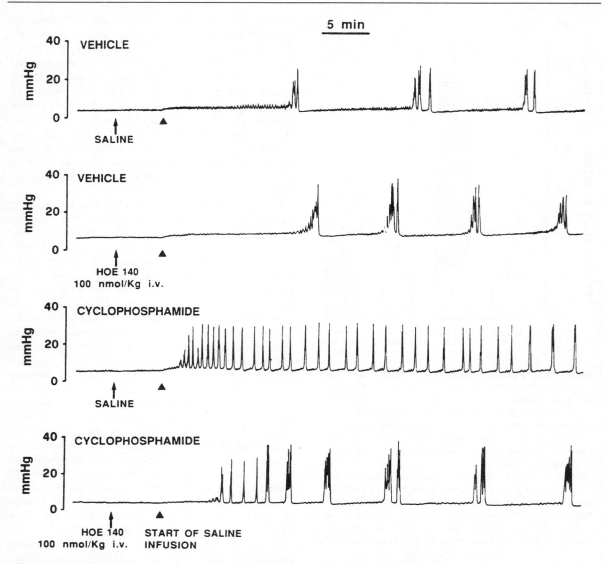

Figure 14.6 Cystometric recordings which illustrate the effect of Hoe 140 (100 nmol/kg i.v.) on the response of the rat urinary bladder *in vivo* (urethane anesthesia) to saline filling (in each tracing the start of saline infusion is marked by triangles). Hoe 140 or saline were injected intravenously 5 min before beginning of cystometric recordings in either vehicle- (upper panels) or cyclophosphamide-treated rats (lower panels). Saline was infused into the urinary bladder at a rate of 0.046 ml/min for 60 min. In each tracing, the initial phase of saline filling was not accompanied by any significant increase of intraluminal pressure, corresponding to the filling phase of the cystometric recording. Filling progressed until a threshold value of saline distention (bladder capacity) was reached, at which time the micturition reflex was triggered. The reflex is evident in tracing as brisk elevation of intravesical pressure, during which part of the infused volume was expelled. During nonstop filling of the urinary bladder, a period of quiescence follows each voiding, until micturition threshold is attained again and a new voiding occurs. Three main parameters were calculated: (1) bladder capacity as the infused volume to trigger the first micturition reflex; (2) the amplitude of reflex bladder contraction; and (3) the number of voiding acts recorded over the 60 min infusion period (micturition frequency). Previous administration of cyclophosphamide induces cystitis, evidenced at cystometry as a detrusor hyperreflexia (reduced bladder capacity and increased micturition frequency). In this model of cystitis the detrusor hyperreflexia induced by cyclophosphamide is abolished by systemic capsaicin pretreatment (Maggi *et al.*, 1992b) indicating that capsaicin-sensitive primary afferent neurons play a central role in its genesis. Hoe 140 (100 nmol/kg i.v.) did not affect urodynamic parameters in normal rats, but increased bladder capacity and reduced micturition frequency in cyclophosphamide-treated rats, indicating the involvement of endogenous kinins, via B_2 receptors in the induced detrusor hyperreflexia. Reproduced with permission from Maggi *et al.* (1993).

6. References

Acevedo, C.G., Lewin, J., Contreras, E. and Huidobro-Toro, J.P. (1990). Bradykinin facilitates the purinergic motor component of the rat bladder neurotransmission. Neurosci. Lett. 113, 227–232.

Ahluwalia, A., Maggi, C.A., Santicioli, P., Lecci, A. and Giuliani, S. (1994). Characterization of the capsaicin-sensitive component of cyclophosphamide-induced inflammation in the rat urinary bladder. Br. J. Pharmacol. 111, 1017–1022.

Ammons, W.S. (1988). Spinoreticular cell responses to intrarenal injections of bradykinin. Am. J. Physiol. 255, R994–R1001.

Andersson, K.E., Hedlund, H. and Stahl, M. (1992). Contractions induced by angiotensin I, angiotensin II and bradykinin in isolated smooth muscle from human detrusor. Acta Physiol. Scand. 145, 253–259.

Armstrong, D., Jepson, J.B., Keele, C.A. and Stewart, J.W. (1957). Pain-producing substance in human inflammatory exudates and plasma. J. Physiol. 135, 350–370.

Au, J.L.S., Dalton, J.T. and Wientjes, M.G. (1991). Evidence of significant absorption of sodium salicylate from urinary bladder of rats. J. Pharmacol. Exp. Ther. 258, 357–364.

Baccaglini, P.T. and Hogan, P.G. (1983). Some rat sensory neurons in culture express characteristics of differentiated pain sensory cells. Proc. Natl Acad. Sci. USA 80, 594–598.

Beck, P.W. and Handwerker, H.O. (1974). Bradykinin and serotonin effects on various types of cutaneous nerve fibres. Pflügers Arch. 347, 209–222.

Bhoola, K.D., Figueroa, C.D. and Worthy, K. (1992). Bioregulation of kinins: kallikreins, kininogens and kininases. Pharmacol. Rev. 44, 4–80.

Burgess, G.M., Mullaney, I., McNeill, M., Dunn, P.M. and Rang, H.P. (1989). Second messengers involved in the mechanism of action of bradykinin in sensory neurons in culture. J. Neurosci. 9, 3314–3325.

Butt, S.K., Dawson, L.G. and Hall, J.M. (1995). Bradykinin B_1 receptors in the rabbit urinary bladder: induction of responses smooth muscle contraction and phosphatidylinositol hydrolysis. Br. J. Pharmacol. 114, 612–617.

Carretero, O.A. and Scicli, A.G. (1990). Kinins as regulators of blood flow and blood pressure. In "Hypertension: Pathophysiology, Diagnosis and Management" (eds. J.H. Laragh and B.M. Breener), pp. 805–817. Raven Press, New York.

Catacutan-Labay, and Boyarsky, S. (1966). Bradykinin: effect on ureteral peristalsis. Science 151, 78–79.

Caverson, M.M. and Ciriello, J. (1987). Effect of stimulation of afferent renal nerves on plasma levels of vasopressin. Am. J. Physiol. 252, R801–R807.

Cervero, F. and Sann, H. (1989). Mechanically evoked responses of afferent fibres innervating the guinea-pig ureter: an in vitro study. J. Physiol. 412, 245–266.

Chelsky, M.J., Rosen, S.I., Knight, L.C., Maurer, A.H., Hanno, P.M. and Ruggieri, M.R. (1994). Bladder permeability in interstitial cystitis is similar to that of normal volunteers: direct measurements by transvesical absorption of ^{99m}technetium-diethylenetriaminepentaacetic acid. J. Urol. 151, 346–349.

Cox, P.J. (1979). Cyclophosphamide cystitis – identification of acrolein as the causative agent. Biochem. Pharmacol. 28, 2045–2049.

Coyne, D.W. and Morrison, A.R. (1991). Kinins: biotransformation and cellular mechanisms of action. In "Hormones, Autacoids and the Kidney" (eds. S. Goldfarb and F. Ziyadeh), pp. 263–280. Churchill Livingstone, New York.

Day, T.A. and Ciriello, J. (1987). Effects of renal receptor activation on neurosecretory vasopressin cells. Am. J. Physiol. 253, R234–R241.

Downie, J.W. and Rouffignac, S. (1981). Response of rabbit detrusor muscle to bradykinin. Life Sci. 28, 603–608.

Falconieri-Erspamer, G., Negri, L. and Piccinelli, D. (1973). The use of preparations of urinary bladder smooth muscle for bioassay of and discrimination between polypeptides. Naunyn Schmiedeberg's Arch. Pharmacol. 279, 61–64.

Fellows, G.J. and Marshall, D.H. (1972). The permeability of human bladder epithelium to water and sodium. Inv. Urol. 9, 339–344.

Foad, B.S. and Hess, E.V. (1976). Urinary bladder complications with cyclophosphamide therapy. Arch. Int. Med. 136, 616–622.

Giuliani, S., Santicioli, P., Lippe, I.Th., Lecci, A. and Maggi, C.A. (1993). Effect of bradykinin and tachykinin receptor antagonist on xylene-induced cystitis in rats. J. Urol. 150, 1014–1017.

Grinberg-Funes, D.J., Sheldon, C. and Weiss, M. (1990). The use of prostaglandin $F_{2\alpha}$ for the prophylaxis of cyclophosphamide-induced cystitis in rats. J. Urol. 144, 1500–1504.

Hial, V., Keiser, H.R. and Pisano, J.J. (1976). Origin and content of methionyl-lysyl-bradykinin lysyl-bradykinin and bradykinin in human urine. Biochem. Pharmacol. 25, 2499–2503.

Janig, W. and Morrison, J.F.B. (1986). Functional properties of spinal visceral afferent supplying abdominal and pelvic organs, with special emphasis on visceral nociception. Prog. Brain Res. 67, 87–114.

Juan, H. and Lembeck, F. (1974). Action of peptides and other algesic agents on paravascular pain receptors of the isolated perfused rabbit ear. Naunyn Schmiedeberg's Arch. Pharmacol. 283, 151–164.

Kopp, U.C. and Smith, L.A. (1993). Role of prostaglandins in renal sensory receptor activation by substance P and bradykinin. Am. J. Physiol. 265, R544–R551.

Koziol, J.A., Clark, D.C., Gittes, R.F. and Tan, E.M. (1993). The natural history of interstitial cystitis: a survey of 374 patients. J. Urol. 149, 465–469.

Lecci, A., Giuliani, S., Meini, S. and Maggi, C.A. (1995). Pharmacological analysis of the local and reflex responses to bradykinin on rat urinary bladder motility in vivo. Br. J. Pharmacol. 114, 708–714.

Lewis, S.A. (1986). The mammalian urinary bladder: it's more than accommodating. News Physiol. Sci. 1, 61–65.

Lundstrom, S., Leissner, K.H., Hahlander, L.A. and Kral, J.G. (1982). Prostaglandin synthetase inhibition with diclofenac sodium in treatment of renal colic: comparison with use of narcotic analgesics. Lancet 1, 1096–1097.

Lunglmayr, G. and Czech, K. (1971). Absorption studies on intraluminal thiothepa for topical cytostatic treatment of low-stage bladder tumors. J. Urol. 106, 72–74.

Maggi, C.A., Abelli, L., Giuliani, S., Santicioli, P., Geppetti, P., Somma, V., Frilli, S. and Meli, A. (1988). The contribution of sensory nerves to xylene-induced cystitis in rats. Neuroscience 26, 709–723.

Maggi, C.A., Patacchini, R., Santicioli, P., Geppetti, P., Cecconi, R., Giuliani, S. and Meli, A. (1989). Multiple mechanisms in the motor responses of the guinea-pig isolated urinary bladder to bradykinin. Br. J. Pharmacol. 98, 619–629.

Maggi, C.A., Santicioli, P., Del Bianco, E. and Giuliani, S. (1992a). Local motor responses to bradykinin and bacterial chemotactic peptide formyl-methionyl-leucyl-phenylalanine (FMLP) in the guinea-pig isolated renal pelvis and ureter. J. Urol. 148, 1944–1950.

Maggi, C.A., Lecci, A., Santicioli, P., Del Bianco, E. and Giuliani, S. (1992b). Cyclophosphamide cystitis in rats: involvement of capsaicin-sensitive primary afferents. J. Autonom. Nerv. Sys. 38, 201–208.

Maggi, C.A., Santicioli, P., Del Bianco, E., Lecci, A. and Giuliani, S. (1993). Evidence for the involvement of bradykinin in chemically-evoked cystitis in anaesthetized rats. Naunyn Schmiedeberg's Arch. Pharmacol. 347, 432–437.

Manning, D.C. and Snyder, S.H. (1989). Bradykinin receptors localized by quantitative autoradiography in kidney, ureter and bladder. Am. J. Physiol. 256, F909–F915.

Marceau, F., Barabé, J., St Pierre, S. and Regoli, D. (1980). Kinin receptors in experimental inflammation. Can. J. Physiol. Pharmacol. 58, 536–542.

Marceau, F., Lussier, A., Regoli, D. and Giroud, J.P. (1983). Pharmacology of kinins: their relevance to tissue injury and inflammation. Gen. Pharmacol. 14, 209–229.

Marin-Grez, M., Speck, G., Hilgenfeldt, U. and Schaechtelin, G. (1983). Inhibition of the ureteral contractions by rat urine with kallikrein antibodies. Experientia 39, 360–362.

Massad, C.A., Kogan, B.A. and Trigo-Rocha, F.E. (1992). The pharmacokinetics of intravesical and oral oxybutynin chloride. J. Urol. 148, 595–597.

Meini, S., Santicioli, P. and Maggi, C.A. (1995). Propagation of impulses in the guinea-pig ureter and its blockade by CGRP. Naunyn Schmiedeberg's Arch. Pharmacol. 351, 79–86.

Mizumura, K., Minagawa, M., Tsujii, Y. and Kumazawa, T. (1990). The effects of bradykinin agonists and antagonists on visceral polymodal receptor activities. Pain 40, 221–227.

Moskowitz, M.O., Byrne, D.S., Callahan, H.J., Parsons, C.L., Valderrama, E. and Moldwin, R.M. (1994). Decreased expression of a glycoprotein component of bladder surface mucin (GP1) in interstitial cystitis. J. Urol. 151, 343–345.

Moss, N. (1982). Renal function and renal afferent and efferent activity. Am. J. Physiol. 243, F425–F433.

Nakahata, N., Ono, T. and Nakanishi, H. (1987). Contribution of prostaglandin E_2 to bradykinin-induced contraction in rabbit urinary detrusor. Jpn. J. Pharmacol. 43, 351–359.

Oosterlink, W., Philip, N.H., Charig, C., Gillies, G., Hetherington, J.W. and Lloyd, J. (1990). A double blind single dose comparison of intramuscular ketorolac tromethamine and pethidine in the treatment of renal colic. J. Clin. Pharmacol. 30, 336–342.

Parsons, C.L. (1990). Interstitial cystitis: clinical manifestations and diagnostic criteria in over 200 cases. Neurourol. Urodyn. 9, 241–250.

Parsons, C.L., Stauffer, C. and Schmidt, J.D. (1980). Bladder surface glycosaminoglycans: an efficient mechanism of environmental adaptation. Science 208, 605–607.

Parsons, C.L., Lilly, J.D. and Stein, P. (1991). Epithelial dysfunction in nonbacterial cystitis. J. Urol. 145, 732–735.

Perlmutter, A., Miller, L., Trimble, L.A., Marion, D.N., Vaughan, E.D. and Felsen, D. (1993). Toradol, an NSAID used for renal colic, decreases renal perfusion and ureteral pressure in a canine model of unilateral ureteral obstruction. J. Urol. 149, 926–930.

Phillips, F.S., Sternberg, S.S., Cronin, A.P. and Vidal, P.N. (1961). Cyclophosphamide and urinary bladder toxicity. Cancer Res. 21, 1577–1586.

Pinna, C., Caratozzolo, O. and Puglisi, L. (1992). A possible role for urinary bladder epithelium in bradykinin-induced contraction in diabetic rats. Eur. J. Pharmacol. 214, 143–148.

Recordati, G.M., Moss, N.G., Genovesi, S. and Rogenes, P.R. (1980). Renal receptors in the rat sensitive to alterations of their environment. Circ. Res. 46, 395–405.

Regoli, D. and Barabé, J. (1980). Pharmacology of bradykinin and related kinins. Pharmacol. Rev. 32, 1–46.

Rhaleb, N.E., Rouissi, N., Drapeau, G., Jukic, D. and Regoli, D. (1991). Characterization of bradykinin receptors in peripheral organs. Can. J. Physiol. Pharmacol. 69, 938–943.

Roslan, R., Campbell, E.A. and Dray, A. (1995). The induction of bradykinin B_1 receptors in the non-inflamed and inflamed rat urinary bladder. Br. J. Pharmacol. 114, 228P.

Santicioli, P., Carganico, G., Meini, S., Giuliani, S., Giachetti, A. and Maggi, C.A. (1995). Modulation by stereoselective inhibition of cyclooxygenase of electromechanical coupling in the guinea-pig isolated renal pelvis. Br. J. Pharmacol. 114, 1149–1158.

Saria, A., Lundberg, J.M., Skofitsch, G. and Lembeck, F. (1983). Vascular protein leakage in various tissues induced by substance P, capsaicin, bradykinin, serotonin, histamine and by antigen challenge. Naunyn Schmiedeberg's Arch. Pharmacol. 324, 212–218.

Scicli, A.G., Gandolfi, R. and Carretero, O.A. (1978). Site of formation of kinins in the dog nephron. Am. J. Physiol. 234, F36–F40.

Shimamoto, K., Ando, T., Nakao, T., Tanaka, S., Sakuma, M. and Miyahara, M. (1978). A sensitive radioimmunoassay method for urinary kallikrein in man. J. Lab. Clin. Invest. 91, 721–728.

Smits, J.F. and Brody, M.J. (1984). Activation of afferent renal nerves by intrarenal bradykinin in conscious rats. Am. J. Physiol. 247, R1003–R1008.

Steers, W.D. (1992). Physiology of the urinary bladder. In "Campbell's Urology" (eds. P.C. Walsh, A.B. Retik, T.A. Stamey and E.D. Vaughan), pp. 145–181. W.B. Saunders, Philadelphia.

Steidle, C.P., Cohen, M.L. and Neubauer, B.L. (1990). Bradykinin-induced contractions of canine prostate and bladder: effect of angiotensin converting enzyme inhibition. J. Urol. 144, 390–392.

Steranka, L.R., Manning, D.C., De Haas, C.J., Ferkany, J.W., Borowsky, S.A., Connor, J.R., Vavrek, R.J., Stewart, J.M. and Snyder, S.H. (1988). Bradykinin as a pain mediator: receptors are localized to sensory neurons and antagonists have analgesic actions. Proc. Natl Acad. Sci. USA 85, 3245–3249.

Stillwell, T.J. and Benson, R.C. (1988). Cyclophosphamide-induced hemorrhagic cystitis – a review of 100 patients. Cancer 61, 451–457.

Szolcsanyi, J. (1984). Capsaicin-sensitive chemoceptive neural system with dual sensory-efferent function. In "Antidromic Vasodilatation and Neurogenic Inflammation" (eds. L.A. Chahl, J. Szolcsanyi and F. Lembeck), pp. 27–52. Akademiai Kiado, Budapest.

Tammela, T., Wein, A.J., Monson, F.C. and Levin, R.M. (1993). Urothelial permeability of the isolated whole bladder. Neurourol. Urodyn. 12, 39–47.

Watts, S.W. and Cohen, M.L. (1991). Effect of bombesin, bradykinin, substance P and CGRP in prostate, bladder body and neck. Peptides 12, 1057–1062.

Yamasu, A., Oh-ishi, S., Hayashi, I., Hayashi, K., Hayashi, M.,

Yamaki, K., Nakano, T. and Sunahara, N. (1989). Differentiation of kinin fractions in ureter urine and bladder urine of normal and kininogen deficient rats. J. Pharmacobio. Dyn. 12, 287–292.

Zawalinski, V.C., Constantinou, C.E. and Burnstock, G. (1975). Ureteral pacemaker potentials recorded with the sucrose gap technique. Experientia S31, 931–933.

15. The Kallikrein–Kinin System in Asthma and Acute Respiratory Distress Syndrome

Stephen G. Farmer

1. Introduction

There is evidence that the kallikrein–kinin system is important in chronic asthma, as well as other manifestations of airway inflammation including allergic and viral rhinitis (see Chapters 16 and 17 of this volume). In addition, for many years, there have been indications that this cascade is important in the pathophysiology of acute lung injury, the acute respiratory distress syndrome (ARDS) and multiple organ failure, associated with conditions such as sepsis, pancreatitis, primary pneumonia or multiple trauma (see Chapters 10 and 12). The present chapter reviews evidence for involvement of kinins in asthma and ARDS.

2. Airway Kinin Receptors

Kinin receptors are expressed by virtually every mammalian cell type (Bhoola *et al.*, 1992; Farmer and Burch, 1992; Hall, 1992) (see Chapter 2) (Fig. 15.1). In addition to exerting direct actions, bradykinin (BK) stimulates the release of many endogenous substances commonly implicated in inflammatory lung and airway diseases. Furthermore, BK synergizes with various cytokines and may act *in vivo* to amplify ongoing inflammatory responses (Kimball and Fisher, 1988; Tiffany and Burch, 1989; Vandekerckhove *et al.*, 1991; Lerner *et al.*, 1992; Ferreira *et al.*, 1993; Tsukagoshi *et al.*, 1994; Amrani *et al.*, 1995; Paegelow *et al.*, 1995). BK is

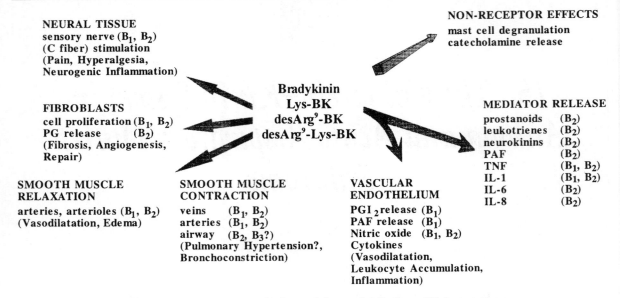

NEURAL TISSUE
sensory nerve (B_1, B_2)
(C fiber) stimulation
(Pain, Hyperalgesia,
Neurogenic Inflammation)

NON-RECEPTOR EFFECTS
mast cell degranulation
catecholamine release

FIBROBLASTS
cell proliferation (B_1, B_2)
PG release   (B_2)
(Fibrosis, Angiogenesis,
Repair)

Bradykinin
Lys-BK
desArg^9-BK
desArg^9-Lys-BK

MEDIATOR RELEASE
prostanoids   (B_2)
leukotrienes   (B_2)
neurokinins   (B_2)
PAF   (B_2)
TNF   (B_1, B_2)
IL-1   (B_1, B_2)
IL-6   (B_2)
IL-8   (B_2)

SMOOTH MUSCLE
RELAXATION
arteries, arterioles (B_1, B_2)
(Vasodilatation, Edema)

SMOOTH MUSCLE
CONTRACTION
veins   (B_1, B_2)
arteries   (B_1, B_2)
airway   (B_2, B_3?)
(Pulmonary Hypertension?,
Bronchoconstriction)

VASCULAR
ENDOTHELIUM
PGI$_2$ release (B_1)
PAF release   (B_1)
Nitric oxide (B_1, B_2)
Cytokines
(Vasodilatation,
Leukocyte Accumulation,
Inflammation)

Figure 15.1   Ubiquitous expression and tissue distribution of kinin receptors.

one of the most promiscuous and noxious substances known.

Kinin receptors are currently classified as B_1 and B_2 subtypes (Farmer and Burch, 1992; Hall, 1992), and their distribution, regulation and pharmacology are discussed in considerable detail by Judith Hall and Ian Morton in Chapter 2. In general, B_2 receptors exhibit much higher affinity for BK and Lys-BK (kallidin) than for the metabolites, desArg^9-BK and Lys-desArg^9-BK, whereas B_1 receptors are more sensitive to the latter. Although a human-lung fibroblast B_1 receptor was recently cloned and expressed (Menke *et al.*, 1994) (see Chapter 3), and bovine tracheal smooth muscle cells may express B_1 receptors (Marsh and Hill, 1994), nothing is known about the role of this kinin receptor subtype, if any, in human airway diseases, and it is not considered further here. The human B_2 receptor has also been cloned, and is a member of the heptahelical transmembrane-spanning family of G protein-coupled receptors (Eggerickx *et al.*, 1992; Hess *et al.*, 1992). Most effects of BK are mediated by B_2 receptors, although there is pharmacological evidence for heterogeneity among this subtype, as well as the conflicting reports of airway smooth muscle B_3 receptors (Farmer *et al.*, 1989, 1991a,b; Perkins *et al.*, 1991; Trifilieff *et al.*, 1991; Field *et al.*, 1992; Pyne and Pyne, 1993, 1994; Farmer and DeSiato, 1994; Pruneau *et al.*, 1995; Scherrer *et al.*, 1995). These aspects of kinin receptor pharmacology are discussed in Chapter 2.

3.   *Bradykinin and Asthma*

In the airways, BK produces multitudinous effects (Farmer, 1991a,b, 1994, 1995) (Fig. 15.1). Animal studies reveal direct and indirect airway actions of BK, and these include: (1) neurogenic inflammation, via stimulation of sensory nerve fibers (Lundberg and Saria, 1983; Saria *et al.*, 1988; Geppetti, 1993; Qian *et al.*, 1993; Nakajima *et al.*, 1994; Yoshihara *et al.*, 1996); (2) arteriolar dilatation, venoconstriction, and plasma protein extravasation, leading to mucosal edema, congestion and swelling (Ichinose and Barnes, 1990; Rogers *et al.*, 1990; Brattsand *et al.*, 1991; Pedersen *et al.*, 1991; Kimura *et al.*, 1992; Bertrand *et al.*, 1993; Qian *et al.*, 1993; O'Donnell and Anderson, 1995; Featherstone *et al.*, 1996); (3) smooth muscle contraction, contributing to bronchoconstriction (Farmer, 1991b, 1995); (4) release of inflammatory mediators such as cytokines (Koyama *et al.*, 1995; Paegelow *et al.*, 1995) and metabolites of arachidonic acid (Greenberg *et al.*, 1979; Bakhle *et al.*, 1985; Gavras and Gavras, 1988; Churchill *et al.*, 1989; Widdicombe *et al.*, 1989; Rogers *et al.*, 1990; Farmer *et al.*, 1991a,b). These, too, can elicit bronchoconstriction, mucus secretion, edema and recruitment of leukocytes into the airways (Fig. 15.1).

3.1   AIRWAY EFFECTS OF BRADYKININ IN HUMANS

The early observation (Herxheimer and Stresemann, 1961), that inhalation of "... 0.5% bradykinin aerosol was very active in asthmatic patients, but not in normal subjects ...", was the first of many studies that observed BK to be a very potent bronchoconstrictor in asthmatics and, yet, essentially inactive in nonasthmatics (Lecomte *et al.*, 1962; Varonier and Panzani, 1968; Fuller *et al.*, 1987; Polosa and Holgate, 1990). In addition to bronchoconstriction, inhaled BK causes cough, wheezing and retrosternal discomfort (Barnes, 1992).

Although there is currently no information regarding BK receptor antagonists in asthmatics, other aspects of human pulmonary pharmacology of BK have been studied.

Note added in proof: A recent report demonstrated that a BK B$_2$ receptor antagonist, Hoe 140, displayed beneficial activity following 4-weeks treatment by inhalation in moderate to severe asthmatic subjects (Akbary *et al.*, 1996).

BK-induced bronchoconstriction in humans is inhibited by muscarinic receptor antagonists (Fuller *et al.*, 1987), indicating that it is mediated via vagal reflexes, but unaffected by inhibitors of cyclooxygenase (Fuller *et al.*, 1987; Polosa *et al.*, 1990). Further evidence that BK acts via neuronal mechanisms was provided by a report that FK 224, an antagonist of neurokinin NK$_1$ and NK$_2$ receptors, inhibited BK-induced bronchoconstriction in asthmatics (Ichinose *et al.*, 1992). BK-stimulated bronchoconstriction in humans is not mediated by histamine or prostaglandins, since H$_1$ receptor antagonists and cyclooxygenase inhibitors are without effect (Fuller *et al.*, 1987; Polosa *et al.*, 1990).

Acute effects of BK on airway caliber in asthmatics, therefore, involve stimulation of pulmonary sensory nerves, which initiates reflex bronchoconstriction via acetylcholine release from parasympathetic nerve endings, as well as tachykinin release via antidromal sensory axon reflexes. Indeed, BK is one of the most potent known stimulants of C-fibers (Dray and Perkins, 1993; Geppetti, 1993) (see Chapter 9). Thus, local release of tachykinins, induced by BK, in addition to eliciting bronchospasm, may contribute to neurogenic inflammation in the airways.

3.2 KININ GENERATION IN ASTHMA

Elevated kinin levels are detected in plasma of asthmatics (Abe *et al.*, 1967; Christiansen *et al.*, 1987), and there may be a significant correlation between circulating kinin levels and symptom severity (Abe *et al.*, 1967). Also, bronchoalveolar (BAL) levels of kinins are significantly elevated in symptomatic asthmatics, but not in asymptomatic subjects (Baumgarten *et al.*, 1992). BAL kinin-generating activity and levels of kinins increase, following allergen challenge, in allergic asthmatics (Christiansen *et al.*, 1987, 1992; Liu *et al.*, 1991). The predominant kinin-generating enzyme is a tissue kallikrein (Christiansen *et al.*, 1987). Tissue kallikrein is relatively resistant to endogenous protease inhibitors such as C1-inhibitor or α$_1$-proteinase inhibitor (α$_1$-PI). This was confirmed with the kallikrein activity found in asthmatic BAL fluid following allergen challenge. Christiansen and colleagues (1989) reported that C1-inhibitor had no effect on BAL kallikrein activity, whereas α$_1$-PI was inhibitory. Inhibition, however, was markedly prolonged in onset. Furthermore, although α$_1$-

PI was detected in BAL fluid, there was no difference in its activity in normal subjects compared with asthmatics even though kallikrein levels were elevated in the latter (Christiansen *et al.*, 1989). These data indicate increased production of kinins in the asthmatic airways, and that bronchial kallikrein activity and hence, local kinin production, are virtually unregulated in this disease.

Averill *et al.* (1992) reported that BAL levels of BK, and prostaglandins E$_2$ and F$_{2\alpha}$, and eosinophil products increased during late asthmatic responses, and that BK concentrations correlated with BAL increases in platelet-derived products. They suggested that prostaglandin-stimulated activation of platelets may be important in late-onset airway obstruction, and that endogenous BK releases epithelial prostaglandins which activate platelets. Therefore, the human data discussed above provide evidence that kinin generation occurs both in the acute phases of an asthmatic attack, as well as during the prolonged inflammatory aspects of this disease. These studies, in conjunction with animal experiments (see Section 3.3) implicate the kallikrein–kinin system in the pathogenesis of asthma.

3.3 BRADYKININ RECEPTOR ANTAGONISTS IN ANIMAL MODELS OF ASTHMA

3.3.1 Sheep

BK receptor antagonists exhibit noteworthy efficacy in some animal models of asthma. For example, in sheep with a natural hypersensitivity to *Ascaris suum*, inhalation of extracts of the nematode elicits acute bronchoconstriction that is associated with an increase in airway responsiveness (Abraham *et al.*, 1991a). Also, eosinophils and neutrophils infiltrate the airways (Solér *et al.*, 1990). Pretreatment of these animals with DArg-[Hyp^3,DPhe^7]-BK (NPC 567), a B$_2$ receptor antagonist, abolished hyperresponsiveness, and reduced airway inflammation (Solér *et al.*, 1990). Since the BK antagonist was effective only when administered prior to antigen inhalation, rather than postantigen, it was suggested that BK released during airway anaphylaxis may have a function in initiating the inflammatory response (Solér *et al.*, 1990). In allergic sheep, which exhibit late-onset bronchial obstruction 6–8 h following allergen challenge, administration of NPC 567 abolishes the late response (Abraham *et al.*, 1991b). Antigen inhalation also increased BAL levels of BK, as well as PGE$_2$, thromboxane, LTB$_4$ and LTD$_4$. Along with its ability to diminish the late response, the kinin antagonist reduced the amounts of inflammatory mediators, and blunted the severity of airway infiltration by granulocytes (Abraham *et al.*, 1991b). Thus, pulmonary generation of kinins in allergic sheep stimulates the synthesis and release of eicosanoids, which, in turn, may contribute to the genesis of airway inflammation, hyperresponsiveness, and late-onset airway obstruction.

3.3.2 Guinea-Pigs

In conscious guinea-pigs, inhalation of BK aerosols causes dose-dependent bronchoconstriction in animals pretreated with inhibitors of BK degradation (M.A. DeSiato and S.G. Farmer, unpublished observations). Similar to observations in human asthmatics, inhaled BK-induced bronchoconstriction in guinea-pigs is inhibited by atropine but unaltered by a cyclooxygenase inhibitor, indomethacin (Fig. 15.2).

Guinea-pigs repeatedly exposed to aerosol antigen exhibit airway hyperresponsiveness to acetylcholine (ACh) and airway infiltration by eosinophils (Ishida *et al.*, 1989). Antigen challenge increases both lung kallikrein activity and circulating kinin levels in this species (Brocklehurst and Lahiri 1962; Jonasson and Becker, 1966). In addition, following intratracheal allergen challenge, BAL kallikrein-like activity increases for as long as 6 h, and this may be related to maintained levels of bronchoconstriction that last this long in these animals (Featherstone *et al.*, 1992).

Inhalation of B_2 receptor antagonists, prevented airway hyperresponsiveness (Fig. 15.3), and dramatically reduced airway eosinophilia (Fig. 15.4) in sensitized guinea-pigs, which had been exposed repeatedly to inhaled challenge

with allergen (Farmer *et al.*, 1992). Interestingly, the B_2 receptor antagonists also abrogated antigen-induced cyanosis and delayed onset of dyspnea, doubling the time taken for onset of respiratory distress (Farmer *et al.*, 1992). The latter observation probably indicates that the kinin antagonists reduce airway hyperresponsiveness to endogenous mediators such as leukotriene C_4 (LTC_4) and histamine. The ability of kinin receptor antagonists to inhibit antigen-induced airway hyperresponsiveness in addition to eosinophilia indicates an important role for endogenous kinins. Few other drugs, including glucocorticosteroids, are consistently reported to inhibit both antigen-induced airway hyperresponsiveness *and* eosinophilia (Murlas and Roum, 1985; Havill *et al.*, 1990; Sanjar *et al.*, 1990; Freedman *et al.*, 1995). Moreover, the inhibition of eosinophil infiltration by the BK receptor antagonists suggests that BK has a significant function in maintaining allergic inflammation in guinea-pig airways.

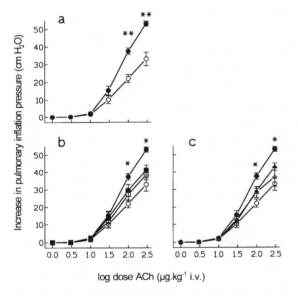

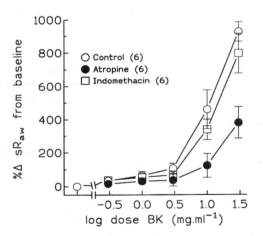

Figure 15.2   Dose-dependent increases in specific airway resistance (sR_{aw}) in conscious guinea pigs in response to inhalation of aerosolized bradykinin (BK). Pulmonary function was determined in a whole-body, double-chambered, constant-volume plethysmograph, and sR_{aw} determined electronically. Aerosols were delivered from a DeVilbiss Pulmosonic Nebulizer, with each dose of BK being administered for 15 s, and 15 min elapsing between successive doses. All animals were pretreated with propranolol (1 mg/kg), thiorphan (10 mg/kg) and captopril (5 mg/kg) i.v. 15 min prior to administration of BK. BK-induced bronchoconstriction was inhibited by pretreatment with atropine (0.1 mg/kg i.v.), but unaffected by indomethacin (5 mg/kg i.v.). Numbers in parentheses represent the numbers of animals in each experiment. From M.A. DeSiato and S.G. Farmer (unpublished data).

Figure 15.3   Dose–response curves for the bronchoconstrictor effect of intravenous acetylcholine (ACh) in anesthetized guinea-pigs sensitized to ovalbumin and challenged twice weekly for 4–5 weeks with inhaled ovalbumin. (a) Data show airway hyperresponsiveness to ACh 24–48 h following the final antigen challenge. ○, controls; ●, antigen-challenged animals. Ovalbumin challenges caused a significant increase in airway responsiveness. (b) Effects of the B_2 receptor antagonist, DArg-[Hyp³,DPhe⁷]-BK (NPC 567), in control animals (□) and in animals chronically challenged with antigen (■). (c) Effects of the B2 receptor antagonist, DArg-[Hyp³,Thi⁵,DTic⁷,Tic⁸]-BK (NPC 16731), in control animals (△) and in animals chronically challenged with antigen (▲). Both kinin receptor antagonists significantly inhibited the effect of ovalbumin challenge. *P < 0.05; **P < 0.01. Data are expressed as mean ± s.e. mean. Reproduced from Farmer et al. (1992).

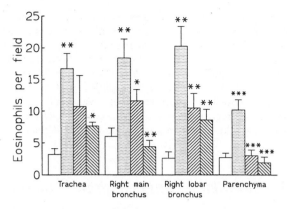

Figure 15.4   NPC 567 and NPC 16731, bradykinin B₂ receptor antagonists, on the numbers of eosinophils present in tissue sections from various airway regions of guinea pigs. Open columns, control tissues; stippled columns, tissues from animals chronically challenged with ovalbumin (OA); left-hatched columns, challenged animals treated with NPC 567; right-hatched columns, challenged animals treated with NPC 16731. In all airway regions examined, OA exposures significantly increased tissue eosinophil numbers. NPC 16731 significantly inhibited or abolished OA-induced eosinophilia in all airway regions. NPC 567 had significant effects in main and lobar bronchi, and in parenchyma. $*P < 0.05$; $**P < 0.01$; $***P < 0.001$. Data are expressed as mean ± s.e. mean. Reproduced from Farmer et al. (1992).

3.3.2.1   Airway Neurogenic Inflammation in Guinea-Pigs

Consistent with the data noted above is the observation that airway plasma protein extravasation, in response to antigen challenge of sensitized guinea-pigs, was reduced by a BK receptor antagonist, Hoe 140, and a Substance

P receptor antagonist, CP 96,345 (Bertrand et al., 1993). The effect of the antagonists, however, was not additive. Hoe 140 or CP 96,345 also inhibited lung microvascular leakage in response to aerosol BK, indicating that antigen challenge releases BK, which in turn stimulates sensory nerves to release Substance P. Indeed, others have demonstrated that BK-induced microvascular leakage is inhibited by an NK₁ receptor antagonist (Qian et al., 1993; Sakamoto et al., 1993). Thus, allergen-induced plasma extravasation has a component mediated by endogenous BK, acting directly as well as through the release of secondary mediators, causing neurogenic inflammation (Geppetti, 1993).

In agreement with the above, we found that inhaled BK-induced bronchoconstriction in conscious guinea pigs is virtually abolished by pretreatment with either capsaicin, which depletes neuropeptide-containing sensory nerve endings, or by an antagonist of neurokinin NK₂ receptors (Fig. 15.5). Thus, the bronchospastic effects of inhaled BK in this species are mediated almost entirely by neuronal mechanisms involving sensory neuropeptides and, to some extent, acetylcholine released from parasympathetic nerve endings (Fig. 15.2).

3.3.2.2   Kallikrein Activity in the Airways

As discussed in Chapters 5, 11 and 16 of this volume, the kinin-generating enzyme, tissue kallikrein, is expressed in a variety of tissues and cells. Similarly, Proud and Vio (1993) localized tissue kallikrein in human trachea, and kallikrein and kinin-forming activities are elevated in the asthmatic airways (Christiansen et al., 1987, 1989, 1992). Poblete et al. (1993) reported that tissue kallikrein is localized in the serous cells of seromucous glands of both human and guinea-pig trachea and in the extrapulmonary and intrapulmonary bronchi of both species. Although no kininogen source has yet been

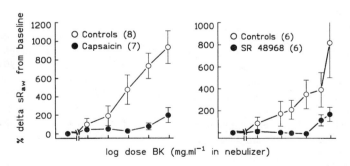

Figure 15.5   Dose-dependent increases in specific airway resistance (sR_{aw}) in conscious guinea pigs in response to inhalation of aerosolized bradykinin (BK). Pulmonary function was determined in a whole-body, double-chambered, constant-volume plethysmograph and sR_{aw} determined electronically. Aerosols were delivered from a DeVilbiss Pulmosonic Nebulizer, with each dose of BK being administered for 15 s and 15 min elapsing between successive doses. All animals were pretreated with propranolol (1 mg/kg), thiorphan (10 mg/kg) and captopril (5 mg/kg) i.v. 15 min prior to administration of BK. BK-induced bronchoconstriction was inhibited by treatment with capsaicin (50 mg/kg s.c.) or an NK₂ receptor antagonist, SR 48968 (10 μM/kg i.v.). Numbers in parentheses represent the numbers of animals in each experiment. From M.A. DeSiato and S.G. Farmer, (unpublished data).

reported in airway tissues, it is possible that this kinin precursor may be expressed by airway and lung endothelial or epithelial cells, as described for the vasculature in Chapter 4 of this volume. Indeed, we presented evidence that ferret isolated trachea has the capacity to generate and respond to endogenous BK *in vitro* (Farmer *et al.*, 1994).

3.3.3   Rats

Limited information has been published regarding the airway effects of kinins in the rat. This is probably due in part to the reputed lack of activity of BK as a broncho-constrictor, either *in vivo* or *in vitro* in rats (Bhoola *et al.*, 1962; Greeff and Moog, 1964). BK induces micro-vascular leakage in rat trachea (Bhoola *et al.*, 1962; Greeff and Moog, 1964; Robbins *et al.*, 1993), and has been reported to cause the release of inflammatory neuropeptides from this tissue preparation (Ray *et al.*, 1991; Hua and Yaksh, 1993).

3.3.3.1   Eosinophils and Airway Hyperresponsiveness

Elevated eosinophil-derived cytotoxic proteins such as major basic protein (MBP) are detected in the lungs of asthmatics (Jaeschke *et al.*, 1991), and their levels correlate with the degree of epithelial damage and the severity of airway hyperresponsiveness (Wardlaw *et al.*, 1988). Furthermore, the degree of epithelial loss in asthma correlates with airway hyperresponsiveness (Jeffery *et al.*, 1989). Inhalation of MBP increases airway responsiveness in rats (Coyle *et al.*, 1993; Uchida *et al.*, 1993) and, since cationic proteins can activate kallikrein (Venge *et al.*, 1979), Coyle and co-investigators (1995) examined the role of kinins. They examined the effects of tracheal instillation of MBP and poly-L-lysine on respon-siveness of rats to methacholine-induced broncho-constriction. They also determined the effects of MBP on lung BK levels and kallikrein activity. In addition, effects of a peptidic B_2 receptor antagonist, DArg-[Hyp^3,DHypE (transpropyl)^7,Oic^8]-BK (NPC 17731), were studied.

Pretreatment with NPC 17731, at a dose sufficient to inhibit effects of exogenous BK, abolished the cationic protein-induced airway hyperresponsiveness (Coyle *et al.*, 1995). Poly-L-lysine increased lung kallikrein activity three-fold, and a concomitant eight-fold increase in lung BK, reaching a maximum level after 15 min. This was still elevated 60 min later when animals exhibited airway hyperresponsiveness (Coyle *et al.*, 1995). Similarly, MBP produced a two-fold increase in lung kallikrein activity and a five-fold increase in BK levels.

These data suggest that the ability of eosinophil granule proteins to induce airway hyperresponsiveness is mediated by BK. Not only did MBP-induced increases in lung kallikrein activity and BK levels correlate with the onset of hyperresponsiveness, but a B_2 antagonist abolished the ability of the cytotoxic proteins to induce airway hyperresponsiveness. Although the mechanisms

underlying hyperresponsiveness are unknown, it is clear that BK's involvement is an early event, as the antagonist was ineffective if administered after the cationic proteins (Coyle *et al.*, 1995). Interestingly, it has been demon-strated that human nasal levels of BK increase following antigen challenge in allergic rhinitics, and that the increase correlated with levels of eosinophil cationic protein, suggesting a link between eosinophilic inflam-mation and kinin generation (Svensson *et al.*, 1990).

4.   Bradykinin Antagonists in Humans

At present, there are no data published on the effects of BK receptor antagonists in asthma (see note on p. 251), although Hoechst are reputed to be conducting clinical trials with Hoe 140 in asthmatic patients. The paucity of such studies is due in no small measure to the lack of metabolically stable, nontoxic, nonpeptide kinin receptor antagonists. The only published nonpeptide B_2 receptor antagonist, Win 64338 (see Chapter 2), is quite toxic in experimental animals (Farmer and DeSiato, 1995; Sawutz *et al.*, 1995; and our unpublished observations).

Nevertheless, the studies in animals discussed above, as well as a recent report in human rhinitis, suggest that BK may be a cardinal mediator of allergic airway inflammation. Austin *et al.* (1994) showed that intranasal administration of Hoe 140 (200 µg), 2 min prior to antigen challenge, essentially abolished nasal blockage in 15 subjects with allergic rhinitis. This dose also inhibited nasal effects of exogenous BK (Austin and Foreman, 1994). The authors noted that their data indicate that the nasal blockage in perennial allergic rhinitis "... is mediated to a significant extent by kinin, formed in the nasal airway following antigen challenge" (Austin *et al.*, 1994). The reader is referred to Chapter 17 for a review of studies with kinin receptor antagonists in human upper airway disease.

Interestingly, it was recently reported that, although Hoe 140 was effective in perennial rhinitis, it is ineffective in seasonal allergic rhinitis (Akbary and Bender 1993; Austin *et al.*, 1994). It is feasible that the pathophysiology of the two types of allergic rhinitis may differ, with BK being a major mediator of perennial, but not seasonal rhinitis. There is perhaps a note of caution here for potential trials in patients with asthma of different etiologies.

5.   Bradykinin in Acute Respiratory Distress Syndrome

Detailed reviews on the immunobiology of acute lung injury (ALI) and ARDS have been published (Schlag and Redl, 1989; Rinaldo and Christman, 1990; Neuhof,

1991; Gadek, 1992; Hasegawa *et al.,* 1994; Plitman and Snapper, 1994; Strieter and Kunkel, 1994; Demling, 1995; Schuster, 1995). Although acute respiratory failure associated with trauma or disease had long been known, Ashbaugh and colleagues (1967) first noted in 1967 that its occurrence, following various pathological insults, was associated with strikingly uniform clinical features. As the lung abnormalities manifested were similar to those in infant respiratory distress syndrome, the condition became known as "adult respiratory distress syndrome" (now referred to as "*acute* respiratory distress syndrome"; Bernard *et al.,* 1994). ARDS represents the most severe and often fulminant form of ALI (Rinaldo and Christman, 1990; Bitterman, 1992; Plitman and Snapper, 1994). This syndrome is a major contributor to the mortality of patients in intensive care units, imparting enormous human and financial costs (Bernard *et al.,* 1994).

The nastiness of ALI and its sequelae is exemplified by vivid literature descriptions. Thus, the nature of the inflammatory response and the myriad mediators released in response to lung injury have been described as "the inferno that is sparked" (Reynolds, 1987), "a biological explosion" (Bitterman, 1992), "overzealous attempts by the body to protect itself" (Gadek, 1992), or "a mediator arsenal" (Neuhof, 1991). Possibly because of the rapid and violent nature of the body's response to ALI, many years of research have failed to produce an effective drug for ARDS. The obduracy of the problem may reflect the multiple biochemical pathways along which the pathogenic cascade can proceed and the brevity of the "window of opportunity" following lung injury, early in ARDS, during which drugs have the greatest chance of success. Novel drugs for ARDS may have to be administered prophylactically to patients predicted to have a high probability of developing the syndrome following ALI. Indeed, in many animal models (predominantly of sepsis), for a drug to improve survival it must be given *prior to* the insult. This is not a luxury usually afforded in the clinic.

Notwithstanding, although multifarious conditions predispose to ARDS, the abnormalities seen in pathophysiology of the syndrome are remarkably uniform (Hyers 1991; Bitterman, 1992; Hooker *et al.,* 1992; Nieuwenhuijzen *et al.,* 1995), and this has led investigators to postulate the existence of a final, common pathway for development of all ARDS. At present, however, a unifying pathophysiological mechanism or mediator for ARDS has not been identified. Certainly, clinical trials with anticytokine or anti-endotoxin approaches have proven disappointing in septic ARDS (Goldstein and Luce, 1990; Fisher *et al.,* 1994; Bone *et al.,* 1995; Quezado *et al.,* 1995; Solomkin and Hill, 1995; Willatts *et al.,* 1995).

Most, if not all, of the complex mammalian biochemical cascades and the mediators subsequently generated have been implicated in the pathogenesis of

ARDS. Coagulation and fibrinolysis, the complement system, lipid mediators and a variety of cytokines have been accused as perpetrators. This section will discuss the activation of the kallikrein–kinin system in ARDS, the role of kinins in the syndrome, and the effects of antagonists of BK receptors in animal models, in addition to a recent study with a peptide kinin antagonist in human sepsis. Pixley and Colman provide a review of the kinin system in sepsis in Chapter 10 of this volume.

5.1   EFFECTS OF KININS CONSISTENT WITH A ROLE IN ARDS

BK and other components of the kinin cascade exert many effects that are consistent with a role in ARDS (Fig. 15.1). Similarly, there is much evidence that the kinin system is important in acute pancreatitis (Chapter 12), multiple trauma and sepsis (Chapter 10), which each can lead to ARDS.

Several of the features of ARDS mimicked by kinins include: (1) lung microvascular leakage, leading to edema and decreased lung compliance; (2) pulmonary hypertension (Hauge *et al.,* 1966; Bönner *et al.,* 1992; Nossaman *et al.,* 1994); (3) systemic vasodilation and hypotension (Regoli *et al.,* 1981; Gavras and Gavras, 1988; Weipert *et al.,* 1988; Katori *et al.,* 1989; Wilson *et al.,* 1989; Gavras and Gavras, 1991; Noronha-Blob *et al.,* 1991; Pixley *et al.,* 1993; Ueno *et al.,* 1995a,b); (4) release of mediators associated with ARDS, including platelet-activating factor (PAF) and various cytokines (see Section 3); (5) evidence for activation of the contact system, and consequent generation of kallikrein following ALI (Saugstad *et al.,* 1982, 1992; Aasen *et al.,* 1983; Kalter *et al.,* 1983; Carvalho *et al.,* 1988; Fuhrer *et al.,* 1989; Wanaka *et al.,* 1990; DeLa Cadena *et al.,* 1993; Khan *et al.,* 1993; Pixley *et al.,* 1993; Brus *et al.,* 1994) (see also Chapters 10 and 16); (6) kallikrein, in addition to generating kinins, is chemotactic for neutrophils and induces release of neutrophil elastase (see Chapter 11).

Many studies demonstrate that BK causes lung edema in animal models of ALI (O'Brodovich *et al.,* 1981; Pang *et al.,* 1982; Ichinose and Barnes, 1990; Sakamoto *et al.,* 1992), and decreases lung compliance (Hauge *et al.,* 1966; Greenberg *et al.,* 1979; Jin *et al.,* 1989). Furthermore, hypoxia is known to decrease lung angiotensin-converting enzyme (ACE) activity. For example, in sheep it has been observed that, although BK has little effect on lung microvascular permeability, the peptide causes microvascular leakage, increased lung lymph flow and edema in hypoxemic animals (Pang *et al.,* 1982). The authors suggest that, under normoxic conditions, the lung may provide a protective function by degrading BK in a single pass through the lungs (see also Chapter 7). If hypoxia, such as occurs in ARDS, is superimposed the lung's protective function is impaired (Pang *et al.,* 1982). Indeed, there is severe depression of lung ACE activity following ALI (Wiedemann *et al.*, 1990) and serum ACE decreases in septicemia-induced ARDS in humans (Casey *et al.*, 1981). Thus, hypoxia may result in depressed degradation of BK formed in the lungs of ARDS patients, thereby potentiating the harmful effects of the peptide.

5.1.1 Effects of Bradykinin on Pulmonary Blood Pressure

BK has several effects on the cardiovascular system that are consistent with a role in ARDS-inducing conditions like acute pancreatitis and sepsis (see Chapters 10 and 12). Thus, although BK is one of the most potent known systemic vasodilators, causing profound hypotension (Bönner *et al.*, 1989, 1992; Gavras and Gavras, 1991), it can increase pulmonary blood pressure in humans (Bönner *et al.*, 1989, 1992). Bönner *et al.* (1992) noted that the pulmonary hypertensive effect in humans was probably due to increased cardiac output as pulmonary vascular resistance was not increased by BK. Although there is a wealth of information on the ability of kinins to relax systemic blood vessels, only a limited number of studies have examined the effects on pulmonary vessels. Nevertheless, BK causes contraction of guinea-pig pulmonary artery *in vitro* (Hock *et al.*, 1991). Similarly, in both rabbit (Seeger *et al.*, 1983) and rat perfused lungs (Pardi *et al.*, 1992), BK causes vasoconstriction in the pulmonary vasculature. Moreover, if pulmonary blood pressure is already elevated, the pulmonary vaso-constrictor effect of BK is enhanced, suggesting that BK may aggravate pre-existing pulmonary hypertension (Pardi *et al.*, 1992). Also in rats, monocrotaline-induced ALI, pulmonary hypertension and lung edema are associated with large increases in kininogen gene expression in the lungs and liver (Chao *et al.*, 1993).

As in the chronic scenario of asthma, BK releases many of the mediators associated with ALI, including PAF (McIntyre *et al.*, 1985; Ricupero *et al.*, 1992), cytokines including interleukin-1 (IL-1), IL-6, IL-8 and tumor necrosis factor (TNF) (Farmer, 1991a; Vandekerckhove *et al.*, 1991; Ferreira *et al.*, 1993; Paegelow *et al.*, 1995), and nitric oxide (Mayhan, 1992; Wiemer and Wirth, 1992). In addition, BK has synergistic effects upon fibroblast proliferation, angiogenesis and prostaglandin biosynthesis in response to IL-1, IL-6 and TNF (Kimball and Fisher, 1988; Burch and Tiffany, 1989; O'Neill and Lewis, 1989; Vandekerckhove *et al.*, 1991; Lerner *et al.*, 1992; Cisar *et al.*, 1993; Hu and Fan, 1993) (Fig. 15.1) (see also Chapter 18).

5.2 KININ GENERATION IN ARDS

The kallikrein–kinin system is activated by tissue injury such as that occurring in ALI (see Chapters 4, 5, 10 and 16). Moreover, generation of kinins has long had an association with sepsis syndrome and ARDS, in that contact activation occurs (Carvalho *et al.*, 1988; Colman,

1994; Hasegawa *et al.*, 1994) and, consequently, decreased kininogen and prekallikrein levels, increased kallikrein activity, and increased levels of kinins occur in animals and humans with experimental endotoxin shock (Traber *et al.*, 1988; Katori *et al.*, 1989; Naess *et al.*, 1991; DeLa Cadena *et al.*, 1993), bacterial sepsis (Aasen *et al.*, 1983; Pixley *et al.*, 1993; Roman *et al.*, 1993), acute pancreatitis (Ofstad, 1970; Griesbacher *et al.*, 1993; Shimizu *et al.*, 1993; Bläckberg and Ohlsson, 1994), and septic and nonseptic ARDS (Carvalho *et al.*, 1988; de Oliveira and de Oliveira, 1988; Saugstad *et al.*, 1992; Koivunen *et al.*, 1994). Indeed, human BAL levels of kinins are significantly elevated in ARDS (Baumgarten *et al.*, 1992).

Utilizing a monoclonal antibody against Factor XII, Pixley and colleagues (1993) examined contact activation during lethal Gram-negative bacteremia in baboons. Untreated septic animals exhibited contact activation manifested by a decrease in high molecular weight kininogen (HMWK) and an increase in circulating levels of α_2-macroglobulin–kallikrein complexes. In contrast, in baboons treated with anti-Factor XII antibody, no contact activation occurred. Whereas untreated baboons developed irreversible hypotension, the treated animals demonstrated a fall in blood pressure that reversed quickly and survived longer than controls. Untreated and treated groups developed disseminating intravascular coagulation (DIC). As discussed in Chapter 10 of this volume, the authors suggest that treatment of septic shock may require a combination of anticoagulants to control DIC concurrent with an inhibitor of the contact system to prevent the action or generation of BK (Pixley *et al.*, 1993). In partial agreement with this study, preliminary data with PKSI-527, a specific inhibitor of plasma kallikrein, showed significant inhibition of endotoxin-induced DIC in rats (Katsuura *et al.*, 1989).

In a rat model of severe head trauma, inhibition of kallikrein with hexadimethrine almost completely abolished the ARDS that occurs in these animals (de Oliveira and de Oliveira, 1988). Similarly, depletion of kininogen with bromelain, 90 min prior to trauma, abolished ARDS. In contrast, administration of bromelain concomitant with trauma resulted in massive kininogen generation and a dramatic potentiation of ARDS. The authors noted that "The drastic changes in the pathologic evolution of ARDS as a result of manipulation of bradykinin generation and availability... reinforces the importance of the kallikrein–kinin system in the induction of the syndrome" (de Oliveira and de Oliveira, 1988).

In humans, it has long been known that contact activation occurs in sepsis and ARDS (McGuire *et al.*, 1982; Schlag and Redl, 1989; Neuhof, 1991; Colman, 1994; Hasegawa *et al.*, 1994). For example, the proteolytic activity in the neutrophil-rich BAL fluid of ARDS patients can cleave Hageman factor, prekallikrein, plasminogen, HMWK and various complement components (McGuire et al., 1982). Similarly, levels of prekallikrein and HMWK are markedly decreased in trauma- and sepsis-induced ARDS (Mulligan et al., 1994). This study also reported a strong correlation between BAL kallikrein activity (i.e., BK-generating activity) and the numbers of neutrophils with the release of elastase in ARDS. In preterm babies with ARDS and also in full-term neonates with sepsis-induced ARDS, the kinin system is activated (Saugstad et al., 1992; Roman et al., 1993; Koivunen et al., 1994). Plasma prekallikrein levels were more markedly reduced in neonates with septic shock than in babies with uncomplicated sepsis (Roman et al., 1993).

5.3   BRADYKININ RECEPTOR ANTAGONISTS IN SEPSIS AND ARDS IN ANIMALS

BK receptor antagonists have beneficial effects, such as reversing hypotension and metabolic defects, and markedly decreasing mortality in animal models of acute pancreatitis, endotoxin shock and overt sepsis (Weipert et al., 1988; Wilson et al., 1989; Noronha-Blob et al., 1991; Whalley et al., 1992; Otterbein et al., 1993; Hirano and Takeuchi, 1994; Szabó et al., 1994; Closa et al., 1995; Farmer et al., 1995; Griesbacher et al., 1995; Ridings et al., 1995; Hoffmann et al., 1996).

Studies with several B_2 receptor antagonists in animals have demonstrated marked attenuation of the sustained hypotension that is associated with mortality in experimental endotoxinemia. Thus, NPC 567 effectively inhibited endotoxin-induced hypotension and increased plasma levels of prostacyclin. Moreover, the kinin antagonist reduced the 24 h mortality rate from 100% to 50% (Wilson et al., 1989). Unfortunately, NPC 567 was effective only when administered 30 min prior to infusion of endotoxin. When co-administered with endotoxin, no beneficial effects were observed (Wilson et al., 1989).

Another B_2 receptor antagonist, CP-0127 (see Chapter 2), protects against hypotension, and markedly reduces mortality in rabbits and rats infused with E. coli endotoxin (Whalley et al., 1992). Similar results were reported in rats and mice, where infusion of NPC 17731 30 min prior to endotoxin inhibited elevation of plasma prostacyclin and TNF-α (Noronha-Blob et al., 1991). Further, NPC 17731 pretreatment delayed the onset and reduced the magnitude of neutropenia, and abolished endotoxin-induced hypoglycemia (Noronha-Blob et al., 1991). CP-0127 was also examined in a rat model of traumatic shock. In this report, CP-0127 abolished trauma-induced hypotension and doubled survival time (Christopher et al., 1994). It was noted that CP-0127 retarded the accumulation of neutrophils in the gut (the rat "shock" organ), and attenuated endothelial damage to mesenteric vessels. We recently reported that Hoe 140, administered as late as 12 h after induction of septic peritonitis in rats, significantly reduced mortality in a model where untreated animals are all dead within 18 h postinoculation (Farmer et al., 1995).

5.4   BRADYKININ RECEPTOR ANTAGONISTS IN SEPSIS AND ARDS IN HUMANS

Analysis of a Phase II trial with Bradycor (CP-0127), a B_2 receptor antagonist (see Section 5.2), in patients with systemic inflammatory response syndrome and presumed sepsis did not demonstrate clinically meaningful reductions in 28-day mortality (Scrip, July 27, 1994). Five hundred patients were administered three doses of Bradycor for 3 days. Similarly, this drug showed a negative effect on 28-day mortality in a Phase II trial in 251 sepsis patients.

Preliminary data were recently reported in a small clinical trial with Bradycor in Hantavirus pulmonary syndrome (HPS), a relatively rare ARDS-like condition, with a mortality of over 50% (Hallin et al., 1996). HPS is characterized by a fulminant pulmonary edema and cardiogenic shock, with death owing to shock. Levy and colleagues (1995) reported activation of the kinin system in two patients with HPS, which returned to normal with disease resolution. Three HPS patients who were eligible for treatment received Bradycor at a dose of 3 mg/kg/min for 3 days, with 1 dying on day 3. One patient was entered under the Phase II sepsis study and survived, but was randomized to placebo. One patient randomized under an ongoing placebo control protocol died prior to the drug study. Further studies with nonpeptide B_2 and perhaps B_1 receptor antagonists in ARDS are awaited.

6.   Conclusions

It is unlikely that the clinical manifestations of severe and often life-threatening conditions such as chronic asthma and acute lung injury are due to a single mediator. Nevertheless, there is little doubt that the kallikrein–kinin system is activated in these disorders, and that local levels of kinins are markedly elevated in the inflammatory milieu. Thus, BK and related kinins may have important roles in initiating and/or maintaining the pathophysiological changes that occur in the lungs. Furthermore, BK has deleterious, pro-inflammatory effects on many cell types that are mediated directly on those cells, as well as by the myriad other potentially injurious substances whose release this peptide can stimulate.

Experiments in animal models of asthma and ARDS, particularly those demonstrating beneficial effects of selective kinin receptor antagonists, provide convincing testimony for a pivotal role of the kinins in airway pathophysiology. Only after conducting clinical trials, utilizing kinin receptor antagonists, will the roles of these endogenous peptides in human disease become apparent.

7. References

Aasen, A.O., Smith-Erichsen, N. and Amundsen, E. (1983). Plasma kallikrein–kinin system in septicemia. Arch. Surg. 118, 343–345.

Abe, K., Watanabe, N., Kumagai, N., Mouri, T., Seki, T. and Yoshinaga, K. (1967). Circulating plasma kinin in patients with bronchial asthma. Experientia 23, 626–627.

Abraham, W.M., Ahmed, A., Cortes, A., Solér, M., Farmer, S.G., Baugh, L.E. and Harbeson, S.L. (1991a). Airway effects of inhaled bradykinin, substance P and neurokinin A in sheep. J. Allergy Clin. Immunol. 87, 557–564.

Abraham, W.M., Burch, R.M., Farmer, S.G., Ahmed, A. and Cortes, A. (1991b). A bradykinin antagonist modifies allergen-induced mediator release and late bronchial responses in sheep. Am. Rev. Respir. Dis. 143, 787–796.

Akbary, M.A. and Bender, N. (1993). Efficacy and safety of Hoe 140 (a bradykinin antagonist) in patients with mild-to-moderate seasonal allergic rhinitis. "Kinins '93", Proceedings, 22.02.

Akbary, A.M., Wirth, K.J. and Schölkens, B.A. (1996). Efficacy and tolerability of Icatibant (Hoe 140) in patients with moderately severe chronic bronchial asthma. Immuno-pharmacology, 33, 238–242

Amrani, Y., Martinet, N. and Bronner, C. (1995). Potentiation by tumour necrosis factor-α of calcium signals induced by bradykinin and carbachol in human tracheal smooth muscle cells. Br. J. Pharmacol. 114, 4–5.

Ashbaugh, D.G., Bigelow, D.B., Petty, T.L. and Levine, B.E. (1967). Acute respiratory distress in adults. Lancet 2, 319–323.

Austin, C.E. and Foreman, J.C. (1994). A study of the action of bradykinin and bradykinin analogues in the human nasal airway. J. Physiol. 478, 351–356.

Austin, C.E., Foreman, J.C. and Scadding, G.K. (1994). Reduction by Hoe 140, the B_2 kinin receptor antagonist, of antigen-induced nasal blockage. Br. J. Pharmacol. 111, 969–971.

Averill, F.J., Hubbard, W.C., Proud, D., Gleich, G.J. and Liu, M.C. (1992). Platelet activation in the lung after antigen challenge in a model of allergic asthma. Am. Rev. Respir. Dis. 145, 571–576.

Bakhle, Y.S., Moncada, S., De Nucci, G. and Salmon, J.A. (1985). Differential release of eicosanoids by bradykinin, arachidonic acid and calcium ionophore A23187 in guinea-pig isolated perfused lung. Br. J. Pharmacol. 86, 55–62.

Barnes, P.J. (1992). Bradykinin and asthma. Thorax 47, 979–983.

Baumgarten, C.R., Lehmkuhl, B., Henning, R., Brunnee, T., Dorow, P., Schilling, W. and Kunkel, G. (1992). Bradykinin and other inflammatory mediators in BAL-fluid from patients with active pulmonary inflammation. Agents Actions 38, (Suppl. III), 475–481.

Bernard, G.R., Artigas, A., Brigham, K.L., Carlet, J., Falke, K., Hudson, L., Lamy, M., Legall, J.R., Morris, A. and Spragg, R. (1994). The American–European consensus conference on ARDS. Definitions, mechanisms, relevant outcomes, and clinical trial coordination. Am. J. Respir. Crit. Care Med. 149, 818–824.

Bertrand, C., Nadel, J.A., Yamawaki, I. and Geppetti, P. (1993). Role of kinins in the vascular extravasation evoked by antigen and mediated by tachykinins in guinea pig trachea. J. Immunol. 151, 4902–4907.

Bhoola, K.D., Collier, H.O.J., Schachter, M. and Shorley, P.G. (1962). Action of some peptides on bronchial muscle. Br. J. Pharmacol. 19, 190–197.

Bhoola, K.D., Figueroa, C.D. and Worthy, K. (1992). Bio-regulation of kinins: kallikreins, kininogens, and kininases. Pharmacol. Rev. 44, 1–80.

Bitterman, P.B. (1992). Pathogenesis of fibrosis in acute lung injury. Am. J. Med. 92 (Suppl. 6A), 39S–43S.

Bläckberg, M. and Ohlsson, K. (1994). Studies on the release of tissue kallikrein in experimental pancreatitis in the pig. Eur. Surg. Res. 26, 116–124.

Bone, R.C., Balk, R.A., Fein, A.M., Perl, T.M., Wenzel, R.P., Reines, D., Quenzer, R.W., Iberti, T.J., Macintyre, N. and Schein, R.M.H. (1995). A second large controlled clinical study of E5, a monoclonal antibody to endotoxin: results of a prospective, multicenter, randomized, controlled study. Crit. Care Med. 23, 994–1006.

Bönner, G., Schunk, U., Preis, S., Wambach, G. and Toussaint, T. (1989). Einflüsse von Bradykinin auf die systemische und die pulmonale Hämodynamik am Menschen. Klin. Wochenschr. 67, 1085–1095.

Bönner, G., Preis, S., Schunk, U., Wagmann, M., Chrosch, R. and Toussaint, C. (1992). Effect of bradykinin on arteries and veins in systemic and pulmonary circulation. J. Cardiovasc. Pharmacol. 20 (Suppl. 9), S21–S27.

Brattsand, R., O'Donnell, S.R., Miller-Larsson, A. and Rauchle, K.L. (1991). Attenuation of bradykinin-induced mucosal inflammation by topical budesonide in rat trachea. Agents Actions 34, 200–202.

Brocklehurst, W.E. and Lahiri, S.C. (1962). The production of bradykinin during anaphylaxis. J. Physiol. 160, 15–16P.

Brus, F., Van Oeveren, W., Okken, A. and Oetomo, S.B. (1994). Activation of the plasma clotting, fibrinolytic, and kinin–kallikrein system in preterm infants with severe idiopathic respiratory distress syndrome. Pediatr. Res. 36, 647–653.

Burch, R.M. and Tiffany, C.W. (1989). Tumor necrosis factor causes amplification of arachidonic acid metabolism in response to interleukin 1, bradykinin, and other agents. J. Cell. Physiol. 141, 85–89.

Carvalho, A.C., Demarinis, S., Scott, C.F., Silver, L.D., Schmaier, A.H. and Colman, R.W. (1988). Activation of the contact system of plasma proteolysis in the adult respiratory distress syndrome. J. Lab. Clin. Med. 112, 270–277.

Casey, J.D., Krieger, B., Kohler, J. et al. (1981). Decreased serum angiotensin converting enzyme in adult respiratory distress syndrome associated with sepsis. Crit. Care Med. 9, 651–654.

Chao, J., Simson, J.A.V., Chung, P., Chen, L.-M. and Chao, L. (1993). Regulation of kininogen gene expression and local-ization in the lung after monocrotaline-induced pulmonary hypertension in rats. Proc. Soc. Exp. Biol. Med. 203, 243–250.

Christiansen, S.C., Proud, D. and Cochrane, C.G. (1987). Detection of tissue kallikrein in the bronchoalveolar lavage fluid of asthmatic subjects. J. Clin. Invest. 79, 188–197.

Christiansen, S.C., Zuraw, B.L., Proud, D. and Cochrane, C.G. (1989). Inhibition of human bronchial kallikrein in asthma. Am. Rev. Respir. Dis. 139, 1125–1131.

Christiansen, S.C., Proud, D., Sarnoff, R.B., Juergens, U., Cochrane, C.G. and Zuraw, B.L. (1992). Elevation of tissue kallikrein and kinin in the airways of asthmatic subjects after endobronchial allergen challenge. Am. Rev. Respir. Dis. 145, 900–905.

Christopher, T.A., Ma, X.-L., Gauthier, T.W. and Lefer, A.M. (1994). Beneficial actions of CP-0127, a novel bradykinin receptor antagonist, in murine traumatic shock. Am. J. Physiol. 266, H867–H873.

Churchill, L., Chilton, F.H., Resau, J.H., Bascom, R., Hubbard, W.C. and Proud, D. (1989). Cyclooxygenase metabolism of endogenous arachidonic acid by cultured human tracheal epithelial cells. Am. Rev. Respir. Dis. 140, 449–459.

Cisar, L.A., Mochan, E. and Schimmel, R. (1993). Interleukin-1 selectively potentiates bradykinin-stimulated arachidonic acid release from human synovial fibroblasts. Cell. Signal. 5, 463–472.

Closa, D., Hotter, G., Prats, N., Gelpí, E. and Roselló-Catafau, J. (1995). A bradykinin antagonist inhibited nitric oxide generation and thromboxane biosynthesis in acute pancreatitis. Prostaglandins 49, 285–294.

Colman, R.W. (1994). Disseminated intravascular coagulation due to sepsis. Semin. Hematol. 31 (Suppl. 1), 10–17.

Coyle, A.J., Ackerman, S.J. and Irvin, C.G. (1993). Cationic proteins induce airway hyperresponsiveness dependent on charge interactions. Am. Rev. Respir. Dis. 147, 896–900.

Coyle, A.J., Ackerman, S.J., Burch, R.M., Proud, D. and Irvin, C.G. (1995). Human eosinophil-granule major basic protein and synthetic polycations induce airway hyperresponsiveness in vivo dependent on bradykinin generation. J. Clin. Invest. 95, 1735–1740.

de Oliveira, G.G. and de Oliveira, A.M.P. (1988). Adult respiratory distress syndrome (ARDS): the pathophysiologic role of catecholamine–kinin interactions. J. Trauma 28, 246–253.

DeLa Cadena, R.A., Suffredini, A.F., Page, J.D., Pixley, R.A., Kaufman, N., Parrillo, J.E. and Colman, R.W. (1993). Activation of the kallikrein–kinin system after endotoxin administration to normal human volunteers. Blood 81, 3313–3317.

Demling, R.H. (1995). The modern version of adult respiratory distress syndrome. Annu. Rev. Med. 46, 193–202.

Dray, A. and Perkins, M. (1993). Bradykinin and inflammatory pain. Trends Neurosci. 16, 99–104.

Eggerickx, D., Raspe, E., Bertrand, D., Vassart, G. and Parmentier, M. (1992). Molecular cloning, functional expression and pharmacological characterization of a human bradykinin B2 receptor gene. Biochem. Biophys. Res. Commun. 187, 1306–1313.

Farmer, S.G. (1991a). Role of kinins in airway diseases. Immunopharmacology 22, 1–20.

Farmer, S.G. (1991b). Airway pharmacology of bradykinin. In "Bradykinin Antagonists. Basic and Clinical Research" (ed. R.M. Burch), pp. 213–236. Marcel Dekker, New York.

Farmer, S.G. (1994). The pharmacology of bradykinin in human airways. In "Drugs and the Lung" (eds. C.P. Page and W.J. Metzger), pp. 449–465. Raven Press, New York.

Farmer, S.G. (1995). Bradykinin. Receptors, signal transduction and effects. In "Airway Smooth Muscle: Peptide Receptors, Ion Channels and Signal Transduction" (eds. D. Raeburn and M. Giembycz), pp. 51–65. Birkhäuser Verlag, Basel.

Farmer, S.G. and Burch, R.M. (1992). Biochemical and molecular pharmacology of kinin receptors. Annu. Rev. Pharmacol. Toxicol. 32, 511–536.

Farmer, S.G. and DeSiato, M.A. (1994). Effects of a novel non-peptide bradykinin B2 receptor antagonist, on intestinal and airway smooth muscle. Further evidence for the tracheal B3 receptor. Br. J. Pharmacol. 112, 461–464.

Farmer, S.G. and DeSiato, M.A. (1995). Effects of Win 64338,

a nonpeptide bradykinin antagonist, in vivo in rodents. Am. J. Respir. Crit. Care Med. 151, A110.

Farmer, S.G., Burch, R.M., Meeker, S.N. and Wilkins, D.E. (1989). Evidence for a pulmonary bradykinin B3 receptor. Mol. Pharmacol. 36, 1–8.

Farmer, S.G., Burch, R.M., Kyle, D.J., Martin, J.A., Meeker, S.N. and Togo, J. (1991a). DArg[Hyp3-Thi5-DTic7-Tic8]-bradykinin, a potent antagonist of smooth muscle BK2 receptors and BK3 receptors. Br. J. Pharmacol. 102, 785–787.

Farmer, S.G., Ensor, J.E. and Burch, R.M. (1991b). Evidence that cultured airway smooth muscle cells contain bradykinin B2 and B3 receptors. Am. J. Respir. Cell Mol. Biol. 4, 273–277.

Farmer, S.G., Wilkins, D.E., Meeker, S., Seeds, E.A.M. and Page, C.P. (1992). Effects of bradykinin receptor antagonists on antigen-induced respiratory distress, airway hyper-responsiveness and eosinophilia in guinea-pigs. Br. J. Pharmacol. 107, 653–659.

Farmer, S.G., Broom, T. and DeSiato, M.A. (1994). Effects of bradykinin receptor agonists, and captopril and thiorphan in ferret isolated trachea: evidence for bradykinin generation in vitro. Eur. J. Pharmacol. 259, 309–313.

Farmer, S.G., DeSiato, M.A. and Miyamoto, Y.J. (1995). Effects of a bradykinin (BK) receptor antagonist in septic peritonitis in rats. Inflamm. Res. 44, Suppl. 3, S283.

Featherstone, R.L., Parry, J.E. and Church, M.K. (1992). Kallikrein-like activity in the bronchoalveolar lavage (BAL) fluid of sensitized, challenged guinea-pigs. Agents Actions 38 (Suppl. III), 462–466.

Featherstone, R.L., Parry, J.E. and Church, M.K. (1996). The effects of a kinin antagonist on changes in lung function and plasma extravasation into the airways following challenge of sensitized guinea-pigs. Clin. Exp. Allergy 26, 235–240.

Ferreira, S.H., Lorenzetti, B.B. and Poole, S. (1993). Bradykinin initiates cytokine-mediated inflammatory hyperalgesia. Br. J. Pharmacol. 110, 1227–1231.

Field, J.L., Hall, J.M. and Morton, I.K.M. (1992). Putative novel bradykinin B3 receptors in the smooth muscle of the guinea-pig taenia caeci and trachea. Agents Actions 38, (Suppl. I), 540–545.

Fisher, C.J., Jr, Dhainaut, J.-F.A., Opal, S.M., Pribble, J.P., Balk, R.A., Slotman, G.J., Iberti, T.J., Rackow, E.C., Shapiro, M.J., Greenman, R.L., Reines, H.D., Shelly, M.P., Thompson, B.W., LaBrecque, J.F., Catalano, M.A., Knaus, W.A. and Sadoff, J.C. (1994). Recombinant human interleukin 1 receptor antagonist in the treatment of patients with sepsis syndrome: results from a randomized, double-blind, placebo-controlled trial. JAMA 271, 1836–1843.

Freedman, J.E., Fabian, A. and Loscalzo, J. (1995). Impaired EDRF production by endothelial cells exposed to fibrin monomer and FDP. Am. J. Physiol. 268, C520–C526.

Fuhrer, G., Heller, W., Junginger, W., Gröber, O. and Roth, K. (1989). Components of the kallikrein–kinin-system in patients with ARDS. Prog. Clin. Biol. Res. 308, 737–742.

Fuller, R.W., Dixon, C.M.S., Cuss, F.M. and Barnes, P.J. (1987). Bradykinin-induced bronchoconstriction in humans: mode of action. Am. Rev. Respir. Dis. 135, 176–180.

Gadek, J.E. (1992). Adverse effects of neutrophils on the lung. Am. J. Med. 92 (Suppl. 6A), 27S–31S.

Gavras, H. and Gavras, I. (1991). Effects of bradykinin antagonists on the cardiovascular system. In "Bradykinin Antagonists. Basic and Clinical Research" (ed. R.M. Burch), pp. 171–189. Marcel Dekker, New York.

Gavras, I. and Gavras, H. (1988). Anti-hormones and blood pressure: bradykinin antagonists in blood pressure regulation. Kidney Intern. 34 (Suppl. 26), S60–S62.

Geppetti, P. (1993). Sensory neuropeptide release by bradykinin: mechanisms and pathophysiological implications. Regul. Pept. 47, 1–23.

Goldstein, G. and Luce, J.M. (1990). Pharmacologic treatment of the adult respiratory distress syndrome. Clin. Chest Med. 11, 773–787.

Greeff, K. and Moog, E. (1964). Vergleichende Untersuchungen über die bronchoconstrictorische und gefäßconstrictorische Wirkung des Bradykinins, Histamins und Serotonins an isolierten Lungenpräparaten. Naunyn-Schmiedeberg's Arch. Exp. Pathol. Pharmakol. 248, 204–215.

Greenberg, R., Osman, G.H., O'Keefe, E.H. and Antonaccio, M.J. (1979). The effects of captopril (SQ 14,225) on bradykinin-induced bronchoconstriction in the anesthetized guinea pig. Eur. J. Pharmacol. 57, 287–294.

Griesbacher, T., Tiran, B. and Lembeck, F. (1993). Pathological events in experimental acute pancreatitis prevented by the bradykinin antagonist, Hoe 140. Br. J. Pharmacol. 108, 405–411.

Griesbacher, T., Kolbitsch, C., Tiran, B. and Lembeck, F. (1995). Effects of the bradykinin antagonist, icatibant (Hoe 140), on pancreas and liver functions during and after caerulein-induced pancreatitis in rats. Naunyn Schmiedeberg's Arch. Pharmacol. 352, 557–564.

Hall, J.M. (1992). Bradykinin receptors: pharmacological properties and biological roles. Pharmacol. Ther. 56, 131–190.

Hallin, G.W., Simpson, S.Q., Crowell, R.E., James, D.S., Koster, F.T., Mertz, G.J. and Levy, H. (1996). Cardiopulmonary manifestations of hantavirus pulmonary syndrome. Crit. Care Med. 24, 252–258.

Hasegawa, N., Husari, A.W., Hart, W.T., Kandra, T.G. and Raffin, T.A. (1994). Role of the coagulation system in ARDS. Chest 105, 268–277.

Hauge, A., Lunde, P.K.M. and Waaler, B.A. (1966). The effect of bradykinin, kallidin and eledoisin upon the pulmonary vascular bed of an isolated blood-perfused rabbit lung preparation. Acta Physiol. Scand, 66, 269–277.

Havill, A.M., Van Valen, R.G. and Handley, D.A. (1990). Prevention of non-specific airway hyperreactivity after allergen challenge in guinea-pigs by the PAF receptor antagonist SDZ 64-412. Br. J. Pharmacol. 99, 396–400.

Herxheimer, H. and Stresemann, E. (1961). The effect of bradykinin aerosol in guinea-pigs and in man. J. Physiol. 158, 38P.

Hess, J.F., Borkowski, J.A., Young, G.S., Strader, C.D. and Ransom, R.W. (1992). Cloning and pharmacological characterization of a human bradykinin (BK-2) receptor. Biochem. Biophys. Res. Commun. 184, 260–268.

Hirano, T. and Takeuchi, S. (1994). A bradykinin antagonist, D-Arg-[Hyp^3,Thi^{5,8},D-Phe^7]-bradykinin, prevents pancreatic subcellular redistribution of lysosomal enzyme in rats with caerulein-induced acute pancreatitis. Med. Sci. Res. 22, 803–805.

Hock, F.J., Wirth, K., Albus, U., Linz, W., Gerhards, H.J., Wiemer, G., St Henke, G., Breipohl, G., König, W., Knolle, J. and Schölkens, B.A. (1991). Hoe 140 a new potent and long acting bradykinin-antagonist: in vitro studies. Br. J. Pharmacol. 102, 769–773.

Hoffmann, T.F., Leiderer, R., Waldner, H. and Messmer, K.

(1996). Bradykinin antagonists HOE-140 and CP-0597 diminish microcirculatory injury after ischaemia–reperfusion of the pancreas in rats. Br. J. Surg. 83, 189–195.

Hooker, T.P., Hammond, M.D. and Salem, A. (1992). Adult respiratory distress syndrome: a review for the clinician. J. Am. Osteopath. Assoc. 92, 886–896.

Hu, D.-E. and Fan, T.-P.D. (1993). [Leu^8]desArg^9-bradykinin inhibits the angiogenic effect of bradykinin and interleukin-1 in rats. Br. J. Pharmacol. 109, 14–17.

Hua, X.-Y. and Yaksh, T.L. (1993). Pharmacology of the effects of bradykinin, serotonin, and histamine on the release of calcitonin gene-related peptide from C-fiber terminals in the rat trachea. J. Neurosci. 13, 1947–1953.

Hyers, T.M. (1991). Adult respiratory distress syndrome: definition, risk factors, and outcome. In "Adult Respiratory Distress Syndrome" (eds. W.M. Zapol and F. Lemaire), pp. 23–36. Marcel Dekker, New York.

Ichinose, M. and Barnes, P.J. (1990). Bradykinin-induced airway microvascular leakage and bronchoconstriction are mediated via a bradykinin B$_2$ receptor. Am. Rev. Respir. Dis. 142, 1104–1107.

Ichinose, M., Nakajima, N., Takahashi, T., Yamauchi, H., Inoue, H. and Takishima, T. (1992). Protection against bradykinin-induced bronchoconstriction in asthmatic patients by neurokinin receptor antagonist. Lancet 340, 1248–1251.

Ishida, K., Kelly, L.J., Thompson, R.J., Beattie, L.L. and Schellenberg, R.R. (1989). Repeated antigen challenge induces airway hyperresponsiveness with tissue eosinophilia in guinea pigs. J. Appl. Physiol. 67, 1133–1139.

Jaeschke, H., Farhood, A.I. and Smith, C.W. (1991). Neutrophil-induced liver cell injury in endotoxin shock is a CD11b/CD18-dependent mechanism. Am. J. Physiol. 261, C1051–C1056.

Jeffery, P.K., Wardlaw, A.J., Nelson, F.C., Collins, J.V. and Kay, A.B. (1989). Bronchial biopsies in asthma. An ultrastructural, quantitative study and correlation with hyperreactivity. Am. Rev. Respir. Dis. 140, 1745–1753.

Jin, L.S., Seeds, E., Page, C.P. and Schachter, M. (1989). Inhibition of bradykinin-induced bronchoconstriction in the guinea-pig by a synthetic B$_2$ receptor antagonist. Br. J. Pharmacol. 97, 598–602.

Jonasson, O. and Becker, E.L. (1966). Release of kallikrein from guinea-pig lung during anaphylaxis. J. Exp. Med. 123, 509–522.

Kalter, E.S., van Dijk, W.C., Timmerman, A., Verhoef, J. and Bouma, B.N. (1983). Activation of purified human prekallikrein triggered by cell wall fractions of E. coli and S. aureus. J. Infect. Dis. 148, 682–691.

Katori, M., Majima, M., Odoi-Adome, R., Sunahara, N. and Uchida, Y. (1989). Evidence for the involvement of a plasma kallikrein–kinin system in the immediate hypotension produced by endotoxin in anaesthetized rats. Br. J. Pharmacol. 98, 1383–1391.

Katsuura, Y., Kimura, Y., Ohno, N., Naito, T., Okamoto, S. and Wanaka, K. (1989). DIC with or without MOF like syndrome; effect of selective plasma kallikrein synthetic-inhibitor. Thromb. Haemostas. 62, 372.

Khan, M.M.H., Yamamoto, T., Araki, H., Shibuya, Y. and Kambara, T. (1993). Role of Hageman factor/kallikrein-kinin system in pseudomonal elastase-induced shock model. Biochim. Biophys. Acta Gen. Subj. 1157, 119–126.

Kimball, E.S. and Fisher, M.C. (1988). Potentiation of IL-1 induced BALB/3T3 fibroblast proliferation by neuropeptides. J. Immunol. 141, 4203-4208.

Kimura, K., Inoue, H., Ichinose, M., Miura, M., Katsumata, U., Takishima, T. and Takahashi, T. (1992). Bradykinin causes airway hyperresponsiveness and enhances maximal airway narrowing. Role of microvascular leakage and airway edema. Am. Rev. Respir. Dis. 146, 1301–1305.

Koivunen, E., Wang, B., Dickinson, C.D. and Ruoslahti, E. (1994). Peptides in cell adhesion research. In "Extracellular Matrix Components" (eds. E. Ruoslahti and E. Engvall), pp. 346–369. Academic Press, San Diego.

Koyama, S., Rennard, S.I. and Robbins, R.A. (1995). Bradykinin stimulates bronchial epithelial cells to release neutrophil and monocyte chemotactic activity. Am. J. Physiol. 269, L38–L44.

Lecomte, J., Petit, J.M., Mélon, J., Troquet, J. and Marcelle, R. (1962). Propriétés bronchoconstrictrices de la bradykinine chez l'homme asthmatique. Arch. Int. Pharmacodyn. Ther. 137, 232–235.

Lerner, U.H., Brunius, G. and Modeer, T. (1992). On the signal transducing mechanisms involved in the synergistic interaction between interleukin-1 and bradykinin on prostaglandin biosynthesis in human gingival fibroblasts. Biosci. Rep. 12, 263–271.

Levy, H., Rodell, T., Hallin, G., Simpson, S., Modafferi, D. and Neidhart, M. (1995). Treatment of Hantavirus pulmonary syndrome (HPS) with CP-0127, a bradykinin antagonist. Crit. Care Med. 23 (Suppl.) A37.

Liu, M.C., Hubbard, W.C., Proud, D., Stealey, B.A., Galli, S.J., Kagey-Sobotka, A., Bleecker, E.R. and Lichtenstein, L.M. (1991). Immediate and late inflammatory responses to ragweed antigen challenge of the peripheral airways in allergic asthmatics. Am. Rev. Respir. Dis. 144, 51–58.

Lundberg, J.M. and Saria, A. (1983). Capsaicin-induced desensitization of airway mucosa to cigarette-smoke, mechanical and chemical irritants. Nature 302, 251–253.

Marsh, K.A. and Hill, S.J. (1994). Des-Arg^9-bradykinin-induced increases in intracellular calcium ion concentration in single bovine tracheal smooth muscle cells. Br. J. Pharmacol. 112, 934–938.

Mayhan, W.G. (1992). Role of nitric oxide in modulating permeability of hamster cheek pouch in response to adenosine 5'-diphosphate and bradykinin. Inflammation 16, 295–305.

McGuire, W.W., Spragg, R.G., Cohen, A.B. and Cochrane, C.G. (1982). Studies on the pathogenesis of the adult respiratory distress syndrome. J. Clin. Invest. 69, 543–553.

McIntyre, T.M., Zimmerman, G.A., Satoh, K. and Prescott, S.M. (1985). Cultured endothelial cells synthesize both platelet-activating factor and prostacyclin in response to histamine, bradykinin and adenosine monophosphate. J. Clin. Endocrinol. 76, 271–280.

Menke, J.G., Borkowski, J.A., Bierilo, K.K., MacNeil, T., Derrick, A.W., Schneck, K.A., Ransom, R.W., Strader, C.D., Linemeyer, D.L. and Hess, J.F. (1994). Expression cloning of a human B$_1$ bradykinin receptor. J. Biol. Chem. 269, 21583–21586.

Mulligan, M.S., Miyasaka, M., Tamatani, T., Jones, M.L. and Ward, P.A. (1994). Requirements for L-selectin in neutrophil-mediated lung injury in rats. J. Immunol. 152, 832–840.

Murlas, C. and Roum, J.H. (1985). Bronchial hyperreactivity occurs in steroid-treated guinea pigs depleted of leukocytes by cyclophosphamide. J. Appl. Physiol. 58, 1630–1637.

Naess, F., Roeise, O., Pillgram-Larsen, J., Ruud, T.E., Stadaas, J.O. and Aasen, A.O. (1991). Plasma kallikrein generation in endotoxemia is abolished by ultra high doses of methylprednisolone: in vivo studies in a pig model. Circ. Shock 34, 349–355.

Nakajima, N., Ichinose, M., Takahashi, T., Yamauchi, H., Igarashi, A., Miura, M., Inoue, H., Takashima, T. and Shirato, K. (1994). Bradykinin-induced airway inflammation. Am. J. Respir. Crit. Care Med. 149, 694–698.

Neuhof, H. (1991). Actions and interactions of mediator systems and mediators in the pathogenesis of ARDS and multiorgan failure. Acta Anesthesiol. Scand. 35 (Suppl. 95), 7–14.

Nieuwenhuijzen, G.A.P., Meyer, R.A., Hendriks, T. and Goris, R.J.A. (1995). Deficiency of complement factor C5 reduces early mortality but does not prevent organ damage in an animal model of multiple organ dysfunction syndrome. Crit. Care Med. 23, 1686–1693.

Noronha-Blob, L., Prosser, J.C., Lowe, V.C., Sullivan, J.P., Kyle, D.J., Martin, J.A. and Burch, R.M. (1991). NPC-17731 delays the onset of hypotension and reduces mortality in response to lethal doses of endotoxin in rats and mice. "Kinins '91", Proceedings, 365.

Nossaman, B.D., Feng, C.J. and Kadowitz, P.J. (1994). Analysis of responses to bradykinin and influence of HOE 140 in the isolated perfused rat lung. Am. J. Physiol. 266, H2452–H2461.

O'Brodovich, H.M., Stalcup, S.A., Pang, L.M., Lipset, J.S. and Mellins, R.B. (1981). Bradykinin production and increased pulmonary endothelial permeability during acute respiratory failure in unanesthetized sheep. J. Clin. Invest. 67, 514–522.

O'Donnell, S.R. and Anderson, G.P. (1995). The effects of formoterol on plasma exudation produced by a localized acute inflammatory response to bradykinin in the tracheal mucosa of rats in vivo. Br. J. Pharmacol. 116, 1571–1576.

Ofstad, E. (1970). Formation and destruction of plasma kinins during experimental acute hemorrhagic pancreatitis in dogs. Scand. J. Gastroenterol. 5 (Suppl. 5), 7–44.

O'Neill, L.A.J. and Lewis, G.P. (1989). Interleukin-1 potentiates bradykinin- and TNFα-induced PGE$_2$ release. Eur. J. Pharmacol. 166, 131–137.

Otterbein, L., Lowe, V.C., Kyle, D.J. and Noronha-Blob, L. (1993). Additive effects of a bradykinin antagonist, NPC 17761, and a leumedin, NPC 15669, on survival in animal models of sepsis. Agents Actions 39, C125–C127.

Paegelow, I., Werner, H., Vietinghoff, G. and Wartner, U. (1995). Release of cytokines from isolated lung strips by bradykinin. Inflamm. Res. 44, 306–311.

Pang, L.M., O'Brodovich, H.M., Mellins, R.B. and Stalcup, S.A. (1982). Bradykinin-induced increase in pulmonary microvascular permeability in hypoxic sheep. J. Appl. Physiol. 52, 370–375.

Pardi, R., Inverardi, L. and Bender, J.R. (1992). Regulatory mechanisms in leukocyte adhesion: flexible receptors for sophisticated travelers. Immunol. Today 13, 224–230.

Pedersen, K.E., Rigby, P.J. and Goldie, R.G. (1991). Quantitative assessment of increased airway microvascular permeability to [^{125}I]-labelled plasma fibrinogen induced by platelet activating factor and bradykinin. Br. J. Pharmacol. 104, 128–132.

Perkins, M.N., Burgess, G.M., Campbell, E.A., Hallett, A., Murphy, R.J., Naeem, S., Patel, I.A., Rueff, A. and Dray, A.

(1991). HOE140: a novel bradykinin analogue that is a potent antagonist at both B₂ and B₃ receptors *in vitro*. Br. J. Pharmacol. 102, 171P.

Pixley, R.A., DeLa Cadena, R., Page, J.D., Kaufman, N., Wyshock, E.G., Chang, A., Taylor, F.B., Jr and Colman, R.W. (1993). The contact system contributes to hypotension but not disseminated intravascular coagulation in lethal bacteremia. *In vivo* use of a monoclonal anti-factor XII antibody to block contact activation in baboons. J. Clin. Invest. 91, 61–68.

Plitman, J.D. and Snapper, J.R. (1994). Drug therapy for the adult respiratory distress syndrome. In "Drugs and the Lung" (eds. C.P. Page and W.J. Metzger), pp. 521–547. Raven Press, New York.

Poblete, M.T., Garces, G., Figueroa, C.D. and Bhoola, K.D. (1993). Localization of immunoreactive tissue kallikrein in the seromucous glands of the human and guinea-pig respiratory tree. Histochem. J. 25, 834–839.

Polosa, R. and Holgate, S.T. (1990). Comparative airway responses to inhaled bradykinin, kallidin, and [des-Arg⁹]bradykinin in normal and asthmatic subjects. Am. Rev. Respir. Dis. 142, 1367–1371.

Polosa, R., Phillips, G.D., Lai, C.K.W. and Holgate, S.T. (1990). Contribution of histamine and prostanoids to bronchoconstriction provoked by inhaled bradykinin. Allergy 45, 174–182.

Proud, D. and Vio, C.P. (1993). Localization of immuno-reactive tissue kallikrein in human trachea. Am. J. Respir. Cell Mol. Biol. 8, 16–19.

Pruneau, D., Luccarini, J.M., Defrène, E., Paquet, J.L. and Bélichard, P. (1995). Pharmacological evidence for a single bradykinin B₂ receptor in the guinea-pig. Br. J. Pharmacol. 116, 2106–2112.

Pyne, S. and Pyne, N.J. (1993). Differential effects of B₂ receptor antagonists upon bradykinin-stimulated phospholipase C and D in guinea-pig cultured tracheal smooth muscle. Br. J. Pharmacol. 110, 477–481.

Pyne, S. and Pyne, N.J. (1994). Bradykinin-stimulated phosphatidate and 1,2-diacylglycerol accumulation in guinea-pig airway smooth muscle: evidence for regulation "downstream" of phospholipases. Cell. Signal. 6, 269–277.

Qian, Y., Emonds-Alt, X. and Advenier, C. (1993). Effects of capsaicin, (±)-CP 96,345 and SR 48968 on the bradykinin-induced airways microvascular leakage in guinea-pigs. Pulm. Pharmacol. 6, 63–67.

Quezado, Z.M.N., Banks, S.M. and Natanson, C. (1995). New strategies for combatting sepsis: the magic bullets missed the mark...but the search continues. Trends Biotechnol. 13, 56–63.

Ray, N.J., Jones, A.J. and Keen, P. (1991). Morphine, but not sodium cromoglycate, modulates the release of substance P from capsaicin-sensitive neurones in the rat trachea. Br. J. Pharmacol. 102, 797–800.

Regoli, D., Marceau, F. and Lavigne, J. (1981). Induction of B₁ receptors for kinins in the rabbit by a bacterial lipopolysaccharide. Eur. J. Pharmacol. 71, 105–115.

Reynolds, H.Y. (1987). Lung inflammation: normal host defense or a complication of some diseases? Annu. Rev. Med. 38, 295–323.

Ricupero, D., Taylor, L., Tlucko, A., Navarro, J. and Polgar, P. (1992). Mechanisms in bradykinin stimulated arachidonate release and synthesis of prostaglandin and platelet activating factor. Mediators Inflamm. 1, 133–140.

Ridings, P.C., Blocher, C.R., Fisher, B.J., Fowler, A.A., III and Sugerman, H.J. (1995). Beneficial effects of a bradykinin antagonist in a model of gram-negative sepsis. J. Trauma Injury Infect. Crit. Care 39, 81–89.

Rinaldo, J.E. and Christman, J.W. (1990). Mechanisms and mediators of the adult respiratory distress syndrome. Clin. Chest Med. 11, 621–632.

Robbins, R.A., Koyama, S., Spurzen, J., Rickard, K.A., Nelson, K.J., Gossman, G.L., Thiele, G.M. and Rennard, S.I. (1993). Modulation of neutrophil and mononuclear cell adherence to bronchial epithelial cells. Am. J. Respir. Cell Mol. Biol. 7, 19–29.

Rogers, D.F., Dijk, S. and Barnes, P.J. (1990). Bradykinin-induced plasma exudation in guinea-pig airways: involvement of platelet-activating factor. Br. J. Pharmacol. 101, 739–745.

Roman, J., Velasco, F., Fernandez, F., Fernandez, M., Villalba, R., Rubio, V., Vicente, A. and Torres, A. (1993). Coagulation, fibrinolytic and kallikrein systems in neonates with uncomplicated sepsis and septic shock. Haemostasis 23, 142–148.

Sakamoto, T., Elwood, W., Barnes, P.J. and Chung, K.F. (1992). Effect of Hoe 140, a new bradykinin receptor antagonist, on bradykinin- and platelet-activating factor-induced bronchoconstriction and airway microvascular leakage in guinea pig. Eur. J. Pharmacol. 213, 367–373.

Sakamoto, T., Barnes, P.J. and Chung, K.F. (1993). Effect of CP-96,345, a non-peptide NK₁ receptor antagonist, against substance P-, bradykinin- and allergen-induced airway microvascular leakage and bronchoconstriction in the guinea pig. Eur. J. Pharmacol. 231, 31–38.

Sanjar, S., Aoki, S., Kristersson, A., Smith, D. and Morley, J. (1990). Antigen challenge induces pulmonary airway eosinophil accumulation and airway hyperreactivity in sensitized guinea-pigs: the effect of anti-asthma drugs. Br. J. Pharmacol. 99, 679–686.

Saria, A., Martling, C.-R., Yan, Z., Theodorsson-Norheim, E., Gamse, R. and Lundberg, J.M. (1988). Release of multiple tachykinins from capsaicin-sensitive sensory nerves in the lung by bradykinin, histamine, dimethylphenyl piperazinium, and vagal nerve stimulation. Am. Rev. Respir. Dis. 137, 1330–1335.

Saugstad, O.D., Harvie, A. and Langslet, A. (1982). Activation of the kallikrein-kinin system in premature infants with respiratory distress syndrome (RDS). Acta Paediatr. Scand. 71, 965–968.

Saugstad, O.D., Buo, L., Johansen, H.T., Roise, O. and Aasen, A.O. (1992). Activation of the plasma kallikrein–kinin system in respiratory distress syndrome. Pediatr. Res. 32, 431–435.

Sawutz, D.G., Salvino, J.M., Dolle, R.E., Seoane, P.R. and Farmer, S.G. (1995). Pharmacology and structure–activity relationships of the non-peptide bradykinin receptor antagonist, Win 64338. Can. J. Physiol. Pharmacol. 73, 805–811.

Scherrer, D., Daeffler, L., Trifilieff, A. and Gies, J.-P. (1995). Effects of WIN 64338, a nonpeptide bradykinin B₂ receptor antagonist, on guinea-pig trachea. Br. J. Pharmacol. 115, 1127–1128.

Schlag, G. and Redl, H. (1989). Pathophysiology of acute respiratory failure. In "The Kallikrein–Kinin System in Health and Disease" (eds. H. Fritz, I. Schmidt and G. Dietze), pp. 331–355. Limbach-Verlag Braunschweig, Munich.

Schuster, D.P. (1995). What is acute lung injury? What is ARDS? Chest 107, 1721–1726.

Seeger, W., Neuhof, H., Graubert, E., Wolf, H. and Roka, L. (1983). Comparative influence of the Ca-ionophore A23187, bradykinin, kallidin and eldoisin on the rabbit pulmonary vasculature with special reference to arachidonate metabolism. Adv. Exp. Med. Biol. 156A, 533–551.

Shimizu, I., Wada, S., Okahisa, T., Kamamura, M., Yano, M., Kodaira, T., Nishino, T., Shima, K. and Ito, S. (1993). Radioimmunoreactive plasma bradykinin levels and histological changes during the course of cerulein-induced pancreatitis in rats. Pancreas 8, 220–225.

Solér, M., Sielczak, M. and Abraham, W.M. (1990). A bradykinin antagonist blocks antigen-induced airway hyperresponsiveness and inflammation in sheep. Pulm. Pharmacol. 3, 9–15.

Solomkin, J.S. and Hill, C.S. (1995). Design problems in clinical sepsis trials. Prog. Surg. 20, 59–66.

Strieter, R.M. and Kunkel, S.L. (1994). Acute lung injury: the role of cytokines in the elicitation of neutrophils. J. Invest. Med. 42, 640–651.

Svensson, C., Andersson, M., Persson, C.G.A., Venge, P., Alkner, U. and Pipkorn, U. (1990). Albumin, bradykinins, and eosinophil cationic protein on the nasal mucosal surface in patients with hay fever during natural allergen exposure. J. Allergy Clin. Immunol. 85, 828–833.

Szabó, C., Battistini, B., Regoli, D., Thiemermann, C. and Vane, J.R. (1994). Role of bradykinin and platelet-activating factor in the immediate release of nitric oxide in response to endotoxin in the anaesthetized rat. Br. J. Pharmacol. 112, 234P.

Tiffany, C.W. and Burch, R.M. (1989). Bradykinin stimulates tumor necrosis factor and interleukin 1 release from macrophages. FEBS Lett. 247, 189–192.

Traber, D.L., Redl, H., Schlag, G., Herndon, D.N., Kimura, R., Prien, T. and Traber, L.D. (1988). Cardiopulmonary responses to continuous administration of endotoxin. Am. J. Physiol. 254, H833–H839.

Trifilieff, A., Haddad, E.-B., Landry, Y. and Gies, J.-P. (1991). Evidence for two high affinity bradykinin binding sites in the guinea-pig lung. Eur. J. Pharmacol. 207, 129–134.

Tsukagoshi, H., Sakamoto, T., Xu, W., Barnes, P.J. and Chung, K.F. (1994). Effect of interleukin-1β on airway hyperresponsiveness and inflammation in sensitized and nonsensitized Brown–Norway rats. J. Allergy Clin. Immunol. 93, 464–469.

Uchida, D.A., Ackerman, S.J., Coyle, A.J., Larsen, G.L., Weller, P.F., Freed, J. and Irvin, C.G. (1993). The effect of human eosinophil granule major basic protein on airway responsiveness in the rat in vivo. A comparison with polycations. Am. Rev. Respir. Dis. 147, 982–988.

Ueno, A., Ishida, H. and Oh-ishi, S. (1995a). Comparative study of endotoxin-induced hypotension in kininogen-deficient rats with that in normal rats. Br. J. Pharmacol. 114, 1250–1256.

Ueno, A., Tokumasu, T., Naraba, H. and Oh-ishi, S. (1995b). Involvement of bradykinin in endotoxin-induced vascular permeability increase in the skin of rats. Eur. J. Pharmacol. 284, 211–214.

Vandekerckhove, F., Opdenakker, G., Van Ranst, M., Lenaerts, J.-P., Put, W., Billiau, A. and Van Damme, J. (1991). Bradykinin induces interleukin-6 and synergizes with interleukin-1. Lymphokine Cytokine Res. 10, 285–289.

Varonier, H.S. and Panzani, R. (1968). The effect of inhalations of bradykinin on healthy and atopic (asthmatic) children. Int. Arch. Allergy. 34, 293–296.

Venge, P., Dahl, R. and Hallgren, R. (1979). Enhancement of factor XII reactions by eosinophil cationic protein. Thromb. Res. 14, 641–649.

Wanaka, K., Okamoto, S., Bohgaki, M., Hijikata-Okunomiya, A., Naito, T. and Okada, Y. (1990). Effect of a highly selective plasma-kallikrein synthetic inhibitor on contact activation relating to kinin generation, coagulation and fibrinolysis. Thromb. Res. 57, 889–895.

Wardlaw, A.J., Dunnette, S., Gleich, G.J., Collins, J.V. and Kay, A.B. (1988). Eosinophils and mast cells in the bronchoalveolar lavage in subjects with mild asthma. Relation to bronchial hyperreactivity. Am. Rev. Respir. Dis. 137, 62–70.

Weipert, J., Hoffmann, H., Siebeck, M. and Whalley, E.T. (1988). Attenuation of arterial blood pressure fall in endotoxin shock in the rat using the competitive bradykinin antagonist Lys-Lys-[Hyp^2,$Thi^{5,8}$,D-Phe^7]-BK (B4148). Br. J. Pharmacol. 94, 282–284.

Whalley, E.T., Soloman, J.A., Modafferi, D.M., Bonham, K.A. and Cheronis, J.C. (1992). CP-1027, a novel bradykinin antagonist increases survival in rat and rabbit models of endotoxin shock. Agents Actions 38 (Suppl. III), 413–420.

Widdicombe, J.H., Ueki, I.F., Emery, D., Margolskee, D., Yergey, J. and Nadel, J.A. (1989). Release of cyclooxygenase products from primary cultures of tracheal epithelia of dog and human. Am. J. Physiol. 257, L361–L365.

Wiedemann, H.P., Matthay, M.A. and Gillis, C.N. (1990). Pulmonary endothelial cell injury and altered lung metabolic function. Early detection of the adult respiratory distress syndrome and possible functional significance. Clin. Chest Med. 11, 723–736.

Wiemer, G. and Wirth, K. (1992). Production of cyclic GMP via activation of B_1 and B_2 kinin receptors in cultured bovine aortic endothelial cells. J. Pharmacol. Exp. Ther. 262, 729–733.

Willatts, S.M., Radford, S. and Leitermann, M. (1995). Effect of the antiendotoxic agent, taurolidine, in the treatment of sepsis syndrome: a placebo-controlled, double-blind trial. Crit. Care Med. 23, 1033–1039.

Wilson, D.D., de Garavilla, L., Kuhn, W., Togo, J., Burch, R.M. and Steranka, L.R. (1989). D-Arg-[Hyp^3-D-Phe^7]-bradykinin, a bradykinin antagonist, reduces mortality in a rat model of endotoxic shock. Circ. Shock 27, 93–101.

Yoshihara, S., Geppetti, P., Hara, M., Lindén, A., Ricciardolo, F.L.M., Chan, B. and Nadel, J.A. (1996). Cold air-induced bronchoconstriction is mediated by tachykinin and kinin release in guinea pigs. Eur. J. Pharmacol. 296, 291–296.

16. Bradykinin Formation in Allergic Diseases and Hereditary Angioedema

Sesha R. Reddigari, Michael Silverberg *and* Allen P. Kaplan

1. Introduction

The plasma kinin-forming system consists of three essential plasma proteins, which interact in a complex fashion once bound to certain negatively charged surfaces or macromolecular complexes. These are coagulation Factor XII (also called Hageman Factor), prekallikrein and high molecular weight kininogen (HMWK). Once Factor XII is activated to Factor XIIa, it converts plasma prekallikrein to kallikrein and kallikrein digests HMWK to liberate bradykinin (BK) Factor XIIa also converts coagulation Factor XI to Factor XIa to continue the intrinsic coagulation cascade. The interactions of all four of these proteins to initiate blood clotting is known as "contact activation", thus the formation of BK is a cleavage product of the initiating step of this cascade (Fig. 16.1).

2. Protein Constituents of the Kallikrein–Kinin System

2.1 FACTOR XII

Factor XII circulates as a single-chain zymogen with no detectable enzymatic activity (Silverberg and Kaplan, 1982). It has a molecular weight of 80,000 by sodium dodecyl sulfate (SDS) gel electrophoresis (Revak *et al.*, 1974) (Table 16.1), is synthesized in the liver and circulates in plasma at a concentration of 30–35 µg/ml. Its primary sequence has been deduced from cDNA analysis (Cool *et al.*, 1985; Que and Davie, 1986) and from direct protein sequence data (Fujikawa and McMullen, 1983; McMullen and Fujikawa, 1985). The 596 amino acids present account for a molecular weight

Copyright © 1997 Academic Press Limited
All rights of reproduction in any form reserved.

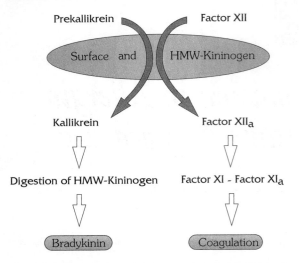

Figure 16.1   Relationship of bradykinin formation to the initiation of the intrinsic coagulation cascade.

of 66,915, the remainder (16.8%) being carbohydrate. The protein has distinct domains homologous to fibronectin, plasminogen and plasminogen activators (Cool *et al.*, 1985; Castellino and Beals, 1987) at its N-terminus while the C-terminus has the catalytic domain. This latter portion is homologous to serine proteases such as pancreatic trypsin and even more so to the catalytic domain of plasminogen activators.

Factor XII is unusual because it is capable of autoactivating once bound to initiating "surfaces" (Silverberg *et al.*, 1980a; Tankersley and Finlayson, 1984). Thus, Factor XII that is bound undergoes a conformational change, which renders it a substrate for Factor XIIa (Griffin, 1978). Gradually, all of the bound Factor XII can be converted to Factor XIIa. Whether plasma normally has a trace of Factor XIIa present is unknown, but if so, its concentration is less than 0.01% of Factor XII. The alternative is that the first molecule of Factor XIIa is formed by the interaction of two Factor XII zymogen molecules on the surface, but this presumes some minimal activity present in the zymogen. If so, it is below our limits of detection and we favor the former scenario (Silverberg and Kaplan, 1982).

Activation of Factor XII is due to cleavage of the

molecule at a critical Arg-Val bond (Fujikawa and McMullen, 1983) contained within a disulfide bridge such that the resultant Factor XIIa is a two-chain, disulfide-linked 80 kDa enzyme consisting of a heavy chain of 50 kDa and a light chain of 28 kDa (Revak *et al.*, 1977). The light chain contains the enzymatic active site (Meier *et al.*, 1977) and is at the carboxy-terminal end, while the heavy chain contains the binding site for the surface and is at the amino terminus (Pixley *et al.*, 1987). Further cleavage can occur at the C-terminal end of the heavy chain to produce a series of fragments of activated Factor XII, which retain enzymatic activity (Kaplan and Austen, 1970, 1971). The most prominent of these is a 30 kDa species termed Factor XII$_f$. Careful examination of Factor XII$_f$ on SDS gels under nonreducing conditions reveals a doublet in which the higher band at 30 kDa is gradually converted to the lower band, which has a molecular weight of 28.5 kDa (Dunn and Kaplan, 1982). Reduced SDS gels demonstrate that these species are comprised of the light chain of Factor XIIa and a very small piece of the original heavy chain. These fragments lack the binding site to the surface and lose much of the ability of Factor XIIa to convert Factor XI to Factor XIa and do not participate in Factor XII autoactivation. However, these fragments remain potent activators of prekallikrein (Kaplan and Austen, 1970). Thus, formation of Factor XII$_f$ allows BK production to continue until the enzyme is inactivated and the reactions can proceed at sites distant from the initiating surface. A diagrammatic representation of the cleavages in Factor XII to generate Factor XIIa and Factor XII$_f$ is shown in Fig. 16.2.

Once Factor XIIa interacts with prekallikrein, rapid conversion to kallikrein ensues, followed by an important positive feedback in which kallikrein digests surface-bound Factor XII to form Factor XIIa and then Factor XII$_f$ (Cochrane *et al.*, 1973; Meier *et al.*, 1977; Dunn and Kaplan, 1982). This reaction is 50–100 times more rapid than the autoactivation reaction (Dunn and Kaplan, 1982; Tankersley and Finlayson, 1984). Thus, quantitatively, most of the Factor XIIa or Factor XII$_f$ activity generated when plasma is activated is a result of kallikrein activation of Factor XII. Yet the autoactivation phenomenon can be demonstrated in plasma that is congenitally deficient in prekallikrein (Fletcher trait) and cannot therefore generate BK (Wuepper, 1973; Saito *et al.*, 1974; Weiss *et*

Table 16.1   Physical and chemical data of contact activation proteins

Protein	Factor XII	Prekallikrein	Factor XI	HMWK
Molecular weight (kDa)	80.4	79.5	140	116.7
% Carbohydrate	16.8	15	5	40
Isoelectric point	6.3	8.7	8.6	4.7
Extinction coefficient, 280 nm ($E^{1\%}$)	14.2	11.7	13.4	7
Plasma concentration, μg/ml (nM)	30–35 (400)	35–50 (534)	4–6 (36)	70–90 (686)

HMWK, high molecular weight kininogen.

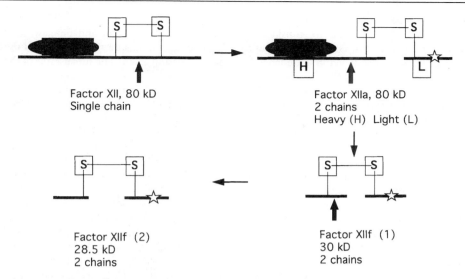

Figure 16.2   Activation of human Hageman factor (Factor XII) to form Factor XIIa and then Factor XII$_f$. Stars, active sites; bold arrows, cleavage sites.

al., 1974). Clotting in Fletcher trait plasma (i.e., conversion of Factor XI to Factor XIa by Factor XIIa) does proceed, albeit at a much slower rate and the partial thromboplastin time (PTT) can be shown to shorten progressively as the time of incubation of the plasma with the surface is increased prior to recalcification. This is possibly due to Factor XII autoactivation on the surface. As more and more Factor XIIa forms, the rate of Factor XI activation increases and the PTT approaches normal.

2.2   PREKALLIKREIN

Prekallikrein is also a zymogen without detectable proteolytic activity, which is converted to kallikrein by cleavage during contact activation (Mandle and Kaplan, 1977). On SDS gels, it has two bands at 88 kDa and 85 kDa. The entire amino-acid sequence of the protein has been determined by a combination of direct protein sequencing and amino-acid sequence prediction from cDNAs isolated from a λgt-11 expression library (Chung *et al.*, 1986). A signal peptide of 19 residues (which is cleaved off prior to secretion) is followed by the sequence of the mature plasma prekallikrein, which has 619 amino acids with a calculated molecular weight of 69,710. There is 15% carbohydrate as well. The heterogeneity on SDS gel electrophoresis is not reflected in the amino-acid sequence, and thus it may be due to variation in glycosylation. Activation of prekallikrein by Factor XIIa or Factor XII$_f$ is due to cleavage of a single Arg-Ile bond within a disulfide bridge such that a heavy chain of 56 kDa is disulfide linked to a light chain of either 33 kDa or 36 kDa, each of which has a di-isofluorophosphate-inhibitable active site (Bouma *et al.*, 1980; Mandle and Kaplan, 1977). This light-chain heterogeneity reflects the two forms of the zymogen.

The amino-acid sequence of the kallikrein heavy chain is unusual and is homologous only to the corresponding portion of Factor XI. It has four tandem repeats, each of which contains approximately 90–91 amino acids. The presence of six cysteines per repeat suggests a repeating structure with three disulfide loops. It is postulated that a gene coding for the ancestor of this repeat sequence duplicated and then the entire segment duplicated again to give the present structure. The light chain, containing the active site is homologous to many of the catalytic domains of other enzymes of the coagulation cascade.

In contrast to Factor XII, prekallikrein does not circulate as a separate protein. It is bound to HMWK in a 1:1 bimolecular complex through a site on its heavy chain. The binding is firm, with a dissociation constant of 12–15 nM (Mandle *et al.*, 1976; Bock *et al.*, 1985) and this is unchanged upon conversion of prekallikrein to kallikrein. Thus, about 80–90% of prekallikrein is normally complexed to HMWK in plasma (Reddigari and Kaplan, 1989b). It is the prekallikrein–HMWK complex that binds to surfaces during contact activation and the binding is primarily through HMWK (Wiggins *et al.*, 1977), although some interaction of prekallikrein with the surface can be inferred (McMillin *et al.*, 1974). The dissociation from the surface of 10–20% of the kallikrein which forms may serve to propagate the formation of BK in the fluid phase and at sites distant from the initiating reaction (Cochrane and Revak, 1980; Silverberg *et al.*, 1980b).

2.3   PROKALLIKREIN AND TISSUE KALLIKREINS

Tissue kallikrein is a single-chain acidic glycoprotein that is physicochemically and immunologically distinct from plasma kallikrein. The reported molecular weight ranges from 25 to 43 kDa, which is due to considerable heterogeneity caused largely by proteolysis subsequent to secretion. Tissue kallikrein is produced in abundance by major exocrine organs such as the pancreas and salivary glands, as well as the kidney (Ole-Moiyoi *et al.*, 1977). However, it is actually widespread and can be found in the colon (Schachter *et al.*, 1983), prostate, the pituitary gland (Powers and Nasjletti, 1982; Figueroa *et al.*, 1989), brain tissue (Chao *et al.*, 1983) and neutrophils (Figueroa *et al.*, 1989).

Within a given species, tissue kallikreins derived from all organs are immunologically identical (Fritz *et al.*, 1977), and cDNA cloning and subsequent analysis reveals the same amino-acid sequence of 238 residues (Baker and Shine, 1985; Fukushima *et al.*, 1985). In rodents, tissue kallikrein is a member of a large multigene family of over 20 members; in humans it is encoded by no more than two or three closely related genes but only one yields bona fide tissue kallikrein (Baker and Shine, 1985; Wines *et al.*, 1989). Tissue kallikrein is synthesized within these cells (tissues) as a pre-proenzyme and in some organs, the proenzyme, i.e., prokallikrein predominates intracellularly where it is converted to tissue kallikrein. Urine, for example, contains a mixture of prokallikrein and tissue kallikrein (Takada *et al.*, 1985). A multitude of isoelectric forms can be found which reflect microheterogeneity owing to variable glycosylation. Intracellular conversion of prokallikrein to kallikrein occurs by enzymes that are not well characterized. Secreted prokallikrein can be converted to tissue kallikrein extracellularly by either plasmin or plasma kallikrein (Takada *et al.*, 1985; Takahashi *et al.*, 1986). The molecular biology of tissue kallikrein is reviewed in Chapter 5.

2.4 FACTOR XI

Coagulation Factor XI is the second substrate of Factor XIIa (Fig. 16.1), but it has no role in BK formation. Factor XI is unique among the clotting factors because the circulating zymogen consists of two identical chains linked by disulfide bonds (Bouma and Griffin, 1977; Kurachi and Davie, 1977). The dimer has an apparent molecular weight of 160 kDa on SDS gel electrophoresis but reveals a single 80 kDa protein upon reduction. Factor XI activation follows the familiar pattern of cleavage of a single peptide bond (Arg-Ile) within a disulfide bridge to yield an amino-terminal heavy chain of 50 kDa and a disulfide-linked light chain of 33 kDa. Since both subunits can be cleaved by Factor XIIa, and each resultant light chain bears a functional active site, Factor XIa is a four-chain protein with two active sites. The concentration in plasma is only 4–8 µg/ml, the lowest among the contact proteins, and its heavy chain(s), like that of kallikrein, binds to the light chain of HMWK. Thus, Factor XI and HMWK also circulate as a complex (Thompson *et al.*, 1977). The dissociation constant is 70 nM (Tait and Fujikawa, 1987), which is high enough to ensure that virtually all the Factor XI in plasma is complexed. The molar ratio of the complex can consist of one or two molecules of HMWK per Factor XI because of the dimeric nature of Factor XI (Warn-Cramer and Bajaj, 1985). The binding site for HMWK on Factor XI has been localized to the first (N-terminal) tandem repeat (Baglia *et al.*, 1989). The Factor XI–HMWK complex binds to the surface and conversion to Factor XIa must occur on the surface. Fluid-phase conversion by Factor XII$_f$ is only 2–4% of that of surface-bound Factor XIIa. The primary function of Factor XIa is to activate Factor IX to IXa, which is the first calcium-dependent reaction in the intrinsic coagulation cascade.

The amino-acid sequence of human Factor XI has been determined by cDNA analysis (Kaplan and Austen, 1971). It has a 19-amino-acid leader peptide followed by a 60% amino-acid sequence for each of the two chains of the mature protein. The amino-acid sequence of the heavy chain of Factor XIa, like that of kallikrein, has four tandem repeats of about 90 amino acids each with six cysteines/repeat implying three disulfide bonds. Unpaired cysteines in the first and fourth repeats are postulated to form the interchain disulfide bridges between monomers to produce the homodimer.

2.5 HIGH MOLECULAR WEIGHT KININOGEN

High molecular weight kininogen circulates in plasma as a 115 kDa nonenzymatic glycoprotein at concentration of 70–90 µg/ml (Proud *et al.*, 1980; Adam *et al.*, 1985; Berrettini *et al.*, 1986; Colman and Müller-Esterl, 1988; Reddigari and Kaplan, 1989b). Its apparent molecular weight by gel filtration is aberrant at about 200,000 indicative of a large partial specific volume owing to its conformation in solution (Mandle *et al.*, 1976). It forms noncovalent complexes with both prekallikrein and Factor XI with dissociation constants of 15 nM (Bock and Shore, 1983; Bock *et al.*, 1985) and 70 nM (Thompson *et al.*, 1979; Tait and Fujikawa, 1987), respectively. There is sufficient HMWK in plasma theoretically to bind both Factor XIIa substrates, and the excess HMWK (about 10–20%) circulates uncomplexed. The complexes of HMWK with prekallikrein or Factor XI are formed with the light-chain region of HMWK, the isolated light chain (after reduction and alkylation) possessing the same binding characteristics as the whole molecule (Thompson *et al.*, 1978, 1979). HMWK functions as a coagulation cofactor and this activity resides in the light chain (Thompson *et al.*, 1978), which consists of a basic (histidine-rich) amino-terminal domain that binds to initiating surfaces (Ikari *et al.*, 1981) and a carboxy-terminal domain that binds prekallikrein or Factor XI (Tait and Fujikawa, 1986). The one cysteine residue in the light chain links it to the heavy chain. The prekallikrein binding site maps to residues 194–224 and the Factor XI site to residues 185–242 (Tait and Fujikawa, 1986, 1987). Since these sites overlap, one molecule of HMWK can only harbor either prekallikrein or Factor XI at a time.

During contact activation, kallikrein cleaves HMWK at two positions within a disulfide bridge, first at the C-terminal Arg-Ser (Mori and Nagasawa, 1981; Mori *et al.*, 1981) and then at the N-terminal Lys-Arg to release the nonapeptide, BK (Arg-Pro-Pro-Gly-Phe-Ser-Pro-Phe-Arg). A two-chain disulfide-linked kinin-free HMWK results, consisting of a heavy chain of 65,000 and a light chain variously reported at molecular weights of 56–62,000. A subsequent further cleavage of the light chain yields a final product of 46–49,000 (Schiffman *et al.*, 1980; Mori and Nagasawa, 1981; Mori *et al.*, 1981; Bock and Shore, 1983; Reddigari and Kaplan, 1988), which retains all light-chain functions. Tissue kallikrein can also digest HMWK to liberate Lys-BK (kallidin) leaving the heavy-chain disulfide linked to the 56–62 kDa light chain. The additional cleavage of the light chain is not made by this enzyme (Reddigari and Kaplan, 1988). It is important to note that tissue kallikrein is immunologically and structurally unrelated to plasma kallikrein. It is secreted by various organs or cells such as salivary glands, kidney, pancreas, prostate, pituitary gland and neutrophils, and is found in high concentrations in saliva, urine and prostatic fluid. Its primary substrate is low molecular weight kininogen (LMWK), but it can release Lys-BK from either HMWK or LMWK. Lys-BK is functionally very similar to BK, albeit slightly less potent. A plasma amino peptidase (Guimaraes *et al.*, 1973) removes the N-terminal Lys to convert it to BK. LMWK is discussed further in Section 2.6.

The very unusual domain structure of HMWK is shown in Fig. 16.3. Domain 5, the histidine-rich region at the N-terminal end of the light chain, binds to initiating surfaces, while the binding of prekallikrein or Factor XI at the C-terminal domain 6 of the light chain account for the cofactor function of HMWK in intrinsic coagulation and kinin generation. The complete amino-acid sequence of HMWK has been determined as translated from the cDNA as well as by direct sequence analysis of the purified protein (Kitamura *et al.*, 1985; Kellermann *et al.*, 1986; Lottspeich *et al.*, 1985; Takagaki *et al.*, 1985). HMWK has 626 amino acids with a calculated molecular weight of 69,896. An unusually high content of carbohydrate accounts for 40% of the observed molecular weight of 115 kDa. The heavy chain of 362 residues is derived from the N-terminus. This is followed by the nine-residue BK (domain 4) sequence and then the light chain of 265 residues. The N-terminal end is blocked with pyroglutamic acid (cyclic glutamine). The carbohydrate is distributed via three N-linked glycosidic linkages on the H chain and nine O-linked glycosidic linkages on the L chain. The H chain has three contiguous and homologous "apple"-type domains consisting of residues 1–116, 117–238 and 239–360 (Fig. 16.3). There are 17 cysteines, one of which is disulfide linked to the L chain and the other from eight disulfide loops within these domains (Kellermann *et al.*, 1986). The three domains on the heavy chain are homologous to the cystatin family of protease inhibitors (of sulfhydryl proteases such as cathepsins). Domains 2 and 3 (but not 1) retain this inhibitory function such that, for example, native HMWK can bind and inactivate two molecules of papain (Gounaris *et al.*, 1984; Müller-Esterl *et al.*, 1985a; Higashiyama *et al.*, 1986; Ishiguro *et al.*, 1987). Limited proteolysis of the heavy chain can occur at susceptible bonds that separate the domains so that individual domains can be isolated. Cleavage at these sites may occur under certain pathologic conditions.

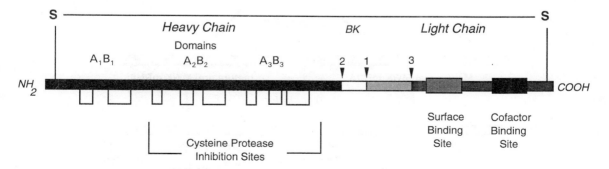

Figure 16.3   The structure of high molecular weight kininogen. The heavy-chain region consists of three homologous domains – A₁B₁, A₂B₂ and A₃B₃ – of which the latter two are sulfhydryl protease inhibition sites. The small rectangles under the heavy chain represent the eight intra-heavy-chain disulfide bonds. The light-chain region contains the surface-binding site and overlapping binding sites (cofactor) for prekallikrein and Factor XI. The heavy and light chains are held together by the ninth disulfide bond. Arrowheads 1 and 2 indicate the sites of cleavage by plasma kallikrein to release bradykinin, and arrowhead 3, secondary cleavage of the light chain.

2.6 Low Molecular Weight Kininogen and Kininogen Genes

Plasma contains another precursor of BK (MW 68 kDa), low molecular weight kininogen. As noted above, its digestion by tissue kallikrein yields Lys-BK and a kinin-free two-chain molecule consisting of a 65 kDa heavy-chain disulfide linked to a light chain of only 4 kDa (Jacobsen and Kritz, 1967; Lottspeich *et al.*, 1985; Müller-Esterl *et al.*, 1985b; Johnson *et al.*, 1987). LMWK is not cleaved by plasma kallikrein. The amino-acid sequences of HMWK and LMWK are identical from the amino terminus through to the BK sequence plus the next 12 residues (Müller-Esterl *et al.*, 1985b) after which the two sequences diverge. Thus, LMWK does not bind to surfaces or to prekallikrein or Factor XI. The kininogens are produced from a single gene thought to have originated by two successive duplications of a primordial cystatin-like gene (Kitamura *et al.*, 1985). As represented in Fig. 16.4, there are 11 exons. The first nine code for the heavy chain and each of the three domains in this portion of the protein is represented by three exons. The tenth exon codes for BK and the light chain of HMWK, while the light chain of LMWK is encoded by exon 11. The mRNAs for HMWK and LMWK are produced by alternative splicing at a point 12 amino acids beyond the BK sequence, thus enabling the two proteins to have different light chains (Fig. 16.4).

3. *Mechanisms of Bradykinin Formation*

3.1 Contact Activation

Contact activation was initially observed by the interaction of blood with glass surfaces (Margolis, 1958); subsequently, finely divided kaolin was used extensively as an experimental surface and for coagulation assays such as the PTT (Proctor and Rapaport, 1961). Ellagic acid (Ratnoff and Crum, 1964), a tannin-like substance used as a component of many commercial assay systems, was purported to be a soluble initiator, but was later shown to form large sedimentable aggregates catalyzed by trace heavy metal ions, so it too is particulate (Bock *et al.*, 1981). More recently, dextran sulfate (Kluft, 1978; Fujikawa *et al.*, 1980) and sulfatide (Fujikawa *et al.*, 1980) have been used to study contact activation. Although sulfatide, a galactose sulfate sphingolipid found in nerve tissue, is an activator, it occurs in quantities too small to be an effective activator. When purified, however, it can form highly charged micelles, which are very efficient initiators (Tans and Griffin, 1982; Griep *et al.*, 1985). Dextran

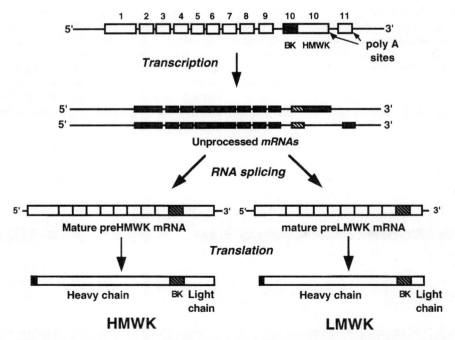

Figure 16.4   The gene for high molecular weight kininogen (HMWK). The boxes labeled 1–9 represent the exon coding for the heavy chain of both HMWK and low molecular weight kininogen (LMWK). Exon 10 codes for the bradykinin (BK) sequence and the light chain of HMWK, whereas exon 11 codes for the light chain of LMWK. The mature mRNAs are assembled by alternative splicing events in which the light-chain sequences are attached to the 3′-end of the 12-amino-acid common sequence C terminal to BK.

sulfate is a truly soluble activator and a close homologue of naturally occurring sulfated mucopolysaccharides. High molecular weight preparations of 500 kDa are typically used (Fujikawa *et al.*, 1980; Silverberg and Kaplan, 1982; Tankersley and Finlayson, 1984), but in a study of Factor XII autoactivation (Silverberg and Diehl, 1987b), much smaller fractions were effective down to as low as 5 kDa. The rate of Factor XII activation increased markedly with dextran sulfate at 10 kDa (or more) where the theoretical number of Factor XII molecules capable of binding per particle increased from one to two, and similar results were seen with heparin. This presumably provides a critical intermolecular interaction required for optimal autoactivation.

Naturally occurring polysaccharides are effective for inducing contact activation if they are highly sulfated, and these include heparin and chondroitin sulfate E (described in rodent mucosal mast cells) (Hojima *et al.*, 1984). Other mucopolysaccharides known to catalyze Factor XII autoactivation are dermatan sulfate, keratin polysulfate or chondroitin sulfate C. The basement membrane of endothelial cell matrix may support contact activation, but this has not been demonstrated *in vivo*. Collagen, long thought to be an initiator, was proven to be ineffective and the activity reported was possibly due to contaminating matrix proteins. One pathophysiologic substance very likely to initiate contact activation *in vivo* is bacterial endotoxin (Pettinger and Young, 1970; Morrison and

Cochrane, 1974; Roeise *et al.*, 1988), and there is good reason to believe that the contact cascade is activated in septic shock and the observed symptoms are due, in part, to the generation of BK (Mason *et al.*, 1970; Kaufman *et al.*, 1991). Crystals of uric acid and pyrophosphate can also initiate kinin formation via this pathway (Kellermeyer and Breckenridge, 1965; Ginsberg *et al.*, 1980).

3.1.1 Regulation of Contact Activation

The various interactions of these constituents are shown in Fig. 16.5, which also includes the steps inhibited by C1 inhibitor. The autoactivation of Factor XII as shown is very slow. However, the reciprocal reactions involving kallikrein contribute to a tremendously fast activation of Factor XII as illustrated by the finding that, if one molecule each of Factor XIIa and kallikrein per milliliter are present in a mixture of Factor XII and prekallikrein at their plasma concentrations, 50% of Factor XII would be activated in 13 s (Tankersley and Finlayson, 1984). This corresponds to $5 \times 10^{-13}\%$ of active enzyme in the preparations. The *in vivo* source of the active enzyme is unknown, but may be formed by other plasma proteases, e.g., plasmin or activation along cell surfaces. In fact, very slow turnover of the cascade may always be occurring (Silverberg and Kaplan, 1982; Bernardo *et al.*, 1993a,b; Reddigari *et al.*, 1993a; Shibayama *et al.*, 1994) and controlled by plasma inhibitors (Weiss *et al.*, 1986). Introduction of a surface or other polyanionic substance

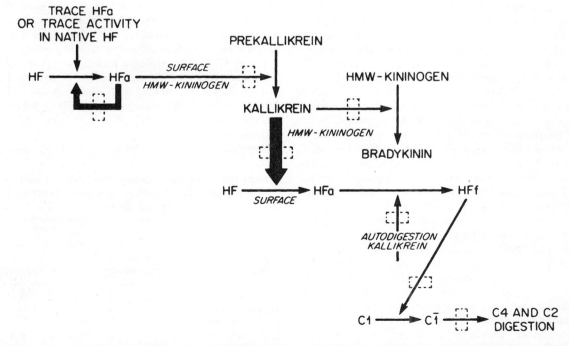

Figure 16.5 Pathway for bradykinin formation indicating the autoactivation of Hageman factor (HF or Factor XII), the positive feedback by which kallikrein activates HF, cleavage of high molecular weight (HMW) kininogen to release bradykinin, formation of Hageman factor fragment (HFf) and enzymatic activation of C1. The steps inhibitable by C1 inhibitor are indicated by the rectangles of broken lines.

could accelerate many thousand-fold the baseline turnover of Factor XII and prekallikrein to ignite the cascade. The addition of the cofactor HMWK (which was not included in the aforementioned kinetic analysis) accelerates these reactions even further, but requires the surface to be present. The surface appears to create a local milieu in the contiguous fluid (Griffin and Cochrane, 1976; Griep *et al.*, 1985; Silverberg and Diehl, 1987a) phase where the local concentrations of reactants are greatly increased, which increases the rates of the reciprocal interaction. In addition, surface-bound Factor XII undergoes a conformational change that renders it more susceptible to cleavage (Griffin, 1978). An alternative idea (McMillin *et al.*, 1974; Ratnoff and Saito, 1979; Kurachi *et al.*, 1980), that binding of Factor XII induces a conformation change that exposes an active site, has essentially been disproved. Inhibitors such as C1 inhibitor are not bound to the surface, and thus the balance between activation and inactivation is upset. The effect of dilution on plasma also diminishes the effect of inhibitors far more than any slowing of enzymatic reaction rates. The net effect is, therefore, a marked augmentation of reaction rate. Using dextran sulfate, the effect of the surface upon Factor XIIa conversion of prekallikrein to kallikrein was 70-fold (Tankersley and Finlayson, 1984), while the effect upon digestion of Factor XII by kallikrein was as much as 3,000–12,000-fold (Tankersley and Finlayson, 1984; Rosing *et al.*, 1985). This latter reaction is about 2,000-fold more rapid than the rate of Factor XII autoactivation, and this kinetic dominance means that prekallikrein must be considered to be a coagulation factor. As indicated earlier, the PTT of prekallikrein-deficient plasma is much prolonged, but does autocorrect as Factor XII autoactivates on the surface. On the other hand, Factor XII-deficient plasma has a markedly abnormal PTT, does not autocorrect, and is essentially devoid of intrinsic clotting or kinin formation. Alternatively, purified Factor XII preparations activate when tested with a surface or polyanion under physiologic conditions (Silverberg *et al.*, 1980b; Tans *et al.*, 1983; Tankersley and Finlayson, 1984), whereas prekallikrein does not. Hence, Factor XII is considered absolutely requisite for intrinsic coagulation while prekallikrein acts as an accelerator.

In plasma, the involvement of HMWK was indicated by the discovery of persons whose plasma had a very prolonged PTT and who generated no BK upon incubation with kaolin, but were not deficient in Factor XII or prekallikrein (Colman *et al.*, 1975; Wuepper *et al.*, 1975; Donaldson *et al.*, 1976). This phenomenon was explained by the identification of HMWK as a nonenzymatic cofactor in contact activation. It appeared to accelerate activation of both Factor XII and prekallikrein as well as Factor XI (Griffin and Cochrane, 1976; Meier *et al.*, 1977; Revak *et al.*, 1977; Wiggins *et al.*, 1977). The discovery that prekallikrein and Factor XII circulate bound to HMWK provided the mechanistic key

to the explanation (Thompson *et al.*, 1977). One function of HMWK is to present the substrates of Factor XIIa in a conformation that facilitates their activation. Thus, prekallikrein that is bound to a surface in the absence of HMWK is not subsequently activated by Factor XIIa (Silverberg *et al.*, 1980b). A synthetic peptide encompassing the HMWK binding site for prekallikrein can interfere with contact activation by competitively interfering with the binding of prekallikrein to the HMWK light chain (Tait and Fujikawa, 1987). Similarly, a monoclonal antibody to this binding site inhibits coagulation and kinin formation in plasma (Reddigari and Kaplan, 1989a). Factor XI activation is almost totally dependent upon the formation of a surface-binding complex with HMWK. HMWK also augments the rate of Factor XII activation in plasma (Revak *et al.*, 1977; Wiggins *et al.*, 1977), although it does not augment the activity of kallikrein against synthetic substrates. The effect seems to be largely indirect. First, it is required for efficient formation of kallikrein in surface-activated plasma. Second, since kallikrein can dissociate from surface-bound HMWK, it can interact with surface-bound Factor XII on an adjacent particle, thereby disseminating the reaction (Wiggins *et al.*, 1977; Cochrane and Revak, 1980; Silverberg *et al.*, 1980b). As a result, the effective kallikrein/Factor XII ratio is increased in the presence of HMWK. Finally, in plasma, HMWK can displace other adhesive glycoproteins such as fibrinogen from binding to the surface (Schmaier *et al.*, 1984) (see also Chapter 4). These data indicate that HMWK must also be considered to be a coagulation cofactor because it is required for the generation of kallikrein (a Factor XII activator) as well as the activation of Factor XI. HMWK-deficient plasma has a profoundly prolonged activated PTT that is almost as abnormal as that of Factor XII deficiency (Colman *et al.*, 1975; Wuepper *et al.*, 1975; Donaldson *et al.*, 1976), although persons with congenital HMWK deficiency have no bleeding diathesis.

Regulation of contact activation occurs via plasma

Table 16.2 Plasma inhibitors of enzymes of contact activation: relative contributions to inhibition in normal plasma

	Enzyme			
Inhibitor	Factors XIIa and XIIf		Kallikrein	Factor XIa
C1 inhibitor	91.3	93	52 (84)[a]	47
Antithrombin III[b]	1.5	4	nd	5
α_2-Macroglobulin	4.3	—	35 (16)[a]	—
α_1-Protease inhibitor	—	—	nd	23.5
α_2-Antiplasmin	3.0	3	nd	24.5

nd, not determined separately.
[a] Data obtained from generation of kallikrein *in situ*.
[b] Data are for results obtained in the absence of added heparin.

protease inhibitors. A summary of the major control proteins of this pathway is given in Table 16.2. The C1 inhibitor is a major inhibitor of Factor XIIa or XII$_f$ (Forbes et al., 1970; Schreiber et al., 1973; de Agostini et al., 1984; Pixley et al., 1985) and it is not active against other coagulation enzymes except for inhibition of Factor XIa. The inhibitor is cleaved by the protease and then binds at the active site of the protease in a 1:1 molar covalent complex that completely inactivates the enzyme (Travis and Salvesen, 1983). Antithrombin III (AT-III), which is a critical control protein for much of the coagulation cascade, makes a minor contribution to Factor XIIa$_f$ inactivation (Stead et al., 1976; de Agostini et al., 1984; Pixley et al., 1985; Cameron et al., 1989). Heparin can augment the inhibition by AT-III, although there is some variance reported as to the magnitude of augmentation. Heparin can also function as an activating polyanion for contact activation (Hojima et al., 1984; Silverberg and Diehl, 1987b). Curiously, α_2-macroglobulin (α_2-M), which is an inhibitor of broad reactivity with enzymes, does not significantly inhibit any of the forms of activated Factor XII.

Kallikrein is inhibited by both C1 inhibitor and α_2-M (Gigli et al., 1970; McConnell, 1972; Harpel, 1974), which together account for over 90% of the inhibitory activity of plasma (Schapira et al., 1982; van der Graaf et al., 1983). α_2-M does not bind to the active site, but traps the protease within its structure so as to interfere sterically with its ability to cleave large protein substrates (Barrett and Starkey, 1973). About one-third of the enzyme's activity on small synthetic substrates is retained by the complex, while the activity on its natural substrates is <1%. Although these two inhibitors contribute roughly equally when kallikrein is added to plasma (Schapira et al., 1982; van der Graaf et al., 1983; Harpel et al., 1985), when a surface such as kaolin is added, 70–80% of the kallikrein formed is bound to C1 inhibitor (Harpel et al., 1985). The reason for this difference is unknown. Conversely, at low temperatures, C1 inhibitor is less effective, and much of the kallikrein inhibition is mediated by α_2-M (Harpel et al., 1985).

Factor XIa is inhibited to a great extent by C1 inhibitor (Wuillemin et al., 1995). When purified Factor XIa (FXIa) was added to plasma and its distribution among various inhibitors was determined, most of the added XIa was found complexed to C1 INH, even in the presence of heparin, followed by FXIa: α_1-antitrypsin (α_1-AT) and FXIa: α_2-antiplasmin. However α_1-AT was found to be the major inhibitor when chromatographic plasma fractions were tested against Factor XIa (Heck and Kaplan, 1974; Scott et al., 1982). C1 inhibitor and AT-III were also previously found to inhibit Factor XIa (Scott et al., 1982; Meijers et al., 1988).

Activation on a surface occurs very quickly while inhibition has a slower reaction rate. In plasma of patients with hereditary angioedema (see Section 5) in which C1 inhibitor is absent or dysfunctional, the amount of surface needed to produce maximal activation is 10–20-fold less than that needed to activate normal plasma (Cameron et al., 1989).

3.2 TISSUE PATHWAY OF BRADYKININ FORMATION

The tissue pathway for BK formation is far simpler because only tissue kallikrein is involved regardless of which cell or tissue produces it. Tissue kallikrein is encoded by a single gene and prokallikrein (Takahashi, 1986), is processed to yield the secreted enzyme. Tissue kallikrein (Nusta et al., 1978) digests HMWK or LMWK to Lys-BK, and a plasma aminopeptidase rapidly converts Lys-BK to BK (Sheikh and Kaplan, 1989). BK and Lys-BK possess identical biologic activities (see below) and act via the same receptor subtype. In most assays, however, Lys-BK has 80–90% the potency of BK. The concentration of LMWK is much greater than that of HMWK, thus LMWK is preferentially cleaved. By contrast, plasma kallikrein is quite specific in utilizing HMWK as substrate.

The regulatory mechanisms of the tissue kallikrein pathway of BK formation are not well known and it was thought for a long time that none of the inhibitors of the contact pathway are involved in controlling tissue kallikrein. However, recently Rahman et al. (1994) detected tissue kallikrein: α_1-AT complexes in synovial fluid from rheumatoid arthritis patients. In porcine plasma exogenously added tissue kallikrein was found, to some extent, to form complexes with α_2-M and α_1-AT (Blackberg, 1994). However, tissue kallikrein activity was still present some 12 h after the addition, suggesting that these macromolecular inhibitors may not provide significant regulation of tissue kallikrein in plasma. Since tissue kallikrein is not a major player in plasma kinin generation, the significance of the above studies is unclear. Other kallikrein-binding proteins, which specifically bind and inhibit tissue kallikrein (but not plasma kallikrein) have been identified in rat and human tissues (Bhoola et al., 1992). However, the functional significance of these binding proteins is still unclear [see Bock and Shore (1983) for a review on tissue kallikreins and also Chapter 5 of this volume]. Peptide sequences analogous to the C-terminal sequence of kinins (Ac-Ser-Pro-Phe-Arg-Ser-Val-Gln-NH$_2$) have been reported to be potent synthetic inhibitors of kallikrein (Burton and Benetos, 1989).

4. Bradykinin: Functions and Control Mechanisms

As discussed throughout this volume, the functions of BK are many, and include venular dilatation, increased vascular permeability, constriction of uterine and gastrointestinal smooth muscle, constriction of coronary and pulmonary vasculature, bronchoconstriction and activation of phospholipase A_2 to augment arachidonic acid metabolism. It acts in most tissues via B_2 receptors, and selective B_2 receptor antagonists have been recently synthesized. In plasma, BK is first digested by carboxypeptidase N (see Chapter 7), which removes the C-terminal Arg leaving desArg^9-BK (Sheik and Kaplan, 1986b). This peptide interacts with B_1 receptors in the vasculature to cause hypotension. It has been reported that B_1 receptors are induced during inflammatory conditions (see Chapter 8), while B_2 receptors are synthesized constitutively. When serum is examined, the rate of removal of the C-terminal Arg is more rapid than can be attributed to carboxypeptidase N (Sheikh and Kaplan, 1989). This may be due to secretion (from cells) or activation of carboxypeptidase U, a newly described exopeptidase (Wang *et al.*, 1994). The next cleavage is by angiotensin converting enzyme (ACE) which digests desArg^9-BK via its tripeptidase activity (Sheikh and Kaplan, 1986a) to yield Arg-Pro-Pro-Gly-Phe + Ser-Pro-Phe. These products are biologically inactive. Further slow digestion leads to the final products of Arg-Pro-Pro plus one mole each of the amino acids Gly, Ser, Pro, and Arg, and two moles of Phe (Sheikh and Kaplan, 1989). The metabolic degradation of kinins is reviewed in Chapter 7.

The cough and angioedema associated with the use of ACE inhibitors for treatment of heart failure, hypertension, diabetes or scleroderma may be due to inhibition of kinin inactivation and accumulation of BK.

All the components of the contact activation cascade have been demonstrated to bind to endothelial cells also and this is described in Chapter 4. Schmaier *et al.* (1988) and Van Iwaarden *et al.* (1988) first described binding of HMWK to human umbilical vein endothelial cells (HUVEC) in a zinc-dependent reaction. This binding was demonstrated *in situ* subsequently by immunochemical staining of umbilical cord segments following incubation with HMWK (Nishikawa *et al.*, 1992). There are 1×10^6 to 1×10^7 binding sites (an unusually large number) with a high affinity ($K_d \approx 40$–50 nM). Binding is seen with either chain of HMWK (i.e., heavy and light chains), thus a complex interaction with subsites within the receptor seems likely (Reddigari *et al.*, 1993a). A similar complex interaction has been observed with platelets, although the binding site number is far less. Since prekallikrein and Factor XI circulate bound to HMWK, these are brought to the endothelial cell surface via HMWK (Berrettini *et al.*, 1992). There are no separate receptor sites for either prekallikrein or Factor XI. When we examined binding of Factor XII, we found binding characteristics of a receptor, which was strikingly similar to that seen with HMWK including a requirement for zinc (Reddigari *et al.*, 1993b). We then demonstrated that HMWK and Factor XII can compete for binding at a comparable molar basis suggesting that they bind to the same receptor (Reddigari *et al.*, 1993b).

We also demonstrated that Factor XII can slowly autoactivate when bound to endothelial cells (Reddigari *et al.*, 1993b) and that addition of kallikrein can digest bound HMWK to liberate BK at a rate proportional to the kallikrein concentration and with a final BK level dependent on the amount of bound HMWK (Nishikawa *et al.*, 1992). Thus, activation of the cascade along the endothelial cell surface is likely as illustrated in Fig. 16.6. BK is liberated and then interacts with B_2 receptors to increase permeability. BK can also stimulate cultured endothelial cells to secrete tissue plasminogen activator (Smith *et al.*, 1985), prostacyclin and thromboxane A_2 (Hong, 1980; Crutchly *et al.*, 1983), and can thereby

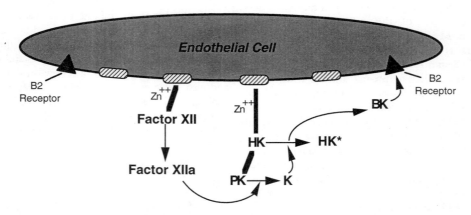

Figure 16.6   Assembly of the plasma cascade on the surface of endothelial cells. Zinc-dependent binding of Factor XII (Hageman factor) and high molecular weight kininogen (HMWK) is shown followed by the enzymatic steps leading to bradykinin (BK) release. Interaction of bradykinin with a B_2 receptor leads to vasodilation and increased vascular permeability. PK, prekallikrein; K, kallikrein; HMWK*, cleaved HMWK.

modulate platelet function and stimulate local fibrinolysis.

Neutrophils also bind HMWK via the C3b$_i$ (CR3) receptor (Wachtfogel *et al.*, 1994), which is absent on endothelial cells and all components of the kinin-forming cascade can interact at the cell surface (Henderson *et al.*, 1994). (See also Chapter 11).

5. *Hereditary Angioedema and Acquired C1 Inhibitor Deficiency*

Hereditary angioedema (HAE) is an autosomal dominant disorder caused by absence of the C1 inactivator (Donaldson and Evans, 1963) in which patients may have attacks of swelling involving almost any portion of the body. A traumatic episode or infection may initiate an attack. However, such a triggering event may not be evident, and the swelling appears to occur spontaneously. It is *not* associated with urticaria, and patients with both urticaria and angioedema without a family history invariably have a normal C1 INH. In addition to the family history, the presence of visceral involvement suggests the hereditary disorder. The most severe complication is laryngeal edema, which has been a major cause of mortality in this disease. Patients can also have abdominal attacks (Donaldson and Rosen, 1966) lasting 1–2 days consisting of vomiting, severe abdominal pain, and guarding in the absence of fever, leukocytosis, or abdominal rigidity. Although this can occasionally be difficult to distinguish from an acute abdominal condition, the attacks are self-limited and have been shown to be caused by edema of the bowel wall (Pearson *et al.*, 1972). The ultrastructural lesion seen in tissues of patients with HAE consists of gaps in the postcapillary venule endothelial cells, edema and virtually no cellular infiltrate (Sheffer *et al.*, 1971), consistent with release of a vasoactive factor such as a kinin.

Patients with HAE have measurable levels of the activated first component of complement (C1), although this protein generally circulates as an unactivated enzyme. The serum level of C4 is diminished even when the patient is free of symptoms and is usually undetectable during an attack (Ruddy *et al.*, 1968). A C4 determination is therefore the simplest test utilized to diagnose HAE. Rocket immunoelectrophoresis for C4d is a very sensitive assay indicating C4 cleavage that is more sensitive than C4 quantification (Zuraw *et al.*, 1986). C2, the other substrate of C1, is usually within normal limits when the patient is asymptomatic, but it also is diminished during an attack (Austen and Sheffer, 1965). When a diminished C4 level is obtained, a direct assay of protein C1 inhibitor level should be performed. A diminished or absent level of C1 inhibitor confirms the diagnosis. However, approximately 15–20% of patients have a normal or elevated C1 inhibitor protein (Rosen *et*

al., 1971). In these cases the protein is not functional and often has an abnormal electrophoretic mobility. Thus, a functional C1 inhibitor assay is then necessary to determine whether the diagnosis is truly HAE.

The pathogenesis of the swelling appears to involve both the complement pathway and the plasma kinin-forming pathway. Intracutaneous injection of C1 into normal individuals causes the formation of a small wheal, whereas injection into a patient with HAE yields localized angioedema (Klemperer *et al.*, 1968). A kinin-like peptide has been isolated from these patients and its formation appears to be inhibited in C2-deficient plasma (Donaldson *et al.*, 1969). However, direct demonstration of such a kinin-like peptide on interaction of C1 with C4 and C2 or C2 alone is lacking. Most recent evidence regarding this peptide suggests that cleavage of C2b by plasmin generates a kinin (Donaldson *et al.*, 1977b), although attempts to confirm this experiment have failed (Curd *et al.*, 1983; Fields *et al.* 1983). The only identifiable kinin seen in these latter studies was BK. On the other hand the amino-acid sequence of C2b is known, and Strang *et al.* (1988) have synthesized peptides of various lengths and tested each for kinin-like activity. One peptide was shown to cause edema when injected intracutaneously, reminiscent of the "C2-kinin" originally described. However, it has not yet been shown to be a cleavage product of C2b, nor has it been shown to be present during attacks of swelling in HAE patients. Thus at this point the data suggest, but do not prove, that a kinin-like molecule is derived from C2b. On the other hand, the presence of BK has been documented, and it is a likely contributor to the swelling that is seen. Perhaps more than one kinin-like moiety will ultimately be shown to be the cause of the angioedema. Twenty-four hour urine histamine excretion is also increased during attacks of angioedema, suggesting that C3a, C4a or C5a is being generated. Although the plasma levels of C3 and C5 are normal in this disorder, C3 turnover is enhanced (Carpenter *et al.*, 1969). The lesions, however, are not pruritic and antihistamines have no effect on the clinical course of the disease.

C1 inhibitor also inhibits activated Factor XII (Schreiber *et al.*, 1973), kallikrein (Gigli *et al.*, 1970) and plasmin (Schreiber *et al.*, 1973). Thus it is also an important modulator of BK generation. Patients appear to be hyperresponsive to injections of kallikrein (Juhlin and Michaelsson, 1969), as they are to C1, and elevated levels of plasma BK have been observed during attacks (Talamo *et al.*, 1969). There is also evidence that the C1 activation observed in HAE may be Factor XII-dependent (Donaldson, 1968; Eisen and Loveday, 1972). Thus some Factor XII-dependent enzyme may be initiating the classic complement cascade. Plasmin has been shown to be capable of activating C1s (Ratnoff and Naff, 1967) and may represent one such enzyme. Ghebrehiwet *et al.* (1981, 1983) demonstrated that Factor XII$_f$ can directly initiate the classic complement cascade by activation of C1 and plasmin can cleave Factor XII$_a$ to form Factor XII$_f$ (Kaplan and Austen, 1971). Plasmin-α_2 antiplasmin levels increase during attacks of swelling, indicating conversion of plasminogen to plasmin. This may represent a critical link between the intrinsic coagulation-kinin cascade and complement activation. The presence of kallikrein-like activity in induced blisters of patients with HAE supports this notion (Curd *et al.*, 1980) as does the progressive generation of kinin upon incubation of HAE plasma in plastic (noncontact-activated tubes) (Fields *et al.*, 1983) and the low prekallikrein and HMWK levels seen during attacks (Schapira *et al.*, 1983).

The treatment for attacks of HAE usually involves intermittent administration of subcutaneous epinephrine. However, there are no studies to support its efficacy, since attacks usually abate in 3–4 days even if no medication is given. A tracheostomy is indicated if laryngeal edema occurs, and mild analgesics may be used to relieve the discomfort of severe swelling and/or abdominal pain. Intravenous fluids may be necessary if the patient is unable to eat or drink. Successful prevention of attacks has been reported with androgens (Davis *et al.*, 1974) and large doses of antifibrinolytic agents, such as ϵ-aminocaproic acid (Lundl *et al.*, 1968; Frank *et al.*, 1972) or tranexamic acid (Sheffer *et al.*, 1972). The precise mechanism of action of these latter agents is unknown. Their primary effect is to inhibit conversion of plasminogen to plasmin and, in higher concentrations, they also inhibit plasmin, which may activate C1 as well as Factor XII (Kaplan and Austen, 1971) and, perhaps, cleave C2b further. These agents also inhibit the activation of C1 by immune complexes, suggesting a direct effect on the C1 molecule (Soter *et al.*, 1975). At the present time the drugs of choice are androgen derivatives, such as danazol or stanozolol, which not only prevent attacks of swelling in patients with HAE but also induce synthesis of normal C1 inhibitor and cause the C4 level to return towards normal (Gelfand *et al.*, 1976).

Yet these drugs may also have other as yet unknown effects, because many patients respond to low doses that are insufficient to raise C4 and C1 inhibitor (Warin *et al.*, 1980; Sheffer *et al.*, 1981, 1987). Alternatively, tissue levels may be more important than plasma levels and be affected first. Patients with HAE are heterozygous and, therefore, possess one normal gene. However, levels may be less than 25% of normal rather than the expected 50% owing to depletion by binding to activated enzymes. Those patients with a functionally abnormal C1 inhibitor also respond to androgen therapy (Gadek *et al.*, 1979). Such patients possess single amino-acid substitutions at the active site of the inhibitor (Skriver *et al.*, 1989). With treatment, normal C1 inhibitor is synthesized along with the abnormal protein, suggesting that such patients possess one gene that manufactures an abnormal protein (structural gene defect) and a second gene that codes for a normal product but is inhibited, or the product is

catabolized excessively (Quastel *et al.*, 1983). A proposed model for the pathogenesis of HAE is shown in Fig. 16.5, indicating the reactions inhibited by C1 inhibitor.

An acquired form of this disease has been described in patients with lymphoma who have circulating low molecular weight immunoglobulin M (IgM) and depressed C1 inhibitor levels. This entity has an unusual complement utilization profile, as C1 levels are low in addition to depletion of C4, C2 and C3. Usually, it is the C1q subcomponent that is quantified. The low C1 distinguishes this acquired disorder from the hereditary form (Caldwell *et al.*, 1972; Hauptmann *et al.*, 1976; Schreiber *et al.*, 1976). The depressed C1 inhibitor level may be caused by depletion secondary to C1 activation by immune complexes or C1 interaction with the tumor cell (Schreiber *et al.*, 1976). In the later instance, C1 fixation and C1 inhibitor depletion have been shown to be caused by an anti-idiotypic antibody bound to Ig on the surface of a B-cell lymphoma (Geha *et al.*, 1985). Other patients with connective tissue disorders (primarily systemic lupus erythematosus) (Donaldson *et al.*, 1977a) and carcinoma (Cohen *et al.*, 1978) have acquired C1 inhibitor deficiency, and respond to androgen therapy, which is helpful in these disorders by enhancing C1 inhibitor synthesis.

A second form of acquired C1 inhibitor deficiency has been described, which is due to synthesis of an autoantibody (IgG) directed to C1 inhibitor (Alsenz *et al.*, 1987). Like the aforementioned acquired forms of C1 inhibitor deficiency, there is low C4 and C1 inhibitor in the absence of a family history, but C1 levels are usually within normal limits. Under normal circumstances, C1 inhibitor is cleaved by the enzymes it inactivates, which exposes the active site in the inhibitor. The cleaved C1 inhibitor then binds stoichiometrically to the enzyme and inactivates it. When antibody to C1 inhibitor is present, C1 inhibitor is cleaved, but the antibody inhibits its ability to inactivate the enzyme. Cleaved, functionless C1 inhibitor then circulates (Zuraw and Curd, 1986) and unopposed activation of the complement and kinin-forming cascades takes place (Malbran *et al.*, 1988). Fibrolysis is also activated based upon increased levels of plasmin–antiplasmin complexes, which appears most prominent in this autoimmune form of C1 inhibitor deficiency (Cugno *et al.*, 1994). Therapeutic modalities other than androgen therapy are required, including the use of inhibitors of fibrinolysis (Cugno *et al.*, 1994), plasmapheresis and immuno-suppressive agents. The complement profile in this autoimmune disorder more closely resembles that of HAE, where a family history is not present (e.g., patient with a new mutation) or is unavailable since C1 or C1q levels are normal. The diagnosis, if suspected, can be corroborated by doing an immunoblot of C1 inhibitor and noting increased levels of a 95 kDa cleaved form, which increases prominently during attacks of swelling (Zuraw and Curd, 1986; Cugno *et al.*, 1994).

5.1 KININ FORMATION IN HEREDITARY ANGIOEDEMA

The pathogenesis of HAE suggests liberation of a kinin that has variously been considered to be a product of the second component of complement or produced by contact activation. As shown earlier in Fig. 16.5, if C1 inhibitor is either absent (Type I HAE) or dysfunctional (Type II HAE), there is insufficient inhibition of all the activated forms of Factor XII, kallikrein or activated C1 (specifically C1r and C1s, each of which is inhibited by C1 inhibitor). Production of BK is markedly augmented under these conditions and it has been shown that addition of dextran sulfate at concentrations insufficient to activate normal plasma leads to complete digestion of HMWK in HAE plasma within a few minutes (Cameron et al., 1989). Thus, seemingly insignificant trauma or infections may be sufficient to initiate an attack in such patients.

Soon after C1 inhibitor deficiency was shown to be the cause of HAE, evidence was presented to suggest that cleavage of C2 (Donaldson, 1968) or C2b (Donaldson et al., 1977b) would generate a kinin that was responsible for the symptoms. Attempts to produce this kinin by cleavage of C2 or C2b have been unsuccessful (Fields et al., 1983), and such a kinin has not been shown to circulate in patients during an attack. On the other hand, synthesis of overlapping peptides within the C2b portion of the molecule revealed a sequence which possessed kinin-like peptides (Strang et al., 1988). However, enzymatic cleavage of the protein to release this peptide has not been achieved.

Activated kallikrein is present in markedly augmented amounts in blister fluids derived from HAE patients and BK has been reported to be the major kinin found when HAE plasma is activated (Curd et al., 1980). The bulk of evidence favors a major role for BK in causing the symptoms of HAE (Fields et al., 1983). However, the presence of an additional kinin-like fragment derived from C2 by a mechanism that is not yet understood is possible. If so, synergy between the two kinins might occur. Use of B_2 receptor antagonists in such patients, once they are available for human use, should help settle the question.

Complement, nevertheless, is clearly activated in HAE and this may be due to the autoactivation of C1r when C1 inhibitor is absent. C4 levels are diminished, presumably due to consumption by C1s in HAE patients even when they are asymptomatic. With attacks of swelling, C4 levels approach zero and C2 levels diminish. As shown in Fig. 16.5, this process may be augmented by Factor XII$_f$, which has been shown to cleave enzymatically and activate C1r, and to a lesser degree, C1s (Ghebrehiwet et al., 1983). Use of androgenic agents such as Danazol and Stanozolol may increase synthesis of C1 inhibitor sufficiently to prevent swelling. Levels of C4 and C1 inhibitor increase; however, the magnitude may not parallel the clinical effect. Use of agents that inhibit

plasminogen activation (ϵ-amino caproic acid, tranexamic acid) are also effective therapies that prevent formation of plasmin, and they may also have direct inhibitory effects on C1 activation (Soter et al., 1975). Plasmin is also an enzymatic activator of Factor XII (Kaplan and Austen, 1971) and might thereby contribute to Factor XII$_f$ production and BK formation, or plasmin might digest C2b to yield a kinin-like fragment (Strang et al., 1988).

5.2 CONTACT ACTIVATION IN ALLERGIC DISEASES

By analogy with observations using dextran sulfate, naturally occurring glycosaminoglycans or proteoglycans may induce contact activation. We have tested heparin proteoglycan from the Furth murine mastocytoma for its ability to activate a mixture of Factor XII and prekallikrein. There is progressive conversion of prekallikrein to kallikrein as the concentration of mast-cell heparin is increased. The potency of heparin proteoglycan equals that of dextran sulfate and its activity is inhibited by heparinase I or II, but not by heparitinase or chondroitinase ABC. Of the glycosaminoglycans we have tested, heparin, dermatan sulfate, keratin polysulfate and chondroitin sulfate C are positive in the assay (in that order), while heparan sulfate and chondroitin sulfate A are negative. Collagen Types I, III, IV and V, laminin, fibronectin and vitronectin are also negative. Activation can then occur by release of heparin and/or other mucopolysaccharides secreted by mast cells and basophils upon exposure to plasma proteins and via interaction of these proteins with exposed connective tissue proteoglycans during tissue injury. The proteins of the kinin-forming system are present in interstitial fluid of rabbit skin, and the source, therefore, may not solely be dependent upon exudation and activation of plasma.

Any aspect of inflammation, which leads to dilution of plasma constituents or exclusion of inhibitors, will augment contact activation since inhibitory functions are very dependent upon concentration. Thus, the activitability of plasma can be shown to be related directly to dilution. Once levels of C1 inhibitors are less than 25% of normal, patients with HAE are prone to attacks of swelling.

5.2.1 Allergic Rhinitis

Activation of the plasma and tissue kinin-forming systems have been observed in allergic reactions in the nose, lungs and skin, and include the immediate reaction as well as the late-phase reaction, although the contributions of the plasma and tissue kallikrein pathways to each aspect of allergic inflammation are likely to be quite different. Antigen challenge to the nose followed by nasal lavage revealed an increase in tosyl arginine methyl ester (TAME) esterase activity, which is largely attributable to kallikrein(s) (Proud et al., 1983). The activation was seen during the immediate response as well as the late phase reaction (Creticos *et al.*, 1984). Both LMWK and HMWK were shown to be present in nasal lavage fluid (Baumgarten *et al.*, 1985), and fractionation of nasal washings demonstrated evidence of both tissue kallikrein (Baumgarten *et al.*, 1986a) and plasma kallikrein (Baumgarten *et al.*, 1986b). Tissue kallikrein can be secreted by glandular tissue as well as by infiltrating cells, such as neutrophils, and will cleave LMWK to yield Lys-BK. Plasma kallikrein will digest HMWK to yield BK directly. High-performance liquid chromatography analysis in nasal washings revealed both BK and Lys-BK, and the former can be formed from Lys-BK by aminopeptidase action. A portion of the BK, however, is also likely to be the direct result of plasma kallikrein activity.

Studies of the allergen-induced late-phase reactions in the skin (Atkins *et al.*, 1987) have demonstrated the presence of kallikrein–C1 inhibitor and activated Factor XII–C1 inhibitor complexes in induced blisters observed for an 8-h period. Elevated levels of these complexes were seen between 3 and 6 h coincident with the late-phase response and were specific for the antigen to which the patient was sensitive (Fig. 16.7).

5.2.2 Asthma

Tissue kallikrein has been found in bronchoalveolar lavage fluids of asthmatics (Christiansen *et al.*, 1987)

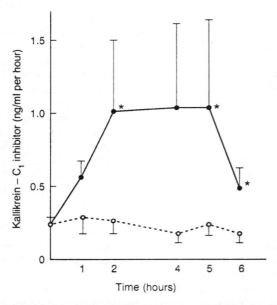

Figure 16.7 Time-course of formation of kallikrein–C1 inhibitor complex during the evolution of a cutaneous late-phase reaction induced by injecting ragweed into a ragweed-sensitive patient. Induced blister chambers are used to obtain fluid for assay. This indirectly reflects contact activation and bradykinin generation.

and, therefore, represents a source of kinins in asthma. The proteins of the plasma cascade have not been assessed in asthma, and the contribution of this cascade is uncertain. The possible functions of BK as a bronchoconstrictor has been assessed and a variety of observations may be relevant to asthma in humans (see Chapter 15). First, the presence of BK, whether during the early or late phase will act as a local vasodilator and increase vascular permeability so as to contribute to submucosal edema and airway narrowing. This effect of BK cannot be antagonized by any medication currently available, with the possible exception of injected epinephrine. The observed effects of BK on bronchial smooth muscle are quite complex (see Chapters 2 and 15), and many of the studies are performed using animal models so that extrapolation of the results to human asthma may be risky.

Studies with cat bronchial arteries that are perfused with BK reveal the expected vasodilation and increase in vascular permeability. The response to acetylcholine is augmented suggesting a contribution to bronchial hyperreactivity (Kimura *et al.*, 1992). The effects of inhaled BK upon pulmonary resistance or airway responsiveness was also tested in a sheep model, and bronchoconstriction was observed (Abraham *et al.*, 1991). However, allergic sheep were six times more sensitive than nonallergic sheep. The response was inhibited by a B_2 receptor antagonist and by atropine. Conversely, BK did not cause airway hyperresponsiveness to carbachol. Thus, in this sheep model, BK appeared to cause bronchoconstriction indirectly by vagal stimulation. This is similar to data obtained in humans in which BK-induced bronchoconstriction was dependent upon the vagus (acetylcholine) but did not produce hyperresponsiveness to histamine (Fuller *et al.*, 1987). A recent study of cross-tachyphylaxis between BK and hypertonic saline challenge in atopic asthmatics, suggesting a shared final pathway; involvement of the vagus nerve, release of acetylcholine, and/or a common release of neuropeptides (Rajakulasingam *et al.*, 1995).

The effects of BK upon bronchoconstriction and even vascular permeability may be due to release of neuropeptides such as substance P and neurokinin A from afferent (sensory) vagal fibers, or noncholinergic nonadrenergic fibers, which act via a series of neurokinin receptors (Saria *et al.*, 1988; Ichinose *et al.*, 1992; Sakamoto *et al.*, 1993). In a rat model, Tsukagoshi *et al.* (1994) demonstrated hyperresponsiveness of airways to BK, but *not* acetylcholine, which is dependent upon interleukin-1, superoxide release and nitric oxide suppression. Thus, hyperresponsiveness can be selective for a particular substance in a complex inflammatory situation. It is clear that the effects of BK in humans and the mechanism(s) by which BK is produced in human asthma are not yet clear.

Once selective, potent BK receptor antagonists become available for use in humans, we may be able to more readily decipher those effects attributable to BK (direct or indirect) from the array of additional inflammatory substances that are released in allergic diseases.

6. References

Abraham, W.M., Ahmed, A., Cortes, A., Solér, M., Farmer, S.G., Baugh, L.E. and Habeson, H.L. (1991). Airway effects of inhaled bradykinin, substance P and neurokinin A in sheep. J. Allergy Clin. Immunol. 87, 557–564.

Adam, A., Albert, A., Calay, G., Closset, J., Damas, J. and Franchimont, P. (1985). Human kininogens of low and high molecular mass: quantification by radioimmunoassay and determination of reference values. Clin. Chem. 31, 423–426.

Alsenz, J., Bork, K. and Loos, M. (1987). Autoantibody-mediated acquired deficiency of C1 inhibitor. N. Engl. J. Med. 316, 1360.

Atkins, P.C., Miragliotta, G., Talbot, S.F., Zweiman, B. and Kaplan, A.P. (1987). Activation of plasma Hageman factor and kallikrein in ongoing allergic reactions in the skin. J. Immunol. 139, 2744–2748.

Austen, K.F. and Sheffer, A.L. (1965). Detection of hereditary angioneurotic edema by demonstration of a reduction in the second component of human complement. N. Engl. J. Med. 272, 649.

Baglia, F.A., Sinha, D. and Walsh, P.N. (1989). Functional domains in the heavy-chain region of factor XI: a high molecular weight kininogen-binding site and a substrate binding site for factor IX. Blood 74, 244–251.

Baker, A.R. and Shine, J. (1985). Human kidney kallikrein: cDNA cloning and sequence analysis. DNA 4, 445.

Barrett, A.J. and Starkey, P.M. (1973). The interaction of α_2-macroglobulin with proteinases. Biochem. J. 133, 709–724.

Baumgarten, C.R., Togias, A.G., Naclerio, R.M., Lichtenstein, L.M., Norman, P.S. and Proud, D. (1985). Influx of kininogens into nasal secretions after antigen challenge of allergic individuals. J. Clin. Invest. 76, 191–197.

Baumgarten, C.R., Nichols, R.C., Naclerio, R.M., Lichtenstein, L.M., Norman, P.S. and Proud, D. (1986a). Plasma kallikrein during experimentally-induced allergic rhinitis: role in kinin formation and contribution to TAME esterase activity in nasal secretions. J. Immunol. 137, 977–982.

Baumgarten, C.R., Nichols, R.C., Naclerio, R.M. and Proud, D. (1986b). Concentrations of glandular kallikrein in human nasal secretions increase during experimentally-induced allergic rhinitis. J. Immunol. 137, 1323–1328.

Bernardo, M.M., Day, D.E., Halvorson, H.R., Olson, S.T. and Shore, J.D. (1993a). Surface-independent acceleration of Factor XII activation by zinc ions. II. Direct binding and fluorescence studies. J. Biol. Chem. 268, 12477–12483.

Bernardo, M.M., Day, D.E. and Olson, S.T. (1993b). Surface-independent acceleration of Factor XII activation by zinc ions. I. Kinetic characterization of the metal ion rate enhancement. J. Biol. Chem. 268, 12468.

Berrettini, M., Lammle, B., White, T., Heeb, M.J., Schwarz, H.P., Zuraw, B., Curd, J. and Griffin, J.H. (1986). Detection of in vitro and in vivo cleavage of high molecular weight kininogen in human plasma immunoblotting with monoclonal antibodies. Blood 68, 455–462.

Berrettini, M., Schleef, R.R., Heeb, M.J., Hopmeier, P. and Griffin, J.H. (1992). Assembly and expression of intrinsic factor IX activator complex on the surface of cultured human endothelial cells. J. Biol. Chem. 267, 19833–19839.

Bhoola, K.D., Figueroa, C.D. and Worthy, K. (1992). Bioregulation of kinins: kallikreins, kininogens, and kininases. Pharmacol. Rev. 44, 1.

Blackberg, M. (1994). Interactions in vitro and in vivo between porcine tissue kallikrein and porcine plasma proteinase inhibitors. Scand. J. Clin. Lab. Invest. 54, 643.

Bock, P.E. and Shore, J.D. (1983). Protein–protein interaction of blood coagulation. Characterization of a fluoroscein-labeled human high molecular weight kininogen light chain as a probe. J. Biol. Chem. 258, 15079–15086.

Bock, P.E., Srinivasan, K.R. and Shore, J.D. (1981). Activation of intrinsic blood coagulation by ellagic acid: insoluble ellagic acid metal ion complexes are the activating species. Biochemistry 20, 7258–7271.

Bock, P.E., Shore, J.D., Tans, G. and Griffin, J.H. (1985). Protein–protein interactions in contact activation of blood coagulation. Binding of high molecular weight kininogen and the 5-(iodoacetamido) fluorescein-labeled kininogen light chain to prekallikrein, kallikrein, and the separated kallikrein heavy and light chains. J. Biol. Chem. 260, 12434–12443.

Bouma, B.N. and Griffin, J.H. (1977). Human blood coagulation factor XI: purification, properties, and mechanism of activation by factor XII. J. Biol. Chem. 252, 6432–6437.

Bouma, B.N., Miles, L.A., Barretta, G. and Griffin, J.H. (1980). Human plasma prekallikrein. Studies of its activation by activated factor XII and of its inactivation by diisopropyl phosphofluoridate. Biochemistry 19, 1151–1160.

Burton, J. and Benetos, A. (1989). The design of specific inhibitors of tissue kallikrein and their effect on the blood pressure of rat. Adv. Exp. Med. Biol. 247B, 9.

Caldwell, J.R., Ruddy, S., Schur, P. and Austen, K.F. (1972). Acquired C1 inhibitor deficiency in lymphosarcoma. Clin. Immunol. Immunopathol. 1, 39.

Cameron, C.L., Fisslthaler, B., Sherman, A., Reddigari, S. and Silverberg, M. (1989). Studies on contact activation: effects of surfaces and inhibitors. Med. Prog. Tech. 15, 53–62.

Carpenter, C.B., Ruddy, S., Shehadeh, I.H. et al. (1969). Complement metabolism in man: hypercatabolism of the fourth (C4) and third (C3) components in patients with renal allograft rejection and hereditary angioedema. J. Clin. Invest. 48, 1495.

Castellino, F.J. and Beals, J.M. (1987). The genetic relationships between the kringle domains of human plasminogen, prothrombin, tissue plasminogen activator, urokinase, and coagulation factor XII. J. Mol. Evol. 26, 358–369.

Chao, J., Woodley-Miller, C. and Chao, L. (1983). Identification of tissue kallikrein in brain and in the cell-free translation product encoded by brain mRNA. J. Biol. Chem. 258, 15173.

Christiansen, S.C., Proud, D. and Cochrane, C.G. (1987). Detection of tissue kallikrein in the bronchoalveolar lavage fluids of asthmatic subjects. J. Clin. Invest. 79, 188–197.

Chung, D.W., Fujikawa, K., McMullen, B.A. and Davie, E.W. (1986). Human plasma prekallikrein, a zymogen to a serine protease that contains four tandem repeats. Biochemistry 25, 2410–2417.

Cochrane, C.G. and Revak, S.D. (1980). Dissemination of contact activation in plasma by plasma kallikrein. J. Exp. Med. 152, 608–619.

Cochrane, C.G., Revak, S.D. and Wuepper, K.D. (1973). Activation of Hageman factor in solid and fluid phases. J. Exp. Med. 138, 1564–1583.

Cohen, S.H., Koethe, S.M., Kozin, F. *et al.* (1978). Acquired angioedema associated with rectal carcinoma and its response to danazol therapy. J. Allergy Clin. Immunol. 62, 217.

Colman, R.W. and Müller-Esterl, W. (1988). Nomenclature of kininogens. Thromb. Haemost. 60, 340–341.

Colman, R.W., Bagdasarian, A., Talamo, R.C., Scott, C.F., Seavey, M., Guimaraes, J.A., Pierce, J.V. and Kaplan, A.P. (1975). Williams trait: human kininogen deficiency with diminished levels of plasminogen proactivator and prekallikrein associated with abnormalities of the Hageman factor dependent pathways. J. Clin. Invest. 56, 1650–1662.

Cool, D.E., Edgell, C.S., Louie, G.V., Zoller, M.J., Brayer, G.D. and MacGillivray, R.T.A. (1985). Characterization of human blood coagulation Factor XII cDNA. J. Biol. Chem. 25, 13666–13676.

Creticos, P.S., Peters, S.P., Adkinson, N.F., Jr, Naclerio, R.M., Hayes, E.C., Norman, P.S. and Lichtenstein, L.M. (1984). Peptide leucotriene release after antigen challenge in patients sensitive to ragweed. N. Engl. J. Med. 310, 1626–1630.

Crutchly, D.J., Ryan, J.W., Ryan, U.S. and Fisher, G.H. (1983). Bradykinin induced release of prostacyclin and thromboxanes from bovine pulmonary artery endothelial cells. Biochim. Biophys. Acta 751, 99–107.

Cugno, M., Cicardi, M. and Agostini (1994). Activation of the contact system and fibroanalysis in autoimmune acquired angioedema: a rationale for prophylactic use of tranexamic acid. J. Allergy Clin. Immunol. 93, 870–876.

Curd, J.G., Prograis, L.F., Jr and Cochrane, C.G. (1980). Detection of active kallikrein in induced blister fluids of hereditary angioedema patients. J. Exp. Med. 152, 742–747.

Curd, J.G., Yelvington, M., Burridge, N. *et al.* (1983). Generation of bradykinin during incubation of hereditary angioedema plasma. Mol. Immunol. 19, 1365.

Davis, P.J., Davis, F.B. and Charache, P. (1974). Long-term therapy of hereditary angioedema (HAE): preventive management with fluoxymesterone and oxymetholone in severely affected males and females. Johns Hopkins Med. J. 135, 391.

de Agostini, A., Lijnen, H.R., Pixley, R.A., Colman, R.W. and Schapira, M. (1984). Inactivation of factor XII active fragment in normal plasma. Predominant role of C1-inhibitor. J. Clin. Invest. 73, 1542–1549.

Donaldson, V.H. (1968). Mechanisms of activation of C1 esterase in hereditary angioneurotic edema plasma *in vitro*: the role of Hageman factor, a clot-promoting agent. J. Exp. Med. 127, 411–429.

Donaldson, V.H. and Evans, R.R. (1963). A biochemical abnormality in hereditary angioneurotic edema. Am. J. Med. 35, 37.

Donaldson, V.H. and Rosen, F.S. (1966). Hereditary angioneurotic edema: a clinical survey. Pediatrics 37, 1017.

Donaldson, V.H., Ratnoff, O.D., DaSilva, W.D. *et al.* (1969). Permeability-increasing activity in hereditary angioneurotic edema plasma. II. Mechanism of formation and partial characterization. J. Clin. Invest. 48, 642.

Donaldson, V.H., Glueck, H.I. and Miller, M.A. (1976). Kininogen deficiency in Fitzgerald trait: role of high molecular weight kininogen in clotting and fibrinolysis. J. Lab. Clin. Med. 89, 327–337.

Donaldson, V.H., Hess, E.V. and McAdams, P.J. (1977a).

Lupus-erythematosus-like disease in three unrelated women with hereditary angioneurotic edema. Ann. Intern. Med. 86, 312.

Donaldson, V.H., Rosen, F.S. and Bing, D.H. (1977b). Role of the second component of complement (C2) and plasmin in kinin release in hereditary angioneurotic edema (H.A.N.E.). Plasma Trans. Assoc. Am. Physicians 40, 174–183.

Dunn, J.T. and Kaplan, A.P. (1982). Formation and structure of human Hageman factor fragments. J. Clin. Invest. 70, 627–631.

Dunn, J.T., Silverberg, M. and Kaplan, A.P. (1982). The cleavage and formation of activated Hageman factor by autodigestion and by kallikrein. J. Biol. Chem. 275, 1779–1784.

Eisen, V. and Loveday, C. (1972). Activation of arginine and tyrosine esterase in serum from patients with hereditary angioedema. Br. J. Pharmacol. 46, 157.

Erdos, E.G. and Sloane, G.M. (1962). An enzyme in human plasma that inactivates bradykinin and kallidins. Biochem. Pharmacol. 11, 585–592.

Fields, T., Ghebrehiwet, B. and Kaplan, A.P. (1983). Kinin formation in hereditary angioedema plasma: evidence against kinin derivation from C2 and in support of "spontaneous" formation of bradykinin. J. Allergy Clin. Immunol. 72, 54–60.

Figueroa, C., Maciver, A.G. and Bhoola, K.D. (1989). Identification of tissue kallikrein in human polymorphonuclear leukocytes. Br. J. Haematol. 72, 321.

Forbes, C.O., Pensky, J. and Ratnoff, O.D. (1970). Inhibition of activated Hageman factor and activated plasma thromboplastin antecedent by purified C1 inactivator. J. Lab. Clin. Med. 76, 809–815.

Frank, M.M., Sergent, J.S., Kane, M.A. and Alling, D.W. (1972). Epsilon aminocaproic and therapy of hereditary angioneurotic edema: a double blind study. N. Engl. J. Med. 286, 808.

Fritz, H., Fiedler, F. and Dietl, T. (1977). On the relationships between porcine pancreatic, submandibular and urinary kallikrein. In "Kininogenases: Kallikrein" (eds. G.L. Haberland, J.W. Rohen and T. Suski), pp. 15. Schattauer-Verlag, New York.

Fujikawa, K. and McMullen, B.A. (1983). Amino acid sequence of human b-Factor XIIa. J. Biol. Chem. 258, 10924–10933.

Fujikawa, K., Heimark, R.L., Kurachi, K. and Davie, E.W. (1980). Activation of bovine factor XII (Hageman factor) by plasma kallikrein. Biochemistry 19, 1322–1330.

Fukushima, D., Kitamura, N. and Nakanishi, S. (1985). Nucleotide sequence of cloned cDNA for human pancreatic kallikrein. Biochemistry 24, 8037.

Fuller, R.W., Dixon, C.S., Cuss, F.M.C. and Barnes, P.J. (1987). Bradykinin-induced bronchoconstruction in humans: mode of action. Am. Rev. Resp. Dis. 135, 176–180.

Gadek, J.E., Hosea, S.W., Gelfand, J.A. and Frank, M.A. (1979). Response of variant hereditary angioedema phenotypes of danazol therapy. J. Clin. Invest. 64, 280.

Geha, R.S., Quinti, I., Austen, K.F. *et al.* (1985). Acquired C1-inhibitor deficiency associated with antiidiotypic antibody to monoclonal immunoglobulin. N. Engl. J. Med. 312, 534.

Gelfand, J.A., Sherms, R.J., Alling, D.W. and Frank, M.M. (1976). Treatment of hereditary angioedema with danazol; reversal of clinical and biochemical abnormalities. N. Engl. J. Med. 295, 1444.

Ghebrehiwet, B., Silverberg, M. and Kaplan, A.P. (1981). Activation of the classical pathway of complement by Hageman factor fragment (HFf). J. Exp. Med. 153, 665.

Ghebrehiwet, B., Randazzo, B.P., Dunn, J.T., Silverberg, M. and Kaplan, A.P. (1983). Mechanism of activation of the classical pathway of complement by Hageman factor fragment. J. Clin. Invest. 71, 1450–1456.

Gigli, I., Mason, J.W., Coleman, R.W. and Austen, K.F. (1970). Interaction of plasma kallikrein with the C1 inhibitor. J. Immunol. 104, 574–581.

Ginsberg, M., Jaques, B., Cochrane, C.G. and Griffin, J.H. (1980). Urate crystal dependent cleavage of Hageman factor in human plasma and synovial fluid. J. Lab. Clin. Med. 95, 497–506.

Gounaris, A.D., Brown, M.A. and Barrett, A.J. (1984). Human plasma α_1-cystein proteinase inhibitor. Purification by affinity chromatography, characterization, and isolation of an active fragment. Biochem. J. 221, 445–452.

Griep, M.A., Fujikawa, K. and Nelsestuen, G.L. (1985). Binding and activation properties of human factor XII, prekallikrein and derived peptides with acidic lipid vesicles. Biochemistry 24, 4124–4130.

Griffin, J.H. (1978). Role of surface in surface-dependent activation of Hageman factor (blood coagulation Factor XII). Proc. Natl Acad. Sci. 75, 1998–2002.

Griffin, J.H. and Cochrane, C.G. (1976). Mechanisms for the involvement of high molecular weight kininogen in surface-dependent reactions of Hageman factor. Proc. Natl Acad. Sci. USA 73, 2554–2558.

Guimaraes, J.A., Borges, D.R., Prado, E.S. and Prado, J.L. (1973). Kinin converting aminopeptidase from human serum. Biochem. Pharmacol. 22, 403–414.

Harpel, P.C. (1974). Circulatory Inhibitors of human plasma kallikrein. In "Chemistry and Biology of the Kallikrein–Kinin System in Health and Disease" (eds. J.J. Pisano and K.F. Austen), pp. 169–177. US Government Printing Office, Washington, DC.

Harpel, P.C., Lewin, M.F. and Kaplan, A.P. (1985). Distribution of plasma kallikrein between C1 inactivator and α_2-macroglobulin-kallikrein complexes. J. Biol. Chem. 260, 4257–4263.

Hauptmann, G., Lang, J.M., North, M.L. et al. (1976). Acquired C1 inhibitor deficiencies in lymphoproliferative diseases with serum immunoglobulin abnormalities. Blut 32, 155.

Heck, L.W. and Kaplan, A.P. (1974). Substrates of Hageman factor: I. Isolation and characterization of human factor XI (PTA) and inhibition of the activated enzyme by α_1-antitrypsin. J. Exp. Med. 140, 1615–1630.

Henderson, L.M., Figueroa, C.D. and Muller-Esterl, W. (1994). Assembly of contact phase factors on the surface of the human neutrophil membrane. Blood 84, 474.

Higashiyama, S., Ohkubo, I., Ishiguro, H., Kunimatsu, M., Sawaki, K. and Sasaki, M. (1986). Human high molecular weight kininogen as a thiol protease inhibitor: presence of the entire inhibition capacity in the native form of heavy chain. Biochemistry 25, 1669–1675.

Hojima, Y., Cochrane, C.G., Wiggins, R.C., Austen, K.F. and Stevens, R.L. (1984). In vitro activation of the contact (Hageman factor) system of plasma by heparin and chondroitin sulfate E. Blood 63, (6) 1453–1459.

Hong, S.L. (1980). Effect of bradykinin and thrombin on prostacyclin synthesis in endothelial cells from calf and pig aorta and human umbilical cord vein. Thromb. Res. 18, 787–795.

Ichinose, M., Nakajima, N., Takahashi, T., Yumamuchi, H., Inoue, H. and Takashima, T. (1992). Protection against bradykinin-induced bronchoconstriction in asthmatic patients by neurokinin receptor antagonists. Lancet 130, 1248–1251.

Ikari, N., Sugo, T., Fujii, S., Kato, H. and Iwanaga, S. (1981). The role of bovine high molecular weight kininogen in contact-mediated activation of bovine factor XII: interaction of HMW kininogen with kaolin and plasma prekallikrein. J. Biochem. 89, 1699–1709.

Ishiguro, H., Higashiyama, S., Ohkubo, I. and Sasaki, M. (1987). Mapping of functional domains of human high molecular weight and high molecular weight kininogens using murine monoclonal antibodies. Biochemistry 26, 7021–7029.

Jacobsen, S. and Kritz, M. (1967). Some data on two purified kininogens from human plasma. Br. J. Pharmacol. 29, 25–36.

Johnson, D.A., Salveson, G., Brown, M. and Barrett, A.J. (1987). Rapid isolation of human kininogens. Thromb. Res. 48, 187–193.

Juhlin, L. and Michaelsson, G. (1969). Vascular reactions in hereditary angioedema. Acta Derm. Venereol. 49, 20.

Kaplan, A.P. and Austen, K.F. (1970). A prealbumin activator of prekallikrein. J. Immunol. 105, 802–811.

Kaplan, A.P. and Austen, K.F. (1971). A prealbumin activator of prekallikrein II: derivation of activators of prekallikrein from active Hageman factor by digestion with plasmin. J. Exp. Med. 133, 696–712.

Kaufman, N., Page, J.D., Pixley, R.A., Schein, R., Schmaier, A.H. and Colman, R.W. (1991). α_2 Macroglobulin-kallikrein complexes detect contact system activation in hereditary angioedema and human sepsis. Blood 77, 2660–2667.

Kellermann, J., Lottspeich, F., Henschen, A. and Muller-Esterl, W. (1986). Completion of the primary structure of human high molecular weight kininogen. The amino acid sequence of the entire heavy chain and evidence for its evolution by gene triplication. Eur. J. Biochem. 154, 471–478.

Kellermeyer, R.W. and Breckenridge, R.T. (1965). The inflammatory process in acute gouty arthritis, I. Activation of Hageman factor by sodium urate crystals. J. Lab. Clin. Med. 63, 307–315.

Kimura, K., Inouo, H., Ichinose, M., Miura, M., Katsumata, V., Takahashi, T. and Takashima, T. (1992). Bradykinin causes airway hyperresponsiveness and enhances maximal airway narrowing: role of microvascular leakage and airway edema. Am. Rev. Resp. Dis. 146, 1301–1305.

Kitamura, N., Kitagawa, H., Fukushima, D., Takagaki, Y., Miyata, T. and Nakanishi, S. (1985). Structural organization of the human kininogen gene and a model for its evolution. J. Biol. Chem. 260, 8610–8617.

Klemperer, M.R., Donaldson, V.H. and Rosen, F.S. (1968). The vasopermeability response in man to purified C1 esterase. J. Clin. Invest. 47, 604 (abstract).

Kluft, C. (1978). Determination of prekallikrein in human plasma: optimal conditions for activating prekallikrein. J. Lab. Clin. Med. 91, 83–93.

Kurachi, K. and Davie, E.W. (1977). Activation of human factor XI (plasma thromboplastin antecedent) by factor XII (activated Hageman factor). Biochemistry 16, 5831–5839.

Kurachi, K., Fujikawa, K. and Davie, E.W. (1980). Mechanism of activation of bovine factor XI by factor XII and factor XII$_a$. Biochemistry 19, 1330–1338.

Lottspeich, F., Kellermann, J., Henschen, A., Foertsch, B. and Muller-Esterl, W. (1985). The amino acid sequence of the light chain of human high molecular mass kininogen. Eur. J. Biochem. 152, 307–314.

Lundh, B., Laurell, A., Wetterqvist, H. *et al.* (1968). A case of hereditary angioneurotic edema successfully treated with ε-aminocaproic acid. Clin. Exp. Immunol. 3, 733.

Malbran, A., Hammer, C.H., Frank, M.M. and Fries, L.F. (1988). Acquired angioedema: observations in the mechanism of action of autoantibodies directed against C1 esterase inhibitor. J. Allergy Clin. Immunol. 81, 1199.

Mandle, R.J.J. and Kaplan, A.P. (1977). Hageman factor substrates. II. Human plasma prekallikrein. Mechanism of activation by Hageman factor and participation in Hageman factor-dependent fibrinolysis. J. Biol. Chem. 252, 6097–6104.

Mandle, R.J., Jr, Colman, R.W. and Kaplan, A.P. (1976). Identification of prekallikrein and HMW-kininogen as a complex in human plasma. Proc. Natl Acad. Sci. USA 73, 4179–4183.

Margolis, J. (1958). Activation of plasma by contact with glass: evidence for a common reaction which releases plasma kinin and initiates coagulation. J. Physiol. 144, 1–22.

Mason, J.W., Kleeberg, U.R., Dolan, P. and Colman, R.W. (1970). Plasma kallikrein and Hageman factor in gram-negative bacteremia. Ann. Intern. Med. 73, 545–551.

McConnell, D.J. (1972). Inhibitors of kallikrein in human plasma. J. Clin. Invest. 51, 1611–1623.

McMillin, C.R., Saito, H., Ratnoff, O.D. and Walton, A.G. (1974). The secondary structure of human Hageman factor (factor XII) and its alteration by activating agents. J. Clin. Invest. 54, 1312–1322.

McMullen, B.A. and Fujikawa, K. (1985). Amino acid sequence of the heavy chain of human a-factor XIIa (activated Hageman factor). J. Biol. Chem. 260, 5328–5341.

Meier, H.L., Pierce, J.V., Colman, R.W. and Kaplan, A.P. (1977). Activation and function of human Hageman factor. The role of high molecular kininogen and prekallikrein. J. Clin. Invest. 60, 18–31.

Meijers, J.C., Vlooswijk, R.A.A. and Bouma, B.N. (1988). Inhibition of human blood coagulation factor XIa by C1 Inhibitor. Biochemistry 27, 959–963.

Mori, K. and Nagasawa, S. (1981). Studies on human high molecular weight (HMW) kininogen. II. Structural change in HMW kininogen by the action of human plasma kallikrein. J. Biochem. 89, 1465–1473.

Mori, K., Sakamoto, W. and Nagasawa, S. (1981). Studies on human high molecular weight (HMW) kininogen III. Cleavage of HMW kininogen by the action human salivary kallikrein. J. Biochem. 90, 503–509.

Morrison, D.C. and Cochrane, C.G. (1974). Direct evidence for Hageman factor (factor XII) activation by bacterial lipo-polysaccharides (endotoxins). J. Exp. Med. 140, 797–811.

Müller-Esterl, W., Fritz, H., Machleidt, I.W., Ritonja, A., Brzin, J., Kotnik, M., Turk, V., Kellermann, J. and Lottspeich, F. (1985a). Human plasma kininogens are identical with α$_2$-cysteine protease inhibitors. Evidence from immunological, enzymological, and sequence data. FEBS Lett. 182, 310–314.

Müller-Esterl, W., Rauth, G., Lottspeich, F., Kellermann, J. and Henschen, A. (1985b). Limited proteolysis of human low-molecular mass kininogen by tissue kallikrein. Isolation and characterization of the heavy and light chains. Eur. J. Biochem. 149, 15–22.

Nishikawa, K., Kuna, P., Calcaterra, E., Kaplan, A.P. and Reddigari, S.R. (1992). Generation of the vasoactive peptide bradykinin from human high molecular weight kininogen bound to human umbilical vein endothelial cells. Blood 80, 1980–1988.

Nustad, K., Orstavik, T.B. and Bautvik, K.M. (1978). Glandular kallikreins. Gen. Pharmacol. 9, 1.

Ole-Moiyoi, O., Spragg, J. and Helbert, S.P. (1977). Immunologic reactivity of purified human urinary kallikrein (urokallikrein) with anti-serum directed against human pancreas. J. Immunol. 118, 667.

Pearson, K.D., Buchignani, J.S., Shimkin, P.M. *et al.* (1972). Hereditary angioneurotic edema of the gastrointestinal tract. Am. J. Roentgenol. Radium Ther. Nucl. Med. 116, 256.

Pettinger, M.A. and Young, R. (1970). Endotoxin-induced kinin (bradykinin) formation: activation of Hageman factor and plasma kallikrein in human plasma. Life Sci. 9, 313–322.

Pixley, R.A., Schapira, M. and Colman, R.W. (1985). The regulation of human factor XII by plasma proteinase inhibitors. J. Biol. Chem. 260, 1723–1729.

Pixley, R.A., Stumpo, L.G., Birkmeyer, K., Silver, L. and Colman, R.W. (1987). A monoclonal antibody recognizing an icosapeptide sequence in the heavy chain of human factor XII inhibits surface-catalyzed activation. J. Biol. Chem. 262, 10140–10145.

Powers, C.A. and Nasjletti, A. (1982). A novel kinin generating protease (kininogenase) in the porcine anterior pituitary. J. Biol. Chem. 257, 5594.

Proctor, R.R. and Rapaport, S.J. (1961). The partial thromboplastin time with kaolin: a simple screening test for first stage clotting deficiencies. Am. J. Clin. Pathol. 35, 212–219.

Proud, D., Pierce, J.V. and Pisano, J.J. (1980). Radioimmunoassay of human high molecular weight kininogen in normal and deficient plasma. J. Lab. Clin. Med. 95, 563–574.

Proud, D., Togias, A.G., Naclerio, R.M., Crush, S.A., Norman, P.S. and Lichtenstein, L.M. (1983). Kinins are generated *in vivo* following nasal airway challenge of allergic individuals with allergen. J. Clin. Invest. 72, 1678–1685.

Quastel, M., Harrison, R., Cicardi, M., Alper, C.A. and Rosen, F.S. (1983). Behavior *in vivo* of normal and dysfunctional C1 inhibitor in normal subjects and patients with hereditary angioedema. J. Clin. Invest. 71, 1041–1046.

Que, B.G. and Davie, E.W. (1986). Characterization of a cDNA coding for human factor XII (Hageman factor). Biochemistry 8, 1525–1528.

Rahman, M.M., Worthy, K. and Elson, C.J. (1994). Inhibitor regulation of tissue kallikrein activity in the synovial fluid of patients with rheumatoid arthritis. Br. J. Rheumatol. 33, 215.

Rajakulasingam, K., Makker, H.K., Howarth, P.H. and Holgate, S.T. (1995). Cross refractoriness between bradykinin and hypertonic saline challenges on asthma. J. Allergy Clin. Immunol. 96, 502–509.

Ratnoff, O.D. and Crum, J.D. (1964). Activation of Hageman factor by solutions of ellagic acid. J. Lab. Clin. Med. 63, 359–377.

Ratnoff, O.D. and Naff, G.B. (1967). The conversion of C'1S to C'1 esterase by plasmin and trypsin. J. Exp. Med. 125, 337.

Ratnoff, O.D. and Saito, H. (1979). Amidolytic properties of single chain activated Hageman factor. Proc. Natl Acad. Sci. USA 76, 1461–1463.

Reddigari, S.R. and Kaplan, A.P. (1988). Cleavage of high molecular weight kininogen by purified kallikreins and upon contact activation of plasma. Blood 71, 1334–1340.

Reddigari, S. and Kaplan, A.P. (1989a). Monoclonal antibody to human high molecular weight kininogen recognizes its prekallikrein binding site and inhibits its coagulant activity. Blood 74, 695–702.

Reddigari, S. and Kaplan, A.P. (1989b). Quantification of human high molecular weight kininogen by immunoblotting with a monoclonal anti-light chain antibody. J. Immunol. Meth. 119, 19–25.

Reddigari, S.R., Kuna, P., Miragliotta, G., Shibayama, Y., Nishikawa, K. and Kaplan, A.P. (1993a). Human high molecular weight kininogen binds to human umbilical vein endothelial cells via its heavy and light chain. Blood 81, 1306–1311.

Reddigari, S.R., Shibayama, Y., Brunnée, T. and Kaplan, A.P. (1993b). Human Hageman factor (FXII) and high molecular weight kininogen compete for the same binding site on human umbilical vein endothelial cells. J. Biol. Chem. 268, 11982–11987.

Revak, S.D., Cochrane, C.G., Johnston, A.R. and Hugli, T.E. (1974). Structural changes accompanying enzymatic activation of human Hageman factor. J. Clin. Invest. 54, 619–627.

Revak, S.D., Cochrane, C.G. and Griffin, J.H. (1977). The binding and cleavage characteristics of human Hageman factor during contact activation: a comparison of normal plasma with plasma deficient in factor XI, prekallikrein or high molecular weight kininogen. J. Clin. Invest. 58, 1167–1175.

Roeise, O., Bouma, B.N., Stadaas, J.O. and Aasen, A.O. (1988). Dose dependence of endotoxin-induced activation of the plasma contact system: an in vitro study. Circ. Shock. 26, 419–430.

Rosen, F.S., Alper, C.A., Pensky, J. et al. (1971). Genetic heterogeneity of the C1 esterase inhibitor in patients with hereditary angioneurotic edema. J. Clin. Invest. 50, 2143.

Rosing, J., Tans, G. and Griffin, J.H. (1985). Surface-dependent activation of human factor XII by kallikrein, and its light chain. Eur. J. Biochemistry 151, 531–538.

Ruddy, S., Gigli, I., Sheffer, A.L. and Austen, K.F. (1968). "The Laboratory Diagnosis of Hereditary Angioedema". Excepta Medica, Amsterdam.

Saito, H., Ratnoff, O.D. and Donaldson, V.H. (1974). Defective activation of clotting, fibrinolytic and permeability-enhancing systems in human Fletcher trait. Circ. Res. 34, 641–651.

Sakamoto, T., Barnes, B.J. and Chung, K.F. (1993). Effects of CP-96,345, a non-peptide NK_1 receptor antagonist against substance P, bradykinin, and allergen-induced airway microvascular leakage and bronchoconstriction in the guinea pig. Eur. J. Pharmacol. 231, 31–38.

Saria, A., Murtling, C.-R., Yan, Z., Theodorsson-Norheim, E., Gamse, R. and Lundberg, J.M. (1988). Release of multiple tachykinins from capsaicin-sensitive sensory nerves in the lung by bradykinin, dimethylphenyl piperazinium and vagal nerve stimulation. Am. Rev. Resp. Dis. 137, 1330–1335.

Schachter, M., Peret, M.W. and Billing, A.G. (1983). Immunolocalization of protease kallikrein in the colon. J. Histochem. Cytochem. 31, 1255.

Schapira, M., Scott, C.F. and Colman, R.W. (1982). Contribution of plasma protease inhibitors to the inactivation of kallikrein in plasma. J. Clin. Invest. 69, 462–468.

Schapira, M., Silver, L.D., Scott, C.F., Schmaier, A.H., Prograis, L.J., Jr, Curd, J.G. and Colman, R.W. (1983). Prekallikrein activation and high molecular weight kininogen consumption in hereditary angioedema. N. Engl. J. Med. 308, 1050–1053.

Schiffman, S., Mannhalter, C. and Tynerk, D. (1980). Human high molecular weight kininogen. Effects of cleavage by kallikrein on protein structure and procoagulant activity. J. Biol. Chem. 255, 6433–6438.

Schmaier, A.H., Silver, L., Adams, A.L., Fischer, G.C., Munoz, P.C., Vroman, L. and Colman, R.W. (1984). The effects of high molecular weight kininogen on surface-adsorbed fibrinogen. Thromb Res. 83, 51–57.

Schmaier, A.H., Kuo, A., Lundberg, D., Murray, S. and Cines, D.B. (1988). The expression of high molecular weight kininogen on human umbilical vein endothelial cells. J. Biol. Chem. 263, 16327–16333.

Schreiber, R.D., Kaplan, A.P. and Austen, K.F. (1973). Inhibition by C1 INH of Hageman factor fragment activation of coagulation, fibrinolysis, and kinin generation. J. Clin. Invest. 52, 1402–1409.

Schreiber, R.D., Zweiman, B., Atkins, P. et al. (1976). Acquired angioedema with lymphoproliferative disorder: association of C1 inhibitor deficiency with cellular abnormality. Blood 48, 567.

Scott, C.F., Schapira, M., James, H.L., Cohen, A.B. and Colman, R.W. (1982). Inactivation of factor XIa by plasma protease inhibitors. Predominant role of α_1-protease inhibitor and protective effect of high molecular weight kininogen. J. Clin. Invest. 69, 844–852.

Sheffer, A.L., Craig, J.M. and Willms-Kretschmer, K. (1971). Histopathological and ultrastructural observations on tissues from patients with hereditary angioneurotic edema. J. Allergy. 47, 292.

Sheffer, A.L., Austen, K.F. and Rosen, F.S. (1972). Tranexamic acid therapy in hereditary angioneurotic edema. N. Engl. J. Med. 287, 452.

Sheffer, A.L., Fearon, D.T. and Austen, K.F. (1981). Clinical and biochemical effects of stanazolol therapy for hereditary angioedema. J. Allergy Clin. Immunol. 68, 181.

Sheffer, A.L., Fearon, D.T. and Austen, K.F. (1987). Hereditary angioedema: a decade of management with stanozolol. J. Allergy Clin. Immunol. 80, 855.

Sheikh, I.A. and Kaplan, A.P. (1986a). Studies of the digestion of bradykinin, lysylbradykinin and des-arg^9 bradykinin by angiotensin converting enzyme. Biochem. Pharmacol. 35, 1951–1956.

Sheikh, I.A. and Kaplan, A.P. (1986b). Studies of the digestion of bradykinin, lysylbradykinin and kinin degradation products by carboxypeptidases A, B and N. Biochem. Pharmacol. 35, 1957–1963.

Sheikh, I.A. and Kaplan, A.P. (1989). The mechanism of digestion of bradykinin and lysylbradykinin (kallidin) in human serum: the role of carboxy-peptidase, angiotensin converting enzyme, and determination of final degradation products. Biochem. Pharmacol. 38, 993–1000.

Shibayama, Y., Brunnee, T. and Kaplan, A.P. (1994). Activation of human Hageman factor (Factor XII) in the presence of zinc and phosphate ions. J. Med. Biol. Res. 27, 1817.

Silverberg, M. and Diehl, S.V. (1987a). The activation of the contact system of human plasma by polysaccharide sulfates. Ann. N. York Acad. Sci. 516, 268–279.

Silverberg, M. and Diehl, S.V. (1987b). The autoactivation of factor XII (Hageman factor) induced by low Mr heparin and dextran sulphate. Biochem. J. 248, 715–720.

Silverberg, M. and Kaplan, A.P. (1982). Enzymatic activities of activated and zymogen forms of human Hageman factor (factor XII). Blood 60, 64–70.

Silverberg, M., Dunn, J.T., Garen, L. and Kaplan, A.P. (1980a). Autoactivation of human Hageman factor. J. Biol. Chem. 255, 7281–7286.

Silverberg, M., Nicoll, J.E. and Kaplan, A.P. (1980b). The mechanism by which the light chain of cleaved HMW-kininogen augments the activation of prekallikrein, factor XI, and Hageman factor. Thromb. Res. 20, 173–189.

Skriver, K., Radziejewska, E., Silbermann, J.A., Donaldson, V.H. and Bock, S.C. (1989). CpG mutations in the reactive site of human C1 inhibitor. J. Biol. Chem. 264, 3066–3071.

Smith, D., Gilbert, M. and Owen, W.G. (1985). Tissue plasminogen activator release in vivo in response to vasoactive agents. Blood 66, 835–839.

Soter, N.A., Austen, K.F. and Gigli, I. (1975). Inhibition by ε-aminocaproic acid of the activation of the first component of the complement system. J. Immunol. 114, 928–932.

Stead, N., Kaplan, A.P. and Rosenberg, R.D. (1976). Inhibition of activated factor XII by antithrombin-heparin cofactor. J. Biol. Chem. 251, 6481–6488.

Strang, C.J., Cholin, S., Spragg, J., Davis, A.E., Schneeberger, E.E., Donaldson, V.H. and Rosen, F.S. (1988). Angioedema induced by a peptide derived from complement component C2. J. Exp. Med. 168, 1685–1698.

Tait, J. and Fujikawa, K. (1986). Identification of the binding site plasma prekallikrein in human high molecular weight kininogen. J. Biol. Chem. 261, 15396–15401.

Tait, J.F. and Fujikawa, K. (1987). Primary structure requirements for the binding of human high molecular weight kininogen to plasma prekallikrein and factor XI. J. Biol. Chem. 262, 11651–11656.

Takada, Y., Skidgel, R.A. and Erdos, E.G. (1985). Purification of human urinary prokallikrein. Identification of the site of activation by the metalloproteinase thermolysin. Biochem. J. 232, 851.

Takagaki, Y., Kitamura, N. and Nakanishi, S. (1985). Cloning and sequence analysis of cDNAs for high molecular weight and low molecular weight prekininogens. J. Biol. Chem. 260, 8601–8609.

Takahashi, S., Irie, A. and Katayama, Y. (1986). N-terminal amino acid sequence of human urinary prokallikrein. J. Biochem. (Tokyo) 99, 989.

Talamo, R.C., Haber, E. and Austen, K.F. (1969). A radioimmunoassay for bradykinin in plasma and synovial fluid. J. Lab. Clin. Med. 74, 816.

Tankersley, D.L. and Finlayson, J.S. (1984). Kinetics of activation and autoactivation of human factor XII. Biochemistry 23, 273–279.

Tans, G. and Griffin, J.H. (1982). Properties of sulfatides in factor XII dependent contact activation. Blood 59, 69–75.

Tans, G., Rosing, J. and Griffin, J.D. (1983). Sulfatide dependent autoactivation of human blood coagulation factor XII (Hageman factor). J. Biol. Chem. 258, 8215–8222.

Thompson, R.E., Mandle, R.J., Jr and Kaplan, A.P. (1977). Association of factor XI and high molecular weight kininogen in human plasma. J. Clin. Invest. 60, 1376–1380.

Thompson, R.E., Mandle, R.J. and Kaplan, A.P. (1978). Characterization of human high molecular weight kininogen. Procoagulant activity associated with the light chain of kinin-free high molecular weight kininogen. J. Exp. Med. 147, 488–499.

Thompson, R.E., Mandle, R.J. and Kaplan, A.P. (1979). Studies of the binding of prekallikrein and factor XI to high molecular weight kininogen and its light chain. Proc. Natl Acad. Sci. USA 76, 4862–4866.

Travis, J. and Salvesen, G.S. (1983). Human plasma proteinase inhibitors. Ann. Rev. Biochem. 52, 655–709.

Tsukagoshi, H., Robbins, R.A., Barnes, P.J. and Chung, K.F. (1994). Role of nitric oxide and superoxide anions in interleukin 1 β-induced airway hyperresponsiveness to bradykinin. A. J. Resp. Crit. Care Med. 150, 1019–1025.

van der Graaf, F., Koedam, J.A. and Bouma, B.N. (1983). Inactivation of kallikrein in human plasma. J. Clin. Invest. 71, 149–158.

Van Iwaarden, F., deGroot, P.G. and Bouma, B.N. (1988). The binding of high molecular weight kininogen to cultured human endothelial cells. J. Biol. Chem. 263, 4698–4703.

Wachtfogel, Y.T., DeLa, C.R., Kunapuli, S.P., Rick, L., Miller, M., Schultze, R.L., Altieri, D.C., Edgington, T.S. and Colman, R.W. (1994). High molecular weight kininogen binds to Mac-1 on neutrophils by its heavy chain (domain 3) and its light chain (domain 5). J. Biol. Chem. 269, 19307–19312.

Wang, W., Hendriks, D.K. and Scharpé, S.S. (1994). Carboxypeptidase U, a plasma carboxypeptidase with high affinity for plasminogen. J. Biol. Chem. 269, 15937–15944.

Warin, A.P., Greaves, M.W., Gatecliff, M. *et al.* (1980). Treatment of hereditary angioedema by low-dose attenuated androgens: dissociation of clinical response from levels of C1 esterase inhibitor and C4. Br. J. Dermatol. 103, 405.

Warn-Cramer, B.J. and Bajaj, S.P. (1985). Stoichiometry of binding of high molecular weight kininogen to factor XI/XIa. Biochem. Biophys. Res. Commun. 133, 417–422.

Weiss, A.S., Gallin, J.I. and Kaplan, A.P. (1974). Fletcher factor deficiency. A diminished rate of Hageman factor activation caused by absence of prekallikrein with abnormalities of coagulation, fibrinolysis, chemotactic activity, and kinin generation. J. Clin. Invest. 53, 622–633.

Weiss, R., Silverberg, M. and Kaplan, A.P. (1986). The effect of C1 inhibitor upon Hageman factor autoactivation. Blood 68, 239–243.

Wiggins, R.C., Bouma, B.N., Cochrane, C.G. and Griffin, J.H. (1977). Role of high molecular weight kininogen in surface-binding and activation of coagulation factor XIa and prekallikrein. Proc. Natl Acad. Sci. USA 77, 4636–4640.

Wines, D.R., Brady, J.M. and Pritchett, D.B. (1989). Organization and expression of the rat kallikrein gene family. J. Biol. Chem. 264, 7653.

Wuepper, K.D. (1973). Prekallikrein deficiency in man. J. Exp. Med. 138, 1345–1355.

Wuepper, K.D., Miller, D.R. and LaCombe, M.J. (1975). Flaujeac trait. Deficiency of human plasma kininogen. J. Clin. Invest. 56, 1663–1672.

Wuillemin, W.A., Minnema, M. and Meijers, J.C. (1995). Inactivation of factor XIa in human plasma assessed by measuring factor XIa-protease inhibitor complexes: major role for C1-inhibitor. Blood 85, 1517.

Zuraw, B.L. and Curd, J.G. (1986). Demonstration of modified inactive first component of complement (C1) inhibitor in the plasmas of C1 inhibitor deficient patients. J. Clin. Invest. 78, 567.

Zuraw, B.L., Sugimoto, S. and Curd, J.G. (1986). The value of rocket immunoelectrophoresis for C4 activation in the evaluation of patients with angioedema or C1-inhibitor deficiency. J. Allergy Clin. Immunol. 78, 1115.

17. The Nasal Airway Pharmacology of Bradykinin

James W. Dear *and* John C. Foreman

1. Introduction

As air passes through the nasal cavity it is effectively filtered, with particles down to a size of around 1 μm being deposited on the mucous blanket lining the nasal mucosa, and cleared. The glandular and vascular elements of the nasal cavity warm and humidify the inspired air so that, together with the filtering, it is delivered to the lungs in the optimal state. Nasal airway function is, in part, controlled by neural reflexes with both afferent and efferent nerve fibers in the mucosa.

One of the cardinal signs of nasal inflammation is a reduced airflow through the nasal airways or nasal blockage (congestion). For nonturbulent flow through a simple tube, the resistance to flow is inversely proportional to the fourth power of the radius of the tube (Poiseuille's Law). Therefore, very small changes in the diameter of the nasal cavity produce large changes in the resistance to airflow. An anatomically important feature of the nasal cavity is the presence of three horizontal bones projecting from the lateral wall of the cavity: the superior, middle and inferior nasal turbinates, or conchae. Swelling of the mucosa surrounding these bones, particularly the inferior turbinate, decreases the caliber of the nasal airways and is a common cause of blockage during inflammation. Blood vessels of the nasal mucosa contain cavernous sinusoids, between the capillaries and venules, which are usually in an empty, contracted state. When they dilate and become engorged with blood, however, the nasal mucosa thickens leading to blockage. The mucosa covering the middle and inferior turbinate bones has the highest concentration of sinusoids and, therefore, these bones form a major site of resistance to nasal airflow. In addition to vasodilatation, plasma exudation in the nasal microvasculature causes edema, which contributes further to nasal blockage. Submucosal glandular secretions into the nasal cavity also reduce the diameter and increases resistance to airflow.

1.1 ALLERGIC RHINITIS

Allergic rhinitis is an allergen-induced inflammatory condition of the nasal mucosa, which presents clinically as nasal congestion, sneezing, rhinorrhea and pruritus of the eyes and nose. The nasal airway of patients with allergic rhinitis shows increased responses to stimuli such as bradykinin (BK), histamine and methacholine, this hyperresponsiveness being a hallmark of allergic rhinitis (Borum, 1979; Druce et al., 1985). Allergic rhinitis may be either seasonal, where allergen exposure occurs at certain times in the year, or perennial, when there is almost constant allergen exposure. The allergens for seasonal allergic rhinitis include tree pollens, grass pollens, molds and certain spores. Seasonal allergic rhinitis is more commonly called "hay fever". Patients presenting with perennial allergic rhinitis are sensitive to allergens that include housedust mite (*Dermatophagoides pteronyssinus*), animal dander and some foods. Patients suffering from perennial allergic rhinitis have a different clinical presentation from patients with seasonal allergic rhinitis. In perennial rhinitis, nasal blockage is the predominant symptom, whereas patients with seasonal allergic rhinitis also present with marked rhinorrhea and sneezing.

Nasal challenge studies with allergens and other pharmacological agents, including BK, have been used to investigate the pathogenesis of allergic rhinitis (Walden et al., 1988). Nasal allergic responses can be induced experimentally by spraying antigen into the nasal cavity of allergic subjects. Symptoms of allergic rhinitis, including nasal blockage, rhinorrhea, increased vascular permeability and sneezing, are provoked immediately after antigen challenge. During this immediate response, the concentrations of several inflammatory mediators in nasal washes increases. These mediators include histamine, kinins, eicosanoids and TAME-esterase. TAME-esterase activity is a measure of the levels of enzymes having arginine esterase activity, and increases are highly correlated with the physiological responses to antigen challenge (Naclerio et al., 1983). In humans, TAME-esterase activity comprises about 75% plasma kallikrein, 20% mast cell tryptase and 5% glandular kallikrein. Three to eleven hours after the immediate response has resolved, many subjects exhibit a second, late response. Again, this is characterized by an increase in mediator release into the nasal cavity, which correlates with the onset of the symptoms of allergic rhinitis. Kinins are generated during both the immediate and the late

reaction. The late-phase reaction is also characterized by cellular infiltration into the nasal airways: eosinophil, basophil and neutrophil numbers increasing significantly. This review will concentrate on the role of the kinins in the pathophysiology of allergic rhinitis.

2. Methods For Studying Kinin Pharmacology in the Nasal Airways

A variety of techniques are available for studying the pharmacology of the human nasal airways. These include: (1) application of substances into the nasal airways; (2) assessment of the patency of the nasal airways; (3) nasal airway lavage to collect released substances; (4) collection of nasal secretions; (5) assessment of the patients' subjective appreciation of nasal symptoms, such as blockage, pain, rhinorrhea, etc.; and (6) brushing the nasal passages to collect superficial cells, biopsy of nasal tissue, or the use of tissue removed during routine surgery. Application of substances may be achieved by instilling or aerosolizing solutions into the nasal airways, or by placing a paper disc containing the substance into the nose at a suitable point such as the inferior turbinate bone.

There are several methods for the assessment of the patency of the nasal airways, and these include rhinomanometry (Kern, 1973), acoustic rhinometry (Jackson et al., 1977), nasal peak flow (Holmstrom et al., 1990) and body plethysmography (Nolte and Luder-Luhr, 1973). Rhinomanometry can be employed in the anterior or the posterior modes. Essentially, the equipment comprises a pressure transducer and a flow transducer.

2.1 ACTIVE POSTERIOR RHINOMANOMETRY

In active (i.e., during breathing) posterior rhinomanometry, a pressure transducer is connected to a tube, which is placed in the oropharynx and held through sealed lips. The flow transducer is connected to a mask, which is held firmly over the nose and mouth. The subject breathes through both nostrils simultaneously, and the total nasal air flow and intranasal pressures are acquired continuously over several breath cycles. From pressure-flow curves, nasal airway resistance is calculated from the airflow at a predetermined reference pressure.

2.2 ANTERIOR RHINOMANOMETRY

In anterior rhinomanometry, the pressure transducer is similarly connected to a tube held in the oropharynx through sealed lips, but the flow transducer is connected separately to each nostril via a tube sealed into the nostril with tape. The major disadvantage of this technique is that each side of the nasal airway is measured separately, and it takes several minutes to change the connection to the flow transducer from one side to the other. Because of nasal cycling (spontaneous cyclical changes of resistance), there may be changes over time in the nasal airways resistance of each side, and these can affect the determination of the effect of applied substances.

2.3 ACOUSTIC RHINOMETRY

In acoustic rhinometry, a sound wave from an impulse generator is directed via a tube ($\approx$ 1 cm wide and 1 m long) into the nasal cavity. The reflected sound waves are collected by a microphone and transmitted to a computer for data analysis. A single measurement takes approximately 10 ms and, as with anterior rhinomanometry, measurements are taken separately from each nostril. However, because the device can be transferred rapidly from one nostril to the other, the impact of nasal cycling is minimized. The program used to analyze the reflected sound waves provides values of nasal cross-sectional area at various distances from the nares into the nasal cavity. Nasal volumes can also be calculated. In a comparison of volumes and cross-sectional areas at various distances into the nasal cavity, we have shown that the parameter most sensitive to drug effects is the minimal cross-sectional area (A_{min}) (Austin and Foreman, 1994b).

Both rhinomanometry and acoustic rhinometry can provide reproducible data on nasal patency. The coefficient of variation for rhinomanometry was found to be 9.6% and that for acoustic rhinometry 2.9%. Apart from less variability in acoustic rhinometry measurements, the method has the advantage of not requiring subject training and is the preferred method in clinical studies. In contrast, rhinomanometry requires a high level of subject cooperation and training for consistent data. Since the patency of the nasal airway varies from subject to subject, even under baseline conditions, it is usual to express changes in either nasal airway resistance or A_{min} as percentage changes from baselines determined from repeated measurements.

2.4 NASAL LAVAGE AND TISSUE SAMPLING

Nasal lavage has been performed in a variety of ways. In one method (Hilding, 1972), a modified Foley catheter is used to produce a seal between the posterior nasal cavity and the pharynx. With the head extended, saline is added into and recovered from the nasal cavity via the catheter. In other methods, a nasal aerosol is applied and then recovered, either by asking the subject to expel it into a receptacle (Linder et al., 1988), or by suction into a trap (Druce et al., 1985). In the method of Naclerio et al. (1983), the subject extends the head and voluntarily holds the tongue against the soft palate to occlude the nasopharynx. Saline is introduced into the nostril and, after 10 s, the subject flexes the head over a collecting funnel. This method yields recoveries of 80–90% and is noninvasive.

Nasal secretions can be measured by placing a preweighed absorbent material into the nasal cavity and weighing it after a given time period (Holmberg et al., 1990). The subject may also blow secretion on to a preweighed absorbent material. Alternatively, mucus glycoprotein can be measured in nasal lavage fluid (Greiff et al., 1993). Symptoms such as pain, itching, running, blockage and sneezing can be recorded using a visual analog scale (Aitken, 1969).

Cells and tissue can be obtained by brushing the nasal cavity (Pipkorn et al., 1988), by nasal biopsy, or by acquiring tissue removed at turbinectomy. Brushing and biopsy can give information about histological changes in the nasal cavity, but tissue from turbinectomy specimens is sufficient for preparations for ligand binding studies, or for preparing explants that can be used to measure mucus secretion in vitro (Mullol et al., 1993).

3. Bradykinin as a Mediator of Allergic Rhinitis

3.1 KININS ARE SYNTHESIZED DURING NASAL ALLERGY

BK and lysylbradykinin (Lys-BK, kallidin) generation are associated with the pathology of allergic rhinitis (Dolovich et al., 1970; Proud et al., 1983). Subjects who are allergic to ragweed or housedust mite release kinins into the nasal cavity during an immediate response to nasal allergen challenge (Fig. 17.1). The concentration of kinin present in nasal lavage fluid increases after challenge in allergic subjects, but not in matched, nonallergic controls. Similarly, there is a significant increase in the nasal level of kinins during the late-phase reaction (Naclerio et al., 1985). The amount of kinin released into the nasal cavity during the late-phase response is significantly lower than in the immediate response, and nasal kinin levels during both the immediate and late-phase reactions correlate with the onset of symptoms.

In addition to increased kinin levels during experimental allergic rhinitis, there is a parallel increase in other components of the kallikrein–kinin system. After allergen challenge, nasal lavage fluid from allergic subjects contains elevated levels of kininogen and albumin (Baumgarten et al., 1985). The albumin in nasal lavage fluid is a common measure of plasma exudation, and increases in kininogen and albumin during nasal allergic reactions indicate that kininogen, passing from the plasma into the nasal cavity, provides a substrate for kinin generation via the action of kallikreins. Both tissue and

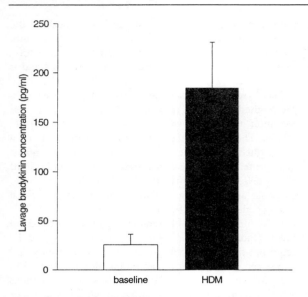

Figure 17.1   Bradykinin is generated during experimentally induced allergic rhinitis. Housedust mite antigen (HDM) administered intranasally to subjects with perennial allergic rhinitis causes an increase in the bradykinin concentration of the nasal lavage.

plasma kallikreins are elevated during experimentally induced allergic rhinitis (Baumgarten et al., 1986a,b). In addition, nasal lavage fluid contains mast cell proteases, which may also generate kinins. Kininases that degrade kinins have been demonstrated in the human nasal mucosa. For example, immunohistochemistry shows staining for neutral endopeptidase in the nasal airway epithelium, submucosal serous glands and the endothelium of small blood vessels (Ohkubo et al., 1993). Similarly, immunoreactivity for angiotensin converting enzyme (ACE or kininase II) and carboxypeptidase N (kininase I) have been demonstrated in the nasal epithelium, endothelium and superficial lamina propria (Ohkubo et al., 1994a,b). Following allergen challenge, nasal lavage fluid contains aminopeptidase and carboxypeptidase enzymes that convert BK to desArg[9]-BK (Proud et al., 1987). Nasal lavage fluid also contains low levels of ACE (Proud et al., 1987), and pretreating subjects with an ACE inhibitor, such as captopril, resulted in a small increase in nasal kinin levels following allergen challenge. Captopril had no significant effects on histamine or TAME-esterase activity, or on sneezing or other symptoms (Proud et al., 1990b). The lack of increased symptoms, despite increased kinin levels, may be the result of experimental design, since an intranasal vasoconstrictor was administered prior to challenge, and this may have functionally antagonized the effects of raised kinin levels. This study was performed using subjects with seasonal allergic rhinitis, although kinins are implicated more strongly in the pathophysiology of

perennial allergic rhinitis (see below). It would be interesting, therefore, to perform similar experiments in perennial allergic rhinitis with a view to investigating whether captopril can also increase symptoms in this condition. Interestingly, there is a report that captopril treatment for hypertension had the side effect of causing rhinorrhea (Balduf et al., 1992), which may be due to elevated nasal kinins.

The evidence presented so far, for kinin formation in nasal allergy, has been obtained from studies where allergic rhinitis was induced experimentally. However, there is also evidence supporting kinin generation in the nasal cavity of atopic individuals during the pollen season. Svensson et al. (1990) demonstrated elevated kinins in nasal secretions from seasonal allergic rhinitics as a result of natural allergen exposure, and the same group later demonstrated that a glucocorticosteroid reduced the nasal kinin levels during the pollen season (Svensson et al., 1994). It was concluded that the steroid drug had reduced plasma exudation into the nasal cavity and, therefore, inhibited kininogen influx, thereby reducing kinin formation. Kinin levels are also elevated during experimental and naturally occurring rhinovirus infections (Naclerio et al., 1988; Proud et al., 1990a).

3.2   KININS MIMIC THE SYMPTOMS OF ALLERGIC RHINITIS

Kinins sprayed into the nasal airways of normal or atopic human subjects induce some of the symptoms of allergic rhinitis, including increased nasal resistance, microvascular leakage and nasal pain (Fig. 17.2). Proud et al. (1988) showed that intranasal BK causes nasal blockage, increased secretions and watery eyes. Furthermore, kinin challenge increased albumin and TAME-esterase activity in nasal lavage fluid. The plasma exudation induced by BK was significantly higher in patients with seasonal allergic rhinitis, challenged out of season, than in normal subjects (Brunnée et al., 1991), and this example of nasal airway hyperresponsiveness is a sign of allergic rhinitis (Borum, 1979; Druce et al., 1985). Platelet-activating factor (PAF) has been implicated in the development of this nonspecific hyperresponsiveness. Intranasally administered PAF induces hyperresponsiveness to histamine and BK, accompanied by an increase in eosinophil cationic protein in the nasal lavage fluid, indicating that activation of eosinophils may be involved (Austin and Foreman, 1993). Vitamin E, an antioxidant, attenuates PAF-induced hyperresponsiveness, suggesting a role for free radicals in the PAF-induced nasal hyperresponsiveness. However, the existence of nasal hyperresponsiveness in allergic rhinitis is debatable, since some studies have failed to demonstrate that nasal allergy produces an enhanced response to histamine, BK or methacholine (Doyle et al., 1990).

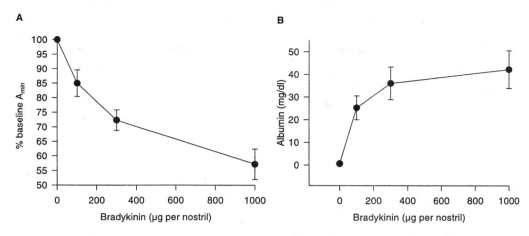

Figure 17.2 The effect of intranasal bradykinin challenge on nasal airway patency and vascular permeability. (A) shows the decrease in the minimum cross-sectional area of the nasal cavity (A_{min}) with increasing bradykinin dose. The nasal patency is expressed as a percentage of the pre-challenge A_{min}. (B) shows the increase in the albumin content of the nasal lavage with increasing bradykinin dose.

Following BK challenge, normal subjects release significantly more albumin and total protein into the nose than subjects with severe perennial allergic rhinitis (Baraniuk et al., 1994a). In normal subjects, BK does not stimulate lysozyme secretion, a marker for submucosal serous cell secretion, into the nasal cavity and stimulates glycoconjugate secretion only at high doses (1,000 nmol) (Baraniuk et al., 1994a). In contrast, BK stimulates significant lysozyme secretion into the nasal cavity of perennial allergic rhinitis patients and also induces significantly more glycoconjugate secretion than in normal subjects (Baraniuk et al., 1994a). Thus, in perennial allergic rhinitis, there are altered glandular responses to BK. The nasal blockage and secretion, watery eyes, sore throat, sneezing, albumin and TAME-esterase activity induced by BK appear not to desensitize with repeated administrations (Churchill et al., 1991). The ability of BK to cause nasal blockage has been confirmed by demonstrating dose-dependent increases in nasal airway resistance, measured by active posterior rhinomanometry (Doyle et al., 1990; Rajakulasingam et al., 1991; Austin and Foreman, 1994a). By using acoustic rhinometry, BK challenge can be shown to reduce the A_{min} of the nasal airway (Austin and Foreman, 1994b) and, compared to histamine, BK is more potent in increasing the resistance and plasma exudation (Rajakulasingam et al., 1993). BK, applied topically to the nasal mucosa, appears to have no effect on mucosal blood flow in normal or atopic subjects as measured by ^{133}Xe washout (Holmberg et al., 1990). However, this may be explained by the topically applied BK only reaching the superficial vessels of the mucosa, whereas the ^{133}Xe washout method determines blood flow in deeper areas. Nasal challenge with Lys-BK also induces some of the symptoms of rhinitis (Rajakulasingam et al.,

1991). However, B_1 kinin receptor agonists such as desArg[9]-BK and Lys-desArg[9]-BK have no effect on nasal airway resistance, albumin release, or on symptom scores (Rajakulasingam et al., 1991; Austin and Foreman, 1994a). The actions of kinins appear, therefore, to be mediated via B_2 receptors (see Chapter 2).

3.3   BRADYKININ RECEPTOR ANTAGONISTS IN HUMAN NASAL AIRWAYS

3.3.1   Normal Subjects

In autoradiography studies, [^{125}I]-BK binding sites have been demonstrated in small muscular arteries, venous sinusoids and submucosal nerve fibers of the mucosa of human inferior turbinate bones (Baraniuk et al., 1990b). Pongracic et al. (1991) tested the effect of intranasal administration of a B_2 receptor antagonist, [DArg[0]-Hyp[3],DPhe[7]]-BK (NPC 567), on various nasal responses to BK. Nevertheless, even at a high dose (1 mg per nostril), NPC 567 did not affect symptom scores or albumin and TAME-esterase activity in nasal lavage fluid in response to BK. The antagonist also had no significant effect on BK-induced nasal blockage measured by active posterior rhinomanometry (Austin and Foreman, 1994a).

In contrast to NPC 567, a relatively labile and weak B_2 receptor antagonist (see Chapter 2), a more recently described potent, stable antagonist, icatibant (more commonly known as "Hoe 140") (Hock et al., 1991; Wirth et al., 1991) blocks the actions of BK in the nasal airway (Fig. 17.3). Icatibant, delivered by pump spray into the nasal cavity, significantly antagonized BK-induced nasal blockage (Austin and Foreman,

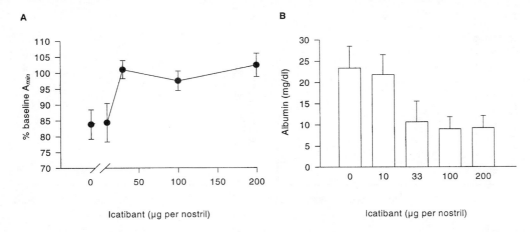

Figure 17.3   The effect of icatibant on bradykinin-induced nasal blockage and increased nasal vascular permeability. (A) shows the reduction in the minimum nasal cross-sectional area (A_{min}) induced by bradykinin (100 µg) in the absence and presence of icatibant, administered 2 min prior to bradykinin challenge. **(B)** shows the albumin content of the nasal lavage following bradykinin challenge (100 µg) in the absence and presence of icatibant, again administered 2 min prior to challenge. Icatibant 33, 100 and 200 µg significantly antagonizes the actions of bradykinin.

1994a). BK-induced albumin release into the nasal airway was also antagonized by icatibant. Another B_2 receptor antagonist, [1-adamantane acetyl-DArg-Hyp[3],Thi[5,8],DPhe[7]]-BK, also antagonized BK-induced nasal blockage and albumin release in humans (Austin and Foreman, 1994a). The ability of these novel compounds to antagonize the nasal effects of BK, in contrast to the inactivity of NPC 567, appears to result from their higher affinity for the B_2 receptor and longer duration of action (Hock *et al.*, 1991) (see Section 5).

3.3.2   Perennial Allergic Rhinitis

Figure 17.4 shows that icatibant dose-dependently antagonizes housedust mite-induced nasal blockage in individuals with allergic rhinitis to this antigen (Austin *et al.*, 1994). The antigen-induced nasal blockage was significantly antagonized by pretreating the nasal airway with icatibant 2 min prior to challenge. It should be noted that icatibant may have effects other than antagonism at kinin receptors. As described earlier, BK levels in the nasal lavage are increased following housedust mite challenge in subjects with perennial allergic rhinitis (Fig. 17.1). However, if the nasal cavity is pretreated with icatibant there is no significant increase in the BK level in the lavage fluid following challenge (Dear *et al.*, 1996a). This may reflect inhibition by icatibant of human tissue kallikrein. Such an effect would lead to a reduction in endogenous kinin formation in the nose. Symptom score measurements after antigen challenge revealed that nasal blockage, but not sneezing, itching or rhinorrhea were increased, and this was antagonized by icatibant. Interestingly, housedust mite

challenge in sensitive subjects does not cause an increase in albumin levels in the nasal lavage fluid. Since BK itself increases albumin release into the nasal cavity of normal subjects, the lack of albumin after antigen challenge is difficult to explain if BK is to be considered as a major mediator of perennial allergic rhinitis. However, subjects with severe perennial allergic rhinitis secrete significantly less albumin into the nasal cavity following BK challenge than normal subjects (Baraniuk *et al.*, 1994a). Hence, this provides further evidence that, in perennial allergic rhinitis, nasal vascular responses are altered compared with normal subjects. The inhibition of nasal blockage by icatibant, together with the lack of effect of housedust mite challenge on nasal vascular permeability, indicates that the main kinin-mediated component in perennial allergic rhinitis is nasal blockage, in which there is little or no contribution from swelling caused by plasma extravasation. Histamine H_1 receptor antagonists, while reducing the pruritus, sneezing and rhinorrhea of perennial allergic rhinitis, have negligible effects on nasal congestion (Meltzer, 1995), indicating that histamine is not a major mediator of the nasal blockage in this disease.

A central role for kinins in the pathogenesis of perennial allergic rhinitis may be explained, in part, by the observation that housedust mite proteases can activate the kallikrein–kinin system and generate BK both *in vivo* and *in vitro* (Stewart *et al.*, 1989, 1991, 1993; Takahashi *et al.*, 1990; Maruo *et al.*, 1991; Matsushima *et al.*, 1992). Nevertheless, in order to determine whether these proteases are responsible for kinin generation in perennial allergic rhinitis, it would be necessary to inhibit their kallikrein-like activity without

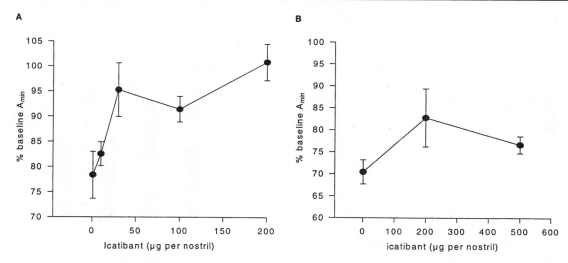

Figure 17.4   The effect of icatibant on the nasal blockage in experimentally induced perennial and seasonal allergic rhinitis. (A) shows the reduction in the minimum cross-sectional area of nasal airway (A_{min}·) induced by housedust mite challenge in sensitive subjects and its reversal by icatibant pretreatment. (B) shows the reduction in the minimum cross-sectional area induced by grass challenge in sensitive subjects and the lack of effect of icatibant pretreatment. In both parts icatibant was delivered 2 min prior to antigen challenge and icatibant, 33, 100 and 200 µg significantly antagonizes housedust mite-induced nasal blockage.

affecting antigenicity. Since highly specific mite protease inhibitors, that affect neither the ability of the mite antigens to cross-link immunoglobulin E (IgE) nor inhibit endogenous plasma and tissue kallikrein are unavailable, this hypothesis remains unresolved.

3.3.3   Seasonal Allergic Rhinitis

Grass pollen administered intranasally by aerosol causes nasal blockage in sensitive subjects. Icatibant, even at a high dose (500 µg), failed significantly to inhibit this response (Dear and Foreman, unpublished) (Fig. 17.4), suggesting that the nasal blockage in seasonal allergic rhinitis may not be mediated by kinins, as in perennial allergic rhinitis. H_1 receptor antagonists do not significantly reduce the nasal blockage in seasonal allergic rhinitis (Rokenes *et al.*, 1988; Pipkorn 1990), indicating that histamine does not play a significant role. Inhibition of 5-lipoxygenase, however, attenuates grass-induced increases in nasal airway resistance, suggesting that leukotriene synthesis may play a role in the nasal blockage of seasonal rhinitis (Howarth *et al.*, 1995). In contrast to antigen challenge in perennial rhinitis, grass pollen challenge of subjects with seasonal allergic rhinitis causes an increase in nasal albumin that is unaffected by icatibant. The failure of icatibant to antagonize the nasal blockage and the increase in vascular permeability, induced by grass pollen, is consistent with observations (Akbary and Bender, 1993) that this antagonist does not block symptoms of pollen-induced seasonal allergic rhinitis.

4.   Mechanisms of Action of Kinins in Human Nasal Airways

4.1   THE ROLE OF NITRIC OXIDE

Nitric oxide (NO) is synthesized from arginine by NO synthases (NOS) (reviewed by Vallance and Collier, 1994). Three isoforms of NOS have been identified: a constitutive "endothelial type" in the vascular endothelium, platelets and heart; a constitutive "neuronal type" in some central and peripheral neurons; and macrophages express an inducible NOS (iNOS). The latter isoform is not constitutively expressed in quiescent cells, but appears after activation by the products of infection or by cytokines. In most tissues and organs, BK-induced vasodilatation is mediated by NO (e.g., Kelm and Schrader, 1988; Cowan and Cohen, 1991; Fulton *et al.*, 1992; Vials and Burnstock, 1992). In the rat mesentery, the effects of BK appear to be mediated by NO released from nonadrenergic, noncholinergic neurons (Yu *et al.*, 1993). In humans, NO mediates BK-induced vasodilatation in the forearm (Cockcroft *et al.*, 1994; O'Kane *et al.*, 1994).

Neuronal NOS immunoreactivity is localized in human nasal extrinsic perivascular and periglandular nerves, suggesting a role in local blood flow and mucus secretion (Kulkarni *et al.*, 1994). NO production has also been demonstrated in the paranasal sinuses (Lundberg *et al.*, 1995). Interestingly, neuronal NOS immunoreactivity has also been located in mast cells of the normal

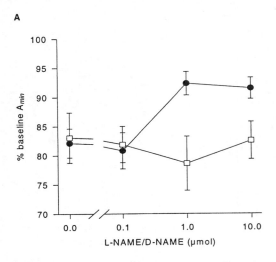

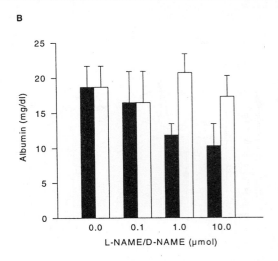

Figure 17.5   The role of nitric oxide in the nasal blockage and vascular permeability increases induced by bradykinin challenge. (A) shows the reduction in the minimum cross-sectional area of the nasal cavity (A_{min}.) induced by bradykinin (100 μg) in the absence and presence of LNAME (filled symbols) and DNAME (open symbols). (B) shows the albumin content of the nasal lavage following bradykinin challenge (100 μg) in the absence and presence of LNAME (filled bars) and DNAME (open bars). LNAME and DNAME were administered into the nasal cavity 30 min prior to bradykinin challenge, and LNAME (1 and 10 μmol) significantly reduced bradykinin-induced nasal blockage and albumin release.

human nasal mucosa (Bacci *et al.*, 1994). NO metabolites are present in the nasal lavage fluid after challenge with housedust mite in perennial allergic rhinitis (Garrelds *et al.*, 1995).

N^G-nitro-Larginine methyl ester (LNAME), a NOS inhibitor, reduces BK-induced nasal blockage in normal subjects in a dose-dependent manner (Fig. 17.5), whereas responses to histamine are unaffected. Although BK-induced nasal blockage is significantly attenuated by LNAME, it is not abolished completely, even at a high dose of inhibitor. These observations are consistent with those in other tissues (Cowan and Cohen, 1991; Weldon *et al.*, 1995), where a mechanism other than NO production may account for a proportion of BK-induced vasodilatation. Further evidence for a role of NO in BK-induced nasal blockage includes the ability of another NOS inhibitor, N^G-monomethyl-Larginine, to antagonize BK-induced nasal blockage.

Figure 17.5 also shows that the increase in human nasal vascular permeability, in response to BK, is significantly attenuated by LNAME, whereas DNAME (the inactive isomer) has no effect, suggesting a specific effect of LNAME. Furthermore, the inhibitory effect of LNAME on the plasma extravasation induced by BK (or histamine) was reversed by adding excess Larginine, the substrate for NOS (Dear *et al.*, 1996b). These data indicate that BK-induced plasma extravasation into the nasal cavity is, at least in part, mediated by NO.

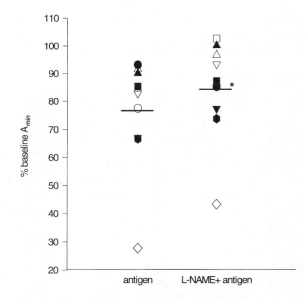

Figure 17.6   The role of nitric oxide in the nasal blockage induced by housedust mite challenge in subjects with perennial allergic rhinitis. Each symbol represents a subject's reduction in minimum nasal cross-sectional area (A_{min}.) following housedust mite challenge without and with LNAME (1 μg) pretreatment. LNAME, administered 30 min prior to challenge, significantly reduces the nasal blockage induced by housedust mite challenge. P < 0.05

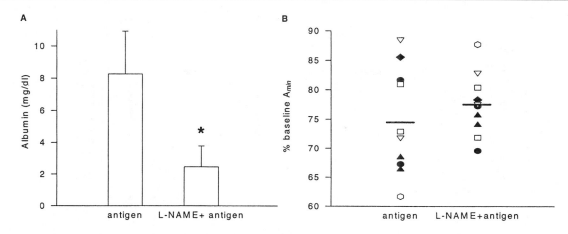

Figure 17.7   The role of nitric oxide in experimentally induced seasonal allergic rhinitis. (A) shows the increase in the albumin content of the nasal lavage induced by grass challenge on subjects with seasonal allergic rhinitis, without and with LNAME (1 μmol) pretreatment. LNAME significantly reduces the albumin content of the nasal lavage. (B) shows the reduction in the minimum nasal cross-sectional area (A_{min}.) induced by grass challenge of subjects with seasonal allergic rhinitis, without and with LNAME (1 μmol) pretreatment. LNAME has no significant effect. Each symbol represents one subject.

Nasal blockage generated by housedust mite antigen in subjects with perennial allergic rhinitis is partly mediated by NO production. This can be demonstrated since pretreating the nasal cavity with LNAME before antigen challenge, attenuates the housedust mite-induced blockage (Fig. 17.6). As BK-induced nasal blockage in normal subjects is also attenuated by LNAME, the effect of the NOS inhibitor on nasal blockage following housedust mite challenge might result from inhibition of allergen-induced BK production causing release of NO. The increase in microvascular permeability, induced by grass pollen challenge in seasonal allergic rhinitis, also appears to be mediated by NO generation, since pretreatment with LNAME reduces the albumin content of nasal lavage fluid following antigen challenge (Fig. 17.7) (Dear *et al.*, 1995). In contrast, nasal blockage induced by grass pollen is unaffected by LNAME.

4.2 THE ROLE OF HISTAMINE

The role of histamine in BK-induced nasal pathology is controversial. Rajakulasingam *et al.* (1992b) reported that an orally administered H_1 receptor antagonist, terfenadine, had no effect on BK-induced increases in nasal resistance, rhinorrhea, nasal pain or albumin in nasal lavage fluid. In contrast, we reported that terfenadine and cetirizine both attenuate BK-induced blockage as measured by active posterior rhino-manometry and acoustic rhinometry at the higher doses of BK. Figure 17.8 shows the reduction by cetirizine in a double-blind, placebo-controlled trial of BK-induced

decreases in A_{min}. Increased nasal vascular permeability, in response to higher doses of BK, were also attenuated by cetirizine. BK does not, however, increase the release of histamine into the nasal lavage fluid (Austin *et al.*, 1996). A possible explanation of this may be that the histamine released in response to BK is cleared into the nasal circulation rather than the nasal cavity. Interestingly, isolated human nasal mast cells release histamine *ex vivo* in response to BK. Human nasal turbinate bones, collected from patients undergoing surgery and enzymatically dispersed, yield cells that release histamine in response to BK (Fig. 17.8). This observation, together with the effect of H_1 antagonists, strongly supports a role for histamine in the nasal actions of high concentrations of BK, although the relevance of such high kinin levels to nasal pathology is questionable.

4.3 THE ROLE OF NERVES

The nasal mucosa contains type C sensory fibers, as well as parasympathetic and sympathetic nerve endings. C-fibers contain the neuropeptides, calcitonin gene-related peptide (CGRP), substance P, neurokinin A and gastrin-releasing peptide (Lundberg *et al.*, 1987; Sunday *et al.*, 1988; Baraniuk *et al.*, 1990a). Parasympathetic fibers contain acetylcholine, vasoactive intestinal peptide (VIP) and peptide histidine methionine (Said and Mutt, 1988; Baraniuk *et al.*, 1990c). Sympathetic neurons contain norepinephrine and neuropeptide Y (Baraniuk *et al.*, 1992b). In addition, there is evidence for the localization of neuronal NOS in nerves of the human nasal mucosa (Kulkarni *et al.*, 1994).

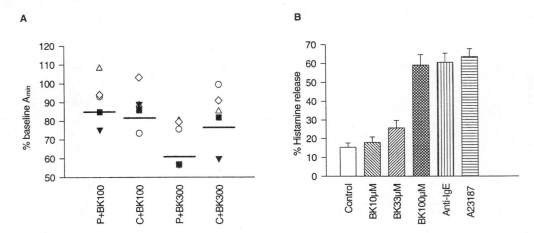

Figure 17.8   The role of histamine in bradykinin-induced nasal blockage. (A) shows the reduction in the minimum nasal cross-sectional area (A_{min}) induced by bradykinin (100 and 300 μg) after oral pretreatment with placebo (P) or cetirizine (10 mg) (C). Cetirizine significantly inhibits the reduction in A_{min} induced by bradykinin (300 μg) **(B)** shows the histamine release induced from dispersed human nasal tissue by bradykinin (BK), anti-IgE and the calcium ionophore, A23187. Bradykinin (33 and 100 μM) causes a significant histamine release.

Neuropeptides may play a role in allergic rhinitis as they can stimulate secretion from nasal submucosal glands (Raphael *et al.*, 1991), and their topical application increases nasal airway resistance, recruitment of inflammatory cells and microvascular leakage in subjects with allergic rhinitis (Lurie *et al.*, 1994). Following nasal allergen challenge in sensitive subjects, substance P, CGRP, VIP and somatostatin levels are raised in the lavage fluid (Walker *et al.*, 1988).

Although BK potently activates C-fibers, the contribution of these fibers to the nasal actions of BK is unclear. Capsaicin stimulates a subpopulation of sensory nerves and, when introduced into the nasal cavity, causes discomfort and rhinorrhea, effects also induced by BK (Phillip *et al.*, 1994; Rajakulasingam *et al.*, 1992a). Capsaicin also induces significant increases in the total protein content of the nasal lavage fluid. These observations suggest that capsaicin-sensitive neurons induce glandular secretion. Alternatively, BK causes nasal blockage and increased vascular permeability, whereas capsaicin has no such effects. Hence, capsaicin-sensitive neurons may mediate BK-induced nasal discomfort and rhinorrhea, but seem not to be involved in nasal blockage and increased vascular permeability. However, repeated applications of capsaicin, to desensitize capsaicin-sensitive neurons, failed to block nasal pain produced by Lys-BK in normal human nasal mucosa (Geppetti *et al.*, 1991). Thus, kinin-induced nasal pain appears to be independent of capsaicin-sensitive neurons. Unfortunately, studies with capsaicin-induced desensitization did not measure quantitatively the nasal blockage, rhinorrhea or vascular permeability changes induced by BK.

Repeated application of capsaicin blocks the algesic response to nasal challenge at low pH with citric acid, indicating that capsaicin-sensitive neurons in the nose mediate the pain response to acidic stimuli (Geppetti *et al.*, 1993). Pretreating the normal human nasal cavity with a muscarinic receptor antagonist, ipratropium bromide, has no significant effect on responses to BK, suggesting that cholinergic pathways do not play a role in normal nonatopic subjects (Rajakulasingam *et al.*, 1992b). However, parasympathetic nerves may play a role in BK-mediated nasal effects in allergic rhinitis. If BK is applied to only one nostril of subjects with perennial allergic rhinitis, levels of total protein and glyco-conjugates increase in the contralateral, unstimulated nostril (Baraniuk *et al.*, 1994a,b). Normal subjects, when treated this way, do not show any changes in contralateral secretions. The contralateral total protein and glycoconjugate secretion induced by BK is abolished by ipratropium bromide, indicating that parasympathetic nerve reflexes mediate this secretion (Baraniuk *et al.*, 1994b). There is considerable evidence that tachykinins, released from sensory nerves, mediate BK-induced increases in microvascular permeability in the guinea-pig and rat nasal mucosa (Bertrand *et al.*, 1993; Ricciardolo *et al.*, 1994). There is, however, no evidence to support a role for nerves in the increased nasal vascular permeability or blockage induced by BK in normal or atopic humans (Dear *et al.*, 1996b).

4.4   THE ROLE OF PROSTAGLANDINS

In many systems in the body, BK mediates its actions via the generation of prostaglandins (Farmer and Burch, 1992; Hall, 1992). In the human nasal airways, aspirin pretreatment has no significant effect on BK-induced increases in albumin or TAME-esterase levels in the lavage fluid, or on symptom scores reported (Churchill *et al.*, 1991). Baraniuk *et al.* (1992a) reported that ibuprofen had no effect on BK-induced total protein or albumin release in the human nasal cavity, whereas the cyclooxygenase inhibitor actually increased the glycoconjugate released by BK provocation. They also measured lysozyme but, as noted above, BK does not cause lysozyme release into the nasal cavity of normal subjects. After ibuprofen treatment, however, 10 nmol BK (but not 100 or 1,000 nmol) caused significant lysozyme release. The lack of a dose-dependent effect by BK on lysozyme release in the presence of ibuprofen is difficult to explain. It appears that ibuprofen enhances BK-induced glycoconjugate production, presumably either by blocking the production of inhibitory prostaglandins or promoting the production of lipoxygenase products. However, an action of ibuprofen, unrelated to cyclooxygenase, may explain the results. Human nasal explants show the opposite response to ibuprofen, where the drug decreased BK-induced glycoconjugate release from nasal tissue *ex vivo* (Baraniuk *et al.*, 1990b).

5. Radioligand Binding Studies of Human Nasal Bradykinin Receptors

Binding of agonists and antagonists to the human nasal B_2 receptor can be studied using $[^{125}I]$-Hoe 140 and nasal membranes prepared from inferior turbinate bones. In saturation binding experiments, at room temperature, $[^{125}I]$-Hoe 140 binds to a single, high-affinity site on the nasal membrane (Dear *et al.*, 1996a). The equilibrium dissociation constant is 0.36 ± 0.08 nM, and the B_{max} value of 0.145 ± 0.012 pmol/mg protein indicates a low receptor density in this tissue. Kinetic binding studies at 4°C produced an association rate constant of 0.24 nM^{-1}min^{-1} and a dissociation rate constant of 0.15 min^{-1}. The equilibrium dissociation constant derived from kinetic binding is 0.62 nM, which agrees well with the value from saturation binding. $[^{125}I]$-Hoe 140 binding is displaced by various kinin receptor ligands (Table 17.1).

NPC 567 displaces $[^{125}I]$-Hoe binding with an affinity approximately 200 times lower than unlabelled Hoe 140 (icatibant). The lowest dose of icatibant which antagonized the nasal actions of BK *in vivo* was 33 μg (25 nmol). Therefore, at least 6.7 mg (5 μmol) of NPC 567 per nostril would, theoretically, be required to antagonize responses to intranasal BK in human subjects. Previous investigations (Pongracic *et al.*, 1991; Austin and Foreman, 1994a; Dear *et al.*, 1996a) used NPC 567 at a maximum dose of 10 mg per nostril and were unable

Table 17.1 Competition binding studies with $[^{125}I]$-Hoe 140, a potent kinin B_2 receptor antagonist, in membranes prepared from human nasal inferior turbinate bones

Displacing ligand	Affinity coefficient, K_i (nM)	Hill coefficient, n_H
Icatibant (Hoe 140)[a]	0.48	1.06
[N-Adamantaneacetyl-DArg-Hyp³,Thi⁵,⁸,DPhe⁷]-BK	1.19	1.36
BK	0.97	1.47
Win 64338[a]	40.87	0.82
NPC 567[a]	90	0.91

[a] For structures and pharmacology of these antagonists, see Chapter 2. BK, bradykinin.

to demonstrate antagonism of the nasal actions of BK. The lack of effect of this antagonist *in vivo* most likely is due to its low affinity for B_2 receptors.

6. Conclusion

There is strong evidence implicating BK, acting via nasal B_2 receptors, as a major mediator of perennial allergic rhinitis but less so in seasonal rhinitis. The actions of BK in human nasal airways occur partly through the generation of NO and partly through the release of histamine. There is little evidence to support a role for eicosanoids or neuronal mechanisms in the actions of BK on the nasal airway except for an involvement of the parasympathetic neurons in BK-induced effects in perennial rhinitis.

7. References

Aitken, R.C.B. (1969). Measurements of feelings using visual analogue scales. Proc. Roy. Soc. Med. 62, 989–993.
Akbary, M.A. and Bender, N. (1993). Efficacy and safety of Hoe 140 (a bradykinin antagonist) in patients with mild-to-moderate seasonal allergic rhinitis. "Kinins, Brazil", Proceedings, Abstract 22.02.
Austin, C.E. and Foreman, J.C. (1993). The effect of platelet-activating factor on the responsiveness of the human nasal mucosa. Br. J. Pharmacol. 110, 113–118.
Austin, C.E. and Foreman, J.C. (1994a). A study of the action of bradykinin and bradykinin analogues in the human nasal airway. J. Physiol. 478, 351–356.
Austin, C.E. and Foreman, J.C. (1994b). Acoustic rhinometry compared with posterior rhinomanometry in the measurement of histamine- and bradykinin-induced changes in nasal airway patency. Br. J. Clin. Pharmacol. 37, 33–37.
Austin, C.E., Foreman, J.C. and Scadding, G.K. (1994). Reduction by Hoe 140, the B2 kinin receptor antagonist, of antigen-induced nasal blockage. Br. J. Pharmacol. 111, 969–971.

Austin, C.E., Dear, J.W., Neighbour, H., Lund, V. and Foreman, J.C. (1996). The contribution of histamine to the action of bradykinin in the human nasal airway. Immunopharmacology 34, 181–189.

Bacci, S., Arbi-Riccardi, R., Mayer, B., Rumio, C. and Borghi-Cirri, M.B. (1994). Localization of NO synthase immuno-reactivity in mast cells of human nasal mucosa. Histochemistry 102, 89–92.

Balduf, M., Steinkraus, V. and Ring, J. (1992). Captopril associated lacrimation and rhinorrhoea. Br. Med. J. 305, 693.

Baraniuk, J.N., Lundgren, J.D., Goff, J., Mullol, J., Castellino, S., Merida, M., Shelhamer, J.H. and Kaliner, M.A. (1990a). Calcitonin gene-related peptide in the human nasal mucosa. Am. J. Physiol. 258, 181–188.

Baraniuk, J.N., Lundgren, J.D., Mizoguchi, H., Peden, D., Gawin, A., Merida, M., Shelhamer, J.H. and Kaliner, M.A. (1990b). Bradykinin and respiratory mucous membranes. Am. Rev. Respir. Dis. 141, 706–714.

Baraniuk, J.N., Lundgren, J.D., Okayama, M., Mullol, J., Merida, M., Shelhamer, J.H. and Kaliner, M.A. (1990c). Vasoactive intestinal peptide in human nasal mucosa. J. Clin. Invest. 86, 825–831.

Baraniuk, J.N., Silver, P.B., Kaliner, M.A. and Barnes, P.J. (1992a). Ibuprofen augments bradykinin-induced glyco-conjugate secretion by human nasal mucosa in vivo. J. Allergy Clin. Immunol. 89, 1032–1039.

Baraniuk, J.N., Silver, P.B., Kaliner, M.A. and Barnes, P.J. (1992b). Neuropeptide Y is a vasoconstrictor in human nasal mucosa. J. Appl. Physiol. 73, 1867–1872.

Baraniuk, J.N., Silver, P.B., Kaliner, M.A. and Barnes, P.J. (1994a). Perennial rhinitis subjects have altered vascular, glandular, and neural responses to bradykinin nasal provocation. Int. Arch. Allergy Immunol. 103, 202–208.

Baraniuk, J.N., Silver, P.B., Kaliner, M.A. and Barnes, P.J. (1994b). Effects of ipratropium bromide on bradykinin nasal provocation in chronic allergic rhinitis. Clin. Exp. Allergy 24, 724–729.

Baumgarten, C.R., Togias, A., Naclerio, R.M., Lichtenstein, L.M., Norman, P.S. and Proud, D. (1985). Influx of kininogens into nasal secretions after antigen challenge of allergic individuals. J. Clin. Invest. 76, 191–197.

Baumgarten, C.R., Nichols, R.C., Naclerio, R.M., Lichtenstein, L.M., Norman, P.S. and Proud, D. (1986a). Plasma kallikrein during experimentally induced allergic rhinitis: role in kinin formation and contribution to TAME-esterase activity in nasal secretions. J. Immunol. 137, 977–982.

Baumgarten, C.R., Nichols, R.C., Naclerio, R.M. and Proud, D. (1986b). Concentrations of glandular kallikrein in human nasal secretions increase during experimentally induced allergic rhinitis. J. Immunol. 137, 1323–1328.

Bertrand, C., Geppetti, P., Baker, J., Petersson, G., Piedmonte, G. and Nadel, J.A. (1993). Role of peptidases and NK1 receptors in vascular extravasation induced by bradykinin in rat nasal mucosa. J. Appl. Physiol. 74, 2456–2461.

Brunnée, T., Nigam, S., Kunkel, G. and Baumgarten, C.R. (1991). Nasal challenge studies with bradykinin: influence upon mediator generation. Clin. Exp. Allergy 21, 425–431.

Borum, P. (1979). Intranasal methacholine challenge: a test for the measurement of nasal reactivity. J. Allergy Clin. Immunol. 63, 253–257.

Churchill, L., Pongracic, J.A., Reynolds, C.J., Naclerio, R.M. and Proud, D. (1991). Pharmacology of nasal provocation with bradykinin: studies of tachyphylaxis, cyclooxygenase inhibition, α-adrenergic stimulation, and receptor subtype. Int. Arch. Allergy Appl. Immunol. 95, 322–331.

Cockcroft, J.R., Chowienczyk, P.J., Brett, S.E. and Ritter, J.M. (1994). Effect of N^G-monomethyl-L-arginine on kinin-induced vasodilation in the human forearm. Br. J. Clin. Pharmacol. 38, 307–310.

Cowan, C.L. and Cohen, R.A. (1991). Two mechanisms mediate relaxation by bradykinin of pig coronary artery: NO-dependent and -independent responses. Am. J. Physiol. 261, H830–H835.

Dear, J.W., Scadding, G.K. and Foreman, J.C. (1995) Reduction by N^G-nitro-arginine methyl ester (L-NAME) of antigen-induced nasal airway plasma extravasation in human subjects in vivo. Br. J. Pharmacol. 116, 1720–1722.

Dear, J.W., Wirth, K., Scadding, G.K. and Foreman, J.C. (1996a). Characterization of the kinin receptor in the human nasal airway using the binding of [^{125}I]-Hoe 140. Br. J. Pharmacol. 119, 1054–1062.

Dear, J.W., Ghali, S. and Foreman, J.C. (1996b). Attenuation of human nasal airway responses to bradykinin and histamine by inhibitors of nitric oxide synthase. Br. J. Pharmacol. 118, 1177–1182.

Dolovich, J., Back, N. and Arbesman, C.E. (1970). Kinin-like activity in nasal secretions of allergic patients. Int. Arch. Allergy 38, 337–344.

Doyle, W.J., Boehm, S. and Skoner, D.P. (1990). Physiologic responses to intranasal dose–response challenges with histamine, methacholine, bradykinin and prostaglandin in adult volunteers with and without nasal allergy. J. Allergy Clin. Immunol. 86, 924–935.

Druce, H.W., Wright, R.H., Kossof, D. and Kaliner, M.A. (1985). Cholinergic nasal hyperreactivity in atopic subjects. J. Allergy Clin. Immunol. 76, 445–452.

Farmer, S.G. and Burch, R.M. (1992). Biochemical and molecular pharmacology of kinin receptors. Annu. Rev. Pharmacol. Toxicol. 32, 511–536.

Fulton, D., McGife, J.C. and Quilley, J. (1992). Contribution of NO and cytochrome P450 to the vasodilator effect of bradykinin in the rat kidney. Br. J. Pharmacol. 107, 722–725.

Garrelds, I.M., van Amsterdam, J.G.C., de Graaf-in't Veld, C., Gerth van Wijk, R. and Zijlstra, F.J. (1995). NO metabolites in nasal lavage fluid of patients with house dust mite allergy. Thorax 50, 275–279.

Geppetti, P., Fusco, B.M., Alessandri, M., Tramontana, M., Maggi, C.A., Drapeau, G., Fanciullacci, M. and Regoli, D. (1991). Kallidin applied to the human nasal mucosa produces algesic response not blocked by capsaicin desensitization. Regul. Pept. 33, 321–329.

Geppetti, P., Tramontana, M., Del Bianco, E. and Fusco, B.M. (1993). Capsaicin-desensitization to the human nasal mucosa selectively reduces pain evoked by citric acid. Br. J. Clin. Pharmacol. 35, 178–183.

Greiff, L., Wollmer, P., Erjefalt, I., Andersson, M., Pipkorn, U. and Persson, C.G. (1993). Effects of nicotine on the human nasal mucosa. Thorax 48, 651–655.

Hall, J.M. (1992). Bradykinin receptors: pharmacological pro-perties and biological roles. Pharmacol. Ther. 56, 131–190.

Hilding, A. (1972). A simple method for collecting near normal nasal secretions. Ann. Otol. Rhinol. Laryngol. 81, 422–423.

Hock, F.J., Wirth, K., Albus, U., Linz, W., Gerhards, H.J., Wiemer, G., Henke, St., Breipohl, G., Konig, W., Knolle, J.

and Scholkens, B.A. (1991). Hoe 140 a new potent and long acting bradykinin-antagonist: *in vitro* studies. Br. J. Pharmacol. 102, 769–773.

Holmberg, K., Bake, B. and Pipkorn, U. (1990). Vascular effects of topically applied bradykinin on the human nasal mucosa. Eur. J. Pharmacol. 175, 35–41.

Holmstrom, M., Scadding, G.K., Lund, V.J. and Darby, Y.C. (1990). A comparison between rhinomanometry and nasal inspiratory peak flow. Rhinology 28, 191–196.

Howarth, P.H., Harrison, K. and Lau, L. (1995). The influence of 5-lipoxygenase inhibition in allergic rhinitis. Int. Arch. Allergy Immunol. 107, 423–424.

Jackson, A.C., Butler, J.P., Mille, E.J., Hoppin, F.G. and Dawson, S.V. (1977). Airway geometry by analysis of acoustic pulse response measurements. J. Appl. Physiol. 43, 523–536.

Kelm, M. and Schrader, J. (1988). NO release from the isolated guinea-pig heart. Eur. J. Pharmacol. 155, 317–321.

Kern, E. (1973). Rhinomanometry. Otolaryngol. Clin. N. Am. 6, 863–874.

Kulkarni, A.P., Getchell, T.V. and Getchell, M.L. (1994). Neuronal NO synthase is localised in extrinsic nerves regulating perireceptor processes in the chemosensory nasal mucosae of rats and humans. J. Comp. Neurol. 345, 125–138.

Linder, A., Strandberge, K. and Deuschl. H. (1988). Variations in histamine concentrations in nasal secretions from patients with allergic rhinitis. Allergy 43, 119–126.

Lundberg, J.M., Lunblad, L., Martling, C.R., Saria, A., Stjarne, P. and Anggard, A. (1987). Coexistence of multiple peptides and classic transmitters in airway neurons: functional and pathophysiological aspects. Am. Rev. Respir. Dis. 136, S16–S22.

Lundberg, J.O.N., Farkas-Szallasi, T., Weitzberg, E., Rinder, J., Lidholm, J., Anggard, A., Hokfelt, T., Lundberg, J.M. and Alving, K. (1995). High NO production in human paranasal sinuses. Nature Med. 1, 370–373.

Lurie, A., Nadel, J.A. Roisman, G., Siney, H. and Dusser, D.J. (1994). Role of neutral endopeptidase and kininase 2 on substance P-induced increase in nasal obstruction in patients with allergic rhinitis. Am. J. Respir. Crit. Care. Med. 149, 113–117.

Maruo, K., Akaike, T., Matsumura, Y., Kohmoto, S., Inada, Y., Ono, T., Arao, T. and Maeda, H. (1991). Triggering of the vascular permeability reaction by activation of the Hageman factor–prekallikrein system by house dust mite proteinase. Biochim. Biophys. Acta 1074, 62–68.

Matsushima, A., Kodera, Y., Ozawa, S., Kobayashi, M., Maeda, H. and Inada, Y. (1992). Inhibition of mite protease (df-protease) with protease inhibitors. Biochem. Int. 28, 717–723.

Meltzer, E.O. (1995). An overview of current pharmacotherapy in perennial rhinitis. J. Allergy Clin. Immunol. 95, 1097–1110.

Mullol, J., Chowdhury, B.A., White, M.V., Ohkibo, K., Rieves, R.D., Baraniuk, J., Hausfeld, J.N., Shelhamer, J.H. and Kaliner, M.A. (1993). Endothelin in human nasal mucosa. Am. J. Resp. Cell Mol. Biol. 8, 393–402.

Naclerio, R.M., Meier, H.L., Kagey-Sobotka, A., Adkinson, N.F., Meyers, D.A., Norman, P.S. and Lichtenstein, L.M. (1983). Mediator release after nasal airway challenge with antigen. Am. Rev. Resp. Dis. 128, 597–602.

Naclerio, R.M., Proud, D., Togias, A.G., Adkinson, N.F., Jr., Meyers, D.A., Kagey-Sobotka, A., Plaut, M., Norman, P.S. and Lichtenstein, L.M. (1985). Inflammatory mediators in late antigen-induced rhinitis. N. Engl. J. Med. 313, 65–70.

Naclerio, R.M., Proud, D., Lichtenstein, L.M., Kagey-Sobotka, A., Hendley, J.O., Sorrentino, J. and Gwaltney, J.M. (1988). Kinins are generated during experimental rhinovirus colds. J. Infect. Dis. 157, 133–142.

Nolte, D. and Luder-Luhr, I. (1973). Comparing measurements of nasal resistance by body plethysmography and rhinomanometry. Respiration 30, 31–38.

O'Kane, K.P.J., Webb, D.J., Collier, J.G. and Vallance, P.J.T. (1994). Local L-N^G-monomethyl-arginine attenuates the vasodilator action of bradykinin in the human forearm. Br. J. Clin. Pharmacol. 38, 311–315.

Ohkubo, K., Baraniuk, J.N., Hohman, R.J., Kaulbach, H.C., Hausfeld, J.N., Merida, M. and Kaliner, M.A. (1993). Human nasal mucosal neutral endopeptidase (NEP): location, quantification and secretion. Am. J. Respir. Cell Mol. Biol. 9, 557–567.

Ohkubo, K., Lee, C.H., Baraniuk, J.N., Merida, M., Hausfeld, J.N. and Kaliner, M.A. (1994a). Angiotensin-converting enzyme in the human nasal mucosa. Am. J. Respir. Cell Mol. Biol. 11, 173–180.

Ohkubo, K., Okuda, M. and Kaliner, M.A. (1994b). Immunological localization of neuropeptide-degrading enzymes in the nasal mucosa. Rhinology 32, 130–133.

Phillip, G., Baroody, F.M., Proud, D., Naclerio, R.M. and Togias, A.G. (1994). The human nasal response to capsaicin. J. Allergy. Clin. Immunol. 94, 1035–1045.

Pipkorn, U. (1990). H_1-antagonists as a tool in clarifying the pathophysiology of nasal allergy. Clin. Exp. Allergy 20 (Suppl. 2), 25–29.

Pipkorn, U., Karlsson, G. and Enerbaeck, L. (1988). A brush method to harvest cells from the nasal mucoas for microscopic and biochemical analysis. J. Immun. Meth. 112, 37–42.

Pongracic, J.A., Naclerio, R.M., Reynolds, C.J. and Proud, D. (1991). A competitive kinin receptor antagonist, [DArg^0,Hyp^3,DPhe^7]-bradykinin, does not affect the response to nasal provocation with bradykinin. Br. J. Clin. Pharmacol. 31, 287–294.

Proud, D., Togias, A., Naclerio, R.M., Crush, S.A., Norman, P.S. and Lichtenstein, L.M. (1983). Kinins are generated *in vivo* following nasal airway challenge of allergic individuals with allergen. J. Clin. Invest. 72, 1678–1685.

Proud, D., Baumgarten, C.R., Naclerio, R.M. and Ward, P.E. (1987). Kinin metabolism in human nasal secretions during experimentally induced allergic rhinitis. J. Immunol. 138, 428–434.

Proud, D., Reynolds, C.J., Lacapra, S., Kagey-Sobotka, A., Lichtenstein, L.M. and Naclerio, R.M. (1988). Nasal provocation with bradykinin induces symptoms of rhinitis and a sore throat. Am. Rev. Respir. Dis. 137, 613–616.

Proud, D., Naclerio, R.M., Gwaltney, J.M. and Hendley, J.O. (1990a). Kinins are generated in nasal secretions during natural rhinovirus colds. J. Infect. Dis. 161, 120–123.

Proud, D., Naclerio, R.M., Meyers, D.A., Kagey-Sobotka, A., Lichtenstein, L.M. and Valentine, M.D. (1990b). Effects of a single-dose pretreatment with captopril on the immediate response to nasal challenge with allergen. Int. Arch. Allergy Appl. Immunol. 93, 165–170.

Rajakulasingam, K., Polosa, R., Holgate, S.T. and Howarth, P.H. (1991). Comparative nasal effects of bradykinin, kallidin and [Des-Arg⁹]-bradykinin in atopic rhinitic and normal volunteers. J. Physiol. 437, 557–587.

Rajakulasingam, K., Polosa, R., Lau, L.C.K., Church, M.K., Holgate, S.T. and Howarth, P.H. (1992a). Nasal effects of bradykinin and capsaicin: influence on plasma protein leakage and role of sensory neurons. J. Appl. Physiol. 72, 1418–1424.

Rajakulasingam, K., Polosa, R., Lau, L.C.K., Church, M.K., Holgate, S.T. and Howarth, P.H. (1992b). The influence of terfenadine and ipratropium bromide alone and in combination on bradykinin-induced nasal symptoms and plasma protein leakage. Clin. Exp. Allergy 22, 717–723.

Rajakulasingam, K., Polosa, R., Lau, L.C.K., Church, M.K., Holgate, S.T. and Howarth, P.H. (1993). Comparative nasal effects of bradykinin and histamine: influence on nasal airways resistance and plasma protein exudation. Thorax 48, 324–329.

Raphael, G.D., Baraniuk, J.N. and Kaliner, M.A. (1991). How and why the nose runs. J. Allergy Clin. Immunol. 87, 457–467.

Ricciardolo, F.L.M., Nadel, J.A., Bertrand, C., Yamawaki, I., Chan, B. and Geppetti, P. (1994). Tachykinins and kinins in antigen-evoked plasma extravasation in guinea-pig nasal mucosa. Eur. J. Pharmacol. 261, 127–132.

Rokenes, H.K., Andersson, B. and Rundcrantz, H. (1988). Effect of terfenadine and placebo on symptoms after nasal allergen provocation. Clin. Allergy 18, 63–69.

Said, S.I. and Mutt, V. (1988). Vasoactive intestinal peptide and related peptides. Ann. N. York Acad. Sci. 527, 283–287.

Stewart, G.A., Thompson, P.J. and Simpson, R.J. (1989). Protease antigens from house dust mite. Lancet 2, 154–155.

Stewart, G.A., Lake, F.R. and Thompson, P.J. (1991). Faecally derived hydrolytic enzymes from dermatophagoides pteronyssinus: physicochemical characterisation of potential allergens. Int. Arch. Allergy Appl. Immunol. 95, 248–256.

Stewart, G.A., Thompson, P.J. and McWilliam, A.S. (1993). Biochemical properties of aeroallergens: contributory factors in allergic sensitization? Pediatr. Allergy Immunol. 4, 163–172.

Sunday, M.E., Kaplan, L.M., Motoyama, E., Chin, W.W. and Spindel, E.R. (1988). Gastrin releasing peptide (mammalian bombesin) gene expression in health and disease. Lab. Invest. 59, 5–24.

Svensson, C., Andersson, M., Persson, C.G.A., Venge, P., Alkner, U. and Pipkorn, U. (1990). Albumin, bradykinin, and eosinophil cationic protein on the nasal mucosal surface in patients with hay fever during natural allergen exposure. J. Allergy Clin. Immunol. 85, 828–833.

Svensson, C., Klementsson, H., Andersson, M., Pipkorn, U. and Alkner, U. (1994). Glucocorticoid-induced attenuation of mucosal exudation of fibrinogen and bradykinins in seasonal allergic rhinitis. Allergy 49, 177–183.

Takahashi, K., Aoki, T., Kohmoto, S., Nishimura, H., Kodera, Y., Matsushima, A. and Inada, Y. (1990). Activation of kallikrein–kinin system in human plasma with purified serine protease from dermatophagoides farinae. Int. Arch. Allergy Appl. Immunol. 91, 80–85.

Vallance, P. and Collier, J. (1994). Biology and clinical relevance of NO. Br. Med. J. 309, 453–457.

Vials, A. and Burnstock, G. (1992). Effects of NO synthase inhibitors, L-N^G-nitroarginine and L-N^G-nitroarginine methyl ester, on responses to vasodilators of the guinea-pig coronary vasculature. Br. J. Pharmacol. 107, 604–609.

Walden, S.M., Proud, D., Bascom, R., Lichtenstein, L.M., Kagey-Sobotka, A., Adkinson, N.F. and Naclerio, R.M. (1988). Experimentally induced nasal allergic responses. J. Allergy Clin. Immunol. 81, 940–949.

Walker, K.B., Serwonska, M.H., Valone, F.H., Harkonen, W.S., Frick, O.L., Scriven, K.H., Ratnoff, W.D., Browning, J.G., Payan, D.G. and Goetzl, E.J. (1988). Distinctive patterns of release of neuroendocrine peptides after nasal challenge of allergic subjects with ryegrass antigen. J. Clin. Immunol. 8, 108–113.

Weldon, S.M., Winquist, R.J. and Madweb, J.B. (1995). Differential effects of L-NAME on blood pressure and heart rate responses to acetylcholine and bradykinin in cronomolgus primates. J. Pharmacol. Exp. Ther. 272, 126–133.

Wirth, K., Hock, F.J., Albus, U., Linz, W., Alpermann, H.G., Anagnostopoulos, H., Henke, St., Breipohl, G., Konig, W., Knolle, J. and Scholkens, B.A. (1991). Hoe 140, a new potent and long acting bradykinin-antagonist: in vivo studies. Br. J. Pharmacol. 102, 774–777.

Yu, X.-Y., Li, Y.-J. and Deng, H.-W. (1993). The regulatory effect of bradykinin on the actions of sensory nerves in the perfused rat mesentery by NO. Eur. J. Pharmacol. 241, 35–40.

18. Bradykinin as a Growth Factor

David A. Walsh *and* Tai-Ping D. Fan

1. Introduction

Cell proliferation is a central process in physiological and pathological conditions, ranging from essential roles in embryogenesis through repair of tissue injury to the unequivocally deleterious effects of neoplasia. Predictably, mitosis is normally tightly regulated through complicated pathways involving a wide range of autocrine, paracrine and humoral factors, as well as through other interactions of the cell with its local environment, including extracellular matrix and neighboring cells. Abnormally high proliferation rates are characteristic of many diseases in which the kallikrein–kinin system is activated, including chronic inflammatory diseases such as human arthritis, psoriasis and neoplasia. Evidence from experiments *in vitro* discussed below indicates that kinin receptor agonists may stimulate cell proliferation, particularly in the presence of other growth factors. However, the role of endogenous bradykinin

(BK) generation in cellular proliferation *in vivo* remains poorly understood. This deserves further study now that specific and metabolically stable BK receptor antagonists, and BK receptor knockout animals are becoming available. In this chapter we review the evidence that BK can regulate cell proliferation, and discuss the possible relevance of this to angiogenesis, chronic inflammatory synovitis, neoplasia and other proliferative conditions.

2. Cell Proliferation in vitro

The kallikrein–kinin system has been implicated in cell growth for many years. Kallikreins have growth-promoting activity, although the relation of this to BK generation has not been fully elucidated (Landsleitner *et al.*, 1974; Catalioto *et al.*, 1987; Korbelik *et al.*, 1989). Certainly enzymatic activity is required for the growth-promoting effects of kallikrein, as both are inhibited by

The Kinin System
ISBN 0–12–249340–0

aprotinin or phenylmethylsulfonylfluoride (Catalioto *et al.*, 1987). However, kallikrein-like arginyl esterases are involved in the processing of other growth factors, including epidermal growth factor and nerve growth factor (Drinkwater *et al.*, 1988).

Proliferative effects of BK were demonstrated on lymphocytes over 25 years ago (Perris and Whitfield, 1969; Whitfield *et al.*, 1970), and since then much data has accumulated on mitogenic actions of kinins in a variety of cell systems. Most authors have used [^{3}H]-thymidine incorporation as an index of cell proliferation, while some have substantiated their observations by documenting changes in cell number, or changes in [^{3}H]-thymidine labeling index. As our understanding of the biology of cell proliferation has increased, it has become clear that each of these methods if used alone may give misleading results. [^{3}H]-Thymidine incorporation is a sensitive measure of proliferation, but also may occur during the development of polyploidy, in the absence of cell division. Changes in cell number represent a balance between proliferation and cell death, typically by apoptosis. Effects of BK on apoptosis and polyploidy have received little attention.

Results of cell-culture studies are further complicated by the multiple components of kinin systems that may be present. Mitogenic effects of exogenous BK may be obscured through the synthesis of endogenous kinins, through kinin metabolism by cellular enzymes, and through the simultaneous expression of multiple subtypes of kinin receptors, and of receptor-coupled second-messenger pathways with different and possibly opposite effects on cell proliferation. Comparisons between studies are further complicated by differences between cell types, changes in phenotype with prolonged culture or subcloning, and the presence of synergistic cofactors in the culture medium. Probably as a result of such heterogeneity, BK has variably been found to be a major growth factor, to stimulate cell proliferation slightly, to have no effect or to inhibit proliferation. Despite these limitations, however, a picture is emerging that BK may act alone as a growth factor, and that its effects may be synergistic or inhibited in the presence of other growth factors.

2.1 STIMULATION OF CELL PROLIFERATION BY BRADYKININ

Fibroblasts and fibroblast cell lines have been used extensively to study the effects of kinins on cellular proliferation (Table 18.1). BK stimulates fibroblast growth (Boucek and Noble, 1973; Straus and Pang, 1984) and Lys-BK also stimulates fibroblast DNA synthesis (Owen and Villereal, 1983).

BK alone, however, appears to be a weak mitogen for fibroblasts. [^{3}H]-Thymidine incorporation is often increased by only 10–50%, compared with the several-fold increases typically observed with other mitogens such as platelet-derived growth factor (PDGF) (Coughlin *et al.*, 1985; Kimball and Fisher, 1988; Godin *et al.*, 1991; Taylor *et al.*, 1992). Furthermore, stimulation of fibroblast proliferation may require incubation with BK for more than 24 h (van Zoelen *et al.*, 1994), and may have a bell-shaped concentration–response curve, with inhibition at concentrations greater than 10 nM (Patel and Schrey, 1992). Other experiments indicated that BK alone had no effect on [^{3}H]-thymidine incorporation in fibroblasts from dental pulp, amniotic fluid, lungs or dermis, or in the Swiss 3T3 fibroblasts (Owen and Villereal, 1983; Straus and Pang, 1984; Marceau and Tremblay, 1986; Olsen *et al.*, 1988; van Corven *et al.*, 1989; Kiehne and Rozengurt, 1994; Sundqvist *et al.*, 1995).

BK has modest stimulatory activity on the proliferation of a range of other cell types, including rat mesenteric arterial smooth muscle cells (Dixon *et al.*, 1994) and smooth muscle-like renal mesangial cells (Ganz *et al.*, 1990; Bascands *et al.*, 1993). BK may stimulate coronary venule endothelial cell proliferation, an essential step in angiogenesis (Ziche *et al.*, 1992). However, mitogenic effects on vascular endothelium from other sources have been more difficult to demonstrate (Porta *et al.*, 1992). BK doubled human keratinocyte growth rates in one study, suggesting a possible role in wound repair (Talwar *et al.*, 1990), but Johnson and colleagues (1992) found no effect on human keratinocyte DNA synthesis. Also, although early studies indicated that BK, at micromolar concentrations, can stimulate rat thymic lymphocyte proliferation (Perris and Whitfield, 1969; Whitfield *et al.*, 1970), the kinin was later found to inhibit lymphocyte proliferation induced by phytohemagglutinin, concanavalin A, or by antigen (Kimura *et al.*, 1979, 1983).

Of clinical interest, BK stimulates the proliferation of a variety of human tumor cell lines, including A431 carcinoma cells and small-cell lung carcinoma cells (Tilly *et al.*, 1987; Roberts and Gullick, 1989; Sethi and Rozengurt, 1991). Transfection of fibroblasts by the proto-oncogene Ha-*ras* increases their susceptibility to BK-stimulated proliferation in the presence of insulin (Roberts and Gullick, 1989). This may indicate that malignant cells are more prone to the mitogenic actions of BK, and that BK may be an important stimulant of tumor growth.

2.2 ANTIMITOGENIC EFFECTS OF KININS

Under some conditions *in vitro*, BK has been found to inhibit, rather than stimulate, proliferation of a number of cell types, including fibroblasts (Newman *et al.*, 1989;

Table 18.1   Proliferative actions of kinins in vitro.

Cell type	Source	Agonist	Antagonist[a]	Percent increase in growth[b]	Second messenger[c]	Synergy[b]	Reference
Fibroblast	Balb/c 3T3 mouse	BK	—	1,100	—	IL-1α, IL-1β, PDGF, insulin	Kimball and Fisher (1988), Godin et al. (1991)
	Human embryo	BK	—	50	—	—	Boucek and Noble (1973)
	IMR-90	BK	—	200	—	Indomethacin	Goldstein and Wall (1984), Straus and Pang (1984)
		desArg9-BK	B1 receptor	200	—	—	
	Human foreskin	Lys-BK	—	≤300	—	Vasopressin, EGF, insulin	Owen and Villereal (1983), Coughlin et al. (1985)
		BK					
	Rabbit R51	desArg9-BK	B1 receptor	60	—	—	Marceau and Tremblay (1986)
	3T3 mouse	BK	B2 receptor	500	PKC, tyrosine kinase	Vasopressin, EGF, insulin	Olsen et al. (1988), Roberts and Gullick (1989), Kiehne and Rozengurt (1994)
	Rat kidney	BK	—	100	—	TGF-β, Retinoic acid	van Zoelen et al. (1994)
Thymocyte	Rat	BK, desArg9-BK	—	100	—	Caffeine	Whitfield et al. (1970), Tiffany and Burch (1989)
Venule endothelium	Bovine coronary	BK	—	130	—	—	Ziche et al. (1992)
Keratinocyte	Human	Lys-BK	—	100	—	Insulin	Talwar et al. (1990)
Endometrium	Human	BK	—	100	—	—	Endo et al. (1991)
Mesangium	Rat	BK	B2 receptor	200	PKC	Insulin	Ganz et al. (1990), Issandou and Darbon (1991), Bascands et al. (1993)
		desArg9-BK	B1 receptor	500	—	Insulin	
Small cell lung cancer	Human	BK	B2 receptor	200	—	—	Sethi and Rozengurt (1991)
Arterial smooth muscle	Rat mesentery	BK	—	≤100	—	—	Dixon (1994)
Human B2 receptor-transfected hamster CCL 39 cells		BK	—	89	—	Insulin	Taylor et al. (1992)

[a] Effect of agonist inhibited by receptor-selective antagonist.

[b] Increased growth given as percentage increase compared with growth in the absence of kinin. Values represent the maximum reported effect, in the presence of synergistic factors as stated.

[c] Second-messenger system mediating effect of agonist indicated by interventional experiment types, e.g., downregulation of protein kinase C (PKC). BK, bradykinin; EGF, epidermal growth factor; IL-1, interleukin-1; PDGF, platelet-derived growth factor; PKC, protein kinase C; TGF-β, transforming growth factor-β.

Patel and Schrey, 1992; McAllister et al., 1993; Lahaye et al., 1994), epithelial cells (Raspe et al., 1992), lymphocytes (Kimura et al., 1979, 1983) and some tumor cell lines (Sukhanova et al., 1993). Boucek and Noble (1973) reported that, while BK alone stimulated fibroblast growth, it inhibited the proliferative action of serotonin. Antimitogenic effects may occur at higher concentrations of BK than does stimulation of proliferation in the same system (Patel and Schrey, 1992), and have been observed most often as an inhibition of the growth-stimulating effects of PDGF, insulin-like growth factor-1 (IGF-1) or epidermal growth factor (EGF) (Newman et al., 1989; McAllister et al., 1993; Lahaye et al., 1994).

The existence both of mitogenic and antimitogenic effects of BK, sometimes observed in a single-cell type under different conditions, indicates interaction with several pathways important in the regulation of cell proliferation. Such multiplicity of action may result from activation of multiple receptor subtypes and second-messenger systems, the biological importance of which varies with cell type and environment.

2.3 KININ RECEPTOR SUBTYPES MEDIATING REGULATION OF CELL PROLIFERATION

Much evidence suggests that the B_2 kinin receptors mediate the proliferative actions of BK on fibroblasts. Fibroblast cell lines susceptible to the growth-stimulating effects of BK express B_2 receptors constitutively as determined by the binding of [³H]-BK, which was inhibited by selective ligands of B_2 receptors, but not by B_1 receptor ligands (Roberts and Gullick, 1989; Bathon et al., 1992; Menke et al., 1994; Kiehne and Rozengurt, 1995). Kinin-stimulated fibroblast proliferation was observed with a half maximal stimulation in the nanomolar range for BK, comparable to the observed K_d for [³H]-BK binding, while a B_1 receptor agonist, desArg⁹-BK, had a potency several orders of magnitude lower (Roberts and Gullick, 1989). Furthermore, BK-stimulated mitogenesis in the Swiss 3T3 fibroblast cell line is blocked by B_2 receptor antagonists (Kiehne and Rozengurt, 1994). The mitogenic potential of B_2 receptor activation has been confirmed by transfection studies. A Chinese hamster lung fibroblast cell line, CCL39, does not display specific [³H]-BK binding, nor show any proliferative response to BK, but develops both when transfected with a vector encoding the B_2 receptor (Taylor et al., 1992).

BK, acting through B_2 receptors, stimulates arachidonic acid metabolism in cultured fibroblasts, resulting in the release of prostanoids, including the potentially antimitogenic agent prostaglandin E_2 (PGE$_2$) (Bathon et al., 1992; McAllister et al., 1993). Cyclooxygenase inhibition by indomethacin enhances the stimulation of fibroblast proliferation by BK, suggesting that prostanoid synthesis may mask or even reverse any direct mitogenic actions of the kinin (Goldstein and Wall, 1984; Straus and Pang, 1984). This prevention of cell proliferation appears also to be mediated by B_2 receptors, since indomethacin enhanced BK-induced fibroblast proliferation even in the presence of B_1 receptor antagonists. It appears, therefore, that activated B_2 receptors may stimulate multiple second-messenger pathways in fibroblasts, with the stimulation of proliferative pathways as well as the generation of arachidonate metabolites which inhibit proliferation.

Other evidence supports the view that B_1 receptors can also mediate mitogenic actions of kinins on fibroblasts. Fibroblasts may express B_1 as well as B_2 receptors (Webb et al., 1994). Indeed, a cDNA clone encoding the human B_1 receptor was recently isolated from a cDNA library derived from the IMR 90 human embryonic lung fibroblast cell line (Menke et al., 1994) (see Chapter 3 of this volume). DesArg⁹-BK stimulated [³H]-thymidine incorporation or increased cell numbers in cultures of IMR-90 fibroblasts, and also in human and rabbit dermal fibroblast cell lines in long-term culture (Owen and Villereal, 1983, Goldstein and Wall, 1984; Straus and Pang, 1984; Marceau and Tremblay, 1986). Mitogenic effects could be blocked by a B_1 receptor antagonist. BK-enhanced T-lymphocyte proliferation may also be mediated by B_1 receptors (Tiffany and Burch, 1989; Marceau et al., 1983).

Although the data discussed above indicate that B_1 receptors may mediate some mitogenic actions of BK, as with B_2 receptors, B_1 receptors may also mediate antimitogenic actions of kinins on fibroblasts under certain conditions. BK-induced inhibition of human breast fibroblast proliferation was mimicked by desArg⁹-BK and inhibited by desArg⁹-[Leu⁸]-BK, a B_1 receptor antagonist, but not by a B_2 receptor antagonist (Patel and Schrey, 1992).

The overall effect of kinins on cellular proliferation depends, in part, on the relative level of B_1 and B_2 receptor expression (Roberts and Gullick, 1989; Taylor et al., 1992). Expression of these two receptor subtypes can be differentially regulated. B_2 receptors are usually expressed constitutively, although both B_1 and B_2 receptors may be upregulated by inflammatory mediators such as interleukin-1 (IL-1) (Bathon et al., 1992; Galizzi et al., 1994). B_1 receptor expression may be upregulated during neoplastic transformation, leading to enhanced mitogenic activity (Roberts and Gullick, 1989). A number of cell types, including fibroblasts and rat mesangial cells, may co-express both B_1 and B_2 receptors. Correspondingly, regulation of proliferation by kinins may be the product of simultaneous and different actions through both B_1 and B_2 receptors (Goldstein and Wall, 1984; Bascands et al., 1993).

2.4   SIGNAL TRANSDUCTION IN BRADYKININ-INDUCED MITOGENESIS

BK can activate several intracellular second-messenger pathways simultaneously, which, in turn, enhance or inhibit cellular proliferation. This may, in part, explain the various mitogenic and antimitogenic effects of BK observed in cell culture systems. Activation of protein kinase C (PKC) has been proposed as central to the mitogenic action of BK on rat mesangial cells and Swiss 3T3 fibroblasts (Issandou and Darbon, 1991; Kiehne and Rozengurt, 1994). BK, binding to B_2 receptors, activates PKC through the generation of diacylglycerol by the phosphoinositide pathway. Downregulation of PKC by prolonged exposure to phorbol esters, or inhibition of the kinase by staurosporine, inhibit BK-induced cell proliferation. DesArg^9-BK-stimulated proliferation may also be mediated by PKC (Issandou and Darbon, 1991).

Other intracellular pathways may also be essential to the proliferative activity of BK, since the degree of activation of PKC correlates poorly with mitogenic activity (Coughlin et al., 1985). Tyrosine kinase activity may play a role, since inhibition of tyrosine phosphorylation by a tyrphostin inhibited BK-stimulated DNA synthesis in 3T3 cells (Kiehne and Rozengurt, 1994). Alternatively, apparent discrepancies between the degree of PKC activation and mitogenic effect may be attributable to selective activation of PKC isoforms.

2.5   SIGNAL TRANSDUCTION IN ANTIMITOGENIC ACTIONS OF BRADYKININ

As discussed above, arachidonic acid metabolites, especially PGE_2, have been implicated in the anti-mitogenic effects of kinins (Goldstein and Wall, 1984; Patel and Schrey, 1992; McAllister et al., 1993; Lahaye et al., 1994). Antiproliferative effects of kinins are mimicked by PGE_2 and inhibited by cyclooxygenase inhibitors such as indomethacin. Other intracellular pathways, however, may also participate in antimitogenic actions of BK. Perhaps surprising in view of the established role of PKC activation in cell proliferation, Newman et al. (1989) found that inhibition of fetal lung fibroblast proliferation by BK was prevented by partial downregulation of PKC. These data may indicate that some isoforms of PKC activated by BK have antimitogenic effects.

2.6   SYNERGY BETWEEN BRADYKININ AND OTHER GROWTH FACTORS

In view of the complexity of intracellular signaling pathways participating in the regulation of cell division, there is considerable potential for synergy between growth factors. In particular, factors which stimulate different components of the second-messenger systems leading to cell proliferation, are likely to have more than additive effects when administered together. Indeed BK, although generally a weak mitogen when administered alone, may enhance proliferation several-fold in the presence of other growth factors (Table 18.1).

BK synergistically enhances proliferative responses of fibroblasts to a variety of growth factors including insulin (Woll and Rozengurt, 1988; Roberts and Gullick, 1989), EGF (Olsen et al., 1988), IL-1 (Kimball and Fisher, 1988), transforming growth factor-β (TGF-β), retinoic acid (Van Zoelen et al., 1994) and PDGF (Godin et al., 1991). Similar synergy has been observed in other cell types such as rat mesangial cells (Issandou and Darbon, 1991) and rat liver epithelial cells (Hill et al., 1988). BK also synergizes with insulin in stimulating proliferation of B_2 receptor-transfected cells (Taylor et al., 1992).

Synergy may result from interactions at several levels. IL-1 stimulates both B_1 and B_2 receptor expression (Bathon et al., 1992; Galizzi et al., 1994; Menke et al., 1994), and has mitogenic properties of its own in the absence of BK (Kimball and Fisher, 1988). PDGF pretreatment enhances BK-stimulated Balb/c 3T3 cell proliferation, even if the two mitogens are not present concurrently (Godin et al., 1991). Conversely, BK primes fibroblasts to the mitogenic action of EGF (Olsen et al., 1988).

Synergy of BK, acting through PKC with growth factors whose effects are mediated by tyrosine kinase-coupled receptors was expected. BK, however, can also synergistically enhance the proliferative actions of other peptides believed to act through stimulation of PKC, such as vasopressin. BK synergized with vasopressin in enhancing [^{3}H]-thymidine incorporation into Swiss 3T3 cells (Kiehne and Rozengurt, 1994). Such evidence may indicate stimulation of mitogenic pathways by BK other than those utilizing PKC, or possible synergy between pathways involving different PKC isoforms.

BK does not synergize with other growth factors under all circumstances. Indeed, BK may inhibit EGF, IGF-1- and PDGF-stimulated proliferation of fibroblasts (Newman et al., 1989; Patel and Schrey, 1992; McAllister et al., 1993). Such effects may depend on activation of B_1 receptors, or on receptor coupling to arachidonate pathways, as discussed above. Conversely, tumor necrosis factor-α (TNF-α) and interferon-β inhibit the growth-stimulating effects of BK in 3T3 cells, and it has been suggested that autocrine factors produced by fibroblasts may limit BK-induced proliferation by reducing B_2 receptor expression on the cell surface (Kiehne and Rozengurt, 1995).

2.7 REGULATION OF GROWTH FACTOR RELEASE BY BRADYKININ

In addition to direct mitogenic actions on a number of cell types, BK may also stimulate the synthesis or release of other growth factors. Some of these may act in an autocrine fashion, perhaps synergizing with the mitogenic activity of BK itself. For example, BK increases basic fibroblast growth factor (FGF) mRNA levels in rat dermal fibroblasts (Lowe et al., 1992). BK-induced growth factors may also have a paracrine action on neighboring cells, which do not, themselves, express kinin receptors.

Several of the mediators released by inflammatory cells in response to BK may have growth-promoting activities, including histamine released from mast cells (Matsuda et al., 1994), and IL-1 released from macrophages (Tiffany and Burch, 1989). Resident cells within tissues may also be stimulated by BK to release growth factors. BK stimulates the release of the neuropeptide Substance P from the peripheral terminals of sensory nerves (Geppetti et al., 1991). Substance P is mitogenic for a number of cell types, including fibroblasts, smooth muscle and endothelial cells (Nilsson et al., 1985; Ziche et al., 1990). Nitric oxide and prostaglandins generated by vascular endothelium exposed to BK may also stimulate cell proliferation (BenEzra, 1978; Form and Auerbach, 1983; Ziche et al., 1994).

BK, however, may also stimulate the release of potentially growth-inhibiting agents, such as TNF from macrophages (Tiffany and Burch, 1989). Furthermore, the synthesis of some growth factors, such as IGF-1, may be inhibited by BK under some circumstances (Lowe et al., 1992). The net effect of BK on cell proliferation in vivo depends, therefore, not only on the expression and second-messenger-coupling of BK receptors, but also on the presence of other growth-factor-producing cells in the tissue.

3. Angiogenesis

BK is generated in tissues undergoing neovascularization, as seen during wound healing, chronic inflammation and tumor growth (see below). The precise role of BK in the angiogenic process itself, however, remains to be elucidated fully. Angiogenesis is a multistep process involving enzymatic degradation of endothelial basement membrane, migration and proliferation of postcapillary venule endothelial cells, capillary tube formation, vessel maturation and reorganization (Folkman and Brem, 1992; Fan et al., 1995). Angiogenic factors may influence any of these steps, either by direct actions on vascular cells, or by stimulating the release of other angiogenic factors by neighboring cells. Furthermore, stimulation of blood flow through vasodilation may in itself enhance angiogenesis.

BK stimulates PKC activity in endothelial cells, as in other cell types where BK is mitogenic (Mackie et al., 1986). Ziche et al. (1992) found that BK stimulated [^{3}H]-thymidine incorporation into coronary venule endothelial cells, with a potency comparable to basic FGF. Another study found no effect on [^{3}H]-thymidine incorporation in bovine retinal or aortic endothelial cells (Porta et al., 1992). Such differences may reflect heterogeneity between endothelial cells from different sources, postcapillary venules being the site of endothelial proliferation during angiogenesis in vivo. Nitric oxide and prostaglandins, released by endothelium in response to BK, are capable under some circumstances of stimulating endothelial proliferation and neo-vascularization (BenEzra, 1978; Form and Auerbach, 1983; Ziche et al., 1993, 1994). Whether such actions are, of themselves, sufficient to induce neovascularization in vivo, however, remains to be determined.

BK may also have a stimulatory effect on vessel growth by enhancing the release of angiogenic agents by nonvascular cells. Generation or release of basic FGF, IL-1, histamine and Substance P, each are stimulated by BK (see above) and each may stimulate angiogenesis (Fan and Brem, 1992).

Studies on the effects of kininase II (angiotensin-converting enzyme, ACE) (see Chapter 7) inhibitors on angiogenesis have implicated BK as an angiogenic factor in vivo. Inhibition of ACE would be expected to increase local tissue concentrations of BK and desArg^9-BK, but may also inhibit the formation of angiotensin II or inhibit the degradation of Substance P, each of which may regulate angiogenesis (Ziche et al., 1990; Stoll et al., 1995). Data from experiments with ACE inhibitors are, therefore, difficult to interpret.

A specific ACE inhibitor, lisinopril, increased capillary densities in the sciatic nerves of diabetic rats, where reduced neuronal blood flow is believed to contribute to the pathogenesis of diabetic neuropathy (Cameron et al., 1992). Similarly, ramipril increased myocardial capillary length density in spontaneously hypertensive rats (Unger et al., 1992), a finding subsequently confirmed by Olivetti et al. (1993). This angiogenic effect was observed at doses of the kininase inhibitor, which were insufficient to prevent myocardial hyper-trophy or to render the animals normotensive, suggesting that it was not mediated by inhibition of angiotensin II generation (Unger et al., 1992). Cardiac, but not plasma kininase activity was inhibited, indicating a possible local action in the heart. Ramipril also increased myocardial blood flow, an effect inhibited by B$_2$ receptor antagonists (Schölkens et al., 1988) (see also Chapter 19). However, the role of BK in increasing vascular densities in these animal models has not been rigorously tested using selective kinin receptor agonists and antagonists. Furthermore, kininase inhibitors do not increase vascular density under all conditions. Wang and Prewitt (1990) found that captopril reduced small arteriolar density in the cremaster muscle of both hypertensive and normotensive rats.

More direct evidence for a role of BK in neovascularization *in vivo* was obtained using subcutaneous sponge implantation in rats (Hu and Fan, 1993). In this model of angiogenesis, polyether sponges were implanted subcutaneously in the dorsa of rats and agents were administered through a permanently indwelling cannula into the sponge. During the following 14 days, sponges were progressively infiltrated by granulation tissue containing newly formed microvessels. This angiogenesis can be quantified by morphometric techniques, measurement of intravascular markers such as hemoglobin and by measures of sponge blood flow, in particular ^{133}Xe clearance (Hu *et al.*, 1995). BK increased ^{133}Xe clearance, cellularity and vascularity in the sponge implants (Fig. 18.1). The angiogenic effect of BK was abolished by a B_1 receptor antagonist, desArg^9-[Leu^8]-BK, but not by Ac-DArg-[Hyp^3,DPhe^7,Leu^8]-BK, a B_2 receptor antagonist. Interestingly, BK acted synergistic-

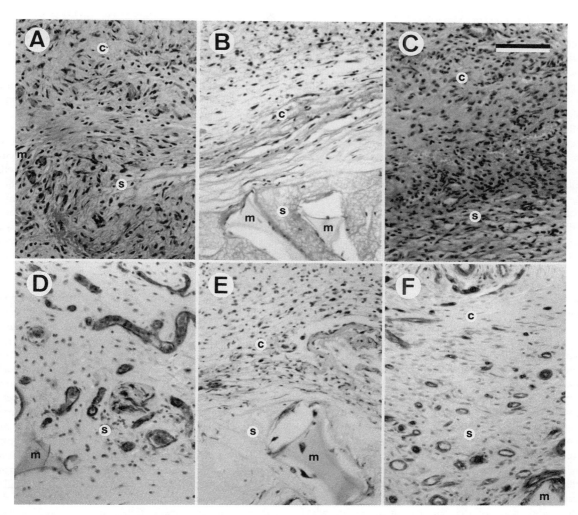

Figure 18.1   Enhanced neovascularization of sponge implants in the presence of bradykinin (BK) and interleukin-1α (IL-1). Polyether sponges were subcutaneously implanted in the dorsa of male Wistar rats and injected daily with 1 nmol of BK (A and D), phosphate-buffered saline (PBS; B and E), or a combination of low-dose BK (10 pmol) and IL-1 (0.3 pmol); C and F. Sponges were removed 10 days after implantation and formalin-fixed, paraffin-wax-embedded sections (10 nm) stained with hematoxylin and eosin to show cellular infiltration (A–C), or with the endothelial marker, *Bandeirea simplicifolia* lectin I, isolectin B_4 to show neovascularization (D–F). Sponges injected with the higher dose of BK (A and D) or with the combination of low-dose BK and IL-1 (C and F), were surrounded by a fibrovascular capsule (c), and stroma (s) between elements of the polyether sponge matrix (m) was infiltrated by a densely vascular granulation tissue. By contrast, sponges injected with the low doses of BK or IL-1 alone (data not shown) or with PBS (B and E), displayed little vascular infiltration. Experimental details are given in Hu and Fan (1993).

ally with IL-1α in enhancing angiogenesis, providing a parallel *in vivo* for their synergistic actions on cell proliferation *in vitro*.

BK may be important in the angiogenesis associated with wound healing, inflammation and neoplasia, where kinin-generation has been identified. The potential for BK receptor antagonists to inhibit angiogenesis, as well as the more acute inflammatory reactions of vasodilation and plasma extravasation, deserves further study.

4. *Proliferative Synovitis*

The synovia lining joints affected by chronic inflammatory arthritis, as in rheumatoid arthritis, are characteristically hyperplastic and hypertrophied. This synovial lining is normally only one or two cells deep, but in chronic arthritis may extend to several hundred micrometers in depth. Synovial expansion arises through influx of macrophage-like cells from the circulation, but also by proliferation of resident synovial fibroblasts (Qu *et al.*, 1994). Furthermore, synovial expansion is associated with angiogenesis, proliferation of endothelial cells and increased microvascular density, particularly in the layers of the synovium beneath the lining region, within a few millimeters of the joint space (Stevens *et al.*, 1991). Such hyperplastic tissue responses are also typical of chronic inflammation in other tissues such as lung and gut. The joint provides a particularly accessible system for the study of chronic inflammation because of our ability to aspirate fluid from, or to administer pharmaceutical agents directly into the joint cavity, which is in direct continuity with synovial connective tissue, with no intervening basement membrane.

Synovial expansion is critical to the pathogenesis of joint damage in rheumatoid arthritis. The hyperplastic synovium encroaches on the articular cartilage, releasing enzymes that degrade cartilage matrix, and infiltrates bone at the margins of the joint, leading to the characteristic bony erosions seen on radiographs. Angiogenesis may contribute to this process, with the neovasculature itself generating matrix-degrading enzymes, as well as providing oxygen and nutrients to maintain the expanded synovium and enhancing access of inflammatory cells. Inhibition of angiogenesis improves outcome in animal models of arthritis (Peacock *et al.*, 1992) and many of the pharmaceutical agents which suppress the activity of rheumatoid arthritis also have antiangiogenic activity (Matsubara and Ziff, 1987; Hirata *et al.*, 1989; Matsubara *et al.*, 1989). Furthermore, suppression of synovial proliferation by radiotherapy or surgical removal of the expanded synovium improve symptoms in the chronically inflamed joint. Below is a summary of the evidence for the presence of kinin systems in chronically inflamed synovium, and a discussion of their possible role in proliferative and angiogenic aspects of the pathogenesis of chronic synovitis.

4.1 KININ SYSTEMS IN CHRONICALLY INFLAMED SYNOVIUM

Jasani *et al.* (1969) described the presence of kinin-forming enzymes in synovial fluid from patients with arthritis. The presence of kallikrein activity in synovial fluids has been confirmed using synthetic substrates and immunoassay (Selwyn *et al.*, 1989; Rahman *et al.*, 1994, 1995). Furthermore, kallikrein-like activity is present in synovial tissue itself (Sharma *et al.*, 1983; Al-Haboubi *et al.*, 1986). Assays of kallikrein activity are, however, complicated by the presence in synovial fluid of inhibitors, including α₂-macroglobulin and C1-esterase inhibitor, which bind and inhibit plasma kallikrein, and α₁-antitrypsin, which inhibits tissue kallikrein (Rahman *et al.*, 1994) (see Chapters 10 and 16).

Synovial kininogenases include both plasma and tissue kallikrein (Nagase *et al.*, 1982; Selwyn *et al.*, 1989). In addition, nonkallikrein kinin-generating enzymes may be released by inflammatory cells, such as mast cells, neutrophils and basophils (Proud and Kaplan, 1988). Synovial kallikrein-like activity may be increased in canine inflamed synovium (Sharma *et al.*, 1983; Al-Haboubi *et al.*, 1986) and such an increase in activity begins early after joint injury (Maslennikov *et al.*, 1985). However, the difficulty in obtaining truly normal human control synovial fluids has hampered the confirmation of this finding in human synovitis, where synovial kallikrein-like activities may be similar in a variety of arthritic conditions (Rahman *et al.*, 1995).

Plasma levels of kininogens, particularly T-kininogen, are persistently elevated in experimental arthritis, behaving as acute phase reactants (van Arman and Nuss, 1969; Pelczarska and Gieldanowski, 1970; Barlas *et al.*, 1985; Okamoto *et al.*, 1989). High and low molecular weight kininogens (HMWK and LMWK, respectively) are both present in synovial tissues and fluids from patients with arthritis (Jasani *et al.*, 1969; Legris *et al.*, 1994). The presence of kininogens need not indicate necessarily a metabolic role for BK in that tissue. These are multifunctional molecules, which also act as cofactors in the activation of serine proteases and can inhibit cysteine proteases, each of which may contribute to the pathogenesis of chronic inflammation. Kinin activity itself has been detected in synovial fluids (Melmon *et al.*, 1967) but rapid degradation by local kininases hinders the accurate determination of BK levels generated *in vivo*.

Synovial fibroblasts, like fibroblasts from other sources, are acutely responsive to BK *in vitro*. Responses are enhanced if the cells are pretreated with other pro-inflammatory cytokines known to be released in the chronically inflamed joint, such as IL-1 or TNF (Bathon *et al.*, 1989; O'Neill and Lewis, 1989). BK increases both phospholipases C and D activities in these cells and enhances PGE_2 production (Bathon *et al.*, 1989; O'Neill and Lewis, 1989; Angel *et al.*, 1994). BK-stimulated release of arachidonic acid was prevented by downregulation of PKC (Cisar *et al.*, 1993).

Many of these activities appear to be mediated by B_2 receptors (Uhl *et al.*, 1992; Bathon *et al.*, 1994). $[^3H]$-BK binds to sites on cultured synovial fibroblasts with the specificity of B_2 receptors, and B_2-like binding has been localized to both lining and sublining regions of synovia from patients with rheumatoid arthritis or osteoarthritis (Bathon *et al.*, 1992). Furthermore, Bathon and colleagues (1992) found a moderate increase in B_2 receptor binding to synovial fibroblasts after pre-incubation with IL-1α, suggesting one possible mechanism by which responsiveness to BK may be enhanced in chronic synovitis.

B_1 receptors are upregulated in animal models of arthritis (see Chapter 8), with enhanced plasma extravasation in response to intra-articular injection of desArg^9-BK, compared with normal joints (Cruwys *et al.*, 1994). Carboxypeptidase N, which converts BK to desArg^9-BK, is present in synovial fluid from inflamed joints, raising the possibility that activation of B_1 receptors by locally generated agonists may be important in human chronic arthritis (Chercuitte *et al.*, 1987). However, the presence of B_1 receptors in chronically inflamed human synovium remains to be confirmed.

4.2 KININS AND SYNOVIAL PROLIFERATION

Much evidence suggests that kinin systems are activated in inflamed joints, and may contribute to acute inflammatory processes (see Chapters 2, 8, 9 and 13). As discussed above, it is likely that locally generated kinins in the synovium may, in addition, contribute to synovial proliferation, through stimulation of fibroblast division and enhancement of angiogenesis. Rheumatoid arthritis is associated with the generation of multiple cytokines and growth factors within the synovium, including IL-1, TGF-β and PDGF (Bucala *et al.*, 1991; Reuterdahl *et al.*, 1991). Each of these growth factors can potentiate mitogenic effects of BK on fibroblasts *in vitro*. The balance between proliferative and antiproliferative effects of kinins *in vivo* will depend on the relative expression of B_1 and B_2 receptors, and their second-messenger coupling, and the local availability of synergistic or inhibitory factors. Elucidation of the consequences of endogenous BK generation on synovial hyperplasia has awaited the development of metabolically stable, selective kinin receptor antagonists, and methods for quantifying cellular proliferation and angiogenesis within the joint.

5. *Neoplasia*

The role of kallikrein–kinin systems in neoplastic disease remains uncertain. Early studies concentrated on possible protection against tumor growth and metastasis afforded by kinin-induced inflammation (Koppelmann *et al.*, 1978), while more recent studies have indicated possible roles in stimulation of tumor cell proliferation (Sethi and Rozengurt, 1991). Angiogenesis is essential for the growth of solid tumors and enhanced angiogenesis correlates with metastasis (Holmgren *et al.*, 1995). BK-induced angiogenesis therefore, may also be detrimental to the survival of the tumor host.

Disturbances of plasma levels of kallikrein–kinin system components have been observed in malignancy by a number of investigators, but are difficult to interpret because of their dynamic nature and the metabolic instability of kinins. Plasma kallikrein activity was increased in fibrosarcoma-bearing hamsters compared with normal animals (Koppelmann *et al.*, 1978), while prekallikrein and kininogen levels were found to be decreased in plasma of patients with malignancy, particularly in the presence of metastases (Deutsch *et al.*, 1983; Matsumura *et al.*, 1991). These findings may indicate increased activation of the kallikrein–kinin system in advanced malignancy, although this may to some extent be counterbalanced by increased levels of endogenous kallikrein inhibitors (Deutsch *et al.*, 1983). Both HMWK and LMWK have been identified in malignant ascites, albeit at lower levels than in plasma (Karlsrud *et al.*, 1991). The hypothesis that kinin systems are activated by tumors is supported by observations that malignant cell lines may activate prekallikrein (Matsumura *et al.*, 1991), and that HMWK in ascitic fluids is extensively cleaved (Karlsrud *et al.*, 1991). However, the association between low plasma kallikrein activity and metastatic disease may alternatively indicate a protective effect of kallikrein against metastasis. An analogous situation is presented by angiostatin, a plasminogen fragment which suppresses angiogenesis. This factor is produced by some primary carcinomas, inhibiting their metastasis (O'Reilly *et al.*, 1994).

Malignant ascites also contains leucine aminopeptidase and carboxypeptidase activities capable of generating (and ultimately inactivating) desArg^9-BK (Matsumura *et al.*, 1990; Rehbock *et al.*, 1992). Generation of B_1 receptor agonists by malignant tissues raises the possibility that B_1 receptor-mediated effects of kinins may be important in tumor biology. In particular, BK-induced angiogenesis is mediated by B_1 receptors, and could contribute, therefore, to tumor growth and metastasis.

As described above, BK stimulates proliferation of a variety of malignant cell lines *in vitro*, including those derived from small-cell lung carcinomata, epidermoid carcinomata and mastocytomata (Roberts and Gullick, 1989; Sethi and Rozengurt, 1991; Sukhanova *et al.*,

1993; Muns *et al.*, 1994). Indeed, transformation of fibroblasts by Ha-ras-transfection increases their proliferative response to BK, possibly by increasing their expression of BK receptors (Parries *et al.*, 1987; Downward *et al.*, 1988; Roberts and Gullick, 1989). This has led to the suggestion that malignant cells may be more susceptible to mitogenic actions of kinins than their normal counterparts. On the other hand, BK had no effect on the proliferation rates of other human lung and breast cancer cell lines (Bunn *et al.*, 1992).

Little has been published on the role of kinins in malignant cell proliferation *in vivo* and studies on the effects of BK receptor antagonists are needed. Inhibition by intralesional BK of the growth of the non-metastasizing SV40 virus-induced fibrosarcoma in hamsters was associated with increased infiltration by lymphocytes, suggesting possible protective effects of BK-induced inflammatory responses (Koppelmann *et al.*, 1978). Whether this is an important mechanism in host-derived tumors, however, or whether mitogenic actions of kinins are more important, remains to be determined.

6.   *Other Proliferative Conditions*

Kinins may have mitogenic roles in other clinical conditions associated with cellular proliferation, including wound healing, reproductive physiology, psoriasis and glomerulonephritis. However, the relative importance of endogenous kallikrein–kinin systems in each of these clinical situations remains to be fully defined.

Tissue repair requires proliferation of connective tissue fibroblasts and of specialized cells such as keratinocytes. Roles for kinins have been postulated in the repair both of skin wounds and of internal organs, for example, the healing of bone fractures and myocardial infarcts. Early reports indicated that systemically administered kallikrein may promote wound healing in X-irradiated rats and in experimental long-bone fractures, although the relevance of local kinin generation to this was not determined (Mandel *et al.*, 1973; Bethge *et al.*, 1979). Nonetheless, BK can stimulate fibroblast and keratinocyte proliferation *in vitro*, and acts synergistically with other growth factors generated during tissue repair, as described above. BK and desArg⁹-BK have other roles in tissue repair, including plasma extravasation, the enhancement of tissue oxygenation and nutrition through vasodilation, and stimulation of collagen synthesis by tissue fibroblasts (Goldstein and Wall, 1984; Weber *et al.*, 1995). Stimulation of angiogenesis by BK may further facilitate healing.

Psoriasis is characterized by excessive keratinocyte proliferation, producing hyperplastic, but noninvasive, skin plaques. Plasma concentrations of kallikreins and kininogens are raised in patients with psoriasis and have been localized to psoriatic skin (Winkelmann, 1984; Hibino *et al.*, 1988; Poblete *et al.*, 1991; Kolosovsky, 1994). BK is capable of simulating human keratinocyte growth *in vitro*, and ACE inhibitors have been associated, in case reports, with the development of psoriaform skin eruptions (Eriksen *et al.*, 1995). Together, these data suggest a possible role for endogenous kinins in keratinocyte proliferation during the creation of psoriatic plaques.

In the kidney, BK and desArg⁹-BK may contribute to the mesangial cell proliferation, characteristic of some forms of glomerulonephritis (Ganz *et al.*, 1990; Issandou and Darbon, 1991; Bascands *et al.*, 1993). BK can stimulate vascular smooth muscle proliferation, and it has been speculated that kinin generation may contribute to the medial and intimal hyperplasia of hypertension and atherosclerosis, respectively (Dixon *et al.*, 1994). Similarly, mitogenic actions of BK may be involved in endometrial proliferation during the menstrual cycle (Endo *et al.*, 1991) and in lymphocyte proliferation, and enhancement of the immune response (Perris and Whitfield, 1969; Whitfield *et al.*, 1970). Other effects of BK on cell growth, as opposed to proliferation, may be important in neuronal development *in utero* and after neuronal injury, since BK synergizes with nerve growth factor to stimulate neurite extension in PC12 cells (Kozlowski *et al.*, 1989). Elucidation of the role of endogenous BK in the pathogenesis of each of these conditions awaits clinical studies with selective antagonists.

7.   *Conclusions*

Kinins, acting through B_1 or B_2 receptors, may stimulate proliferation of a number of normal and neoplastic cell types, particularly in the presence of other polypeptide growth factors. Furthermore, through stimulating neovascularization of proliferating tissues, BK may enhance tissue oxygenation and nutrition, and thereby further facilitate cell growth. Local kinin-generating systems have been characterized in a range of conditions associated with cellular proliferation, including chronic synovitis and neoplasia. It seems probable that locally generated kinins, acting through specific receptors, may contribute to the pathogenesis of at least some of these diseases through their growth-promoting activities. The development of metabolically stable kinin receptor antagonists will allow the more rigorous investigation of the roles of endogenous BK and its analogues in cellular hyperplasia *in vivo*, and may offer novel therapeutic strategies against human proliferative diseases.

8. References

Al-Haboubi, H.A., Bennett, D., Sharma, J.N., Thomas, G.R. and Zeitlin, I.J. (1986). A synovial amidase acting on tissue kallikrein-selective substrate in clinical and experimental arthritis. Adv. Exp. Med. Biol. 198, 405–411.

Angel, J., Audubert, F., Bismuth, G. and Fournier, C. (1994). IL-1 beta amplifies BK-induced prostaglandin E_2 production via a phospholipase D-linked mechanism. J. Immunol. 152, 5032–5040.

Barlas, A., Okamoto, H. and Greenbaum, L.M. (1985). T-kininogen – the major plasma kininogen in rat adjuvant arthritis. Biochem. Biophys. Res. Commun. 129, 280–286.

Bascands, J.L., Pecher, C., Rouaud, S., Emond, C., Tack, J.L., Bastie, M.J., Burch, R., Regoli, D. and Girolami, J.P. (1993). Evidence for existence of two distinct bradykinin receptors on rat mesangial cells. Am. J. Physiol. 264, F548–556.

Bathon, J.M., Proud, D., Krackow, K. and Wigley, F.M. (1989). Preincubation of human synovial cells with IL-1 modulates prostaglandin E_2 release in response to bradykinin. J. Immunol. 143, 579–586.

Bathon, J.M., Manning, D.C., Goldman, D.W., Towns, M.C. and Proud, D. (1992). Characterization of kinin receptors on human synovial cells and upregulation of receptor number by interleukin-1. J. Pharmacol. Exp. Ther. 260, 384–392.

Bathon, J.M., Croghan, J.C., MacGlashan, D.W., Jr and Proud, D. (1994). Bradykinin is a potent and relatively selective stimulus for cytosolic calcium elevation in human synovial cells. J. Immunol. 153, 2600–2608.

BenEzra, D. (1978). Vasculogenic ability of prostaglandins, growth factors, and synthetic chemoattractants. Am. J. Ophthalmol. 86, 455–461.

Bethge, J.F., Babayan, R., Borm, H.P., von Fehrentheil, R., ten Hoff, H., Hose, H., Mangels, P., Piening, H., Reimers, C. and Wider, U. (1979). Experimental acceleration of fracture repair by biochemical media. Res. Exp. Med. 175, 197–222.

Boucek, R.J. and Noble, N.L. (1973). Histamine, norepinephrine, and bradykinin stimulation of fibroblast growth and modification of serotonin response. Proc. Soc. Exp. Biol. Med. 144, 929–933.

Bucala, R., Ritchlin, C., Winchester, R. and Cerami, A. (1991). Constitutive production of inflammatory and mitogenic cytokines by rheumatoid synovial fibroblasts. J. Exp. Med. 173, 569–574.

Bunn, P.A., Jr, Chan, D., Dienhart, D.G., Tolley, R., Tagawa, M. and Jewett, P.B. (1992). Neuropeptide signal transduction in lung cancer: clinical implications of bradykinin sensitivity and overall heterogeneity. Cancer Res. 52, 24–31.

Cameron, N.E., Cotter, M.A. and Robertson, S. (1992). Angiotensin converting enzyme inhibition prevents development of muscle and nerve dysfunction and stimulates angiogenesis in streptozotocin-diabetic rats. Diabetologia 35, 12–18.

Catalioto, R.M., Negrel, R., Gaillard, D. and Ailhaud, G. (1987). Growth-promoting activity in serum-free medium of kallikrein-like arginylesteropeptidases from rat submaxillary gland. J. Cell. Physiol. 130, 352–360.

Chercuitte, F., Beaulieu, A.D., Poubelle, P. and Marceau, F. (1987). Carboxypeptidase N (kininase I) activity in blood and synovial fluid from patients with arthritis. Life Sci. 41, 1225–1232.

Cisar, L.A., Mochan, E. and Schimmel, R. (1993). Interleukin-1 selectively potentiates bradykinin-stimulated arachidonic acid release from human synovial fibroblasts. Cell. Signal. 5, 463–472.

Coughlin, S.R., Lee, W.M., Williams, P.W., Giels, G.M. and Williams, L.T. (1985). c-myc gene expression is stimulated by agents that activate protein kinase C and does not account for the mitogenic effect of PDGF. Cell 43, 243–251.

Cruwys, S.C., Garrett, N.E., Perkins, M.N., Blake, D.R. and Kidd, B.L. (1994). The role of bradykinin B_1 receptors in the maintenance of intra-articular plasma extravasation in chronic antigen-induced arthritis. Br. J. Pharmacol. 113, 940–944.

Deutsch, E., Dragosics, B., Kopsa, H., Mannhalter, C. and Rainer, H. (1983). Prekallikrein, HMW-kininogen and factor XII in various disease states. Thrombosis Res. 31, 351–364.

Dixon, B.S., Sharma, R.V., Dickerson, T. and Fortune, J. (1994). Bradykinin and angiotensin II: activation of protein kinase C in arterial smooth muscle. Am. J. Physiol. 266, C1406–1420.

Downward, J., de Gunzburg, J., Riehl, R. and Weinberg, R.A. (1988). p21ras-induced responsiveness of phosphatidyl-inositol turnover to bradykinin is a receptor number effect. Proc. Natl Acad. Sci. USA 85, 5774–5778.

Drinkwater, C.C., Evans, B.A. and Richards, R.I. (1988). Kallikreins, kinins and growth factor biosynthesis. Trends Biochem. Sci. 13, 169–172.

Endo, T., Fukue, H., Kanaya, M., Mizunuma, M., Fujii, H., Yamamoto, H., Tanaka, S. and Hashimoto, M. (1991). Bombesin and bradykinin increase inositol phosphates and cytosolic free Ca^{2+}, and stimulate DNA synthesis in human endometrial stromal cells. J. Endocrinol. 131, 313–318.

Eriksen, J.G., Christiansen, J.J. and Asmussen, I. (1995). Pustulosis palmoplantaris caused by angiotensin-converting enzyme inhibitors. Ugeskrift Laeger 157, 3335–3336.

Fan, T.-P.D. and Brem, S. (1992). Angiosuppression. In "The Search for New Anticancer Drugs" (eds. W.J. Waring and B.A.J. Ponder), pp. 185–229. Kluwer Academic Publications, Dordrecht.

Fan, T.-P.D., Jaggar, R. and Bicknell, R. (1995). Controlling the vasculature: angiogenesis, anti-angiogenesis and vascular targeting of gene therapy. Trends. Pharmacol. Sci. 16, 57–66.

Folkman, J. and Brem, H. (1992). Angiogenesis and inflammation. In "Inflammation: Basic Principles and Clinical Correlates" (eds. J.I. Gallin, I.M. Goldstein and R. Snyderman), pp. 821–839. Raven Press, New York.

Form, D.M. and Auerbach, R. (1983). PGE_2 and angiogenesis. Proc. Soc. Exp. Biol. Med. 172, 214–218.

Galizzi, J.P., Bodinier, M.C., Chapelain, B., Ly, S.M., Coussy, L., Giroud, S., Neliat, G. and Jean, T. (1994). Up-regulation of [^{3}H]-des-Arg^{10}-kallidin binding to the bradykinin B_1 receptor by interleukin-1β in isolated smooth muscle cells: correlation with B_1 agonist-induced PGI_2 production. Br. J. Pharmacol. 113, 389–394.

Ganz, M.B., Perfetto, M.C. and Boron, W.F. (1990). Effects of mitogens and other agents on rat mesangial cell proliferation, pH, and Ca^{2+}. Am. J. Physiol. 255, F269–278.

Geppetti, P., Del Bianco, E., Tramontana, M., Vigano, T., Folco, G.C., Maggi, C.A., Manzini, S. and Fanciullacci, M. (1991). Arachidonic acid and bradykinin share a common pathway to release neuropeptide from capsaicin-sensitive sensory nerve fibres of the guinea-pig heart. J. Pharmacol. Exp. Ther. 259, 759–765.

Godin, C., Smith, A.D. and Riley, P.A. (1991). Bradykinin stimulates DNA synthesis in competent Balb/c 3T3 cells and enhances inositol phosphate formation induced by platelet-derived growth factor. Biochem. Pharmacol. 42, 117–122.

Goldstein, R.H. and Wall, M. (1984). Activation of protein formation and cell division by bradykinin and des-Arg^9-bradykinin. J. Biol. Chem. 259, 9263–9268.

Hibino, T., Izaki, S., Kimura, H., Izaki, M. and Kon, S. (1988). Partial purification of plasma and tissue kallikreins in psoriatic epidermis. J. Invest. Dermatol. 90, 505–510.

Hill, T.D., Kindmark, H. and Boynton, A.L. (1988). Epidermal growth factor-stimulated DNA synthesis requires an influx of extracellular calcium. J. Cell. Biochem. 38, 137–144.

Hirata, S., Matsubara, T., Saura, R., Tateishi, H. and Hirohata, K. (1989). Inhibition of in vitro vascular endothelial cell proliferation and in vivo neovascularization by low-dose methotrexate. Arthritis Rheum. 32, 1065–1073.

Holmgren, L., O'Reilly, M.S. and Folkman, J. (1995). Dormancy of micrometastases: balanced proliferation and apoptosis in the presence of angiogenesis suppression. Nature Med. 1, 149–153.

Hu, D.-E. and Fan, T.-P.D. (1993). [Leu^8]des-Arg^9-bradykinin inhibits the angiogenic effect of bradykinin and interleukin-1 in rats. Br. J. Pharmacol. 109, 14–17.

Hu, D.-E., Hiley, C.R., Smither, R.L., Gresham, G.A. and Fan, T.-P.D. (1995). Correlation of ^{133}Xe clearance, blood flow and histology in the rat sponge model for angiogenesis; further studies with angiogenic modifiers. Lab. Invest. 72, 601–610.

Issandou, M. and Darbon, J.M. (1991). Des-Arg^9 bradykinin modulates DNA synthesis, phospholipase C, and protein kinase C in cultured mesangial cells. Distinction from effects of bradykinin. J. Biol. Chem. 266, 21037–21043.

Jasani, M.K., Katori, M. and Lewis, G.P. (1969). Intracellular enzymes and kinin enzymes in synovial fluid in joint diseases: origin and relation to disease category. Ann. Rheum. Dis. 28, 497–511.

Johnson, R.M., King, K.L. and Morhenn, V.B. (1992). Comparison of second messenger formation in human keratinocytes following stimulation with epidermal growth factor and bradykinin. Second Messengers Phosphoproteins 14, 21–37.

Karlsrud, T.S., Buo, L., Aasen, A.O. and Johansen, H.T. (1991). Characterization of kininogens in human malignant ascites. Thrombosis Res. 63, 641–650.

Kiehne, K. and Rozengurt, E. (1994). Synergistic stimulation of DNA synthesis by bradykinin and vasopressin in Swiss 3T3 cells. J. Cell. Physiol. 160, 502–510.

Kiehne, K. and Rozengurt, E. (1995). Down-regulation of bradykinin receptors and bradykinin-induced Ca^{2+} mobilization, tyrosine phosphorylation, and DNA synthesis by autocrine factors, tumor necrosis factor alpha, and interferon beta in Swiss 3T3 cells. J. Cell. Physiol. 162, 367–372.

Kimball, E.S. and Fisher, M.C. (1988). Potentiation of IL-1-induced BALB/3T3 fibroblast proliferation by neuropeptides. J. Immunol. 141, 4203–4208.

Kimura, Y., Fujihira, T., Kato, K., Furuya, M., Onda, M. and Shirota, A. (1979). Effect of bradykinin to cyclic AMP levels and response of murine lymphocytes. Adv. Exp. Med. Biol. 120A, 393–407.

Kimura, Y., Saiga, T., Furuya, M., Fujiwara, H., Norose, Y., Okabe, T. and Hida, M. (1983). Suppressive effect of bradykinin to cellular immune responses in vivo and in vitro. Adv. Exp. Med. Biol. 156B, 755–766.

Kolosovsky, E.D. (1994). Evaluation of different treatment methods in patients with psoriasis and content of kallikrein and kallikreinogen in blood plasma. Dermatology 188, 140–141.

Koppelmann, L.E., Moore, T.C. and Porter, D.D. (1978). Increased plasma kallikrein activity and tumour growth suppression associated with intralesional bradykinin injections in hamsters. J. Pathol. 126, 1–10.

Korbelik, M., Skrk, J., Poljak-Blazi, M., Suhar, A. and Boranic, M. (1989). The effects of human kallikrein and aprotinin on nonmalignant and malignant cell growth. Adv. Exp. Med. Biol. 247B, 675–680.

Kozlowski, M.R., Rosser, M.P., Hall, E. and Longden, A. (1989). Effects of bradykinin on PC-12 cell differentiation. Peptides 10, 1121–1126.

Lahaye, D.H., Afink, G.B., Bleijs, D.A., van Alewijk, D.C. and van Zoelen, E.J. (1994). Effect of bradykinin on loss of density-dependent growth inhibition of normal rat kidney cells. Cell. Mol. Biol. 40, 717–721.

Landsleitner, B., Geldmacher, J., Feustel, H. and Ittner, F. (1974). Effect of kallikrein on connective tissue proliferation, animal experiments. Medizinische Welt 25, 1752–1755.

Legris, F., Martel-Pelletier, J., Pelletier, J.P., Colman, R. and Adam, A. (1994). An ultrasensitive chemiluminoenzyme immunoassay for the quantification of human tissue kininogens: application to synovial membrane and cartilage. J. Immunol. Meth. 168, 111–121.

Lowe, W.L., Jr, Yorek, M.A., Karpen, C.W., Teasdale, R.M., Hovis, J.G., Albrecht, B. and Prokopiou, C. (1992). Activation of protein kinase C differentially regulates insulin-like growth factor-I and basic fibroblast growth factor messenger RNA levels. Mol. Endocrinol. 6, 741–752.

Mackie, K., Lai, Y., Nairn, A.C., Greengard, P., Pitt, B.R. and Lazo, J.S. (1986). Protein phosphorylation in cultured endothelial cells. J. Cell. Physiol. 128, 367–374.

Mandel, P., Rodesch, J. and Mantz, J.M. (1973). The treatment of experimental radiation lesions by kallikrein. In "Kininogenases 1" (eds. G.L. Haberland and J.W. Rohen), pp. 171–188. Schattauer, Stuttgart.

Marceau, F. and Tremblay, B. (1986). Mitogenic effect of bradykinin and of des-Arg^9-bradykinin on cultured fibroblasts. Life Sci. 39, 2351–2358.

Marceau, F., Lussier, A., Regoli, D. and Giroud, J.P. (1983). Pharmacology of kinins: their relevance to tissue injury and inflammation. A review. Gen. Pharmacol. 14, 209–229.

Maslennikov, E.I., Vilkov, G.A., Perechai, L.D. and Kozlova, L.S. (1985). Kallikrein–kinin system of the synovial fluid in closed injury to the knee joint. Vopr. Med. Khim. 31, 59–62.

Matsubara, T. and Ziff, M. (1987). Inhibition of human endothelial cell proliferation by gold compounds. J. Clin. Invest. 79, 1440–1446.

Matsubara, T., Saura, R., Hirohata, K. and Ziff, M. (1989). Inhibition of human endothelial cell proliferation in vitro and neovascularization in vivo by D-Penicillamine. J. Clin. Invest. 83, 158–167.

Matsuda, K., Niitsuma, A., Uchida, M.K. and Suzuki-Nishimura, T. (1994). Inhibitory effects of sialic acid- or N-acetyl-glucosamine-specific lectins on histamine release induced by compound 48/80, bradykinin and a polyethylenimine in rat peritoneal mast cells. Jpn J. Pharmacol. 64, 1–8.

Matsumura, Y., Maeda, H. and Kato, H. (1990). Degradation pathway of kinins in tumor ascites and inhibition by kininase inhibitors: analysis by HPLC. Agents Actions 29, 172–180.

Matsumura, Y., Maruo, K., Kimura, M., Yamamoto, T., Konno, T. and Maeda, H. (1991). Kinin-generating cascade in advanced cancer patients and *in vitro* study. Jpn J. Cancer Res. 82, 732–741.

McAllister, B.S., Leeb-Lundberg, F., Olson, M.S. (1993). Bradykinin inhibition of EGF- and PDGF-induced DNA synthesis in human fibroblasts. Am. J. Physiol. 265, C477–484.

Melmon, K.L., Webster, M.E., Goldfinger, S.E. and Seegmiller, J.E. (1967). The presence of a kinin in inflammatory synovial effusion from arthritides of varying etiologies. Arthritis Rheum. 10, 13–20.

Menke, J.G., Borkowski, J.A., Bierilo, K.K., MacNeil, T., Derrick, A.W., Schneck, K.A., Ransom, R.W., Strader, C.D., Linemeyer, D.L. and Hess, J.F. (1994). Expression cloning of a human B_1 bradykinin receptor. J. Biol. Chem. 269, 21583–21586.

Muns, G., Vishwanatha, J.K. and Rubinstein, I. (1994). Effects of smokeless tobacco on chemically transformed hamster oral keratinocytes: role of angiotensin I-converting enzyme. Carcinogenesis 15, 1325–1327.

Nagase, H., Cawston, T.E., De Silva, M. and Barrett, A.J. (1982). Identification of plasma kallikrein as an activator of latent collagenase in rheumatoid synovial fluid. Biochim. Biophys. Acta 702, 133–142.

Newman, E.L., Hyldahl, L., Larsson, O., Engstrom, W. and Rees, A.R. (1989). Bradykinin blocks the action of EGF, but not PDGF, on fibroblast division. FEBS Lett. 251, 225–229.

Nilsson, J., von Euler, A.M. and Dalsgaard, C.-J. (1985). Stimulation of connective tissue cell growth by substance P and substance K. Nature 315, 61–63.

Okamoto, H., Hanaoka, M., Yayama, K., Ohtani, R. and Itoh, N. (1989). Acute-phase response of angiotensinogen in rat adjuvant arthritis. Int. J. Tissue React. 11, 123–127.

Olivetti, G., Cigola, E., Lagrasta, C., Ricci, R., Quaini, F., Monopoli, A. and Ongini, E. (1993). Spirapril prevents left ventricular hypertrophy, decreases myocardial damage and promotes angiogenesis in spontaneously hypertensive rats. J. Cardiovasc. Pharmacol. 21, 362–370.

Olsen, R., Santone, K., Melder, D., Oakes, S.G., Abraham, R. and Powis, G. (1988). An increase in intracellular free Ca^{2+} associated with serum-free growth stimulation of Swiss 3T3 fibroblasts by epidermal growth factor in the presence of bradykinin. J. Biol. Chem. 263, 18030–18035.

O'Neill, L.A. and Lewis, G.P. (1989). Interleukin-1 potentiates bradykinin- and TNF alpha-induced PGE_2 release. Eur. J. Pharmacol. 166, 131–137.

O'Reilly, M.S., Holmgran, L., Shing, Y., Chen, C., Rosenthal, R.A., Moses, M., Lane, W.S., Cao, Y., Sage, E.H. and Folkman, J. (1994). Angiostatin; a novel angiogenesis inhibitor that mediates the suppression of metastases by Lewis lung carcinoma. Cell 79, 315–328.

Owen, N.E. and Villereal, M.L. (1983). Lys-bradykinin stimulates Na^+ influx and DNA synthesis in cultured human fibroblasts. Cell 32, 979–985.

Parries, G., Hoebel, R. and Racker, E. (1987). Opposing effects of a ras oncogene on growth factor-stimulated phosphoinositide hydrolysis: desensitization to platelet-derived growth factor and enhanced sensitivity to bradykinin. Proc. Natl Acad. Sci. USA 84, 2648–2652.

Patel, K.V. and Schrey, M.P. (1992). Inhibition of DNA synthesis and growth in human breast stromal cells by bradykinin: evidence for independent roles of B_1 and B_2 receptors in the respective control of cell growth and phospholipid hydrolysis. Cancer Res. 52, 334–340.

Peacock, D.J., Banquerigo, M.L. and Brahn, E. (1992). Angiogenesis inhibition suppresses collagen arthritis. J. Exp. Med. 175, 1135–1138.

Pelczarska, A. and Gieldanowski, J. (1970). Treatment of adjuvant arthritis with antibradykinin drugs. J. Pharmac. Pharmacol. 22, 617–619.

Perris, A.D. and Whitfield, J.F. (1969). The mitogenic action of bradykinin on thymic lymphocytes and its dependence on calcium. Proc. Soc. Exp. Biol. Med. 130, 1198–1202.

Poblete, M.T., Reynolds, N.J., Figueroa, C.D., Burton, J.L., Müller-Esterl, W. and Bhoola, K.D. (1991). Tissue kallikrein and kininogen in human sweat glands and psoriatic skin. Br. J. Dermatol. 124, 236–241.

Porta, M., Dosso, A.A., Williams, F.M., Kanse, S. and Kohner, E.M. (1992). A study of the effects of angiotensins 1, 2, 3 and bradykinin on the replication of bovine retinal capillary endothelial cells and pericytes. Eur. J. Ophthalmol. 2, 21–26.

Proud, D. and Kaplan, A.P. (1988). Kinin formation: mechanisms and role in inflammatory disorders. Annu. Rev. Immunol. 6, 49–83.

Qu, Z., Garcia, C.H., O'Rourke, L.M., Planck, S.R., Kohli, M. and Rosenbaum, J.T. (1994). Local proliferation of fibroblast-like synoviocytes contributes to synovial hyperplasia. Arthritis Rheum. 37, 212–220.

Rahman, M.M., Worthy, K., Elson, C.J., Fink, E., Dieppe, P.A., Bhoola, K.D. (1994). Inhibitor regulation of tissue kallikrein activity in the synovial fluid of patients with rheumatoid arthritis. Br. J. Rheumatol. 33, 215–223.

Rahman, M.M., Bhoola, K.D., Elson, C.J., Lemon, M. and Dieppe, P.A. (1995). Identification and functional importance of plasma kallikrein in the synovial fluids of patients with rheumatoid, psoriatic, and osteoarthritis. Ann. Rheum. Dis. 54, 345–350.

Raspe, E., Reuse, S., Roger, P.P. and Dumont, J.E. (1992). Lack of correlation between the activation of the calcium phosphatidylinositol cascade and the regulation of DNA synthesis in the dog thyrocyte. Exp. Cell Res. 198, 17–26.

Rehbock, J., Steinhauser, C. and Hermann, A. (1992). Quantification of Ile-Ser-Bradykinin degradation in human serum and ascites. Agents Actions 38(Suppl. 1), 449–453.

Reuterdahl, C., Tingstrom., A., Terracio, L., Funa, K., Heldin, C.H. and Rubin, K. (1991). Characterization of platelet-derived growth factor beta-receptor expressing cells in the vasculature of human rheumatoid synovium. Lab. Invest. 64, 321–329.

Roberts, R.A. and Gullick, W.J. (1989). Bradykinin receptor number and sensitivity to ligand stimulation of mitogenesis is increased by expression of a mutant ras oncogene. J. Cell Sci. 94, 527–535.

Schölkens, B.A., Linz, W. and König, W. (1988). Effects of the angiotensin converting enzyme inhibitor ramipril, in isolated ischaemic rat heart are abolished by a bradykinin antagonist. J. Hypertens. 6(Suppl. 4), 25–28.

Selwyn, B., Figeroa, C.D., Fink, E., Swan, A., Dieppe, P.A. and Bhoola, K.D. (1989). A tissue kallikrein in the synovial fluid of patients with rheumatoid arthritis. Ann. Rheum. Dis. 48, 128–133.

Sethi, T. and Rozengurt, E. (1991). Multiple neuropeptides stimulate clonal growth of small cell lung cancer: effects of bradykinin, vasopressin, cholecystokinin, galanin, and neurotensin. Cancer Res. 51, 3621–3623.

Sharma, J.N., Zeitlin, I.J., Deodhar, S.D. and Buchanan, W.W. (1983). Detection of kallikrein-like activity in inflamed synovial tissue. Arch. Int. Pharm. Ther. 262, 279–286.

Stevens, C.R., Blake, D.R., Merry, P., Revell, P.A. and Levick, J.R. (1991). A comparative study by morphometry of the microvasculature in normal and rheumatoid synovium. Arthritis Rheum. 34, 1508–1513.

Stoll, M., Steckelings, U.M., Paul, M., Bottari, S.P., Metzger, R. and Unger, T. (1995). The angiotensin AT2-receptor mediates inhibition of cell proliferation in coronary endothelial cells. J. Clin. Invest. 95, 651–657.

Straus, D.S. and Pang, K.J. (1984). Effects of bradykinin on DNA synthesis in resting NIL8 hamster cells and human fibroblasts. Exp. Cell Res. 151, 129–135.

Sukhanova, G.A., Potanova, G.V. and Narbutovich, S.A. (1993). Activity of 5'-nucleotidase from thymus and tumors upon administering bradykinin to tumor-bearing mice. Vopr. Med. Khim. 39, 13–15.

Sundqvist, G., Rosenquist, J.B. and Lerner, U.H. (1995). Effects of bradykinin and thrombin on prostaglandin formation, cell proliferation and collagen biosynthesis in human dental-pulp fibroblasts. Arch. Oral Biol. 40, 247–256.

Talwar, H.S., Fisher, G.J. and Voorhees, J.J. (1990). Bradykinin induces phosphoinositide turnover, 1,2-diglyceride formation, and growth in cultured adult human keratinocytes. J. Invest. Dermatol. 95, 705–710.

Taylor, L., Ricupero, D., Jean, J.C., Jackson, B.A., Navarro, J. and Polgar, P. (1992). Functional expression of the bradykinin-B_2 receptor cDNA in Chinese hamster lung CCL39 fibroblasts. Biochem. Biophys. Res. Commun. 188, 786–793.

Tiffany, C.W. and Burch, R.M. (1989). Bradykinin stimulates tumour necrosis factor and interleukin-1 release from macrophages. FEBS Lett. 247, 189–192.

Tilly, B.C., Van Paridon, P.A., Verlaan, I., Wirtz, K.W.A., de Laat, S.W. and Moolenaar, W.H. (1987). Inositol phosphate metabolism in bradykinin-stimulated human A431 carcinoma cells. Biochem. J. 244, 129–135.

Uhl, J., Singh, S., Brophy, L., Faunce, D. and Sawutz, D.G. (1992). Role of bradykinin in inflammatory arthritis: identification and functional analysis of bradykinin receptors on human synovial fibroblasts. Immunopharmacology 23, 131–138.

Unger, T., Mattfeldt, T., Lamberty, V., Bock, P., Mall, G., Linz, W., Schölkens, B.A. and Gohlke, P. (1992). Effect of early onset angiotensin converting enzyme inhibition on myocardial capillaries. Hypertension 20, 478–482.

van Arman, C.G. and Nuss, G.W. (1969). Plasma bradykininogen levels in adjuvant arthritis and carrageenan inflammation. J. Pathol. 99, 245–250.

van Corven, E.J., Groenink, A., Jalink, K., Eichholtz, T. and Moolenaar, W.H. (1989). Lysophosphatidate-induced cell proliferation: identification and dissection of signaling pathways mediated by G proteins. Cell 59, 45–54.

van Zoelen, E.J.J., Peters, P.H.J., Afink, G.B., van Genesen, S., de Roos, A.D.G., van Rotterdam, W. and Theuvenet, A.P.R. (1994). Bradykinin-induced growth inhibition of normal rat kidney (NRK) cells is paralleled by a decrease in epidermal-growth-factor receptor expression. Biochem. J. 298, 335–340.

Wang, D.-H. and Prewitt, R.L. (1990). Captopril reduces aortic and microvascular growth in hypertensive and normotensive rats. Hypertension 15, 68–77.

Webb, M., McIntyre, P. and Phillips, E. (1994). B_1 and B_2 bradykinin receptors encoded by distinct mRNAs. J. Neurochem. 62, 1247–1253.

Weber, K.T., Sun, Y., Katwa, L.C., Cleujens, J.P. and Zhou, G. (1995). Connective tissue and repair in the heart. Potential regulatory mechanisms. Ann. N. York Acad. Sci. 752, 286–299.

Whitfield, J.F., MacManus, J.P. and Gillan, D.J. (1970). Cyclic AMP mediation of bradykinin-induced stimulation of mitotic activity and DNA synthesis in thymocytes. Proc. Soc. Exp. Biol. Med. 133, 1270–1274.

Winkelmann, R.K. (1984). Total plasma kininogen in psoriasis and atopic dermatitis. Acta Derm. Venereol. 64, 261–263.

Woll, P.J. and Rozengurt, E. (1988). Two classes of antagonist interact with receptors for the mitogenic neuropeptides bombesin, bradykinin and vasopressin. Growth Factors 1, 75–83.

Ziche, M., Morbidelli, L., Pacini, M., Geppetti, P., Alessandri, G. and Maggi, C.A. (1990). Substance P stimulates neovascularization in vivo and proliferation of endothelial cells. Microvasc. Res. 40, 264–278.

Ziche, M., Parenti, A., Morbidelli, L., Meininger, C.J., Granger, H.J. and Ledda, F. (1992). The effect of vasoactive factors on the growth of coronary endothelial cells. Cardiologia 37, 573–575.

Ziche, M., Morbidelli, L., Granger, H., Geppetti, P. and Ledda, F. (1993). Nitric oxide promotes DNA synthesis and cyclic GMP formation in endothelial cells from postcapillary venules. Biochem. Biophys. Res. Commun. 192, 1198–1203.

Ziche, M., Morbidelli, L., Masini, E., Granger, H.J., Geppetti, P. and Ledda, F. (1994). Nitric oxide mediates angiogenesis in vivo and endothelial cell growth and migration in vitro promoted by substance P. J. Clin. Invest. 94, 2036–2044.

19. The Role of Kinins in the Cardiac Effects of ACE Inhibitors and Myocardial Ischemia

Wolfgang Linz, Gabriele Wiemer, Klaus Wirth *and* Bernward A. Schölkens

1. Introduction

The kinins, bradykinin (BK) and Lys-BK (kallidin), are potent vasoactive and inflammatory peptides derived from plasma precursors under conditions of tissue injury and ischemia (Bhoola *et al.*, 1992; Carretero *et al.*, 1993). BK and Lys-BK have qualitatively identical biological actions and are ligands at B_2 receptors (see Chapters 2 and 3 of this volume for detailed discussions of the pharmacology and molecular biology of kinin receptors and subtypes). The vasodilator effects of kinins are mediated largely through the release of autacoids generated by the vascular endothelium, and recent evidence suggests that the endothelium itself can release kinins (Wiemer *et al.*, 1994) (see also Chapter 4). Activation of endothelial B_2 kinin receptors leads to the formation of the potent vasodilators and platelet function modulators nitric oxide (NO) and prostacyclin (PGI_2) (Wiemer *et al.*, 1994) by stimulation of phospholipases C and A_2.

In the last decade, two classes of drugs have contributed greatly to the understanding of the cardio-vascular effects of endogenous kinins. First, the use of angiotensin-converting enzyme (ACE) inhibitors has led to growing evidence that kinins contribute to the beneficial cardiovascular effects of these drugs. Second, with the use of specific kinin receptor antagonists, it could be shown that the beneficial effects induced by increased endogenous kinins during ACE inhibition were abolished. Since ACE inhibitors not only attenuate the formation of angiotensin II, but also inhibit kinin degradation, inhibition of ACE results in accumulation of endogenous kinins. Thus, ACE inhibitors have a dual action in that they attenuate or prevent the systemic and local actions of angiotensin II, which are often disadvantageous, and potentiate the beneficial cardiovascular and metabolic effects of kinins (Linz *et al.*, 1995). With regard to their cardiovascular actions angiotensin II and kinins are physiological antagonists.

In blood vessels ACE is located mainly at the luminal surface of the endothelial cell membrane and appears to be largely responsible for the local proteolytic breakdown of vascular kinins (Erdös, 1990; Nolly *et al.*, 1994b) (see also Chapter 7 of this volume). Thus, under physiological conditions the effect of endothelium-derived kinins is limited by the activities of endothelial ACE and other enzymes in deeper layers of the vascular wall (Gohlke *et al.*, 1992). On the other hand, if the breakdown of kinins is reduced during ACE inhibition, or the synthesis and/or release of kinins is activated as suggested for ischemic myocardial conditions (Baumgarten *et al.*, 1993; Lamontagne *et al.*, 1995) enhanced production of NO and PGI$_2$ has been observed (Wiemer *et al.*, 1994; Rubin and Levi, 1995). Both autacoids released from cardiac endothelial cells can diffuse to the underlying smooth muscle cells, exerting dilator and mediating the anti-ischemic effects of kinins (Linz *et al.*, 1995).

In this chapter, we illustrate the role of kinins in myocardial ischemia, and discuss several lines of evidence supporting the hypothesis that kinins exert protective effects in the ischemic heart. Thus: (1) administered locally, in doses similar to those that might be released endogenously, kinins exert beneficial effects; (2) drugs that inhibit breakdown of kinins exert beneficial effects; (3) drugs that antagonize kinin receptors abolish the beneficial cardioprotective effects of kinins or ACE inhibitors during ischemia; (4) kinins are released under conditions of ischemia, compensating the effects which result from ischemia; (5) kinin-generating pathways are present in the heart.

2. Kinins Administered Locally Exert Beneficial Cardiac Effects

As early as in 1966, a study in dogs showed that locally and systemically administered kinins increased coronary blood flow, and improved myocardial metabolism (Lochner and Parratt, 1966). Further investigations on the metabolic status of rat ischemic hearts showed a significant decrease of glycogen, ATP and creatine phosphate, and a significant increase of lactate in myocardial tissue samples determined during the pre-ischemic (20 min before occlusion of the left coronary artery), as well as the ischemic period (15 min after occlusion of the left coronary artery) in comparison to freshly prepared hearts. Perfusion of ischemic hearts with BK (1×10^{-10} M) improved all of these parameters to values measured in fresh hearts (Schölkens and Linz, 1992; Linz *et al.*, 1996), an effect probably due to an increase in myocardial glucose uptake and utilization as well as increased rate of glycolytic flux induced by kinins (Rösen *et al.*, 1983; Rett *et al.*, 1986). Beside the improved metabolic state, perfusion of fibrillating hearts during reperfusion with BK (1×10^{-12} to 1×10^{-8} M) reduced the incidence and/or duration of ventricular fibrillation (Linz *et al.*, 1986, 1992; van Gilst *et al.*, 1986). Likewise, cardiodynamics were improved as indicated by an increase in left ventricular pressure,

contractility and coronary flow without changes in heart rate (Linz *et al.*, 1992, 1996). This agrees with observations of Zhu *et al.* (1995) that BK improved postischemic myocardial recovery in isolated rat hearts when given before ischemia or with the onset of reperfusion. Similar findings were reported in anesthetized dogs. BK profoundly reduced the severity of ischemia induced arrhythmias after myocardial infarction (Vegh *et al.*, 1991).

Furthermore, BK (1 ng/kg per minute), infused into the coronary artery of anesthetized dogs, during ischemia–reperfusion, reduced lactate concentrations after 90 min occlusion as well as lactate dehydrogenase activities in the coronary sinus blood. The tissue levels of high energy-rich phosphates and glycogen stores in the ischemic area were preserved and lactate content reduced (Linz *et al.*, 1996). In another study (Martorana *et al.*, 1990), the effect of locally administered BK on the limitation of infarct size was investigated in dogs. The left descending coronary artery was ligated for 6 h. Animals received BK in a subhypotensive dose of 1 ng/kg per minute. The intracoronary route and the very low dose of BK were chosen to obtain a local cardiac effect with no or minimal effects on systemic hemodynamics. The size of the infarction of saline-treated dogs averaged 56% of the area at risk. BK significantly reduced infarct size to 32%. The observation that BK limited infarct size provided evidence for the cardioprotective involvement of kinins in ischemic events.

Additional evidence for a beneficial role of BK during myocardial ischemia comes from studies in pigs, where BK also reduced infarct size (Tio *et al.*, 1991) and improved electrical stability 2 weeks after myocardial infarction (Tobé *et al.*, 1991). It is possible that cyclic guanosine monophosphate (cGMP) improves the energy state in the ischemic heart (Vuorinen *et al.*, 1984). After addition of BK, enhanced release of NO and PGI$_2$ followed by increased cGMP, and cyclic adenosine monophosphate (cAMP) occurred in cultured endothelial cells and in the coronary effluent of isolated ischemic rat hearts (Wiemer *et al.*, 1994), suggesting that the protective effect of kinins during myocardial ischemia is mediated, at least\ partly, by the release of vascular endothelial kinins.

Excessive stimulation of the ischemic heart by catecholamines has been recognized as a deleterious mechanism, and which is also favorably influenced by BK, adding a further mechanism of protection. Thus, in isolated rat hearts, BK reduced the release of noradrenaline that occurs in this preparation during reperfusion after a period of global ischemia (Carlsson and Abrahamsson, 1989). Similarly, noradrenaline overflow following reperfusion of ischemic myocardium in anesthetized dogs was reduced by intracoronary infusion of BK (Ribout *et al.*, 1994). Table 19.1 summarizes beneficial cardiac effects induced by locally administered kinins.

Table 19.1   Kinins administered locally exert beneficial cardiac effects

Cardiac effect(s)	Abolition by kinin antagonist	References
Increase in coronary and/or capillary nutritional flow	Yes	Rett et al. (1986), Linz et al. (1992)
Preservation of high energy-rich phosphates	Yes	Linz et al. (1992, 1996)
Increased myocardial glucose uptake and utilization	Not tested	Rösen et al. (1983)
Increased rate of glycolytic influx	Not tested	Rösen et al. (1983)
Decrease in cytosolic enzyme leakage	Yes	Linz et al. (1992, 1995, 1996)
Abolition of reperfusion-induced arrhythmias	Yes	Vegh et al. (1991), Linz et al. (1995, 1996)
Improved myocardial electrical stability	Not tested	Tio et al. (1991), Tobé et al. (1991)
Improvement in cardiac performance	Yes	Schölkens and Linz (1992), Linz et al. (1992)
Improved postischemic function and/or recovery	Yes	Linz et al. (1996), Zhu et al. (1995)
Reduction of myocardial infarct size	Yes	Martorana et al. (1990)
Reduction of ischemia-induced noradrenaline overflow	Not tested	Carlsson and Abrahamsson (1989), Ribout et al. (1994)
Increased release of NO and PGI_2 from endothelial cells	Yes	Wiemer et al. (1994), Linz et al. (1996)

3.   Inhibition of Kinin Breakdown Induces Beneficial Effects in Myocardial Ischemia

The hypothesis that ACE inhibitors exert their effects at least partially through kinins was based on the knowledge that ACE is a major kininase responsible for the degradation of kinins, demonstrated cardioprotective actions of kinins, and the often surprising similar or even identical experimental findings obtained with either exogenous kinins or ACE inhibitors. In this section, examples are given to demonstrate the beneficial cardiovascular effects of ACE inhibitors. In Section 4, a number of examples will be given that the effects of ACE inhibitors can be abolished by B_2 receptor antagonists, providing evidence for the participation of kinins in these models.

After acute coronary artery occlusion, captopril reduced the extent of cellular necrosis at the end of a 6 h occlusion period (Ertl et al., 1982). The authors ascribed this reduction in ischemic injury to an increase in regional myocardial blood flow. Enalapril, another ACE inhibitor, has also been demonstrated to reduce myocardial infarct size in rats subjected to a 24 h complete coronary artery occlusion, but without reperfusion (Hock et al., 1985). This treatment also significantly blunted creatine kinase depletion. Enalapril, given 30 min after the onset of ischemia, also beneficially modified plasma creatine kinase changes and ST-segment elevation in cats subjected to a 5 h coronary artery occlusion (Lefer and Peck, 1984).

The contribution of locally formed cardiac kinins in the infarct limiting effects of ramiprilat was investigated in anesthetized dogs subject to ligation of the left descending coronary artery (Martorana et al., 1990). The intracoronary route and a very low dose (40 ng/kg

per minute) was chosen to obtain a local cardiac effect with no or minimal effects on systemic hemodynamics. The ACE inhibitor significantly reduced infarct size. Thus, at a dose that had no effect on systemic hemodynamics ramiprilat effectively limited infarct size following coronary occlusion.

Comparable results were reported in anesthetized rabbits with myocardial infarction (Hartman et al., 1993b). Ramiprilat given intravenously just prior to reperfusion (coronary artery occlusion 30 min, reperfusion 2 h), reduced the infarct size from 41% to 20%. The reduction in myocardial infarct size by ramiprilat was independent of the inhibition of angiotensin II synthesis (Hartman et al., 1993a). In anesthetized, bilaterally nephrectomized dogs, this observation could be confirmed. When captopril was administered intravenously, both prior to and following coronary artery occlusion, infarct size was reduced. However, suppression of angiotensin II formation by a chymotrypsin inhibitor did not reduce myocardial infarct size (Noda et al., 1993). In addition, augmentation of coronary blood flow by ACE inhibition was enhanced by endogenous kinins but not by an angiotensin II receptor antagonist (Ruocco et al., 1992). A role for kinins, NO and prostaglandins in the protective effect of ACE inhibitors on ischemic/reperfusion myocardial infarction was found in rats (Liu et al., 1994).

Studies with BK and ramiprilat in isolated rat hearts with post-ischemic reperfusion arrhythmias led to an almost identical profile of changes in cardiodynamics and metabolism, including reduction in ventricular fibrillation, indicating that local ACE inhibition attenuates degradation of BK and related kinins (Schölkens et al., 1988).

Furthermore, BK accounted for improved post-ischemic function and decreased glutathione release of guinea-pig hearts treated with ramiprilat (Massoudy et *al.*, 1994). The underlying protective mechanism was attributed to a direct oxygen radical scavenger action of NO. Similar observations on improved postischemic function were made by others in isolated rat hearts with ischemic damage and reperfusion arrhythmias (Fleetwood *et al.*, 1991), and on metabolic recovery in postischemic rat hearts (Werrmann and Cohen, 1994) as well as in isolated rabbit hearts (Rump *et al.*, 1993).

In rats, the beneficial effects of ACE inhibitors on remodeling, which is a significant chronic reorganization process of cardiac tissue after myocardial infarction, were studied. An ACE inhibitor, moexipril, reduced infarct size by half (Stauss *et al.*, 1994b). The same group investigated the effects of ACE inhibitor treatment on cardiac remodeling in kinin-deficient Brown–Norway Katholiek rats and in kinin-replete Brown–Norway Hannover rats (Stauss *et al.*, 1994a). Animals were pretreated with ramipril (1 mg/kg per day) for 1 week before the induction of myocardial infarction, followed by continuous treatment for further 6 weeks. Ramipril reduced infarct size and end-diastolic pressure only in kinin-replete but not in kinin-deficient animals, suggesting that the effects of the ACE inhibitor were mediated by the potentiation of endogenous kinins (Stauss *et al.*, 1994a).

In dogs, BK and PGI_2 contributed to the attenuation of myocardial stunning by the ACE inhibitor ramipril (Ehring *et al.*, 1994). There was also a protective role of kinins in cardiac anaphylaxis. Guinea pigs were sensitized by an intravenous injection of homologous cytotropic IgG1. Twelve hours later, the hearts of the sensitized animals were isolated and perfused according to the Langendorff method. Antigen challenge with dinitro-

phenyl bovine serum albumin was associated with a 30% increase in kinin overflow. Increasing the half-life of BK and related kinins with ACE inhibitors, captopril or enalapril, attenuated and reversed vasoconstriction and alleviated arrhythmias (Rubin and Levi, 1995). Furthermore, an inhibition of an early increase in left ventricular mass produced by transmyocardial direct-current shock was observed in a canine model of ventricular remodeling when the animals were treated with ramipril (McDonald *et al.*, 1995).

In patients with type II diabetes, the effects of insulin were improved by ACE inhibition (Rett *et al.*, 1988). Indeed, there was also an increase of plasma kinins in normotensive and hypertensive persons when treated with an ACE inhibitor (Pellacani *et al.*, 1994, 1995). In addition, ACE inhibition increased flow-mediated dilatation in the forearm of normal subjects (Jeserich *et al.*, 1995). Table 19.2 summarizes beneficial cardiac effects induced by inhibition of breakdown of kinins.

4. Antagonism of B_2 Receptors Reverses Cardioprotective Effects of Bradykinin or ACE Inhibitors During Ischemia

Dissection of the relative roles of a reduction of (deleterious) angiotensin II levels and enhancement of protective kinins by ACE inhibitors requires nearly complete inhibition of the action of kinins to enable definitive conclusions on the participation of kinins in the

Table 19.2   Inhibition of kinin breakdown induces beneficial cardiac effects

Cardiac effect(s)	Abolition by kinin antagonist	References
Augmentation of coronary blood flow	Yes	Ruocco *et al.* (1992)
Inhibition of early increase in left ventricular mass by myocardial direct-current shock	Yes	McDonald *et al.* (1995)
Reduction of infarct-induced heart failure (remodeling)	Yes	Stauss *et al.* (1994a,b)
Improved postischemic function (arrhythmias)	Yes	Rösen *et al.* (1983)
Decrease in cytosolic enzyme leakage	Yes	Linz *et al.* (1986), van Gilst *et al.* (1986), Fleetwood *et al.* (1991), Rump *et al.* (1993), Massoudy *et al.* (1994), Werrmann and Cohen (1994)
Attenuation of myocardial stunning	Yes	Ehring *et al.* (1994)
Reduction of myocardial infarct size in		
Rats	Yes	Liu *et al.* (1994)
Rabbits	Yes	Hartman *et al.* (1993a,b)
Dogs	Yes	Martorana *et al.* (1990), Noda *et al.* (1993)
Protective role in cardiac anaphylaxis	Yes	Rubin and Levi (1995)
ACE inhibition has same profile as kinins	Yes	Schölkens *et al.* (1988), Linz *et al.* (1995, 1996)
Increase in plasma kinins in hypertensive humans	Not tested	Pellacani *et al.* (1994, 1995)

effects of ACE inhibitors. Only the advent of potent and specific B_2 receptor antagonists has enabled this pharmacological approach.

Vavrek and Stewart (1985) discovered that substitution of D-phenylalanine for proline at position 7 of the primary sequence of BK converted it into a specific antagonist for B_2 kinin receptors. Later, Hoe 140 (DArg-[Hyp^3,Thi^5,DTic^7,Oic^8]-BK, icatibant), one of the most potent and stable B_2 receptor antagonists, was discovered (Henke et al., 1990; Bao et al., 1991; Hock et al., 1991; Wirth et al., 1991) (see Chapter 2).

Hoe 140 is useful to evaluate the role of kinins in myocardial ischemia. Its sequence contains two unnatural amino acids, D-tetrahydro-isoquinoline-3-carboxylic acid (D-Tic), and octahydroindol-2-carboxylic acid (Oic), replacing a proline residue at position 7 of the BK sequence, and a phenylalanine residue at position 8. Hoe 140, binds tightly to the B_2 but not B_1 receptors with a K_d of less than 0.05 nM (Hock et al., 1991), thereby exceeding the K_d of BK by a factor of at least 10 (Hess et al., 1992).

Hoe 140 reversed most of the cardioprotective effects induced by BK or ACE inhibitors during ischemia (Tables 19.1 and 19.2). Hoe 140, administered alone, aggravated ischemia induced effects in isolated rat heart preparations (Linz et al., 1992). On the other hand, when Hoe 140 (10^{-10} M) was perfused with Krebs–Henseleit buffer in isolated rat hearts without ischemia, no changes in cardiovascular parameters (left ventricular pressure, dP/dt_{max} as a measure for contractility, heart rate, coronary flow) as well as myocardial metabolism (release of lactate dehydrogenase, creatine kinase, lactate in the venous effluent) and in the myocardium (glycogen, ATP and creatine phosphate values) were found. In rat studies in toxicology, where Hoe 140 was given intravenously or subcutaneously in large doses of 10 mg/kg per day for 4 weeks, no adverse effects on cardiodynamics were observed. This indicates that B_2 receptor blockade with Hoe 140 may only interfere with kinin-induced cardiac effects when acute severe myocardial ischemia occurs.

Kinins also seem to participate in the regulation of coronary blood flow in humans as demonstrated by intracoronary infusion of Hoe 140. Endogenous kinins may mediate basal and flow stimulated endothelium-dependent vasodilation in the human coronary circulation (Groves et al., 1994), as both were found significantly reduced by the B_2 receptor antagonist. These findings suggest that kinins are produced under basal conditions and their production may influence vascular shear stress (Malek and Izumo, 1994).

In addition, there was also found a direct role of endogenous kinins in insulin sensitivity and blood pressure regulation during hyperinsulinemia in normotensive rats. Inhibition of the B_2 receptors significantly reduced sensitivity to insulin and resulted in an increase in blood pressure (Kohlman et al., 1995).

In conclusion, potent and specific B_2 kinin receptor antagonists have contributed to identify a significant participation of kinins to the cardioprotective effects of ACE inhibitors. The favorable outcome of large clinical studies with ACE inhibitors after myocardial infarction in patients becomes plausible from the amount of very consistent experimental data that kinins were involved [The Acute Infarction Efficacy (AIRE) Study Investigators, 1993].

5. Kinins are Released Under Conditions of Ischemia

Although it has been demonstrated that exogenous kinins exert beneficial effects and kinins participate in the effects of ACE inhibitors in a number of models, demonstration of increased kinin levels and possible protective effects of kinins during myocardial ischemia in the absence of ACE inhibitors would very much support the contention that kinins may play a role in myocardial ischemia. By use of a specific radioimmunoassay for kinins (Proud et al., 1983), it was demonstrated that in isolated normoxic rat hearts, kinins are released into the perfusate (Baumgarten et al., 1993). In the same model, perfusion with distilled water, a procedure that destroys the function of the endothelium, markedly attenuated basal kinin release pointing to the myocardial endothelium as a source of kinins (Wiemer et al., 1994). During ischemia, the respective kinin outflow increased more than five-fold. Thus, kinins are continuously formed in the isolated rat heart and, moreover, ischemia is a stimulus for an enhanced kinin release, which may contribute to reduce the sequelae of myocardial ischemia (Baumgarten et al., 1993; Koide et al., 1993; Lamontagne et al., 1995).

In bilaterally nephrectomized, anesthetized dogs immediately after coronary occlusion, a significant increase of kinins in the anterior interventricular vein was observed (Noda et al., 1993). These results are in line with other findings showing an activation of kinin-generating pathways in cardiac (Kimura et al., 1973; Hashimoto et al., 1977), and other ischemic tissues (Pitt et al., 1970; Wilkens et al., 1970; Hashimoto et al., 1978; Matsuki et al., 1987; Zeitlin et al., 1989; Poucher et al., 1993; Pan et al., 1994).

A phenomenon of the heart with possible participation of kinins is the methodological approach of ischemic preconditioning. Ischemic preconditioning can be defined as a protective adaptive mechanism produced by short periods of ischemic stress resulting in a marked, albeit temporary, resistance of the heart to a subsequent more prolonged period of that same stress. This protection includes reductions in ischemic cellular damage (infarct size), in left ventricular dysfunction and in life-threatening ventricular arrhythmias (Parratt, 1994b).

In isolated rat hearts, preconditioning induced by five subsequent 1-min cycles of global ischemia, each followed by 4 min of reperfusion, protected against reperfusion arrhythmias that occurred during a subsequent 15 min local ischemia and a 30 min reperfusion period. In comparison to nonpreconditioned hearts, preconditioned hearts were protected against ventricular fibrillation. Similar effects were observed by five short-term infusions of BK (1×10^{-10} M). Thus, BK mimicked the effect of preconditioning. Coperfusion with Hoe 140 (1×10^{-9} M) abolished the cardioprotective effects of both preconditioning and BK infusion proving significant contribution of endogenous kinins to ischemic preconditioning in this model. Indeed, outflow of kinins and PGI_2 into the venous effluent was enhanced even during short-term global ischemia (Linz *et al.*, 1996).

Further evidence for a role of kinins in ischemic preconditioning comes from studies *in vivo* in anesthetized open-chest rabbits (Wall *et al.*, 1994). Prior to 30 min of coronary occlusion, rabbits were subject to ischemic preconditioning (5 min occlusion followed by 10 min reperfusion). Systemic hemodynamic responses were similar in both treatment groups. Compared to nonpreconditioned controls, preconditioning reduced infarct size significantly. Pretreatment with Hoe 140 abolished the cardioprotective effect. In addition, BK infusion without preconditioning reduced infarct size significantly. This effect was also prevented by Hoe 140, and the antagonist alone did not exacerbate the degree of myocardial necrosis. Myocardial area at risk as a percent of total ventricular mass was not different between the treatment groups. These results in rabbits suggest that endogenous kinins mediate the cardioprotective events associated with ischemic preconditioning. This is supported by a study in anesthetized dogs where the antiarrhythmic effect of ischemic preconditioning was blocked by Hoe 140 (Vegh *et al.*, 1994).

NO may mediate the cardioprotective effects induced by preconditioning or BK since, in dogs, intracoronary administration of methylene blue prevents the pronounced antiarrhythmic effect of ischemic preconditioning (Vegh *et al.*, 1993). This raises the possibility that kinins may act as "primer" mediators in preconditioning (Vegh *et al.*, 1993, 1994; Parrat, 1994a) together or among other mediators like adenosine, acetylcholine and stimulators of protein kinase C, for which already some evidence of an involvement in preconditioning exists.

6. Kinin-generating Pathways are Present in the Heart

In humans, circulating concentrations of immunoreactive kinins are low (Pellacani *et al.*, 1994). Human tissue kallikrein induced hypotension in transgenic mice (Wang *et al.*, 1994), and direct gene delivery of human tissue kallikrein reduced blood pressure in spontaneously hypertensive rats (Wang *et al.*, 1995). The kinin system is considered to operate mainly at a local tissue level. This is supported by the finding that in the rat kinin tissue concentrations were about ten-fold higher than circulating plasma levels (Campbell *et al.*, 1993). Increased levels of bradykinin and its metabolites were found in tissues including hearts of young spontaneously hypertensive rats when compared to normotensive animals, suggesting increased kallikrein activity in the 6-week-old hypertensive animals. However, the difference was not present in older animals aged 10 and 20 weeks (Campbell *et al.*, 1995a). Interestingly, in a transgenic TGR(mRen-2)27 rat strain, in which the Ren-2 mouse gene is transferred into the genome of the Sprague–Dawley rat, severe hypertension develops at a young age, and plasma and heart tissue levels of angiotensin II are increased significantly. In contrast, kinin levels remained unchanged providing direct support for an angiotensin II-dependent mechanism of hypertension in these animals (Campbell *et al.*, 1995b).

In a demonstration of the existence of kinin-generating pathways in the heart, kallikrein was increased in tissue and in the incubation medium of rat heart slices, suggesting that the heart most probably contains the capacity to generate kinins (Nolly *et al.*, 1994a). These findings were corroborated by other studies showing that kallikrein gene expression takes place in rat and human hearts (Clements and Mukhtar, 1994).

The existence of BK binding sites in rat cardiomyocytes (Minshall *et al.*, 1994) and Ca^{2+}-independent NO-synthase activity in adult rat cardiomyocytes were demonstrated (Schulz *et al.*, 1992). Furthermore, BK receptor binding in cardiac fibroblasts was increased with the appearance of myocardial fibrosis following chronic angiotensin II, or aldosterone administration or after myocardial infarction. This coincided with increased ACE binding suggesting that, under pathophysiological conditions, altered receptor expression may modulate fibroblast collagen turnover and thereby govern fibrogenesis (Sun *et al.*, 1994).

Additional evidence for the actions of endogenous kinins derives from Brown–Norway rats (May/Pfd/f), which are kinin-deficient. Brown–Norway rats completely lack one of the kinin precursors, the high molecular weight kininogen and are deficient in low molecular weight kininogen as well as plasma prekallikrein (Damas and Adams, 1980). Compared to control Brown–Norway Hannover rats, which have a normal kallikrein system, ischemic working hearts from May/Pfd/f rats showed increased heart rate, decreased coronary flow, and an impaired myocardial metabolism via decreased glycogen stores and decreased energy-rich phosphates. This suggests that a myocardial kinin-generating system is important for the improved function (e.g., reduction in reperfusion arrhythmias) and metabolism (e.g., increases in energy-rich phosphates) of the heart during ischemia (Schölkens and Linz, 1992; Linz et al., 1996).

7.   Conclusion

A potential protective role of endogenous kinins in myocardial ischemia seems to be evident. Their cardioprotective profile resembles that of ACE inhibitors and is abolished by specific B_2 receptor antagonists. Kiningenerating pathways are present in the heart and, consequently, kinins are released during ischemia with subsequent formation of PGI_2 and NO, probably derived from the coronary vascular endothelium.

8.   References

The AIRE Study Investigators (1993). Effect of ramipril on mortality and morbidity of survivors of acute myocardial infarction with clinical evidence of heart failure. Lancet 342, 821–828.

Bao, G., Qadri, F., Stauss, B., Stauss, H., Gohlke, P. and Unger, T. (1991). HOE 140, a new highly potent and long-acting bradykinin antagonist in conscious rats. Eur. J. Pharmacol. 200, 179–182.

Baumgarten, C.R., Linz, W., Kunkel, G., Schölkens, B.A. and Wiemer, G. (1993). Ramiprilat increases bradykinin outflow from isolated rat hearts. Br. J. Pharmacol. 108, 293–295.

Bhoola, K.D., Figueroa, C.D. and Worthy, K. (1992). Bioregulation of kinins, kallikreins, kininogens, and kininases. Pharmacol. Rev. 44, 1–80.

Campbell, D.J., Kladis, A. and Duncan, A.-M. (1993). Bradykinin peptides in kidney, blood, and other tissues of the rat. Hypertension 21, 155–165.

Campbell, D.J., Duncan, A.-M., Kladis, A. and Harrap, S.B. (1995a). Increased levels of bradykinin and ist metabolites in tissues of young spontaneously hypertensive rats. J. Hypertens. 13, 739–746.

Campbell, D.J., Rong, P., Kladis, A., Rees, B., Ganten, D. and Skinner, S. (1995b). Angiotensin and bradykinin peptides in the TGR(mRen-2)27 rat. Hypertension 25, 1014–1020.

Carlsson, L. and Abrahamsson, T. (1989). Ramiprilat attenuates local ischemia-induced release of noradrenaline in the ischemic myocardium. Eur. J. Pharmacol. 166, 157–164.

Carretero, O.A., Carbini, L.A. and Scicli, A.G. (1993). The molecular biology of the kallikrein–kinin system: I. General description, nomenclature and the mouse gene family. J. Hypertens. 11, 693–697.

Clements, J. and Mukhtar, A. (1994). Kallikrein gene expression in the heart and kidney. Kallikreins and kinins and cardiovascular function. Official Satellite to the 15th Scientific Meeting of the International Society of Hypertension, Melbourne, 25 March, Abstract book, p. 7.

Damas, J. and Adams, A. (1980). Congenital deficiency in plasma kallikrein and kininogen on the Brown Norway rat. Experientia 36, 586–587.

Ehring, T., Baumgart, D., Krajcar, M., Hümmelgen, M., Kompa, S. and Heusch, G. (1994). Attenuation of myocardial stunning by the ACE inhibitor ramiprilat through a signal cascade of bradykinin and prostaglandins but not nitric oxide. Circulation 90, 1368–1385.

Erdös, E.G. (1990). Some old and some new ideas on kinin metabolism. J. Cardiovasc. Pharmacol. 15 (Suppl. 6), S20–S24.

Ertl, G., Kloner, R.A., Alexander, R.W. and Braunwald, E. (1982). Limitation of experimental infarct size by angiotensin converting enzyme inhibition. Circulation 65, 40–48.

Fleetwood, G., Boutinet, M. and Wood, J.M. (1991). Involvement of the renin–angiotensin system in ischemic damage and reperfusion arrhythmias in the isolated perfused rat heart. J. Cardiovasc. Pharmacol. 17, 351–356.

Gohlke, P., Bünning, P., Bönner, G. and Unger, T. (1992). ACE inhibitor effect on bradykinin metabolism in the vascular wall. Agents Actions Suppl. 38/III, 178–185.

Groves, P.H., Kurz, S. and Drexler, H. (1994). The role of bradykinin in basal and flow-mediated endothelium-dependent vasodilation in the human coronary circulation. Circulation 90(4/2), abstract 0184.

Hartman, J.C., Hullinger, T.G., Wall, T.M. and Shebuski, R.J. (1993a). Reduction of myocardial infarct size by ramiprilat is independent of angiotensin II synthesis inhibition. Eur. J. Pharmacol. 234, 229–236.

Hartman, J.C., Wall, T.M., Hullinger, T.G. and Shebuski, R.J. (1993b). Reduction of myocardial infarct size in rabbits by ramiprilat: reversal by the bradykinin antagonist HOE 140. J. Cardiovasc. Pharmacol. 21, 996–1003.

Hashimoto, K., Hirose, M., Furukawa, H. and Kimura, E. (1977). Changes in hemodynamics and bradykinin concentration in coronary sinus blood in experimental coronary occlusion. Jpn Heart J. 18, 679–689.

Hashimoto, K., Hamamoto, H., Honda, Y., Hirose, M., Furukawa, S. and Kimura, E. (1978). Changes in components of kinin system and hemodynamics in acute myocardial infarction. Am. Heart J. 95, 619–626.

Henke, S., Anagnostopoulos, H., Breipohl, G., Gerhards, H., Knolle, J., Schölkens, B.A. and Lembeck, F. (1990). Novel highly potent bradykinin antagonists and their impact on allergic diseases. In "IUPAC: International Union of Pure and Applied Chemistry. Trends in Medicinal Chemistry '90" (eds. S. Sarel, R. Mechoulam and I. Agranat), pp. 229–231. Blackwell, Scientific Publications, London.

Hess, J.F., Borkowski, J.A., Young, G.S., Strader, C.D. and Ransom, R.W. (1992). Cloning and pharmacological characterization of a human bradykinin (BK-2) receptor. Biochem. Biophys. Res. Commun. 184, 260–268.

Hock, C.E., Ribeiro, G.T. and Lefer, A.M. (1985). Preservation of ischemic myocardium by a new converting enzyme inhibitor, enalaprilic acid, in acute myocardial infarction. Am. Heart J. 109, 222–228.

Hock, F.J., Wirth, K., Albus, U., Linz, W., Gerhards, H.J., Wiemer, G., Henke, St., Breipohl, G, König, W., Knolle, J. and Schölkens, B.A. (1991). HOE 140 a new and long acting bradykinin-antagonist: in vitro studies. Br. J. Pharmacol. 102, 769–773.

Jeserich, M., Hornig, B., Lohmann, A. and Drexler, H. (1995). ACE-inhibition increases flow-mediated dilation in the forearm of normal subjects. Eur. Soc. Meeting, Amsterdam, August, book of abstracts.

Kimura, E., Hashimoto, K., Furukawa, S. and Hayakawa, H. (1973). Changes in bradykinin level in coronary sinus blood after the experimental occlusion of a coronary artery. Am. Heart J. 85, 635–647.

Kohlman, O., de Assis Rocha Neves, F., Ginoza, M., Tavares, A., Cezaretti, M.L., Zanella, M.T., Ribeiro, A.B., Gavras, I. and Gavras, H. (1995). Role of bradykinin in insulin sensitivity and blood pressure regulation during hyperinsulinemia. Hypertension 25, 1003–1007.

Koide, A., Zeitlin, I.J. and Parratt, J.R. (1993). Kinin formation in ischemic heart and aorta of anesthetized rats. J. Physiol. 467, abstract 125P.

Lamontagne, D., Nadeau, R. and Adam, A. (1995). Effect of enalaprilat on bradykinin and des-Arg^9-bradykinin release following reperfusion of the ischaemic rat heart. Br. J. Pharmacol. 115, 476–478.

Lefer, A.M. and Peck, R.C. (1984). Cardioprotective effects of enalapril in acute myocardial ischemia. Pharmacology 29, 61–69.

Linz, W., Schölkens, B.A. and Han, Y.-F. (1986). Beneficial effects of the converting enzyme inhibitor, ramipril, in ischemic rat hearts. J. Cardiovasc. Pharmacol. 8 (Suppl. 10), S91–S99.

Linz, W., Wiemer, G. and Schölkens, B.A. (1992). ACE-inhibition induces NO-formation in cultured bovine endothelial cells and protects isolated ischemic rat hearts. J. Mol. Cell. Cardiol. 24, 909–919.

Linz, W., Wiemer, G., Gohlke, P., Unger, T. and Schölkens, B.A. (1995). Contribution of kinins to the cardiovascular actions of angiotensin converting enzyme inhibitors. Pharmacol. Rev. 47, 25–49.

Linz, W., Wiemer, G. and Schölkens, B.A. (1996). Role of kinins in the pathophysiology of myocardial ischemia: *in vitro* and *in vivo* studies. Diabetes 45 (Suppl. 1), 551–558.

Liu, Y.-H., Yang, X.-P., Sharov, V.G., Sabbah, H.N., Scicli, A.G. and Carretero, O.A. (1994). Role of kinins, nitric oxide and prostaglandins in the protective effect of ACE inhibitors on ischemia/reperfusion myocardial infarction in rats. Hypertension 24, 380.

Lochner, W. and Parratt, J.R. (1966). A comparison of the effects of locally and systemically administered kinins on coronary blood flow and myocardial metabolism. Br. J. Pharmacol. 26, 17–26.

Malek, A.M. and Izumo, S. (1994). Molecular aspects of signal transduction of shear stress in the endothelial cell. J. Hypertens. 12, 989–999.

Martorana, P.A., Kettenbach, B., Breipohl, G., Linz, W. and Schölkens, B.A. (1990). Reduction of infarct size by local angiotensin-converting enzyme inhibition is abolished by a bradykinin antagonist. Eur. J. Pharmacol. 182, 395–396.

Massoudy, P., Becker, B.F. and Gerlach, E. (1994). Bradykinin accounts for improved postischemic function and decreased glutathione release of guinea pig heart treated with the angiotensin-converting enzyme inhibitor ramiprilat. J. Cardiovasc. Pharmacol. 23, 632–639.

Matsuki, T., Shoji, T., Yoshida, S., Kudoh, Y., Motoe, M., Inoue, M., Nakata, T., Hosada, S., Shimamoto, K., Yellon, D. and Imura, O. (1987). Sympathetically induced myocardial ischemia causes the heart to release plasma kinin. Cardiovasc. Res. 21, 428–432.

McDonald, K.M., Mock, J., D'Aloia, A., Parrish, T., Hauer, K., Francis, G., Stillman, A. and Cohn, J.N. (1995). Bradykinin

antagonism inhibits the antigrowth effect of converting enzyme inhibition in the dog myocardium after discrete transmural myocardial necrosis. Circulation 91, 2043–2048.

Minshall, R.D., Miletich, D.J., Vogel, S.M., Becker, R.P., Erdös, E.G. and Rabito, S.T. (1994). Existence of bradykinin (BK) binding sites in rat cardiomyocytes. FASEB Meeting, Abstract 5122.

Noda, K., Sasaguri, M., Ideishi, M., Ikeda, M. and Arakawa, K. (1993). Role of locally formed angiotensin II and bradykinin in the reduction of myocardial infarct size in dogs. Cardiovasc. Res. 27, 334–340.

Nolly, H., Carbini, L.A., Scicli, G., Carretero, O.A. and Scicli, A.G. (1994a). A local kallikrein–kinin system is present in rat hearts. Hypertension 23, 919–923.

Nolly, H., Damiani, M.T. and Miatello, R. (1994b). Vascular-derived kinins and local control of vascular tone. Braz. J. Med. Biol. Res. 27, 1995–2011.

Pan, H.-L., Stahl, G.L., Rendig, S.V., Carretero, O. and Longhurst, J.C. (1994). Endogenous BK stimulates ischemically sensitive abdominal visceral C fiber afferents through kinin B$_2$ receptors. Am. J. Physiol. 267, H2398–H2406.

Parratt, J.R. (1994a). Cardioprotection by angiotensin converting enzyme inhibitors – the experimental evidence. Cardiovasc. Res. 28, 183–189.

Parratt, J.R. (1994b). Protection of the heart by ischemic preconditioning: mechanisms and possibilities for pharmacological exploitation. Trends Pharmacol. Sci. 15, 19–25.

Pellacani, A., Brunner, H.R. and Nussberger, J. (1994). Plasma kinins increase after angiotensin-converting enzyme inhibition in human subjects. Clin. Sci. 87, 567–574.

Pellacani, A., Cavallone, S., Veglio, F., Chiandussi, L. and Nussberger, J. (1995). Plasma kinins increase after acute ACE-inhibition with ramipril in patients with essential hypertension. 17th European Meeting on Hypertension, Milan 9–12 June, abstract 647.

Pitt, B., Mason, J., Conti, C.R. and Colman, R.W. (1970). Activation of the plasma kallikrein system during myocardial ischemia. Adv. Exp. Med. Biol. 8, 403–410.

Poucher, S.M., Garcia, S. and Brooks, R. (1993). The effect of the bradykinin antagonist HOE 140 upon skeletal muscle blood flow in anaesthetized cats. J. Physiol. 467, 315P.

Proud, D., Togias, A., Naclerio, R.M., Crush, S.A., Norman, P.S. and Lichtenstein, L.M. (1983). Kinins are generated *in vivo* following nasal airway challenge of allergic individuals with allergen. J. Clin. Invest. 72, 1678–1685.

Rett, K., Maerker, E., Lodri, C., Wicklmayr, M. and Dietze, G. (1986). Effects of kallikrein, bradykinin and insulin on substrate metabolism in isolated perfused rat heart. In "Kinins IV, Part B" (eds. L.M. Greenbaum and H.S. Margolius), pp. 379–384. Plenum Press, New York.

Rett, K., Lotz, N., Wickelmayer, M., Fing, E., Jauch, K.W., Günther, B. and Dietze, G. (1988). Verbesserte Insulinwirkung durch ACE-Hemmung beim Typ II-Diabetiker. Dtsch Med. Wochenschr. 113, 243–249.

Ribuot, C., Yamaguchi, N., Godin, D., Jett, L., Adam, A. and Nadeau, J. (1994). Intracoronary infusion of bradykinin: effects on noradrenaline overflow following reperfusion of ischemic myocardium in the anesthetized dog. Fundam. Clin. Pharmacol. 8, 532–538.

Rösen, P., Eckel, J. and Reinauer, H. (1983). Influence of bradykinin on glucose uptake and metabolism studied in isolated cardiac myocytes and isolated perfused rat hearts. Hoppe Seyler's Z. Physiol. Chem. 364, 431–438.

Rubin, L.E. and Levi, R. (1995). Protective role of bradykinin in cardiac anaphylaxis. Coronary-vasodilating and anti-arrhythmic activities mediated by autocrine/paracrine mechanisms. Circ. Res. 76, 434–440.

Rump, A.F.E., Koreuber, D., Rösen, R. and Klaus, W. (1993). Cardioprotection by ramiprilat in isolated rabbit hearts. Eur. J. Pharmacol. 241, 201–207.

Ruocco, N.A., Yu, T.-K., Bergelson, B.A., Cannistra, A.J., Cody, C. and Ryan, T.J. (1992). Augmentation of coronary blood flow by ACE inhibition enhanced by endogenous bradykinin but not by angiotensin II receptor blockade. Circulation 86 (Suppl. I), I640.

Schölkens, B.A. and Linz, W. (1992). Bradykinin-mediated metabolic effects in isolated perfused rat hearts. Agents Actions Suppl. 38/II, 36–42.

Schölkens, B.A., Linz, W. and König, W. (1988). Effects of the angiotensin converting enzyme inhibitor, ramipril, in isolated ischemic rat heart are abolished by a bradykinin antagonist. J. Hypertens. 6 (Suppl. 4), S25–S28.

Schulz, R., Nava, E. and Moncada, S. (1992). Induction and potential biological relevance of a Ca^{2+}-independent nitric oxide synthase in the myocardium. Br. J. Pharmacol. 105, 575–580.

Stauss, H.M., Adamiak, D., Zhu, Y.C., Redlich, T. and Unger, T. (1994a). ACE inhibition following myocardial infarction (MI) in kinin-deficient Brown–Norway Katholiek rats (BNK). Council of High Blood Pressure Research, 48th Annual Fall Conference and Scientific Sessions, Chicago, September 27–30, book of abstracts.

Stauss, H.M., Zhu, Y.C., Redlich, T., Adamiak, D., Mott, A., Kregel, K.C. and Unger, T. (1994b). Angiotensin-converting enzyme inhibition in infarct-induced heart failure in rats: bradykinin versus angiotensin II. J. Cardiovasc. Risk 1, 255–262.

Sun, Y., Cleutjens, J.P.M., Diaz-Arias, A. and Weber, K. (1994). Cardiac angiotensin converting enzyme and myocardial fibrosis in the rat. Cardiovasc. Res. 281, 1423–1432.

Tio, R.A., Tobé, T.J.M., Bel, K.J., de Langen, C.D.J., van Gilst, W.H. and Wesseling, H. (1991). Beneficial effects of bradykinin on porcine ischemic myocardium. Basic Res. Cardiol. 86, 107–116.

Tobé, T.J.M., de Langen, C.D.J., Tio, R.A., Bel, K.J., Mook, P.H. and Wesseling, H. (1991). In vivo effect of bradykinin during ischemia and reperfusion: improved electrical stability two weeks after myocardial infarction in the pig. J. Cardiovasc. Pharmacol. 17, 600–607.

van Gilst, W.H., de Graeff, P.A., Wesseling, H. and de Langen, C.D.J. (1986). Reduction of reperfusion arrhythmias in the ischemic isolated rat heart by angiotensin converting enzyme inhibitors: a comparison of captopril, enalapril, and HOE 498. J. Cardiovasc. Pharmacol. 8, 722–728.

Vavrek, R.J. and Stewart, J.M. (1985). Competitive antagonists of bradykinin. Peptides 6, 161–164.

Vegh, A., Szekeres, L. and Parratt, J.R. (1991). Local intra-coronary infusions of bradykinin profoundly reduce the severity of ischemia-induced arrhythmias in anaesthetized dogs. Br. J. Pharmacol. 104, 294–295.

Vegh, A., Papp, J.G., Szekeres, L. and Parratt, J.R. (1993). Prevention by an inhibitor of the L-arginine-nitric oxide pathway of the antiarrhythmic effects of bradykinin in anesthetized dogs. Br. J. Pharmacol. 110, 18–19.

Vegh, A., Papp, J.G. and Parratt, J. (1994). Attenuation of the antiarrhythmic effects of ischemic preconditioning by blockade of bradykinin B_2 receptors. Br. J. Pharmacol. 113, 1167–1172.

Vuorinen, P., Laustiola, K. and Metsä-Ketelä, T. (1984). The effects of cyclic AMP and cyclic GMP on redox state and energy state in hypoxic rat atria. Life Sci. 35, 155–161.

Wall, T.M., Sheehy, R. and Hartman, J.C. (1994). Role of bradykinin in myocardial preconditioning. J. Pharmacol. Exper. Ther. 270, 681–689.

Wang, J., Xiong, W., Yang, Z., Davis, T., Dewey, M.J., Chao, J. and Chao, L. (1994). Human tissue kallikrein induces hypotension in transgenic mice. Hypertension 23, 236–243.

Wang, C., Chao, L. and Chao, J. (1995). Direct gene delivery of human tissue kallikrein reduces blood pressure in spontaneously hypertensive rats. J. Clin. Invest. 95, 1710–1716.

Werrmann, J.G. and Cohen, S.M. (1994). Comparison of effects of angiotensin-converting enzyme inhibition with those of angiotensin II receptor antagonism on functional and metabolic recovery in postischemic working rat heart as studied by ^{31}P nuclear magnetic resonance. J. Cardiovasc. Pharmacol. 24, 573–586.

Wiemer, G., Schölkens, B.A. and Linz, W. (1994). Endothelial protection by converting enzyme inhibitors. Cardiovasc. Res. 28, 166–172.

Wilkens, H., Back, N., Steger, R. and Karn, J. (1970). The influence of blood pH on peripheral vascular tone: possible role of proteases and vaso-active polypeptides. In "Shock. Biochemical, Pharmacological and Clinical Aspects" (eds. A. Bertelli and N. Back), pp. 201–214. Plenum Press, New York.

Wirth, K., Hock, F.J., Albus, U., Linz, W., Alpermann, H.G., Anagnostopoulos, H., St. Henke, G., Breipohl, G., König, W., Knolle, J. and Schölkens, B.A. (1991). HOE 140 a new potent and long acting bradykinin-antagonist: in vivo studies. Br. J. Pharmacol. 102, 774–777.

Zeitlin, I.J., Fagbemi, S.O. and Parratt, J.R. (1989). Enzymes in normally perfused and ischemic dog hearts which release a substance with kinin like activity. Cardiovasc. Res. 23, 91–97.

Zhu, P., Zaugg, C.E., Simper, D., Hornstein, P., Allegrini, P.R. and Buser, P.T. (1995). Bradykinin improves postischemic recovery in the rat heart: role of high energy phosphates, nitric oxide, and prostacyclin. Cardiovasc. Res. 29, 658–663.

Glossary

Notes: This glossary is up to date for the current volume only and will be supplemented with each subsequent volume.

α_1, α_2 **receptors** Adrenoceptor subtypes
α_1-**ACT** α_1-Antichymotrypsin
α_1-**AP** α_1-antiproteinase *also known as* α_1-antitrypsin and α_1-proteinase inhibitor
α_1-**AT** α_1-Antitrypsin inhibitor *also known as* α_1-antiproteinase and α_1-proteinase inhibitor
α_1-**PI** α_1-Proteinase inhibitor *also known as* α_1-antitrypsin and α_1-antiproteinase
α_2-**AP** α_2-antiplasmin
α_2-**M** α_2-macroblobulin
A Absorbance
AI, AII Angiotensin I, II
Å Angstrom
AA Arachidonic acid
aa Amino acids
AAb Autoantibody
ABAP 2′,2′-azobis-2-amidino propane
Ab Antibody
Ab1 Idiotype antibody
Ab2 Anti-idiotype antibody
Ab2α Anti-idiotype antibody which binds outside the antigen binding region
Ab2β Anti-idiotype antibody which binds to the antigen binding region
Ab3 Anti-anti-idiotype antibody
Abcc Antibody dependent cellular cytotoxicity
ABA-L-GAT Arsanilic acid conjugated with the synthetic polypeptide L-GAT
AC Adenylate cyclase
ACAT Acyl-co-enzyme-A acyltransferase
ACAID Anterior chamber-associated immune deviation
ACE Angiotensin-converting enzyme
ACh Acetylcholine
ACTH Adrenocorticotrophin hormone
ADCC Antibody-dependent cellular cytotoxicity
ADH Alcohol dehydrogenase
Ado Adenosine

ADP Adenosine diphosphate
ADPRT Adenosine diphosphate ribosyl transferase
AES Anti-eosinophil serum
Ag Antigen
AGE Advanced Glycosylation end-product
AGEPC 1-*O*-alkyl-2-acetyl-*sn*-glyceryl-3-phosphocholine; *also known as* PAF and APRL
AH Acetylhydrolase
AHP after-hyperpolarization
AID Autoimmune disease
AIDS Acquired immune deficiency syndrome
A/J A Jackson inbred mouse strain
ALI Acute lung injury
ALP Anti-leukoprotease
ALS Amyotrophic lateral sclerosis
cAMP Cyclic adenosine monophosphate *also known as* adenosine 3′,5′-phosphate
AM Alveolar macrophage
AML Acute myelogenous leukaemia
AMP Adenosine monophosphate
AMVN 2,2′-azobis (2,4-dimethylvaleronitrile)
ANAb Anti-nuclear antibodies
ANCA Anti-neutrophil cytoplasmic auto antibodies
cANCA Cytoplasmic ANCA
pANCA Perinuclear ANCA
AND Anaphylactic degranulation
ANF Atrial natriuretic factor
ANP Atrial natriuretic peptide
Anti-I-A, Anti-I-E Antibody against class II MHC molecule encoded by I-A locus, I-E locus
anti-Ig Antibody against an immunoglobulin
anti-RTE Anti-tubular epithelium
AP-1 Activator protein-1
APA B-azaprostanoic acid
APAS Antiplatelet antiserum
APC Antigen-presenting cell
APD Action potential duration
apo-B Apolipoprotein B
APTT Activated partial thromboplastin times
APRL Anti-hypertensive polar renal

lipid *also known as* PAF
APUD Amine precursor uptake and decarboxylation
AR Aldose reductase
AR-CGD Autosomal recessive form of chronic granulomatous disease
ARDS Adult respiratory distress syndrome
AS Ankylosing spondylitis
ASA Acetylsalicylic acid *also known as* aspirin
4-ASA, 5-ASA 4-, 5-aminosalicylic acid
ATIII antithrombin III
ATHERO-ELAM A monocyte adhesion molecule
ATL Adult T cell leukaemia
ATP Adenosine triphosphate
ATPase Adenosine triphosphatase
ATPγs Adenosine 3′ thiotriphosphate
AITP Autoimmune thrombocytopenic purpura
AUC Area under curve
AVP Arginine vasopressin

β_1, β_2 **receptors** Adrenoceptor subtypes
β_2 **(CD18)** A leucocyte integrin
β_2**M** β_2-Microglobulin
β-**TG** β-Thromboglobulin
B$_7$/BB$_1$ *Known to be* expressed on B cell blasts and immunostimulatory dendritic cells
BAF Basophil-activating factor
BAL Bronchoalveolar lavage
BALF Bronchoalveolar lavage fluid
BALT Bronchus-associated lymphoid tissue
B cell Bone marrow-derived lymphocyte
BCF Basophil chemotactic factor
B-CFC Basophil colony-forming cell
BCG Bacillus Calmette-Guérin
BCNU 1,3-bis (2-chloroethyl)-1-nitrosourea
bFGF Basic fibroblast growth factor
Bg Birbeck granules
BHR Bronchial hyperresponsiveness
BHT Butylated hydroxytoluene
b.i.d. *Bis in die* (twice a day)

BK Bradykinin

BK₁, BK₂ receptors Bradykinin receptor subtypes *also known as* B₁ and B₂ receptors

B1-CFC Blast colony-forming cells

B-lymphocyte Bursa-derived lymphocyte

BM Bone marrow

BMCMC Bone marrow cultured mast cell

BMMC Bone marrow mast cell

BOC-FMLP Butoxycarbonyl-FMLP

BAEC Bovine aortic endothelial cells

bp Base pair

BPAEC Bovine pulmonary artery endothelial cells

BPB Para-bromophenacyl bromide

BPI Bacterial permeability-increasing protein

BSA Bovine serum albumin

BSS Bernard-Soulier Syndrome

^{51}**Cr** Chromium^{51}

C1, C2 . . . C9 The 9 main components of complement

C1 inhibitor A serine protease inhibitor which inactivates C1r/C1s

C1q Complement fragment 1q

C1qR Receptor for C1w; facilitates attachment of immune complexes to mononuclear leucocytes and endothelium

C3a Complement fragment 3a (anaphylatoxin)

C3a₇₂₋₇₇ A synthetic carboxyterminal peptide C3a analogue

C3aR Receptor for anaphylatoxins, C3a, C4a, C5a

C3b Complement fragment 3b (anaphylatoxin)

C3bi Inactivated form of C3b fragment of complement

C4b Complement fragment 4b (anaphylatoxin)

C4BP C4 binding protein; plasma protein which acts as co-factor to factor I inactivate C3 convertase

C5a Complement fragment 5a (anaphylatoxin)

C5aR Receptor for anaphylatoxins C3a, C4a and C5a

C5b Complement fragment 5b (anaphylatoxin)

C$_ε$2, C$_ε$3, C$_ε$4 Heavy chain of immunoglobulin E: domains 2, 3 and 4

Ca *The chemical symbol for* calcium

[CA²⁺]ᵢ Intracellular free calcium concentration

CAH Chronic active hepatitis

CALLA Common lymphoblastic leukaemia antigen

CALT Conjunctival associated lymphoid tissue

CaM Calmodulin

CAM Cell adhesion molecule

cAMP Cyclic adenosine monophosphate *also known as* adenosine 3′,5′-phosphate

CaM-PDE Ca²⁺/CaM-dependent PDE

CAP57 Cationic protein from neutrophils

CAT Catalase

CatG Cathepsin G

CB Cytochalasin B

CBH Cutaneous basophil hypersensitivity

CBP Cromolyn-binding protein

CCK Cholecystokinin

CCR Creatinine clearance rate

CD Cluster of differentiation (a system of nomenclature for surface molecules on cells of the immune system); cluster determinant

CD1 Cluster of differentiation 1 *also known as* MHC class I-like surface glycoprotein

CD1a Isoform a *also known as* non-classical MHC class I-like surface antigen; present on thymocytes and dendritic cells

CD1b *Known to be* present on thymocytes and dendritic cells

CD1c Isoform c *also known as* non-classical MHC class I-like surface antigen; present on thymocytes

CD2 Defines T cells involved in antigen non-specific cell activation

CD3 *Also known as* T cell receptor-associated surface glycoprotein on T cells

CD4 Defines MHC class II-restricted T cells subsets

CD5 *Known to be* present on T cells and a subset of B cells; *also known as* Lyt 1 in mouse

CD7 Cluster of differentiation 7; present on most T cells and NK cells

CD8 Defines MHC class I-restricted T cell subset; present on NK cells

CD10 *Known to be* common acute leukaemia antigen

CD11a *Known to be* an α chain of LFA-1 (leucocyte function antigen-1) present on several types of leucocyte and which mediates adhesion

CD11c *Known to be* a complement receptor 4 α chain

CD13 Aminopeptidase N; present on myeloid cells

CD14 *Known to be* a lipid-anchored glycoprotein; present on monocytes

CD15 *Known to be* Lewis X, fucosyl-*N*-acetyllactosamine

CD16 *Known to be* Fcγ receptor III

CD16-1, CD16-2 Isoforms of CD16

CD19 Recognizes B cells and follicular dendritic cells

CD20 *Known to be* a pan B cell

CD21 C3d receptor

CD23 Low affinity FcεR

CD25 Low affinity receptor for interleukin-2

CD27 Present on T cells and plasma cells

CD28 Present on resting and activated T cells and plasma cells

CD30 Present on activated B and T cells

CD31 *Known to be* on platelets, monocytes, macrophages, granulocytes, B-cells and endothelial cells; *also known as* PECAM

CD32 Fcγ receptor II

CD33⁺ *Known to be* a monocyte and stem cell marker

CD34 *Known to be* a stem cell marker

CD35 C3b receptor

CD36 *Known to be* a macrophage thrombospondin receptor

CD40 Present on B cells and follicular dendritic cells

CD41 *Known to be* a platelet glycoprotein

CD44 *Known to be* a leucocyte adhesion molecule; *also known as* hyaluronic acid cell adhesion molecule (H-CAM), Hermes antigen, extracellular matrix receptor III (ECMIII); present on polymorphonuclear leucocytes

CD45 *Known to be* a pan leucocyte marker

CD45RO *Known to be* the isoform of leukosialin present on memory T cells

CD46 *Known to be* a membrane cofactor protein

CD49 Cluster of differentiation 49

CD51 *Known to be* vitronectin receptor alpha chain

CD54 *Known to be* Intercellular adhesion molecule-1 *also known as* ICAM-1

CD57 Present on T cells and NK subsets

CD58 A leucocyte function-associated antigen-3, *also known to be* a member of the β-2 integrin family of cell adhesion molecules

CD59 *Known to be* a low molecular weight HRf present to many haematopoietic and non-haematopoietic cells

CD62 *Known to be present on* activated platelets and endothelial cells; *also known as* P-selectin

CD64 *Known to be* Fcγ receptor I

CD65 *Known to be* fucoganglioside

CD68 Present on macrophages

CD69 *Known to be* an activation inducer molecule; present on activated lymphocytes

CD72 Present on B-lineage cells

CD74 An invariant chain of class II B cells
CDC Complement-dependent cytotoxicity
cDNA Complementary DNA
CDP Choline diphosphate
CDR Complementary-determining region
CD$_{xx}$ Common determinant xx
CEA Carcinoembryonic antigen
CETAF Corneal epithelial T cell activating factor
CF Cystic fibrosis
Cf Cationized ferritin
CFA Complete Freund's adjuvant
CFC Colony-forming cell
CFU Colony-forming unit
CFU-Mk Megakaryocyte progenitors
CFU-S Colony-forming unit, spleen
CGD Chronic granulomatous disease
CHAPS 3-[(3-cholamidopropyl)-dimethylammonio]-1-propane sulphonate
cGMP Cyclic guanosine monophosphate *also known as* guanosine 3′,5′-phosphate
CGRP Calcitonin gene-related peptide
CH2 Hinge region of human immunoglobulin
CHO Chinese hamster ovary
CI Chemical ionization
CIBD Chronic inflammatory bowel disease
CK Creatine phosphokinase
CKMB The myocardial-specific isoenzyme of creatine phosphokinase
Cl *The chemical symbol for* chlorine
CL Chemiluminescent
CLA Cutaneous lymphocyte antigen
CL18/6 Anti-ICAM-1 monoclonal antibody
CLC Charcot–Leyden crystal
CMC Critical micellar concentration
CMI Cell mediated immunity
CML Chronic myeloid leukaemia
CMV Cytomegalovirus
CNS Central nervous system
CO Cyclooxygenase
CoA Coenzyme A
CoA-IT Coenzyme A – independent transacylase
Con A Concanavalin A
COPD Chronic obstructive pulmonary disease
COS Fibroblast-like kidney cell line established from simian cells
CoVF Cobra venom
CP Creatine phosphate
Cp Caeruloplasmin
c.p.m. Counts per minute
CPJ Cartilage/pannus junction
Cr *The chemical symbol for* chromium
CR Complement receptor

CR1, CR2 & CR4 Complement receptor types 1, 2 and 4
CR3-α Complement receptor type 3-α
CRF Corticotrophin-releasing factor
CRH Corticotrophin-releasing hormone
CRI Cross-reactive idiotype
CRP C-reactive protein
CSA Cyclosporin A
CSF Colony-stimulating factor
CSS Churg–Strauss syndrome
CT Computed tomography
CTAP-III Connective tissue-activating peptide
CTD Connective tissue diseases
C terminus Carboxy terminus of peptide
CThp Cytotoxic T lymphocyte precursors
CTL Cytotoxic T lymphocyte
CTLA-4 *Known to be* co-expressed with CD20 on activated T cells
CTMC Connective tissue mast cell
CVF Cobra venom factor

2D Second derivative
Da Dalton (the unit of relative molecular mass)
DAF Decay-accelerating factor
DAG Diacylglycerol
DAO Diamine oxidase
D-Arg D-Arginine
DArg-[Hyp^3,DPhe^7]-BK A bradykinin B$_2$ receptor antagonist. Peptide derivative of bradykinin
DArg-[Hyp^3,Thi^5,DTic^7,Tic^8]-BK A bradykinin B$_2$ receptor antagonist. Peptide derivative of bradykinin
DBNBS 3,5-dibromo-4-nitroso-benzenesulphonate
DC Dendritic cell
DCF Oxidized DCFH
DCFH 2′,7′-dichlorofluorescin
DEC Diethylcarbamazine
DEM Diethylmaleate
desArg^9-BK Carboxypeptidase N product of bradykinin
desArg^{10}KD Carboxypeptidase N product of kallidin
DETAPAC Diethylenetriaminepentaacetic acid
DFMO α1-Difluoromethyl ornithine
DFP Diisopropyl fluorophosphate
DFX Desferrioxamine
DGLA Dihomo-γ-linolenic acid
DH Delayed hypersensitivity
DHA Docosahexaenoic acid
DHBA Dihydroxybenzoic acid
DHR Delayed hypersensitivity reaction
DIC Disseminated intravascular coagulation
DL-CFU Dendritic cell/Langerhans cell colony forming

DLE Discoid lupus erythematosus
DMARD Disease-modifying anti-rheumatic drug
DMF N,N-dimethylformamide
DMPO 5,5-dimethyl-l-pyrroline N-oxide
DMSO Dimethyl sulfoxide
DNA Deoxyribonucleic acid
D-NAME D-Nitroarginine methyl erster
DNase Deoxyribonuclease
DNCB Dinitrochlorobenzene
DNP Dinitrophenol
Dpt4 *Dermatophagoides pteronyssinus* allergen 4
DGW2, DR3, DR7 HLA phenotypes
DREG-56 (Antigen) L-selectin
DREG-200 A monoclonal antibody against L-selectin
ds Double-stranded
DSCG Disodium cromoglycate
DST Donor-specific transfusion
DTH Delayed-type hypersensitivity
DTPA Diethylenetriamine pentaacetate
DTT Dithiothreitol
dv/dt Rate of change of voltage within time

ε Molar absorption coefficient
EA Egg albumin
EACA Epsilon-amino-caproic acid
EAE Experimental autoimmune encephalomyelitis
EAF Eosinophil-activating factor
EAR Early phase asthmatic reaction
EAT Experimental autoimmune thyroiditis
EBV Epstein-Barr virus
EC Endothelial cell
ECD Electron capture detector
ECE Endothelin-converting enzyme
E-CEF Eosinophil cytotoxicity enhancing factor
ECF-A Eosinophil chemotactic factor of anaphylaxis
ECG Electrocardiogram
ECGF Endothelial cell growth factor
ECGS Endothelial cell growth supplement
E. coli *Escherichia coli*
ECP Eosinophil cationic protein
EC-SOD Extracellular superoxide dismutase
EC-SOD C Extracellular superoxide dismutase C
ED$_{35}$ Effective dose producing 35% maximum response
ED$_{50}$ Effective dose producing 50% maximum response
EDF Eosinophil differentiation factor
EDL Extensor digitorum longus
EDN Eosinophil-derived neurotoxin
EDRF Endothelium-derived relaxing factor

EDTA Ethylenediamine tetraacetic acid *also known as* etidronic acid

EE Eosinophilic eosinophils

EEG Electroencephalogram

EET Epoxyeicosatrienoic acid

EFA Essential fatty acid

EFS Electrical field stimulation

EG1 Monoclonal antibody specific for the cleaved form of eosinophil cationic peptide

EGF Epidermal growth factor

EGTA Ethylene glycol-bis(β-aminoethyl ether) N,N,N',N'-tetraacetic acid

EHNA Erythro-9-(2-hydroxy-3-nonyl)-adenine

EI Electron impact

EIB Exercise-induced bronchoconstriction

eIF-2 Subunit of protein synthesis initiation factor

ELAM-1 Endothelial leucocyte adhesion molecule-1

ELF Respiratory epithelium lung fluid

ELISA Enzyme-linked immunoabsorbent assay

EMS Eosinophilia-myalgia syndrome

ENS Enteric nervous system

EO Eosinophil

EO-CFC Eosinophil colony-forming cell

EOR Early onset reaction *also known as* EAR

EPA Eicosapentaenoic acid

EpDIF Epithelial-derived inhibitory factor *also known as* epithelium-derived relaxant factor

EPO Eosinophil peroxidase

EPOR Erythropoietin receptor

EPR Effector cell protease

EPX Eosinophil protein X

ER Endoplasmic reticulum

ERCP Endoscopic retrograde cholangiopancreatography

E-selectin Endothelial selectin *formerly known as* endothelial leucocyte adhesion molecule-1 (ELAM-1)

ESP Eosinophil stimulation promoter

ESR Erythrocyte sedimentation rate

e.s.r. Electron spin resonance

ET, ET-1 Endothelin, -1

ETYA Eicosatetraynoic acid

FA Fatty acid

FAB Fast-electron bombardment

Fab Antigen binding fragment

F(ab')2 Fragment of an immunoglobulin produced pepsin treatment

FACS Fluorescence activated cell sorter

factor B Serine protease in the C3 converting enzyme of the alternative pathway

factor D Serine protease which cleaves factor B

factor H Plasma protein which acts as a co-factor to factor I

factor I Hydrolyses C3 converting enzymes with the help of factor H

FAD Flavine adenine dinucleotide

FapyAde 5-formamido-4,6-diamino-pyrimidine

FapyGua 2,6-diamino-4-hydroxy-5-formamidopyrimidine

FBR Fluorescence photobleaching recovery

Fc Crystallizable fraction of immunoglobulin molecule

Fcγ Receptor for Fc portion of IgG

FcγRI Ig Fc receptor I *also known as* CD64

FcγRII Ig Fc receptor II *also known as* CD32

FcγRIII Ig Fc receptor III *also known as* CD16

Fc$_\varepsilon$RI High affinity receptor for IgE

Fc$_\varepsilon$RII Low affinity receptor for IgE

FcR Receptor for Fc region of antibody

FCS Foetal calf (bovine) serum

FEV$_1$ Forced expiratory volume in 1 second

Fe-TPAA Fe(III)-tris[N-(2-pyridylmethyl)-2-aminoethyl]amine

Fe-TPEN Fe (II)-tetrakis-N,N,N',N'-(2-pyridyl methyl-2-aminoethyl)amine

FFA Free fatty acids

FGF Fibroblast growth factor

FID Flame ionization detector

FITC Fluorescein isothiocyanate

FKBP FK506-binding protein

FLAP 5-lipoxygenase-activating protein

FMLP N-Formyl-methionyl-leucyl-phenylalanine

FNLP Formyl-norleucyl-leucyl-phenylalanine

FOC Follicular dendritic cell

FPLC Fast protein liquid chromatography

FPR Formyl peptide receptor

FS cell Folliculo-stellate cell

FSG Focal sequential glomerulosclerosis

FSH Follicle stimulating hormone

FX Ferrioxamine

5-FU 5-fluorouracil

Ga G-protein

G6PD Glucose 6-phosphate dehydrogenase

GABA γ-Aminobutyric acid

GAG Glycosaminoglycan

GALT Gut-associated lymphoid tissue

GAP GTPase-activating protein

GBM Glomerular basement membrane

GC Guanylate cyclase

GC-MS Gas chromatography mass spectroscopy

G-CSF Granulocyte colony-stimulating factor

GDP Guanosine 5'-diphosphate

GEC Glomerular epithelial cell

GEMSA guanidinoethylmercapto-succinic acid

GF-1 An insulin-like growth factor

GFR Glomerular filtration rate

GH Growth hormone

GH-RF Growth hormone-releasing factor

Gi Family of pertussis toxin sensitive G-proteins

GI Gastrointestinal

GIP Granulocyte inhibitory protein

GlyCam-1 Glycosylation-dependent cell adhesion molecule-1

GMC Gastric mast cell

GM-CFC Granulocyte-macrophage colony-forming cell

GM-CSF Granulocyte-macrophage colony-stimulating factor

GMP Guanosine monophosphate (guanosine 5'-phosphate)

Go Family of pertussis toxin sensitive G-proteins

GP Glycoprotein

gp45–70 Membrane co-factor protein

gp90^{MEL} 90 kD glycoprotein recognized by monoclonal antibody MEL-14; *also known as* L-selectin

GPIIb-IIIa Glycoprotein IIb-IIIa *known to be* a platelet membrane antigen

GppCH$_2$P Guanyl-methylene diphosphanate *also known as* a stable GTP analogue

GppNHp Guanylyl-imidiodiphosphate *also known as* a stable GTP analogue

GRGDSP Glycine–arginine–glycine–aspartic acid–serine–proline

Gro Growth-related oncogene

GRP Gastrin-related peptide

Gs Stimulatory G protein

GSH Glutathione (reduced)

GSHPx Glutathione perioxidase

GSSG Glutathione (oxidized)

GT Glanzmann Thrombasthenia

GTP Guanosine triphosphate

GTP-γ-S Guanosine5'O-(3-thiotriphosphate)

GTPase Guanosine triphosphatase

GVHD Graft-versus-host-disease

GVHR Graft-versus-host-reaction

H Histamine

H$_1$, H$_2$, H$_3$ Histamine receptor types 1, 2 and 3

H$_2$O$_2$ *The chemical symbol for* hydrogen peroxide

HAE Hereditary angiodema
Hag Haemagglutinin
Hag-1, Hag-2 Cleaved haemagglutinin subunits-1, -2
H & E Haematoxylin and eosin
hIL Human interleukin
Hb Haemoglobin
HBBS Hank's balanced salt solution
HCA Hypertonic citrate
H-CAM Hyaluronic acid cell adhesion molecule
HDC Histidine decarboxylase
HDL High-density lipoprotein
HEL Hen egg white lysozyme
HEPE Hydroxyeicosapentanoic acid
HEPES N-2-Hydroxylethylpiperazine-N'-2-ethane sulphonic acid
HES Hypereosinophilic syndrome
HETE 5,8,9,11,12 and 15 Hydroxyeicosatetraenoic acid
5(S)HETE A stereo isomer of 5-HETE
HETrE Hydroxyeicosatrienoic acid
HEV High endothelial venule
HF Hageman factor
HFN Human fibronectin
HGF Hepatocyte growth factor
HHTrE 12(S)-Hydroxy-5,8,10-heptadecatrienoic acid
HIV Human immunodeficiency virus
HL60 Human promyelocytic leukaemia cell line
HLA Human leucocyte antigen
HLA-DR2 Human histocompatability antigen class II
HMG CoA Hydroxylmethylglutaryl coenzyme A
HMT Histidine methyltransferase
HMVEC Human microvascular endothelial cell
HMW High molecular weight
HMWK Higher molecular weight kininogen
HNC Human neutrophil collagenase (MMP-8)
HNE Human neutrophil elastase
HNG Human neutrophil gelatinase (MMP-9)
HODE Hydroxyoctadecanoic acid
HO· Hydroxyl radical
HO$_2$· Perhydroxyl radical
HPETE, 5-HPETE & 15-HPETE 5 and 15 Hydroperoxyeicosatetraenoic acid
HPETrE Hydroperoxytrienoic acid
HPODE Hydroperoxyoctadecanoic acid
HPLC High-performance liquid chromatography
HPS Hantavirus pulmonary syndrome
HRA Histamine-releasing activity
HRAN Neutrophil-derived histamine-releasing activity
HRf Homologous-restriction factor

HRF Histamine-releasing factor
HRP Horseradish peroxidase
HSA Human serum albumin
HSP Heat-shock protein
HS-PG Heparan sulphate proteoglycan
HSV, HSV-1 Herpes simplex virus, -1
^{3}HTdR Tritiated thymidine
5-HT 5-Hydroxytryptamine *also known as* Serotonin
HTLV-1 Human T-cell leukaemia virus-1
HUVEC Human umbilical vein endothelial cell
[Hyp^3]-BK Hydroxyproline derivative of bradykinin
[Hyp^4]-KD Hydroxyproline derivative of kallidin

^{111}In Indium^{111}
i.a. intra-arterial
Ia immune reaction-associated antigen
Ia+ Murine class II major histocompatibility complex antigen
IB4 Anti-CD18 monoclonal antibody
IBD Inflammatory bowel disease
IBMX 3-isobutyl-1-methylxanthine
IBS Inflammatory bowel syndrome
iC3 Inactivated C3
iC4 Inactivated C4
IC$_{50}$ Concentration producing 50% inhibition
I$_{Ca}$ Calcium current
ICAM Intercellular adhesion molecules
ICAM-1, ICAM-2, ICAM-3 Intercellular adhesion molecules-1, -2, -3
cICAM-1 Circulating form of ICAM-1
ICE IL-1β-converting enzyme
i.d. Intradermal
ID$_{50}$ Dose of drug required to inhibit response by 50%
IDC Interdigitating cell
IDD Insulin-dependent (type 1) diabetes
IEL Intraepithelial leucocyte
IELym Intraepithelial lymphocytes
IFA Incomplete Freund's adjuvant
IFN Interferon
IFNα, IFNβ, IFNγ Interferons α, β, γ
Ig Immunoglobulin
IgA, IgE, IgG, IgM Immunoglobulins A, E, G, M
IgG1 Immunoglobulin G class 1
IgG$_{2a}$ Immunoglobulin G class 2a
IGF-1 Insulin-like growth factor
Ig-SF Immunoglobulin supergene family
IGSS Immuno-gold silver stain
IHC Immunohistochemistry

IHES Idiopathic hypereosinophilic syndrome
IκB NFκB inhibitor protein
IL Interleukin
IL-1, IL-2 . . . IL-8 Interleukins-1, 2 . . . -8
IL-1α, IL-1β Interleukin-1α, -1β
ILR Interleukin receptor
IL-1R, IL-2R; IL-3R-IL-6R Interleukin 1–6 receptors
IL-1Ra Interleukin-1 receptor antagonist
IL-2Rβ Interleukin-2 receptor β
IMF Integrin modulating factor
IMMC Intestinal mucosal mast cell
iNOS inducible NOS
i.p. Intraperitoneally
IP$_1$ Inositol monophosphate
IP$_2$ Inositol biphosphate
IP$_3$ Inositol 1,4,5-trisphosphate
IP$_4$ Inositol tetrakisphosphate
IPF Idiopathic pulmonary fibrosis
IPO Intestinal peroxidase
IpOCOCq Isopropylidene OCOCq
I/R Ischaemia-reperfusion
IRAP IL-1 receptor antagonist protein
IRF-1 Interferon regulatory factor 1
I$_{sc}$ Short-circuit current
ISCOM Immune-stimulating complexes
ISGF3 Interferon-stimulated gene Factor 3
ISGF3α, ISGFγ α, γ subunits of ISGF3
IT Immunotherapy
ITP Idiopathic thrombocytopenic purpura
i.v. Intravenous

K *The chemical symbol for* potassium
K$_a$ Association constant
kb Kilobase
20KDHRF A homologous restriction factor; binds to C8
65KDHRF A homologous restriction factor, also known as C8 binding protein; interferes with cell membrane pore-formation by C5b-C8 complex
Kcat Catalytic constant; a measure of the catalytic potential of an enzyme
K$_d$ dissociation constant
kD Kilodalton
KD Kallidin
K$_i$ Antagonist binding affinity
Ki67 Nuclear membrane antigen
KLH Keyhole limpet haemocyanin
K$_m$ Michaelis constant
KOS KOS strain of herpes simplex virus

λ$_{max}$ Wavelength of maximum absorbance
LAD Leucocyte adhesion deficiency
LAK Lymphocyte-activate killer (cell)

LAM, LAM-1 Leucocyte adhesion molecule, -1
LAR Late-phase asthmatic reaction
L-Arg L-Arginine
LBP LPS binding protein
LC Langerhans cell
LCF Lymphocyte chemoattractant factor
LCR Locus control region
LDH Lactate dehydrogenase
LDL Low-density lipoprotein
LDV Laser Doppler velocimetry
Le^x (Lewis X) Leucocyte ligand for selectin
LFA Leucocyte function-associated antigen
LFA-1 Leucocyte function-associated antigen-1; *also known to be* a member of the β-2 integrin family of cell adhesion molecules
LG β-Lactoglobulin
LGL Large granular lymphocyte
LH Luteinizing hormone
LHRH Luteinizing hormone-releasing hormone
LI Labelling index
LIF Leukaemia inhibitory factor
LIS Lateral intercellular spaces
LMP Low molecular mass polypeptide
LMW Low molecular weight
LMWK Low molecular weight kininogen
L-NOARG L-Nitroarginine
LO Lipoxygenase
5-LO, 12-LO, 15-LO 5-, 12-, 15-Lipoxygenases
LP(a) Lipoprotein (a)
LPS Lipopolysaccharide
L-selectin Leucocyte selectin, *formerly known as* monoclonal antibody that recognizes murine L-selectin (MEL-14 antigen), leucocyte cell adhesion molecule-1 (LeuCAM-1), lectin cell adhesion molecule-1 (LeCAM-1 or LecCAM-1), leucocyte adhesion molecule-1 (LAM-1)
LT Leukotriene
LTA$_4$, LTB$_4$, LTC$_4$, LTD$_4$, LTE$_4$ Leukotrienes A$_4$, B$_4$, C$_4$, D$_4$, and E$_4$
L$_y$-1$^+$ (Cell line)
LX Lipoxin
LXA$_4$, LXB$_4$, LXC$_4$, LXD$_4$, LXE$_4$ Lipoxins A$_4$, B$_4$, C$_4$, D$_4$ and E$_4$
Lys-BK Kallidin

M Monocyte
M3 Receptor Muscarinic receptor subtype 3
M-540 Merocyanine-540
mAb Monoclonal antibody
mAb IB4, mAb PB1.3, mAb R 3.1, mAb R 3.3, mAb 6.5, mAb 60.3 Monoclonal antibodies IB4, PB1.3, R 3.1, R 3.3, 6.5, 60.3

MABP Mean arterial blood pressure
MAC Membrane attack molecule
Mac Macrophage (also abbreviated to MΦ)
Mac- Macrophage-1 antigen; a member of the β-2 integrin family of cell adhesion molecules (also abbreviated to MΦ1), *also known as* monocyte antigen-1 (M-1), complement receptor-3 (CR3), CD11b/CD18
MAF Macrophage-activating factor
MAO Monoamine oxidase
MAP Monophasic action potential
MAPTAM An intracellular Ca^{2+} chelator
MARCKS Myristolated, alanine-rich C kinase substrate; specific protein kinase C substrate
MBP Major basic protein
MBSA Methylated bovine serum albumin
MC Mesangial cells
MCAO Middle cerebral artery occlusion
M cell Microfold or membranous cell of Peyer's patch epithelium
MCP Membrane co-factor protein
MCP-1 Monocyte chemotactic protein-1
M-CSF Monocyte/macrophage colony-stimulating factor
MC$_T$ Tryptase-containing mast cell
MC$_{TC}$ Tryptase- and chymase-containing mast cell
MDA Malondialdehyde
MDCK Madin Darby Canine kidney
MDGF Macrophage-derived growth factor
MDP Muramyl dipeptide
MEA Mast cell growth-enhancing activity
MEL Metabolic equivalent level
MEM Minimal essential medium
MG Myasthenia gravis
MGSA Melanoma-growth-stimulatory activity
MGTA DL2-mercaptomethyl-3-guanidinoethylthio-propanoic acid
MHC Major histocompatibility complex
MI Myocardial ischaemia
MIF Migration inhibition factor
mIL Mouse interleukin
MIP-1α Macrophage inflammatory protein 1α
MI/R Myocardial ischaemia/reperfusion
MIRL Membrane inhibitor of reactive lysis
mix-CFC Colony-forming cell mix
Mk Megakaryocyte
MLC Mixed lymphocyte culture
MLymR Mixed lymphocyte reaction
MLR Mixed leucocyte reaction

mmLDL Minimally modified low-density lipoprotein
MMC Mucosal mast cell
MMCP Mouse mast cell protease
MMP, MMP1 Matrix metalloproteinase, -1
MNA 6-Methoxy-2-napthylacetic acid
MNC Mononuclear cells
MΦ Macrophage (also abbreviated to Mac)
MPG *N*-(2-mercaptopropionyl)-glycine
MPO Myeloperoxidase
MPSS Methyl prednisolone
MPTP *N*-methyl-4-phenyl-1,2,3,6-tetrahydropyridine
MRI Magnetic resonance imaging
mRNA Messenger ribonucleic acid
MS Mass spectrometry
MSAP Mean systemic arterial pressure
MSS Methylprednisolone sodium succinate
MT Malignant tumour
MW Molecular weight

Na *The chemical symbol for* sodium
NA Noradrenaline *also known as* norepinephrine
NAAb Natural autoantibody
NAb Natural antibody
NAC *N*-acetylcysteine
NADH Reduced nicotinamide adenine dinucleotide
NADP Nicotinamide adenine diphosphate
NADPH Reduced nicotinamide adenine dinucleotide phosphate
NAF Neutrophil activating factor
L-NAME L-Nitroarginine methyl ester
NANC Non-adrenergic, non-cholinergic
NAP Neutrophil-activating peptide
NAPQI *N*-acetyl-*p*-benzoquinone imine
NAP-1, NAP-2 Neutrophil-activating peptides -1 and -2
NBT Nitro-blue tetrazolium
NC1 Non-collagen 1
N-CAM Neural cell adhesion molecule
NCEH Neutral cholesteryl erster hydrolase
NCF Neutrophil chemotactic factor
NDGA Nordihydroguaretic acid
NDP Nucleoside diphosphate
Neca 5′-(*N*-ethyl carboxamido)-adenosine
NED Nedocromil sodium
NEP Neutral endopeptidase (EC 3.4.24.11)
NF-AT Nuclear factor of activated T lymphocytes
NF-κB Nuclear factor-κB

NgCAM Neural-glial cell adhesion molecule
NGF Nerve growth factor
NGPS Normal guinea-pig serum
NIH 3T3 (fibroblasts) National Institute of Health 3T3-Swiss albino mouse fibroblast
NIMA Non-inherited maternal antigens
NIRS Near infrared spectroscopy
Nk Neurokinin
NK Natural killer
Nk-1, Nk-2, Nk-3 Neurokinin receptor subtypes 1, 2 and 3
NkA Neurokinin A
NkB Neurokinin B
NLS Nuclear location sequence
NMDA N-methyl-D-aspartic acid
L-NMMA L-Nitromonomethyl arginine
NMR Nuclear magnetic resonance
NNA Nω-nitro-L-arginine
1,N^2-NET β-(2-Naphthyl-1,N^2-etheno
1,N^2-PET β-Phenyl-1,N^2-etheno
NO *The chemical symbol for* nitric oxide
NOD Non-obese diabetic
NOS Nitric oxide synthase
c-NOS Ca^{2+}-dependent constitutive form of NOS
i-NOS Inducible form of NOS
NPK Neuropeptide K
NPY Neuropeptide Y
NRS Normal rabbit serum
NSAID Non-steroidal anti-inflammatory drug
NSE Nerve-specific enolase
NT Neurotensin
N terminus Amino terminus of peptide

$^1\Delta O_2$ Singlet Oxygen (Delta form)
$^1\Sigma O_2$ Singlet Oxygen (Sigma form)
O_2^- *The chemical symbol for* the superoxide anion radical
OA Osteoarthritis
OAG Oleoyl acetyl glycerol
OD Optical density
ODC Ornithine decarboxylase
ODFR Oxygen-derived free radical
ODS Octadecylsilyl
OH$^-$ *The chemical symbol for* hydroxyl ion
·OH *The chemical symbol for* hydroxyl radical
8-OH-Ade 8-hydroxyadenine
6-OHDA 6-hydroxyguanine
8-OH-dG 8-hydroxydeoxyguanosine *also known as* 7,8-dihydro-8-oxo-2′-deoxyguanosine
8-OH-Gua 8-hydroxyguanine
OHNE Hydroxynonenal
4-OHNE 4-hydroxynonenal
OT Oxytocin

OVA Ovalbumin
ox-LDL Oxidized low-density lipoprotein
OZ Opsonized zymosan

Ψa Apical membrane potential
P Probability
P Phosphate
P_aO_2 Arterial oxygen pressure
P_i Inorganic phosphate
p150,95 A member of the β-2-integrin family of cell adhesion molecules; *also known as* CD11c
PA Phosphatidic acid
pA_2 Negative logarithm of the antagonist dissociation constant
PAEC Pulmonary artery endothelial cells
PAF Platelet-activating factor *also known as* APRL and AGEPC
PAGE Polyacrylamide gel electrophoresis
PAI Plasminogen activator inhibitor
PA-IgG Platelet associated immunoglobulin G
PAM Pulmonary alveolar macrophages
PAS Periodic acid-Schiff reagent
PBA Polyclonal B cell activators
PBC Primary biliary cirrhosis
PBL Peripheral blood lymphocytes
PBMC Peripheral blood mononuclear cells
PBN N-*tert*-butyl-α-phenylnitrone
PBS Phosphate-buffered saline
PC Phosphatidylcholine
PCA Passive cutaneous anaphylaxis
pCDM8 Eukaryotic expression vector
PCNA Proliferating cell nuclear antigen
PCR Polymerase chain reaction
PCT Porphyria cutanea tarda
p.d. Potential difference
PDBu 4α-phorbol 12,13-dibutyrate
PDE Phosphodiesterase
PDGF Platelet-derived growth factor
PDGFR Platelet-derived growth factor receptor
PE Phosphatidylethanolamine
PECAM-1 Platelet endothelial cell adhesion molecule-1; *also known as* CD31
PEG Polyethylene glycol
PET Positron emission tomography
PEt Phosphatidylethanolamine
PF_4 Platelet factor 4
PG Prostaglandin
PGAS Polyglandular autoimmune syndrome
PGD_2 Prostaglandin D_2
PGE_1, PGE_2, PGF_2, $PGF_{2\alpha}$, PGG_2, PGH_2 Prostaglandins E_1, E_2, F_2, $F_{2\alpha}$, G_2, H_2
PGF, PGH Prostaglandins F and H

PGI_2 Prostaglandin I_2 *also known as* prostacyclin
P_aO_2 Arterial oxygen pressure
PGP Protein gene-related peptide
Ph^1 Philadelphia (chromosome)
PHA Phytohaemagglutinin
PHD PHD[8(1-hydroxy-3-oxo-propyl)-9,12-dihydroxy-5,10 heptadecadienic acid]
PHI Peptide histidine isoleucine
PHM Peptide histidine methionine
P_i Inorganic phosphate
pI Isoelectric point
PI Phosphatidylinositol
PI-3,4-P2 Phosphatidylinositol 3, 4-biphosphate
PI-3,4,5-P3 Phosphatidylinositol 3, 4, 5-trisphosphate
PI-3-kinase Phosphatidylinositol-3-kinase
PI-4-kinase Phosphatidylinositol-4-kinase
PI-3-P Phosphatidylinositol-3-phosphate
PI-4-P Phosphatidylinositol-4-phosphate
PI-4,5-P2 Phosphatidylinositol 4,5-biphosphate
PIP Phosphatidylinositol monophosphate
PIP_2 Phosphatidylinositol biphosphate
PIPES piperazine-N, N′-bis(2-ethanesulfonic acid)
PK Protein kinase
PKA, PKC, PKG Protein kinases A, C and G
PL Phospholipase
PLA, PLA_2, PLC, PLD Phospholipases A, A_2, C and D
PLN Peripheral lymph node
PLNHEV Peripheral lymph node HEV
PLP Proteolipid protein
PLT Primed lymphocyte typing
PMA Phorbol myristate acetate
PMC Peritoneal mast cell
PMN Polymorphonuclear neutrophil
PMSF Phenylmethylsulphonyl fluoride
PNAd Peripheral lymph node vascular addressin
PNH Paroxysmal nocturnal hemoglobinuria
PNU Protein nitrogen unit
p.o. *Per os* (by mouth)
POBN α-4-pyridyl-oxide-N-t-butyl nitrone
PPD Purified protein derivative
PPME Polymeric polysaccharide rich in mannose-6-phosphate moieties
PQ Phenylquinone
PRA Percentage reactive activity
PRD, PRDII Positive regulatory domain, -II
PR3 Proteinase-3

PRBC Parasitized red blood cell
proET-1 Proendothelin-1
PRL Prolactin
PRP Platelet-rich plasma
PS Phosphatidylserine
P-selectin Platelet selectin *formerly known as* platelet activation-dependent granule external membrane protein (PADGEM), granule membrane protein of MW 140 kD (GMP-140)
PT Pertussis toxin
PTCA Percutaneous transluminal coronary angioplasty
PTCR Percutaneous transluminal coronary recanalization
Pte-H$_4$ Tetrahydropteridine
PTH Parathyroid hormone
PTT Partial thromboplastin times
PUFA Polyunsaturated fatty acid
PUMP-1 Punctuated metalloproteinase *also known as* matrilysin
PWM Pokeweed mitogen
Pyran Divinylether maleic acid

q.i.d. *Quater in die* (four times a day)
QRS Segment of electrocardiogram

·R Free radical
R15.7 Anti-CD18 monoclonal antibody
RA Rheumatoid athritis
RANTES A member of the IL8 supergene family (*R*egulated on *a*ctivation, *n*ormal *T* *e*xpressed and *s*ecreted)
RAST Radioallergosorbent test
R$_{aw}$ Airways resistance
RBC Red blood cell
RBF Renal blood flow
RBL Rat basophilic leukaemia
RC Respiratory chain
RE RE strain of herpes simplex virus type 1
REA Reactive arthritis
REM Relative electrophoretic mobility
RER Rough endoplasmic reticulum
RF Rheumatoid factor
RFL-6 Rat foetal lung-6
RFLP Restriction fragment length polymorphism
RGD Arginine–glycine–asparagine
rh- Recombinant human – (prefix usually referring to peptides)
RIA Radioimmunoassay
RMCP, RMCPII Rat mast cell protease, -II
RNA Ribonucleic acid
RNase Ribonuclease
RNHCl *N*-Chloramine
RNL Regional lymph nodes
ROM Reactive oxygen metabolite
RO· *The chemical symbol for* alkoxyl radical

ROO· *The chemical symbol for* peroxy radical
ROP Retinopathy of prematurity
ROS Reactive oxygen species
R-PIA *R*-(1-methyl-1-phenyltheyl)-adenosine
RPMI 1640 Roswell Park Memorial Institute 1640 medium
RS Reiter's syndrome
RSV Rous sarcoma virus
RTE Rabbit tubular epithelium
RTE-a-5 Rat tubular epithelium antigen a-5
r-tPA Recombinant tissue-type plasminogen activator
RT-PCR Reverse transcriptase/ polymerase chain reaction
RW Ragweed

S Svedberg (unit of sedimentation density)
SALT Skin-associated lymphoid tissue
SaR Sacroplasmic reticulum
SAZ Sulphasalazine
SC Secretory component
SCF Stem cell factor
SCFA Short-chain fatty acid
SCG Sodium cromoglycate *also known as* DSCG
SCID Severe combined immunodeficiency syndrome
sCR1 Soluble type-1 complement receptors
SCW Streptococcal cell wall
SD Stranded deviation
SDS Sodium dodecyl sulphate
SDS-PAGE Sodium dodecyl sulphate-polyacrylamide gel electrophoresis
SEM Standard error of the mean
SERPIN Serine protease inhibitor
SGAW Specific airway conductance
SHR Spontaneously hypertensive rat
SIM Selected ion monitoring
SIN-1 3-Morpholinosydnonimine
SIRS Systemic inflammatory response syndrome
SIV Simian immunodeficiency virus
SK Streptokinase
SLE Systemic lupus erythematosus
SLe^x Sialyl Lewis X antigen
SLO Streptolysin-O
SLPI Secretory leucocyte protease inhibitor
SM Sphingomyelin
SNAP *S*-Nitroso-*N*-acetylpenicillamine
SNP Sodium nitroprusside
SOD Superoxide dismutase
SOM Somatostatin *also known as* somatotrophin release-inhibiting factor
SOZ Serum-opsonized zymosan
SP Sulphapyridine
SR Systemic reaction
sr Sarcoplasmic reticulum

sR$_{aw}$ Specific airways resistance
SRBC Sheep red blood cells
SRS Slow-reacting substance
SRS-A Slow-reacting substance of anaphylaxis
STZ Streptozotocin
Sub P Substance P

T Thymus-derived
α-TOC α-Tocopherol
t$_{1/2}$ Half-life
T84 Human intestinal epithelial cell line
TAME Tosyl-L-arginine methyl ester
TauNHCl Taurine monochloramine
TBA Thiobarbituric acid
TBAR Thiobarbituric acid-reactive product
TBM Tubular basement membrane
TBN di-*tert*-Butyl nitroxide
tBOOH *tert*-Butylhydroperoxide
TCA Trichloroacetic acid
T cell Thymus-derived lymphocyte
TCR T cell receptor α/β or γ/δ heterodimeric forms
TDI Toluene diisocyanate
TEC Tubular epithelial cell
TF Tissue factor
Tg Thyroglobulin
TGF Transforming growth factor
TFGα, TGFβ, TGFβ$_1$ Transforming growth factors α, β, and β$_1$
T$_H$ T helper cell
T$_H$o T helper o
T$_H$p T helper precursor
T$_H$0, T$_H$1, T$_H$2 Subsets of helper T cells
THP-1 Human monocytic leukaemia
Thy 1 + Murine T cell antigen
t.i.d. Ter in die (three times a day)
TIL Tumour-infiltrating lymphocytes
TIMP Tissue inhibitors of metalloproteinase
TIMP-1, TIMP-2 Tissue inhibitors of metalloproteinases 1 and 2
Tla Thymus leukaemia antigen
TLC Thin-layer chromatography
TLCK Tosyl-lysyl-CH$_2$Cl
TLP Tumour-like proliferation
Tm T memory
TNF, TNF-α Tumour necrosis factor, -α
tPA Tissue-type plasminogen activator
TPA 12-*O*-tetradeconylphorbol-13-acetate
TPCK Tosyl-phenyl-CH$_2$Cl
TPK Tyrosine protein kinases
TPP Transpulmonary pressure
TRAP Thrombospondin related anomalous protein
Tris Tris (hydroxymethyl)-aminomethane
TSH Thyroid-stimulating hormone
TSP Thrombospondin
TTX Tetrodotoxin

TX Thromboxane
TXA$_2$, TXB$_2$ Thromboxane A$_2$, B$_2$
Tyk2 Tyrosine kinase

U937 (cells) Histiocytic lymphoma, human
UC Ulcerative colitis
UCR Upstream conserved region
UDP Uridine diphosphate
UPA Urokinase-type plasminogen activator
UTP Uridine triphosphate
UV Ultraviolet
UVA Ultraviolet A
UVB Ultraviolet B
UVR Ultraviolet irradiation
UW University of Wisconsin (preserving solution)

VAP Viral attachment protein
VC Veiled cells
VCAM, VCAM-1 Vascular cell adhesion molecule, -1, *also known as* inducible cell adhesion molecule MW 110 kD (INCAM-110)
VF Ventricular fibrillation
V/GSH Vanadate/glutathione complex
VIP Vasoactive intestinal peptide
VLA Very late activation antigen beta chain; *also known as* CD29
VLA α2 Very late activation antigen alpha 2 chain; *also known as* CD49b
VLA α4 Very late activation antigen alpha 4 chain; *also known as* CD49d
VLA α6 Very late activation antigen alpha 6 chain; *also known as* CD49f
VLDL Very low-density lipoprotein
***V* max** Maximal velocity
***V* min** Minimal velocity
VN Vitronectrin
VO$_4^-$ *The chemical symbol for* vanadate
vp Viral protein
VP Vasopressin
VPB Ventricular premature beat
VT Ventricular tachycardia

vWF von Willebrand factor

W Murine dominant white spotting mutation
WBC White blood cell
WGA Wheat germ agglutinin
WI Warm ischaemia

XD Xanthine dehydrogenase
XO Xanthine oxidase

Y1/82A A monoclonal antibody detecting a cytoplasmic antigen in human macrophages

ZA Zonulae adherens
ZAP Zymosan-activated plasma
ZAS Zymosan-activated serum
zLYCK Carboxybenzyl-Leu-Tyr-CH$_2$,Cl
ZO Zonulae occludentes

Key to Illustrations

Helper lymphocyte

Suppressor lymphocyte

Killer lymphocyte

Plasma cell

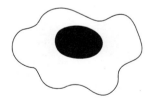
Bacterial or Tumour cell

Blood vessel lumen

Eosinophil passing through vessel wall

Neutrophil passing through vessel wall

 Resting neutrophil

 Activated neutrophil

 Resting eosinophil

 Activated eosinophil

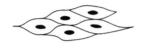 Smooth muscle

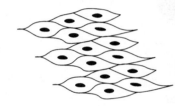 Smooth muscle thickening

 Smooth muscle contraction

 Normal blood vessel

 Endothelial cell permeability

 Resting macrophage

 Activated macrophage

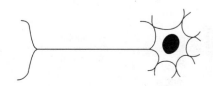 Nerve

 Intact
epithelium

 Damaged
epithelium

 Intact
epithelium
with
submucosal
gland

 Normal
submucosal
gland

 Hypersecreting
submucosal
gland

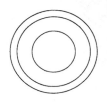 Normal
airway

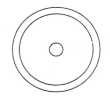 Oedema

 Bronchospasm

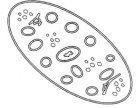 Resting platelet

 Activated
platelet

 Airway
hypersecreting
mucus

Resting
basophil

Activated
basophil

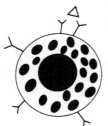

Resting
mast cell

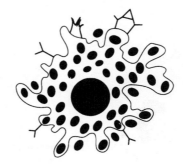

Activated
mast cell

Resting
chondrocyte

Activated
chondrocyte

Cartilage

Fibroblast

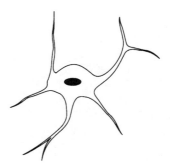

Dendritic cell/
Langerhans cell

Arteriole

Venule

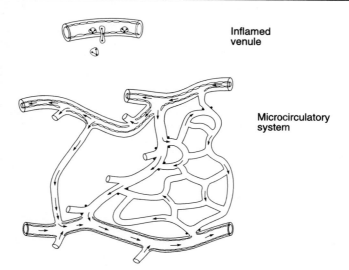

Inflamed
venule

Microcirculatory
system

Index